CONVERSION FACTORS*

Length
1 m = 3.281 ft
1 m = 3.937 × 10 in.

Area
1 m² = 1.550 × 10³ in.²
1 m² = 1.076 × 10 ft²

Volume
1 m³ = 6.102 × 10⁴ in.³
1 m³ = 3.532 × 10 ft³
1 m³ = 2.642 × 10² U.S. gallons

Mass
1 kg = 2.205 lbm

Force
1 N = 2.248 × 10⁻¹ lbf

Energy
1 J = 9.478 × 10⁻⁴ Btu
= 7.376 × 10⁻¹ ft-lbf
1 kW-hr = 3.412 × 10³ Btu
= 2.655 × 10⁶ ft-lbf

Power
1 W = 3.412 Btu/hr
1 W = 1.341 × 10⁻³ hp
1 W = 2.844 × 10⁻⁴ tons of
refrigeration

Pressure
1 Pa = 1.450 × 10⁻⁴ lbf/in.²
1 Pa = 2.088 × 10⁻² lbf/ft²
1 Pa = 9.869 × 10⁻⁶ std atm
1 Pa = 2.961 × 10⁻⁴ in. mercury wg
1 Pa = 4.019 × 10⁻³ in. water

Temperature
1 deg R difference = 1 deg F
difference = 5/9 deg C difference
= 5/9 deg K difference
deg F = 9/5 (deg C) + 32

Velocity
1 m/s = 1.969 × 10² ft/min
1 m/s = 3.281 ft/sec

Acceleration
1 m/s² = 3.281 ft/sec²

Mass Density
1 kg/m³ = 6.243 × 10⁻² lbm/ft³

Mass Flow Rate
1 kg/s = 2.205 lbm/sec
1 kg/s = 7.937 × 10³ lbm/hr

Volume Flow Rate
1 m³/s = 2.119 × 10³ ft³/min
1 m³/s = 1.585 × 10⁴ gal/min

Thermal Conductivity

$$1 \frac{\text{W}}{\text{m-C}} = 5.778 \times 10^{-1} \frac{\text{Btu}}{\text{hr-ft-F}}$$

$$1 \frac{\text{W}}{\text{m-C}} = 6.934 \frac{\text{Btu-in}}{\text{hr-ft}^2\text{-F}}$$

Heat-transfer Coefficient

$$1 \frac{\text{W}}{\text{m}^2\text{-C}} = 1.761 \times 10^{-1} \frac{\text{Btu}}{\text{hr-ft}^2\text{-F}}$$

Specific Heat

$$1 \frac{\text{J}}{\text{kg-C}} = 2.389 \times 10^{-4} \frac{\text{Btu}}{\text{lbm-F}}$$

Viscosity, Absolute

$$1 \frac{\text{N-s}}{\text{m}^2} = 10^3 \text{ centipoise} = 1 \text{ Pa-s}$$

$$1 \frac{\text{N-s}}{\text{m}^2} = 5.720 \times 10^{-1} \frac{\text{lbm}}{\text{ft-sec}}$$

Viscosity, Kinematic
1 m²/s = 1.076 × 10 ft²/sec
1 m²/s = 10⁶ centistoke

*All factors have been rounded off to four significant figures.

Heating, Ventilating, and Air Conditioning
Analysis and Design

Heating, Ventilating, and Air Conditioning
ANALYSIS and DESIGN

Fourth Edition

Faye C. McQuiston

Oklahoma State University

Jerald D. Parker

Oklahoma Christian University
of
Science and Arts

John Wiley & Sons, Inc.
New York • Chichester • Brisbane • Toronto • Singapore

ACQUISITIONS EDITOR	Cliff Robichaud
MARKETING MANAGER	Susan Elbe
SENIOR PRODUCTION EDITOR	Ingrao Associates and Nancy Prinz
MANUFACTURING MANAGER	Andrea Price
ILLUSTRATION COORDINATOR	Jamie Perea

This book was set in Times Roman by General Graphic Services and printed and bound by Hamilton Printing Company. The cover was printed by New England Book Components, Inc.

Library of Congress Cataloging-in-Publication Data

McQuiston, Faye C.
    Heating, ventilating, and air conditioning: analysis and design/
    Faye C. McQuiston, Jerald D. Parker—4th ed.
        p.   cm.
    Includes index.
    ISBN 0-471-58107-0
    1. Heating.   2. Ventilation.   3. Air conditioning.
    I. Parker, Jerald D.
    TH7222.M38   1994
    697—dc20   93-28394
                                                    CIP

Printed in the United States of America

10 9 8 7 6 5 4 3 2

Preface

Advances in the areas of load calculations, indoor air quality (IAQ), and the requirements for environmentally acceptable refrigerants have prompted this revision of the third edition. The revisions reflect primarily the result of research sponsored by the American Society of Heating, Refrigerating and Air-Conditioning Engineers (ASHRAE) and the continued development of ASHRAE standards related to IAQ, comfort, and refrigerants. The original objective of this book, to produce an up-to-date, convenient classroom teaching aid based on ASHRAE literature has not changed. It is intended for use by engineering students at the undergraduate and graduate level as well as practicing engineers. Mastery of the material should enable a person to effectively participate in the design of all types of HVAC systems.

Basic courses in thermodynamics, heat transfer, fluid mechanics, and dynamics are desirable prerequisites. There is sufficient material for two-semester length courses with considerable latitude in course makeup. Although the book is intended to be primarily a teaching device, it should also be useful as a reference and as an aid in studying new procedures.

A number of revisions have been made based on suggestions from users of the previous editions. Data and references have been updated throughout the text; however, in a few instances, useful material from older sources has been retained. New problems have been added, existing problems have been revised, and the problems have been rearranged to fit the new order of the material in the chapters. In all major areas there are problems that can be solved either manually or using available computer software. A solution manual is available from the publisher. Instructors should provide some examples and problems that emphasize their own design philosophies and the requirements of local geographical regions.

Chapter 2, Air-Conditioning Systems, has been revised in an effort to improve understanding for the beginning student. Increased emphasis has been placed on the control of HVAC systems by adding more explanation of control theory and more detailed illustrations of typical systems and controls.

Chapter 4, Indoor Air Quality—Comfort and Health, has been completely revised to place greater emphasis on indoor air quality and health and the use of ANSI/ASHRAE Standard 62. The sections dealing with comfort have also been updated to agree with the latest comfort standard, ASHRAE Standard 55.

Chapter 8, The Cooling Load, has undergone extensive revision based on the latest ASHRAE research in this area. The Transfer Function Method is more fully explained with sufficient data and examples to enhance understanding of this method. The new CLTD/CLF/SCL manual calculation method is also presented with adequate data and examples. Recent literature and software for load calculations is referenced.

All the material related to energy calculations has been collected together in Chapter 9.

Chapter 12, Fans and Building Air Distribution, has been revised to reference the new

ASHRAE duct fitting database with a sampling of those data. Discussion of evolving duct design optimizing procedures is now included.

Mass transfer and direct contact heat transfer material has been condensed and placed in Chapter 13.

Chapter 15, Refrigeration, has been revised to include discussion related to ozone depletion, the safety and environmental effects of refrigerants, and the selection of replacement refrigerants.

Instructors using this text are encouraged to involve students in the use of personal computers and the many programs available for use. The authors may be contacted for information related to procurement of software.

Uncertainty exists as to when a complete conversion from English to the international system of units (SI) will occur in the United States. However, engineers should be comfortable with both systems of units when they enter practice. Therefore, this book continues to use a dual system of units, with some emphasis placed on the English system. Instructors should blend the two systems of units as they see fit.

We are deeply indebted to ASHRAE for providing much support in the production of this book. Many companies and individuals, too numerous to list, contributed suggestions, ideas, photographs, and commentary. Thank you every one.

FAYE C. MCQUISTON
JERALD D. PARKER

About the Authors

Faye C. McQuiston is Professor Emeritus of Mechanial and Aerospace Engineering at Oklahoma State University, Stillwater, Oklahoma. He received B.S. and M.S. degrees in mechanical engineering from Oklahoma State University in 1958 and 1959 and a Ph.D. in mechanical engineering from Purdue University in 1970. Dr. McQuiston joined the Oklahoma State faculty in 1962 after a three-year period in industry. He was a National Science Foundation Faculty Fellow from 1967 to 1969. Dr. McQuiston is an active member of the American Society of Heating, Refrigerating and Air-Conditioning Engineers (ASHRAE), recently completing a term as a Vice-President. He has served on the Board of Directors, the Technology, Education, Member, and Publishing councils, is a past member of the Research and Technical, Education, and Standards Committees. He was honored with the Best Paper Award in 1979, the Region VIII Award of Merit in 1981, the Distinguished Service Award in 1984, and the E. K. Campbell Award in 1986. He was also elected to the grade of Fellow in 1986. Dr. McQuiston is a registered professional engineer and a consultant to several system design and equipment manufacturing firms. He is active in research related to the design of heating and air-conditioning systems, particularly in the areas of heat-exchanger design and simulation and load calculations. He has written extensively in the area of heating and air conditioning and is the coauthor of a basic fluid mechanics and heat transfer text.

Jerald D. Parker is a Professor of Mechanical Engineering at Oklahoma Christian University of Science and Arts after serving 33 years on the Mechanical Engineering faculty at Oklahoma State University. He received B.S. and M.S. degrees in mechanical engineering from Oklahoma State University in 1955 and 1958 and a Ph.D. in mechanical engineering from Purdue University in 1961. During his tenure at Oklahoma State, he spent one year on leave with the engineering department of duPont in Newark, Delaware. He has been active at both the local and national level in ASME, where he is a fellow. In ASHRAE he has served as Chairman of the Technical Committee on Fluid Mechanics and Heat Transfer, Chairman of a Standards Project Committee, and a member of the Continuing Education Committee. He is a registered professional engineer and a member of the Oklahoma and National Societies of Professional Engineers. He is coauthor of a basic text in fluid mechanics and heat transfer and has contributed articles for handbooks, technical journals, and magazines. His research has been involved with ground-coupled heat pumps, solar-heated asphalt storage systems, and chilled-water storage and distribution. He has served as a consultant in cases involving performance and safety of heating, cooling, and process systems.

Contents

Symbols

English Letter Symbols

A	area, ft^2 or m^2
A	apparent solar irradiation for zero air mass, Btu/(hr-ft^2) or W/m^2
ADPI	air distribution performance index, dimensionless
$ASHGF$	absorbed solar heat gain factor
B	atmospheric extinction coefficient.
b	transfer function coefficient, Btu/(hr-ft^2-F) or W/(m^2-C)
b	bypass factor, dimensionless
C	concentration, lbm/ft^3 or kg/m^3
C	unit thermal conductance, Btu/(hr-ft^2-F) or W/(m^2/C)
C	discharge coefficient, dimensionless
C	loss coefficient, dimensionless
C	fluid capacity rate, Btu/(hr-F) or W/C
C	clearance factor, dimensionless
C_d	overall flow coefficient, dimensionless
C_d	draft coefficient, dimensionless
C_p	pressure coefficient, dimensionless
C_v	flow coefficient, dimensionless
CLF	cooling load factor, dimensionless
CLTD	cooling load temperature difference, F or C
COP	coefficient of performance, dimensionless
c	specific heat, Btu/(lbm-F) or J/(kg-C)
c	transfer function coefficient, Btu/(hr-ft^2-F) or W/(m^2-C)
cfm	volume flow rate, ft^3/min
clo	clothing thermal resistance, (ft^2-hr-F)/Btu or (m^2-C)/W
D	diameter, ft or m
D	diffusion coefficient, ft^2/sec or m^2/s
DD	degree days, F-day or C-day
db	dry bulb temperature, F or C
DR	daily range of temperature, F or C
d	bulb diameter, ft or m
d	sun's declination, degrees
d	transfer function coefficient, dimensionless
E	effective emittance, dimensionless
EDT	effective draft temperature, or C
ET	effective temperature, F or C
F	configuration factor, dimensionless
F	quantity of fuel, ft^3 or m^3
F_c	fraction of heat gain, dimensionless
F(s)	wet surface function, dimensionless
f	friction factor, dimensionless
f_t	Darcy friction factor with fully turbulent flow, dimensionless

FP	correlating parameter, dimensionless
G	irradiation, Btu/(hr-ft^2) or W/m^2
G	mass velocity, lbm/(ft^2-sec) or kg/m^2-s
g	local acceleration due to gravity, ft/sec^2 or m/s^2
g	transfer function coefficient, Btu/(hr-ft) or W/C
g_c	dimensional constant, 32.17 (lbm-ft)/(lbf-sec^2) or 1.0 (kg-m)/(N-s^2)
H	heating value of fuel, Btu or J per unit of volume
H	head, ft or m
h	height or length, ft or m
h	heat-transfer coefficient, Btu/(hr-ft^2-F) or W/(m^2-C) (also used for mass-transfer coefficient with subscripts m, d, and i)
h	hour angle, degrees
hp	horsepower
i	enthalpy, Btu/lbm or J/kg
J	Joule's equivalent, 778.28 (ft-lbf)/Btu
JP	correlating parameter, dimensionless
$J(s)$	wet surface function, dimensionless
$J_i(s)$	wet surface function, dimensionless
j	Colburn j-factor, dimensionless
K	color correction factor, dimensionless
K	resistance coefficient, dimensionless
K_t	unit length conductance, Btu/(ft-hr-F) or W/(m-C)
k	thermal conductivity, (Btu-ft)/(ft^2-hr-F), (Btu-in.)/(ft^2-hr-F) or (W-m)/(m^2-C)
k	isentropic exponent, c_p/c_v, dimensionless
L	fin dimension, ft or m
L	total length, ft or m
Le	Lewis number, Sc/Pr, dimensionless
LMTD	log mean temperature difference, F or C
l	latitude, degrees
l	lost head, ft or m
M	molecular mass, lbm/(lb mole) or kg/(kg mole)
M	fin dimension, ft or m
MRT	mean radiant temperature, F or C
m	mass, lbm or kg
$\dot{m}$	mass flow rate or mass tranfer rate, lbm/sec or kg/s
N	number of hours or other integer
N	fraction of absorbed solar heat gain
Nu	Nusselt number, hx/k, dimensionless
NC	noise criteria, dimensionless
NTU	number of transfer units, dimensionless
P	pressure, lb/ft^2 or psia or N/m^2 or Pa
P	heat exchanger parameter, dimensionless
P	circumference, ft or m
Pr	Prandtl number, $\mu c_p/k$, dimensionless
PD	piston displacement, ft^3/min or m^3/s
p	partial pressure, lbf/ft^2 or psia or Pa
p	transfer function coefficient, dimensionless

$\dot{Q}$	volume flow rate, ft^3/sec or m^3/s
q	heat transfer, Btu/lbm or J/kg
$\dot{q}$	heat transfer rate, Btu/hr or W
R	gas constant, (ft-lbf)/(lbm-R) or J/(kg-K)
R	unit thermal resistance, (ft^2-hr-F)/Btu or (m^2-C)/W
R	heat exchanger parameter, dimensionless
R	fin radius, ft or m
R'	thermal resistance, (hr-F)/Btu or C/W
$\overline{R}$	gas constant, (ft-lbf)/(lb mole-R) or J/(kg mole-K)
Re	Reynolds number $\rho\overline{V}D/\mu$, dimensionless
R_f	unit fouling resistance, (hr-ft^2-F)/Btu, or (m^2-C)/W
r	radius, ft or m
rpm	revolutions per minute
S	fin spacing, ft or m
S	equipment characteristic, Btu/(hr-F) or W/C
Sc	Schmidt number, ν/D, dimensionless
Sh	Sherwood number, $h_m x/D$, dimensionless
SC	shading coefficient, dimensionless
SCL	solar cooling load, Btu/(hr-ft^2) or W/m^2
SHF	sensible heat factor, dimensionless
SHGF	solar heat gain factor, Btu/(hr-ft^2) or W/m^2
s	entropy, Btu/(lbm-R) or J/(kg-K)
T	absolute temperature, R or K
TSCL	total solar cooling load, Btu/ft^2 or W-hr/m^2
TSHGF	transmitted solar heat gain factor
t	temperature, F or C
t^*	thermodynamic wet bulb temperature, F or C
U	overall heat transfer coefficient, Btu/(hr-ft^2-F) or W/(m^2-C)
u	velocity in x direction, ft/sec or m/s
V	volume, ft^3 or m^3
$\overline{V}$	velocity, ft/sec or m/s
v	specific volume, ft^3/lbm or m^3/kg
v	transfer function coefficient, dimensionless
v	velocity in y direction, ft/sec or m/s
W	humidity ratio, lbmv/lbma or kgv/kga
W	equipment characteristics, Btu/hr or W
$\dot{W}$	power, Btu/hr or W
WBGT	wet bulb globe temperature, F or C
w	skin wettedness, dimensionless
w	work, Btu, or ft-lbf, or J
w	transfer function coefficient, dimensionless
X	normalized input, dimensionless
X	fraction of daily range
x	mole fraction
x	quality, lbmv/lbm or kgv/kg
x, y, z	length, ft or m
Y	normalized capacity, dimensionless

Subscripts

a	transverse dimension
a	air
a	average
a	attic
as	adiabatic saturation
as	denotes change from dry air to saturated air
avg	average
B	barometric
b	branch
b	longitudinal dimension
b	base
c	cool or coil
c	convection
c	ceiling
c	cross section or minimum free area
c	cold
c	condenser
c	Carnot
c	collector
cl	center line
D	direct
D	diameter
d	dewpoint
d	total heat
d	diffuse
d	design
d	downstream
dry	dry surface
e	equivalent
e	sol-air
e	equipment
e	evaporator
f	film
f	friction
f	fin
fg	refers to change from saturated liquid to saturated vapor
fl	fluorescent light
fl	floor
fr	frontal
g	refers to saturated vapor
g	globe
H	horizontal
HD	on horizontal surface, daily
h	heat
h	hydraulic
h	head

h	heat transfer
h	hot
i	j-factor for total heat transfer
i	inside or inward
i	instantaneous
l	latent
l	liquid
m	mean
m	mass transfer
m	mechanical
ND	direct normal
n	integer
o	outside
o	total or stagnation
o	initial condition
oh	humid operative
P	pressure
p	constant pressure
p	pump
R	reflected
R	refrigerating
r	radiation
r	room air
s	stack effect
s	sensible
s	saturated vapor or saturated air
s	supply air
s	shaft
s	static
s	surface
sc	solar constant
sh	shade
sl	sunlit
t	temperature
t	total
t	contact
t	tube
u	unheated
u	upstream
V	vertical
v	vapor
v	ventilation
v	velocity
w	wind
w	wall
w	liquid water
wet	wet surface
x	length

x	extraction
1, 2, 3	state of substance at boundary of a control volume
1, 2, 3	a constituent in a mixture
∞	free stream condition

Greek Letter Symbols

α	angle of tilt from horizontal, degrees
α	absorptivity or absorptance, dimensionless
α	total heat-transfer area over total volume, ft^{-1} or m^{-1}
α	thermal diffusivity, ft^2/sec or m^2/s
β	fin parameter, dimensionless
β	altitude angle, degrees
γ	wall solar azimuth angle, degrees
Δ	change in a quantity or property
δ	boundary layer thickness, ft or m
ϵ	heat exchanger effectiveness, dimensionless
ϵ	emittance or emissivity, dimensionless
η	efficiency, dimensionless
θ	angle of incidence, degrees
θ	time, sec
μ	degree of saturation, percent or fraction
μ	dynamic viscosity, lbm/(ft-sec) or (N-s)/m^2
ν	kinematic viscosity, ft^2/sec or m^2/s
ρ	mass density, lbm/ft^3 or kg/m^3
ρ	reflectivity or reflectance, dimensionless
Σ	angle of tilt from horizontal, degrees
σ	Stefan-Boltzmann constant, Btu/(hr-ft^2-R^4) or J/(s-m^2-K^4)
σ	free flow over frontal area, dimensionless
τ	transmissivity or transmittance, dimensionless
ϕ	fin parameter, dimensionless
ϕ	solar azimuth angle, degrees
ϕ	relative humidity, percent or fraction
ψ	wall azimuth angle, degrees
Ψ	fin parameter, dimensionless
Ψ	zenith angle, degrees

List of Charts*

Chart 1a	ASHRAE Psychrometric Chart No. 1 (IP) (reprinted by permission of ASHRAE).
Chart 1b	ASHRAE Psychrometric Chart No. 1 (SI) (reprinted by permission of ASHRAE).
Chart 1Ha	ASHRAE Psychrometric Chart No. 4 (IP) (reprinted by permission of ASHRAE).

* For the convenience of the reader all charts have been folded into a pocket located on the inside back cover.

Chapter 1

Introduction

We all appreciate the relief from discomfort afforded by a modern air-conditioning system. Many of our homes and most offices and commercial facilities would not be comfortable without year-round control of the indoor environment. The "luxury label" attached to air-conditioning prior to World War II has given way to an appreciation of its practicality in making our lives healthier and more productive. Along with that rapid development in improving human comfort came the realization that goods could be produced better, faster, and more economically in a properly controlled environment. In fact, many products of today could not be produced at all were the temperature, humidity, and air quality not controlled within very narrow limits. The development and industrialization of the United States, especially the Southern states, would never have been possible without year-round control of the indoor environment. One has only to look for a manufacturing or printing plant, electronics laboratory, or other high-technology facility and the vast office complexes associated with our economy to understand the truth of that statement. Indeed, virtually every residential, commercial, industrial, and institutional building in the United States, Canada, and other industrial countries of the world have controlled environments the year round.

Most systems installed prior to the 1970s were designed with little attention to energy conservation since fuels were abundant and inexpensive. Escalating energy costs since that decade have caused increased interest in efficiency of operation. During the same period the need for closely controlled environments in laboratories, hospitals, and industrial facilities continued to grow. A third factor of expanding awareness was the importance of comfort and indoor air quality on both health and performance. Practitioners of the arts and sciences of HVAC system design and simulation were challenged as they never had been before. Developments in electronics, controls, and computers have furnished the tools allowing HVAC to become a high-technology industry. Although tools and methods have changed, and a better understanding of the parameters that define comfort and indoor air quality have been accomplished, many of the basics of good system design have not changed. These basic elements of HVAC system design are the emphasis of this text and furnish a basis for presenting recent developments of importance and procedures for designing functional, well-controlled, and energy-efficient systems to maintain human comfort and health as well as industrial productivity.

1-1 HISTORICAL NOTES

Historically *air conditioning* has implied cooling or otherwise improving the indoor environment during the warm months of the year. In modern times the term has taken on a more literal meaning that can be applied to year-round environmental situations. That is, air conditioning refers to the control of temperature, moisture content, cleanliness, air quality, and air circulation as required by occupants, a process, or a product in the space. This definition was first proposed by Willis Carrier (1).

There is evidence of the use of evaporative effects and ice for cooling in very early times; however, it was not until the middle of the nineteenth century that a practical refrigerating machine was built. By the end of the nineteenth century, the concept of central heating was fairly well developed, and early in the twentieth century cooling for comfort got its start. Carrier is credited with the first successful attempt in 1902 to reduce the humidity of air and maintain it at a specified level (1). This marked the birth of true environmental control as we know it today. Developments since that time have been rapid.

More detailed discussions of the history of heating, refrigeration, and air conditioning are given in articles by Willis R. Woolrich, John W. James, Walter A. Grant, and William L. McGrath (2). Because of the wide scope and diverse nature of the Heating, Ventilating, and Air Conditioning (HVAC) field, literally thousands of engineers, their names too numerous to list, have developed the industry. The accomplishments of all these unnamed persons are summarized in the *ASHRAE* Handbook* consisting of four volumes entitled: *Fundamentals, Refrigeration, HVAC Systems and Equipment,* and *HVAC Applications.* Research designed to improve the handbooks is sponsored by ASHRAE and monitored by ASHRAE members. The principles presented in this textbook follow the handbooks closely.

At the beginning we will review the units and dimensions common to HVAC analysis and design.

1-2 UNITS AND DIMENSIONS

In HVAC computations as in all engineering work, consistent units must be employed. A *unit* is a specific quantitative measure of a physical characteristic in reference to a standard. Examples of a unit are the foot and the meter, which are used to measure the physical characteristic length. A physical characteristic, such as length, is called a *dimension.* Other dimensions of interest in HVAC computations are force, time, temperature, and mass.

In this text two systems of units will be employed. The first is called the *English Engineering System* and is most commonly used in HVAC work in the United States with some modification such as use of inches instead of feet. The system is sometimes referred to as the inch-pound or IP system. The second is the *International System* (or SI for *Systeme International d'Unites*), the system in use in engineering practice through-out most of the world and widely adopted in the United States.

Equipment designed using U. S. conventional units will be operational for years and even decades. For the foreseeable future, then, it will be necessary for many engineers

---

* ASHRAE is an abbreviation for the American Society of Heating, Refrigerating and Air-Conditioning Engineers, Incorporated.

Table 1-1

	Units	
Dimensions	English Engineering System	International System
Mass	pound mass (lbm)	kilogram (kg)
Length	foot (ft)	meter (m)
Time	second (sec)	second (s)
Temperature	degree Fahrenheit (F)	degree Kelvin (K)

to work in either system of units and to be able to make conversion from one system to another. The base SI units important in HVAC work are given in Table 1-1 with the comparable English Engineering units.

The SI system is described in several documents: (3), (4), and (5). The system consists of seven base units (the four base units given in Table 1-1 plus units of electric current, amount of substance, and luminous intensity). In addition there are two supplementary units for the plane angle (rad), the solid angle (r), and a list of derived units. The derived units in the SI system that are frequently used in HVAC are given in Table 1-2.

Because there is only one unit for each dimension in the SI system, there is a carefully defined set of multiple and submultiple prefixes to avoid the use of a large number of

Table 1-2 Derived Units Frequently Used in HVAC

Dimension	SI Unit	Special SI Name & Symbol	English Unit
Acceleration, angular	rad/s^2		rad/sec^2
Acceleration, linear	m/s^2		ft/sec^2
Area	m^2		ft^2
Density	kg/m^3		lbm/ft^3
Energy	N-m	joule (J)	Btu, ft-lb
Force	$(kg\text{-}m)/s^2$	newton (N)	lbf
Frequency	l/s	hertz (Hz)	l/sec
Power	J/s	watt (W)	hp
Pressure	N/m^2	pascal (Pa)	psi
Specific heat capacity	J/(kg-C)		Btu/(lbm-F)
Stress	N/m^2	pascal (Pa)	psi
Thermal conductivity	W/(m-C)		Btu/(hr-ft-F)
Thermal flux density	W/m^2		$Btu/(hr\text{-}ft^2)$
Velocity, angular	rad/s		rad/sec
Velocity, linear	m/s		ft/sec
Viscosity, dynamic	$(N\text{-}s)/m^2$		$lbf\text{-}sec/ft^2$
Viscosity, kinematic	m/s^2		ft/sec^2
Volume	m^3		ft^3

Table 1-3 Multiple and Submultiple SI Prefixes

Multiplying Factor	Prefix	Symbol
1 000 000 000 000 = 10^{12}	tera	T
1 000 000 000 = 10^9	giga	G
1 000 000 = 10^6	mega	M
1 000 = 10^3	kilo	k
100 = 10^2	hecto[a]	h
10 = 10^1	deka[a]	da
0.1 = 10^{-1}	deci	d
0.01 = 10^{-2}	centi[a]	c
0.001 = 10^{-3}	milli	m
0.000 001 = 10^{-6}	micro	μ
0.000 000 001 = 10^{-9}	nano	n
0.000 000 000 001 = 10^{-12}	pico	p

[a]There prefixes are to be avoided if possible.

zeros for very large or very small quantities. These prefix names and symbols are given in Table 1-3.

Conversion factors between IP and SI units are given inside the front cover.

EXAMPLE 1-1

Describe the quantity of pressure equal to 1000 N/m^2 in suggested SI terminology.

SOLUTION

1000 can be described by the prefix kilo (k). The N/m^2 has the special name of pascal. Therefore,

$$1000 \, \text{N/m}^2 = 1 \text{ kilo pascal} = 1 \text{ kPa}$$

EXAMPLE 1-2

A certain corkboard has a thermal conductivity of 0.025 Btu/(hr-ft-F). Convert this quantity to the equivalent SI value.

SOLUTION

According to the conversion factor given in the inside front cover, to convert from Btu/(hr-ft-F) to W/(m-C) you must divide the given value by 0.5778.

$$\frac{(0.025)\dfrac{\text{Btu}}{\text{hr-ft-F}}}{0.5778\dfrac{\text{Btu/(hr-ft-F)}}{\text{W/(m-C)}}} = 0.043 \text{ W/(m-C)}$$

Because the prefix centi is to be avoided if possible (Table 1-3), the quantity is expressed as shown. Notice that the final quantity is not expressed to any more significant figures than the original quantity, in this case two.

In the SI system the temperature is given in degrees Kelvin (K), which is a thermodynamic absolute temperature. It is quite common to specify temperature in degrees Celsius or Centigrade even where all other units might be SI units. One degree change on the Celsius scale is identical to one degree change on the Kelvin scale. For this reason the value of thermal conductivity obtained in Example 1-1 would be no different if it had been expressed in the units of W/(m-K). In this text quantities involving temperature *change* may be expressed in either absolute or nonabsolute units.

The relationship between temperature on the Kelvin Scale and temperature on the Celsius scale is

$$K = C + 273.15 \qquad\qquad \textbf{(1-1)}$$

For example, 100 C is identical to 373.15 K.

In the English Engineering system the unit of temperature is the degree Fahrenheit. When the thermodynamic absolute temperature is needed, the temperature is specified in degrees Rankine (R). The relationship between the Fahrenheit scale and the Rankine scale is

$$R = F + 459.67 \qquad\qquad \textbf{(1-2)}$$

Since both the Rankine and Kelvin scales are absolute scales, absolute zero is identical on each scale. The relationship between the Rankine scale and the Kelvin scale is

$$K = \tfrac{5}{9}R \qquad\qquad \textbf{(1-3)}$$

For example, 900 R is identical in temperature to 500 K. The relationship between the four temperature scales is given in Fig. 1-1.

Consistent units must always be employed in physical computations. For example, thermal and mechanical energy may be interchangeable in a given situation. Thermal energy traditionally has been specified in terms of British Thermal Units (Btu) in engineering work in English-speaking countries. Mechanical energy is frequently specified in the units of foot pounds force (ft-lbf). If the net energy (the sum of the thermal and mechanical energy) is to be computed, it is necessary to specify both types of energy in the same unit. The relationship between the Btu and the ft-lbf is

$$1 \text{ Btu} = 778.28 \text{ ft-lbf} \qquad\qquad \textbf{(1-4)}$$

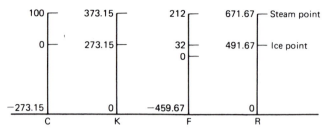

Figure 1-1 Relationship between temperatures scales.

EXAMPLE 1-3

A system is known to contain 100 Btu of thermal energy and 30,000 ft-lbf of mechanical energy. What is the total energy (mechanical plus thermal) contained by the system in Btu?

SOLUTION

Use Eq. 1-4 as a conversion factor.

$$100\,\text{Btu} + \frac{(30,000)\,\text{ft-lbf}}{(778.28)\,\text{ft-lbf/Btu}} = (100 + 38.5)\,\text{Btu} = 139\,\text{Btu}$$

In this example notice that an equation (Eq. 1-4) was changed to a conversion factor by simple algebra. Also note that the answer was rounded off to three significant figures to match the accuracy of the given data.

$$1\,\text{Btu} = 778.28\,\text{ft-lbf}$$
$$1 = 778.28\,\text{ft-lbf/Btu}$$

Since 778.28 (ft-lbf)/Btu is equivalent to unity, it can be placed in the appropriate term in such a way as to cancel the undesired units, and yet not change the true value of the physical quantity represented by the term. In Example 1-3, 30,000 ft-lbf of energy is identical to 38.5 Btu of energy.

From Table 1-2 we see that the derived unit of energy in the SI system is the joule, which is equivalent to one newton-meter. The derived unit of power in the SI system is the watt, which is equivalent to one joule per second.

EXAMPLE 1-4

The specific heat of air at normal conditions is approximately equal to 0.241 Btu/(lbm-F). Express this value of the specific heat of air in SI units, joule per kilogram, and degree.

SOLUTION

From the conversion factor given in the inside cover, the following relationship is true

$$\frac{(0.241)\dfrac{\text{Btu}}{\text{lbm-F}}}{(2.389 \times 10^{-4})\dfrac{\text{Btu/lbm-F}}{\text{J/kgC}}} = 1009 \text{ J(kg-C)} = 1.01 \text{ kJ/(kg-C)}$$

Conversion factors from SI to English are given on the inside front cover. Additional useful conversion factors, primarily English units, are given on the inside back cover.

1-3   FUNDAMENTAL CONCEPTS

Generally speaking, background preparation for a study of HVAC system design is covered in the typical thermodynamics, fluid mechanics, and heat transfer curriculum. The general concepts taught in dynamic systems are also important to understanding and analyzing any HVAC system.

The most important concept in this area is the First Law of Thermodynamics or energy balance. In some cases the balance will be on a *closed system* or fixed mass. More often the balance will involve a *control volume,* with mass flowing in and out.

The principles dealing with the behavior of liquids and gases flowing in pipes and ducts are most important, especially the relationship between flow and pressure loss. The concepts are closely related and used in conjunction with thermodynamic concepts. Emphasis will be placed on complete fluid distributing systems instead of a single element. This is a significant extension of basic fluid mechanic concepts. Most problems will be of a steady-flow nature even though changes in flow rate and fluid properties may occur from time to time.

Generally the most simple concepts of heat transfer, dealing with conduction, convection, and radiation, are used in typical system design. Again the concepts of thermodynamics and fluid mechanics are intertwined with most heat-transfer processes. In most cases, steady state can be assumed for design purposes. In those cases where transient effects are important, computer routines are usually used to obtain the required results. In this connection a basic understanding of a dynamic system is essential.

Some of the more important functions required to air condition a space completely are briefly described below.

Heating

Heating is the transfer of energy to a space or to the air in a space by virtue of a difference in temperature between the source and the space or air. This process may take different forms such as direct radiation and free convection to the space, direct heating of forced circulated air, or through heating of water that is circulated to the vicinity of the space and used to heat the circulated air. Heat transfer, which is manifested in a rise in temperature of the air, is called *sensible heat transfer.*

The rate of sensible heat transfer can be related to the rise in temperature of an air stream being heated by

$$\dot{q}_s = \dot{m}\, c_p(t_e - t_i) = \frac{\dot{Q}\, c_p}{v}(t_e - t_1) \tag{1-5}$$

where:

$\dot{q}_s$ = rate of sensible heat transfer, Btu/hr or W
$\dot{m}$ = mass rate of air flow, lbm/hr or kg/s
c_p = constant pressure specific heat of air, Btu/(lbm-F) or J/(kg-K)
$\dot{Q}$ = volume flow rate of air flow, ft^3/hr or m^3/s
v = specific volume of air, ft^3/lbm or m^3/kg
t_e = temperature of air at exit, F or C
t_i = temperature of air at inlet, F or C

The specific volume and the volume flow rate of the air are usually specified at the inlet conditions. Note that the mass flow rate of the air $\dot{m}$, equal to the volume flow rate divided by the specific volume, is considered not to change between inlet and outlet. The specific heat is assumed to be an average value.

EXAMPLE 1-5

Determine the rate at which heat must be added in Btu/hr to a 3000 cfm air stream to change its temperature from 70 to 120 F. Assume an inlet air specific volume of 13.5 ft^3/lbm and a specific heat of 0.24 Btu/(lbm-F).

SOLUTION

The heat being added is sensible as it is contributing to the temperature change of the air stream. Eq. 1-5 applies.

$$\dot{q}_s = \frac{\dot{Q}\, c_p}{v}(t_e - t_i) = \frac{(3000)\dfrac{\text{ft}^3}{\text{min}}(0.24)\dfrac{\text{Btu}}{\text{lbm-F}}(120\text{-}70)\text{F}\,(60)\dfrac{\text{min}}{\text{hr}}}{(13.5)\dfrac{\text{ft}^3}{\text{lbm}}}$$

$$\dot{q}_s = 160{,}000 \text{ Btu/hr}$$

Humidifying

The transfer of water vapor to atmospheric air is referred to as humidification. Heat transfer is associated with this mass transfer process; however, the transfer of mass and energy are manifested in an increase in the concentration of water in the air–water vapor mixture. Here the term *latent heat transfer* is used. This process is usually accomplished by introducing water vapor or by spraying fine droplets of water that evaporate into the circulating air stream. Wetted mats or plates may also be used.

The latent energy required in a humidifying process can be calculated if the rate at which water is being vaporized and the enthalpy of vaporization (latent enthalpy) are known. The relation is

$$\dot{q}_l = i_{fg}\,\dot{m}_w \tag{1-6}$$

where:

$\dot{q}_l$ = rate of latent heat addition, Btu/hr or W
i_{fg} = enthalpy of vaporization, Btu/lbm or J/kg
$\dot{m}_w$ = rate at which water is vaporized, lbm/hr or kg/s

Equation 1-6 does not necessarily give the total energy exchanged with the air stream in a humidification process. If the fluid being injected is at any condition other than dry saturated vapor at the air stream temperature, the temperature of the air stream will change and there will be some sensible heating or cooling. This concept is developed in Chapter 3.

EXAMPLE 1-6

It is desired to add 0.01 lbm of water vapor to each pound of perfectly dry air flowing at the rate of 3000 cfm using saturated (liquid) water in the humidifier. Assuming a value of 1061 Btu/lbm for the enthalpy of vaporization of water, estimate the rate of latent energy input necessary to perform this humidification of the air stream.

SOLUTION

Since the rate of water addition is tied to the mass of the air, we must determine the mass flow rate of the air stream. Let us assume that the specific volume of the air given in Example 1-1 is a suitable value to use in this case; then

$$\dot{m}_{air} = \frac{\dot{Q}}{v} = \frac{3000\ \text{ft}^3/\text{min}}{13.5\ \text{ft}^3/\text{lbm}} = 222\ \text{lbm/min}$$

and the latent heat transfer

$$\dot{q}_e = (1061\ \text{Btu/lbm}_w)\,(222\ \text{lbm}_a/\text{min})\,(0.01\ \text{lbm}_w/\text{lbm}_a)\,(60\ \text{min/hr})$$

$$= 141{,}000\ \text{Btu/hr}$$

The use of psychrometric charts and psychrometric formulas to determine sensible and latent heat transfers will be discussed in Chapter 3.

Cooling

Cooling is the transfer of energy from the space or air supplied to the space by virtue of a difference in temperature between the source and the space or air. In the usual cooling process air is circulated over a surface maintained at a low temperature. The surface may be in the space to be cooled or at some remote location from it, the air being ducted to and from the space. Usually water or a volatile refrigerant is the cooling medium. Cooling usually denotes sensible heat transfer, with a decrease in the air temperature. Equation 1-5 is valid in this case, and a negative value for sensible heat rate will be obtained, showing heat transfer is from the air stream.

Dehumidifying

The transfer of water vapor from atmospheric air is called dehumidification. Latent heat transfer is associated with this process. The transfer of energy is from the air; as a consequence, the concentration of water in the air–water vapor mixture is lowered. This process is most often accomplished by circulating the air over a surface maintained at a sufficiently low temperature to cause the condensation of water vapor from the mixture. It is also possible to dehumidify by spraying cold water into the air stream. Equation 1-6 can be used to predict the latent heat transfer with a negative sign showing the transfer of energy from the air stream. The use of psychrometric processes for this case will be discussed in Chapter 3.

Cleaning

The cleaning of air usually implies filtering; additionally it may be necessary to remove contaminant gases from the air. Filtering is most often done by a process in which dirt particles are captured in a porous medium. Electrostatic cleaners are also used, especially to remove very small particles; in some cases water sprays may be used. Contaminant gases may be removed by absorption, by physical adsorption, and by other means. Air cleaning will be discussed in more detail in Chapter 4.

Air Motion

The motion of air in the vicinity of the occupant should be sufficiently strong to remove energy generated by the body but gentle enough to be unnoticed. The desired air motion is achieved by the proper placement of air inlets to the space and by the use of various air-distributing devices. The importance of air motion especially where occupant comfort is required cannot be underestimated.

These functions of the air-conditioning system may not be active all of the time. Residences and many commercial establishments have inactive cooling and dehumidifying sections during the winter months and keep the heating and humidifying sections inactive during the summer. In large commercial installations, however, it is not uncommon to have all of the functions under simultaneous control for the entire year. Obviously this requires elaborate controls and sensing devices. Precise control of the moisture content of the air is difficult, and humidifying and dehumidifying are usually not done during both winter and summer unless they are absolutely necessary for process control or product preservation, even though the heating and cooling functions may be active all of the time. Because cleaning of the air and good air circulation are always necessary, these functions are used continuously except for some periods when the space may not be occupied.

At appropriate places in the chapters that follow, the concepts discussed herein will be introduced and reviewed in a context useful in HVAC systems design. In the next chapter, HVAC systems are described and their general functions discussed. This will give an indication of how all the basic concepts enter the design problem.

REFERENCES

1. Willis Carrier, *Father of Air Conditioning,* Fetter Printing Company, Louisville, KY, 1991.
2. *Principles of Heating, Ventilating, and Air Conditioning,* American Society of Heating, Refrigerating and Air-Conditioning Engineers, Inc., Atlanta, GA, 1990.
3. *ASHRAE SI for HVAC and R,* 6th ed., American Society of Heating, Refrigerating and Air-Conditioning Engineers, Inc., New York, 1986.
4. W. F. Stoecker, *Using SI Units in Heating, Air-Conditioning, and Refrigeration,* Business News Publishing Co., Birmingham, MI, 1975.
5. *Metric Practice Guide E380-72,* American Society for Testing and Materials, Philadelphia, 1972 (also ANSI Standard Z210.1).

PROBLEMS

1-1. Write the following quantities in the suggested SI terminology.
  (a) 11,000 newton per square meter   (e) 12,000 kilowatt
  (b) 12,000 watts   (f) 0.012 meter
  (c) 1600 joule per kilogram   (g) 0.0000012 second
  (d) 101,101 pascal

1-2. Convert the following quantities from English to SI units.
  (a) 98 Btu/(hr-ft-F)   (d) 1050 Btu/lbm
  (b) 0.24 Btu/(lbm-F)   (e) 1.0 ton (cooling)
  (c) 0.04 lbm/(ft-hr)   (f) 14.7 lbf/in.^{2}

1-3. Convert the following quantities from SI to English units.
  (a) 120 kPa   (d) 10^{-6} (N-s)/m^2
  (b) 100 W/(m-C)   (e) 1200 kW
  (c) 0.8 W/(m^2-C)   (f) 1000 kJ/kg

1-4. The kinetic energy of a flowing fluid is proportional to the velocity squared divided by two. Compute the kinetic energy per unit mass for the following velocities in the units indicated.
  (a) velocity of 100 ft/sec; English units
  (b) velocity of 60 m/s; SI units
  (c) velocity of 400 ft/sec; SI units
  (d) velocity of 500 m/s; English units

1-5. The potential energy of a fluid is proportional to the elevation of the fluid. Compute the potential energy for the elevations given below per unit mass.
  (a) elevation of 200 ft; English units
  (b) elevation of 70 m; SI units
  (c) elevation of 120 ft; SI units
  (d) elevation of 38 m; English units

1-6. A gas is contained in a vertical cylinder with a frictionless piston. Compute the pressure in the cylinder for the following cases.

(a) piston mass of 20 lbm and area of 7 in.^{2}

(b) piston mass of 10 kg and diameter of 100 mm

1-7. A pump develops a total head of 50 ft of water under a given operating condition. What pressure is the pump developing in SI units and terminology?

1-8. A fan is observed to operate with a pressure difference of 4 in. of water. What is the pressure difference in SI units and terminology?

1-9. Compute the Reynolds number (Re $= \rho \bar{V} D/\mu$) for 21 C water flowing in a standard 2-in. pipe at a velocity of 1 m/s using SI units. Tables B-3b and D-1 will be useful. What is the Reynolds number in English units?

1-10. Compute the thermal diffusivity ($\alpha = k/\rho c_p$) of the following substances in SI units.

(a) saturated liquid water at 38 C

(b) air at 47 C and 101.3 kPa

(c) saturated liquid refrigerant 22 at 38 C

(d) saturated liquid refrigerant 12 at 38 C

1-11. Compute the Prandtl number (Pr $= \mu c_p/k$) for the following substances in SI units.

(a) saturated liquid water at 10 C

(b) saturated liquid water at 60 C

(c) air at 20 C

(d) saturated liquid refrigerant 22 at 38 C

1-12. Compute the heat transferred from water as it flows through a heat exchanger at a steady rate of 1 m^3/s. The decrease in temperature of the water is 5 C and the mean bulk temperature is 60 C. Use SI units.

1-13. Make the following volume and mass flow rate calculations in SI units. (a) Water flowing at an average velocity of 2 m/s in nominal $2\frac{1}{2}$-in., type L copper tubing. (b) Standard air flowing at an average velocity of 4 m/s in a 0.3 m diameter duct.

1-14. A room with dimensions of 3 × 10 × 20 m is estimated to have outdoor air brought in at an infiltration rate of $\frac{1}{4}$ volume change per hour. Determine the infiltration rate in m^3/s.

1-15. Air enters a heat exchanger at a rate of 5000 cubic feet per minute at a temperature of 50 F and pressure of 14.7 psia. The air is heated by hot water flowing in the same exchanger at a rate of 11,200 pounds per hour with a decrease in temperature of 10 F. At what temperature does the air leave the heat exchanger?

1-16. Water flowing at a rate of 1.5 kg/s through a heat exchanger heats air from 20 C to 30 C flowing at a rate 2.4 m^3/s. The water enters at a temperature of 90 C and the air is at 0.1 MPa. At what temperature does the water leave the exchanger?

1-17. Air at a mean temperature of 50 F flows over a thin-wall 1-in. O.D. tube, 10 feet in length, which has condensing water vapor flowing inside at a pressure of 14.7 psia. Compute the heat transfer rate if the average heat transfer coefficient between the air and tube surface is 10 Btu/(hr-ft^2-F).

1-18. Repeat Problem 1-17 for air at 10 C, a tube with diameter 25 mm, a stream pressure of 101 kPa, and a tube length of 4 m, and find the heat transfer coefficient in SI units if the heat transfer rate is 1250 W.

1-19. Air at 1 atm and 76 F is flowing at the rate of 5000 cfm. At what rate must energy be removed, in Btu/hr, to change the temperature to 58 F assuming that no dehumidification occurs?

1-20. Air flowing at the rate of 1000 cfm and with a temperature of 80 F is mixed with 600 cfm of air at 50 F. Use Eq. 1-5 to estimate the final temperature of the mixed air. Assume $c_p = 0.24$ Btu/(lbm-F) for both streams.

Chapter 2

Air-Conditioning Systems

Air-conditioning systems generally have common basic elements; however, they may differ dramatically in physical appearance and arrangement. Even with the same elements present, the manner in which systems are controlled and operated may also be quite different. This chapter discusses some of the more common basic elements and the types of systems that are used to meet the requirements of different building types and uses, load variations, and economic considerations. Some basic elements of control theory are also introduced along with examples of how several types of HVAC systems might be controlled

The earliest systems for air conditioning from centrally located equipment supplied tempered air through ducts for heating and ventilating. The addition of cooling and dehumidification equipment permitted year-round comfort in spaces where the heat gains and losses were relatively uniform throughout the conditioned area. Because this was often not the case, the conditioned area was divided into zones with separate requirements. This led to the need for supplementing the central system with additional equipment and more sophisticated controls. More recently, emphasis on indoor air quality and on energy conservation and economics have influenced selection, design, and control of HVAC systems.

Air-conditioning systems are categorized according to the means by which the heating and cooling is controlled and by their special equipment arrangement to accomplish specific purposes (1).

2-1 THE COMPLETE SYSTEM

Commercial air-conditioning systems all have about the same general arrangement of components, with variations to accommodate local requirements. Figure 2-1 is a schematic showing the major elements of such a system. The air-conditioning and distribution system, shown in the upper right portion of Fig. 2-1, may be of several types and will be discussed later. However, this part of the overall system will generally have means to heat, cool, humidify, dehumidify, clean, and distribute air to the various conditioned spaces in a zone. Note also that the system has means to admit outdoor air and to exhaust air as well as filter the mixed air.

A cooling fluid must be supplied to the cooling coil (heat exchanger) in the air handler. The fluid may be a liquid or a volatile substance, commonly called a

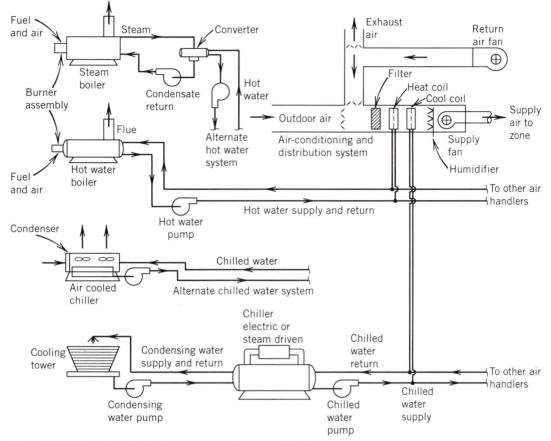

Figure 2-1 Schematic of a typical commercial air-conditioning system.

refrigerant. However, a liquid is generally used in commerical applications. Exceptions are discussed later. As shown in Fig. 2-1, the liquid is usually cooled with devices referred to as *chillers*. The chillers are refrigerating machines operating on the vapor compression or absorption cycles (Chapter 15). Heat is rejected to the atmosphere by use of a *cooling tower,* or the chiller may have an air-cooled condenser as shown. Pumps are required to circulate the liquid through piping, and the liquid cooling equipment is remote from the air-conditioning equipment.

To provide heat, a heating fluid must be supplied to the heating coil in the air handler. The fluid is usually hot water, but steam can also be used. Hot water or steam is provided by a boiler at some remote location and circulated by pump and piping. Water may be heated using steam with a heat exchanger, called a *converter*. The fuel for the boilers may be natural gas, liquified petroleum gas (LPG), or fuel oil.

The humidifier is supplied with water vapor or an atomized water spray. Water vapor is most desirable and can be supplied by a steam boiler or a small special steam-generating device.

The components of a complete system will be further discussed below.

2-2 THE AIR-CONDITIONING AND DISTRIBUTION SYSTEM

Important first choices in HVAC central system design involve determination of the individual zones to be conditioned and the type and location of the HVAC equipment. Normally, the equipment is located outside the conditioned area in a basement, on the roof, or in a service area at the core of a commercial building.

A *zone* is a conditioned space under the control of a single thermostat. The thermostat is a control device that senses the space temperature and sends a correcting signal if that temperature is not within some desired range. In some special cases the zone humidity may also be controlled by a humidistat. It is most important that the temperature within the area conditioned by a central system be uniform if a single-zone, constant-air-volume duct system is to be used because air temperature is sensed only at that single location where the thermostat is located. Because conditions vary some in most typical zones, it is important that the thermostat be located carefully, in a spot free from local disturbances and where the temperature is likely to be most nearly the average of the occupied space.

Uniform loads are generally experienced in spaces with relatively large open areas and small external loads such as theaters, auditoriums, department stores, and public areas of most buildings. In large commerical buildings the interior zones are usually fairly uniform if provisions are made to take care of local heat sources such as large equipment or computers. Variations of temperature within a zone can be reduced by adjusting local air flows or changing supply air temperatures.

Spaces with stringent requirements for cleanliness, humidity, temperature control, and/or air distribution are usually isolated as separate zones within the larger building and served with separate systems and furnished with precision controls.

For applications requiring close aseptic or contamination control of the environment, such as surgical operating rooms, all-air-type systems generally are used to provide adequate dilution of the controlled space. These applications usually involve careful control of the diluting air through the controlled environment space.

In spaces such as large office buildings, factories, and large department stores, practical considerations require not only multiple zones but also multiple installation of central systems. In the case of tall buildings, each central system may serve several floors.

Large installations such as college campuses, military bases, and research facilities may best be served by a *central station* or *central plants,* where chillers and boilers provide chilled water and hot water or steam through a piping system to the entire facility, often through underground piping. For these large installations central plants provide higher diversity, greater efficiency, less maintenance cost, and lower labor costs compared to individual central facilities in each building.

The choices described above are usually controlled by economic factors, involving a trade-off between first costs and operating costs for the installation. As the distance over which energy must be transported is increased, the cost of moving that energy tends to become more significant when compared to the costs of operating the chillers and boilers. As a general rule the smaller systems tend to be the most economical if they move the energy as directly as possible. For example, in a small heating system the air would most likely be heated directly in a furnace and transported through ducts to the controlled space. Likewise in the smaller units the refrigerating system would likely involve a direct exchange between the refrigerant and the supply air (a D-X system). In installations were the energy must be moved over greater distances, a liquid (or steam) transport system would likely be used, but this involves extra heat exchange steps. Most commercial systems, such as the one shown in Fig. 2-1, utilize a dual or combination method of transporting energy. In such systems energy is carried from the boiler (or converter) to the air handler heating coil by a liquid, usually water. The energy is then carried to the conditioned space by ducted air. In cooling, energy is carried by return air from the conditioned space to the air handler cooling coil and then transported to the chiller evaporator by a liquid. This energy is rejected by the refrigerator condenser to the ambient air. Where a cooling tower is utilized, energy is carried from the chiller condenser to the cooling tower by a liquid.

2-3   CENTRAL MECHANICAL EQUIPMENT

Once the user's needs have been appraised, zones defined, and loads and air requirements calculated and the type of overall system has been determined, the designer can start the process of selection and arrangement of the various system components. It is important that the equipment be suited for the particular application, sized properly, accessible for easy maintenance, and no more complex in arrangement and control than is necessary to produce the conditions required to meet the design criteria. The economic trade-off between initial investment and operating costs must always be kept in mind.

Consideration of the type of fuel or energy source must be made at the same time as the selection of the energy-consuming equipment to assure the least lifecycle cost for the owner. Chapter 15 of the *ASHRAE Handbook* (2) gives the types and properties of fuels and energy sources and guidance in their proper use. This selection is not only important from an economic standpoint but is also important in making the best use of our natural resources.

Ductwork and piping make up a significant part of the design of an HVAC system. These will be described in detail in later chapters. The remaining components can be generally grouped into five categories:

Air handlers and fans
Heating sources
Refrigeration
Pumps
Controls and instrumentation

Familarity with some of the components of HVAC systems will make the design and analysis material that follows more meaningful as well as more interesting. A quick look at some of the components in the categories above will now be given. This will be followed by descriptions of some common arrangements of these components in modern HVAC systems.

Air-Handling Equipment

The general arrangement of a commercial central system for year-round air-conditioning was given in Fig. 2-1. Most of the components are available in subassembled sections ready for bolting together in the field or completely assembled by the manufacturer. In the upper-right-hand corner of Fig. 2-1 can be seen the simplified schematic of an air handler, showing the fans, heating and cooling coils, filter, humidifier, and controlling dampers. The fan is located downstream of the coils and is referred to as a *draw-through configuration*. A photograph of an actual air handler of this type is shown in Fig. 2-2.

In cases where several zones are to be serviced by a single air handler, the heating and

Figure 2-2 Single-zone, draw-through air handler. (Reproduced by permission of the Trane Company, LaCrosse, WI)

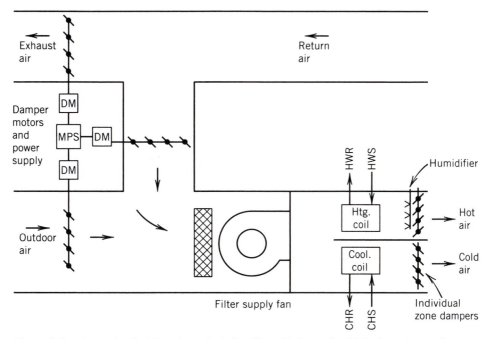

Figure 2-3 Schematic of a blow-through air handler with hot and cold decks and zone dampers.

cooling coils may be placed in a side-by-side or parallel arrangement. A schematic of such an arrangement is shown in Fig. 2-3. The heating and cooling coils are referred to as the *hot deck* and the *cold deck,* respectively. The arrangement with the fan upstream of the coils is called a *blow-through configuration*. Figure 2-4 shows a photograph of a multizone, blow-through configuration. The discharge area of the air handler may be divided so that several zones may be served with separate temperature control in each zone, or the air handler may be used without the dampers in a dual-duct system. Both of these arrangements will be discussed later in this chapter.

Internal components of the central air handler include the cooling, heating, and, for some units, the preheat coils. These are usually of the finned-tube type such as are shown in Fig. 14-10. Coil design and selection will be considered in Chapters 3 and 14. The humidifier in a commercial air handler is usually a type utilizing steam such as is shown in Fig. 2-5. The fans are usually centrifugal types, shown in Fig. 2-6. These and other fan types will be looked at in more detail in Chapter 12. A unit-type filter is shown in Fig. 2-7; the various filter types and design procedures will be considered in Chapter 4. Dampers, which can be seen in Fig. 2-4, will be discussed in Chapter 12.

The ductwork to deliver air is usually a unique design to fit a particular building. The air ducts should deliver conditioned air to an area as quietly and economically as possible. In some installations the air delivery system consumes a significant part of the total energy, making good duct design and fan selection a very important part of the engineering process. Design of the duct system must be coordinated with the building design to avoid last-minute changes. Chapter 12 explains this part of the system design.

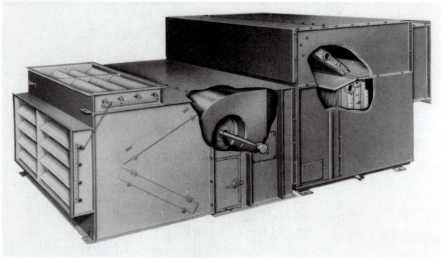

Figure 2-4 A multizone, blow-through air handler showing the coils, fan, filters, and mixing box. (Reproduced by permission of the Trane Company, LaCrosse, WI)

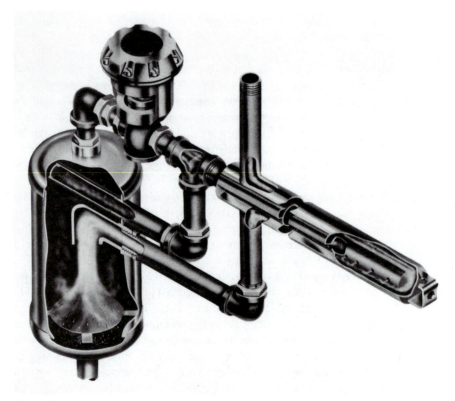

Figure 2-5 A commercial-type Steam Humidifier. (Courtesy of Spirax Sarco, Inc.)

Figure 2-6 A centrifugal fan. (Reproduced by permission of the Trane Company, LaCrosse, WI)

Figure 2-7 A unit-type air filter. (Courtesy of the Farr Company, Los Angeles, CA)

Heating Equipment

Steam and hot water boilers for heating are manufactured for high or low pressure, using coal, oil, electricity, gas, or waste material for fuel.

Low-pressure boilers are rated for a working pressure of 15 psig or 103 kPa for steam, 160 psig or 1.1 MPa for water, with a maximum temperature limitation of 250 F or 121 C. High-rise buildings usually use high-pressure boilers. However, a low-pressure steam boiler may be used with a heat exchanger, called a converter, to heat water, which is then pumped to the various floors in a high-rise building. This eliminates the need for an expensive high-pressure boiler. Packaged boilers with all components and controls assembled as a unit are available. Figure 2-8 shows such a unit.

Heating of the space may be accomplished by heating and circulating air with the air handler as shown in Fig. 2-1, or a radiant system may be used where a single enclosed finned tube is installed along the outer wall near the floor. This will be discussed later.

Refrigeration Equipment

Understanding the basics of the refrigeration system is important in selection of that and other components of the HVAC system and in the proper design of controls to assure optimum performance of the overall system. The subject of refrigeration is often presented as one or more courses separate from and in addition to the subject of HVAC analysis and design. This text presents a basic discussion of refrigeration and refriger-

Figure 2-8 Packaged fire-tube hot water boiler. (Courtesy of Federal Corporation, Oklahoma City, OK)

ants in Chapter 15. Additional information can be found in the *ASHRAE Handbook, Refrigeration* (3).

The basic components of the typical commercial HVAC refrigeration system are the compressor, the condenser, the evaporator, the expansion valve, and the control system. In many cases the cooling tower is also utilized as a means of removing the heat rejected by the condenser (see Fig. 2-1).

The compressor is the energy-consuming component of the refrigeration system, and its performance and reliability are significant to the overall performance of the HVAC system.

Three major types of compressors are used in systems:

1. Reciprocating: $\frac{1}{16}$ to 150 hp or 50 W to 112 kW.
2. Helical rotary: 100 to 1000 tons or 350 to 3500 kW.
3. Centrifugal: 100 tons or 350 kW to an upper limit of capacity determined only by physical size.

Compressors come with many types of drives including electric, gas, and diesel engines, and gas and steam turbines. Many compressors are purchased as part of a condensing unit; they consist of compressor, drive, condenser, and all necessary safety and operating controls. Reciprocating condensing units are available with an air-cooled condenser as part of the unit or arranged for remote installation. A commericial-size, air-cooled condensing unit is shown in Fig. 2-9. Compressors and refrigeration equipment are discussed fully in Chapter 15.

As the name implies, chillers cool water or other fluid that is circulated to a remote

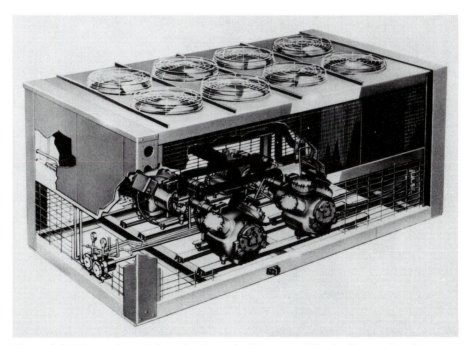

Figure 2-9 A large-air-cooled condensing unit. (Courtesy of Carrier Corporation, Syracuse, NY)

location where it is used to cool air with a cooling coil in an air handler (see Fig. 2-1). Chillers operating on the vapor compression cycle take many forms, ranging in size from about three tons to more than a thousand tons. Smaller units usually use reciprocating compressors with air-cooled condensers, whereas large units use centrifugal compressors and reject heat to water from cooling towers. Figure 2-10 shows a large centrifugal chiller.

Absorption chillers are also used. These are available in large units from 50 to 1500 tons or 176 kW to 5 MW capacity. Their generator sections are heated with low-pressure steam, hot water, or other hot liquids. In large installations the absorption chiller is frequently combined with centrifugal compressors driven by steam turbines. Steam from the noncondensing turbine is taken to the generator of the absorption machine. When the centrifugal unit is driven by a gas turbine or an engine, the absorption machine generator is heated with the exhaust-jacket water. Because of rising energy costs many absorption chillers that use primary energy have been replaced with more energy-efficient reciprocating or centrifugal equipment.

To remove the heat from the water-cooled condensers of air-conditioning systems, the water is usually cooled by contact with the atmosphere. This is accomplished by natural draft or mechanical draft cooling towers or by spray ponds. Of these, the mechanical draft tower can be designed for more exacting conditions because it is independent of the wind.

Figure 2-10 A large centrifugal chiller. (Reproduced by permission of the Trane Company, LaCrosse, WI)

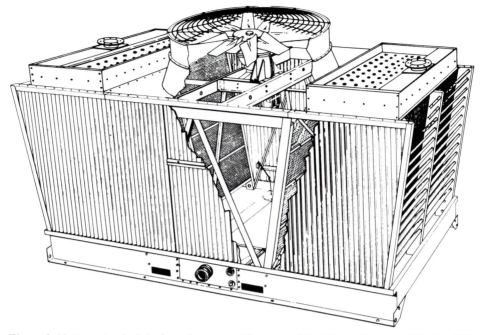

Figure 2-11 A mechanical-draft cooling tower. (Courtesy of the Marley Company, Mission, KS)

Air-conditioning systems use towers ranging from small package towers of 5 to 500 tons or 17.5 to 1760 kW or intermediate-size towers of 2000 to 4000 tons or 7 to 14 MW. A mechanical draft cooling tower is shown in Fig. 2-11. Water treatment is a definite requirement for satisfactory operation. Units that operate during the entire year must be protected against freezing.

Pumps and Piping

The pumps used in air-conditioning systems are usually single-inlet centrifugal pumps. The pumps for the larger system and for heavy duty have a horizontal split case with double-suction impeller for easier maintenance and high efficiency. End-suction pumps, either close coupled or flexible connected, are used for smaller tasks. Figure 2-12a shows a medium-sized direct-coupled centrifugal pump, and Fig. 2-12b shows an inline centrifugal pump. The major applications for pumps in the equipment room are primary and secondary chilled water, hot water, condenser water, steam condensate return, boiler feed water, and fuel oil.

Air-conditioning pipe systems can be divided into two parts, the piping in the main equipment room and the piping required for the air-handling systems throughout the building. The air-handling system piping follows procedures developed in detail in Chapter 10. The major piping in the main equipment room consists of fuel lines, refrigerant piping, steam, and water connections.

Figure 2-12*a* A single-inlet direct-coupled centrifugal pump. (Courtesy of Pacific Pump Company, Oakland, CA)

Controls and Instrumentation

Because the loads in the various zones of a building will vary with time, there must be controls to match the output of the HVAC system to the loads. An HVAC system is designed to meet the extremes in the demand, but most of the time it will be operating at part load conditions. A properly designed control system will maintain good indoor air quality and comfort under all anticipated conditions with the lowest possible cost of operation.

Controls may be energized in a variety of ways: pneumatic, electric, electronic, or they may even be self-contained, where no external power is required. Some HVAC systems have combination systems, for example, pneumatic and electronic. The trend in recent times is more and more toward the use of electronic controllers, and interest is especially high regarding digital control, sometimes called direct digital control or DDC. Developments in both analog and digital electronics and in computers have allowed control systems to become much more sophisticated and permit an almost limitless variety of control sequences within the physical capability of the HVAC equipment. Along with better control has come additional monitoring capability, and

Figure 2-12*b* A small in-line centrifugal pump. (Courtesy of Taco, Incorporated, Cranston, RI)

energy management systems (EMS) have become quite common. This has permitted a better determination of unsafe operating conditions and better control of the spread of contamination or fire. By minimizing human intervention in the operation of the system, the possibility of human error has been reduced. More detailed information on HVAC controls can be found in the automatic control chapter of *ASHRAE Handbook, HVAC Applications* (4) and in Haines (5). Some common control methods and systems will be discussed in later sections of this text. A brief review of control fundamentals might be helpful before proceeding further.

All control systems, including even the simplest ones, have three necessary elements: sensor, controller, and controlled device. Consider the control of the air temperature downstream of a heating coil, as in Fig. 2-13. The position of the control valve determines the rate at which hot water circulates through the heating coil. As hot water passes through the coil at a higher rate, the air (presumed to be flowing at a constant rate) will be heated to a higher temperature. A temperature sensor is located at a position downstream of the coil so as to measure the temperature of the air leaving the coil. The temperature sensor sends a signal (voltage, current, or resistance) to the controller that corresponds to the sensor's temperature. The controller has been given a *set point* equal to the desired downstream air temperature and compares the signal from the sensor with the set point temperature. If the temperature described by the signal from the sensor is greater than the set point temperature, the controller will send a signal to partially close the control valve. This is a *closed loop* system because the change in the controlled device (the control valve) results in a change in the downstream air temperature (the controlled variable), which in turn is detected by the sensor. The process by which the

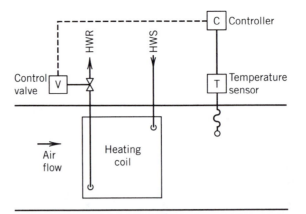

Figure 2-13 Elementary air-temperature control system.

change in output is sensed is called *feedback*. In an *open loop* system the sensor is not directly affected by the action of the controlled device. An example of an open loop system is the sensing of outdoor temperature to set the water temperature in a heating loop. In this case adjustment of the water temperature has no effect on the outdoor temperature sensor.

Different types of *control actions* are available to meet the need for a variety of kinds of control response. Most may be classified into one of the following:

Two-position or on–off action
Timed two-position action
Floating action
Modulating action

The two-position or on–off action is the simplest and is common in all types of equipment. An example is an electric heater turned on and off by a thermostat, or a pump turned on and off by a pressure switch. To prevent rapid cycling when this type action is used, there must be a difference between the setting at which the controller operates to one position and the setting at which it changes to the other. Figure 2-14 illustrates how the controlled variable might change with time with two-position action. Note that there is a time lag in the response of the controlled variable resulting in the actual operating differential being greater than the set or control differential. This

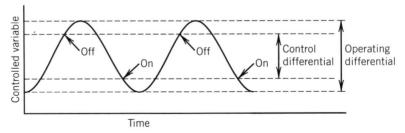

Figure 2-14 Two position (on-off) control action.

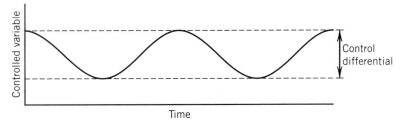

Figure 2-15 Floating control action.

difference can be reduced by artificially shortening *on* or *off* time in anticipation of the system response. For example, a thermostat in the heating mode may have a small internal heater activated during the *on* period, causing the *off* signal to occur sooner than it would otherwise. With this device installed the thermostat would be said to have an *anticipator* or *heat anticipation.*

Figure 2-15 illustrates the controlled variable behavior when the control action is floating. With this action the controlled device can stop at any point in its stroke and be reversed. The controller has a neutral range in which no signal is sent to the controlled device, which is allowed to *float* in a partially open position. The controlled variable must have a relatively rapid response to the controlling signal for this type of action to operate properly.

Modulating action is illustrated in Fig. 2-16. With this action the output of the controller can vary infinitely over its range. The controlled device will seek a position corresponding to its own range and the output of the controller. Figure 2-16 helps in the definition of three terms that are important in modulating control and that have not been previously defined. *Throttling range* is the amount of change in the controlled variable required to run the actuator of the controlled device from one end of its stroke to the other. Figure 2-17 shows the throttling range for a typical cooling system controlled by a thermostat; in this case it is the temperature at which the thermostat calls for maximum cooling minus the temperature at which the thermostat calls for minimum cooling. The actual value of the controlled variable is called the *control point.* The system is said to be in control if the control point is inside the throttling range, and out of control if the control point is outside that range. The difference between the set point and the control point is said to be the *offset* or *control point shift* (sometimes called drift, droop, or deviation). The action represented by the solid line in Fig. 2-17 is called direct action (DA) since an increase in temperature causes an increase in the heat extraction or

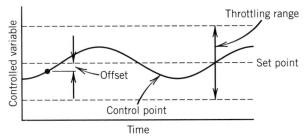

Figure 2-16 Modulating control action.

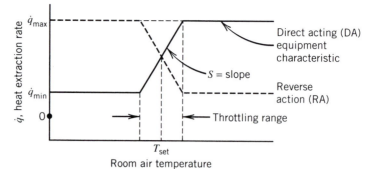

Figure 2-17 Typical equipment characteristic for thermostat control of room temperature.

cooling. The dashed line represents reverse action (RA), where an increase in temperature would cause a decrease in the controlled variable, for example, heat input.

The simplest modulating action is referred to as *proportional control,* the name sometimes used to describe the modulating control system. This is the control action used in most pneumatic and older electrical HVAC control systems. The output of a proportional controller is equal to a constant plus the product of the error and the gain:

$$O = A + e K_p \qquad (2\text{-}1)$$

where:

O = controller output
A = the controller output with no error, a constant
e = the error, equal to the set point minus the measured value of the controlled variable
K_p = proportional gain constant

The gain is usually an adjustable quantity, set to give a desired response. High gain makes the system more responsive but may make the system unstable. Lowering the gain decreases responsiveness but makes the system more stable. The gain of the control system shown in Fig. 2-17 is given by the slope of the equipment characteristic line S in the throttling range. For this case the units of gain would be those of heat rate per degree, for example Btu/hr-F or W/C.

Figures 2-18a and 2-18b illustrate stability. In Fig. 2-18a the controlled variable is shown with maximum error at time zero and a response that brings the control point quickly to a stable value with a small offset. Figure 2-18b illustrates an unstable system, where the control point continues to oscillate about the set point, never settling down to a constant, low-offset value as with the stable system.

There will always be some offset with proportional control, the magnitude for a given HVAC system increasing with decreases in the control system gain and the load. System performance, comfort, and energy consumption are affected, usually adversely, by this offset. Offset can be eliminated by the use of a refinement to proportional

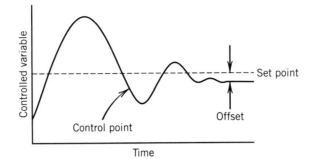

Figure 2-18*a* A stable system under proportional control.

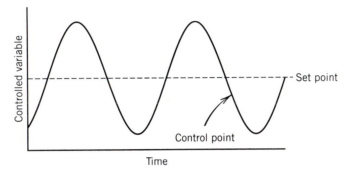

Figure 2-18*b* An unstable system under proportional control.

control, referred to as *proportional plus integral* (PI) *control*. The controller is designed to behave in the following manner:

$$O = A + e\,K_p + K_i \int e\,dt \tag{2-2}$$

where K_i = integral gain constant.

In this mode the output of the controller is additionally affected by the error integrated over time. This means that the error or offset will eventually be reduced for all practical purposes to zero. The integral gain constant K_i is equal to x/t, where x is the number of samples of the measured variable taken per unit time, somtimes called the *reset rate*. In much of the HVAC industry, PI control has been referred to as *proportional with reset,* but the correct term *proportional plus integral* is becoming more widely used. Most electronic controllers and many pneumatic controllers use PI, and computers can be easily programmed for this mode.

An additional correction involving the derivative of the error is used in the proportional plus integral derivative (PID) mode. This increases the rate of correction as the error increases, giving rapid response where needed. Most HVAC systems are relatively slow in response to changes in controller output, and PID systems may overcontrol.

Although many electronic controllers are available with PID mode, the extra derivative feature is usually not helpful to good HVAC control.

The control of particular types of HVAC systems will be described throughout this text. To simplify the diagrams and discussions, a set of typical symbols and abbreviations will be used (Appendix F). These symbols and abbreviations are the same as used by Haines (5) and generally are easily interpreted.

System monitoring is closely related to system control, and it is important to provide adequate instrumentation for this purpose. At the time of installation all equipment should be provided with adequate gages, thermometers, flow meters, and balancing devices so that system performance is properly established. In addition, capped thermometer wells, gage cocks, capped duct openings, and volume dampers should be provided at strategic points for system balancing. A central control center to monitor and control a large number of control points should be considered for any large and complex air-conditioning system. Fire detection and security systems are often integrated with HVAC monitoring and control system.

2-4   ALL-AIR SYSTEMS

An all-air system provides complete sensible heating and humidification and sensible and latent cooling by supplying air to the conditioned space. In such systems there may be piping connecting the chillers and the heating devices to the air-handling device. No additional cooling is required at the zone. Figure 2-1 is an example of an all-air system.

The all-air system may be adapted to all types of air-conditioning systems for comfort or process work. It is applied in buildings requiring individual control of conditions and having a multiplicity of zones such as office buildings, schools and universities, laboratories, hospitals, stores, hotels, and ships. Air systems are also used for any special applications where a need exists for close control of temperature and humidity, including clean rooms, computer rooms, hospital operating rooms, textile and tobacco factories.

Heating may be accomplished by the same duct system used for cooling, by a separate perimeter air system, or by a separate perimeter baseboard system using hot water, steam, or electric-resistance heat. Many commercial buildings need no heating in interior spaces but only around the perimeter. It is the function of the perimeter heating system to offset the heat losses at the exterior envelope of the building. During those times when heat is required only in perimeter zones served by baseboard systems, the air system provides the necessary ventilation but not cooling. The most common application of perimeter heating is with variable-air-volume (VAV) systems, which will be described later in this section.

Single-zone System

The simplest all-air system is a supply unit (air handler) serving a single zone. The unit can be installed either within a zone or remote from the space it serves and may operate with or without ductwork. A single-zone system responds to only one set of space conditions. Thus it is limited in application to where reasonably uniform temperatures can be maintained throughout the zone. Figure 2-19 shows a schematic of the air handler and associated dampers and controls for a single-zone constant-volume all-air system. In this particular system the room thermostat maintains the desired temperature

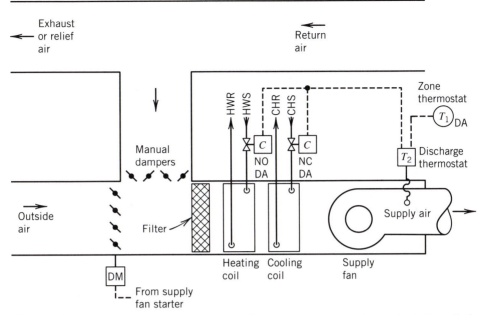

Figure 2-19 Air handler and associated controls for a simple constant-volume, single-duct all-air system.

in the zone by control of the temperature of the air being supplied to the zone. The discharge thermostat takes a signal from the zone thermostat and opens or closes the appropriate valve on the heating or cooling coil to maintain the desired room temperature. Because the heating valve is normally open (NO) and direct acting and the zone thermostat is direct acting, an increase in room temperature will cause the hot water valve to close to a lower flow condition. The cold water valve will be closed as long as there is a call for heat. When cooling is required the hot water valve will be closed and the cooling water valve will respond in the proper direction to the thermostat. The discharge thermostat could be eliminated from the circuit and the zone thermostat control the valves directly, but response would be slower.

It this case, where the air delivered by the fan is constant, the rate of outside air intake is determined by the setting of the dampers. The outside dampers have a motor to drive them from a closed position when the fan is off to the desired full open position with the fan running. In this case the dampers in the recirculated air stream are manually adjustable. They are often set to operate in tandem with the outside air dampers and with the exhaust or relief dampers should they be present.

Reheat Systems

The reheat system is a modification of the single-zone constant-volume system. Its purpose is to permit zone or space control for areas of unequal loading, or to provide heating or cooling of perimeter areas with different exposures, or for process or comfort applications where close control of space conditions is desired. As the word *reheat* implies, the application of heat is a secondary process, being applied to either precondi-

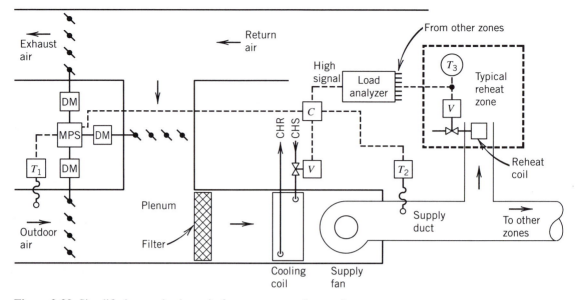

Figure 2-20 Simplified control schematic for a constant-volume reheat system.

tioned primary air or recirculated room air. A single low-pressure reheat system is produced when a heating coil is inserted in the duct system. The more sophisticated systems utilize higher pressure duct designs and pressure-reduction devices to permit system balancing at the reheat zone. The medium for heating may be hot water, steam, or electricity.

Conditioned air is supplied from a central unit at a fixed cold air temperature designed to offset the maximum cooling load in the space. The control thermostat activates the reheat unit when the temperature falls below the upper limit of the controlling instrument's setting. A schematic arrangement of the components for a typical reheat system is shown in Fig. 2-20. To conserve energy reheat should not be used unless absolutely necessary. At the very least, reset control should be provided to maintain the cold air at the highest possible temperature to satisfy the space cooling requirement.

Figure 2-20 also shows an *economizer* arrangement where outdoor air is used to provide cooling when outdoor temperatures are sufficiently low. Sensor T_1 determines the damper positions and thus the outdoor air intake. The outdoor damper must always be open sufficiently to provide the minimum outdoor air required for maintaining good indoor air quality. Since humidity may be a problem, many designers provide a humidistat on the outdoor air intake to assure that air is not used for cooling when outdoor humidities are too high.

Variable-volume System

The variable-volume system compensates for varying load by regulating the volume of air supplied through a single duct. Special zoning is not required because each space supplied by a controlled outlet is a separate zone. Figure 2-21 is a schematic of a single-duct variable-air-volume (VAV) system.

Significant advantages of the variable-volume system are low initial cost and low

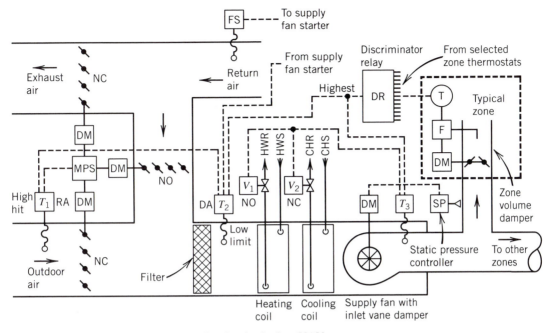

Figure 2-21 Simplified control schematic of a single-duct VAV system.

operating costs. The first cost of the system is far lower in comparison with other systems that provide individual space control because it requires only single runs of duct and a simple control at the air terminal. Where diversity of loading occurs, smaller equipment can be used and operating costs are generally the lowest among all the air systems. Because the volume of air is reduced with a reduction in load, the refrigeration and fan horsepower follow closely the actual air-conditioning load of the building. During intermediate and cold seasons, outdoor air can be used for economy in cooling. In addition, the system is virtually self-balancing.

Until recently there were two reasons why variable-volume systems were not recommended for applications with loads varying more than 20 percent. First, throttling of conventional outlets down to 50 or 60 percent of their maximum design volume flow might result in the loss of control of room air motion with noticeable drafts resulting. Second, the use of mechanical throttling dampers produces noise, which increases proportionally with the amount of throttling.

Improvements in volume-throttling devices and aerodynamically designed outlets have helped overcome these problems and extended the potential application of variable-volume systems. Present systems can now handle building perimeter areas where load variations are greatest, and where throttling to 10 percent of design volume flow is often necessary.

Although some heating may be done with a variable-volume system, it is primarily a cooling system and should be applied only where cooling is required the major part of the year. Buildings with internal spaces with large internal loads are the best candidates. A secondary heating system should be provided for boundary surfaces during the heating season. Baseboard perimeter heat is often used. During the heating season, the VAV system simply provides tempered ventilation air to the exterior spaces.

Reheat may be used in conjunction with the VAV system. In this case the air is throttled to some predetermined ratio and then reheat takes over to temper the air.

An important aspect of VAV system design is fan control. There are significant fan power savings where fan speed is reduced in relation to the volume of air being circulated. This topic is discussed in detail in Chapter 12.

Single-duct variable-volume systems should be considered in applications where full advantage can be taken of their low cost of installation and operation. Applications exist for office buildings, hotels, hospitals, apartments, and schools. Further details of the use of VAV systems may be obtained from reference 1.

Dual- or Double-duct System

In the dual-duct system the central station equipment supplies warm air through one duct run and cold air through the other. The temperature in an individual space is controlled by mixing the warm and cool air in proper proportions. Variations of the dual-duct system are possible, with one form shown in Fig. 2-22.

For best performance some form of constant volume regulation should be incorporated into the system to maintain a constant flow of air. Without this the system is difficult to control because of the wide variations in system static pressure that occur from the normal demand from loading changes.

Many double-duct systems are installed in office buildings, hotels, hospitals, schools,

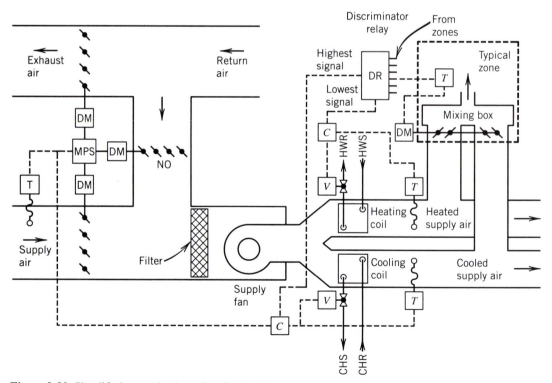

Figure 2-22 Simplified control schematic of a dual-duct system.

and large laboratories. A common characteristic of these multiroom buildings is their highly variable sensible heat load. This system provides great flexibility in satisfying multiple loads and in providing prompt and opposite temperature response as required.

Space or zone thermostats may be set once to control year-round temperature conditions. All outdoor air can be used when the outdoor temperature is low enough to handle the cooling load. A dual-duct system should be provided with control that will automatically reset the cold air supply to the highest temperature acceptable and the hot air supply to the lowest temperature acceptable.

Variable air volume may be incorporated into the dual-duct system, with various arrangements possible. Two supply fans are usually used in this case, one for the hot deck and one for the cold deck, with each controlled by the static pressure downstream in each duct.

From the energy conservation viewpoint, the dual-duct system has the same disadvantage as reheat. Although many of these systems are in operation, few are now being designed and installed.

Multizone System

The multizone central station units provide a single supply duct for each zone and obtain zone control by mixing hot and cold air at the central unit in response to room or zone thermostats. For a comparable number of zones, this system provides greater flexibility than the single duct and involves lower cost than the dual-duct system, but it is physically limited by the number of zones that may be provided at each central unit.

Typical multizone equipment is similar in some respects to the dual-duct system, but the two air streams are proportioned within the equipment instead of being mixed at each space served, and air of the proper temperature is provided as it leaves the equipment. Figure 2-23 shows a sketch of a multizone system. The system conditions groups of rooms or zones by means of a blow-through central apparatus having heating and cooling coils in parallel downstream from the fan.

The multizone, blow-through system is applicable to locations and areas having high sensible heat loads and limited ventilation requirements. The use of many duct runs and control systems can make initial costs of this system high compared to other all-air systems. Also to obtain very fine control this system might require larger refrigeration and air-handling equipment, which should be considered in estimating both initial and operating costs.

The use of these systems with simultaneous heating and cooling is now discouraged for energy conservation. However, through the use of outdoor air and either heating or cooling, satisfactory control may be attained in many applications.

2-5   AIR-AND-WATER SYSTEMS

In the all-air systems discussed in the previous section, the spaces within the building are cooled solely by air supplied to them from the central air-conditioning equipment. In contrast, in an air-and-water system both air and water are distributed to each space to perform the cooling function. In virtually all air–water systems both cooling and heating functions are carried out by changing the air or water temperatures (or both) to permit control of space temperature during all seasons of the year.

There are several basic reasons for the use of this type of system. Because of the

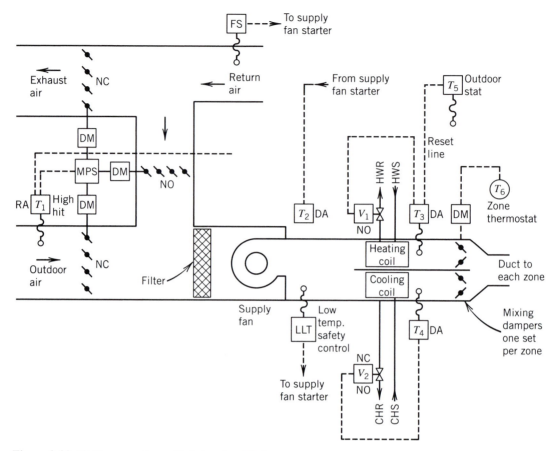

Figure 2-23 Multizone system with hot and cold plenum reset.

greater specific heat and much greater density of water compared to air, the cross-sectional area required for the distribution pipes is markedly less than that required for ductwork to accomplish the same cooling task. Consequently, the quantity of air supplied can be low compared to an all-air system, and less building space need be allocated for the cooling distribution system.

The reduced quantity of air is usually combined with a high-velocity method of air distribution to minimize the space required. If the system is designed so that the air supply is equal to the air needed to meet outside air requirements or that required to balance exhaust (including exfiltration) or both, the return air system can be eliminated for the areas conditioned in this manner.

The pumping horsepower necessary to circulate the water throughout the building is usually significantly less than the fan horsepower to deliver and return the air. Thus not only space but also operating cost savings can be realized.

Systems of this type have been commonly applied to office buildings, hospitals, hotels, schools, better apartment houses, research laboratories, and other buildings. Space saving has made these systems beneficial in high-rise structures.

The air side of air-and-water systems is comprised of central air-conditioning equipment, a duct distribution system, and a room terminal. The air is supplied at constant volume and is often referred to as primary air to distinguish it from room air that is recirculated over the room coil.

The water side in its basic form consists of a pump and piping to convey water to the heat transfer surface within each conditioned space. The heat exchange surface may be a coil that is an integral part of the air terminal (as with induction units), a completely separate component within the conditioned space (radiant panel), or either of these (as is true of fan–coil units).

Individual room temperature control is obtained by varying the capacity of the coil (or coils) within the room by regulation of either the water flow through it or the air flow over it. The coil may be converted to heating service during the winter, or a second coil or a heating device within the space may provide heating capacity, depending on the type of system.

Air-and-water systems are categorized as two-pipe, three-pipe, and four-pipe systems. They are basically similar in function and all incorporate both cooling and heating capabilities for all-season air conditioning. However, arrangements of the secondary water circuits and control systems differ greatly.

Air–Water Induction System

The basic arrangement for air–water induction units is shown in Fig. 2-24. Centrally conditioned primary air is supplied to the unit plenum at high pressure. The plenum is acoustically treated to attenuate part of the noise generated in the duct system and in the unit. A balancing damper is used to adjust the primary air quantity within limits.

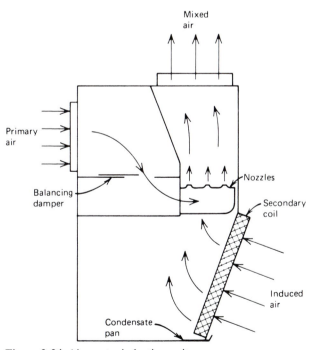

Figure 2-24 Air–water induction unit.

The high-pressure air flows through the induction nozzles and induces secondary air from the room and over the secondary coil. This secondary air is either heated or cooled at the coil depending on the season, the room requirement, or both. Ordinarily no latent cooling is accomplished at the room coil, but a drain pan is provided to collect condensed moisture resulting from unusual latent loads of short duration. The primary and secondary air are mixed and discharged to the room.

Induction units are usually installed at a perimeter wall under a window, but units designed for overhead installation are available. During the heating season the floor-mounted induction unit can function as a convector during off hours with hot water to the coil and without a primary air supply.

Fan–Coil Conditioner System

The fan–coil conditioner unit is a versatile room terminal that is applied to both air–water and water-only systems.

The basic elements of fan–coil units are a finned-tube coil and a fan section, as in Fig. 2-25. The fan section recirculates air continuously from within the perimeter space through the coil, which is supplied with either hot or chilled water. In addition, the unit may contain an auxiliary heating coil, which is usually of the electric resistance type but which can be of the steam or hot water type. Thus the recirculated room air is either heated or cooled. Primary air made up of outdoor air sufficient to maintain air quality is supplied by a separate central system usually discharged at ceiling level. The primary air is normally tempered to room temperature during the heating season, but is cooled and dehumidified in the cooling season. The primary air may be shut down during unoccupied periods to conserve energy.

2-6 ALL-WATER SYSTEMS

All-water systems are those with fan–coil, unit ventilator, or valance-type room terminals, with unconditioned ventilation air supplied by an opening through the wall or

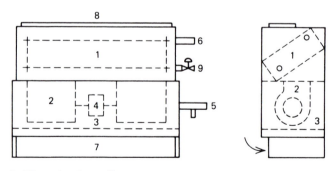

1. Finned tube coil
2. Fan scrolls
3. Filter
4. Fan motor
5. Auxiliary condensate pan
6. Coil connections
7. Return air opening
8. Discharge air opening
9. Water control valve

Figure 2-25 Typical fan–coil unit.

by infiltration. Cooling and dehumidification are provided by circulating chilled water or brine through a finned coil in the unit. Heating is provided by supplying hot water through the same or a separate coil using water distribution from central equipment. Electric heating or a separate steam coil may also be used. Humidification is not practical in all-water systems unless a separate package humidifier is provided in each room.

The greatest advantage of the all-water system is its flexibility for adaptation to many building module requirements.

A fan–coil system applied without provision for positive ventilation or one taking ventilation air through an aperture is one of the lowest first-cost central station–type perimeter systems in use today. It requires no ventilation air ducts, it is comparatively easy to install in existing structures, and, as with any central station perimeter system utilizing water in pipes instead of air ducts, its use results in considerable space savings throughout the building. However, it must be understood that this type system may not meet today's stringent indoor air quality (IAQ) requirements.

All-water systems have individual room control with quick response to thermostat settings and freedom from recirculation of air from other conditioned space. These systems have remote chilling and heating equipment. When fan–coil units are used, each is its own zone with a choice of heating or cooling at all times, and no seasonal changeover is required. All-water systems can be installed in existing buildings with a minimum of interference in the use of occupied space.

There is no positive ventilation unless openings to the outside are used, and then ventilation can be affected by wind pressures and stack action on the building. Special precautions are required at each unit with an outside air opening to prevent freezing of coil and potential water damage from rain. Because of these problems and energy conservation considerations, it is becoming standard practice to eliminate the outdoor air feature of these systems and rely on other means to provide outdoor air. This type of system is not recommended for applications requiring high indoor air quality.

Maintenance and service work has to be done in the occupied areas; as the units become older, the fan noise can become objectionable. Each unit requires a condensate drain line. It is difficult to limit bacterial growth in the unit. In extremely cold weather it is often necessary to close the outside air dampers to prevent freezing of coils, reducing ventilation air to that obtained by infiltration. Filters are small and inefficient and require frequent changing to maintain air volume.

Figure 2-26 illustrates a typical unit ventilator, used in all water systems, with two separate coils, one used for heating and the other for cooling with a four-pipe system. In some cases the unit ventilator may have only one coil, such as the fan–coil of Fig. 2-25.

The heating coil may use hot water, steam, or electricity. The cooling coil can be either a chilled water coil or a direct expansion refrigerant coil. Heating and cooling coils are sometimes combined in a single coil by providing separate tube circuits for each function. In such cases the effect is the same as having two separate coils.

Unit ventilator capacity control is essentially the same as described for fan coils in the previous section. In most cases, however, the control system will be slightly more sophisticated.

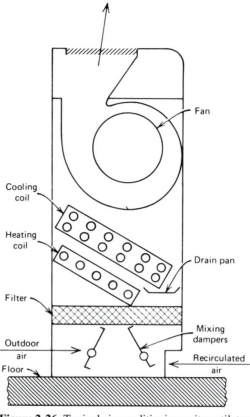

Figure 2-26 Typical air-conditioning unit ventilator
with separate coils.

2-7 UNITARY AIR CONDITIONERS

Unitary air-conditioning equipment consists of factory-matched components for inclusion in air-conditioning systems that are field designed to meet the needs of the user.

The following list of variations is indicative of the vast number of types of unitary air conditioners available.

1. Arrangement: single or split (evaporator connected in the field).
2. Heat rejection: air cooled, evaporative condenser, water cooled.
3. Unit exterior: decorative for in-space application, functional for equipment room and ducts, weatherproofed for outdoors.
4. Placement: floor standing, wall mounted, ceiling suspended, roof mounted.
5. Indoor air: vertical upflow, counterflow, horizontal, 90 and 180 degree turns, with fan, or for use with forced air furnace.
6. Locations: *indoor*—exposed with plenums or furred in ductwork, concealed in closets, attics, crawl spaces, basements, garages, utility rooms, or equipment rooms; *wall*—built in, window, transom; *outdoor*—rooftop, wall mounted, or on ground.

7. Heat: intended for use with upflow, horizontal, or counterflow forced-air furnace, combined with furnace, combined with electrical heat, combined with hot water or steam coil.

Unitary air conditioners are designed with fan capability for duct work.

Heat pumps (Sec. 2-8) are also offered in many of the same types and capacities as unitary air conditioners.

Packaged reciprocating and centrifugal water chillers can be considered as unitary air conditioners, particularly when applied with unitary-type chilled water blower coil units.

Figure 2-27 depicts a small single package air conditioner while Fig. 2-28 shows a large commercial packaged system.

The nearly infinite combination of coil configurations, evaporator temperatures, air-handling arrangements, refrigerating capacities, and other variations that are available in central systems are not possible with unitary systems. Consequently, in many respects a higher level of design ingenuity and performance is required to develop superior system performance using unitary equipment than for central systems.

The need for zoning is widely recognized because it is desirable to give the occupant some control over his or her individual environment. Unitary equipment tends to be a single-zone system; however, for large single spaces the use of multiple units can achieve the same objectives as a large multiple-zone system.

Room Air Conditioners

A room air conditioner is an encased assembly designed as a unit primarily for mounting in a window, through a wall, or as a console. The basic function of a room air conditioner is to provide comfort by cooling, dehumidifying, filtering or cleaning, and circulating the air in one room. It may also provide ventilation by introducing outdoor air into the room, and by exhausting the room air to the outside. The conditioner may be designed to provide heating by reverse cycle (heat pump) operation or by electric resistance elements. Room air conditioners do not generally provide the best all-around environment and are considered to be appliances. Figure 2-29 shows a schematic view of a typical room air conditioner.

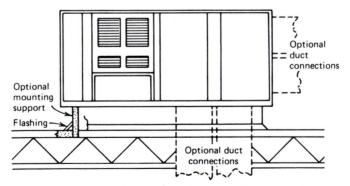

Figure 2-27 Rooftop installation of air-cooled single-package or single-package year-round unit.

Figure 2-28 A large commercial packaged air-conditioning system. (Courtesy of Carrier Corporation, Syracuse, NY)

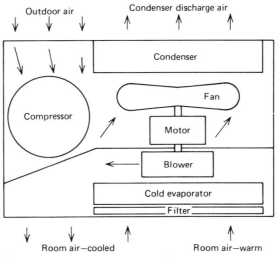

Figure 2-29 Schematic view of room air conditioner.

Through-the-wall Conditioner Systems

A through-the-wall system uses an air-cooled room air conditioner designed for mounting through the wall and in many cases is capable of providing both the heating and cooling function. Design and manufacture parameters vary widely. Specification grades range from appliance grade through heavy-duty commercial grade. The latter is called a *packaged terminal air conditioner*.

The through-the-wall concept incorporates a complete air-cooled refrigeration and air-handling system in an individual package; it utilizes space normally occupied by the building wall for equipment with the remainder projecting inside the room.

Each packaged terminal air conditioner has a self-contained, direct-expansion cool-

Figure 2-30 Packaged terminal air conditioner with combination heating and cooling chasis.

ing system, heating coil (electric, hot water, or steam), and packaged controls. Figure 2-30 shows a typical unit; it consists of a combination wall sleeve and room cabinet, combination heating and cooling chassis, and outdoor louver.

Through-the-wall systems are best applied in multizone applications. The initial cost of the through-the-wall system of air conditioning in multiroom applications is considerably less than central systems adapted to provide simultaneous functions of either heating or cooling in each room under control of the room occupants. With a through-the-wall system of heating and air conditioning, considerable space is saved by eliminating both ductwork and equipment rooms.

The through-the-wall system is limited to multizone systems and generally cannot be used economically in large spaces. One of the most popular applications has been in hotels and motels. This system is also applicable in renovation of existing buildings because there is less disruption and construction-forced sacrifice of rentable space than with alternative systems.

2-8 HEAT PUMP SYSTEMS

The term *heat pump* as applied to HVAC systems is a system in which refrigeration equipment is used such that heat is taken from the condenser and given up to the conditioned space when heating service is wanted and is removed from the space and taken to the evaporator when cooling and dehumidification are desired. The thermal cycle is identical with that of ordinary refrigeration, except that heating is provided when required. In some applications both the heating and cooling effects obtained in the cycle are utilized at the same time.

Unitary heat pumps (as opposed to applied heat pumps) are shipped from the factory as a complete preassembled unit including internal wiring, controls, and piping. Only the ductwork, external power wiring, and piping (for water-source heat pumps) are required to complete the installation. For the split unit it is also necessary to connect the refrigerant piping between the indoor and outdoor sections. In appearance and dimensions, casings of unitary heat pumps closely resemble those of conventional air-conditioning units having equal capacity.

Capacities of unitary heat pumps range from about 1½ to 25 tons or 5 to 90 kW, although there is no specific limitation. This equipment is almost universally used in residential and the smaller commercial and industrial installations. The multiunit type of installation with a number of individual units used in a closed-loop system are discussed below.

Large central heat pumps of modern design within the capacity range of about 30 to 1000 horsepower or 20 to 750 kW of compressor-motor rating are now opearting in a substantial number of buildings. A single or central system is generally used throughout the building, but in some instances the total capacity is divided among several separate heat pump systems to facilitate zoning.

Heat Pump Types

The air-to-air heat pump is the most common type. It is particularly suitable for factory-built unitary heat pumps and has been widely used for residential and light commercial applications. Outdoor air offers a universal heat-source, heat-sink medium for the heat pump. Extended-surface, forced-convection heat transfer coils are normally employed to transfer the heat between the air and the refrigerant.

The capacity of an air-to-air heat pump is highly dependent on the outdoor temperature. Therefore, it is usually necessary to provide supplemental heat at low outdoor temperature. The most common type of supplemental heat for heat pumps in the United States is electrical-resistance heat. This is usually installed in the air-handler unit and is designed to turn on automatically, sometimes in stages, as the indoor temperature drops. Heat pumps that have fossil fuel–fired supplemental heat are referred to as hybrid or bivalent heat pumps.

Air-to-water heat pumps are sometimes used in large buildings where zone control is necessary and are also sometimes employed for the production of hot or cold water in industrial applications as well as heat reclaiming. Heat pumps for water heating are commercially available in residential sizes.

A water-to-water heat pump uses water as the heat source and sink for both cooling and heating operation. Heating–cooling changeover may be accomplished in the refrigerant circuit, but in many cases it is more convenient to perform the switching in the water circuits.

A water-to-air pump uses water as a heat source and sink and uses air to transmit heat to or from the conditioned space. Water is a satisfactory and in many cases an ideal heat source. Well water is particularly attractive because of its relatively high and nearly constant temperature, generally about 50 F or 10 C in northern areas and 60 F or 16 C and higher in the south. However, abundant sources of suitable water are not always available and the application of this type of system is limited. Frequently sufficient water may be available from wells, but the condition of the water may cause corrosion in heat exchangers or it may induce scale formation. Other considerations are the costs of drilling, piping, and pumping and the means for disposing of used water. Surface or stream water may be utilized, but under reduced winter temperatures the cooling spread between inlet and outlet must be limited to prevent freeze-up in the water chiller, which is absorbing the heat.

Under certain industrial circumstances waste process water such as spent warm water in laundries and warm condenser water may be a source for specialized heat pump operations.

Closed-loop Systems

In many cases a building may require cooling in interior zones while needing heat in exterior zones. The needs of the north zones of a building may also be different from those of the south. In some cases a closed-loop heat pump system is a good choice. Closed-loop systems may be solar assisted. A closed-loop system is shown in Fig. 2-31.

Individual water-to-air heat pumps in each room or zone accept energy from or reject energy to a common water loop, depending on whether that area has a call for heating or for cooling. In the ideal case the loads will balance and there will be no surplus or deficiency of energy in the loop. If cooling demand is such that more energy is rejected to the loop than is required for heating, the surplus is rejected to the atmosphere by a cooling tower. In the other case an auxiliary boiler furnishes any deficiency.

The ground has been used successfully as a source-sink for heat pumps, using either a vertical or a horizontal run of plastic pipe as the ground-coupling device. Water from the heat pump exchanges heat with the surrounding earth before being returned back to

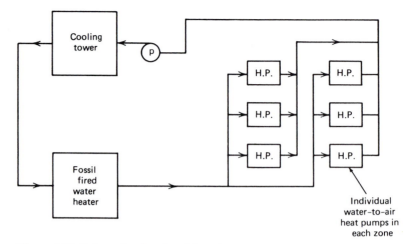

Figure 2-31 Schematic of a closed-loop heat pump system.

the heat pump. Proper sizing of the well depends on the nature of the soil surrounding the well, the water table level, and the efficiency of the heat pump. Reference 4 deals with the design of ground-coupled systems.

2-9 HEAT RECOVERY SYSTEMS

In large commercial applications considerable heat energy is generated internally and may require removal even during the coldest weather. This condition usually occurs within the central spaces, which do not have exterior walls. It is necessary to exhaust

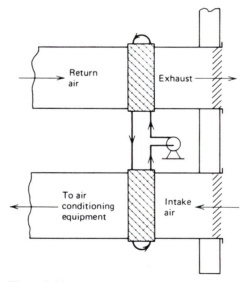

Figure 2-32 Air-to-water type of heat-recovery system.

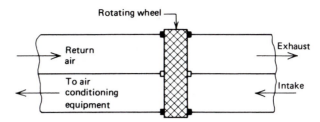

Figure 2-33 Rotating-type heat exchanger used for heat recovery.

considerable quantities of air from large commercial structures because of the introduction of outdoor ventilation air. Considerable savings in energy can be realized if the heat energy from the interior spaces and the exhaust air can be recovered and used in heating the exterior parts of the structure. Heat energy may also be recovered from waste water.

Redistribution of heat energy within a structure can be accomplished through the use of heat pumps of the air-to-air or water-to-water type. Another approach is the use of the dual-path systems described earlier.

Recovery of heat energy from exhaust air is accomplished through the use of air-to-air heat exchangers, rotating (periodic type) heat exchangers, and air-to-water heat exchangers connected by a circulating water loop. Sometimes spray systems are used; they may contain desiccants to enhance latent heat transfer. Figures 2-32 and 2-33

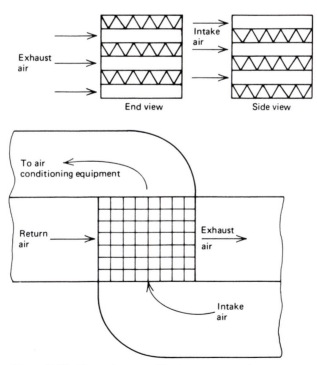

Figure 2-34 Air-to-air type of heat-recovery system.

illustrate the air-to-water and rotating heat recovery systems while Fig. 2-34 shows how an air-to-air system might be arranged. The air-to-air and rotating type systems are the most effective in recovering energy but require that the intake and exhaust to the building be at the same location, whereas the air-to-water system may have the exhaust and intake at widely separated locations; however, the air-to-water system is not as effective because of the added thermal resistance of the water. To prevent freezing, a brine must also be introduced, which further reduces the effectiveness of the air-to-water system.

All of the previously described systems may also be effective during the cooling season, when they function to cool and dehumidify outdoor ventilation air.

2-10 SUMMARY

It is apparent from the preceding sections that the entire subject of HVAC systems is complex and that the types and combinations of equipment are numerous. The design engineer must constantly keep acquainted with the literature from the various manufacturers to stay abreast of new innovations. It is the business of the sales or application engineer to be particularly well versed on available equipment and to give the best advice possible to the design engineer.

REFERENCES

1. *ASHRAE Handbook, Systems and Equipment Volume,* American Society of Heating, Refrigerating and Air-Conditioning Engineers, Inc., Atlanta, GA, 1992.
2. *ASHRAE Handbook, Fundamentals Volume,* American Society of Heating, Refrigerating and Air-Conditioning Engineers, Inc., Atlanta, GA, 1989.
3. *ASHRAE Handbook, Refrigeration Volume,* American Society of Heating, Refrigerating and Air-Conditioning Engineers, Inc., Atlanta, GA, 1990.
4. *ASHRAE Handbook, HVAC Applications,* American Society of Heating, Refrigerating and Air-Conditioning Engineers, Inc., Atlanta, GA, 1991.
5. Roger W. Haines, *Control Systems for Heating, Ventilating and Air Conditioning,* 3rd ed., Van Nostrand Reinhold, New York, 1983.

PROBLEMS

2-1. Consider the small single-story office building in Fig. 2-35. Lay out an all-air central system using an air handler with two zones. There is space between the ceiling and roof for ducts. The air handler is equipped with a direct expansion cooling coil and a hot water heating coil. Show all associated equipment schematically. Describe how the system might be controlled.

2-2. Suppose the building of Problem 2-1 is to use a combination air–water system where fan–coil units in each room are used for heating. Schematically lay out this part of the system with related equipment. Discuss the general method of control for (a) the supplied air and (b) the fan–coil units.

2-3. Lay out a year-round all-water system for the building of Problem 2-1. Show all equipment schematically. Discuss the control and operation of the system in the summer, winter, and between seasons.

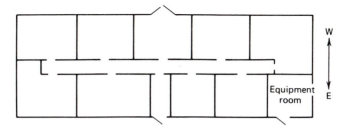

Figure 2-35 Floor plan of small office building.

2-4. Apply single-package year-round rooftop type unit(s) to the single story building in Figure 2-35.

2-5. Suppose a variable-air-volume (VAV) all-air system is to be used to condition the space shown in Figure 2-36. Assume that the space is the ground floor of a multistory office building. Describe the system using a schematic diagram. The lighting and occupant load are variable. Discuss the general operation of the system during the (a) colder months and (b) the warmer months.

2-6. Devise a central equipment arrangement for the system of Problem 2-5 that would save energy during the winter months. Sketch the system schematically.

2-7. Suppose an air-to-water heat pump is used to condition each space of Figure 2-36 where the water side of each heat pump is connected to a common water circuit. Sketch this system schematically showing all necessary additional equipment. Discuss the operation of this system during the (a) colder months, (b) warmer months, and (c) intermediate months.

2-8. A building such as that shown in Figure 2-36 requires some outdoor air. Explain and show schematically how this may be done with the system of Problem 2-5. Incorporate some sort of heat recovery device in the system. What controls would be necessary?

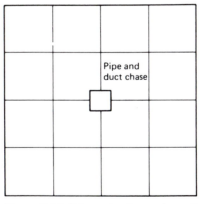

Figure 2-36 Schematic floor plan of one floor of a large building.

2-9. How can an economizer be used to advantage during the (a) winter months, (b) the summer months, and (c) the intermediate seasons?

2-10. The system proposed in Problem 2-7 requires the distribution of outdoor ventilation air to each space. Sketch a central air-handler system for this purpose that has energy recovery equipment and an economizer. Do not sketch the air distribution system. Discuss the control of this system assuming that the air will always be delivered at 72 F.

2-11. Make a single-line, block diagram of an all-water cooling system. The system has unit ventilators in each room with a packaged water chiller, and pumps. Explain how the system will be controlled.

2-12. Sketch a diagram of an air–water system that uses fan–coils around the perimeter and an overhead air distribution system from a central air handler. Show a hot water boiler, chiller, and water distribution pumps. Explain the operation of the system in the summer and in the winter. What kind of controls does the system need?

2-13. Make a sketch of a variable-volume system with a secondary perimeter heating system for a perimeter zone. Discuss the operation and control of the system for the different seasons of the year.

2-14. Diagram a combination air-to-air heat recovery and economizer system. Describe the operation and control of the system for various times of the year.

2-15. A large manufacturing facility requires hot and chilled water and electricity in its operation. Describe how internal combustion engines operating on natural gas could provide part or all of these needs using heat recovery and generating electricity. The objective would be to save energy.

2-16. Thermal storage is often used to smooth the demand for cooling in large buildings. Imagine that the chiller can also make ice during the nighttime hours for use later when the peak cooling demand is high. Make a sketch of such a central plant and describe its operation for a typical daily cycle. How would this system benefit the building owner? Describe the control system.

2-17. Make a sketch of a single-zone system for a small building that uses a ground-coupled heat pump. Show all the major parts of the system including the ground heat exchanger. Discuss operation of the system in summer and winter.

2-18. Sketch a variable-air-volume reheat system that has four zones. Discuss the operation of a typical zone.

2-19. Sketch a dual-duct VAV system. Show the fans and a typical zone. Describe a way to control the speed of the fans as the terminal devices reduce the air flow to the various zones.

2-20. It is desirable for the water leaving a cooling coil to be at a fixed temperature for return to the chiller. Sketch a coil, control valve, etc., to accomplish this action and describe operation of the system.

2-21. A small system has one cooling coil which is supplied water from a chiller. The chiller must have a constant volume flow rate of water but the flow rate of water through the coil must be varied to maintain a constant leaving water temperature. Sketch and describe a piping and control system to accomplish this action.

Chapter 3

Moist Air Properties and Conditioning Processes

A thorough understanding of psychrometrics, which deals with the properties of moist air, and the ability to analyze the various processes involving air, is basic to the HVAC engineer. Atmospheric air makes up the environment in almost every design situation.

In 1911, Willis H. Carrier made a significant contribution to the air-conditioning field when he published relations for moist air properties together with a psychrometric chart. These formulas became fundamental to the industry.

In 1983, formulas of Hyland and Wexler were published (1, 2). These formulas, developed at the National Bureau of Standards (now NIST) and based on the thermodynamic temperature scale, are the basis for the thermodynamic properties of moist air given in the 1989 *ASHRAE Handbook, Fundamentals Volume* (3) and Table A-2a of this text. Threlkeld (4) has shown that errors in calculation of the major properties will be less than 0.7 percent when perfect gas relations are used. This chapter emphasizes the use of the perfect gas relations.

Material in this chapter involves primarily the thermodynamic analysis. That is, only the states at the beginning and end of a process are considered. In the final analysis rate processes (heat transfer, fluid mechanics, and mass transfer) must also be included. This important part of the problem is covered in Chapters 13 and 14.

3-1 MOIST AIR AND THE STANDARD ATMOSPHERE

Atmospheric air is a mixture of many gases plus water vapor and countless pollutants. Aside from the pollutants, which may vary considerably from place to place, the composition of the dry air is relatively constant, varying slightly with time, location, and altitude. In 1949, a standard composition of dry air was fixed by the International Joint Committee on Psychrometric Data as shown in Table 3-1.

The ideal gas relation

$$Pv = \frac{P}{\rho} = R_a T \qquad (3\text{-}1)$$

Table 3-1 Composition of Dry Air

Constituent	Molecular Mass	Volume Fraction
Oxygen	32.000	0.2095
Nitrogen	28.016	0.7809
Argon	39.944	0.0093
Carbon dioxide	44.010	0.0003

has been shown to produce small errors when used to make psychrometric calculations. Based on the composition of air in Table 3-1, the molecular mass M_a of dry air is 28.965, and the gas constant R_a is

$$R_a = \frac{\overline{R}}{M_a} = \frac{1545.32}{29.965} = 53.352 \,(\text{ft-lbf})/(\text{lbm-R}) \qquad \text{or} \qquad 287 \,\text{J/(kg-K)} \qquad \textbf{(3-2)}$$

$*\,28.965$

where $\overline{R}$ is the universal gas constant; $\overline{R} = 1545.32$ (ft-lbf)/(lb mole-R) or 8314 J/(kg mole-K).

Most air-conditioning processes involve a mixture of dry air and water vapor. The amount of water vapor may vary from zero to a maximum determined by the temperature and pressure of the mixture. The latter case is called saturated air, a state of neutral equilibrium between the moist air and the liquid or solid phases of water. The molecular mass of water is 18.015 and the gas constant for water vapor is

$$R_v = \frac{1545.32}{18.015} = 85.78 \,(\text{ft-lbf})/(\text{lbm-R}) \qquad \text{or} \qquad 462 \,\text{J/(kg-K)} \qquad \textbf{(3-3)}$$

The *ASHRAE Handbook* (3) gives the following definition of the U. S. Standard Atmosphere:

1. Acceleration due to gravity is constant at 32.174 ft/sec^2 or 9.807 m/s^2.
2. Temperature at sea level is 59.0 F, 15 C or 288.1 K.
3. Pressure at sea level is 29.921 in. of mercury or 101.039 kPa.*
4. The atmosphere consists of dry air, which behaves as a perfect gas.

Standard sea level density computed using Eq. 3-1 with the standard temperature and pressure is 0.0765 lbm/ft^3 or 1.115 kg/m^3. The *ASHRAE Handbook* (3) summarizes standard atmospheric data for altitudes up to 60,000 ft or 18,291 m. Atmospheric pressure may be computed as a function of elevation by the following relation:

$$P = a + bH \qquad \textbf{(3-4)}$$

where the constants a and b are given in Table 3-2 and H is the elevation above sea level

---

* Standard atmospheric pressure is also commonly taken to be 14.696 lbf/in.^{2} or 101.325 kPa, which corresponds to 30.0 in. of mercury and standard atmospheric temperature is sometimes assumed to be 70 F (21 C).

Table 3-2 Constants for Eq. 3-4

Constant	$H \leq 4000$ ft or 1220 m		$H > 4000$ ft or 1220 m	
	IP	SI	IP	SI
a	29.92	101.325	29.42	99.436
b	-0.001025	-0.01153	-0.0009	-0.010

in feet or meters. The pressure P is in in. Hg or kPa. Elevation above sea level is given in Table C-1 for many locations in the United States and Canada.

3-2 FUNDAMENTAL PARAMETERS

Moist air up to about three atmospheres pressure obeys the perfect gas law with sufficient accuracy for engineering calculations. The Gibbs Dalton law for a mixture of perfect gases states that the mixture pressure is equal to the sum of the partial pressures of the constituents:

$$P = p_1 + p_2 + p_3 \tag{3-5}$$

For moist air

$$P = p_{N_2} + p_{O_2} + p_{CO_2} + p_A + p_v \tag{3-6}$$

Because the various constituents of the dry air may be considered to be one gas, it follows that the total pressure of moist air is the sum of the partial pressures of the dry air and the water vapor:

$$P = p_a + p_v \tag{3-7}$$

Each constituent in a mixture of perfect gases behaves as if the others were not present. To compare values for moist air assuming ideal gas behavior with actual table values, consider a saturated mixture of air and water vapor at 80 F. Table A-1a gives the saturation pressure p_s of water as 0.5073 lbf/in.^{2}. For saturated air this is the partial pressure p_v of the vapor. The mass density is $1/v = 1/632.8$ or 0.001580 lbm/ft^3. By using Eq. 3-1 we get

$$\frac{1}{v} = \rho = \frac{p_v}{R_v T} = \frac{0.5073(144)}{85.78(459.67 + 80)} = 0.001578 \text{ lbm/ft}^3$$

This result is accurate within about 0.1 percent. For nonsaturated conditions water vapor is superheated and the agreement is generally better. Several useful terms will now be defined.

Humidity ratio W is the ratio of the mass of the water vapor m_v to the mass of the dry air m_a in the mixture:

$$W = \frac{m_v}{m_a} \tag{3-8}$$

Relative humidity ϕ is the ratio of the mole fraction of the water vapor x_v in a mixture to the mole fraction x_s of the water vapor in a saturated mixture at the same temperature and pressure:

$$\phi = \left. \left| \frac{x_v}{x_s} \right| \right|_{t,P} \tag{3-9}$$

For a mixture of perfect gases, the mole fraction is equal to the partial pressure ratio of each constituent. The mole fraction of the water vapor is

$$x_v = \frac{p_v}{P} \tag{3-10}$$

Using Eq. 3-9 and letting p_s stand for the partial pressure of the water vapor in a saturated mixture, we may express the relative humidity as

$$\phi = \frac{p_v/P}{p_s/P} = \left. \left| \frac{p_v}{p_s} \right| \right|_{t,P} \tag{3-11}$$

Since the temperature of the dry air and the water vapor are assumed to be the same in the mixture

$$\phi = \frac{p_v/R_v T}{p_s/R_v T} = \left[\frac{\rho_v}{\rho_s} \right]_{t,P} \tag{3-12}$$

where the densities ρ_v and ρ_s are referred to as the absolute humidities of the water vapor (mass of water per unit volume of mixture). Values of ρ_s may be obtained from Tables A-1, Tables A-2, or calculated by formula such as the ones given by Hyland and Wexler (1) in reference 3.

Using the perfect gas law we can derive a relation between the relative humidity ϕ and the humidity ratio W:

$$m_v = \frac{p_v V}{R_v T} = \frac{p_v V M_v}{\overline{R} T} \tag{3-13a}$$

and

$$m_a = \frac{p_a V}{R_a T} = \frac{p_a V M_a}{\overline{R} T} \tag{3-13b}$$

and

$$W = \frac{M_v p_v}{M_a p_a} \qquad \text{(3-14a)}$$

For the air–water vapor mixture, Eq. 3-14a reduces to

$$W = \frac{18.015 \, p_v}{28.965 \, p_a} = 0.6219 \frac{p_v}{p_a} \qquad \text{(3-14b)}$$

Combining Eqs. 3-11 and 3-14b gives

$$\phi = \frac{W p_a}{0.6129 p_s} \qquad \text{(3-15)}$$

$$0.6219 \, p_s$$

Degree of saturation μ is the ratio of the humidity ratio W to the humidity ratio W_s of a saturated mixture at the same temperature and pressure:

$$\mu = \left[\frac{W}{W_s} \right]_{t,P} \qquad \text{(3-16)}$$

This parameter is most useful when the data of Tables A-2 are used.

Dewpoint temperature t_d is the temperature of saturated moist air at the same pressure and humidity ratio as the given mixture. As a mixture is cooled at constant pressure, the temperature at which condensation first begins is the dewpoint. At a given mixture (total) pressure, the dewpoint temperature is fixed by the humidity ratio W or by the partial pressure of the water vapor. Thus t_d, W, and p_v are not independent properties.

The *enthalpy i* of a mixture of perfect gases is equal to the sum of the enthalpies of each constituent and for the air–water vapor mixture is usually referenced to the mass of dry air. This is because the amount of water vapor may vary during some processes but the amount of dry air typically remains constant:

$$i = i_a + W i_v \qquad \text{(3-17)}$$

Each term has the units of energy per unit mass of dry air. With the assumption of perfect gas behavior, the enthalpy is a function of temperature only. If zero Fahrenheit or Celsius is selected as the reference state where the enthalpy of dry air is zero, and if the specific heat c_{pa} and c_{pv} are assumed to be constant, simple relations result

$$i_a = c_{pa} t \qquad \text{(3-18)}$$

$$i_v = i_g + c_{pv} t \qquad \text{(3-19)}$$

where the enthalpy of saturated water vapor i_g at 0 F is 1061.2 Btu/lbm and 2501.3 kJ/kg at 0 C.

Using Eqs. 3-17, 3-18, and 3-19 with c_{pa} and c_{pv} taken as 0.240 and 0.444 Btu/(lbm-F), respectively,

$$i = 0.240t + W(1061.2 + 0.444t) \text{ Btu/lbma} \qquad \text{(3-20a)}$$

$$(i_g @ \ 0°F)$$

In SI units Eq. 3-20a becomes

$$i = 1.0t + W(2501.3 + 1.86t) \text{ kJ/kga} \qquad \text{(3-20b)}$$

where c_{pa} and c_{pv} are 1.0 and 1.86 kJ/(kg-C), respectively.

EXAMPLE 3-1

Compute the enthalpy of saturated air at 60 F and standard atmospheric pressure.

SOLUTION

Equation 3-20a will be used to compute enthalpy; however, the humidity ratio W_s must first be determined from Eq. 3-14b.

$$W_s = 0.6219 \frac{p_s}{p_a} = 0.6219 \frac{p_s}{P - p_s}$$

From Table A-1a, $p_s = 0.2563$ psia and

$$W_s = 0.6219 \left[\frac{0.2563}{14.696 - 0.2563} \right] = 0.01104 \text{ lbmv/lbma}$$

$$i_s = (0.24)60 + 0.01104[1061.2 + (0.444)60] = 26.41 \text{ Btu/lbma}$$

The result may be compared with the precise data of Table A-2a. The enthalpy calculated using ideal gas relations is about 0.25 percent low but quite satisfactory for engineering calculations.

3-3 ADIABATIC SATURATION

The equations discussed in the previous section show that at a given pressure and dry bulb temperature of an air–water vapor mixture, one additional property is required to completely specify the state, except at saturation. Any of the parameters discussed (ϕ, W, or i) would be acceptable; however, there is no practical way to measure any of them. The concept of adiabatic saturation provides a convenient solution.

Consider the device shown in Fig. 3-1. The apparatus is assumed to operate such that the air leaving at point 2 is saturated. The temperature t_2, where the relative humidity is 100 percent, is then defined as the *adiabatic saturation temperature* t_2^*, or *thermodynamic wet bulb temperature*. If we assume that the device operates in a steady-flow–steady-state manner, an energy balance on the control volume yields

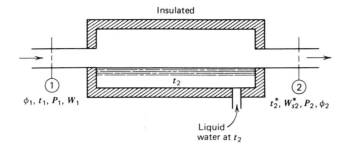

Figure 3-1 Schematic of adiabatic saturation device.

$$i_{a1} + W_1 i_{v1} + (W_{s2}^* - W_1)i_w^* = W_{s2}^* i_{v2}^* + i_{a2}^*$$ (3-21a)

or

$$W_1(i_{v1} - i_w^*) = c_{pa}(t_2^* - t_1) + W_{s2}^*(i_{v2}^* - i_w^*)$$ (3-21b)

where the * superscript refers to the adiabatic saturation temperature, and

$$W_1(i_{v1} - i_w^*) = c_{pa}(t_2^* - t_1) + W_{s2}^* i_{fg2}^*$$ (3-21c)

Solving for W_1 yields

$$W_1 = \frac{c_{pa}(t_2^* - t_1) + W_{s2}^* i_{fg2}^*}{(i_{v1} - i_w^*)}$$ (3-21d)

It is clear from Eq. 3-21d that W_1 is a function of t_1, t_2^*, P_1, P_2:

$$W_{s2}^* = 0.6219 \frac{p_{v2}}{(p_2 - p_{v2})}$$ (3-14b)

and $p_{v2} = p_{s2}$ at t_2^*. The enthalpy of vaporization i_{fg2}^* depends only on t_2^*; the enthalpy of the vapor i_{v1} is a function of t_1 and i_w^* is a function of t_2^*. Therefore, the humidity ratio of an air–water vapor mixture can be determined from the entering and leaving temperatures and pressures of the adiabatic saturator. Consider the following example.

EXAMPLE 3-2

The pressure entering and leaving an adiabatic saturator is 14.696 lbf/in.[2], the entering temperature is 80 F, and the leaving temperature is 64 F. Compute the humidity ratio W_1 and the relative humidity ϕ_1.

SOLUTION

Because the mixture leaving the device is saturated, $p_{v2} = p_{s2}$, and W_2 can be calculated using Eq. 3-14b.

$$W_{s2}^* = 0.6219 \left| \frac{0.2952}{14.696 - 0.2952} \right| = 0.0127 \text{ lbmv/lbma}$$

Now using Eq. 3-21d and data from Table A-1a, we get

$$W_1 = \frac{c_{pa}(t_2^* - t_1) + W_{s2}^* i_{fg2}^*}{(i_{v1} - i_w^*)}$$

$$W_1 = \frac{0.24(64 - 80) + (0.0127 \times 1057.3)}{(1096.4 - 32.09)} = 0.0090 \text{ lbmv/lbma}$$

Then using Eq. 3-14b

$$W_1 = 0.6219 \frac{p_{v1}}{14.696 - p_{v1}} = 0.0090 \text{ lbmv/lbma}$$

$$p_{v1} = 0.2098 \text{ psia}$$

Finally, from Eq. 3-11

$$\phi_1 = \frac{p_{v1}}{p_{s1}} = \frac{0.2098}{0.5073} = 0.414$$

It seems that the state of moist air could be completely determined from pressure and temperature measurements. However, the adiabatic saturator is not a practical device because it would have to be infinitely long in the flow direction.

3-4   WET BULB TEMPERATURE AND THE PSYCHROMETRIC CHART

A practical device used in place of the adiabatic saturator is the *psychrometer*. This apparatus consists of two thermometers, or other temperature-sensing elements, one of which has a wetted cotton wick covering the bulb (Fig. 3-2). The temperatures indicated

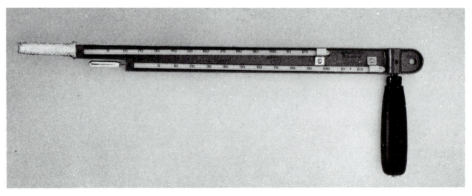

Figure 3-2 Typical hand-held psychrometer. (Courtesy of Central Scientific Company, Chicago, IL)

by the psychrometer are called the *wet bulb* and the *dry bulb* temperatures. The dry bulb temperature corresponds to t_1 in Fig. 3-1 and the wet bulb temperature is an approximation to t_2^* in Fig. 3-1, whereas P_1 and P_2 are equal to barometric or total mixture pressure. The combination heat-and-mass-transfer process from the wet bulb thermometer is not the same as the adiabatic saturation process; however, the error is relatively small when the wet bulb thermometer is used under suitable conditions. When the wet bulb temperature appears in psychrometric equations and charts, it is really the adiabatic saturation temperature that is being considered.

Threlkeld (4) has analyzed the problem and correlated wet bulb temperature with the adiabatic saturation temperature. The results are shown in Figs. 3-3 and 3-4 for thermometers with diameters of 0.3 and 0.1 in. or 7.5 and 2.5 mm. The data are shown as percent deviation of $(t_{wb} - t_2^*)$ from the *wet bulb depression* $(t - t_{wb})$ for shielded and unshielded wet bulbs. Three different temperature combinations are given and several general conclusions may be drawn. It appears that the unshielded wet bulb will generally more closely approximate the adiabatic saturation temperature. Temperatures taken with air velocities greater than about 200 ft/min (1 m/s) generally have the least deviation. Lower air velocities may be used with smaller wet bulbs. Threlkeld drew the following general conclusion: For atmospheric temperature above freezing, where the wet bulb depression does not exceed about 20 F (11 C), and where no unusual radiation

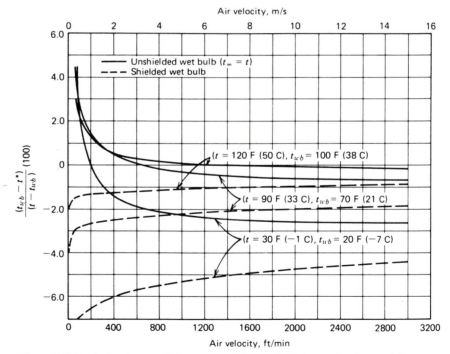

Figure 3-3 Deviation $(t_{wb} - t^*)$ in percent of the wet bulb depression $(t - t_{wb})$ for a wet bulb diameter of 0.3 in. (7.5 mm) and a barometric pressure of 14.696 psia (101.325 kPa). (James L. Threlkeld, *Thermal Environmental Engineering,* 2nd ed. 1970, p. 208. Reprinted by permission of Prentice-Hall, Inc., Englewood Cliffs, NJ).

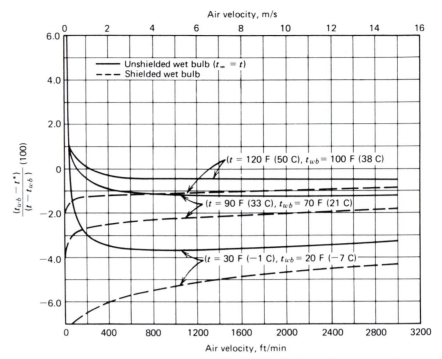

Figure 3-4 Deviation $(t_{wb} - t^*)$ in percent of the wet bulb depression $(t - t_{wb})$ for a wet bulb diameter of 0.1 in. (2.5 mm) and a barometric pressure of 14.696 psia (101.325 kPa). (James L. Threlkeld, *Thermal Environmental Engineering*, 2nd ed. 1970, p. 209. Reprinted by permission of Prentice-Hall, Inc., Englewood Cliffs, NJ)

circumstances exist, $(t_{wb} - t_2^*)$ should be less than about 0.5 F (0.27 C) for an unshielded mercury-in-glass thermometer as long as the air velocity exceeds about 100 ft/min (0.5 m/s). If thermocouples are used, the velocity may be somewhat lower with similar accuracy. Figure 3-5 shows a psychrometer properly installed to meet the foregoing conditions.

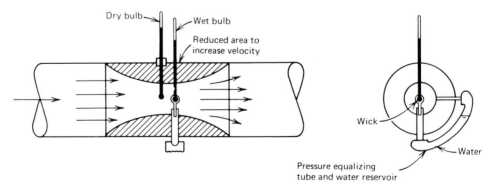

Figure 3-5 A psychrometer installed in a duct.

Thus, for most engineering problems the wet bulb temperature obtained from a properly operated, unshielded psychrometer may be used directly in Eq. 3-21d in place of the adiabatic saturation temperature.

To facilitate engineering computations, a graphical representation of the properties of moist air has been developed and is known as a psychrometric chart. Richard Mollier was the first to use such a chart with enthalpy as a coordinate. Modern-day charts are somewhat different but still retain the enthalpy coordinate feature. ASHRAE has developed five of the Mollier-type charts to cover the necessary range of variables. Figure 3-6 is an abridgment of ASHRAE Chart 1 that covers the normal range of variables at standard atmospheric pressure. The charts are based on the precise data of Tables A-2; within the readability of the charts, however, agreement with the perfect gas relations is very good. Details of the actual construction of the charts may be found in references 3 and 6. Charts for both English and SI units are provided in the packet in the back of the book.

In Fig. 3-6 dry bulb temperature is plotted along the horizontal axis. The dry bulb temperature lines are straight but not exactly parallel and incline slightly to the left. Humidity ratio is plotted along the vertical axis on the right-hand side of the chart. The scale is uniform with horizontal lines. The saturation curve slopes upward from left to right. Dry bulb, wet bulb, and dew point temperatures all coincide on the saturation curve. Relative humidity lines with shapes similar to the saturation curve appear at regular intervals. The enthalpy scale is drawn obliquely on the left of the chart with parallel enthalpy lines inclined downward to the right. Although the wet bulb temperature lines appear to coincide with the enthalpy lines, they diverge gradually in the body of the chart and are not parallel to one another. The spacing of the wet bulb lines is not uniform. Specific volume lines appear inclined from the upper left to the lower right and are not parallel. A protractor with two scales appears at the upper left of ASHRAE Charts 1. One scale gives the sensible heat ratio and the other the ratio of enthalpy difference to humidity ratio difference. *The enthalpy, specific volume, and humidity ratio scales are all based on a unit mass of dry air and not a unit mass of the moist air.*

EXAMPLE 3-3

Read the properties of moist air at 75 F db, 60 F wb, and standard sea level pressure from ASHRAE Psychrometric Chart 1*a*.

SOLUTION

The intersection of the 75 F db and 60 F wb lines defines the given state. This point on the chart is the reference from which all the other properties are determined.

Humidity Ratio, W. Move horizontally to the right and read $W = 0.0077$ lbmv/lbma on the vertical scale.

Relative Humidity, ϕ. Interpolate between the 40 and 50 percent relative humidity lines and read $\phi = 41$ percent.

Enthalpy, i. Follow a line of constant enthalpy upward to the left and read $i = 26.4$ Btu/lbma on the oblique scale.

Specific Volume, v. Interpolate between the 13.5 and 14.0 specific volume lines and read $v = 13.65$ ft^3/lbma.

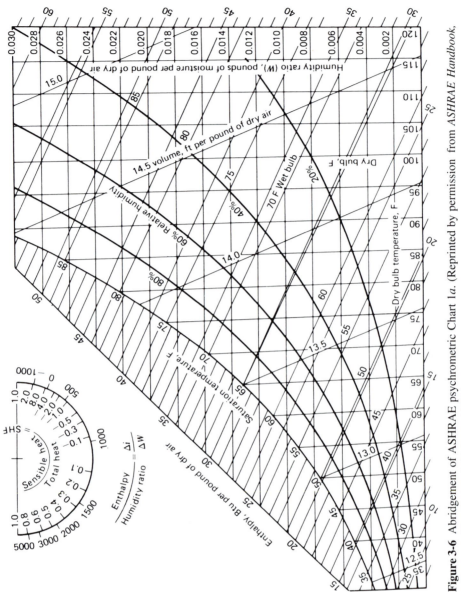

Figure 3-6 Abridgement of ASHRAE psychrometric Chart 1*a*. (Reprinted by permission from *ASHRAE Handbook, Fundamentals Volume,* 1989.)

Dew Point Temperature, t_d. Move horizontally to the left from the reference point and read $t_d = 50$ F on the saturation curve.

Enthalpy, i (alternate method). The nomograph in the upper left-hand corner of Chart 1a gives the difference D between the enthalpy of unsaturated moist air and enthalpy of saturated air at the same wet bulb temperature. Then $i = i_s + D$. For this example $i_s = 26.5$ Btu/lbma, $D = -0.1$ Btu/lbma, and $i = 26.5 - 0.1 = 26.4$ Btu/lbma.

Although psychrometric charts are useful in several aspects of HVAC design, the availability of computer programs to determine moist air properties have made some of these steps easier to carry out. These programs may be easily constructed from the basic equations of this chapter. Computer programs give the additional convenience of choice of units and arbitrary mixture pressures.

3-5   CLASSIC MOIST AIR PROCESSES

Two powerful analytical tools of the HVAC design engineer are the first law of thermodynamics or energy balance, and the conservation of mass or mass balance. These conservation laws are the basis for the analysis of moist air processes. In actual practice the properties may not be uniform across the flow area especially at the outlet, and a considerable length may be necessary for complete mixing. It is customary to analyze these processes by using the bulk average properties at the inlet and outlet of the device being studied.

In this section we will consider the basic processes that are a part of the analysis of most systems.

Heating or Cooling of Moist Air

When air is heated or cooled without the loss or gain of moisture, the process yields a straight horizontal line on the psychrometric chart because the humidity ratio is constant. Such processes can occur when moist air flows through a heat exchanger. In cooling, if part of the surface of the heat exchanger is below the dew point temperature of the air, condensation and the consequent dehumidification will occur. Figure 3-7 shows a schematic of a device used to heat or cool air. Under steady-flow–steady-state conditions the energy balance becomes

$$\dot{m}_a i_2 + \dot{q} = \dot{m}_a i_1 \tag{3-22}$$

An energy balance yields a positive number for $\dot{q}$ for both cooling and heating, and the direction of the heat transfer is implied by the terms *heating* and *cooling*. The enthalpy of the moist air, per unit mass of dry air, at sections 1 and 2 is given by:

$$i_1 = i_{a1} + W_1 i_{v1} \tag{3-23}$$

and

$$i_2 = i_{a2} + W_2 i_{v2} \tag{3-24}$$

Alternatively i_1 and i_2 may be obtained directly from the psychrometric chart. The

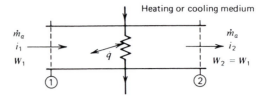

Figure 3-7 Schematic of a heating or cooling device.

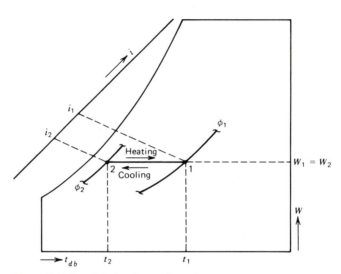

Figure 3-8 Sensible heating and cooling process.

convenience of the chart is evident. Figure 3-8 shows heating and cooling processes. Because the moist air has been assumed to be a perfect gas, Eq. 3-22 may be arranged and written

$$\dot{q}_s = \dot{m}_a c_p (t_1 - t_2) \quad \text{(cooling and heating)} \quad (3\text{-}25)$$

where

$$c_p = c_{pa} + W c_{pv} \quad (3\text{-}26)$$

In the temperature range of interest, $c_{pa} = 0.24$ Btu/(lbma-F) or 1.0 kJ/(kga-C), $c_{pv} = 0.45$ Btu/(lbmv-F) or 1.86 kJ/(kga-C), and W is the order of 0.01. Then c_p is about 0.245 Btu/(lbma-F) or 1.02 kJ/(kga-C).

EXAMPLE 3-4

Find the heat transfer rate required to warm 1500 cfm (ft³/min) of air at 60 F and 90 percent relative humidity to 120 F without the addition of moisture.

SOLUTION

Equation 3-22 or 3-25 may be used to find the required heat transfer rate. First it is necessary to find the mass flow rate of the dry air.

$$\dot{m}_a = \frac{\overline{V}_1 A_1}{v_1} = \frac{\dot{Q}}{v_1} \tag{3-27}$$

The specific volume is read from Chart 1*a* at $t_1 = 60$ F and $\phi = 90$ percent as 13.31 ft^3/lbma:

$$\dot{m}_a = \frac{1500(60)}{13.31} = 6762 \text{ lbma/hr}$$

Also from Chart 1*a*, $i_1 = 25.3$ Btu/lbma and $i_2 = 40$ Btu/lbma. Then by using Eq. 3-22, we get

$$\dot{q} = 6762(40.0 - 25.3) = 99{,}400 \text{ Btu/hr}$$

or if we had chosen to use Eq. 3-25

$$\dot{q} = 6762(0.245)(120 - 60) = 99{,}400 \text{ Btu/hr}$$

Locating point 2 at the end of the heating process on the chart, we can see that the relative humidity decreases when the moist air is heated. The reverse process of cooling results in an increase in relative humidity.

Cooling and Dehumidifying of Moist Air

When moist air is cooled to a temperature below its dew point, some of the water vapor will condense and may leave the air stream. Figure 3-9 shows a schematic of a cooling and dehumidifying device, and Fig. 3-10 shows the process on the psychrometric chart. Although the actual process path may vary considerably depending on the type of surface, surface temperature, and flow conditions, the net heat and mass transfer can be expressed in terms of the initial and final states. By referring to Fig. 3-9, we see that the energy balance gives

$$\dot{m}_a i_1 = \dot{q} + \dot{m}_a i_2 + \dot{m}_w i_w \tag{3-28}$$

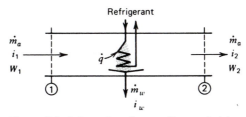

Figure 3-9 Schematic of a cooling and dehumidifying device.

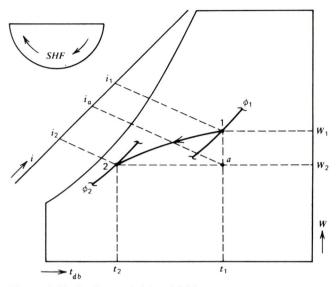

Figure 3-10 Cooling and dehumidifying process.

and the mass flow balance for the water in the air is

$$\dot{m}_a W_1 = \dot{m}_w + \dot{m}_a W_2$$

(3-29)

Combining Eqs. 3-28 and 3-29 yields

$$\dot{q} = \dot{m}_a(i_1 - i_2) - \dot{m}_a(W_1 - W_2)i_w$$

(3-30)

Equation 3-30 gives the total rate of heat transfer from the moist air. The last term on the right-hand side of Eq. 3-30 is usually small compared to the others and is often neglected. The following example illustrates this point.

EXAMPLE 3-5

Moist air at 80 F db and 67 F wb is cooled to 58 F db and 80 percent relative humidity. The volume flow rate is 2000 cfm and the condensate leaves at 60 F. Find the heat transfer rate.

SOLUTION

Equation 3-30 applies to this process, which is similar to Fig. 3-10. The following properties are read from Chart 1a: $v_1 = 13.85$ ft³/lbma, $i_1 = 31.6$ Btu/lbma, $W_1 = 0.0112$ lbmv/lbma, $i_2 = 22.9$ Btu/lbma, $W_2 = 0.0082$ lbmv/lbma. The enthalpy of the condensate is obtained from Table A-1a, $i_w = 28.08$ Btu/lbmw. The mass flow rate $\dot{m}_a$ is obtained from Eq. 3-27.

$$\dot{m}_a = \frac{2000(60)}{13.85} = 8664 \text{ lbma/hr}$$

Then

$$\dot{q} = 8664[(31.6 - 22.9) - (0.0112 - 0.0082)28.08]$$

$$\dot{q} = 8664[(8.7) - (0.084)]$$

The last term, which represents the energy of the condensate, is seen to be small. Neglecting the condensate term, $\dot{q} = 75{,}000$ Btu/hr $= 6.25$ tons.

The cooling and dehumidifying process involves both sensible and latent heat transfer, where sensible heat transfer rate is associated with the decrease in dry bulb temperature and the latent heat transfer rate is associated with the decrease in humidity ratio. These quantities may be expressed as

$$\dot{q}_s = \dot{m}_a c_p (t_1 - t_2) \tag{3-31}$$

and

$$\dot{q}_l = \dot{m}_a (W_1 - W_2) i_{fg} \tag{3-32}$$

By referring to Fig. 3-10 we may also express the latent heat transfer rate as

$$\dot{q}_l = \dot{m}_a (i_1 - i_a) \tag{3-33}$$

and the sensible heat transfer rate is given by

$$\dot{q}_s = \dot{m}_a (i_a - i_2) \tag{3-34}$$

The energy of the condensate has been neglected. Obviously

$$\dot{q} = \dot{q}_s + \dot{q}_l \tag{3-35}$$

The *sensible heat factor* SHF is defined as $\dot{q}_s/\dot{q}$. This parameter is shown on the semicircular scale of Fig. 3-10. The use of this feature of the chart is shown later.

Heating and Humidifying Moist Air

A device to heat and humidify moist air is shown schematically in Fig. 3-11. This process is generally required during the cold months of the year. An energy balance on the device yields

$$\dot{m}_a i_1 + \dot{q} + \dot{m}_w i_w = \dot{m}_a i_2 \tag{3-36}$$

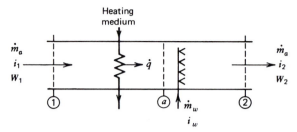

Figure 3-11 Schematic of a heating and humidifying device.

and a mass balance on the water gives

$$\dot{m}_a W_1 + \dot{m}_w = \dot{m}_a W_2 \qquad (3\text{-}37)$$

Equations 3-36 and 3-37 may be combined to obtain

$$\frac{i_2 - i_1}{W_2 - W_1} = \frac{\dot{q}}{\dot{m}_a(W_2 - W_1)} + i_w \qquad (3\text{-}38a)$$

or

$$\frac{i_2 - i_1}{W_2 - W_1} = \frac{\dot{q}}{\dot{m}_w} + i_w \qquad (3\text{-}38b)$$

Equation 3-38a or 3-38b describes a straight line that connects the initial and final states on the psychrometric chart. Figure 3-12 shows a typical combined heating and humidifying process, states 1–2.

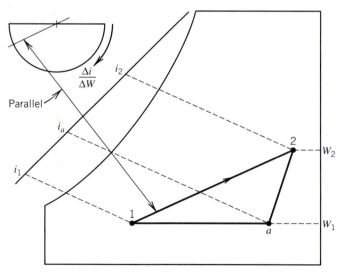

Figure 3-12 Typical heating and humidifying process.

A graphical procedure makes use of the semicircular scale on Charts 1 to locate the process line. The ratio of the change in enthalpy to the change in humidity ratio is

$$\frac{\Delta i}{\Delta W} = \frac{i_2 - i_1}{W_2 - W_1} = \frac{\dot{q}}{\dot{m}_w} + i_w \qquad (3\text{-}39)$$

Figure 3-12 shows the procedure where a straight line is laid out parallel to the line on the protractor through state 1. Although the process may be represented by one line from state 1 to state 2, it is not practical to do this. The heating and humidification processes are usually carried out separately, as shown in Fig. 3-11 and Fig. 3-12 as processes 1–a and a–2.

Adiabatic Humidification of Moist Air

When moisture is added to moist air without the addition of heat, Eq. 3-38b becomes

$$\frac{i_2 - i_1}{W_2 - W_1} = i_w = \frac{\Delta i}{\Delta W} \qquad (3\text{-}40)$$

The direction of the process on the psychrometric chart can vary considerably. Figure 3-13 shows these processes. If the injected water is saturated vapor at the dry bulb temperature, the process will proceed at a constant dry bulb temperature. If the water enthalpy is greater than the enthalpy of saturated vapor at the dry bulb temperature, the air will be heated and humidified. If the water enthalpy is less than the enthalpy of saturated vapor at the dry bulb temperature, the air will be cooled and humidified. One other situation is worthy of mention. When liquid water at the wet bulb temperature is injected, the process follows a line of constant wet bulb temperature.

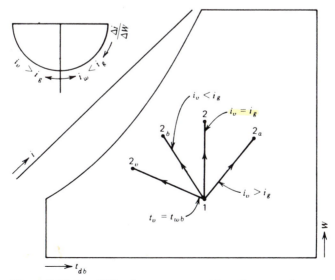

Figure 3-13 Humidification processes without heat transfer.

EXAMPLE 3-6

Moist air at 60 F db and 20 percent relative humidity enters a heater and humidifier at the rate of 1600 cfm. Heating of the air is followed by adiabatic humidification so that it leaves at 115 F db and a relative humidity of 30 percent. Saturated water vapor at 212 F in injected. Determine the required heat transfer rate and mass flow rate of water vapor.

SOLUTION

Figure 3-11 is a schematic of the apparatus. Locate the states as shown in Fig. 3-12 from the given information and Eq. 3-40 using the protractor feature of the psychrometric chart. Process $1-a$ is sensible heating; therefore, a horizontal line to the right of state 1 is constructed. Process $a-2$ is determined from Eq. 3-40 and the protractor:

$$\frac{\Delta i}{\Delta W} = i_w = 1151 \text{ Btu/lbm}$$

where i_w is read from Table A-1a. A parallel line is drawn from state 2 as shown in Fig. 3-12. State a is determined by the intersection on lines $1-a$ and $a-2$. The heat transfer rate is then given by

$$\dot{q} = \dot{m}_a(i_a - i_1)$$

where

$$\dot{m}_a = \frac{\dot{Q}(60)}{v_1} = \frac{(1600)(60)}{13.15} = 7300 \text{ lbma/hr}$$

and i_1 and i_a are read from Chart 1a as 16.8 and 29.2 Btu/lbma, respectively. Then

$$\dot{q} = 7300(29.2 - 16.8) = 90{,}500 \text{ Btu/hr}$$

The mass flow rate of the water vapor is given by

$$\dot{m}_v = \dot{m}_a(W_2 - W_1)$$

where W_2 and W_1 are read from Chart 1a as 0.0194 and 0.0022 lbmv/lbma, respectively. Then

$$\dot{m}_v = 7300(0.0194 - 0.0022) = 125.6 \text{ lbmv/hr}$$

Adiabatic Mixing of Two Streams of Moist Air

The mixing of air streams is quite common in air-conditioning systems. The mixing process usually occurs under steady, adiabatic flow conditions. Figure 3-14 illustrates the mixing of two air streams. An energy balance gives

$$\dot{m}_{a1}i_1 + \dot{m}_{a2}i_2 = \dot{m}_{a3}i_3 \tag{3-41}$$

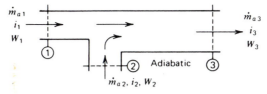

Figure 3-14 Schematic of the adiabatic mixing of two air streams.

The mass balance on the dry air is

$$\dot{m}_{a1} + \dot{m}_{a2} = \dot{m}_{a3} \tag{3-42}$$

and the mass balance on the water vapor is

$$\dot{m}_{a1}W_1 + \dot{m}_{a2}W_2 = \dot{m}_{a3}W_3 \tag{3-43}$$

Combining Eqs. 3-41, 3-42, and 3-43 and eliminating $\dot{m}_{a3}$ yields

$$\frac{i_2 - i_3}{i_3 - i_1} = \frac{W_2 - W_3}{W_3 - W_1} = \frac{\dot{m}_{a1}}{\dot{m}_{a2}} \tag{3-44}$$

The state of the mixed streams lies on a straight line between states 1 and 2 (Fig. 3-15). From Eq. 3-44 the lengths of the various line segments are proportional to the masses of dry air mixed:

$$\frac{\dot{m}_{a1}}{\dot{m}_{a2}} = \frac{\overline{32}}{\overline{13}}; \qquad \frac{\dot{m}_{a1}}{\dot{m}_{a3}} = \frac{\overline{32}}{\overline{12}}; \qquad \frac{\dot{m}_{a2}}{\dot{m}_{a3}} = \frac{\overline{13}}{\overline{12}} \tag{3-45}$$

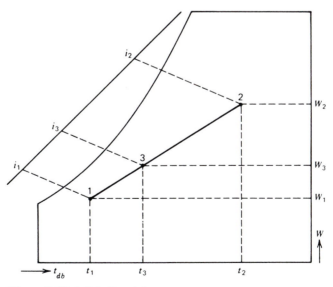

Figure 3-15 Adiabatic mixing process.

This is most easily shown by solving Eq. 3-44 for i_3 and for W_3.

$$i_3 = \frac{\dfrac{\dot{m}_{a1}}{\dot{m}_{a2}} i_{i1} + i_2}{1 + \dfrac{\dot{m}_{a1}}{\dot{m}_{a2}}} \tag{3-44a}$$

$$W_3 = \frac{\dfrac{\dot{m}_{a1}}{\dot{m}_{a2}} W_1 + W_2}{1 + \dfrac{\dot{m}_{a1}}{\dot{m}_{a2}}} \tag{3-44b}$$

Clearly for given states 1 and 2, a straight line will be generated when any constant value of $\dot{m}_{a1}/\dot{m}_{a2}$ is used and the result plotted on the psychometric chart. It is also clear that the location of state 3 on the line is proportional to $\dot{m}_{a1}/\dot{m}_{a3}$. Consider the case when $\dot{m}_{a1}/\dot{m}_{a2} = 1$, for example. This fact provides a very convenient graphical procedure for solving mixing problems in contrast to the use of Eqs. 3-41, 3-42, and 3-43.

Although the mass flow rate is used when the graphical procedure is employed, the volume flow rates may be used to obtain approximate results.

EXAMPLE 3-7

Two thousand cubic feet per minute (cfm) of air at 100 F db and 75 F wb are mixed with 1000 cfm of air at 60 F db and 50 F wb. The process is adiabatic, at a steady flow rate and at standard sea level pressure. Find the condition of the mixed streams.

SOLUTION

A combination graphical and analytical solution is first obtained. The initial states are first located on Chart 1*a* as illustrated on Fig. 3-15 and connected with a straight line. Equations 3-42 and 3-43 are combined to obtain

$$W_3 = W_1 + \frac{\dot{m}_{a2}}{\dot{m}_{a3}} (W_2 - W_1) \tag{3-46}$$

By using the property values from Chart 1*a*, we obtain

$$\dot{m}_{a1} = \frac{1000(60)}{13.21} = 4542 \text{ lbma/hr}$$

$$\dot{m}_{a2} = \frac{2000(60)}{14.4} = 8332 \text{ lbma/hr}$$

$$W_3 = 0.0053 + \frac{8332}{(4542 + 8332)}(0.013 - 0.0053)$$

$$W_3 = 0.0103 \text{ lbmv/lbma}$$

The intersection of W_3 with the line connecting states 1 and 2 gives the mixture state 3. The resulting dry bulb temperature is 86 F and the wet bulb temperature is 68 F.

The complete graphical procedure could also be used where

$$\frac{\overline{13}}{\overline{12}} = \frac{\dot{m}_{a2}}{\dot{m}_{a3}} = \frac{8332}{(8332 + 4542)} = 0.65$$

or

$$\overline{13} = 0.65(\overline{12})$$

The length of line segments $\overline{12}$ and $\overline{13}$ depend on the scale of the psychrometric chart used. However, when the length $\overline{13}$ is laid out along $\overline{12}$ from state 1, state 3 is accurately determined.

3-6 SPACE AIR CONDITIONING—DESIGN CONDITIONS

The complete air-conditioning system may involve two or more of the processes just considered. For example, in the air conditioning of space during the summer, the air supplied must have a sufficiently low temperature and moisture content to absorb the total heat gain of the space. Therefore, as the air flows through the space, it is heated and humidified. If the system is closed loop, the air is then returned to the conditioning equipment, where it is cooled and dehumidified and supplied to the space again. Outdoor air may be mixed with the return air. During the winter months the same general processes occur but in reverse. Systems described in Chapter 2 carry out these conditioning processes with some variations.

Sensible Heat Factor

The *sensible heat factor* (SHF) was defined in Section 3-5 as the ratio of the sensible heat transfer to the total heat transfer for a process:

$$\text{SHF} = \frac{\dot{q}_s}{\dot{q}_s + \dot{q}_l} = \frac{\dot{q}_s}{\dot{q}} \qquad (3\text{-}47)$$

If we recall Eqs. 3-33 and 3-34 and refer to Fig. 3-10, it is evident that the SHF is related to the parameter $\Delta i/\Delta W$. The SHF is plotted on the inside scale of the protractor on Charts 1. The following examples will demonstrate the usefulness of the SHF.

EXAMPLE 3-8

Conditioned air is supplied to a space at 15 C db and 14 C wb at the rate of 0.5 m³/s. The sensible heat factor for the space is 0.70 and the space is to be maintained at 24 C db. Determine the sensible and latent cooling loads for the space.

SOLUTION

Chart 1*b* can be used to solve this problem conveniently. A line is drawn on the protractor through a value of 0.7 on the SHF scale. A parallel line is then drawn from the initial state (15 C db and 14 C wb) to the intersection of the 24 C db line, which defines the final state. Figure 3-16 illustrates the procedure. The total heat transfer rate for the process is given by

$$\dot{q} = \dot{m}_a(i_2 - i_1)$$

and the sensible heat transfer rate is given by

$$\dot{q}_s = (SHF)\dot{q}$$

and

$$\dot{m}_a = \dot{Q}/v_1 = \frac{0.5}{0.827} = 0.605 \text{ kg/s}$$

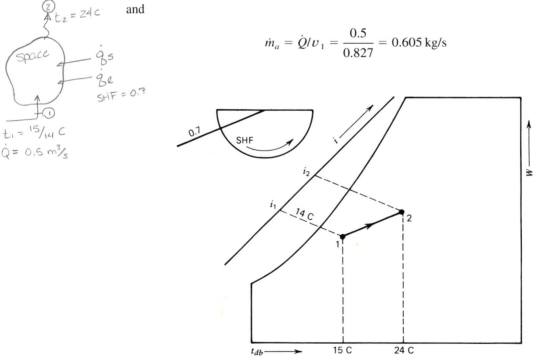

Figure 3-16 The condition line for the space in Example 3-8.

where v_1 is read from Chart 1b. Also from Chart 1b, $i_1 = 39.3$ kJ/kg dry air and $i_2 = 52.6$ kJ/kg dry air. Then

$$\dot{q} = 0.605(52.6 - 39.3) = 8.04 \text{ kJ/s} = 8.04 \text{ kW}$$

$$\dot{q}_s = \dot{q}(\text{SHF}) = 8.04(0.7) = 5.63 \text{ kW}$$

and

$$\dot{q}_l = \dot{q} - \dot{q}_s = 2.41 \text{ kW}$$

The process 1–2 and its extension to the left is called the *condition line* for the space. Assuming that state 2, the space condition, is fixed, air supplied at any state on the condition line will satisfy the load requirements. However, as that state is changed different quantities of air must be supplied to the space. The closer point 1 is to point 2, the more air is required, and the converse is also true.

We will now consider several examples of single-path, constant-flow systems. Heat losses and gains to the ducts and fan power will be neglected for the time being.

EXAMPLE 3-9

A given space is to be maintained at 78 F db and 65 F wb. The total heat gain to the space has been determined to be 60,000 Btu/hr of which 42,000 Btu/hr is sensible heat transfer. The outdoor air requirement of the occupants is 500 cfm. The outdoor air has a temperature and relative humidity of 90 F and 55 percent, respectively. Determine the quantity and the state of the air supplied to the space and the required capacity of the cooling and dehumidifying equipment.

SOLUTION

A simplified schematic is shown in Fig. 3-17. The given quantities are shown and stations are numbered for reference. Losses in connecting ducts will be neglected. By Eq. 3-47 the sensible heat factor for the conditioned space is

$$\text{SHF} = \frac{42,000}{60,000} = 0.7$$

The state of the air entering the space lies on the line defined by the SHF on psychrometric Chart 1a. Therefore, state 3 is located as shown on Fig. 3-18a and a line drawn through the point parallel to the SHF = 0.7 line on the protractor. State 2 may be any point on the line and is determined by the operating characteristics of the equipment, desired indoor air quality, and by what will be comfortable for the occupants. These aspects of the problem will be developed later. For now assume that the dry bulb temperature of the entering air t_2 is 20 F less than the space temperature t_3. Then $t_2 = 58$ F and state 2 is determined. The air quantity required may now be found from an energy balance on the space.

$$\dot{m}_{a2} i_2 + \dot{q} = \dot{m}_{a3} i_3$$

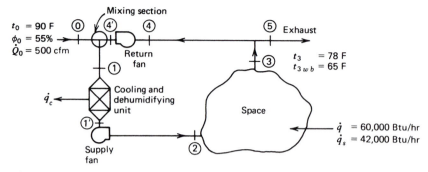

Figure 3-17 Single-line sketch of cooling and dehumidifying system for Example 3-9.

or

$$\dot{q} = \dot{m}_{a2}(i_3 - i_2)$$

and

$$\dot{m}_{a2} = \frac{\dot{q}}{(i_3 - i_2)}$$

From Chart 1a, $i_3 = 30$ Btu/lbma, $i_2 = 23$ Btu/lbma, and

$$\dot{m}_{a2} = \dot{m}_{a3} = \frac{60,000}{(30 - 23)} = 8570 \text{ lbma/hr}$$

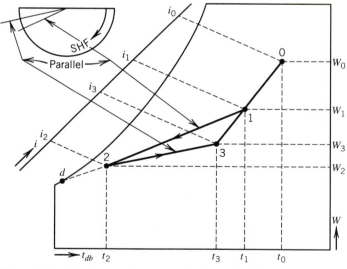

Figure 3-18a Psychrometric process for Example 3-9.

Also from Chart 1a, $v_2 = 13.21$ ft^3/lbma and the air volume flow rate required is

$$\dot{Q}_2 = \dot{m}_{a2}v_2 = \frac{8570(13.21)}{60} = 1890 \, \text{cfm}$$

Before attention is directed to the cooling and dehumidifying process, state 1 must be determined. A mass balance on the mixing section yields

$$\dot{m}_{a0} + \dot{m}_{a4} = \dot{m}_{a1} = \dot{m}_{a2}$$

$$\dot{m}_{a0} = \frac{\dot{Q}_0}{v_0}; \qquad v_0 = 14.23 \, \text{ft}^3/\text{lbma}$$

$$\dot{m}_{a0} = \frac{(500 \times 60)}{14.23} = 2110 \, \text{lbma/hr}$$

Then

$$\dot{m}_{a4} = \dot{m}_{a2} - \dot{m}_{a0} = 8570 - 2110 = 6460 \, \text{lbma/hr}$$

By using the graphical technique discussed in Example 3-7 and referring to Fig. 3-18a, we see that

$$\frac{\overline{31}}{\overline{30}} = \frac{\dot{m}_{a0}}{\dot{m}_{a1}} = \frac{2110}{8570} = 0.246$$

$$\overline{31} = 0.246(\overline{30})$$

State 1 is located at 81 F db and 68 F wb. A line constructed from state 1 to state 2 on Chart 1a then represents the process taking place in the conditioning equipment. An energy balance gives

$$\dot{m}_{a1}i_1 = \dot{q}_c + \dot{m}_{a2}i_2$$

Solving for the rate at which energy is removed in the cooling coil

$$\dot{q}_c = \dot{m}_{a1}(i_1 - i_2)$$

From Chart 1, $i_1 = 32.4$ Btu/lbma and

$$\dot{q}_c = 8570(32.4 - 23) = 80,560 \, \text{Btu/hr} = 6.7 \, \text{tons}$$

The sensible heat factor (SHF) for the cooling coil is found to be 0.6 using the protractor of Chart 1a (Fig. 3-18a). Then

$$\dot{q}_{cs} = 0.6(80,560) = 48,340 \, \text{Btu/hr}$$

and

$$\dot{q}_{cl} = 80,560 - 48,340 = 32,220 \, \text{Btu/hr}$$

The sum of $\dot{q}_{cs}$ and $\dot{q}_{cl}$ is known as the *coil refrigeration load,* which because of outdoor air cooling is different from the *space cooling load.*

In an actual system fans are required to move the air and some energy may be gained from this. Referring to Fig. 3-17, the supply fan is located just downstream of the cooling unit and the return fan is just upstream of the mixing box. All of the power input to the fans is manifested as a sensible energy input to the air, just as if heat were transferred. Heat may also be gained in the supply and return ducts. The power input to the supply air fan and the heat gain to the supply air duct may be summed as shown on Chart 1a, Fig. 3-18b, as process 1′–2. It is assumed that all of the supply fan power input is transformed to internal energy by the time the air reaches the space, state 2. Likewise, heat is gained from point 3 to point 4 and the return fan power input occurs between 4 and 4′, as shown in Fig. 3-18b. The condition line for the space, 2–3, is the same as it was before, when the fans and heat gain were neglected. However, the requirements of the cooling unit have changed. Process 1–1′ now shows that the capacity of the coil must be greater to offset the fan power input and duct heat gain.

An alternate approach to the analysis of the cooling coil in Example 3-9 uses the so-called *coil bypass factor.* Note that when line 1–2 of Fig. 3-18a is extended, it intersects the saturation curve at point *d.* This point represents the *apparatus dewpoint temperature* (t_d) of the cooling coil. The coil cannot cool all of the air passing through it to the coil surface temperature. This fact makes the coil perform in a manner similar to what would happen if a portion of the air were brought to saturation at the coil temperature and the remainder bypassed the coil unchanged. Using Eq. 3-44 and the concept of mixing described in the previous section, the resulting mixture is unsaturated air at point 2. In terms of the length of the line *d*–1, the length *d*–2 is proportional to the mass of air bypassed, and the length 1–2 is proportional to the mass of air not bypassed. Because

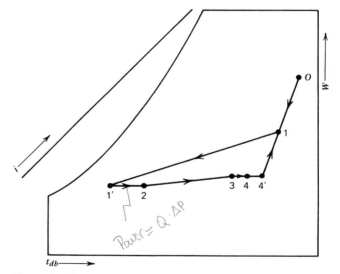

Figure 3-18b Psychrometric processes for Example 3-9, showing the effect of fans and heat gain.

dry bulb lines are not parallel and inclined, it is only approximately true that

$$b = \frac{t_2 - t_d}{t_1 - t_d} \quad \text{by-passes coil} \quad \text{(3-48)}$$

and

$$1 - b = \frac{t_1 - t_2}{t_1 - t_d} \quad \text{through the coil} \quad \text{(3-49)}$$

where b is the fraction of air bypassed, or coil bypass factor, expressed as a decimal, and where the temperatures are dry bulb values. The coil sensible heat transfer rate is

$$\dot{q}_{cs} = \dot{m}_{a1} c_p (t_1 - t_2) \quad \text{(3-50a)}$$

or

$$\dot{q}_{cs} = \dot{m}_{a1} c_p (t_1 - t_d)(1 - b) \quad \text{(3-50b)}$$

EXAMPLE 3-10

Find the bypass factor for the coil of Example 3-9 and compute the sensible and latent heat transfer rates.

SOLUTION

The apparatus dewpoint temperature obtained from Chart 1a as indicated in Fig. 3-18a is 46 F. Then from Eq. 3-48 the bypass factor is

$$b = \frac{58 - 46}{81 - 46} = 0.343 \quad \text{and} \quad 1 - b = 0.657$$

Equation 3-50b expresses the sensible heat transfer rate as

$$\dot{q}_{cs} = 8570(0.245)(81 - 46)(0.657) = 48,280 \text{ Btu/hr}$$

The coil sensible heat factor is used to compute the latent heat transfer rate. From Example 3-9, the SHF is 0.6, then the total heat transfer rate is

$$\dot{q}_t = \dot{q}_{cs}/\text{SHF} = 48,280/0.6 = 80,470 \text{ Btu/hr}$$

and

$$\dot{q}_{cl} = \dot{q}_t - \dot{q}_{cs} = 80,470 - 48,280 = 32,190 \text{ Btu/hr}$$

It should be noted that the bypass factor approach results in values somewhat high in this case.

In Example 3-9 the outdoor air was hot and humid. This is not always the case and

Read

state 0 can be almost anywhere on Charts 1. For example, the southwestern part of the United States is hot and dry during the summer, and evaporative cooling can often be used to advantage under these conditons. A simple system of this type is shown in Fig. 3-19. The dry outdoor air flows through an adiabatic spray chamber and is cooled and humidified. An energy balance on the spray chamber will show that the enthalpies i_o and i_1 are equal; therefore, the process is as shown in Fig. 3-20. Ideally the cooling process terminates at the space condition line. The air then flows through the space and is exhausted. Large quantities of air are required, and this system is not satisfactory where the outdoor relative humidity is high. If W_o is too high the process 0–1 cannot intersect the condition line.

Evaporative cooling can be combined with a conventional system as shown in Fig. 3-21 when outdoor conditions are suitable. There are a number of possibilities. First, if the outdoor air is just mixed with return air without evaporative cooling, the ideal result would be state 1 in Fig. 3-22. The air would require only sensible cooling to state 2 on the condition line. Second, outdoor air could be cooled by evaporation to state 0', Fig. 3-22, and then mixed with return air resulting in state 1'. Sensible cooling would then be

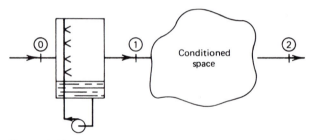

Figure 3-19 A simple evaporative cooling system.

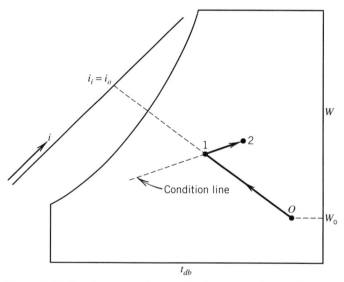

Figure 3-20 Psychrometric diagram for the evaporative cooling system of Figure 3-19.

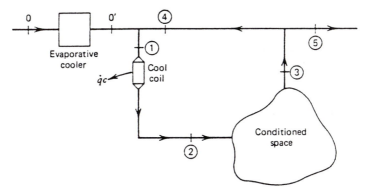

Figure 3-21 Combination evaporative and regular cooling system.

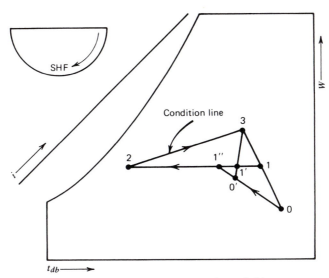

Figure 3-22 Psychrometric diagram for Figure 3-21.

required from state 1′ to state 2. Finally, the outdoor air could ideally be evaporatively cooled all the way to state 1″. This would require the least power for sensible cooling, and the air supplied to the space would be 100 percent outdoor air.

Examples of other single-path systems such as VAV or multizone could be presented; however, under the full-flow design condition, these systems operate the same as the simple system of Figs. 3-17 and 3-18. They will be discussed further in the section on part-load operation.

EXAMPLE 3-11   Read

A space is to be maintained at 75 F and 50 percent relative humidity. Heat losses from the space are 225,000 Btu/hr sensible and 56,250 Btu/hr latent. The latent heat transfer is due to the infiltration of cold dry air. The outdoor air required is 1000 cfm and is at 35 F and 80 percent relative humidity. Determine the quantity of air supplied at 120 F, the state of the supply air, the size of the furnace or heating coil, and the humidifier characteristics.

SOLUTION

Figure 3-23 is a schematic for the problem; it contains the given information and reference points. First consider the conditioned space

$$\text{SHF} = \frac{225{,}000}{(225{,}000 + 56{,}250)} = 0.80$$

The state of the supply air lies on a line drawn through state point 3 parallel to the SHF = 0.8 line on the protractor of Chart 1a. Figure 3-24 shows this construction. State 2 is located at 120 F dry bulb and the intersection of this line. An energy balance on the space gives

$$\dot{m}_{a2} i_2 = \dot{q} + \dot{m}_{a3} i_3$$

or

$$\dot{q} = \dot{m}_{a2}(i_2 - i_3)$$

From Chart 1a, $i_2 = 42$ Btu/lbma, $i_3 = 28.2$ Btu/lbma, and

$$\dot{m}_{a2} = \frac{\dot{q}}{(i_2 - i_3)} = \frac{281{,}250}{(42 - 28.2)} = 20{,}400 \text{ lbma/hr}$$

From Chart 1a, $v_2 = 14.89$ ft³/lbma, and

$$\dot{Q}_2 = \frac{20{,}400}{60} \times 14.89 = 5060 \text{ cfm}$$

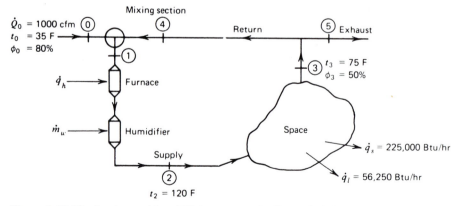

Figure 3-23 The heating and humidifying system for Example 3-11.

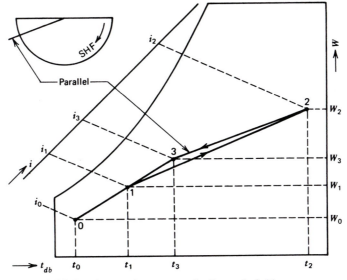

Figure 3-24 Psychrometric processes for Example 3-11.

To find the conditions at state 1, the mixing process must be considered. A mass balance on the mixing section yields

$$\dot{m}_{a0} + \dot{m}_{a4} = \dot{m}_{a1} = \dot{m}_{a2}$$

or

$$\dot{m}_{a4} = \dot{m}_{a2} - \dot{m}_{a0}$$

$$\dot{m}_{a0} = \frac{\dot{Q}_0}{v_0} \quad \text{and} \quad v_0 = 12.53 \text{ ft}^3/\text{lbma}$$

$$\dot{m}_{a4} = 20,400 - \frac{1000 \times 60}{12.53} = 15,600 \text{ lbma/hr}$$

Using the graphical technique and referring to Fig. 3-22, we obtain

$$\overline{31} = \frac{m_{a0}}{m_{a1}}(\overline{30}) = \frac{4790}{20,400}(\overline{30}) = 0.235\,(\overline{30})$$

State 1 is then located at 66 F db and 57 F wb. The line $\overline{12}$ constructed on Chart 1a, Fig. 3-22, repesents the heating and humidifying process that must take place in the heating and humidifying unit. Equation 3-30 is applicable for this situation:

$$\dot{q}_h = \dot{m}_{a1}(i_2 - i_1) - \dot{m}_{a1}(W_2 - W_1)i_w$$

Assume that water at 55 F is used in the humidifier. Then $i_w = 23.08$ Btu/lbm from

Table A-1a. From Chart 1a, i_1 = 24.6 Btu/lbma, W_1 = 0.008 lbmw/lbma, and W_2 = 0.0119 lbmw/lbma. Therefore

$$\dot{q}_h = 20,400[(42 - 24.6) - (0.0119 - 0.008)23.08] = 353,000 \text{ Btu/hr}$$

The amount of water supplied to the humidifier is

$$\dot{m}_w = \dot{m}_{a1}(W_2 - W_1) = 20,400(0.0119 - 0.008)$$
$$\dot{m}_w = 79.6 \text{ lbm/hr} = 1.33 \text{ lbm/min}$$

It is usually necessary to use a preheat coil to heat the outdoor air to a temperature

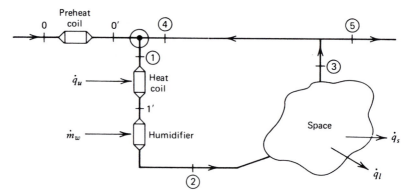

Figure 3-25 Heating system with preheat of outdoor air.

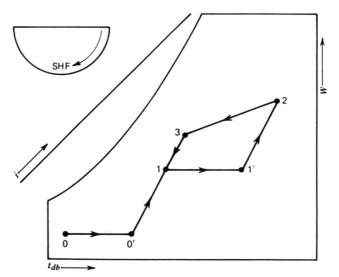

Figure 3-26 Psychrometric diagram for Figure 3-25.

above the dewpoint of the air in the equipment room so that condensation will not form on the air ducts upstream of the regular heating coil. Figure 3-25 shows this arrangement. The outdoor air is heated to state $0'$ where it is mixed with return air, resulting in state 1. The mixed air is then heated to state $1'$, where it is humidified to state 2 on the condition line for supply to the space. Figure 3-26 shows the states on Charts 1.

3-7   SPACE AIR CONDITIONING—OFF-DESIGN CONDITIONS

The previous section treated the common space air conditioning problem assuming that the system was operating steadily at the design condition. Actually the space requires only a part of the designed capacity of the conditioning equipment most of the time. A control system functions to match the required cooling or heating of the space to the conditioning equipment by varying one or more system parameters. For example, the quantity of air circulated through the coil and to the space may be varied in proportion to the space load, as in the *variable-air-volume* (VAV) system. Another approach is to circulate a constant amount of air to the space, but some of the return air is diverted around the coil and mixed with air coming off the coil to obtain a supply air temperature that is proportional to the space load. This is known as face and bypass control, because face and bypass dampers are used to divert the flow. Another possibility is to vary the coil surface temperature with respect to the required load by changing the temperature or the amount of heating or cooling fluid entering the coil. This technique is usually used in conjunction with VAV and face and bypass systems. However, control of the coolant temperature or quantity may be the only variable in some small systems.

These and other control methods will be considered separately later. Figure 3-27a illustrates what might occur when the load on a variable-air-volume system decreases. The solid lines represent the full-load design condition, whereas the broken lines illustrate a part-load condition where the amount of air circulated to the space and across the coil has decreased. Note that the state of the outdoor air $0'$ has changed and could be

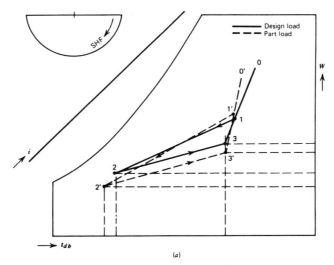

Figure 3-27 Psychrometric processes for off-design conditions.
(a) Variable air-volume control.

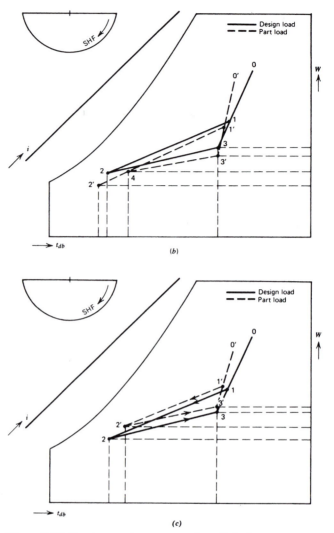

Figure 3-27 Psychrometric processes for off-design conditions. (*b*) Face and bypass control. (*c*) Water-temperature control.

almost anyplace on the chart under part-load conditions. Due to the lower air flow rate through the coil, the air is cooled to a lower temperature and humidity. The thermostat maintains the space temperature, but the humidity in the space will decrease. This explains why control of the water temperature or flow rate is desirable. Decreasing the water flow rate will cause point 2′ to move upward and to the right to a position where the room process curve may terminate at point 3.

The behavior of a constant-air-volume, face and bypass system is shown in Fig. 3-27*b*. The total design air flow rate is flowing at state 2, 3, and 3′ but a lower flow rate occurs at state 2′, leaving the coil. Air at states 2′ and 1′ is mixed downstream of the coil to obtain state 4. The total design flow rate and the enthalpy difference, $i_3′ - i_4$, then match the space load. Note that the humidity at state 4 is lower than necessary, which lowers state 3′ below the design value. At very small space loads, point 4 may be located very near state 1′ on the condition line. In this case the humidity in the space may become high. This is a disadvantage of a multizone face and bypass system. Control of the coil water temperature can help to correct this problem.

A constant air volume system with either water temperature or flow rate control is shown in Fig. 3-27c. In this case both the temperature and humidity of the air leaving the coil increase and the room process curve 2′–3′ may not terminate at state 3. In fact, there will be cases when state 3′ will lie above state 3, causing an uncomfortable condition in the space. For this reason, water control alone is not usually used in commercial applications, but is used in conjunction with VAV and face and bypass systems as discussed before. In fact, all water coils should have control of the water flow rate. This is also important to the operation of the water chiller and piping system. The following example illustrates the analysis of a VAV system with variable water temperature.

EXAMPLE 3-12

A variable-air-volume system operates as shown in Fig. 3-28. The solid lines show the full-load design condition of 100 tons with a room SHF of 0.75. At the estimated minimum load of 15 tons with SHF of 0.9, the air flow rate is decreased to 20 percent of the design value and all outdoor air is shut off. Estimate the supply air and apparatus dewpoint temperatures of the cooling coil for minimum load, assuming that state 3 does not change.

SOLUTION

The solution is best carried out using Chart 1a, as shown in Fig. 3-28. Because the outdoor air is off during the minimum load condition, the space condition and coil

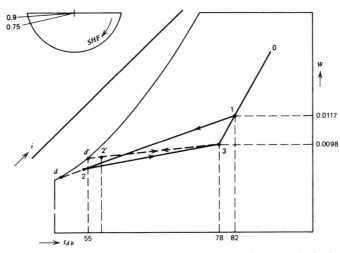

Figure 3-28 Schematic psychrometric processes for Example 3-12.

process lines will coincide as shown by line $3-2'-d'$. This line is constructed by using the protractor of Chart 1a with a SHF of 0.9. The apparatus dewpoint is seen to be 55 F, as compared with 50 F for the design condition. The air flow rate for the design condition is given by

$$\dot{m}_2 = \dot{q}/(i_3 - i_2)$$
$$\dot{m}_2 = 100(12,000)/(29.4 - 23.2) = 193,550 \text{ lbma/hr}$$

or

$$\dot{Q}_2 = \dot{m}_2 v_2/60 = 193,550(13.25)/60 = 42,740 \text{ cfm}$$

Then, the minimum volume flow rate is

$$\dot{Q}_m = 0.2(42,740) = 8,550 \text{ cfm}$$

and the minimum mass flow rate may be estimated by assuming a value for $v_{2'}$:

$$\dot{m}_m = 8550(60)/13.28 = 38,630 \text{ lbma/hr}$$

State point $2'$ may then be determined by computing $i_{2'}$:

$$i_{2'} = i_3 - \frac{\dot{q}_m}{\dot{m}_m} = 29.4 - 15(12,000)/38,630 = 24.7 \text{ Btu/lbma}$$

Then from Chart 1a, the air condition leaving the coil is 60.5 F db and 57.5 F wb. Calculation of the coil water temperature is beyond the scope of this analysis; however, the water temperture would be increased by about 7 degrees from the design to the minimum load condition.

Reheat was mentioned as a variation on the simple constant flow and VAV systems to obtain control under part-load conditions. Figure 3-29 shows how this affects the psychrometric analysis for a typical zone. After the air leaves the cooling coil at state 2, it is then heated to state $2'$ and enters the zone at a higher temperature to accommodate the part-load condition. A VAV reheat system operates similarly.

The economizer cycle is a system used during part-load conditions when outdoor temperature and humidity are favorable to saving operating energy. One must be cautious in the application of such a system, however, if the desired space conditions are to be maintained. Once the cooling equipment and especially the coil have been selected, there are limitations on the quantity and state of the outdoor air. The coil apparatus dewpoint can be used as a guide to avoid impossible situations. For example, a system is designed to operate as shown by the solid process lines in Fig. 3-30. Assume that the condition line 2–3, does not change, but state 0 changes to state $0'$. Theoretically a mixed state $1'$ located anyplace on the line $0'-3$ could occur, but the air must be cooled and dehumidified to state 2. To do this the coil apparatus dewpoint must be reasonable. Values below about 48 F are not economical to attain. Therefore, state $1'$ must be controlled to accommodate the coil. It can be seen in Fig. 3-30 that moving of

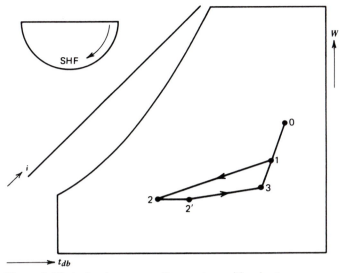

Figure 3-29 A simple constant flow system with reheat.

state 1′ closer to state 0′ lowers the coil apparatus dewpoint rapidly and soon reaches the condition where the coil process line will not intersect the saturation curve, indicating an impossible condition. It is obvious in Fig. 3-30 that less energy is required to cool the air from state 1′ to 2 than from state 1 to 2. There are many other possibilities, which must be analyzed on their own merits. Some may require more or less outdoor air, humidification, or reheat to reach state 2 in Fig. 3-30.

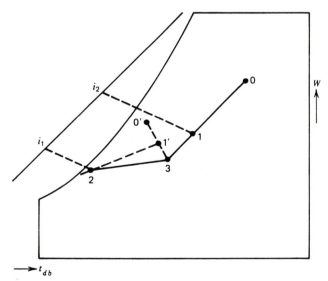

Figure 3-30 Psychrometric processes for an economizer cycle.

REFERENCES

1. R. W. Hyland and A. Wexler, "Formulations for the Thermodynamic Properties of the Saturated Phases of H_2O from 173.15 K to 473.15 K," *ASHRAE Transactions,* Vol. 89, Part 2A, 1983.
2. R. W. Hyland and A. Wexler, "Formulations for the Thermodynamic Properties of Dry Air from 173.15 K to 473.15 K, and of Saturated Moist Air from 173.15 K to 372.15 K, at Pressures to 5 MPa," *ASHRAE Transactions,* Vol. 89, Part 2, 1983.
3. *ASHRAE Handbook, Fundamentals Volume,* American Society of Heating, Refrigerating and Air-Conditioning Engineers, Inc., Atlanta, GA, 1989.
4. James L. Threlkeld, *Thermal Environmental Engineering,* 2nd ed., Prentice-Hall, Englewood Cliffs, NJ, 1970.
5. J. A. Goff, "Standardization of Thermodynamic Properties of Moist Air," *Transactions ASHVE,* 55, 1949.
6. R. B. Stewart, R. J. Jacobsen, and J. H. Becker, "Formulations for Thermodynamic Properties of Moist Air at Low Pressures as Used for Construction of New ASHRAE SI Unit Psychrometric Charts," *ASHRAE Transactions,* Vol. 89, Part 2, 1983.

PROBLEMS

3-1. Calculate values of humidity ratio, enthalpy, and specific volume for saturated air at one standard atmosphere using perfect gas relations for temperatures of (a) 70 F (20 C) and (b) 20 F (-6.7 C).

3-2. Moist air exists at a dewpoint temperature of 18 C, a relative humidity of 60 percent, and a pressure of 96.5 kPa. Determine (a) the humidity ratio and (b) the volume in m^3/kga.

3-3. The temperature of a certain room is 22 C and the relative humidity is 50 percent. The barometric pressure is 100 kPa. Find (a) the partial pressures of the air and water vapor, (b) the vapor density, and (c) the humidity ratio of the mixture.

3-4. Compute the local atmospheric pressure at elevations ranging from sea level to 6000 ft (1830 m) in (a) inches of mercury and (b) kPa.

3-5. Rework Problem 3-1 for an atmospheric pressure corresponding to an elevation of (a) 5280 ft and (b) 1600 m.

3-6. Compute the enthalpy of moist air at 60 F (16C) and 80 percent relative humidity for an elevation of (a) sea level and (b) 5000 ft (1525 m).

3-7. The condition within a room is 70 F db, 50 percent relative humidity, and 14.696 psia pressure. The inside surface temperature of the windows is 40 F. Will moisture condense on the window glass?

3-8. Assume that the dimensions of the room of Problem 3-7 are 30 $\times$ 15 $\times$ 8 ft. Calculate the mass of water vapor in the room.

3-9. A duct has moist air flowing at a rate of 5000 ft^3/min (2.36 m^3/s). What is the mass flow rate of the dry air, where the dry bulb temperature is 60 F (16 C), the relative humidity is 80 percent and where the pressure inside the duct corresponds to (a) sea level, and (b) 6000 ft (1830 m)?

3-10. Compute the dewpoint temperature for moist air at 80 F (27 C) and 50 percent relative humidity for pressures corresponding to (a) sea level and (b) 5000 ft (1225 m).

3-11. Air with a dry bulb temperature of 70 F and a wet bulb temperature of 65 F is at a barometric pressure of 29.92 in. Hg. Without making use of the psychrometric chart find (a) the relative humidity of the air, (b) the vapor density, (c) the dewpoint temperature, (d) the humidity ratio, and (e) the volume occupied by the mixture associated with a pound mass of dry air.

3-12. Air is supplied to a room at 72 F db and 68 F wb from outside air at 40 F db and 37 F wb. The barometric pressure is 29.92 in. Hg. Find (a) the dewpoint temperatures of the inside and outside air, (b) the moisture added to each pound of dry air, (c) the enthalpy of the outside air, and (d) the enthalpy of the inside air.

3-13. Air is cooled from 75 F db and 70 F wb until it is saturated at 55 F. Find (a) the moisture removed per pound of dry air, (b) the heat removed to condense the moisture, (c) the sensible heat removed, and (d) the total amount of heat removed.

3-14. The dry bulb and thermodynamic wet bulb temperature are measured to be 75 F and 62 F, respectively, in a room. Compute the humidity ratio and relative humidity for the air at (a) sea level and (b) 5000 ft (1225 m).

3-15. To what temperature must atmospheric air at standard sea level pressure be cooled to be saturated with a humidity ratio of 0.001 lbv/lba? What is the temperature if the pressure is 5 atmospheres?

3-16. Complete Table 3-3 using the ASHRAE psychrometric chart for (a) sea level, (b) 5000 ft (1500 m) elevation; (c) compare parts (a) and (b).

3-17. To save energy, the environmental conditions in a room are to be regulated so that the dry bulb temperature will be greater than or equal to 78 F (24 C) and the dewpoint temperature will be less than or equal to 64 F (17 C). Find the maximum relative humidity that can occur for standard barometric pressure.

3-18. Atmospheric air at 85 F (29 C), 60 percent relative humidity is compressed to 60 psia (414 kPa) and then cooled in an intercooler before entering a second stage of compression. What is the minimum temperature to which the air can be cooled without condensation?

3-19. A chilled water cooling coil receives 2.5 m³/s of air at 25 C db, 20 C wb. It is necessary for the air to leave the coil at 13 C db, 12 C wb. Assume sea level pressure.
(a) Determine the SHF and the apparatus dewpoint temperature.
(b) Compute the bypass factor.
(c) Compute total and sensible heat transfer rates from the air using enthalpy difference and the SHF.
(d) Compute the total and sensible heat transfer rate from the air using the bypass factor and the SHF.
(e) Compare the results of parts (c) and (d).

3-20. The dry bulb and wet bulb temperatures are measured to be 78 F and 65 F, respectively, in an air duct. Make use of psychrometric Charts 1*a* and 1*b* to find the enthalpy, specific volume, humidity ratio, and relative humidity in (a) English units and (b) SI units.

3-21. The air in Problem 3-20 is heated to a temperature of 110 F. Make use of Charts 1*a* and 1*b* and compute the heat transfer rate if 4000 ft³/min (1.9 m³s) is flowing at state 1 in (a) English units and (b) SI units.

Table 3-3 Psychrometric Properties for Problem 3-16

Dry Bulb F (C)	Wet Bulb F (C)	Dewpoint F (C)	Humid. Ratio, W lbv/lba (kga/kgv)	Enthalpy, i Btu/lba (kJ/kga)	Rel. Humid, Percent	Spec. Vol, v ft³/lba (m³/kga)
85 (29)	60 (16)	50 (10)				
75 (24)	70 (21)	82 (28)	0.01143	30 (70) 50 (116)	60	

3-22. The air in Problem 3-20 is cooled to 50 F (10 C) dry bulb and 90 percent relative humidity. Make use of Charts 1*a* and 1*b* to compute the total heat transfer rate, the sensible heat transfer rate, and the sensible heat factor (SHF) if 4000 ft³/min (1.9 m³/s) is flowing at state 1 in (a) English units and (b) SI units.

3-23. Saturated air at 45 F (7 C) is first heated and then saturated adiabatically. This saturated air is then heated to 105 F (41 C) and 30 percent relative humidity. To what temperature must the air initially be heated?

3-24. Air at 100 F (38 C) db and 65 F (18 C) wb is humidified adiabatically with steam. The steam supplied contains 20 percent moisture (quality of 0.80) at 14.7 psia (101.3 kPa). If the air is humidified to 60 percent relative humidity, what is the dry bulb temperature of the humidified air? Assume sea level pressure.

3-25. Air at 84 F (29 C) db and 60 F (16 C) wb is humidified with the dry bulb temperature remaining constant. Wet steam is supplied for humidification at 20 psia (138 kPa). What quality must the steam have (a) to provide saturated air and (b) to provide air at 70 percent relative humidity? Assume sea level pressure.

3-26. Air at 38 C db and 20 C wb and 101.325 kPa is humidified adiabatically with liquid water supplied at 60 C, in such proportions that the mixture has a relative humidity of 80 percent. Find the dry bulb temperature of the mixture.

3-27. It is desired to heat and humidify 2000 cfm (1.0 m³/s) of air from an initial state defined by a temperature of 60 F (16 C) dry bulb and relative humidity of 30 percent to a final state of 110 F (43 C) dry bulb and 30 percent relative humidity. The air will first be heated by a hot water coil followed by adiabatic humidification using saturated vapor at 5 psig (34.5 kPa). Using the psychrometric chart, find the heat transfer rate for the heating coil and the mass flow rate of the water vapor, and sketch the processes on a skeleton chart showing pertinent data. Use (a) English units and (b) SI units. Assume sea level pressure.

3-28. Air at 40 F (5 C) db and 35 F (2 C) wb is mixed with warm air at 100 F (38 C) db and 77 F (25 C) wb in the ratio of 2 lbm (kga) cool air to 1 lbm (kga) of warm air. (a) Compute the humidity ratio and enthalpy of the mixed air and (b) find the humidity ratio and enthalpy using psychrometric chart 1.

3-29. Air at 10 C db and 5 C wb is mixed with air at 25 C db and 18 C wb in a steady-flow process at standard atmospheric pressure. The volume flow rates are 10 m³/s and 6 m³/2, respectively. (a) Compute the mixture conditions. (b) Find the mixture conditions using Chart 1*b*.

3-30. Rework Problem 3-29, using Chart 1*b*, assuming that the mixture condition can be computed on the basis of the volume flow rates rather than mass flow rate. What is the percent error in the mixture enthalpy and humidity ratios?

3-31. A zone in a building has a design cooling load of 136,000 Btu/hr (40 kW) of which 110,000 Btu/hr (32 kW) is sensible cooling load. The space is to be maintained at 76 F (24 C) dry bulb temperature and 50 percent relative humidity. Locate the space condition line on Charts 1*a* and 1*b*.

3-32. Refer to Problem 3-31 and assume that the air can be supplied to the space at 55 F (13 C). Compute the volume flow rate of the air required in (a) English units and (b) SI units.

3-33. Rework Problem 3-31 using Charts 1H*a* and 1H*b* for (a) 5000 ft and (b) 1500 m elevation, respectively.

3-34. Rework Problem 3-32 using Charts 1H*a* and 1H*b* for (a) 5000 and (b) 1500 m elevation, respectively.

3-35. A meeting hall is to be maintained at 25 C db and 18 C wb. The barometric pressure is 101.3 kPa. The space has a load of 58.6 kW sensible and 58.6 kW latent. The temperature of the supply air to the space cannot be lower than 18 C db. (a) How many kg/s of air must be supplied? (b) What is the required wet bulb temperature of the supply air? (c) What is the sensible heat ratio?

3-36. A space is to be maintained at 72 F (22 C) and 30 percent relative humidity during the winter months. The sensible heat loss from the space is 500,000 Btu/hr (146 kW) and the latent heat loss due to infiltration is 50,000 Btu/hr (14.6 kW). Construct the condition line on (a) Chart 1a and (b) Chart 1b.

3-37. What is the maximum relative humidity to prevent condensation on pipes carrying water at 50 F through a room that has an air temperature of 70 F?

3-38. A space is to be maintained at 21 C dry bulb. It is estimated that the inside wall surface temperature could be as low as 7 C. What maximum relative and specific humidities can be maintained without condensation on the walls?

3-39. Outdoor air with a temperature of 40 F db and 35 F wb and with a barometric pressure of 29 in. Hg is treated and humidified under steady-flow conditions to a final temperature of 70 F db and 40 percent relative humidity. (a) Find the mass of water vapor added to each pound mass of dry air. (b) If the water is supplied at 50 F, how much heat is added per pound mass of dry air?

3-40. Outdoor air at 95 F db and 79 F wb and at a barometric pressure of 29.92 in. Hg is cooled and dehumidified under steady conditions until it becomes saturated at 60 F. (a) Find the mass of water condensed per pound of dry air. (b) If the condensate is removed at 60 F, what quantity of heat is removed per pound of dry air?

3-41. Outdoor air at 38 C db and 26 C wb is cooled to 15 C db and 14 C wb. The process occurs at 101.3 kPa barometric pressure. Determine (a) the mass of water condensed per kilogram of dry air and (b) the sensible and latent heat transfer per kilogram of dry air and the sensible heat factor.

3-42. Moist air enters a refrigeration coil at 89 F db and 75 F wb at a rate of 1400 cfm. The apparatus dewpoint temperature of the coil is 55 F. (a) If 3.5 tons of refrigeration are available, find the dry bulb temperature of the air leaving the coil. (b) Compute the bypass factor. Assume sea level pressure.

3-43. Saturated steam at a pressure of 25 psia is sprayed into a stream of moist air. The initial condition of the air is 55 F db and 45 F wb temperature. The mass rate of air flow is 2000 lbma/min. Barometric pressure is 14.696 psia. Determine (a) how much steam must be added in lbm/min to produce a saturated air condition and (b) the resulting temperature of the saturated air.

3-44. Saturated water vapor at 100 C is used to humidify a stream of moist air. The air enters the humidifier at 13 C db and 2 C wb at a flow rate of 2.5 m³/s. The pressure is 101.35 kPa. Determine (a) the mass flow rate of the steam required to saturate the air and (b) the temperature of the saturated air.

3-45. Moist air at 70 F db and 45 percent relative humidity is recirculated from a room and mixed with outdoor air at 97 F db and 83 F wb. Determine the mixture dry bulb and wet bulb temperatures if the volume of recirculated air is three times the volume of outdoor air. Assume sea level pressure.

3-46. A structure has a calculated cooling load of 10 tons of which 2.5 tons is latent load. The space is to be maintained at 76 F db and 50 percent relative humidity. Ten percent by volume of the air supplied to the space is outdoor air at 100 F db and 50 percent relative humidity. The air supplied to the space cannot be less that 56 F db. Assume sea level pressure and find (a) the minimum amount of air supplied to the space in cfm, (b) the amounts of return air and outdoor air in cfm, (c) the conditions and volume flow rate of the air entering the cooling coil, and (d) the capacity and SHF for the cooling coil.

3-47. Rework Problem 3-46 for an elevation of 5000 feet.

3-48. A building has a calculated cooling load of 410 kW. The latent portion of the load is 100 kW. The space is to be maintained at 25 C db and 50 percent relative humidity. Outdoor air is at 38 C and 50 percent relative humidity, and 10 percent by mass of the air supplied to the space is outdoor air. Air is to be supplied to the space at not less than 18 C. Assume sea level pressure and find (a) the minimum amount of air supplied to the space in m^3/s, (b) the volume flow rates of the return air, exhaust air, and outdoor air, (c) the condition and volume flow rate of the air entering the cooling coil, and (d) the capacity, apparatus dewpoint temperature, bypass factor, and SHF of the cooling coil.

3-49. Rework Problem 3-48 for an elevation of 1500 m.

3-50. A building has a total heating load of 200,000 Btu/hr. The sensible heat factor for the space is 0.8. The space is to be maintained at 72 F db and 40 percent relative humidity. Outdoor air at 40 F db and 20 percent relative humidity in the amount of 1000 cfm is required. Air is supplied to the space at 120 F db. Find (a) the conditions and amount of air supplied to the space, (b) the temperature rise of the air through the furnace, (c) the amount of water at 50 F required by the humidifier, and (d) the capacity of the furnace. Assume sea level pressure.

3-51. Rework Problem 3-50 for an elevation of 5000 feet.

3-52. The system of Problem 3-46 has a supply air fan located just downstream of the cooling coil. The total power input to the fan is 2.0 hp. It is also estimated that heat gain to the supply duct system is 500 Btu/hr. Rework Problems 3-46 taking the fan and duct heat gain into account. Make a sketch of the processes.

3-53. The system of Problem 3-48 has a supply air fan located just downstream of the coil and a return air fan just upstream of the mixing box. The power input to the supply fan is 18 kW and the power input to the return fan is 12 kW. Rework Problem 3-48 taking the fan power into account. Make a sketch of the system processes on a skeleton psychrometric chart.

3-54. A large warehouse located in Denver, Colorado (elevation = 5000 ft or 1500 m), is to be conditioned using an evaporative cooling system. Assume that the space is to be maintained at 80 F (27 C) and 50 percent relative humidity by a 100 percent outdoor air system. Outdoor design conditions are 91 F (33 C) db and 59 F (15 C) wb. The cooling load is estimated to be 100 tons (352 kW) with a sensible heat factor of 0.9. The supply air fan is located just downstream of the spray chamber and is estimated to require a power input of 20 hp (15 kW). Determine the volume flow rate of air to the space and sketch the processes on a skeleton psychrometric chart in (a) English units and (b) SI units.

3-55. The summer design conditions for Tucson, Arizona, are 102 F (39 C) dry bulb and 66 F (19 C) wet bulb temperature. In Shreveport, Louisiana, the design conditions are 96 F (36 C) dry bulb and 76 F (24 C) wet bulb temperature. What is the lowest air temperature that can theoretically be attained in an evaporative cooler for these design conditions in each city?

3-56. Consider a conventional cooling system designed for use at high elevation (5000 ft or 1500 m). The space is to be maintained at 75 F (24 C) db and 40 percent relative humidity, and outdoor design conditions are 100 F (38 C) and 10 percent relative humidity. Outdoor air is to be mixed with return air such that it can be cooled sensibly to 50 F (10 C), where it crosses the condition line. The air is then supplied to the space. Sketch the processes on Chart 1H*a* or 1H*b* and compute the volume flow rate of the supply air and the percent outdoor air by mass per ton of cooling load for a SHF of 0.7 in (a) English units and (b) SI units.

3-57. A space heating system is designed as shown in Fig. 3-23 for a large zone in a building. Under design conditions for Kansas City, Missouri, air enters the preheat coil at 6 F (-14 C) and essentially 0 percent relative humidity. The outdoor air is heated to 60 F (16 C) and mixed with return air. It is then heated and humidified in a separate process to 105 F (40 C) and 30 percent relative humidity (RH) for supply to the space. Saturated vapor at 2.0 psig is used for humidification. Twenty-five percent of the supply air is outdoor air by mass. The total space heating load is 500,000 Btu/hr (145 kW) and the space design conditions are 70 F (21 C) and 30 percent RH. Sketch the psychrometric processes and compute the supply air volume flow rate, the heat transfer rates in both coils, and the steam flow rate in (a) English units and (b) SI units.

3-58. A space is to be maintained at 78 F (26 C) db and 68 F (20 C) wb. The cooling system is a variable-air-volume (VAV) type where the quantity of air supplied and the supply air temperature are controlled. Under design conditions, the total cooling load is 150 tons (530 kW) with a sensible heat factor of 0.6 and the supply air temperature is 60 F (16 C) db. At minimum load, about 18 tons (63 kW) with SHF of 0.8, the air quantity may be reduced no more than 80 percent by volume of the full load design value. Determine the supply air conditions for minimum load. Show all the conditions on a psychrometric chart for (a) English units and (b) SI units. Assume sea level pressure.

3-59. Rework Problem 3-58 for an elevation of 5000 feet (1500 m).

3-60. Investigate the feasibility of conditioning a space with 100 percent outdoor air with a direct expansion cooling coil that requires a fixed volume flow rate of air per ton of cooling capacity. Explain. Assuming that this is not feasible with a fixed flow rate, describe a different kind of coil and cooling medium to accomplish the process.

3-61. A 50-ton constant-volume space air-conditioning system uses face and bypass and water temperature control. At the design condition the space is to be maintained at 77 F (25 C) db and 50 percent relative humidity with 55 F (13 C) db supply air at 90 percent relative humidity. Outdoor air is supplied at 95 F (35 C) db, 60 percent relative humidity with a ratio of 1 lbm (kgm) to 5 lbm (kgm) return air. A part-load condition exists where total space load decreases by 50 percent and the SHF increases to 90 percent. The outdoor air condition changes to 85 F (29 C) db and 70 percent relative humidity. Assume sea level pressure.

  (a) At what temperature must the air be supplied to the space under the part-load condition?

  (b) If the air leaving the coil has a dry bulb temperature of 60 F (15 C), what is the ratio of the air bypassed to that flowing through the coil?

  (c) What is the apparatus dewpoint temperature for both the design and part-load conditions?

    (d) Show all the processes on a psychrometric chart.

3-62. Rework Problem 3-61 for an elevation of 5000 feet (1500 m).

3-63. A condition exists where it is necessary to cool and dehumidify air from 80 F db and 67 F wb to 60 F db and 54 F wb.

    (a) Discuss the feasibility of doing this in one process with a cooling coil. (*Hint:* Determine the apparatus dewpoint temperature for the process.)

    (b) Describe a practical method of achieving the required process and sketch it on a psychrometric chart.

3-64. Consider one zone of a dual-duct conditioning system. Conditions in the zone are to be maintained at 75 F (24 C) and 50 percent relative humidity (RH). Cold deck air is at 51 F (11 C) and 90 percent RH while the hot deck air is outdoor air at 90 F (32 C) and 30 percent RH. The sensible heat factor for the zone is 0.65. Assume sea level pressure. In what proportion must the warm and cold air be mixed to satisfy the space condition? If the total space load is 60 tons (210 kW), what is the total volume flow rate of air supplied to the zone? Sketch the states and processes on a psychrometric chart. (a) Use English units and (b) SI units.

3-65. Rework Problem 3-64 for an elevation of 5000 ft (1500 m).

3-66. In Problem 3-64 return air is cooled by a water coil to the cold deck condition. Sketch the processes for the entire system on a psychrometric chart and determine the coil load (for the one zone) and volume flow rate in (a) English units and (b) SI units.

3-67. During the winter months it is possible to cool and dehumidify a space using outdoor air. Suppose an interior zone of a large building is designed to have a supply air flow rate of 5000 cfm, which can be all outdoor air. The cooling load is constant at 10 tons with a SHF of 0.8 the year round. Indoor conditions are 78 F db and 67 F wb.

    (a) What is the maximum outdoor air dry bulb temperature and humidity ratio that would satisfy the load condition?

    (b) Consider a different time when the outdoor air has a temperature of 40 F db and 20 percent relative humidity. Return air and outdoor air may be mixed to cool the space, but humidification will be required. Assume that saturated water vapor at 14.7 psia is used to humidify the mixed air and compute the amounts of outdoor and return air in cfm.

    (c) At another time, outdoor air is at 70 F db with a relative humidity of 90 percent. The cooling coil is estimated to have a minimum apparatus dewpoint of 50 F. What amount of outdoor and return air should be mixed before entering the coil to satisfy the given load condition?

    (d) What is the refrigeration load for the coil of part (c) above?

3-68. An economizer mixes outdoor air with room return air to reduce the refrigeration load on the cooling coil.

    (a) For a space condition of 25 C db and 20 C wb, describe the maximum wet bulb and dry bulb temperatures that will reduce the coil load.

    (b) Suppose a system is designed to supply 5 m³/s at 18 C db and 17 C wb to a space maintained at the conditions given in part (a) above. What amount of outdoor air at 20 C db and 90 percent relative humidity can be mixed with the return air if the coil SHF is 0.6?

    (c) What is the apparatus dewpoint and the bypass factor in part (b) above?

    (d) Compare the coil refrigeration load in part (b) above with the outdoor air to that without outdoor air.

Chapter 4

Indoor Air Quality—
Comfort and Health

Air conditioning is used to maintain temperature, humidity, air circulation, and air quality within an indoor environment. In the past, comfort at reasonable cost was the primary concern of HVAC designers and those responsible for building construction and maintenance. A comfortable environment was generally taken to be a healthy one, but an indoor environment may be comfortable without being healthy. Some overzealous attempts to save energy through reduction of outdoor air have caused problems with the health of building occupants. Today health as well as comfort are both concerns when humans are occupants of an indoor space. The effect of the indoor environment on quality and process efficiency are additional important factors in many industrial settings. In this chapter we shall be concerned with the conditions that provide a comfortable and healthful indoor environment for humans. Industrial ventilation, as well as specialized environments for laboratories, industry, and health facilities, will not be emphasized, although some of the methods discussed here may have application to some of those cases.

Dilution of air contaminants by introduction of outdoor air is the primary means of assuring acceptable indoor air quality. In recent years the increased tightness of buildings and the concern with energy conservation have often caused reductions in the amount of outdoor air entering buildings. Most modern, occupied buildings have carpeting, furniture, draperies, and other accessories that out-gas a variety of materials, especially volatile organic compounds (VOCs) with potential for harming human health. New terms, such as "sick building syndrome" have been coined and the acronym for indoor air quality (IAQ) has become a familiar term in the HVAC technical literature.

Although health, safety, and economics are being given increased attention, comfort is still a major concern of the HVAC industry. Experience has shown that not everyone can be made completely comfortable by one set of conditions. Although all factors affecting comfort are not completely understood, it is clear that comfort is directly influenced by the air temperature, humidity, and motion as well as thermal radiation from surrounding surfaces. Odor, dust, and noise are additional factors that might cause one to not feel "comfortable." A well-designed HVAC system attempts to keep these variables within specified limits, set by the customer, building codes, and good engineering practice. These complex problems present a challenge to engineers responsible for the design, construction, and/or operation of HVAC systems. This chapter attempts to provide a background to help meet that challenge. ASHRAE Standard 62, ''Ventilation for Acceptable Indoor Air Quality'' (1), is the basis for most building codes and has a direct effect on most HVAC designs. ASHRAE Standard 55, ''Thermal Environmental Conditions for Human Occupancy'' (2), and the *ASHRAE Handbook, Fundamentals Volume* (3) give a basis to produce healthy, comfortable systems. These three documents form the basis for much of the material in this chapter.

4-1 THE BASIC CONCERNS

ASHRAE Standard 62 defines acceptable indoor air quality as air in which there are no known contaminants at harmful concentrations as determined by cognizant authorities and with which a substantial majority (80% or more) of the people exposed do not express dissatisfaction. With acceptable indoor air quaility, not only are occupants comfortable but their environment is free of bothersome odors and harmful levels of contaminants. Maintaining thermal comfort, which involves primarily the control of temperature, humidity, air movement, and the temperature of surrounding surfaces, is not just desirable and helpful in assuring a productive work environment, but in many cases also has a direct effect on the health of the building occupants. Thermal comfort will be discussed in a later section. Factors other than thermal comfort that are controlled by the HVAC system involve maintenance of a clean, healthy, and odor-free indoor environment. These factors are often what is intended by the term *indoor air quality* or IAQ. Maintaining good indoor air quality involves keeping gaseous and particulate contaminants below some acceptable level in the indoor environment. The contaminants include such things as carbon dioxide, carbon monoxide, other gases and vapors, radioactive materials, microorganisms, viruses, allergens, and suspended particulate matter.

Contamination of indoor spaces is caused by human and animal occupancy, by the release of contaminants in the space from the furnishings and accessories or from processes taking place inside the space, and by the introduction of contaminated outdoor air. The contaminants may be apparent, as in the case of large particulate matter or where odors are present, or they may be discernible only by instruments or by the effect that they have on the occupants. Symptoms such as headaches, nausea, and irritations of the eyes or nose may be a clue that indoor air quality in a building is poor. Buildings with an unusual number of occupants having physical problems have come to be described as having ''Sick Building Syndrome.'' Emphasis on comfort and health in the workplace and increased litigation in this area place a great responsibility on contractors, building owners, employers, and even HVAC engineers to be well informed, technically competent, and totally ethical in any actions affecting indoor air quality. Good indoor air quailty usually costs and the economic pressure to save on initial and operating costs can sometimes cause poor decisions that lead to both human suffering and even greater economic consequences.

4-2   COMMON CONTAMINANTS

Carbon Dioxide and Other Common Gases

Carbon dioxide is an exhaled byproduct of human (and all mammal) metabolism, and therefore CO_2 levels are typically higher in occupied spaces than for outdoor air. In heavily occupied spaces such as auditoriums, CO_2 levels will often be a major concern. This is not because of any direct health risk but is due to the fact that CO_2 is an easily measurable indicator of the effectiveness of ventilation of the space. As such it gives at least an indirect indication of potentially unacceptable levels of more harmful gases. The Environmental Protection Agency (EPA) recommends a maximum level of 1000 ppm (1.8 gm/m^3) for continuous CO_2 exposure, specifically for school and residential occupancy, and as a guideline for other building types.

Incomplete combustion of hydrocarbon fuels and tobacco smoking are two significant sources of carbon monoxide. Buildings with internal or nearby parking garages and loading docks are more likely to have high levels of CO. HVAC outdoor air intakes at ground level where heavy street traffic occurs can draw unacceptable levels of CO into the building's air system. Improperly vented and leaking furnaces, chimneys, water heaters, and incinerators are often the source of difficulty. Carbon monoxide is a toxic gas, and levels near 15 ppm can significantly affect body chemistry. The reaction of humans to different CO levels varies significantly, and the effects can be cumulative. Headaches and nausea are common symptoms of those exposed to quantities of CO above their tolerance.

Sulfur oxides are the result of combustion of fuels containing sulfur and may enter a building through outdoor air intakes or from leaks in combustion systems within the building. When hydrolyzed with water, sulfur oxides can form sulfuric acid, creating problems in the moist mucus membranes that may cause upper respiratory tract irritation and induce episodic attacks in individuals with asthmatic tendencies.

Nitrous oxides are produced by combustion of fuel with air at high temperatures. Normally, these contaminants would be brought in with outdoor air that has been contaminated by internal combustion engines and industrial effluents, however indoor combustion sources frequently contribute significant amounts. Opinions seem to differ regarding the health effects of different levels of nitrous oxides. Until this is determined more precisely, it would be wise to minimize indoor levels of nitrous oxide concentrations to the extent practical.

Radon, a naturally occurring radioactive gas resulting from the decay of radium, has received a great deal of attention recently, especially in areas where concentration levels have been found to be very high. The primary concern with radon is the potential for causing lung cancer. In many areas of the United States the indoor radon levels and therefore the risks are typically low. Radon gas may enter a building from the soil through cracks in slab floors and basement walls, or through the water supply, or from building materials containing uranium or thorium. The rate of entry from the soil depends on pressure differences, and therefore pressurization of a space is one means of reducing radon levels in that space. Other preventive measures include the ventilating of crawl spaces and under-floor areas and the sealing of floor cracks. For safety, radon levels should be kept low enough to keep the exposure of occupants below 4 picocuries/liter of air.

Volatile Organic Compunds (VOCs)

A variety of organic chemical species occur in a typical modern indoor environment resulting from combustion sources, pesticides, building materials and finishes, cleaning agents and solvents, and plants and animals. Fortunately, they usually exist in levels that are below recommended standards. Some occupants, however, are hypersensitive to particular chemicals, and for them many indoor environments create problems. Formaldehyde gas, one of the more common VOCs, is irritating to the eyes and the mucus membranes. It seems to have caused a diversity of problems in asthmatic and immunoneurological reactions and is considered to be a potential cancer hazard. Formaldehyde, used in the manufacture of carpets, pressed board, insulations, textiles, paper products, cosmetics, shampoos, and phenolic plastics, seems to enter buildings primarily in building products. These products continue to off-gas formaldehyde for long periods of time with much of this occurring during the first year. Acceptable limits are in the range of 1 ppm as a time-weighted 8-hour average. For homes, levels of 0.1 ppm seem to be a more prudent upper limit.

Particulate Matter

A typical sample of outdoor air might contain soot and smoke, silica, clay, decayed animal and vegetable matter, lint and plant fibers, metallic fragments, mold spores, bacteria, plant pollens, and other living material. The sizes of these particles may range from less than 0.01 micron (10^{-6} m) to the dimensions of leaves and insects. Fig. 4-1 shows the very wide range of sizes of particles and particle dispersoids along with types of gas cleaning equipment that might be effective in each case (1).

When suspended in the air, the mixture is called an *aerosol.* As outdoor air is brought into an indoor environment, it may be additionally contaminated by human sources and activities, interior furnishings and equipment, and pets. Microbial and infectious organisms can persist and even multiply when indoor conditions are favorable. Environmental tobacco smoke (ETS) has been one of the major problems in maintaining good indoor air quality, and concern has been heightened by increased evidence of its role in lung diseases, particularly cancer. Allergies are a common problem in a modern society, and the indoor environment may contain many of the particulates found outdoors. In addition, some occupants may be sensitive to the particulates found primarily indoors such as fibers, molds, and dust from carpets and bedding.

4-3 METHODS TO CONTROL CONTAMINANTS

There are four basic methods to maintain good IAQ in buildings:

1. Source elimination or modification
2. Use of outdoor air
3. Space air distribution
4. Air cleaning

Source Elimination or Modification

Of the four basic methods listed above, source elimination or modification very often is the most effective method for reducing contaminants not generated directly by the

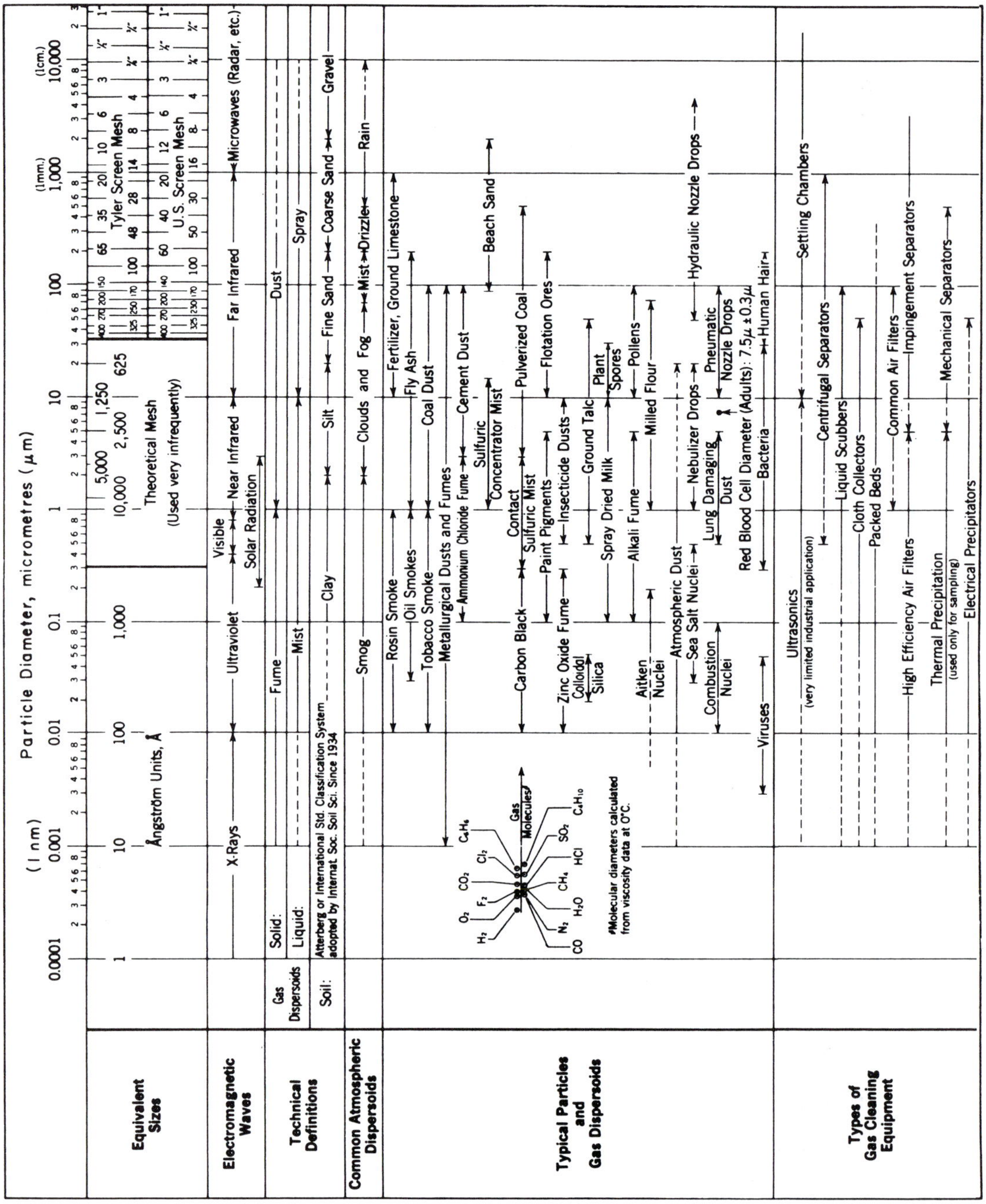

Figure 4-1 Characteristics of particles and particle dispersoids. (Reprinted by permission from ANSI/ASHRAE Standard 62-89, 1989)

human occupants or the necessary activities in the space. In new building design or with retrofitting, this method involves specifying exactly what building materials and furnishings are to be allowed within the building. In existing buildings it involves finding and removing any undesirable contaminants not essential to the functions taking place in the building. Elimination of smoking within a building is an acceptable approach to improving IAQ in both public and private buildings. Many states and cities have laws that prohibit smoking within certain types of facilities. Some employers and building operators have provided special areas for smoking, where the impact can be limited.

Storage of paints, solvents, cleaners, insecticides, and volatile compounds within a building or near the outdoor air intakes can often lead to impairment of the IAQ of the building. Removal or containment of these materials has been found to be necessary in some cases in order to make the indoor environment acceptable.

Use of Outdoor Air

Figure 4-2 is used to help define the various terms involved in the air flow of a typical HVAC system. *Supply air* is that air delivered to the conditioned space and used for ventilation, heating, cooling, humidification, or dehumidification. *Ventilation air* is that portion of supply air that is outdoor air plus any recirculated air that has been treated for the purpose of maintaining acceptable indoor air quality. Indoor spaces occupied for any reasonable length of time require the intake of some outdoor air to maintain air quality. Since outdoor air must normally be conditioned before it enters a space,

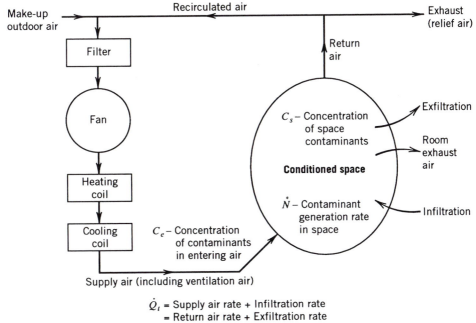

$$\dot{Q}_i = \text{Supply air rate + Infiltration rate}$$
$$= \text{Return air rate + Exfiltration rate}$$
$$+ \text{Room exhaust rate}$$

Figure 4-2 A typical HVAC ventilation system.

economics usually requires the use of a minimum amount of outdoor air to meet the air quality requirements. With economizers and buildings requiring cooling during mild or cold weather, outdoor air is often used to meet the cooling load. In some cases the amount of ventilation air required to maintain good indoor air quality may be less than the supply air actually delivered to the space because of the requirement to maintain comfort. In other situations the minimum rate of supply air may be fixed by the requirements of ventilation to maintain acceptable indoor air quality.

Outdoor air is air taken from the external atmosphere and, therefore, not previously circulated through the system. Some outdoor air may enter a space by *infiltration* through cracks and interstices and through ceilings, floors, and walls of a space or building, but generally in air-conditioned buildings most outdoor air is brought into a space by the supply air. It is usually assumed that outdoor air is free of contaminants that might cause discomfort or harm to humans, but this is not always so. In some localities and where strong contaminant sources might exist near a building, the air surrounding a building may not be free of the contaminants for which there are concerns. The United States Environmental Protection Agency (EPA) has published National Primary and Secondary Ambient-Air Quality Standards for outdoor air (4). These values are listed in ASHRAE Standard 62 and are shown in Table 4-1. Unless otherwise stated, examples and problems in this text will assume that the outdoor air meets the EPA ambient air-quality standards.

Recirculated air is that air removed from the conditioned space and intended for reuse as supply air. It differs from *return air* only in that some of the return air may be exhausted or relieved through dampers or by fans. *Makeup air* is outdoor air supplied to replace exhaust air and exfiltration. *Exfiltration* is air leakage outward through cracks and interstices and through ceilings, floors, and walls of a space or building. Some air

Table 4-1 National Primary Ambient-Air Quality Standards for Outdoor Air as Set by the U. S. Environmental Protection Agency (4)

Contaminant	Long term			Short term		
	Concentration	Averaging		Concentration	Averaging	
	$\mu g/m^3$	ppm		$\mu g/m^3$	ppm	
Sulfur dioxide	80	0.03	1 year	365^a	0.14^a	24 hours
Particles (PM 10)	50^b	—	1 year	150^a	—	24 hours
Carbon monoxide				$40{,}000^a$	35^a	1 hour
Carbon monoxide				$10{,}000^a$	9^a	8 hours
Oxidants (ozone)				235^c	0.12^c	1 hour
Nitrogen dioxide	100	0.055	1 year			
Lead	1.5	—	3 months^d			

[a] Not to be exceeded more than once per year.

[b] Arithmetic mean

[c] Standard is attained when expected number of days per calendar year with maximal average concentrations above 0.12 ppm (235 $\mu g/m^3$) is equal to or less than 1.

[d] Three-month period is a calendar quarter.

Source: Reprinted by permission from ANSI/ASHRAE Standard 62–89, 1989 (1).

may be removed from a space directly by room exhaust, usually with exhaust fans. There must always be a balance between the amount of air mass entering and the amount leaving a space as well as between the amount of air mass entering and leaving the entire air supply system. Likewise there must be a balance on the mass of any single contaminant entering and leaving a space and entering and leaving the entire air supply system.

The basic equation for contaminant concentration in a space is obtained using Figure 4-2, making a balance on the concentrations entering and leaving the conditioned space assuming complete mixing, a uniform rate of generation of the contaminant, and uniform concentration of the contaminant within the space and in the entering air. All balances should be on a mass basis; however, if densities are assumed constant, then volume flow rates may be used. For the steady state case:

$$\dot{Q}_t C_e + \dot{N} = \dot{Q}_t C_s \tag{4-1}$$

where:

$\dot{Q}_t$ = rate at which air enters or leaves the space
C_s = average concentration of a contaminant within the space
$\dot{N}$ = rate of contaminant generation within the space
C_e = concentration of the contaminant of interest in the entering air

Equation 4-1 can be solved for the concentration level in the space C_s or for the necessary rate $\dot{Q}_t$ at which air must enter the space to maintain the desired concentration level of a contaminant within the space. This fundamental equation may be used as the basis for deriving more complex equations for more realistic cases.

EXAMPLE 4-1

A person breathes out carbon dioxide at the rate of 0.30 L/min. The concentration of CO_2 in the incoming ventilation air is 300 ppm or 0.03 percent. It is desired to hold the concentration in the room below 1000 ppm or 0.1 percent. Assuming that the air in the room is perfectly mixed, what is the minimum rate of flow of air required to maintain the desired level?

SOLUTION

Solving Equation 4-1 for $\dot{Q}_t$:

$$\dot{Q}_t = \dot{N}/(C_s - C_e) = [(0.30)\,\text{L/min}]/[0.001 - 0.0003)\,(60)\,\text{s/min}]$$
$$= 7.1\,\text{L/s} = 15\,\text{cfm}$$

It can be seen from this calculation that the ASHRAE Standard 62 requirement of a maximum indoor level for CO_2 of 1000 ppm is equivalent to a minimum outdoor air requirement of 15 cfm/person, assuming that the normal CO_2 production of a person is approximately that given in the example problem.

In most HVAC systems emphasis is placed on maintaining the occupied zone at a nearly uniform condition. The *occupied zone* is the region within an occupied space between the floor and 72 in. (1800 mm) above the floor and more than 2 ft (600 mm) from the wall or fixed air conditioning equipment (2). In most cases perfect mixing of the supply air with the room air does not occur, and some fraction S of the supply air rate $\dot{Q}_s$ bypasses and does not enter the occupied zone, as shown in Figure 4-3. Because of this some of the outdoor air in the room supply air is exhausted without having performed any useful reduction in the contaminants of the occupied zone. The effectiveness E_{oa} with which outdoor air is used can be expressed as the fraction of the outdoor air entering the system that is utilized:

$$E_{oa} = [\dot{Q}_o - \dot{Q}_{oe}]/\dot{Q}_o \tag{4-2}$$

where:

$\dot{Q}_o$ = rate at which outdoor air is taken in
$\dot{Q}_{oe}$ = rate at which unused outdoor air is exhausted

From Figure 4-3, with R equal to the fraction of return air $\dot{Q}_r$ that is recirculated, the rate at which outdoor air is supplied to the space $\dot{Q}_{os}$ is

$$\dot{Q}_{os} = \dot{Q}_o + [R \times S \times \dot{Q}_{os}] \tag{4-3}$$

The amount of unused outdoor air that is exhausted $\dot{Q}_{oe}$ is

$$\dot{Q}_{oe} = (1 - R) \times S \times \dot{Q}_{os} \tag{4-4}$$

Combining Eqs. 4-2, 4-3, and 4-4 yields

$$E_{oa} = [1 - S]/[1 - RS] \tag{4-5}$$

Equation 4-5 gives the effectiveness with which the outdoor air is circulated to the occupied space in terms of the stratification factor S and the recirculation factor R. S is

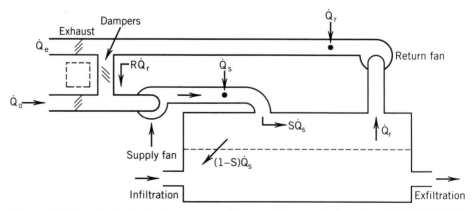

Figure 4-3 Typical air distribution system. (Reprinted by permission from ANSI/ASHRAE Standard 62-89, 1989)

sometimes called the *occupied zone bypass factor*. Using this simple model, with no stratification S would equal zero and there would be total mixing of air, and the effectiveness E_{oa} would be 1.0. Note also that as the exhaust flow becomes small, R approaches 1.0 and the effectiveness again approaches 1.0. This simple model neglects the effect of infiltration and assumes that the occupied space is perfectly mixed air. Appropriate equations for the more general case, where air cleaning occurs, will be developed in a following section.

EXAMPLE 4-2

For a given space it is determined that due to poor location of inlet diffusers relative to the inlet for the air return, and due to partitions around each work space, about 50 percent of the supply air for a space is bypassed around the occupied zone. What fractions of the outdoor air provided for the space are effectively utilized as the recirculation rate is changed from 0.4 to 0.8?

SOLUTION

This is an application of Eq. 4-5, for which each term is dimensionless:

$$E_{oa} = [1 - S]/[1 - RS]$$

For $R = 0.4$

$$E_{oa} = [1 - 0.5]/[1 - (0.4)(0.5)] = 0.625$$

For $R = 0.8$

$$E_{oa} = [1 - 0.5]/[1 - (0.8)(0.5)] = 0.833$$

Standard 62 describes two methods by which acceptable indoor air quailty can be achieved. The first of these procedures, the *Ventilation Rate Procedure* prescribes the rate at which outdoor air must be delivered to a space and various means to condition that air. A sample of these rates is given in Table 4-2, from Standard 62, and are derived from physiological considerations, subjective evaluations, and professional judgments. The Ventilation Rate Procedure prescribes:

- The outdoor air quality acceptable for ventilation
- Outdoor air treatment when necessary
- Ventilation rates for residential, commercial, institutional, vehicular, and industrial spaces
- Criteria for reduction of outdoor air quantities when recirculated air is treated by contaminant-removal equipment
- Criteria for variable ventilation when the air volume in the space can be used as a reservoir to dilute contaminants.

Table 4-2 Outdoor Air Requirements for Ventilation[a]—Commerical Facilities (offices, stores, shops, hotels, sports facilities)

Application	Estimated Maximum[b] Occupancy P/1000 ft² or 100 m²	Outdoor Air Requirements				Comments
		cfm/ person	L/s· person	cfm/ft²	L/s·m²	
Dry Cleaners, Laundries						Dry-cleaning processes may require more air.
Commercial laundry	10	25	13			
Commercial dry cleaner	30	30	15			
Storage, pick up	30	35	18			
Coin-operated laundries	20	15	8			
Coin-operated dry cleaner	20	15	8			
Food and Beverage Service						
Dining rooms	70	20	10			
Cafeteria, fast food	100	20	10			
Bars, cocktail lounges	100	30	15			Supplementary smoke-removal equipment may be required.
Kitchens (cooking)	20	15	8			Makeup air for hood exhaust may require more ventilating air. The sum of the outdoor air and transfer air of acceptable quality from adjacent spaces shall be sufficient to provide an exhaust rate of not less than 1.5 cfm/ft² (7.5 L/s·m²).

Garages, Repair, Service Stations

	cfm/ft²	L/s·m²	Comments
Enclosed parking garage	1.50	7.5	Distribution among people must consider worker location and concentration of running engines; stands where engines are run must incorporate systems for positive engine exhaust withdrawal. Contaminant sensors may be used to control ventilation.
Auto repair rooms	1.50	7.5	

Hotels, Motels, Resorts, Dormitories

	cfm/room	L/s·room	Comments
Bedrooms	30	15	Independent of room size.
Living rooms	30	15	
Baths	35	18	Installed capacity for intermittent use.

		cfm/person	L/s·person	Comments
Lobbies	30	15	8	
Conference rooms	50	20	10	
Assembly rooms	120	15	8	
Dormitory sleeping areas	20	15	8	See also food and beverage services, merchandising, barber and beauty shops, garages.
Gambling casinos	120	30	15	Supplementary smoke-removal equipment may be required.

Offices

		cfm/person	L/s·person	Comments
Office space	7	20	10	
Reception areas	60	15	8	Some office equipment may require local exhaust.
Telecommunication centers and data entry areas	60	20	10	
Conference rooms	50	20	10	Supplementary smoke-removal equipment may be required.

Table 4-2 Outdoor Air Requirements for Ventilation^a—Commerical Facilities (offices, stores, shops, hotels, sports facilities) *(Continued)*

Application	Estimated Maximum^b Occupancy P/1000 ft^2 or 100 m^2	Outdoor Air Requirements				Comments
		cfm/ person	L/s· person	cfm/ft^2	L/s·m^2	
Public Spaces						
Corridors and utilities				0.05	0.25	
Public restrooms, cfm/wc or cfm/urinal		50	25			Normally supplied by transfer air.
Locker and dressing rooms				0.5	2.5	Local mechanical exhaust with no recirculation recommended.
Smoking lounge	70	60	30			
Elevators				1.00	5.0	Normally supplied by transfer air.
Retail Stores, Sales Floors, and Show Room Floors						
Basement and street	30			0.30	1.50	
Upper floors	20			0.20	1.00	
Storage rooms	15			0.15	0.75	
Dressing rooms				0.20	1.00	
Malls and arcades	20			0.20	1.00	
Shipping and receiving	10			0.15	0.75	
Warehouses	5			0.05	0.25	
Smoking lounge	70	60	30			Normally supplied by transfer air, local mechanical exhaust; exhaust with no recirculation recommended.
Specialty Shops						
Barber	25	15	8			
Beauty	25	25	13			
Reducing salons	20	15	8			

Florists	8	15	8			Ventilation to optimize plant growth may dictate requirements.
Clothiers, furniture	8	15	8	0.30	1.50	
Hardware, drugs, fabric	8	15	8			
Supermarkets	8	15	8			
Pet shops				1.00	5.00	
Sports and Amusement						
Spectator areas	150	15	8			
Game rooms	70	25	13			
Ice arenas (playing areas)				0.50	2.50	When internal combustion engines are operated for maintenance of playing surfaces, increased ventilation rates may be required.
Swimming pools (pool and deck area)				0.50	2.50	Higher values may be required for humidity control.
Playing floors (gymnasium)	30	20	10			
Ballrooms and discos	100	25	13			
Bowling alleys (seating areas)	70	25	13			
Theaters						
Ticket booths	60	20	10			
Lobbies	150	20	10			
Auditorium	150	15	8			
Stages, studios	70	15	8			Special ventilation will be needed to eliminate special stage effects (e.g., dry ice vapors, mists, etc.)
Transportation						
Waiting rooms	100	15	8			
Platforms	100	15	8			
Vehicles	150	15	8			Ventilation within vehicles may require special considerations.

Table 4-2 Outdoor Air Requirements for Ventilation[a]—Commercial Facilities (offices, stores, shops, hotels, sports facilities) (Continued)

Application	Estimated Maximum[b] Occupancy P/1000 ft² or 100 m²	Outdoor Air Requirements				Comments
		cfm/ person	L/s· person	cfm/ft²	L/s·m²	
Workrooms						
Meat processing	10	15	8			Spaces maintained at low temperatures (−10°F to +50°F, or −23°C to +10°C) are not covered by these requirements unless the occupancy is continuous. Ventilation from adjoining spaces is permissible. When the occupancy is intermittent, infiltration will normally exceed the ventilation requirement.
Photo studios	10	15	8			
Darkrooms	10			0.50	2.50	
Pharmacy	20	15	8			
Bank vaults	5	15	8			
Duplicating, printing				0.50	2.50	Installed equipment must incorporate positive exhaust and control (as required) of undesirable contaminants (toxic or otherwise).
Institutional Facilities						
Education						
Classroom	50	15	8			
Laboratories	30	20	10			Special contaminant control systems may be required for
Training shop	30	20	10			

					Comments
Music rooms	50	15	8			processes or functions including laboratory animal occupancy.
Libraries	20	15	8			
Locker rooms				0.50	2.50	
Corridors				0.10	0.50	
Auditoriums	150	15	8			
Smoking lounges	70	60	30			Normally supplied by transfer air. Local mechanical exhaust with no recirculation recommended.
Hospitals, Nursing and Convalescent Homes						
Patient rooms	10	25	13			Special requirements or codes and pressure relationships may determine minimum ventilation rates and filter efficiency. Procedures generating contaminants may require higher rates.
Medical procedure	20	15	8			
Operating rooms	20	30	15			
Recovery and ICU	20	15	8			
Autopsy rooms				0.50	2.50	Air shall not be recirculated into other spaces.
Physical Therapy	20	15	8			
Correctional Facilities						
Cells	20	20	10			
Dining halls	100	15	8			
Guard stations	40	15	8			

[a] Table 4-2 prescribes supply rates of acceptable outdoor air required for acceptable indoor air quality. These values have been chosen to control CO_2 and other contaminants with an adequate margin of safety and to account for health variations among people, varied activity levels, and a moderate amount of smoking.

[b] Net occupiable space.

Source: Reprinted by permission from ANSI/ASHRAE Standard 62–89, 1989.

Standard 62 gives procedures by which the outdoor air can be evaluated for acceptability. Table 4-1, taken from Standard 62, lists the EPA standards (4) as the contaminant concentrations allowed in outdoor air. Outdoor air treatment is prescribed where the technology is available and feasible for any concentrations exceeding the values recommended. Where the best available, demonstrated, and proven technology does not allow the removal of contaminants, outdoor air rates may be reduced during periods of high contaminant levels, but recognizing the need to follow local regulations.

Indoor air quality is considered acceptable by the Ventilation Rate Procedure if the required rates of acceptable outdoor air listed in Table 4-2 are provided for the occupied space. Unusual indoor contaminants or sources should be controlled at the source or the Air Quality Procedure, described below, should be followed. Areas within industrial facilities not covered by Table 4-2 should use threshold limit values of references 5 and 6.

For most of the cases in Table 4-2, outdoor air requirements are assumed to be in proportion to the number of space occupants and are given in cfm (L/s) per person. In the rest of the cases the outdoor air requirements are given in cfm/ft^2 (L/s-m^2) and the contamination is presumed to be primarily due to other factors. Although estimated maximum occupancy is given where appropriate for design purposes, the anticipated occupancy should be used. For cases where more than one space is served by a common supply system, the Ventilation Rate Procedure provides a means for calculating the outdoor air requirements for the system. Rooms provided with exhaust air systems, such as toilet and bathrooms, kitchens, and smoking lounges, may be furnished with makeup air from adjacent occupiable spaces providing the quantity of air supplied meets the requirements of Table 4-2.

Except for intermittent or variable occupancy, outdoor air requirements of Table 4-2 must be met under the Ventilation Rate Procedure. Rules for intermittent or variable occupancy are described in Standard 62. If cleaned, recirculated air is to be used to reduce the outdoor air rates below these values then the Air Quality Procedure, described below, must be used.

The second procedure of Standard 62, the *Indoor Air Quality Procedure,* provides a direct solution to acceptable IAQ by restricting the concentration of all known contaminants of concern to some specified acceptable levels. Both quantitative and subjective evaluations are involved. The quantitative evaluation involves the use of acceptable indoor contaminant levels from a variety of sources, some of which are tabulated in tables in Standard 62. The subjective evaluation involve the response of impartial observers to odors that might be present in the indoor environment, and which can obviously occur only after the building is complete and operational.

Air cleaning may be used to reduce outdoor air requirements below those given in Table 4-2 and still maintain the indoor concentration of troublesome contaminants below the levels needed to provide a safe environment. However, there may be some contaminants that are not appreciably reduced by the air-cleaning system and that may be the controlling factor in determining the minimum outdoor air rates required. For example, the standard specifically requires a maximum of 1000 ppm of CO_2, a gas not commonly controlled by air cleaning. The rationale for this requirement on CO_2 is shown in Example 4-3 and documented in Appendix D of the Standard. The calculations show that for assumed normal conditions, this maximum concentration would require a minimum of 15 cfm of outdoor air per person. Notice that there are no values below 15 cfm (or 8 L/s) in Table 4-2. A more active person would produce more CO_2

and would require even higher rates of outdoor air for dilution. In the absence of CO_2 removal by air cleaning, CO_2 levels would need to be monitored in order to permit operation below the 15 cfm/person level for outdoor air. The Standard describes the documentation required of the design criteria and assumptions made when using the Indoor Air Quality Procedure.

Space Air Distribution

Where contaminant sources can be localized, the offending gas can be removed from the conditioned space before it spreads into the occupied zone. This involves control of the local air motion by the creation of pressure differentials, by exhaust fans, or by careful location of inlet diffusers and air return inlets. Care is required in designing for this method of control, and one should recognize that air is not easily directed by suction alone. Simply locating an air return inlet or exhaust fan near a contaminating source may not remove the contaminants before they have moved past some of the occupants.

Air Cleaning

Some outdoor air is necessary in buildings to replenish the oxygen required for breathing and to dilute the carbon dioxide and other wastes produced by the occupants. In many cases it is desirable to clean or filter the incoming outdoor air. In combination with the introduction of outdoor air, source reduction, and good air distribution, cleaning, or filtration of the recirculated air can often provide a cost-effective approach to the control of indoor air contaminants. Design of a proper system for gas cleaning is often the final step to assuring that an HVAC system will provide a healthy and clean indoor environment.

Gas Removal

The 1991 *ASHRAE Handbook, HVAC Applications* (7) has a detailed discussion of the control of gaseous contaminants for indoor air. Industrial gas cleaning and air pollution control is discussed in the 1992 *ASHRAE Handbook, HVAC Systems and Equipment* (8).

Contaminants may be removed from an airstream by *absorption,* by *physical adsorption,* by *chemisorption,* by *catalysis,* and by *combustion.* In some cases particulate matter may also be removed as these processes take place.

Absorbers are commonly used in the life support systems of space vehicles and submarines. Both solid and liquid absorbers may be used to reduce carbon dioxide and carbon monoxide to carbon, returning the oxygen to the conditioned space. Air washers, whose purpose may be to control temperature and humidity in buildings, not only remove contaminant gases from an airstream by absorption, but can also remove particulate matter as well. Contaminant gases are absorbed in liquids when the partial pressure of the contaminant in the air stream is greater than the solution vapor pressure with or without additive for that contaminant.

Although water, sometimes improved by the addition of reagents, is a common liquid for washing, other liquids may be used. The liquids must be maintained with a sufficiently low concentration of contaminants and must not transfer undesirable odors to the air. New or regenerated liquid must be continuously added to avoid these problems. Generally, large quantities of air must be moved through the water without an excessive airstream pressure drop.

Adsorption is the adhesion of molecules to the surface of a solid (the adsorbent) in contrast to absorption, in which the molecules are dissolved into or react with a substance. Good adsorbents must have large surface areas exposed to the gas being adsorbed and therefore typically have porous surfaces. Activated charcoal is the most widely used adsorbent because of its superior adsorbing properties. It is least effective with the lighter gases such as ammonia and ethylene and most effective with gases having high molecular mass. The charcoal may be impregnated with other substances to permit better accommodation of chemically active gases.

Chemisorption is similar in many ways to physical adsorption. It differs in that surface binding in chemisorption is by chemical reaction and therefore only certain pollutant compounds will react with a given chemisorber. In contrast to physical adsorption, chemisorption improves as temperature increases, does not generate heat (but may require heat input), is not generally reversible, is helped by the presence of water vapor, and is a monomolecular layer phenomenon.

Catalysis is closely related to chemisorption since chemical reactions occur at the surface of the catalyst; however, the gaseous pollutant does not react stoichiometrically with the catalyst itself. Because the catalyst is not used up in the chemical reactions taking place, this method of air purification has the potential for longer life than with adsorbers or chemisorbers, assuming that an innocuous product is produced in the reaction. The chemical reactions may involve a breakdown of the contaminant into smaller molecules or it may involve combining the contaminant gas with the oxygen available in the air stream or with a supplied chemical. Only a few catalysts appear to be effective for air purification at ambient temperatures. Catalytic combustion permits the burning of the offending gas at temperatures lower than with unassisted combustion and is widely used in automobiles to reduce urban air pollution.

In some cases odor rather than health may be a concern, or odors may persist even when the levels of all known contaminants are reduced to otherwise acceptable levels. In such cases odor masking or odor counteraction may be last resorts. This involves introducing a pleasant oder to cover or mask an unpleasant one, or the mixing of two odorous vapors together so that both odors tend to be diminished.

Particulate Removal—Filtering

The wide variety of suspended particles in both the outdoor and indoor environments has been described previously. With such a wide range of particulate sizes, shapes, and concentrations, it is impossible to design one type of air particulate cleaner that would be suitable for all applications. Clean rooms in an electronic assembly process require entirely different particulate removal systems than an office or a hospital. Air cleaners for particulate contaminants are covered in more detail in the *ASHRAE Systems and Equipment Handbook* (8). A brief outline of this material is presented here.

The most important characteristics of the aerosol affecting the performance of a particulate air cleaner include the particle's

- Size and shape
- Specific gravity

- Concentration
- Electrical properties

Particulate air cleaners vary widely in size, shape, initial cost, and operating cost. The major factor influencing filter design and selection is the degree of air cleanliness required. Generally, the cost of the filter system will increase as the size of the particles to be removed decreases. The three operating characteristics that can be used to compare various types are

- Efficiency
- Air-flow resistance
- Dust-holding capacity

Efficiency measures the ability of the air cleaner to remove particulate matter from an air stream. Figure 4-4 shows the efficiency of four different high-performance filters as a function of particle size. It can be seen that smaller particles are the most difficult to filter. In applications with dry-type filters and with low dust concentrations, the initial or clean filter efficiency should be considered for design since the efficiency in such cases increases with dust load. Average efficiency over the life of the filter is the most meaningful for most types and applications.

The *air-flow resistance* is the loss in total pressure at a given air flow rate. This is an important factor in operating costs for the system. *Dust-holding capacity* defines the amount of a particular type of dust that an air cleaner can hold when it is operated at a specified air-flow rate to some maximum resistance value or before its efficiency drops

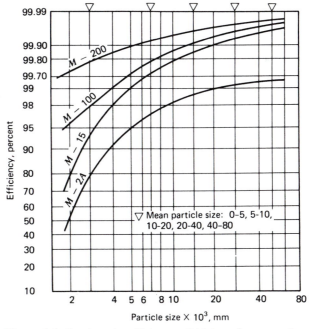

Figure 4-4 Gravimetric efficiency of high-performance dry-media filters.

seriously as a result of the collected dust. Methods for testing and rating air cleaners are given in reference 9. Typical engineering data (physical size, flow rate at a stated pressure drop) for the four filters shown in Fig. 4-4 are given in Table 4-3. The design requirements will rarely be exactly one of the air-flow rates or the pressure losses shown in Table 4-3. In these cases one can assume that the pressure loss across a filter element is proportional to the square of the flow rate. Thus, letting the subscript r stand for rated conditions, the pressure loss Δp at any required rate of flow $\dot{Q}$ can be determined by:

$$\Delta p = \Delta p_r [\dot{Q}/\dot{Q}_r]^2 \qquad (4\text{-}6)$$

The mechanism by which particulate air filters operate include

- Straining
- Direct interception
- Inertial deposition
- Diffusion
- Electrostatic effects

The common types of particulate air cleaners may be put in one of four groups:

- Fibrous media unit filters
- Renewable media filters
- Electronic air cleaners
- Combination air cleaners

Air cleaning has been used for many years to improve the quality of air entering a building, to protect components such as heat exchanger coils from particulate contamination, and to remove contaminants introduced into the recirculated air from the conditioned space. In more recent times with the combined emphasis on indoor air quality and economy of operation, there is an increased interest in air cleaning as a means to satisfy these requirements. Properly designed HVAC systems utilize air cleaning along with source modification, dilution with outdoor air, and space air distribution to give optimum performance with lowest cost.

The performance of an air cleaning system can be studied by using a model shown in Figure 4-5. This is a simplified model in which infiltration, exfiltration, and room exhaust are ignored and the air cleaner is assumed to be located either in the recirculated air stream (location A) or in the supply air stream (location B). *Ventilation effectiveness* E_v, the fraction of supply air delivered to the occupied zone, depends on the room shape, as well as on the location and design of the supply diffusers and the location of the return inlets. These factors will be discussed in more detail in the chapter on room air distribution. Notice the subtle difference between E_v and E_{oa} (defined by Eq. 4-2). E_v is equivalent to $(1 - S)$ in Eq. 4-2. Both terms have been referred to in the literature as ventilation effectiveness.

Assuming that densities do not vary significantly, volume balances can be used in place of mass balances. This seems to be a common assumption in air cleaning calculations, but care should always be exercised to be sure significant errors are not introduced. Making volume balances on the overall air-flow rates, and on any one contaminant of interest, Figure 4-5 can be used to obtain equations for the required constant outdoor air rates for constant air volume systems:

Table 4-3 Engineering Data—High-Performance Dry-Media Filters (Corresponds to Efficiency Data of Figure 4-4)

Standard Size	Meter:	0.3 × 0.6 × 0.2		0.3 × 0.6 × 0.3		0.6 × 0.6 × 0.2		0.6 × 0.6 × 0.3		Pressure Loss	
	Inch:	12 × 24 × 8		12 × 24 × 12		24 × 24 × 8		24 × 24 × 12		Inches of Water	Pa
Rated Capacity[a]		ft³/min	m³/s	ft³/min	m³/s	ft³/min	m³/s	ft³/min	m³/s		
Media Type	M-2A[b]	900	0.42	1025	0.48	1725	0.81	2000	0.94	0.15	37.4
	M-15	900	0.42	1025	0.48	1725	0.81	2000	0.94	0.35	87.2
	M-100	650	0.30	875	0.41	1325	0.62	1700	0.80	0.40	100.0
	M-200	450	0.21	630	0.29	920	0.43	1200	0.56	0.40	100.0
Effective filtering area All media types		ft² 14.5	m² 1.35	ft² 20.8	m² 1.93	ft² 29.0	m² 2.69	ft² 41.7	m² 3.87		

[a]Filters may be operated from 50 to 120 percent of the rated capacities with corresponding changes in pressure drop.

[b]The M-2A is available in 2-in. thickness and standard sizes with a nominal rating of 0.28 in. wg at 500 fpm face velocity.

(handwritten notes: 275.5 fpm; A = 4.356 ft³)

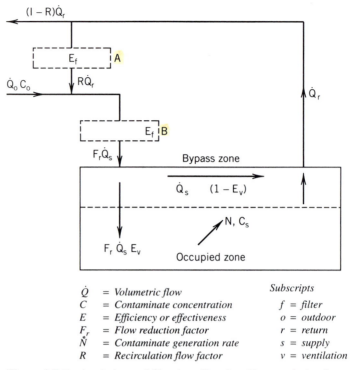

$\dot{Q}$ = Volumetric flow
C = Contaminate concentration
E = Efficiency or effectiveness
F_r = Flow reduction factor
$\dot{N}$ = Contaminate generation rate
R = Recirculation flow factor

Subscripts
f = filter
o = outdoor
r = return
s = supply
v = ventilation

Figure 4-5 Recirculation and filtration. (Reprinted by permission from
ANSI/ASHRAE Standard 62-89, 1989)

Filter location		Required outdoor air rate	
Return Air recirculated	A	$$\dot{Q}_o = \frac{\dot{N} - E_v R \dot{Q}_r E_f C_s}{E_v(C_s - C_o)}$$	(4-7)
Out/IN Air mix	B	$$\dot{Q}_o = \frac{\dot{N} - E_v R \dot{Q}_r E_f C_s}{E_v[C_s - (1 - E_f)C_o]}$$	(4-8)

Standard 62 gives five additional equations for variable air volume systems. Equations 4-7 and 4-8 can be used as an engineering basis for air cleaner (filter) selection. A typical computation might be to determine the required outdoor air that must be taken in by a system to maintain the desired air quality, assuming air cleaning to occur. The equations can also be used to solve for space contaminant concentration, required recirculation rate, or required filter efficiency.

An example of this type of analysis is shown as Fig. 4-6, where the required air cleaner efficiency is plotted versus supply air ventilation rate with outdoor air requirement/person as a parameter. The reduced outdoor air/person and the ventilation effectiveness are assumed fixed. This figure assumes that both particulate and gas phase filtration are used. The particulate filters are rated according to their efficiency on 0.3-micron particles, the typical mass mean diameter of respirable particulate matter in office buildings; and carbon adsorbers are rated on their performance with toulene, a

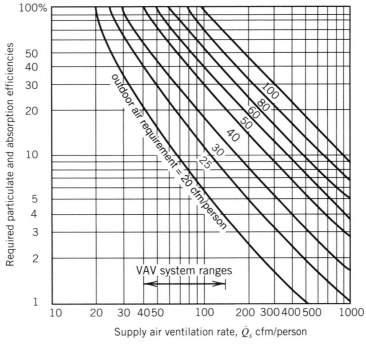

Figure 4-6 Required system filter efficiency with reduced outdoor air of 15 cfm/person for $E_v = 1.0$.

representative volatile organic compound. The computation of Fig. 4-6 further assumes that there is a filter in the incoming outdoor air stream as well as in the supply (mixed) stream.

EXAMPLE 4-3 ✳ See Notes

example okay · gct

A constant air volume system having a filter located in the supply duct (location *B*, Figure 4-5) and a filter efficiency of 70 percent for environmental tobacco smoke (ETS) is to be used to assist in holding the particulate level of the ETS in an occupied zone to below 220 $\mu g/m^3$. Assume that an average occupant (including smokers and nonsmokers) produces about 125 $\mu g/min$ of ETS, and that 20 cfm of outdoor air/person is to be supplied. For a ventilation effectiveness of 0.65 for the space, determine the necessary rate of recirculation assuming no ETS in the incoming outdoor air.

SOLUTION

Solving Eq. 4-8 for $R\dot{Q}_r$

$$R\dot{Q}_r = \{\dot{N} + E_v \dot{Q}_o[(1 - E_f)C_o - C_s]\}/E_v E_f C_s$$

for each person this is:

$$R\dot{Q}_r = \frac{125\,\mu g/min + (0.65)(20)cfm\left[(1 - 0.7)(0) - 220)\mu g/m^3\right](0.0283)m^3/ft^3}{(0.65)(0.7)(220)\mu g/m^3(0.0283)m^3/ft^3}$$

$$R\dot{Q}_r = 15.6\,cfm/person$$

The total rate of supply air to the room $\dot{Q}_t = \dot{Q}_o + R\dot{Q}_r = 20 + 15.6 = 35.6$ cfm/person. If we assumed that there were about 7 persons per 1000 square feet as typical for an office (Table 4-2), the air flow to the space would be

$$\dot{Q}/A = (35.6)\,cfm/person\,(7)\,persons/(1000)\,ft^2 = 0.25\,cfm/ft^2$$

This would probably be less than the supply air-flow rate typically required to meet the cooling load. A less efficient filter might be considered. If the above filter were used with the same rate of outdoor air but with increased supply and recirculation rates, the air in the space would be better than the assumed level.

EXAMPLE 4-4

For Example 4-3 assume that the cooling load requires that $1.0\,cfm/ft^2$ be supplied to the space, determine the recirculation rate per person $\dot{Q}_rR$, and the concentration level of the ETS in the space. Assume that the rate of outdoor air/person and the filter efficiency remain unchanged.

SOLUTION

$$R\dot{Q}_r/A = \dot{Q}_t/A - \dot{Q}_o/A = 1.0 - (7)(20)/1000 = 0.86\,cfm/ft^2$$

$$R\dot{Q}_r = (0.86)\,cfm/ft^2\,(1000)ft^2/(7)\,persons = 123\,cfm/person$$

Solving Eq. 4-8 for C_s

$$C_s = \frac{N + E_v\dot{Q}_o(1 - E_f)C_o}{E_v(\dot{Q}_o + R\dot{Q}_rE_f)} = \frac{125\,\mu g/min\text{-}person}{0.65(20 + (123)(0.7))cfm/person(0.0283)m^3/ft^3}$$

$$C_s = 64\,\mu g/m^3$$

The extra recirculation of the air through the filter has reduced the space concentration level of the tobacco smoke considerably with no use of extra outdoor air.

EXAMPLE 4-5

Assume that the office in Example 4-4 is occupied by 70 persons and that a suitably efficient filter was the M-15 filter of Fig. 4-4 and Table 4-3. Using this filter, design a system that has a pressure loss of no more than 0.30 in. wg in the clean condition.

SOLUTION

Table 4-3 gives the application data needed. There are four sizes of M-15 filters to choose from and the rated cfm at 0.35 in. wg pressure loss is given for each size. We must choose an integer number of filter elements. The total supply cfm required for 70 persons is

$$\dot{Q}_s = (123 + 20)\text{cfm/person} \ (70) \text{ persons} = 10,000 \text{ cfm}$$

It is desirable for the complete filter unit to have a reasonable geometric shape and be as compact as possible. Therefore choose the $24 \times 24 \times 12$ elements for a trial design. The rated cfm will first be adjusted to obtain a pressure loss of 0.30 in. wg using Eq. 4-6:

$$\dot{Q}_n = \dot{Q}_r (\Delta p_n/\Delta p_r)^{1/2} = 2000(0.3/0.35)^{1/2} = 1852 \text{ cfm/element}$$

Then the required number of elements is

$$n = \dot{Q}_s/\dot{Q}_n = 10,000/1852 = 5.40 \text{ elements}$$

Since *n* must be an integer, use 6 elements and the complete filter unit will have dimensions of 48×72 in., a reasonable shape. However, the filter unit will have a pressure loss less than the specified 0.30 in. wg. Again using Eq. 4-6 the actual pressure loss will be approximately

$$\Delta p = \Delta p_r [\dot{Q}/\dot{Q}_r]^2 = 0.35 \, [(10,000/6)/2000]^2 = 0.24 \text{ in. wg}$$

This is not an undesirable result and can be taken into account in the design of the air distribution system.

4-4   COMFORT—PHYSIOLOGICAL CONSIDERATIONS

The *ASHRAE Handbook of Fundamentals* (3) gives a detailed discussion of the physiological principles of human thermal comfort. Only brief, essential details will be given here.

   The amount of heat generated and dissipated by the human body varies considerably with activity and age as well as with size and gender. The body has a complex regulating system acting to maintain the deep body temperature of 98.6 F (36.9 C) regardless of the environmental conditions. A normal, healthy person generally feels most comfortable when the environment is maintained at conditions where the body can easily maintain a thermal balance with the surroundings. ASHRAE Standard 55 specifies conditions in which 80 percent or more of the occupants will find the environment thermally acceptable.

   The environmental factors that affect a person's thermal balance and therefore which influence thermal comfort are

- The *dry bulb temperature* of the surrounding air
- The *humidity* of the surrounding air
- The *relative velocity* of the surrounding air

- The temperature of any surfaces that can directly view any part of the body and thus exchange *radiation*

In addition the two *personal variables* that influence thermal comfort are *activity* and *clothing*.

The basic mechanisms that the body uses to control body temperatures are metabolism, blood circulation near the surface of the skin (cutaneous blood circulation), respiration, and sweating. Metabolism determines the rate at which energy is converted from chemical to thermal form within the body, and blood circulation controls the rate at which the thermal energy is carried to the surface of the skin. In respiration, air is taken in at ambient conditions but leaves saturated with moisture and very near the body temperature. Sweating has a significant effect on the rate at which energy can be carried away from the skin by heat and mass transfer.

The energy generated by a person's metabolism varies considerably with that person's activity. A unit to express the metabolic rate per unit of body surface area is the *met,* defined as the metabolic rate of a sedentary person (seated, quiet), 1 met = 18.4 But/(hr-ft^2) (58.2 W/m^2). Typical metabolic heat generation for various activities are given in Table 4-4. (3) The average adult is assumed to have an effective surface area for heat transfer of 19.6 ft^2 and would therefore dissipate approximately 360 Btu/hr (106 Watts) when functioning in a quiet, seated manner. A table of heat generation by various categories of persons is given for purposes of making cooling load calculations in Chapter 8.

The other personal variable that affects comfort is the type and amount of clothing that a person is wearing. Clothing insulation is usually described as a single equivalent uniform layer over the whole body. Its insulating value is expressed in terms of "clo" units, defined as 1 clo = 0.880 (F-ft^2-hr)/Btu (0.155 m^2-C/W). Typical insulation values for clothing ensembles are given in reference 3. A heavy two-piece business suit and accessories has an insulation value of about 1 clo, whereas a pair of shorts is about 0.05 clo. The operative temperatures and the clo values corresponding to the optimum comfort and the 80 percent acceptability limits are given in Fig. 4-7 from Standard 55 (2).

4-5   ENVIRONMENTAL COMFORT INDICES

In the previous section it was pointed out that, in addition to the personal factors of clothing and activity that affect comfort, there are four environmental factors: temperature, humidity, air motion, and radiation. The first of these, temperature, is easily measured and is alternatively called the air temperature or the dry bulb temperature. The second factor, humidity, can be described, for a given pressure and dry bulb temperature, using some of the terms defined in psychrometrics (see Chapter 3). These include wet bulb and dewpoint temperatures, which can be measured directly, and relative humidity and humidity ratio, which must be determined indirectly from measurement of the directly measurable variables.

The third environmental comfort factor, air motion, can be determined from measurement and, to a certain extent, predicted from the theories of fluid mechanics. Air flow in occupied spaces and air velocity measurements will be discussed in Chapters 10 and 11.

The fourth environmental comfort factor involves the amount of radiant exchange between a person and the surroundings. Cold walls or windows may cause a person to

Table 4-4 Typical Metabolic Heat Generation for Various Activities

	Btu/(hr-ft^2)	met
Resting		
Sleeping	13	0.7
Reclining	15	0.8
Seated, quiet	18	1.0
Standing, relaxed	22	1.2
Walking (on the level)		
0.89 m/s	37	2.0
1.34 m/s	48	2.6
1.79 m/s	70	3.8
Office Activities		
Reading, seated	18	1.0
Writing	18	1.0
Typing	20	1.1
Filing, seated	22	1.2
Filing, standing	26	1.4
Walking about	31	1.7
Lifting/packing	39	2.1
Driving/Flying		
Car	18–37	1.0–2.0
Aircraft, routine	22	1.2
Aircraft, instrument landing	33	1.8
Aircraft, combat	44	2.4
Heavy vehicle	59	3.2
Miscellaneous Occupational Activities		
Cooking	29–37	1.6–2.0
House cleaning	37–63	2.0–3.4
Seated, heavy limb movement	41	2.2
Machine work		
sawing (table saw)	33	1.8
light (electrical industry)	37–44	2.0–2.4
heavy	74	4.0
Handling 50-kg bags	74	4.0
Pick and shovel work	74–88	4.0–4.8
Miscellaneous Leisure Activities		
Dancing, social	44–81	2.4–4.4
Calisthenics/exercise	55–74	3.0–4.0
Tennis, singles	66–74	3.6–4.0
Basketball	90–140	5.0–7.6
Wrestling, competitive	130–160	7.0–8.7

Source: Reprinted by permission from *ASHRAE Handbook, Fundamentals Volume,* 1989.

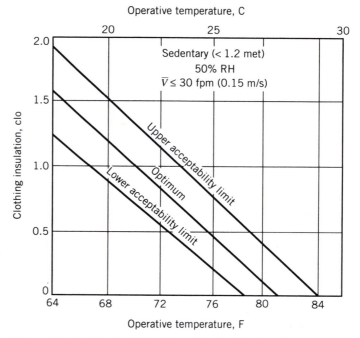

Figure 4-7 Clothing insulation for various levels of comfort at a given temperature during light and primarily sedentary activities (<1.2 met). (Reprinted by permission from ASHRAE Standard 55-92, 1992)

feel cold even though the surrounding air may be at a comfortable level. Likewise warm surfaces such as stoves or fireplaces or ceilings may cause a person to feel warmer than the surrounding air temperature would indicate. Usually these surfaces do not surround a person but occur only on one or two sides. Exact description of the condition is difficult and involves not only the surface temperature but how well that surface is "seen" by parts of one's body. Computation involves the "angle factor" or "configuration factor" used in radiation heat transfer. The basic index used to describe the radiative conditions in a space is the *mean radiant temperature,* the mean temperature of individual exposed surfaces in the environment. The most commonly used instrument to determine the mean radiant temperature is Vernon's *globe thermometer,* which consist of a hollow sphere 6 in. in diameter, flat black paint coating, and a thermocouple or thermometer bulb at its center. The equilibrium temperature assumed by the globe (the *globe temperature*) results from a balance in the convective and radiative heat exchanges between the globe and its surroundings. Measurements of the globe thermometer, air temperature, and air velocity can be combined as a practical way to estimate values of the mean radiant temperature:

$$T_{mrt}^4 = T_g^4 + C\overline{V}^{1/2}(T_g - T_a) \qquad (4\text{-}9)$$

where:

T_{mrt} = mean radiant temperature, R or K

T_g = globe temperature, R or K
T_a = ambient air temperature, R or K
$\overline{V}$ = air velocity, fpm or m/s
C = 0.103 × 10^9 (English units)
    = 0.247 × 10^9 (SI units)

Other indices have been developed to simplify description of the thermal environment and to take into account the combined effects of two or more of the environmental factors controlling human comfort: air temperature, humidity, air movement, and thermal radiation. These indices fall into two categories, depending on how they were developed. *Rational indices* depend on theoretical concepts already developed. *Empirical indices* are based on measurements with subjects or on simplified relationships that do not necessarily follow theory. The rational indices have the least direct use in design, but they form a basis from which we can draw useful conclusions about comfort conditions.

Considered to be the most common environmental index with the widest range of application, the effective temperature ET* is the temperature of an environment at 50 percent relative humidity that results in the same total heat loss from the skin as in the actual environment. It combines temperature and humidity into a single index so that two environments with the same effective temperature should produce the same thermal response even though the temperatures and the humidities may not be the same. Effective temperature depends on clothing and activity; therefore, it is not possible to generate a universal chart utilizing the parameter. Calculations of ET* are tedious and usually involve computer routines, and a *standard effective temperature* SET has been defined for typical indoor conditions. These conditions are

clothing insulation = 0.6 clo
moisture permeability index = 0.4
metabolic activity level = 1.0 met
air velocity < 20 fpm
ambient temperature = mean radiant temperature

The *operative temperature* is the average of the mean radiant and ambient air temperatures, weighted by their respective heat transfer coefficients. For the usual practical applications, it is the mean of the radiant and dry bulb temperatures and is sometimes referred to as the *adjusted dry bulb temperature*. It is the uniform temperature of an imaginary enclosure with which an individual exchanges the same heat by radiation and convection as in the actual environment. The effective temperature and the operative temperature are used in defining comfort conditions in ASHRAE Standard 55, shown in Fig. 4-8. Other indices, defined later, are useful in studies of comfort, stress caused by heat and cold, and reaction of the human body to various environmental conditions.

The *humid operative temperature* is the temperature of a uniform environment at 100 percent relative humidity in which a person loses the same total amount of heat from the skin as in the actual environment. It takes into account all three of the external transfer mechanisms that the body uses to lose heat: radiation, convection, and mass transfer. A similar index is the *adiabatic equivalent temperature,* the temperature of a uniform environment at 0 percent relative humidity in which a person loses the same total amount of heat from the skin as in the actual environment. Notice that these two indices have definitions similar to the effective temperature except for the relative humidities.

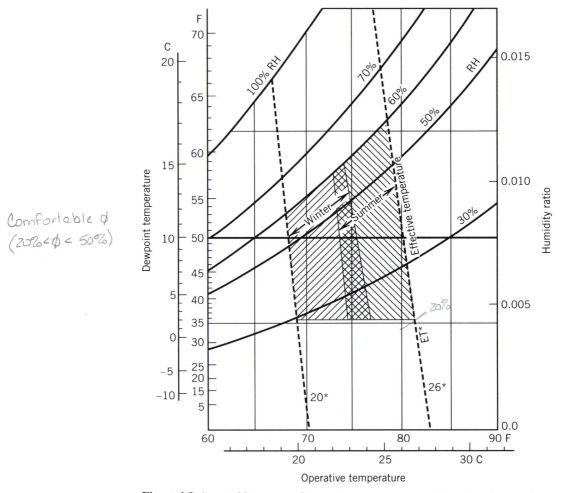

Figure 4-8 Acceptable ranges of operative temperature and humidity for people in typical summer and winter clothing during light and primarily sedentary activity (<1.2 met). (Reprinted by permission from ASHRAE Standard 55-92, 1992)

The *heat stress index* is the ratio of the total evaporative heat loss required for thermal equilibrium to the maximum evaporative heat loss possible for the environment, multiplied by 100, for steady-state conditions, and with the skin temperature held constant at 95 F. Table 4-5 gives the implications of an 8-hour exposure to various heat stress index levels. Except for the factor of 100, the *skin wettedness* is essentially the same as the heat stress index. It is the ratio of observed skin sweating to the maximum possible sweating for the environment as defined by the skin temperature, air temperature, humidity, air motion, and clothing. Skin wettedness is more closely related to the sense of discomfort or unpleasantness than to temperature sensation.

The *wet bulb globe temperature* t_{wbg} is an environmental heat stress index that combines dry bulb temperature t_{db}, a naturally ventilated wet bulb temperature t_{nwb}, and

Table 4-5 Evaluation of Index of Heat Stress

Index of Heat Stress	Physiological and Hygienic Implications of 8-hr Exposures to Various Heat Stresses
0	No thermal strain.
10 20 30	Mild to moderate heat strain. If job involves higher intellectual functions, dexterity, or alertness, subtle to substantial decrements in performance may be expected. In peforming heavy physical work, little decrement is expected unless ability of individuals to perform such work under no thermal stress is marginal.
40 50 60	Severe heat strain, involving a threat to health unless men are physically fit. Break-in period required for men not previously acclimatized. Some decrement in performance of physical work is to be expected. Medical selection of personnel desirable because these conditions are unsuitable for those with cardiovascular or respiratory impairment or with chronic dermatitis. These working conditions are also unsuitable for activities requiring sustained mental effort.
70 80 90	Very severe heat strain. Only a small percentage of the population may be expected to qualify for this work. Personnel should be selected: (a) by medical examination; and (b) by trial on the job (after acclimatization). Special measures are needed to assure adequate water and salt intake. Amelioration of working conditions by any feasible means is highly desirable, and may be expected to decrease the health hazard while increasing job efficiency. Slight ''indisposition'' which in most jobs would be insufficient to affect performance may render workers unfit for this exposure.
100	The maximum strain tolerated daily by fit, acclimatized young men.

Source: Reprinted by permission from *ASHRAE Handbook, Fundamentals Volume,* 1989.

the globe temperature t_g. It is a parameter that combines the effect of all four environmental factors affecting comfort. The equation that defines this index is

$$t_{wbg} = 0.7t_{nwb} + 0.2t_g + 0.1t_{db} \tag{4-10}$$

Equation 4-10 is usually used where solar radiation is significant. In enclosed environments the index is calculated from

$$t_{wbg} = 0.7t_{nwb} + 0.3t_g \tag{4-11}$$

Equations 4-10 and 4-11 are valid for any consistent unit of temperature.

The *wind chill index* WCI is an empirical index for the combined effect of wind and low temperature. For wind velocities less than 50 mph (80 km/hr), the index seems to reliably express subjective discomfort to cold. An index derived from the WCI is the *equivalent wind chill temperature,* the ambient temperature that would produce, in a calm wind, the same WCI as the actual combination of air temperature and wind velocity.

EXAMPLE 4-6

Determine the operative temperature for a work station in a room near a large window where the dry bulb and globe temperatures are measured to be 75 F and 85 F, respectively. The air velocity is estimated to be 30 ft/min at the station.

SOLUTION

The operative temperature depends on the mean radiant temperature, which is given by Eq. 4-9:

$$T_{mrt}^4 = T_g^4 + C\overline{V}^{1/2}(T_g - T_a)$$

or

$$T_{mrt} = [T_g^4 + C\overline{V}^{1/2}(T_g - T_a)]^{1/4}$$

$$T_{mrt} = [(81 + 460)^4 + (0.103 \times 10^9)(30)^{1/2}(81 - 75)]^{1/4} = 546R = 86\,F$$

A good estimate of the operative temperature is

$$t_o = (t_{mrt} + t_a)/2 = (75 + 86)/2 = 80.5 \qquad \text{or} \qquad t_o = 81\,F$$

The operative temperature shows the combined effect of the environment's radiation and air motion, which for this case gives a value 6 degrees F greater than the surrounding air temperature. It will be seen in the next section (Fig. 4-8) that this is probably an uncomfortable environment. The discomfort is caused by thermal radiation from surrounding warm surfaces, not from the air temperature. The humidity has not been taken into account, but at this operative temperature a person would likely be uncomfortable at any level of humidity.

Notice that in Eq. 4-9 absolute temperature must be used in the terms involving fourth power, but that temperature differences can be expressed in Fahrenheit units.

Comfort Conditions

ASHRAE Standard 55 gives the conditions for an acceptable thermal environment. Acceptable ranges of operative temperature and humidity for people in typical summer and winter clothing during light and primarily sedentary activity (≤ 1.2 met) are given in Figure 4-8. The ranges are based on a 10 percent dissatisfaction criterion.

The coordinates of the comfort zones are:

Winter: Operative temperature $t_o = 68 - 74$ F ($20 - 23.5$ C) at 60 percent relative humidity and $t_o = 69 - 76$ F ($20.5 - 24.5$ C) at 36 F (2 C) dewpoint. The slanting side boundaries are loci of constant comfort or thermal sensations and correspond to effective temperatures ET* of 68 and 74 F (20 and 23.5 C).

Summer: Operative temperature $t_o = 73 - 79$ F ($22.5 - 26$ C) at 60 percent relative humidity and $t_o = 74 - 80$ F ($23.5 - 27$ C) at 36 F (2 C) dewpoint. The slanting side boundaries correspond to effective temperatures ET* of 73 and 79 F (23 and 26 C).

In Figure 4-8 the upper and lower humidity limits are based on considerations of dry skin, eye irritation, respiratory health, microbial growth, and other moisture-related phenomena. Care must also be taken to avoid condensation on building surfaces and materials by controlling those surface temperatures.

It can be seen that the winter and summer comfort zones overlap. In this region people in summer dress tend to approach a slightly cool sensation, but those in winter clothing would be near a slightly warm sensation. In reality the boundaries shown in Figure 4-8 should not be thought of as sharp since individuals differ considerably in their reactions to given conditions. Table 4-6 gives the operative temperature range for sedentary persons in minimal clothing such as briefs. The ranges of the table are for air speeds less than 30 fpm (0.15 m/s) and 50 percent relative humidity.

For sedentary persons it is necessary to avoid the discomfort of drafts, but active persons are less sensitive. Figure 4-9 shows the combined effect of air speed and temperature on the comfort zone of Figure 4-8. It can be seen that air temperatures can be raised in the summer if air velocities can also be increased.

Table 4-6 Operative Temperatures for Thermal Acceptability of Sedentary or Slightly Active Persons (≤ 1.2 mets) at 50% Relative Humidity[a]

Season	Description of Typical Clothing	I_{cl} (clo)	Optimum Operative Temperature	Operative Temperature Range for 90% Thermal Acceptability[b]
Winter	Heavy slacks, long sleeve shirt, and sweater	0.9	22.0 C 71.0 F	20–23.5 C 68–75.0 F
Summer	Light slacks and short sleeve shirt	0.5	24.5 C 76.0 F	23–26.0 C 73–79.0 F
	Minimal	0.05	27.0 C 81.0 F	26–29.0 C 79–84.0 F

[a]Other than clothing, there are no adjustments for season or sex to the temperatures. For infants, certain elderly people, and individuals who are physically disabled, the lower limits should be avoided.

[b]Ranges used for low air speed: ≤ 30 fpm (0.15 m/s) and 50% RH.

Source: Reprinted by permission from ASHRAE Standard 55–92, 1992.

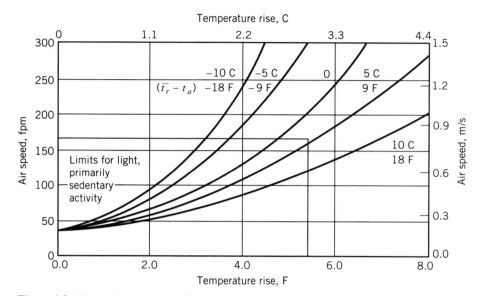

$$\bar{t_r} = t_{mrt}$$

Figure 4-9 Air speed required to offset increased temperature. (Reprinted by permission from ASHRAE Standard 55-92, 1992)

The acceptable operative temperature for active persons can also be calculated (for $1.2 < \text{met} < 3$) from:

$$t_{o,\text{active}} = t_{o,\text{sedentary}} - 5.4(1 + \text{clo})(\text{met} - 1.2) \qquad \textbf{(4-12a)}$$

in degrees F or

$$t_{o,\text{active}} = t_{o,\text{sedentary}} - 3.0(1 + \text{clo})(\text{met} - 1.2) \qquad \textbf{(4-12b)}$$

in degrees C. The minimum allowable operative temperature for these equations to apply is 59 F (15 C). Met levels can be obtained from Table 4-1. The combined effect of operative temperature, activity level, and clothing is shown in Fig. 4-10. One might expect people to remove a part of their clothing when exercising vigorously.

Figure 4-11 shows the allowable mean air speed permitted by Standard 55 as a function of air dry bulb temperature and the turbulence intensity of the air stream. The turbulence intensity may vary between 30 and 60 percent in conventionally ventilated spaces. In rooms with displacement ventilation or without ventilation, the turbulence intensity may be lower. The figure is based on a 15 percent acceptable level and the sensation at the head/feet level, where people are most sensitive. Higher air speeds may be acceptable if the affected occupants have control of local air speed.

People at higher activity levels are assumed to be able to accept higher degrees of temperature nonuniformity than people with light, primarily sedentary activity.

ASHRAE Standard 55 defines allowable rates of temperature change and also describes acceptable measuring range, accuracy, and response time of the instruments used for measuring the thermal parameters as well as locations where measurements should be taken. Procedures for determining air speed and temperature variations in building spaces are given in ASHRAE Standard 113 (10).

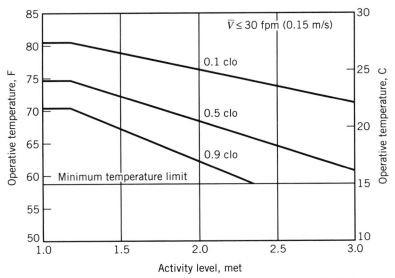

Figure 4-10 Optimum operative temperatures for active people in low air movement environments ($\overline{V} < 30$ fpm or 0.15 m/s). (Reprinted by permission from ASHRAE Standard 55-92, 1992.)

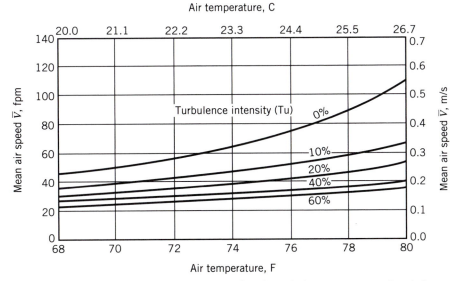

Figure 4-11 Allowable mean air speed as a function of air temperature and turbulence intensity. (Reprinted by permission from ASHRAE Standard 55-92, 1992.)

REFERENCES

1. ANSI/ASHRAE Standard 62-89, ''Ventilation for Acceptable Indoor Air Quality,'' American Society of Heating, Refrigerating and Air-Conditioning Engineers, Inc., Atlanta, GA, 1989.
2. ASHRAE Standard 55-92, ''Thermal Environmental Conditions for Human Occupancy,'' American Society of Heating, Refrigerating and Air-Conditioning Engineers, Inc., Atlanta, GA, 1992.
3. *ASHRAE Handbook, Fundamentals Volume,* American Society of Heating, Refrigerating and Air-Conditioning Engineers, Inc., Atlanta, GA, 1989.
4. National Primary and Secondary Ambient-Air Quality Standards, Code of Federal Regulations, Title 40 Part 50 (40 CFR50), as amended July 1, 1987. U. S. Environmental Protection Agency.
5. *ACGIH Industrial Ventilation—A Manual of Recommended Practice,* American Conference of Governmental Industrial Hygienists, Committee on Industrial Ventilation, Lansing, MI, 1986.
6. *ACGIH Threshold Limit Values and Biological Exposure Indices for 1986–87,* American Conference of Governmental Industrial Hygienists, Cincinnati, OH, 1987.
7. *ASHRAE Handbook, HVAC Applications,* American Society of Heating, Refrigerating and Air-Conditioning Engineers, Inc., Atlanta, GA, 1991.
8. *ASHRAE Handbook, HVAC Systems and Equipment Volume,* American Society of Heating, Refrigerating and Air-Conditioning Engineers, Inc., Atlanta, GA, 1992.
9. ASHRAE Standard 52-76, ''Method of Testing Air Cleaning Devices Used in General Ventilation for Removing Particulate Matter,'' American Society of Heating, Refrigerating and Air-Conditioning Engineers, Inc., Atlanta, GA, 1976.
10. ANSI/ASHRAE Standard 113-1990, ''Method of Testing for Room Air Diffusion,'' ''American Society of Heating, Refrigerating and Air-Conditioning Engineers, Inc., Atlanta, GA, 1990.

PROBLEMS

4-1. What level, in ppm, would the CO_2 concentration come to in a space in steady-state, if CO_2 were being released into the space at the rate of 0.25 cfm (0.118 L/s) and outdoor air with a CO_2 concentration of 200 ppm is being supplied to the space at the rate of 1000 cfm (0.472 m^3/s)? Assume complete mixing in the room.

4-2. How many people could occupy a room where the concentration level of carbon dioxide is to be kept below 1000 ppm if air with a concentration of 300 ppm CO_2 is being supplied to the room at the rate of 3.00 m^3/s (6400 cfm)? Assume that each person is producing carbon dioxide at the average rate of 5.00 mL/s (0.0107 cfm) and that the incoming air is completely mixed with the room air.

4-3. Select a filter system that will have a gravimetric efficiency of at least 95 percent in the particle size range of $0–5 \times 10^{-3}$ mm. The system must handle 2200 cfm and the maximum pressure drop across the filters must be less than or equal to 0.25 in. of water. Space limits the depth of the filters to 8 in.

4-4. An M-100, $24 \times 24 \times 12$ filter described in Table 4-3 is suggested for use with 1200 cfm of air. What pressure loss can be expected?

4-5. A filter system to handle 5600 cfm of air is to be designed using the M-2A media (Table 4-3) and the 12 × 24 × 8 modules. The lost pressure must be 0.10 in. wg or less when the filters are clean. How many modules are required?

4-6. The M-200, 0.6 × 0.6 × 0.2 filters of Table 4-3, are to be used with a volume flow rate of 0.38 m^3/s. What minimum pressure loss can be expected?

4-7. Design a filter system to handle 20 m^3/s of air using the M-15, 0.6 × 0.6 × 0.3 filters of Table 4-3. The pressure loss in the clean condition must be 70 Pa or less.

4-8. It is necessary to design a relatively high-efficiency filter system to handle 1200 cfm of air with a maximum pressure loss of 0.085 in. water when the filters are clean. (a) Consider the M-2A media of Table 4-3 in the 2-in. thickness and determine the face area required. (b) The physical layout of the equipment would permit the use of either 12 × 24 × 2-in. or 16 × 20 × 2-in. sizes. Select the size and number of modules to best match the requirements.

4-9. A laboratory involving animal research must use either 100 percent outdoor air or 25 percent outdoor air with high-performance filters for the return air. Gravimetric efficiency must be at least 99 percent in the size 0–5 × 10^{-6} m. The computed heat gain of the laboratory is 3 tons with a sensible heat factor (SHF) of 0.7. The cooling and dehumidifying unit requires a fixed air-flow rate of 350 cfm per ton. Inside conditions are 78 F db and 40 percent relative humidity, whereas outdoor conditions are 95 F db and 50 percent relative humidity. (a) Investigate the feasibility, and find the required amount of air and the size of the cooling unit when 100 percent outdoor air is used. (b) Find the required amount of air and the size of the cooling unit when 25 percent outdoor air is used and the remainder is recirculated through the high-performance filters. (c) Design the filter system using the data of Fig. 4-4 and Table 4-3 so that the maximum pressure loss is 0.125 in. water with clean filters.

4-10. A smoking area in a factory is designed for 75 people. Design a filter and air-circulation system whereby the outdoor air rate can be reduced to 15 cfm/person. Outdoor and recirculated air should be mixed before filtering. Outdoor air contaminants are negligible. The filter media must have a gravimetric efficiency of about 80 percent in the 0–5 × 10^{-3} mm particle range and the pressure loss must not exceed 0.10 in. water. Refer to Table 4-2.

4-11. Consider a game room where smoking is permitted. Suppose a filter system is available with an efficiency of 80 percent in removing tobacco smoke. Compute the air supply rates assuming the outdoor air rate is set at the minimum level and contaminants in the outdoor air are negligible.

4-12. It is desired to hold the particulate level of environmental tobacco smoke below 180 μg/m^3 in an occupied indoor space using filters with an effective efficiency of 80 percent and an intake of "clean" outdoor air at the rate of 20 cfm. Ten smokers are anticipated in the space and each smoker is expected to contribute about 150 μg/min of ETS to the space. What is the required rate of supply air to the room assuming a ventilation effectiveness of 0.85 and the filter located downstream of the mixed recirculated and outdoor air?

4-13. Solve Example 4-3 assuming that the filter is in location A in Figure 4-3.

4-14. Suppose the game room of Problem 4-11 is designed for 40 people. The cited filter media is rated at 0.2 in. water pressure loss with a face velocity of 600 ft/

min. What net filter face area is required to obtain a pressure loss of 0.15 in. water when clean?

4-15. For a 3000 ft^2 combination gymnasium, recreation, and exercise operation, it is desirable to reduce the outdoor air rate to a minimum by use of filtering and air recirculation. (a) Design a system using filters having $\eta = 0.5$ and a pressure loss of 0.14 in. wg at 350 ft/min face velocity. Pressure loss should not exceed about 0.20 in. wg. Outdoor air contaminants are negligible. (b) Suppose the cooling load dictates a larger supply air rate than the ventilation rate. Discuss how this would influence your choice of a filter.

4-16. A zone consisting of a large classroom isolated from the outdoor environment except for ventilation air has a capacity of 225 people. The cooling load is 125,000 Btu/hr (37 kW) with a sensible heat factor of 0.7. The outdoor air required for this case is the minimum, 15 cfm (7.5 L/s) per person. (a) Compute the required amount of ventilation air (supply air) on the basis of the cooling load assuming that the space dry bulb is 75 F (24 C) and 50 percent relative humidity (RH) and that the air is supplied at 90 percent RH. (b) What is the minimum amount of supply air based on air quality? (c) Compare parts (a) and (b) and resolve any inconsistency between the two.

4-17. A mixed group of men and women occupy a space maintained at 76 F db and 62 F wb. All are lightly clothed, with sedentary activity. (a) By using Fig. 4-8, draw a conclusion about the general comfort of the group. The MRT is 78 F. (b) Suppose the group is moving around (light activity) instead of sitting. What is your conclusion about their comfort? (c) Suppose the group is composed of retired people, 65 years of age or more, playing cards. How would you change the room conditions to ensure their comfort?

4-18. An air-conditioning system is to be designed for a combination shop and laboratory facility. The space will be occupied by men engaged in active work such as hammering, sawing, walking, climbing, and so on. The men will be dressed in short sleeves. (a) Select comfort conditions for the space, assuming that the occupants are to be comfortable. (b) It is expected that the toolroom keeper will work in a fenced-in area in the space with light activity. What is your recommendation about how he or she should dress to be comfortable?

4-19. Suppose that a space is used for both athletic events and classroom space at different times. The relative humidity can be maintained at 40 percent. At what temperature should the thermostat be set when the space is used for (a) classroom? (b) basketball practice?

4-20. A space used for sedentary activity by occupants in light clothing has a dry bulb temperature of 25 C and the air motion is about 0.1 m/s. The mean radiant temperature is about 29 C. (a) Can the occupants be expected to be comfortable? Why? (b) If not, what changes to the room air can be made to improve the comfort?

4-21. For sedentary activity, light clothing, and an air temperature equal to the mean radiant temperature, what is the approximate relationship between air motion (m/s) and the air temperature to maintain comfort? Consider a velocity range of 0.2 to 0.8 m/s.

4-22. Compute the operative temperature in a space where the mean air velocity is 30 ft/min (0.15 m/s) and the dry bulb and globe temperatures are 74 F (23 C) and 78 F (26 C), respectively, in (a) English units and (b) SI units.

4-23. You are asked to investigate a complaint that an office space being maintained at 76 F (24 C) and 40 percent relative humidity is uncomfortably warm. You measure the globe temperature to be 80 F (27 C) and estimate the mean air velocity to be 40 ft/min (0.2 m/s). What is your recommendation to correct the situation? Assume that you cannot change the building or fixtures.

4-24. Consider the work environment of a model shop where people are standing, walking, lifting, and so on, around various types of shop equipment. Air motion is about 20 ft/min (0.1 m/s) and a typical globe temperature measurement is 70 F (21 C). The shop technicians are dressed in shirts and trousers (clo = 0.5 (ft^2-hr-F)/Btu or 0.078 (m^2-C)/W). Estimate the air dry bulb temperature for comfortable conditions in (a) F and (b) C. Assume the relative humidity to be normal (40 to 50 percent).

4-25. To save energy, an office building cooling system is controlled so that the minimum dry bulb temperature in the space is 78 F with a dewpoint temperature of 63 F. (a) Comment on the comfort expectations for the space. (b) What general type of clothing would you recommend? (c) Assuming that the dry bulb temperature is fixed, what could be done to improve comfort conditions?

4-26. During an emergency energy shortage thermostats were ordered set at 65 F to conserve heating fuel. Using Figs. 4-7 and 4-8 discuss the comfort expectations of the following: (a) an executive dressed in a vested business suit, (b) a secretary sitting and typing, (c) a factory worker, (d) a customer in a grocery store, and (e) a patient in a doctor's examining or waiting room.

4-27. Air motion is a parameter that may be adjusted to improve comfort when space temperatures are fixed. Discuss the general effect of increased or decreased air motion when the space temperature is low in winter and high in summer.

4-28. A method of saving energy in large, chilled water cooling systems is to increase the temperature of the water circulating in the system. (a) How will this affect the space comfort conditions? (b) Would there be certain periods of time when this would be feasible? Discuss.

4-29. The fan power required to circulate air to and from the conditioned space is often minimized by reducing air flow to conserve energy. (a) Is this consistent with high space temperatures in the summer months? Explain. (b) Would the same be true for low space temperatures in the winter? Explain.

4-30. The productivity of people is related to their comfort. Give an example where an overzealous attempt to save energy by raising or lowering the space temperature might result in an increase in overall energy use and greater overall cost.

4-31. A classroom in a school is designed for 100 people. (a) What is the minimum amount of clean outdoor air required? (b) The floor area is 1500 ft^2. What is the outdoor air ventilation requirement on the basis of floor area (see Table 4-2)?

Chapter 5

Heat Transmission in Building Structures

The design of an acceptable air-conditioning system is dependent on a good estimate of the heat gain or loss in the space to be conditioned. In the usual structure the walls and roofs are rather complex assemblies of materials. Windows are often made of two or more layers of glass with air spaces between them and usually have drapes or curtains. In basements, floors and walls are in contact with the ground. Because of these conditions precise calculation of heat transfer rates is difficult, but experience and experimental data make reliable estimates possible. Because most of the calculations require a great deal of repetitive work, tables that list coefficients and other data for typical situations are used. Thermal resistance is a very useful concept and will be used extensively in the following sections. Thermal capacitance is an important concept in all transient analysis computations; however, discussion will be included in another chapter.

Generally all three modes of heat transfer—conduction, convection, and radiation—are important in building heat gain and loss. Solar radiation will be treated in a separate section, in Chapter 6, because it creates some unusual complexities in computation. Long wavelength radiation, such as occurs in air gaps, will be considered in this chapter.

5-1   BASIC HEAT-TRANSFER MODES

In the usual situation all three modes of heat transfer occur simultaneously. In this section, however, they will be considered separately for clarity and ease of presentation.

Thermal conduction is the mechanism of heat transfer between parts of a continuum because of the transfer of energy between particles or groups of particles at the atomic level. The Fourier equation expresses steady-state conduction in one dimension as follows:

$$\dot{q} = -kA\frac{dt}{dx} \tag{5-1}$$

where

$\dot{q}$ = heat transfer rate, Btu/hr or W
k = thermal conductivity, Btu/(hr-ft-F) or W/(m-C)
A = area normal to heat flow, ft^2 or m^2
$\dfrac{dt}{dx}$ = temperature gradient, F/ft or C/m

Equation 5-1 incorporates a negative sign because $\dot{q}$ flows in the positive direction of x when dt/dx is negative.

Consider the flat wall of Fig. 5-1a, where uniform temperatures t_1 and t_2 are assumed to exist on each surface. If the thermal conductivity, the heat transfer rate, and the area are constant, Eq. 5-1 may be integrated to obtain

$$\dot{q} = \frac{-kA(t_2 - t_1)}{(x_2 - x_1)} \tag{5-2a}$$

A very useful form of Eq. 5-2a is

$$\dot{q} = \frac{-(t_2 - t_1)}{R'} \tag{5-2b}$$

where R' is the thermal resistance defined by

$$R' = \frac{x_2 - x_1}{kA} = \frac{\Delta x}{kA} \tag{5-3a}$$

The thermal resistance for a unit area of material is very commonly used in handbooks and in the HVAC literature. In this book this quantity, sometimes called the "R-factor," is referred to as the *unit thermal resistance*, or simply the *unit resistance, R*. For a plane wall the unit resistance is

$$R = \frac{\Delta x}{k} \tag{5-3b}$$

Thermal resistance R' is analogous to electrical resistance, and $\dot{q}$ and $(t_2 - t_1)$ are analogous to current and potential difference in Ohm's law. This analogy provides a very convenient method of analyzing a wall or slab made up of two or more layers of dissimilar material. Figure 5-1b shows a wall constructed of three different materials. The heat transferred by conduction is given by Eq. 5-2b, where

$$R' = R'_1 + R'_2 + R'_3 = \frac{\Delta x_1}{k_1 A} + \frac{\Delta x_2}{k_2 A} + \frac{\Delta x_3}{k_3 A} \tag{5-4}$$

Although the foregoing discussion is limited to a plane wall where the cross-sectional area is a constant, a similar procedure applies to a curved wall. Consider the long,

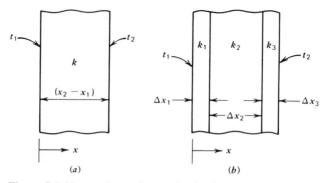

Figure 5-1 Nomenclature for conduction in plane walls.

hollow cylinder shown in cross section in Fig. 5-2. The surface temperatures t_i and t_o are assumed to be uniform and steady over each surface. The material is assumed to be homogeneous with a constant value of thermal conductivity. Integration of Eq. 5-1 with k and $\dot{q}$ constant but A a function of r yields

$$\dot{q} = \frac{2\pi kL}{\ln\left(\dfrac{r_o}{r_i}\right)}(t_i - t_o) \tag{5-5}$$

where L is the length of the cylinder. Here the thermal resistance is

$$R' = \frac{\ln\left(\dfrac{r_o}{r_i}\right)}{2\pi kL} \tag{5-6}$$

Cylinders made up of several layers may be analyzed in a manner similar to the plane wall where resistances in series are summed as shown in Eq. 5-4 except that the individual resistances are given by Eq. 5-6.

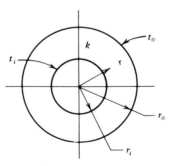

Figure 5-2 Radial heat flow in a hollow cylinder.

Table 5-1 summarizes the thermal resistances for several common situations, and Tables 5-2a and 5-2b give the thermal conductivity k for a wide variety of building and insulating materials. Other useful data are also given in Tables 5-2a and 5-2b; for example, the reciprocal of the unit thermal resistance and the *unit thermal conductance, C*. Note that k has the units of Btu-in./(ft²-hr-F) or W/(m-C). With Δx given in inches or meters, respectively, the unit thermal conductance C is given by

$$C = \frac{1}{R} = \frac{k}{\Delta x} \text{ Btu/(hr-ft}^2\text{-F)} \qquad \text{or} \qquad \text{W/(m}^2\text{-C)} \qquad (5\text{-}7)$$

Thermal convection is the transport of energy by mixing in addition to conduction. Convection is associated with fluids in motion generally through a pipe or duct or along a surface. In the very thin layer of fluid next to the surface, the transfer of energy is by conduction. In the main body of the fluid, mixing is the dominant energy transfer mechanism. A combination of conduction and mixing exists between these two regions. The transfer mechanism is complex and highly dependent on whether the flow is laminar or turbulent.

The usual, simplified approach in convection is to express the heat transfer rate as

$$\dot{q} = hA(t - t_w) \qquad (5\text{-}8a)$$

where

$\dot{q}$ = heat transfer rate from fluid to wall, Btu/hr or W
h = film coefficient, Btu/(hr-ft²-F) or W/(m²-s)
t = bulk temperature of the fluid, F or C
t_w = wall temperature, F or C

The film coefficient h is sometimes called the *unit surface conductance* or alternatively the *convective heat transfer coefficient*. Equation 5-8a may also be expressed in terms of thermal resistance:

$$\dot{q} = \frac{t - t_w}{R'} \qquad (5\text{-}8b)$$

where

$$R' = \frac{1}{hA} \text{ (hr-F)/Btu} \qquad \text{or} \qquad \text{C/W} \qquad (5\text{-}9a)$$

or

$$R = \frac{1}{h} = \frac{1}{C^*} \text{ (hr-ft}^2\text{-F)/Btu} \qquad \text{or} \qquad \text{(m}^2\text{-C)/W} \qquad (5\text{-}9b)$$

---

* Note that the symbol for conductance is C, in contrast to the symbol for the temperature in Celsius degrees, C.

Table 5-1 Thermal Resistances for Some Steady-State Conduction Problems

Number	System	Expressions for the Resistance R' $\dot{q} = \Delta t/R'$ (Btu/hr) or W
1.	Cylinder buried in a semiinfinite medium having a temperature at a great distance t_∞. The ground surface is assumed adiabatic.	

$$R' = \frac{\left(\ln\dfrac{2L}{D}\right)\left(1 + \dfrac{\ln(L/2z)}{\ln(2L/D)}\right)}{2\pi kL}$$

| 2. | A vertical cylinder placed in a semiinfinite medium having an adiabatic surface and temperature t_∞ at a great distance | |

$$R' = \frac{\ln(4L/D)}{2\pi kL}$$

| 3. | Conduction between inside and outside surfaces of a rectangular box having uniform inside and outside temperatures. Wall thickness Δx is less than any inside dimension. | |

$$R' = \frac{1}{k\left[\dfrac{A}{\Delta x} + 0.54\Sigma L + 1.2\Delta x\right]}$$

ΣL = sum of all 12 inside lengths
Δx = thickness of wall
A = inside surface area

Table 5-2a Thermal Properties of Building and Insulating Materials at a Mean Temperature of 75 F (English Units)

Material	Description	Density ρ $\dfrac{\text{lbm}}{\text{ft}^3}$	Thermal Conductivity k $\dfrac{\text{Btu-in.}}{\text{ft}^2\text{-hr-F}}$	Unit Conductance C $\dfrac{\text{Btu}}{\text{hr-ft}^2\text{-F}}$	Unit Resistance		Specific Heat $\dfrac{\text{Btu}}{\text{lbm-F}}$
					Per Inch Thickness $1/k$ $\dfrac{\text{ft}^2\text{-hr-F}}{\text{Btu-in.}}$	For Listed Thickness $1/C$ $\dfrac{\text{hr-ft}^2\text{-F}}{\text{Btu}}$	
Building Board	Asbestos-cement board						
Boards, panels,	$\frac{1}{4}$ in. or 6 mm	120	—	16.5	—	0.06	0.24
subflooring, sheathing,	Gypsum or plasterboard						
wood-based panel	$\frac{3}{8}$ in. or 10 mm	50	—	3.10	—	0.32	0.26
products	$\frac{1}{2}$ in. or 13 mm	50	—	2.22	—	0.45	—
	Plywood	34	0.80	—	1.25	—	0.29
	$\frac{1}{4}$ in. or 6 mm	34	—	3.20	—	0.31	0.29
	$\frac{3}{8}$ in. or 10 mm	34	—	2.13	—	0.47	0.29
	$\frac{1}{2}$ in. or 13 mm	34	—	1.60	—	0.62	0.29
	$\frac{2}{3}$ in. or 20 mm	34	—	1.07	—	0.93	0.29
	Insulating board and sheathing, regular density						
	$\frac{1}{2}$ in. or 13 mm	18	—	0.76	—	1.32	0.31
	$\frac{25}{32}$ in. or 20 mm	18	—	0.49	—	2.06	0.31
	Hardboard, high density, standard tempered	63	1.00	—	1.00	—	0.32
	Particle board						
	Medium density	50	0.94	—	1.06	—	0.31
	Underlayment						
	$\frac{5}{8}$ in. or 16 mm	40	—	1.22	—	0.82	0.29
	Wood subfloor						
	$\frac{3}{4}$ in. or 20 mm	—	—	1.06	—	0.94	0.33

Table 5-2a Thermal Properties of Building and Insulating Materials at a Mean Temperature of 75 F (English Units) (*Continued*)

Material	Description	Density ρ $\dfrac{lbm}{ft^3}$	Thermal Conductivity k $\dfrac{Btu\text{-}in.}{ft^2\text{-}hr\text{-}F}$	Unit Conductance C $\dfrac{Btu}{hr\text{-}ft^2\text{-}F}$	Unit Resistance Per Inch Thickness $1/k$ $\dfrac{ft^2\text{-}hr\text{-}F}{Btu\text{-}in.}$	Unit Resistance For Thickness Listed $1/C$ $\dfrac{hr\text{-}ft^2\text{-}F}{Btu}$	Specific Heat $\dfrac{Btu}{lbm\text{-}F}$
Building Paper	Vapor—permeable felt	—	—	16.7	—	0.06	—
	Vapor—seal, two layers of mopped 15-lb felt	—	—	8.35	—	0.12	—
Finish Flooring Materials	Carpet and fibrous pad	—	—	0.48	—	2.08	0.34
	Carpet and rubber pad	—	—	0.81	—	1.23	0.33
	Tile—asphalt, linoleum, vinyl, or rubber	—	—	20.0	—	0.05	0.30
Insulating Materials Blanket and batt	Mineral fiber—fibrous form processed from rock, slag, or glass						
	Approximately 2–$2\frac{3}{4}$ in. or 50–70 mm	0.3–2.0	—	0.143	—	7	0.17–0.23
	Approximately 3–$3\frac{1}{2}$ in. or 75–90 mm	0.3–2.0	—	0.091	—	11	0.17–0.23
	Approximately $5\frac{1}{4}$–$6\frac{1}{2}$ in. or 135–165 mm	0.3–2.0	—	0.053	—	19	0.17–0.23
Board and slabs	Cellular glass	8.5	0.38	—	2.63	—	0.24
	Glass fiber, organic bonded	4–9	0.25	—	4.00	—	0.23
	Expanded polystyrene—molded beads	1.0	0.28	—	3.57	—	0.29
	Expanded polyurethane—R-11 expanded	1.5	0.16	—	6.25	—	0.38
	Mineral fiber with resin binder	15	0.29	—	3.45	—	0.17

Material	Density	k	C	R (per inch, 1/k)	R (for thickness)	Specific heat
Loose Fill						
Mineral fiber—rock, slag, or glass						
Approximately 3.75–5 in. or 75–125 mm	0.6–2.0	—	—	—	11	0.17
Approximately 6.5–8.75 in. or 165–222 mm	0.6–2.0	—	—	—	19	0.17
Approximately 7.5–10 in. or 191–254 mm	—	—	—	—	22	0.17
Approximately $7\frac{1}{4}$ in. or 185 mm	—	—	—	—	30	0.17
Silica aerogel	7.6	0.17	5.88	—	—	—
Vermiculite (expanded)	7–8	0.47	2.13	—	—	—
Roof Insulation						
Preformed, for use above deck						
Approximately $\frac{1}{2}$ in. or 13 mm	—	—	0.72	—	1.39	—
Approximately 1 in. or 25 mm	—	—	0.36	—	2.78	—
Approximately 2 in. or 50 mm	—	—	0.19	—	5.56	—
Cellular glass	9	0.4	2.5	—	—	0.24
Masonry Materials						
Concretes						
Lightweight aggregates including expanded shale, clay, or slate; expanded slags; cinders; pumice; vermiculite; also cellular concretes	200	5.2	—	0.19	—	—
	100	3.6	—	0.28	—	—
	80	2.5	—	0.40	—	—
	40	1.15	—	0.86	—	—
	20	0.70	—	1.43	—	—
Sand and gravel or stone aggregate (not dried)	140	12.0	—	0.08	—	—
Masonry Units						
Brick, common	120	5.0	—	0.20	—	—
Brick, face	130	9.0	—	0.11	—	—
Concrete blocks, three-oval core—sand and gravel aggregate						
4 in. or 100 mm	—	—	1.4	—	0.71	—
8 in. or 200 mm	—	—	0.9	—	1.11	—
12 in. or 300 mm	—	—	0.78	—	1.28	—

Table 5-2a Thermal Properties of Building and Insulating Materials at a Mean Temperature of 75 F (English Units) (Continued)

Material	Description	Density ρ $\dfrac{\text{lbm}}{\text{ft}^3}$	Thermal Conductivity k $\dfrac{\text{Btu-in.}}{\text{ft}^2\text{-hr-F}}$	Unit Conductance C $\dfrac{\text{Btu}}{\text{hr-ft}^2\text{-F}}$	Unit Resistance Per Inch Thickness $1/k$ $\dfrac{\text{ft}^2\text{-hr-F}}{\text{Btu-in.}}$	Unit Resistance For Thickness Listed $1/C$ $\dfrac{\text{hr-ft}^2\text{-F}}{\text{Btu}}$	Specific Heat $\dfrac{\text{Btu}}{\text{lbm-F}}$
	Lightweight aggregate (expanded shale, clay slate or slag; pumice)						
	3 in. or 75 mm	—	—	0.79	—	1.27	—
	4 in. or 100 mm	—	—	0.67	—	1.50	—
	8 in. or 200 mm	—	—	0.50	—	2.00	—
	12 in. or 300 mm	—	—	0.44	—	2.27	—
Plastering Materials	Cement plaster, sand, aggregate	116	5.0	—	0.20	—	—
	Gypsum plaster:						
	Lightweight aggregate						
	$\frac{1}{2}$ in. or 13 mm	45	—	3.12	—	0.32	—
	$\frac{5}{8}$ in. or 16 mm	45	—	2.67	—	0.39	—
	Lightweight aggregate on metal lath						
	$\frac{3}{4}$ in. or 20 mm	—	—	2.13	—	0.47	—
Roofing	Asbestos-cement shingles	120	—	4.76	—	0.21	—
	Asphalt roll roofing	70	—	6.50	—	0.15	—
	Asphalt shingles	70	—	2.27	—	0.44	—
	Built-in roofing						
	$\frac{3}{8}$ in. or 10 mm	70	—	3.00	—	0.33	0.35
	Slate, $\frac{1}{2}$ in. or 13 mm	—	—	20.00	—	0.05	—
	Wood shingles—plain or plastic film faced	—	—	1.06	—	0.94	0.31

Siding Materials (on flat surface)

Shingles						
Asbestos-cement	120	—	4.76	—	0.21	—
Siding						
Wood, drop, 1 in. or 25 mm	—	—	1.27	—	0.79	0.31
Wood, plywood, $\frac{3}{8}$ in. or 10 mm, lapped	—	—	1.59	—	0.59	0.29
Aluminum or steel, over sheathing, hollowbacked	—	—	1.61	—	0.61	—
Insulating board—backed nominal, $\frac{3}{8}$ in. or 10 mm	—	—	0.55	—	1.82	—
Insulating board—backed nominal, $\frac{3}{8}$ in. or 10 mm, foil-backed	—	—	0.34	—	2.96	—
Architectural glass	—	—	10.00	—	0.10	—

Woods

Maple, oak, and similar hardwoods	45	1.10	—	0.91	—	0.30
Fir, pine, and similar softwoods	32	0.80	—	1.25	—	0.33

Metals

Aluminum (1100)	171	1536	—	0.00065	—	0.214
Steel, mild	489	314	—	0.00318	—	0.120
Steel, stainless	494	108	—	0.00926	—	0.109

Source: Abstracted by permission from *ASHRAE Handbook, Fundamentals Volume,* 1985.

Table 5-2b Thermal Properties of Building and Insulating Materials at a Mean Temperature of 24 C (SI UNITS)

Material	Description	Density ρ $\frac{kg}{m^3}$	Thermal Conductivity k $\frac{W}{m\text{-}C}$	Unit Conductance C $\frac{W}{m^2\text{-}C}$	Unit Resistance Per Meter Thickness $1/k$ $\frac{m\text{-}C}{W}$	For Listed Thickness $1/C$ $\frac{m^2\text{-}C}{W}$	Specific Heat $\frac{kJ}{kg\text{-}C}$
Building Board Boards, panels, subflooring, sheathing, woodbased panel products	Asbestos-cement board $\frac{1}{4}$ in. or 6 mm	1922	—	93.7	—	0.011	1.00
	Gypsum or plasterboard						
	$\frac{3}{8}$ in. or 10 mm	800	—	17.6	—	0.057	1.09
	$\frac{1}{2}$ in. or 13 mm	800	—	12.6	—	0.078	—
	Plywood	545	0.12	—	8.70	—	1.21
	$\frac{1}{4}$ in. or 6 mm	545	—	18.2	—	0.055	1.21
	$\frac{3}{8}$ in. or 10 mm	545	—	12.1	—	0.083	1.21
	$\frac{1}{2}$ in. or 13 mm	545	—	9.09	—	0.110	1.21
	$\frac{3}{4}$ in. or 20 mm	545	—	6.08	—	0.165	1.21
	Insulating board and sheathing, regular density						
	$\frac{1}{2}$ in. or 13 mm	288	—	4.32	—	0.232	1.30
	$\frac{25}{32}$ in. or 20 mm	288	—	2.78	—	0.359	1.30
	Hardboard, high density, standard tempered	1010	0.14	—	6.94	—	1.34
	Particleboard						
	Medium density	800	0.14	—	7.35	—	1.30
	Underlayment $\frac{5}{8}$ in. or 16 mm	640	—	6.93	—	0.144	1.21
	Wood subfloor $\frac{3}{4}$ in. or 20 mm	—	—	6.02	—	0.166	1.38

Material						
Building Paper						
Vapor—permeable felt	—	—	94.8	—	0.011	—
Vapor—seal, two layers of mopped 15-lb felt	—	—	47.4	—	0.021	—
Finish Flooring Materials						
Carpet and fibrous pad	—	—	2.73	—	0.367	1.42
Carpet and rubber pad	—	—	4.60	—	0.217	1.38
Tile—asphalt, linoleum, vinyl, or rubber	—	—	113.0	—	0.009	1.26
Insulating Materials						
Blanket and batt						
Mineral fiber—fibrous form processed from rock, slag, or glass. Approximately 2-2¾ in. or 50–70 mm	4.8–32	—	0.812	—	1.23	0.71–0.96
Approximately 3-3½ in. or 75–90 mm	4.8–32	—	0.517	—	1.94	0.71–0.96
Approximately 5¼-6½ in. or 135–165 mm	4.8–32	—	0.301	—	3.32	0.71–0.96
Board and slabs						
Cellular glass	136	0.0548	—	18.2	—	1.0
Glass fiber, organic bonded	64–144	0.036	—	27.8	—	0.96
Expanded polystyrene—molded beads	16	0.040	—	25.0	—	1.2
Expanded polyurethane—R-11 expanded	24	0.023	—	43.5	—	1.6
Mineral fiber with resin binder	240	0.042	—	23.9	—	0.71
Loose Fill						
Mineral fiber—rock, slag, or glass						
Approximately 3.75–5 in. or 75–125 mm	9.6–32	—	—	—	1.94	0.71
Approximately 6.5–8.75 in. or 165–22 mm	9.6–32	—	0.45	—	3.35	0.71
Approximately 7.5–10 in. or 191–254 mm	—	—	0.28	—	3.87	0.71
Approximately 7¼ in. or 185 mm	—	—	0.23	—	5.28	0.71
Silica aerogel	122	0.025	—	40.8	—	—
Vermiculite (expanded)	122	0.068	—	14.8	—	—

Table 5-2b Thermal Properties of Building and Insulating Materials at a Mean Temperature of 24 C (SI UNITS) (*Continued*)

Material	Description	Density ρ $\frac{kg}{m^3}$	Thermal Conductivity k $\frac{W}{m\text{-}C}$	Unit Conductance C $\frac{W}{m^2\text{-}C}$	Unit Resistance — Per Meter Thickness $1/k$ $\frac{m\text{-}C}{W}$	Unit Resistance — For Listed Thickness $1/C$ $\frac{m^2\text{-}C}{W}$	Specific Heat $\frac{kJ}{kg\text{-}C}$
Roof Insulation	Preformed, for use above deck						
	Approximately $\frac{1}{2}$ in. or 13 mm	—	—	4.1	—	0.24	1.0
	Approximately 1 in. or 25 mm	—	—	2.0	—	0.49	2.1
	Approximately 2 in. or 50 mm	—	—	1.1	—	0.93	3.9
	Cellular glass	144	0.058	—	17.3	—	1.0
Masonry Materials	Lightweight aggregates including	3200	0.75	—	1.32	—	—
Concretes	expanded shale, clay, or slate;	1600	0.52	—	1.94	—	—
	expanded slags; cinders;	1280	0.36	—	2.77	—	—
	pumice; vermiculite; also	640	0.17	—	6.03	—	—
	cellular concretes	320	0.10	—	10.0	—	—
	Sand and gravel or stone	2242	1.73	—	0.58	—	—
	aggregate (not dried)						
Masonry Units	Brick, common	1922	0.72	—	1.39	—	—
	Brick, face	2082	1.30	—	0.77	—	—
	Concrete blocks, three-oval						
	core—sand and gravel aggregate						
	4 in. or 100 mm	—	—	8.0	—	0.13	—
	8 in. or 200 mm	—	—	5.1	—	0.20	—
	12 in. or 300 mm	—	—	4.4	—	0.23	—

Material							
Plastering Materials							
lightweight aggregate (expanded shale, clay slate or slag; pumice)							
3 in. or 75 mm	—	—	—	4.5	—	0.22	—
4 in. or 100 mm	—	—	—	3.8	—	0.26	—
8 in. or 200 mm	—	—	—	2.8	—	0.35	—
12 in. or 300 mm	—	—	—	2.5	—	0.40	—
Cement plaster, sand aggregate,	1858	0.72	—	—	1.39	—	—
Gypsum plaster:							
Lightweight aggregate							
$\frac{1}{2}$ in. or 13 mm	721	—	—	17.7	—	0.056	—
$\frac{5}{8}$ in. or 16 mm	721	—	—	15.2	—	0.066	—
Lightweight aggregate on metal lath $\frac{3}{4}$ in. or 20 mm	—	—	—	12.1	—	0.83	—
Roofing							
Asbestos-cement shingles	1922	—	—	27.0	—	0.037	—
Asphalt roll roofing	1121	—	—	36.9	—	0.027	—
Asphalt shingles	1121	—	—	12.9	—	0.078	—
Built-up roofing $\frac{3}{8}$ in. or 10 mm	1121	—	—	17.0	—	0.059	—
Slate, $\frac{1}{2}$ in. or 13 mm	—	—	—	113.6	—	0.009	—
Wood shingles—plain or plastic film faced	—	—	—	6.02	—	0.166	—
Siding Materials (on flat surface)							
Shingles							
Asbestos-cement	1922	—	—	27.0	—	3.70	—
Siding							
Wood, drop, 1 in. or 25 mm	—	—	—	7.21	—	0.139	1.30
Wood, plywood, $\frac{3}{8}$ in. or 10 mm, lapped	—	—	—	9.03	—	0.111	1.21

Table 5-2b Thermal Properties of Building and Insulating Materials at a Mean Temperature of 24 C (SI UNITS) (*Continued*)

Material	Description	Density ρ $\dfrac{kg}{m^3}$	Thermal Conductivity k $\dfrac{W}{m\text{-}C}$	Unit Conductance C $\dfrac{W}{m^2\text{-}C}$	Unit Resistance Per Meter Thickness $1/k$ $\dfrac{m\text{-}C}{W}$	For Listed Thickness $1/C$ $\dfrac{m^2\text{-}C}{W}$	Specific Heat $\dfrac{kJ}{kg\text{-}C}$
	Aluminum or steel, over sheathing, hollowbacked	—	—	9.14	—	0.109	—
	Insulating board—backed nominal $\frac{3}{8}$ in. or 10 mm	—	—	3.12	—	0.320	—
	Insulating board backed nominal $\frac{3}{8}$ in. or 10 mm, foil-backed	—	—	1.93	—	0.518	—
	Architectural glass	—	—	56.8	—	0.018	—
Woods	Maple, oak, and similar hardwoods	721	0.159	—	6.3	—	1.26
	Fir, pine, and similar softwoods	513	0.115	—	8.67	—	1.38
Metals	Aluminum (1100)	2739	221.5	—	0.0045	—	0.896
	Steel, mild	7833	45.3	—	0.022	—	0.502
	Steel, stainless	7913	15.6	—	0.064	—	0.456

Source: Adapted by permission from *ASHRAE Handbook, Fundamentals Volume*, 1985.

The thermal resistance given by Eq. 5-9a may be summed with the thermal resistances arising from pure conduction given by Eq. 5-3a or 5-6.

The film coefficient h appearing in Eqs. 5-8a and 5-9a depends on the fluid, the fluid velocity, the flow channel, and the degree of development of the flow field (that is, whether the fluid has just entered the channel or is relatively far from the entrance). Many correlations exist for predicting the film coefficient under various conditions. Correlations for forced convection are given in Chapter 3 of the *ASHRAE Handbook* (1) and in textbooks on heat transfer.

In convection it is important to recognize the mechanism that is causing the fluid motion to occur. When the bulk of the fluid is moving relative to the heat transfer surface, the mechanism is called *forced convection,* because such motion is usually caused by a blower, fan, or pump that is forcing the flow. In forced convection buoyancy forces are negligible. In *free convection,* on the other hand, the motion of the fluid is due entirely to buoyancy forces, usually confined to a layer near the heated or cooled surface. The surrounding bulk of the fluid is stationary and exerts a viscous drag on the layer of moving fluid. As a result inertia forces in free convection are usually small. Free convection is often referred to as *natural convection.*

Natural or free convection is an important part of HVAC applications. However, it is one of the most difficult physical situations to describe, and the predicted film coefficients have a greater uncertainty than those of forced convection. Various empirical relations for natural convection film coefficients can be found in reference 1 and in heat transfer textbooks.

Most building structures have forced convection along outer walls or roofs, and natural convection occurs inside narrow air spaces and on the inner walls. There is considerable variation in surface conditions, and both the direction and magnitude of the air motion on outdoor surfaces are very unpredictable. The film coefficient for these situations usually ranges from about 1.0 Btu/(hr-ft²-F) or 6 W/(m²-C) for free convection up to about 6 Btu/(hr-ft²-F) or 35 W/(m²-C) for forced convection with an air velocity of about 15 miles per hour, 20 ft/sec, or 6 m/s. Because of the low film coefficients, especially with free convection, the amount of heat transferred by thermal radiation may be equal to or larger than that transferred by convection.

Thermal radiation is the transfer of thermal energy by electromagnetic waves and is an entirely different phenomenon from conduction and convection. In fact thermal radiation can occur in a perfect vacuum and is actually impeded by an intervening medium. The direct net transfer of energy by radiation between two surfaces that see only each other and that are separated by a nonabsorbing medium is given by

$$\dot{q}_{12} = \frac{\sigma(T_1^4 - T_2^4)}{\dfrac{1 - \epsilon_1}{A_1\epsilon_1} + \dfrac{1}{A_1 F_{12}} + \dfrac{1 - \epsilon_2}{A_2\epsilon_2}} \tag{5-10}$$

where

σ = Boltzmann constant, 0.1713×10^{-8} Btu/(hr-ft²-R⁴) or 5.673×10^{-8} W/(m²-K⁴)

T = absolute temperature, R or K
ϵ = emittance of each surface
A = surface area, ft^2 or m^2
F = configuration factor, a function of geometry only (Chapter 6)

In Eq. 5-10 it is assumed that both surfaces are "gray" (where the emittance ϵ equals the absorptance α). This assumption can often be justified. The student is referred to Parker and McQuiston (2) or other textbooks on heat transfer for a more complete discussion of thermal radiation. Figure 5-3 shows situations where radiation may be a significant factor. For the wall

$$\dot{q}_i = \dot{q}_w = \dot{q}_r + \dot{q}_o$$

and for the air space

$$\dot{q}_i = \dot{q}_r + \dot{q}_c = \dot{q}_o$$

The resistances can be combined to obtain an equivalent overall resistance R' with which the heat transfer rate can be computed using Eq. (5-2b):

$$\dot{q} = \frac{-(t_o - t_i)}{R'} \tag{5-2b}$$

The thermal resistance for radiation is not easily computed, however, because of the fourth power temperature relationship of Eq. 5-10. For this reason and because of the inherent uncertainty in describing the physical situation, theory and experiment have been combined to develop combined or effective unit thermal resistances and unit thermal conductances for many typical surfaces and air spaces. Table 5-3a gives effective film coefficients and unit thermal resistances as a function of wall position, direction of heat flow, air velocity, and surface emittance for exposed surfaces such as outside walls. Table 5-3b gives representative values of emittance ϵ for some building and insulating materials. For example, a vertical brick wall in still air has an effective

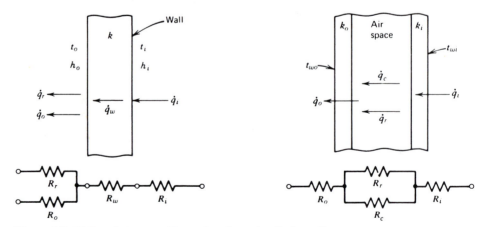

Figure 5-3 Wall and air space illustrating thermal radiation effects.

$$\left(T_1 + T_2\right)\left(T_1{}^2 + T_2{}^2\right)\Delta t \cong 4T_1{}^3\Delta t \quad : \text{ for small } \Delta t$$

Table 5-3a Surface Unit Conductances and Unit Resistances for Air[a]

		Surface Emittances											
		ε = 0.9				ε = 0.2				ε = 0.05			
		h		R		h		R		h		R	
Position of Surface	Direction of Heat Flow	Btu / hr-ft²-F	W / m²-C	hr-ft²-F / Btu	m²-C / W	Btu / hr-ft²-F	W / m²-C	hr-ft²-F / Btu	m²-C / W	Btu / hr-ft²-F	W / m²-C	hr-ft²-F / Btu	m²-C / W
Still Air													
Horizontal	Upward	1.63	9.26	0.61	0.11	0.91	5.2	1.10	0.194	0.76	4.3	1.32	0.232
Sloping—45 degrees	Upward	1.60	9.09	0.62	0.11	0.88	5.0	1.14	0.200	0.73	4.1	1.37	0.241
Vertical	Horizontal	1.46	8.29	0.68	0.12	0.74	4.2	1.35	0.238	0.59	3.4	1.70	0.298
Sloping—45 degrees	Downward	1.32	7.50	0.76	0.13	0.60	3.4	1.67	0.294	0.45	2.6	2.22	0.391
Horizontal	Downward	1.08	6.13	0.92	0.16	0.37	2.1	2.70	0.476	0.22	1.3	4.55	0.800
Moving Air													
Wind is 15 mph or 6.7 m/s (for winter)	Any (any position)	6.0	34.0	0.17	0.029								
Wind is 7½ mph or 3.4 m/s (for summer)	Any	4.0	22.7	0.25	0.044								

[a]Conductances are for surfaces of the stated emittance facing virtual blackbody surroundings at the same temperature as the ambient air. Values are based on a surface-air temperature difference of 10 deg F and for surface temperature of 70 F.

Source: Adapted by permission from *ASHRAE Handbook, Fundamentals Volume*, 1989.

Table 5-3b Reflectance and Emittance of Various Surfaces and Effective Emittances of Air Space

Surface	Reflectance in Percent	Average Emittance ϵ	Effective Emittance E of Air Space	
			With One Surface Having Emittance ϵ and Other 0.90	With Both Surfaces of Emittance ϵ
Aluminum foil, bright	92–97	0.05	0.05	0.03
Aluminum sheet	80–95	0.12	0.12	0.06
Aluminum coated paper, polished	75–84	0.20	0.20	0.11
Steel, galvanized, bright	70–80	0.25	0.24	0.15
Aluminum paint	30–70	0.50	0.47	0.35
Building materials—wood, paper, glass, masonry, nonmetallic paints	5–15	0.90	0.82	0.82

Source: Adapted by permission from *ASHRAE Handbook, Fundamentals Volume,* 1989.

emittance E of about 0.8 to 0.9. In still air the average film coefficient, from Table 5-3a, is about 1.46 Btu/(hr-ft^2-F) or 8.29 W/(m^2-C) and the unit thermal resistance is 0.68 (hr-ft^2-F)/Btu or 0.12 (m^2-C)/W.

If the surface were highly reflective, $E = 0.05$, the film coefficient would be 0.59 Btu/(hr-ft^2-F) [3.4 W/(m^2-C)] and the unit thermal resistance would be 1.7 (hr-ft^2-F)/Btu or 0.298 (m^2-C)/W. It is evident that thermal radiation is a large factor when natural convection occurs. If the air velocity were to increase to 15 mph or about 7 m/s, the average film coefficient would increase to about 6 Btu/(hr-ft^2-F) or 34 W/(m^2-C). With higher air velocities the relative effect of radiation diminishes. Radiation appears to be very important in the heat gains through attic spaces.

Tables 5-4, 5-5, and 5-6 give conductances and resistances for air spaces as a function of position, direction of heat flow, air temperature, and the effective emittance of the space. The effective emittance E is given by

$$\frac{1}{E} = \frac{1}{\epsilon_1} + \frac{1}{\epsilon_2} - 1 \qquad (5\text{-}11)$$

where ϵ_1 and ϵ_2 are for each surface of the air space. The effect of radiation is quite apparent in Tables 5-4, 5-5, and 5-6, where the thermal resistance may be observed to decrease by a factor of two or three as E varies from 0.05 to 0.82.

The preceding paragraphs cover thermal resistances arising from conduction, convection, and radiation. Equation 5-4 may be generalized to give the equivalent resistance of n resistors in series:

$$R_e' = R_1' + R_2' + R_3' + \cdots + R_n' \qquad (5\text{-}12)$$

Figure 5-4 is an example of a wall being heated or cooled by a combination of convection and radiation on each surface and having five different resistances through which the heat must be conducted. The equivalent thermal resistance R_e' for the wall is given by Eq. 5-12 as

$$R_e' = R_i' + R_1' + R_2' + R_3' + R_o' \qquad (5\text{-}13)$$

Each of the resistances may be expressed in terms of fundamental variables using Eqs. 5-3a and 5-9a.

$$R_e' = \frac{1}{h_i A_i} + \frac{\Delta x_1}{k_1 A_1} + \frac{\Delta x_2}{k_2 A_2} + \frac{\Delta x_3}{k_3 A_3} + \frac{1}{h_o A_o} \qquad (5\text{-}14)$$

The film coefficients may be read from Table 5-3a and the thermal conductivities may be obtained from Table 5-2a. For this case, a plane wall, the areas in Eq. 5-14 are all equal.

In the general case the area that is properly a part of the resistance may vary and unit thermal resistances may have to be adjusted. Consider the insulated pipe shown in Fig.

Table 5-4 Unit Thermal Resistance of a Plane $\frac{3}{4}$ In. (20 mm) Air Space[a]

Position of Air Space	Direction of Heat Flow	Mean Air Temperature		Temperature Difference		E = 0.05		E = 0.2		E = 0.82	
		F	C	F	C	ft²-hr-F / Btu	m²-C / W	ft²-hr-F / Btu	m²-C / W	ft²-hr-F / Btu	m²-C / W
Horizontal	Up	90	32	10	6	2.22	0.391	1.61	0.284	0.75	0.132
		50	10	30	17	1.66	0.292	1.35	0.238	0.77	0.136
		50	10	10	6	2.21	0.389	1.70	0.299	0.87	0.153
		0	−18	20	11	1.79	0.315	1.52	0.268	0.93	0.164
		0	−18	10	6	2.16	0.380	1.78	0.313	1.02	0.180
45° Slope	Up	90	32	10	6	2.78	0.490	1.88	0.331	0.81	0.143
		50	10	30	17	1.92	0.338	1.52	0.268	0.82	0.144
		50	10	10	6	2.75	0.484	2.00	0.352	0.94	0.166
		0	−18	20	11	2.07	0.364	1.72	0.303	1.00	0.176
		0	−18	10	6	2.62	0.461	2.08	0.366	1.12	0.197
Vertical	Horizontal	90	32	10	6	3.24	0.571	2.08	0.366	0.84	0.148
		50	10	30	17	2.77	0.488	2.01	0.354	0.94	0.166
		50	10	10	6	3.46	0.609	2.35	0.414	1.01	0.178
		0	−18	20	11	3.02	0.532	2.32	0.408	1.18	0.210
		0	−18	10	6	3.59	0.632	2.64	0.465	1.26	0.222
45° Slope	Down	90	32	10	6	3.27	0.576	2.10	0.370	0.84	0.148
		50	10	30	17	3.23	0.569	2.24	0.394	0.99	0.174
		50	10	10	6	3.57	0.629	2.40	0.423	1.02	0.180
		0	−18	20	11	3.57	0.629	2.63	0.463	1.26	0.222
		0	−18	10	6	3.91	0.689	2.81	0.495	1.30	0.229

[a]Effective emittance of the space E is given by Eq. 5-11. Credit for an air space resistance value cannot be taken more than once and only for the boundary conditions established. Resistances of horizontal spaces with heat flow downward are substantially the same as those for 45° slope with heat flow down and independent of temperature difference.

Source: Adapted by permission from *ASHRAE Handbook, Fundamentals Volume,* 1989.

Table 5-5 Unit Thermal Resistances of a Plane 3.5 In. (89 mm) Air Space[a]

Position of Air Space	Direction of Heat Flow	Mean Air Temperature		Temperature Difference		E = 0.05		E = 0.2		E = 0.82	
		F	C	F	C	ft²-hr-F / Btu	m²-C / W	ft²-hr-F / Btu	m²-C / W	ft²-hr-F / Btu	m²-C / W
Horizontal	Up	90	32	10	6	2.66	0.468	1.83	0.322	0.80	0.141
		50	10	30	17	2.01	0.354	1.58	0.278	0.84	0.148
		50	10	10	6	2.66	0.468	1.95	0.343	0.93	0.164
		0	−18	20	11	2.18	0.384	1.79	0.315	1.03	0.181
		0	−18	10	6	2.62	0.461	2.07	0.365	1.12	0.197
45° Slope	Up	90	32	10	6	2.96	0.521	1.97	0.347	0.82	0.144
		50	10	30	17	2.17	0.382	1.67	0.294	0.86	0.151
		50	10	10	6	2.95	0.520	2.10	0.370	0.96	0.169
		0	−18	20	11	2.35	0.414	1.90	0.335	1.06	0.187
		0	−18	10	6	2.87	0.505	2.23	0.393	1.16	0.204
Vertical	Horizontal	90	32	10	6	3.40	0.598	2.15	0.479	0.85	0.150
		50	10	30	17	2.55	0.449	1.89	0.333	0.91	0.160
		50	10	10	6	3.40	0.598	2.32	0.409	1.01	0.178
		0	−18	20	11	2.78	0.490	2.17	0.382	1.14	0.201
		0	−18	10	6	3.33	0.586	2.50	0.440	1.23	0.217
45° Slope	Down	90	32	10	6	4.33	0.763	2.49	0.438	0.90	0.158
		50	10	30	17	3.30	0.581	2.28	0.402	1.00	0.176
		50	10	10	6	4.36	0.768	2.73	0.481	1.08	0.190
		0	−18	20	11	3.63	0.639	2.66	0.468	1.27	0.224
		0	−18	10	6	4.32	0.761	3.02	0.532	1.34	0.236

[a]Effective emittance of the space E is given by Eq. 5-11. Credit for an air space resistance value cannnot be taken more than once and only for the boundary conditions established. Resistances of horizontal spaces with heat flow downward are substantially independent of temperature differences.

Source: Adapted by permission from *ASHRAE Handbook, Fundamentals Volume,* 1989.

Table 5-6 Unit Thermal Resistances of Plane Horizontal Air Spaces with Heat Flow Downward, Temperature Difference 10 Deg F (6 Deg C)

Air Space				$E = 0.05$		$E = 0.2$		$E = 0.82$	
Thickness		Mean Temperature		$\frac{\text{ft}^2\text{-hr-F}}{\text{Btu}}$	$\frac{\text{m}^2\text{-C}}{\text{W}}$	$\frac{\text{ft}^2\text{-hr-F}}{\text{Btu}}$	$\frac{\text{m}^2\text{-C}}{\text{W}}$	$\frac{\text{ft}^2\text{-hr-F}}{\text{Btu}}$	$\frac{\text{m}^2\text{-C}}{\text{W}}$
Inch	mm	F	C						
$\frac{3}{4}$	20	90	32	3.29	0.579	2.10	0.370	0.85	0.150
		50	10	3.59	0.632	2.41	0.424	1.02	0.180
		0	−18	4.02	0.708	2.87	0.505	1.31	0.231
$1\frac{1}{2}$	38	90	32	5.35	0.942	2.79	0.491	0.94	0.166
		50	10	5.90	1.04	3.27	0.576	1.15	0.203
		0	−18	6.66	1.17	4.00	0.707	1.51	0.266
$3\frac{1}{2}$	89	90	32	8.19	1.44	3.41	0.601	1.00	0.176
		50	10	9.27	1.63	4.09	0.720	1.24	0.218
		0	−18	10.32	1.82	5.08	0.895	1.64	0.289

Source: Adapted by permission from *ASHRAE Handbook, Fundamentals Volume*, 1989.

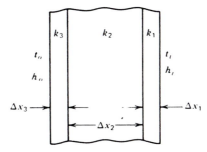

Figure 5-4 Wall with thermal resistances in series.

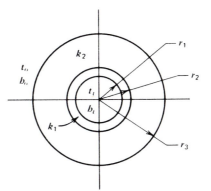

Figure 5-5 Insulated pipe in convective environment.

5-5. Convection occurs on the inside and outside surfaces while heat is conducted through the pipe wall and insulation. The overall thermal resistance for the pipe of Fig. 5-5 is

$$R'_e = R'_o + R'_2 + R'_1 + R'_i \tag{5-15}$$

or using Eqs. 5-6 and 5-9a

$$R'_e = \frac{1}{h_o A_o} + \frac{\ln\left(\dfrac{r_3}{r_2}\right)}{2\pi k_2 L} + \frac{\ln\left(\dfrac{r_2}{r_1}\right)}{2\pi k_1 L} + \frac{1}{h_i A_i} \tag{5-16a}$$

Equation 5-16a may also be written as

$$R'_e = \frac{1}{h_o A_o} + \frac{(r_3 - r_2)}{k_2 A_{m2}} + \frac{(r_2 - r_1)}{k_1 A_{m1}} + \frac{1}{h_i A_i} \tag{5-16b}$$

where

$$A_m = \frac{2\pi L(r_o - r_i)}{\ln\left(\dfrac{r_o}{r_i}\right)} \tag{5-17}$$

Equation 5-16b has a form quite similar to Eq. 5-14; however, the areas are all unequal. The thermal resistance on the outside surface is reduced by the increasingly large area. Where area changes occur in the direction of heat flow, unit resistances or conductances can be used only with appropriate area weighting factors.

Thermal resistances may also occur in parallel. In theory the parallel resistances can be combined into an equivalent thermal resistance in the same way as electrical resistances.

$$\frac{1}{R'_e} = \frac{1}{R'_1} + \frac{1}{R'_2} + \frac{1}{R'_3} + \cdots \frac{1}{R'_n} \tag{5-18}$$

In most heat transfer situations with parallel heat flow paths, however, lateral flow occurs, which may invalidate Eq. 5-18. The effect of lateral heat transfer between two thermal conductors is to lower the equivalent resistance of the system. However, when the ratio of the larger to the smaller of the thermal resistances is less than about 5, Eq. 5-18 gives a reasonable approximation of the equivalent thermal resistance. For large variations in the thermal resistance of parallel conduction paths, numerical techniques using the computer are suggested.

The concept of thermal resistance is very useful and convenient in the analysis of complex arrangements of building materials. After the equivalent thermal resistance has been determined for a configuration, however, the overall unit thermal conductance, usually called the *overall heat transfer coefficient U,* is frequently used.

$$U = \frac{1}{R'A} = \frac{1}{R} \text{ Btu/(hr-ft}^2\text{-F)} \quad \text{or} \quad \text{W/(m}^2\text{-C)} \tag{5-19}$$

The heat transfer rate is then given by

$$\dot{q} = UA\Delta t \tag{5-20}$$

where

UA = conductance, Btu/(hr-F) or W/C
A = surface area, ft^2 or m^2
Δt = overall temperature difference, F or C

For a plane wall the area A is the same at any position through the wall. In dealing with a curved wall, a particular area, such as the outside surface area, is selected for convenience of calculation. For example, in the problem of heat transfer through the ceiling–attic–roof combination, it is usually most convenient to use the ceiling area. The area selected is then used to determine the appropriate value of U from Eq. 5-19.

5-2 TABULATED OVERALL HEAT-TRANSFER COEFFICIENTS

For convenience of the designer, tables have been constructed that give overall coefficients for many common building sections including walls and floors, doors, windows, and skylights. The tables used in the *ASHRAE Handbook* have a great deal of flexibility and are summarized in the following pages.

Walls and Roofs

Walls and roofs vary considerably in the materials from which they are constructed. Therefore, the thermal resistance or the overall heat transfer coefficient is usually computed on an individual basis using Eqs. 5-14 and 5-19. This procedure is demonstrated for a wall and a roof in Tables 5-7a and 5-7b. Note that in each case an element has been changed. The tabular presentation makes it simple to recalculate the thermal resistance. In each case the unit thermal resistance and the overall heat-transfer coefficient have been computed for one set of conditions.

EXAMPLE 5-1

A frame wall is modified to have $3\frac{1}{2}$ in. of mineral fiber insulation between the studs. Compute the overall heat-transfer coefficient U if the unit thermal resistance without the insulation is 4.44 (hr-ft^2-F)/Btu.

SOLUTION

Total unit resistance given	4.44
Deduct the air space unit resistance, Table 5-5	−1.01
Add insulation unit resistance given in Table 5-2a	11.00
Total R in (hr-ft^2-F)/Btu	14.43

Then based on one square foot, we see that

$$U = \frac{1}{R} = \frac{1}{14.43} = 0.07 \text{ Btu/(hr-ft}^2\text{-F)}$$

Equation 5-18 may be used to correct for framing (2 × 4 studs on 16 in. centers)

$$\frac{1}{R_c'} = \frac{1}{R'} + \frac{1}{R_f'} \qquad \text{or} \qquad U_c A_t = U A_b + U_f A_f$$

where

A_t = total area
A_b = area between studs
A_f = area occupied by the studs

The unit thermal resistance of a section through the 2 × 4 stud is equal to the total resistance less the resistance of the air gap plus the resistance of the stud from Table 5-2a. A 2 × 4 stud is only $3\frac{1}{2}$ in. deep and $1\frac{1}{2}$ in. wide.

Table 5-7a Coefficients of Transmission U of Masonry Cavity Walls, Btu/(hr-ft²-F)[a]

Replace Cinder Aggregate Block with 6-in. Lightweight Aggregate Block with Cores Filled (New Item 4 in col. 2)

Construction	Resistance (R) 1 Between Furring	Resistance (R) 1 At Furring	Resistance (R) 2 Between Furring	Resistance (R) 2 At Furring
1. Outside surface (15 mph wind)	0.17	0.17	0.17	0.17
2. Face brick, 4 in.	0.44	0.44	0.44	0.44
3. Cement mortar, 0.5 in.	0.10	0.10	0.10	0.10
4. Concrete block, cinder aggregate, 8 in.	1.72	1.72	2.99	2.99
5. Reflective air space, 0.75 in. (50 F mean; 30 deg F temperature difference)	2.77	—	2.77	—
6. Nominal 1-in. × 3-in. vertical furring	—	0.94	—	0.94
7. Gypsum wallboard, 0.5 in, foil backed	0.45	0.45	0.45	0.45
8. Inside surface (still air)	0.68	0.68	0.68	0.68
Total Thermal Resistance (R)............	$R_i = 6.33$	$R_s = 4.50$	$R_i = 7.60$	$R_s = 5.77$

Construction No. 1: $U_i = 1/6.33 = 0.158$; $U_s = 1/4.50 = 0.222$. With 20% framing (typical of 1-in. × 3-in. vertical furring on masonry @ 16-in. o.c.), $U_{av} = 0.8\ (0.158) + 0.2\ (0.222) = 0.171$

Construction No. 2: $U_i = 1/7.60 = 0.132$ $U_s = 1/5.77 = 0.173$.
With framing unchanged, $U_{av} = 0.8(0.132) + 0.2(0.173) = 0.140$

[a] U factor may be converted to W/(m²-C) by multiplying by 5.68.

Source: Adapted by permission from *ASHRAE Handbook, Fundamentals Volume*, 1977.

Table 5-7b Coefficients of Transmission U of Flat Built-up Roofs,[a] Btu/(hr-ft²-F)[b]

Construction (Heat Flow Up)	1 Resistance (R)	2 Resistance (R)
1. Outside surface (15 mph wind)	0.17	0.17
2. Built-up roofing, 0.375 in.	0.33	0.33
3. Rigid roof deck insulation (none)	—	4.17
4. Concrete slab, lightweight aggregate, 2 in.	2.22	2.22
5. Corrugated metal deck	0	0
6. Metal ceiling suspension system with metal hanger rods	0[c]	0[c]
7. Nonreflective air space, greater than 3.5 in. (50 F mean; 10 deg F temperature difference)	0.93[a]	0.93[a]
8. Metal lath and lightweight aggregate plaster, 0.75 in.	0.47	0.47
9. Inside surface (still air)	0.61	0.61
Total Thermal Resistance (R)	4.73	8.90

Add Rigid Roof Deck Insulation, $C = 0.24$ ($R = 1/C$) (New Item 3 in col. 2)

Construction No. 1: $U_{avg} = 1/4.73 = 0.211$
Construction No. 2: $U_{avg} = 1/8.90 = 0.112$

[a]Use largest air space (3.5 in.) value shown in Table 5-5.
[b]U factor may be converted to W/m^2-C by multiplying by 5.68.
[c]Area of hanger rods is negligible in relation to ceiling area.

Source: Adapted by permission from *ASHRAE Handbook, Fundamentals Volume,* 1977.

$$R_f = \frac{1}{U_f} = 4.4 - 0.97 + (1.25)(3.5) = 7.81$$

or

$$U_f = 0.128 \text{ Btu/(hr-ft}^2\text{-F)}$$

Then using Eq. 5-18 we get

$$U_c = \frac{(0.07)(14.5) + (0.128)(1.5)}{16} = 0.075 \text{ Btu/(hr-ft}^2\text{-F)}$$

EXAMPLE 5-2

Compute the overall average coefficient for the roof-ceiling combination shown in Table 5-7b with 3.5 in. of mineral wool insulation in the ceiling space rather than the rigid roof deck insulation.

SOLUTION

Total unit resistance of the ceiling floor combination in Table 5-7b with no insulation is 4.73 (hr-ft^2-F)/Btu. Assume air space greater than 3.5 in.

Total resistance without insulation	4.73
Add mineral fiber insulation, 3.5 in.	11.00
Total R [(hr-ft^2-F)/Btu]	15.73
Total U [Btu/(ft^2-hr-F)]	0.064

The data given in Tables 5-7a and 5-7b and Examples 5-1 and 5-2 are based on

1. Steady-state heat transfer.
2. Ideal construction methods.
3. Surrounding surfaces at ambient air temperature.
4. Variation of thermal conductivity with temperature negligible.

Some caution should be exercised in applying calculated overall heat transfer coefficients such as those of Tables 5-7 because the effects of poor workmanship and materials are not included. Although a safety factor is not usually applied, a moderate increase in U may be justified in some cases.

The overall heat-transfer coefficients obtained for walls and roofs should always be adjusted for framing, as shown in Tables 5-7 using Eq. 5-18. This adjustment will normally be 5 to 12 percent of the unadjusted coefficient.

The coefficients of Tables 5-7 have all been computed for a 15 mph wind velocity on outside surfaces and should be adjusted for other velocities. The data of Table 5-3a may be used for this purpose.

The following example illustrates the calculation of an overall heat-transfer coefficient for an unvented roof-ceiling system.

EXAMPLE 5-3

Compute the overall heat-transfer coefficient for the roof–ceiling combination shown in Fig. 5-6. The wall assembly is similar to Table 5-7a with an overall heat-transfer coefficient of 0.16 Btu/(hr-ft^2-F). The roof assembly is similar to Table 5-7b without the ceiling and has a conductance of 0.13 Btu/(hr-ft^2-F) between the air space and the outdoor air. The ceiling has a conductance of 0.2 Btu/(hr-ft^2-F) between the conditioned space and the ceiling air space. The air space is 2.0 ft in the vertical direction. The ceiling has an area of 15,000 ft^2 and a perimeter of 500 ft.

SOLUTION

It is customary to base the overall heat-transfer coefficient on the ceiling area horizontal to the floor. Note that heat can enter or leave the air space through the roof or around the perimeter through the wall enclosing the space. The thermal resistances of the roof and the wall are in parallel and together are in series with the resistance of the ceiling. Then for roof and wall

$$A_c \, C_{rw} = C_w A_w + C_r A_r$$

and

$$R'_{rw} = \frac{1}{\left(C_{rw} A_c \right)} = \frac{1}{C_w A_w + C_r A_r}$$

Further

$$R'_o = R'_{rw} + R'_c$$

and

$$R'_o = \frac{1}{C_w A_w + C_r A_r} + \frac{1}{C_c A_c}$$

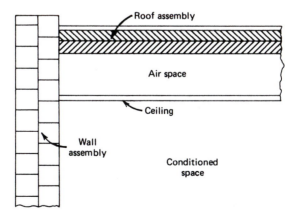

Figure 5-6 Section of a roof–ceiling combination.

Table 5-8a Overall Coefficients of Heat Transmission of Various Fenestration Products[a]

Part A: U-Values for Vertical Installation,[b] Btu/(h·ft²·°F)[e]

Glazing Type[d]	Glass Only		Aluminum Frame No Thermal Break ($U_f =$ 1.9)	Aluminum Frame Thermal Break ($U_f =$ 1.0)	Wood or Vinyl Frame ($U_f =$ 0.4)
	Center of Glass	Edge[e] of Glass			
Single glazing glass	1.11	n/a	1.23	1.10	0.98
$\frac{1}{8}$-in. acrylic	1.03	n/a	1.16	1.03	0.92
Double glass					
$\frac{1}{4}$-in. air space	0.57	0.66	0.78	0.65	0.55
$\frac{3}{8}$-in. air space	0.52	0.62	0.74	0.60	0.51
$\frac{1}{2}$-in. and greater air space	0.49	0.59	0.72	0.59	0.49
Double glass, $\epsilon = 0.15$ on surface 2 or 3					
$\frac{1}{4}$-in. air space	0.45	0.56	0.68	0.55	0.46
$\frac{3}{8}$-in. air space	0.36	0.51	0.62	0.48	0.39
$\frac{1}{2}$-in. and greater air space	0.34	0.50	0.60	0.46	0.37
Double glass					
$\frac{1}{4}$-in. argon space	0.52	0.62	0.74	0.61	0.51
$\frac{3}{8}$-in. argon space	0.48	0.59	0.71	0.57	0.48
$\frac{1}{2}$-in. and greater argon space	0.46	0.57	0.69	0.56	0.47
Double glass, $\epsilon = 0.15$ on surface 2 or 3					
$\frac{1}{4}$-in. argon space	0.36	0.51	0.62	0.48	0.39
$\frac{3}{8}$-in. argon space	0.30	0.48	0.57	0.43	0.34
$\frac{1}{2}$-in. and greater argon space	0.28	0.47	0.55	0.42	0.33
Triple glass					
$\frac{1}{4}$-in. air space	0.38	0.52	0.64	0.50	0.41
$\frac{3}{8}$-in. air space	0.34	0.50	0.60	0.46	0.38
$\frac{1}{2}$-in. air space	0.32	0.49	0.58	0.45	0.36

Triple glass, $\epsilon = 0.40$ on surface 2, 3, 4, or 5

$\frac{1}{4}$-in. air spaces	0.35	0.50	0.61	0.48	0.39
$\frac{3}{8}$-in. air spaces	0.30	0.48	0.57	0.44	0.35
$\frac{1}{2}$-in. and greater air spaces	0.28	0.47	0.55	0.41	0.33

Triple glass

$\frac{1}{4}$-in. argon spaces	0.34	0.50	0.60	0.46	0.38
$\frac{3}{8}$-in. argon spaces	0.31	0.48	0.57	0.44	0.35
$\frac{1}{2}$-in. and greater argon spaces	0.29	0.47	0.56	0.42	0.34

Triple glass, $\epsilon = 0.40$ on surface, 2, 3, 4, or 5

$\frac{1}{4}$-in. argon spaces	0.30	0.48	0.57	0.44	0.35
$\frac{3}{8}$-in. air spaces	0.26	0.46	0.54	0.41	0.32
$\frac{1}{2}$-in. and greater argon spaces	0.25	0.46	0.53	0.39	0.31

Part B: *U*-Value Conversion Table for Sloped and Horizontal Glazing for Upward Heat Flow

Slope	*U*-Value, Btu/(h·ft²·F)												
	0.10	0.20	0.30	0.40	0.50	0.60	0.70	0.80	0.90	1.00	1.10	1.20	1.30
90° (vertical)	0.10	0.20	0.30	0.40	0.50	0.60	0.70	0.80	0.90	1.00	1.10	1.20	1.30
45°	0.14	0.25	0.36	0.47	0.57	0.68	0.79	0.90	1.00	1.11	1.22	1.33	1.44
0 (horizontal)	0.19	0.29	0.40	0.51	0.61	0.72	0.82	0.93	1.04	1.14	1.25	1.35	1.46

[a] Product type is Commercial.

[b] All *U*-factors are based on standard ASHRAE winter conditions of 70 F indoor and 0 F outdoor air temperature, with 15 mph outdoor air velocity and zero solar flux. The outside surface coefficient at these conditions is approximately 5.1 Btu/(h·ft²· F), depending on the glass surface temperature. With the exception of single glazing, small changes in the interior and exterior temperatures do not significantly affect overall U-factors.

[c] *U*-factor may be converted to W/(m²·C) by multiplying by 5.68.

[d] Glazing layer surfaces are numbered from the outside to the inside. Double and triple refer to the number of glazing lites. All data are based on $\frac{1}{8}$-in. glass unless otherwise noted. Thermal conductivities are: 0.53 Btu/(h·ft· F) for glass, and 0.11 Btu/(h·ft· F) for acrylic and polycarbonate.

[e] Based on aluminum spacers data. Edge of glass effect assumed to extend over the 2.5-in. band around perimeter of each glazing unit.

Source: Reprinted by permission *ASHRAE Cooling and Heating Load Calculation Manual,* 2nd ed., 1992.

Substitution yields

$$R'_o = \frac{1}{(0.16)(2)(500) + (0.13)(15,000)} + \frac{1}{(0.2)(15,000)}$$

$$R'_o = 0.000807 = \frac{1}{U_o A_c}$$

then

$$U_o = \frac{1}{(0.000807)(15,000)} = 0.083 \text{ Btu/(hr-ft}^2\text{-F)}$$

Ceiling spaces should be vented to remove moisture from the insulation, but only moderate ventilation rates are required. The effect of ventilation on the transfer of heat through the air space is not significant provided the ceiling is insulated with a unit thermal resistance of about 11 or more. This is true for both winter and summer conditions. It once was thought that increased ventilation during the summer would dramatically reduce the heat gain to the inside space; however, this is apparently incorrect (3). It is generally not economically feasible to use power ventilation. The main reason for the ineffectiveness of ventilation is the fact that most of the heat transfer through the attic is by thermal radiation between the roof and the ceiling insulation. The use of reflective surfaces is therefore much more useful in reducing heat transfer. It is recommended that calculation of the overall transmission coefficient for ceiling spaces be computed using the approach of Example 5-3 with appropriate unit resistances and assuming no ventilation.

Windows

Table 5-8a contains overall heat-transfer coefficients for a range of fenestration products for vertical to horizontal installation. The values given are for winter design conditions; however, when corrected for wind velocity using Table 5-8b, the data are appropriate for estimating design loads for summer conditions. The U-factors may also be corrected using Part B of Table 5-8a. Using a value found in Part A, enter the table on the line for vertical glass and read down to the line for the appropriate orientation of the glass and read the corrected U-factor. The U-factors are based on the rough opening area and accounts for the effect of the frame. Transmission coefficients are given for the center and edge of the glass. Table 5-8a applies only for air-to-air heat transfer and does not account for solar radiation, which will be discussed in Chapter 6.

Doors

Table 5-9 gives overall heat-transfer coefficients for common doors. The values are for winter design conditions; however, they are appropriate for estimating design loads for summer conditions. Solar radiation has not been included.

Table 5-8*b* Glazing *U*-Value Conversion from 15 mph Wind to 7.5 mph Wind and Still Air

Wind Speed, mph			Wind Speed, mph		
15	7.5	0	15	7.5	0
U-Value, Btu/ (h·ft^2· F)			*U*-Value Btu/ (h·ft^2· F)		
0.10	0.10	0.10	0.80	0.74	0.69
0.20	0.20	0.19	0.90	0.83	0.78
0.30	0.29	0.28	1.00	0.92	0.86
0.40	0.38	0.37	1.10	1.01	0.94
0.50	0.47	0.45	1.20	1.10	1.02
0.60	0.56	0.53	1.30	1.19	1.10
0.70	0.65	0.61			

Source: Reprinted by permission *ASHRAE Cooling and Heating Load Calculation Manual,* 2nd ed., 1992.

Concrete Floors and Walls Below Grade

The heat transfer through basement walls and floors depends on the temperature difference between the inside air and the ground, the wall or floor material, and the conductivity of the ground. All of these factors involve considerable uncertainty. Mitalas (4) has studied the below-grade heat-transfer problem and developed methods that accurately predict heat losses for basement walls and floors 3 ft (0.9 m) or more below grade. When the design conditions across the United States are considered, the average coefficients in Tables 5-10 and 5-11, derived from reference 4, give satisfactory results for design load calculations. The heat loss is given by

$$\dot{q} = UA(t_i - t_g) \tag{5-21}$$

where

U = overall heat-transfer coefficient from Table 5-10 or 5-11, Btu/(hr-ft^2-F) or W/(m^2-C)

A = wall or floor surface area below 3 ft (0.9 m), ft^2 or m^2

t_i = inside air temperature, F or C

and

$$t_g = t_{avg} - Amp \tag{5-22}$$

where

t_g = design ground surface temperature, F or C

t_{avg} = average winter temperature, F or C (Table C-2)

Amp = amplitude of ground temperature variation about t_{avg}, F or C (Fig. 5-7)

Studies have shown that the heat losses from below-grade walls and floors are far more dependent on the ground temperature near the surface than the deep ground temperature. Ground surface temperature is known to vary about a mean value by an amplitude

Table 5-9 Transmission Coefficients U for Wood and Steel Doors, Btu/(h·ft²·F)[a]

Nominal Door Thickness, in.	Description	No Storm Door	Wood Storm Door[b]	Metal Storm Door[c]
Wood Doors[d,e]				
$1\frac{3}{8}$	Panel door with $\frac{7}{16}$-in. panels[f]	0.57	0.33	0.37
$1\frac{3}{8}$	Hollow core flush door	0.47	0.30	0.32
$1\frac{3}{8}$	Solid core flush door	0.39	0.26	0.28
$1\frac{3}{8}$	Panel door with $\frac{7}{16}$-in. panels[f]	0.57	0.33	0.36
$1\frac{3}{4}$	Hollow core flush door	0.46	0.29	0.32
$1\frac{3}{4}$	Panel door with $1\frac{1}{8}$-in. panels[f]	0.39	0.26	0.28
$1\frac{3}{4}$	Solid core flush door	0.33	0.28	0.25
$2\frac{1}{4}$	Solid core flush door	0.27	0.20	0.21
Steel Doors[e]				
$1\frac{3}{4}$	Fiberglass or mineral wool core with steel stiffeners, no thermal break[g]	0.60	—	—
$1\frac{3}{4}$	Paper honeycomb core without thermal break[g]	0.56	—	—
$1\frac{3}{4}$	Solid urethane foam core without thermal break[d]	0.40	—	—
$1\frac{3}{4}$	Solid fire rated mineral fiberboard core without thermal break[g]	0.38	—	—
$1\frac{3}{4}$	Polystyrene core without thermal break (18 gage commercial steel)[g]	0.35	—	—
$1\frac{3}{4}$	Polystyrene core without thermal break (18 gage commercial steel)[g]	0.35	—	—
$1\frac{3}{4}$	Polyurethane core without thermal break (18 gage commercial steel)[g]	0.29	—	—
$1\frac{3}{4}$	Polyurethane core without thermal break (24 gage commercial steel)[g]	0.29	—	—
$1\frac{3}{4}$	Polyurethane core with thermal break and wood perimeter (24 gage residential steel)[g]	0.20	—	—
$1\frac{3}{4}$	Solid urethane foam core with thermal break[d]	0.19	0.16	0.17

Note: All U-factors for exterior doors are for doors with no glazing, except for the storm doors that are in addition to the main exterior door. Any glazing area in exterior doors should be included with the appropriate glass type and analyzed. Interpolation and moderate extrapolation are permitted for door thicknesses other than those specified.

[a] U-factor may be converted to W/(m²·C) by multiplying by 5.68.

[b] Values for wood storm door are for approximately 50% glass area.

[c] Values for metal storm door are for any percent glass area.

[d] Values are based on a nominal 32×80 in. door size with no glazing.

[e] Outside air conditions: 15 mph wind speed; 0 F air temperature; inside air temperature; natural convection, 70 F air temperature.

[f] 55% panel area.

[g] ASTM C 236 hotbox data on a nominal 3×7 ft door with no glazing.

Source: Reprinted by permission from *ASHRAE Cooling and Heating Load Calculation Manual*, 2nd Ed. 1992.

Table 5-10 Average Overall Heat-Transfer Coefficients for Basement Walls (Ref. 4)

Depth Below Grade		Wall Uninsulated		Wall Insulated to a Depth of 2 Ft Below Grade				Wall Insulated Over Full Inside Surface			
				R-5	R-0.88	R-10	R-1.76	R-5	R-0.88	R-10	R-1.76
Feet	Meter	$\dfrac{\text{Btu}}{\text{hr-ft}^2\text{-F}}$	$\dfrac{\text{W}}{\text{m}^2\text{-C}}$	$\dfrac{\text{Btu}}{\text{hr-ft}^2\text{-F}}$	$\dfrac{\text{W}}{\text{m}^2\text{-C}}$	$\dfrac{\text{Btu}}{\text{hr-ft}^2\text{-F}}$	$\dfrac{\text{W}}{\text{m}^2\text{-C}}$	$\dfrac{\text{Btu}}{\text{hr-ft}^2\text{-F}}$	$\dfrac{\text{W}}{\text{m}^2\text{-C}}$	$\dfrac{\text{Btu}}{\text{hr-ft}^2\text{-F}}$	$\dfrac{\text{W}}{\text{m}^2\text{-C}}$
4	1.22	0.200	0.035	0.140	0.025	0.130	0.023	0.088	0.015	0.054	0.010
5	1.52	0.180	0.032	0.140	0.025	0.130	0.023	0.082	0.014	0.050	0.009
6	1.83	0.170	0.030	0.140	0.025	0.130	0.023	0.079	0.014	0.048	0.008
7	2.13	0.160	0.028	0.140	0.025	0.130	0.023	0.076	0.013	0.047	0.008

$$\frac{\text{Btu}}{\text{hr ft}^2\text{F}} \times 5.68 = \frac{\text{W}}{\text{m}^2\text{C}}$$

Table 5-11 Average Overall Heat-Transfer Coefficients for Uninsulated Basement Floors, 3 ft or More Below Grade (ref. 4)

Wall Un Insulated		Wall Insulated to a Depth of 2 Ft Below Grade		Wall Insulated Over Full Inside Surface	
$\dfrac{\text{Btu}}{\text{hr-ft}^2\text{-F}}$	$\dfrac{\text{W}}{\text{m}^2\text{-C}}$	$\dfrac{\text{Btu}}{\text{hr-ft}^2\text{-F}}$	$\dfrac{\text{W}}{\text{m}^2\text{-C}}$	$\dfrac{\text{Btu}}{\text{hr-ft}^2\text{-F}}$	$\dfrac{\text{W}}{\text{m}^2\text{-C}}$
0.025	0.004	0.036	0.006	0.045	0.008

Amp that varies with geographic location (Fig. 5-7). The mean ground surface temperature is assumed to be the average winter temperature.

When below-grade spaces are conditioned as living space, the walls should be furred and finished with a vapor barrier, insulating board, and some type of finish layer such as paneling. This will add thermal resistance to the wall. The basement floor should also be finished by laying down an insulating barrier and floor tile or carpet. The overall coefficients for the finished wall or floor may be computed as

$$R'_a = R' + R'_f = \frac{1}{UA} + R'_f = \frac{1}{U_a A} \tag{5-23}$$

Floor Slabs at Grade Level

When considering the heat losses for floor slabs at or near grade level, we must take into account two situations. The first is the unheated slab where heat is supplied to the space

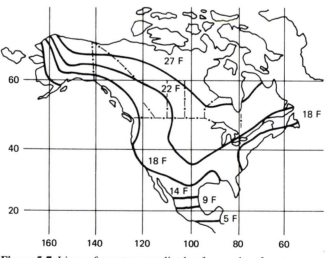

Figure 5-7 Lines of constant amplitude of ground surface temperature. (Reprinted by permission from *ASHRAE Handbook, Fundamentals Volume,* 1989)

from above. The second situation results when the air duct system is installed beneath the slab with air discharged at the perimeter of the structure. In both cases most of the heat loss is from the edge of the slab. When compared with the total heat losses of the structure, this loss may not be significant; however, from the viewpoint of comfort the temperature is important. Proper insulation around the perimenter of the slab is essential in severe climates to ensure a warm floor.

Figure 5-8 shows typical placement of edge insulation for a floor slab. Insulation may also be placed on the outside of the foundation wall extending down to the footing. Table 5-12 contains heat-loss factors for slabs. Note that the heat-loss factors are expressed as heat-transfer rate per unit length of perimeter per degree temperature difference. For summer conditions the heat transfer to the floor slab is negligible.

The heat loss from the slab is expressed as

$$\dot{q} = U'P(t_i - t_o) \tag{5-24}$$

where

U' = heat loss coefficient, Btu(hr-ft-F) or W/(m-C) (Table 5-12)
P = perimeter of slab, ft or m
t_i = inside air temperature, F or C
t_o = outdoor design temperature, F or C

Crawl Spaces

The most important factor in calculating heat losses from a crawl space is its tempera-ture. A heat balance on the crawl space taking into account the various gains and losses will yield the temperature. Heat is transferred to the crawl space through the floor and lost through the foundation wall and the ground much like a slab on grade. The inside or outside of the foundation wall may be insulated, and insulation may extend inward from the base of the foundation wall. The problem of crawl space temperature and heat loss is treated in Chapter 7.

Horizontal Pipes and Flat Surfaces

It is often necessary to compute the heat transfer from pipes, ducts, or flat surfaces. Typically these surfaces are in a free convection environment where radiation may also be important. This problem is complicated because the free convection heat-transfer

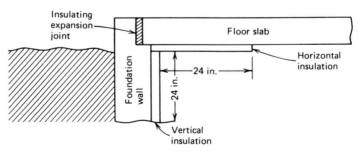

Figure 5-8 Edge insulation for a floor slab.

Table 5-12 Heat-Loss Coefficients for Slab Floor Construction

| Construction | Insulation | Degree Days (Base 65 F)[1] | | | | | | | | | | |
| | | 2950 | | 5350 | | 7433 | | | | | | |
		Btu/(hr-ft-F)	W/(m-C)	Btu/(hr-ft-F)	W/(m-C)	Btu/(hr-ft-F)	W/(m-C)
8-in. (200-mm) block wall, brick face	Uninsulated	0.62	1.07	0.68	1.18	0.72	1.25
	R-5.4 (R-0.95) slab to footing	0.48	0.83	0.50	0.87	0.56	0.97
4-in. (100-mm) block wall, brick face	Uninsulated	0.80	1.39	0.84	1.45	0.93	1.61
	R-5.4 (R-0.95) slab to footing	0.47	0.81	0.49	0.85	0.54	0.94
Metal stud wall, stucco	Uninsulated	1.15	1.99	1.20	2.08	1.34	2.32
	R-5.4 (R-0.95) slab to footing	0.51	0.88	0.53	0.92	0.58	1.00
Poured concrete wall, with perimeter ducts	Uninsulated	1.84	3.19	2.12	3.67	2.73	4.73
	R-5.4 (R-0.95) slab to footing	0.64	1.11	0.72	1.25	0.90	1.56

[1]Degree Day data is given in Table C-2.

Source: Reprinted by permission from *ASHRAE Handbook, Fundamentals Volume,* 1989.

Table 5-13 Transmission Coefficients U for Horizontal Bare Steel Pipes and Flat Surfaces,[a] Btu/(hr-ft²-F)[b]

Pipe Size Inches	Temperature Difference F Between Pipe Surface and Surrounding Air, Air at 80 F[c]									
	50	100	150	200	250	300	350	400	450	500
$\frac{1}{2}$	2.12	2.48	2.80	3.10	3.42	3.74	4.07	4.47	4.86	5.28
$\frac{3}{4}$	2.08	2.43	2.74	3.04	3.35	3.67	4.00	4.40	4.79	5.21
1	2.04	2.38	2.69	2.99	3.30	3.61	3.94	4.33	4.72	5.14
$1\frac{1}{4}$	2.00	2.34	2.64	2.93	3.24	3.55	3.88	4.27	4.66	5.07
$1\frac{1}{2}$	1.98	2.31	2.61	2.90	3.20	3.52	3.84	4.23	4.62	5.03
2	1.95	2.27	2.56	2.85	3.15	3.46	3.78	4.17	4.56	4.97
$2\frac{1}{2}$	1.92	2.23	2.52	2.81	3.11	3.42	3.74	4.12	4.51	4.92
3	1.89	2.20	2.49	2.77	3.07	3.37	3.69	4.08	4.46	4.87
$3\frac{1}{2}$	1.87	2.18	2.46	2.74	3.04	3.34	3.66	4.05	4.43	4.84
4	1.85	2.16	2.44	2.72	3.01	3.32	3.64	4.02	4.40	4.81
$4\frac{1}{2}$	1.84	2.14	2.42	2.70	2.99	3.30	3.61	4.00	4.38	4.79
5	1.83	2.13	2.40	2.68	2.97	3.28	3.59	3.97	4.35	4.76
Vertical surface	1.84	2.14	2.42	2.70	3.00	3.30	3.62	4.00	4.38	4.79
Horizontal surface Facing upward	2.03	2.37	2.67	2.97	3.28	3.59	3.92	4.31	4.70	5.12
Horizontal surface Facing downward	1.61	1.86	2.11	2.36	2.64	2.93	3.23	3.60	3.97	4.37

[a]Values are for flat surfaces greater than 4 ft² or 0.4 m².

[b]U in W/(m²-C) equals Btu/(hr-ft²-F) times 5.678.

[c]The temperature difference in C equals F divided by 1.8. An air temperature of 27 C corresponds to 80 F.

Source: Adapted by permission from ASHRAE Handbook, Fundamentals Volume, 1985.

coefficient and the thermal radiation both depend on the surface temperature, which may not be accurately known. Other variables are the surface finish, ambient air temperature, and air motion. Therefore it is customary to present overall heat-transfer coefficients as shown in Table 5-13, which is for bare steel pipe and flat surfaces. The values given are based on experimental data and analytical solutions for combined free convection and thermal radiation. They are reliable when the surface temperature is known and the surrounding air is still. The average bulk temperature of the fluid flowing in the pipe is a good estimate of the surface temperature because of the low thermal resistance of the pipe wall.

Buried Pipe

To make calculations of the heat transfer to or from buried pipes it is necessary to know the thermal properties of the earth. The thermal conductivity of soil varies considerably with the analysis and moisture content. Typically the range is 0.33 to 1.33 Btu/(ft-hr-F) or 0.58 to 2.3 W/(m-C). A reasonable estimate can be made using the equation for thermal resistance given in Table 5-1 with a known value of k. Thermal conductivity data for various soils and moisture contents are given in reference 1.

5-3 MOISTURE TRANSMISSION

The transfer of moisture through building materials and between the building surfaces and moist air follows theory directly analogous to conductive and convective heat transfer. Fick's law, which has the same form as Eq. 5-1,

$$\dot{m}_w = -DA\frac{dC}{dx} \tag{5-25}$$

governs the diffusion of moisture in a substance. Convective transport of moisture may be expressed as

$$\dot{m}_w = h_m A(C - C_w) \tag{5-26}$$

which is similar to Eq. 5-4. This subject is discussed in Chapter 13. The important point here is that moisture moves from a location where the concentration is high to one where it is low. This movement and accumulation of moisture can cause severe damage to the structure if not controlled.

During the coldest months, the moisture concentration tends to be greatest in the interior space. Moisture is transferred to the walls and ceilings and, if not retarded, diffuses outward into the insulation. The moisture reduces the thermal resistance of the insulation and in some cases it may freeze. Ceilings may fail structurally due to an accumulation of ice.

During the summer months, the moisture transfer process is reversed. This case is not as severe as that for the winter; however, the moisture is still harmful to the insulation and condensation may occur on some inside surfaces.

The transfer of moisture and the resulting damage is controlled through the use of barriers or retardants such as aluminum foil, thin plastic film, or other such material, and through the use of ventilation. Analysis of the problem shows that the moisture retarder should be near the warmest surface to prevent moisture from entering the insulation. Because the winter months are the most critical time, the barrier is usually installed between the inside finish layer and the insulation. During the summer months, the problem can usually be controlled by natural ventilation. This is the most important reason for ventilating an attic in both summer and winter. About 0.5 cfm/ft^2 or 0.15 m^3/(m^2-min) is required to remove the moisture from a typical attic. This can usually be accomplished through natural effects. Walls sometimes have provisions for a small amount of ventilation, but usually the outer wall covering is adequate to retard moisture.

REFERENCES

1. *ASHRAE Handbook, Fundamentals Volume,* American Society of Heating, Refrigerating and Air-Conditioning Engineers, Inc., Atlanta, GA, 1989.
2. J. D. Parker and F. C. McQuiston, *Introduction to Fluid Mechanics and Heat Transfer,* Kendall/Hunt, Dubuque, IA, 1988.
3. NBS Special Publication 548 ''Summer Attics and Whole-House Ventilation,'' U. S. Department of Commerce/National Bureau of Standards, Washington, DC, 1978.
4. G. P. Mitalas, ''Basement Heat Loss Studies at DBR/NRC,'' National Research Council of Canada, Division of Building Research, Ottawa, 1982.

PROBLEMS

5-1. Compute the Unit Conductance C for a 6 in. (150 mm) of glass fiberboard with a thermal conductivity of 0.25 Btu-in./(ft^2-hr-F) (0.43 W/(m-C)) in (a) English units and (b) SI units.

5-2. Refer to Problem 5-1 and compute the unit thermal resistance and the thermal resistance for 100 ft^2 (9.3 m^2) of the glass fiber board in (a) English units and (b) SI units.

5-3. Determine the thermal conductivity of 2 in. (50 mm) of roof insulation with a unit conductance of 0.19 Btu/(hr-ft^2-F) (1.08 W/m^2-C) in (a) English units and (b) SI units.

5-4. Compute the overall thermal resistance for a 2-in. steel pipe with 1 in. of insulation. The inside and outside film coefficients are 500 and 2 Btu/(hr-ft^2-F), respectively, and the insulation has a thermal conductivity of 0.2 Btu-in./(ft^2-hr-F).

5-5. What is the unit thermal resistance for an inside partition made up of $\frac{3}{8}$-in. gypsum board on each side of 2 $\times$ 4 in. studs? (Neglect the studing.)

5-6. Compute the overall thermal resistance of a wall made up of 100-mm face brick and 100-mm common brick with a 20-mm air gap between. There is 13 mm of gypsum plaster on the inside. Assume a 7 m/s wind velocity on the outside and still air inside.

5-7. Assuming that the 2 $\times$ 4 studs form a parallel heat transfer path, compute the unit thermal resistance for the partition of Problem 5-5. The studs are on 16 in. centers.

5-8. The pipe of Problem 5-4 has water flowing inside with a heat-transfer coefficient of 500 Btu/(hr-ft²-F) and is exposed to air on the outside with a film coefficient of 2.0 Btu/(hr-ft²-F). Compute the overall heat-transfer coefficient.

5-9. The partition of Problem 5-5 has still air on each side. Compute the overall heat-transfer coefficient.

5-10. Compute the overall heat-transfer coefficient for a frame construction wall made of brick veneer with 3-in. insulation bats between the studs; the wind velocity is $7\frac{1}{2}$ mph.

5-11. In an effort to save energy it is proposed to change the standard frame wall construction from 2 × 4 studs on 16 in. centers to 2 × 6 studs on 24 in. centers. Make a table similar to Table 5-7a showing both constructions. Use $5\frac{1}{2}$-in. and $3\frac{1}{2}$-in. fibrous glass insulation. Compare the two different constructions.

5-12. Use the data in Table 5-3a for still air and estimate what fraction of the heat transfer for a vertical wall is pure convection. Explain.

5-13. Consider a ceiling space formed by an infinite flat roof and horizontal ceiling. The inside surface of the roof has a temperature of 135 F and the top side of the ceiling insulation has a temperature of 115 F. Estimate the heat transferred by radiation and convection separately and compare them. Assume both surfaces have an emittance of 0.9.

5-14. Rework Problem 5-13 assuming that both surfaces have an emittance of 0.05.

5-15. Estimate the unit thermal resistance for a vertical $\frac{3}{4}$-in. (20-mm) air space. The air space is near the outside surface of a wall that has a large thermal resistance near the inside surface. The outdoor temperature is 5 F (-15 C). Assume nonreflective surfaces.

5-16. Refer to Problem 5-15 and estimate the unit thermal resistance assuming the air space is horizontal with heat flow up.

5-17. A wall is 12 ft wide and 8 ft high and has an overall heat transfer coefficient of 0.1 Btu/(hr-ft²-F). It contains a solid wood door, 80 × 32 × $1\frac{1}{2}$ in., and a single glass window, 60 × 30 in. Assuming parallel heat flow paths for the wall, door, and window, find the overall thermal resistance and overall heat transfer coefficient for the combination. Assume winter conditions. The window is metal sash, no thermal break.

5-18. A wall exactly like the one described in Table 5-7a has dimensions of 10 × 3 m. The wall has a total window area of 5 m² made of double-insulating glass with a 6-mm air space in an aluminum frame with thermal break. There is a steel door, 2 × 1 m, that has a mineral fiber core. Assuming winter conditions, compute the effective overall heat-transfer coefficient for the combination.

5-19. Compute the heat-transfer rate per square foot through a flat, built-up roof-ceiling combination similar to that shown in Table 5-7b. The ceiling is metal lath and plaster with 4-in. fibrous glass batts above. Indoor and outdoor temperatures are 70 and 5 F, respectively.

5-20. Compute the overall heat transfer for a single glass window and compare with the values given in Table 5-8a for the center of the glass.

5-21. Compute the overall heat-transfer coefficient for a $1\frac{3}{4}$-in. (45-mm) solid core wood door and compare to the value given in Table 5-9.

5-22. Determine the overall heat-transfer coefficient for a single glass window with an aluminum frame, no thermal break. Compare this with double insulating glass ($\frac{1}{4}$-in. air space) in a wood frame.

5-23. Refer to Table 5-7a and compute the overall transmission coefficient for the same construction with 1 in. sheathing of expanded polystyrene molded beads and wood siding in place of the brick.

5-24. Compute the overall heat-transfer coefficient for (a) an ordinary vertical single glass window with thermal break. (b) Assume the window has a roller shade. Compute the overall heat-transfer coefficient assuming a $3\frac{1}{2}$-in. (89-mm) air space between the shade and the glass. (c) Compare the results of parts (a) and (b).

5-25. Estimate the overall heat-transfer coefficient for a basement floor that has been covered with carpet and fibrous pad. The wall is fully insulated.

5-26. A basement is 20×20 ft (6×6 m) and 7 ft (2.13 m) below grade. The walls are insulated to a depth of 2 ft (0.6 m) below grade with R-5 (R-0.88) insulation. (a) Estimate the overall heat-transfer coefficients for the walls and floor. (b) Assuming the basement is located in Chicago, Illinois, compute the heat loss from the basement.

5-27. Rework Problem 5-26 assuming that the walls are finished on the inside with R-11 insulation and $\frac{3}{8}$-in. gypsum board and the floor has a carpet and pad.

5-28. Estimate heat loss from the floor slab of a building with a perimeter heating system. The building has dimensions of 15×30 m with a block and brick wall. The insulation is made of 40 mm of expanded polystyrene. The building is located in Kansas City.

5-29. A basement wall extends 6 ft below grade, is fully insulated (R-10), and is 6 in. thick. The inside is finished with $\frac{1}{2}$-in. insulating board, plastic vapor seal, and $\frac{1}{4}$-in. plywood paneling. Compute the overall heat-transfer coefficient for the wall.

5-30. A 24×40 ft (7.3×12.2 m) building has a full basement with walls extending 8 ft (2.4 m) below grade. The inside of the walls is finished with R-5 (R-0.7) insulation, a thin vapor barrier, and $\frac{1}{2}$-in. (12.7-mm) gypsum board. Estimate an overall heat-transfer coefficient for the walls.

5-31. The floor of the basement described in Problem 5-30 is finished with a thin vapor barrier, $\frac{5}{8}$-in. (16-mm) particle board underlayment, and carpet with rubber pad. Estimate an overall heat-transfer coefficient for the floor.

5-32. Assume that the ground temperature t_g is 40 F (10 C) and that the inside temperature is 68 F (20 C) in Problem 5-30 and estimate the temperatures between the wall and insulation and between the gypsum board and insulation.

5-33. Use the temperatures given in Problem 5-32 and compute the temperature between the underlayment and the carpet pad in Problem 5-31.

5-34. A floor slab with heating ducts below has R-5 (R-0.88), insulation from the slab down to the footing. Estimate the heat loss per unit length of perimeter for a location where the degree days are about 5400 F-day.

5-35. A small single-story frame office building in Syracuse, New York, is constructed with a concrete slab floor with an overhead type heating system. Estimate the heat loss per unit length of perimeter. Assume (a) R-5 (R-0.88) edge insulation; (b) no edge insulation. (c) Give a good reason to use the edge insulation other than to save energy.

5-36. Two 3-in. steel pipes cross a basement room near the ceiling. One pipe supplies hot water to a heating unit at 150 F, and the other pipe returns water from the unit at 130 F. If the room temperature is about 75 F, estimate the heat transfer from each pipe per linear foot.

5-37. Estimate the heat transfer rate to a 100-ft length of buried steel pipe carrying chilled water at 40 F. The pipe is 30 in. deep and has a 4-in. diameter. Assume the temperature of the ground is 50 F. The thermal conductivity of the earth is about 8 Btu-in./(hr-ft^2-F).

5-38. Estimate the heat loss from 100 m of buried hot water pipe. The mean water temperature is 49 C. The copper pipe with 10 mm of insulation, $k = 0.05$ W/(m-C), is buried 1 m below the surface and is 100 mm in diameter. Assume a thermal conductivity of the earth of 1.4 W/(m-C) and a ground temperature of 10 C.

5-39. A large beverage cooler resembles a small building and is to be maintained at about 35 F (2 C). The walls and ceiling are well insulated and are finished on the inside with plywood. Assume that the outdoor temperature drops below 35 F (2 C) for only short periods of time during the winter. Where should the vapor retardant be located? Explain what might happen if the retardant is improperly located.

5-40. Consider the wall section shown in Fig. 5-9. (a) Compute the temperatures of surfaces 1 and 2. (b) Assuming that the moist air can diffuse through the gypsum and insulation from the inside, would you expect moisture to condense on surface 1? Explain. (c) Would moisture condense on surface 2? Explain. (d) Where should a vapor retardant be placed?

5-41. Refer to the roof-ceiling assembly shown in Table 5-7b and compute the temperature of the metal roof deck when the outdoor temperature is 0 F (-18 C) and the indoor temperature is 72 F (22 C) with RH of 40 percent (a) with the rigid insulation and (b) without the insulation. (c) Would you expect any condensation problems on the underside of the metal deck? Explain.

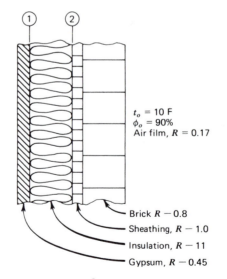

$t_o = 10$ F
$\phi_o = 90\%$
Air film, $R = 0.17$

Brick $R - 0.8$
Sheathing, $R - 1.0$
Insulation, $R - 11$
Gypsum, $R - 0.45$

R in units of (hr-ft^2-F)/Btu

Figure 5-9 Wall section for Problem 5-40.

5-42. Consider the wall section shown in Fig. 5-7*a* and estimate the temperature of the inside surface of the concrete block at the furring. The outdoor temperature is −5 F (−21 C) and the inside temperature is 70 F (21 C) with a relative humidity of 40 percent. Would you recommend a vapor retardant? If so where would you place it? Explain.

Chapter 6

Solar Radiation

Solar radiation has important effects on both the heat gain and heat loss of a building. This effect depends to a great extent on both the location of the sun in the sky and the clearness of the atmosphere as well as on the nature and orientation of the building. It is useful at this point to discuss ways of predicting the variation of the sun's location in the sky during the day and with the seasons for various locations on the earth's surface. It is also useful to know how to predict, for specified weather conditions, the solar irradiation of a surface at any given time and location on the earth, as well as the total radiation striking a surface over a specified period of time.

6-1 THERMAL RADIATION

Solar radiation is made up of several broad classes of electromagnetic radiation, all of which have some common characteristics, but which differ in the effect they produce, primarily because of their wavelength. These broad classes of the solar spectrum include ultraviolet, visible light, and infrared. Overlapping the wavelengths of most of the infrared, all of the visible light, and a part of the ultraviolet spectrum is a range referred to as *thermal radiation* since it is this part of the electromagnetic spectrum that primarily creates a heating effect. Conversely, when a substance has its thermal energy level (temperature) increased, the electromagnetic radiation produced by this temperature increase is primarily in the thermal radiation band. Thermal radiation is that portion of the electromagnetic spectrum with wavelengths of from 0.1×10^{-6} m up to approximately 100×10^{-6} m. In both the IP and the SI systems the common unit for wavelength is the *micron* ($1 \ \mu m = 10^{-6}$ m); therefore, the approximate range of thermal radiation is from 0.1 to 100 microns. A portion of the shorter wavelengths in this range is visible to the human eye. To better understand the heating effect of solar energy on a building we will review briefly the general characteristics of all thermal radiation. This review may yield additional benefits, since it will be shown later that aside from solar effects, thermal radiation plays an important role in heat exchanges in attics and enclosed spaces as well as in the energy exchanges that occur in occupied spaces of a building. For this discussion the terms *radiant energy* or *radiation* should be understood to mean thermal radiation.

The total thermal radiation that impinges on a surface from all directions and from all sources is called the *total* or *global irradiation* (G). Its units are Btu/(hr-ft^2) or W/m^2.

The thermal radiation energy that falls on a surface is subject to absorption and reflection as well as transmission through transparent bodies. *Absorption* is the transformation of the radiant energy into thermal energy stored by the molecules. *Reflection* is the return of radiation by a surface without change of frequency. In effect the radiation is "bounced" off of the surface. Transmission is the passage of radiation through a medium without change of frequency. Energy falling on a surface must be subject to one of these three actions; therefore:

$$\alpha + \rho + \tau = 1 \qquad (6\text{-}1)$$

where

   α = the absorptance, the fraction of the total incident thermal radiation absorbed
   ρ = the reflectance, the fraction of the total incident thermal radiation reflected
   τ = the transmittance, the fraction of the total incident radiation transmitted through the body

When the material is optically smooth and of sufficient thickness to show no change of reflectance or absorptance with increasing thickness, the terms *reflectivity* and *absorptivity* are used to describe the reflectance and absorptance, respectively. In much of the literature there is no distinction between these terms.

   Radiant energy originates at a surface or from the interior of a medium because of the temperature of the material. The rate of emission of energy is stated in terms of the *total emissive power (E)*. Its value depends only on the temperature of the system and the characteristics of the material of the system. Some surfaces emit more energy than others at the same temperature. The units of E may be expressed in Btu/(hr-ft^2) or W/m^2. E is the total energy emitted by the surface into the space and is a multidirectional, total quantity.

   It follows that radiant energy leaving an opaque surface ($\tau = 0$) comes from two sources: (1) the emitted energy and (2) the reflected irradiation.

   A surface that reflects no radiation ($\rho = 0$) is said to be a *blackbody,* since in the absence of emitted or transmitted radiation it puts forth no radiation visible to the eye and thus appears black. A blackbody is a perfect absorber of radiation and is a useful concept and standard for study of the subject of radiation heat transfer. It can be shown that the perfect absorber of radiant energy is also a perfect emitter; thus the perfect radiant emitter is also given the name *blackbody.* For a given temperature T in degrees R, a black emitter exhibits a maximum monochromatic emissive power at wavelength λ_{max}, given by

$$\lambda_{max} = \frac{5215.6}{T} \text{ microns} \qquad (6\text{-}2)$$

This equation is known as *Wien's displacement law.* The maximum amount of radiation is emitted in the wavelengths around the value of λ_{max}. According to Wien's displacement law, as the temperature of a black emitter increases, the major part of the radiation that is being emitted shifts to shorter wavelengths. This is an important concept in engineering, since the concept may be applied to approximate behavior of many nonblack emitters. It implies that higher temperature surfaces are primarily emitters of short wavelength radiation and lower temperature surfaces are primarily emitters of long wavelength radiation. The sun, which has a surface temperature of approximately 10,000 F or 6000 K, emits radiation with a maximum in the visible range. Building surfaces, which are at a much lower temperature, emit radiation primarily at much longer wavelengths.

Most surfaces are not blackbodies, but reflect some incoming radiation and emit less radiation than a blackbody at the same temperature. For such real surfaces we define one additional term, the *emittance* ϵ. The emittance is the fraction of the blackbody energy that a surface would emit at the same temperature, or

$$E = \epsilon E_B \qquad (6\text{-}3)$$

The emittance can vary with the temperature of the surface and with its conditions, such as roughness, degree of contamination, and the like. Although the emittance ϵ and the absorptance α of a given surface are identical for radiation at a given wavelength, the emittance of a building surface is most often quite different from its absorptance for solar radiation. The sun, being at a much higher temperature than a building surface, emits a predominance of radiation having relatively short wavelength compared to that of the building surface. The ratio of absorptance for sunlight to the emittance of a surface, combined with convection effects, controls the outer surface temperature of a building in sunlight. Sunlight has an additional important effect in transmitting energy into a building through openings (fenestrations) such as windows, doors, and skylights.

6-2   THE EARTH'S MOTION ABOUT THE SUN

The sun's position in the sky is a major factor in the effect of solar energy on a building. Equations for predicting the sun's position are best understood by considering the earth's motion about the sun. The earth moves in a slightly elliptical orbit about the sun (Fig. 6-1). The plane in which the earth rotates around the sun (approximately once every $365\frac{1}{4}$ days) is called the *ecliptic plane* or *orbital plane*. The mean distance from the center of the earth to the center of the sun is approximately 92.9×10^6 miles or 1.5×10^8 km. The *perihelion distance,* when the earth is closest to the sun, is 98.3 percent of the mean distance and occurs on January 4. The *aphelion distance,* when the earth is farthest from the sun, is 101.7 percent of the mean distance and occurs on July 5. Because of this the earth receives about 7 percent more total radiation in January than in July.

As the earth moves it also spins about its own axis at the rate of one revolution each 24 hours. There is an additional motion because of a slow wobble or gyroscopic precession of the earth. The earth's axis of rotation is tilted 23.5 deg with respect to the orbital plane. As a result of this dual motion and tilt, the position of the sun in the sky, as seen by an observer on earth, varies with the observer's location on the earth's surface and with the time of day and the time of year. For practical purposes the sun is so small as seen by an observer on earth that it may be treated as a point source of radiation.

At the time of the vernal equinox (March 21) and of the autumnal equinox (September 22 or 23), the sun appears to be directly overhead at the equator and the earth's poles are equidistant from the sun. Equinox means ''equal nights'' and during the time of the two equinoxes all points on the earth (except the poles) have exactly 12 hours of darkness and 12 hours of daylight.

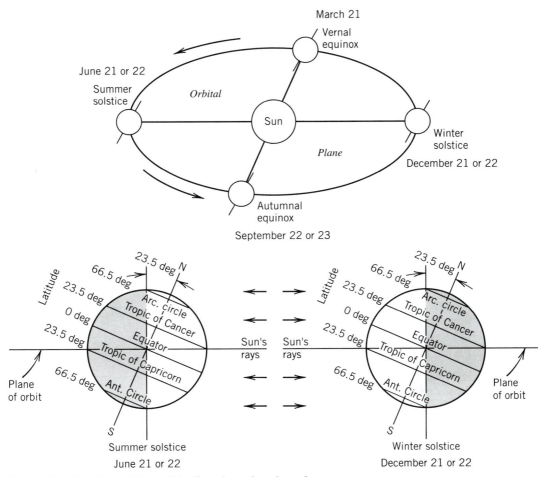

Figure 6-1 The effect of the earth's tilt and rotation about the sun.

During the summer solstice (June 21 or 22) the north pole is inclined 23.5 deg toward the sun. All points on the earth's surface north of 66.5 deg N latitude (the Arctic Circle) are in continuous daylight, whereas all points south of 66.5 deg S latitude (the Antarctic Circle) are in continuous darkness. Relatively warm weather occurs in the northern hemisphere and relatively cold weather occurs in the southern hemisphere. The word solstice means *sun standing still*.

During the summer solstice the sun appears to be directly overhead along the Tropic of Cancer, whereas during the winter solstice it is overhead along the Tropic of Capricorn. The *torrid zone* is the region between, where the sun is at the zenith, directly overhead, at least once during the year. In the *temperature zones* (between 23.5 and 66.5 deg latitude in each hemisphere) the sun is never directly overhead but always appears above the horizon each day. The *frigid zones* are those zones with latitude greater than 66.5 deg where the sun is below the horizon for at least one full day (24 hours) each year. In these two zones the sun is also above the horizon for at least one full day each year.

6-3 TIME

Because of the earth's rotation about its own axis, a fixed location on the earth's surface goes through a 24-hour cycle in relation to the sun. The earth is divided into 360 deg of circular arc by longitudinal lines passing through the poles. Thus, 15 deg of longitude corresponds to $\frac{1}{24}$ of a day or 1 hour of time. A point on the earth's surface exactly 15 deg west of another point will see the sun in exactly the same position as the first point after one hour of time has passed. *Universal Time* or *Greenwich Civil Time* (GCT) is the time along the zero longitude line passing through Greenwich, England. *Local Civil Time* (LCT) is determined by the longitude of the observer, the difference being four minutes of time for each degree of longitude, the more advanced time being on meridians further east. Thus when it is 12:00 noon GCT, it would be 7:00 A.M. LCT along the seventy-fifth deg W longitude meridian.

Clocks are usually set for the same reading throughout a zone covering approximately 15 deg of longitude, although the borders of the time zone may be irregular to accommodate local geographical features. The Local Civil Time for a selected meridian near the center of the zone is called the *Standard Time*. The four standard times zones in the lower 48 states and their standard meridians are

Eastern Standard Time, EST 75 deg
Central Standard Time, CST 90 deg
Mountain Standard Time, MST 105 deg
Pacific Standard Time, PST 120 deg

In much of the United States clocks are advanced one hour during the late spring, summer, and early fall season, leading to *Daylight Savings Time.*

Whereas civil time accounts for days that are precisely 24 hours in length, solar time has slightly variable days because of nonsymmetry of the earth's orbit, irregularities of the earth's rotational speed, and other factors. Time measured by the position of the sun is called *solar time.*

The Local Solar Time (LST) can be calculated from the Local Civil Time (LCT) by a quantity called the *equation of time*, LST = LCT + (equation of time). Values of the equation of time are given in Table 6-1 for the twenty-first day of each month (1). Values for any day may be obtained by linear interpolation.

EXAMPLE 6-1

Determine the local solar time (LST) corresponding to 11:00 A.M. CDST on February 21 in the United States at 95 deg W longitude.

SOLUTION

It is first necessary to convert central daylight savings time to central standard time:

$$CST = CDST - 1 \, \text{hour} = 11:00 - 1 = 10:00 \, \text{A.M.}$$

Then CST is local civil time at 90 deg W longitude. Now local civil time (LCT) at 95 deg W is $5 \times 4 = 20$ minutes less advanced than LCT at 90 deg W. Then

$$LCT = CST - 20 \, \text{min} = 9:40 \, \text{A.M.}$$

Table 6-1 Solar Data for 21st Day of Each Month[a]

	Equation of Time, min.	d Declination, degrees	A $\dfrac{Btu}{hr\text{-}ft^2}$	B (dimensionless ratios)	C
Jan	− 11.2	− 20.0	381.2	0.141	0.103
Feb	− 13.9	− 10.8	376.4	0.142	0.104
Mar	− 7.5	0.0	369.1	0.149	0.109
Apr	1.1	11.6	358.3	0.164	0.120
May	3.3	20.0	350.7	0.177	0.130
June	− 1.4	23.45	346.3	0.185	0.137
July	− 6.2	20.6	346.6	0.186	0.138
Aug	− 2.4	12.3	351.0	0.182	0.134
Sep	7.5	0.0	360.2	0.165	0.121
Oct	15.4	− 10.5	369.7	0.152	0.111
Nov	13.8	− 19.8	377.3	0.142	0.106
Dec	1.6	− 23.45	381.8	0.141	0.103

[a]A, B, C coefficients are based on research by Machler and Iqbal, 1985.

Source: Reprinted by permission from *ASHRAE Cooling and Heating Load Calculation Manual*, 2nd ed., 1992.

From Table 6-1 the equation of time is − 13.9 min. Then

$$LST = LCT + \text{equation of time}$$
$$LST = 9:40 - 0:14 = 9:26 \text{ A.M}$$

6-4 SOLAR ANGLES

The direction of the sun's rays can be described if three fundamental quantities are known:

1. Location on the earth's surface
2. Time of day
3. Day of the year

It is convenient to describe these three quantities by giving the latitude, the hour angle, and the sun's declination, respectively. Figure 6-2 shows a point P located on the surface of the earth in the northern hemisphere. The *latitude l* is the angle between the line OP and the projection of OP on the equatorial plane. This is the same latitude that is commonly used on globes and maps to describe the location of a point with respect to the equator.

The *hour angle h* is the angle between the projection of P on the equatorial plane and the projection on that plane of a line from the center of the sun to the center of the earth. Fifteen degrees of hour angle corresponds to one hour of time. The hour angle varies from zero at local solar noon to a maximum at sunrise or sunset. Solar noon occurs when

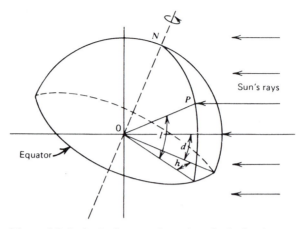

Figure 6-2 Latitude, hour angle, and sun's declination.

the sun is at the highest point in the sky, and hour angles are symmetrical with respect to solar noon. Thus, the hour angles of sunrise and sunset on a given day are identical.

The sun's *declination d* is the angle between a line connecting the center of the sun and earth and the projection of that line on the equatorial plane. Figure 6-3 shows how the sun's declination varies throughout a typical year. On a given day in the year, the declination varies slightly from year to year, but for typical HVAC calculations the values from any year are sufficiently accurate. Table 6-1 shows typical values of the sun's declination for the twenty-first day of each month. More precise values may be found in reference 2.

It is convenient in HVAC computations to define the sun's position in the sky in terms of the solar altitude β, and the solar azimuth ϕ, which depend on the fundamental quantities l, h, and d.

The *solar altitude* β (sun's altitude angle) is the angle between the sun's ray and the projection of that ray on a horizontal surface (Fig. 6-4). It is the angle of the sun above the horizon. It can be shown by analytic geometry that the following relationship is true (2):

$$\sin \beta = \cos l \cos h \cos d + \sin l \sin d \tag{6-4}$$

The *sun's zenith* angle Ψ is the angle between the sun's rays and a perpendicular to the horizontal plane at point P (Fig. 6-4). Obviously

$$\beta + \Psi = 90 \text{ degrees} \tag{6-5}$$

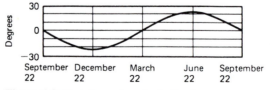

Figure 6-3 Variation of sun's declination.

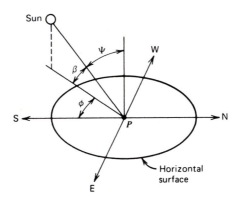

Figure 6-4 The solar altitude β, zenith angle Ψ, and azimuth angle ϕ.

The daily maximum altitude (solar noon) of the sun at a given location can be shown to be

$$\beta_{\text{noon}} = 90 - |(l - d)| \text{ degrees} \qquad \textbf{(6-6)}$$

where $|(l - d)|$ is the absolute value of $(l - d)$.

The *solar azimuth* angle ϕ is the angle in the horizontal plane measured between south and the projection of the sun's rays on that plane (Fig. 6-4). Again by analytic geometry (2) it can be shown that

$$\cos \phi = (\sin \beta \sin l - \sin d)/(\cos \beta \cos l) \qquad \textbf{(6-7)}$$

For a vertical surface the angle measured in the horizontal plane between the projection of the sun's rays on that plane and a normal to the vertical surface is called the *wall solar azimuth* γ. Figure 6-5 illustrates this quantity.

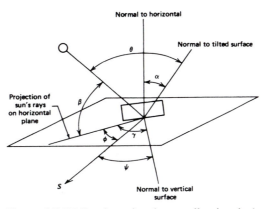

Figure 6-5 Wall solar azimuth γ, wall azimuth ψ, and angle of tilt α for an arbitrary tilted surface.

If ψ is the wall azimuth measured east or west from south, then obviously

$$\gamma = \phi \pm \psi \tag{6-8}$$

The *angle of incidence* θ is the angle between the sun's rays and the normal to the surface, as shown in Fig. 6-5. The *angle of tilt* α is the angle between the normal to the surface and the normal to the horizontal surface. It may be shown that

$$\cos\theta = \cos\beta\cos\gamma\sin\alpha + \sin\beta\cos\alpha \tag{6-9a}$$

Then for a vertical surface

$$\cos\theta = \cos\beta\cos\gamma \tag{6-9b}$$

and for a horizontal surface

$$\cos\theta = \sin\beta \tag{6-9c}$$

EXAMPLE 6-2

Find the solar altitude and azimuth at 10:00 A.M. Central Daylight Savings Time on July 21 at 40 deg N latitude and 85 deg W longitude.

SOLUTION

The Local Civil Time (LCT) is

$$10:00 - 1:00 + 4(90 - 85) = 9:20 \text{ A.M.}$$

The equation of time is -6.2 min; therefore the Local Solar Time (LST) to the nearest minute is

$$\text{LST} = 9:20 - 0:06 = 9:14 \text{ A.M.}$$

β is calculated from Eq. 6-4:

$$h = 2\,\text{hr}\,46\,\text{min} = 2.767\,\text{h} = 41.5\,\text{deg}$$

$$\beta = \sin^{-1}(\cos 40 \cos 41.5 \cos 20.6 + \sin 40 \sin 20.6)$$

$$\beta = 49.7\,\text{deg}$$

ϕ is calculated from Eq. 6-7:

$$\phi = \cos^{-1}\left[\frac{\sin 40 \sin 49.7 - \sin 20.6}{\cos 49.7 \cos 40}\right] = 73.7\,\text{deg}$$

6-5   SOLAR IRRADIATION

The *mean solar constant* G_{sc} is the rate of irradiation on the surface normal to the sun's rays beyond the earth's atmosphere and at the mean earth–sun distance. According to a fairly recent study (2), the mean solar constant is

$$G_{sc} = 433.4 \text{ Btu/(hr-ft}^2)$$

$$= 1367 \text{ W/m}^2$$

The irradiation from the sun varies about ± 3.5 percent because of the variation in distance between the sun and earth. Because of the large amount of atmospheric absorption of this radiation, and because this absorption is so variable and difficult to predict, a precise value of the solar constant is not used in most HVAC calculations.

The radiant energy emitted by the sun closely resembles the energy that would be emitted by a blackbody (an ideal radiator) at about 10,800 F or 5982 C. Figure 6-6 (3) shows the spectral distribution of the radiation from the sun as it arrives at the outer

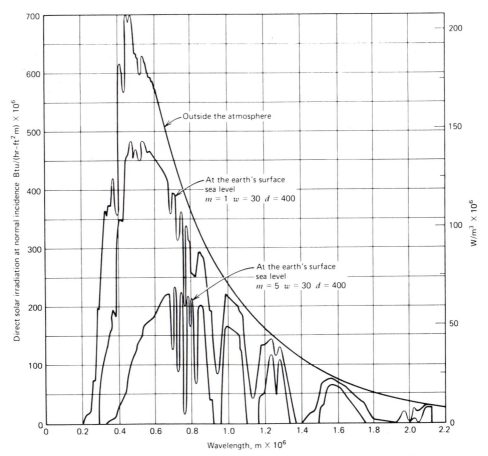

Figure 6-6 Spectral distribution of direct solar irradiation at normal incidence during clear days. (Adapted by permission from *ASHRAE Transactions,* Vol. 64, p. 50.)

edge of the earth's atmosphere (the upper curve). The peak radiation occurs at a wavelength of about 0.48×10^{-6} in the green portion of the visible spectrum. Forty percent of the total energy emitted by the sun occurs in the visible portion of the spectrum, between 0.4 and 0.7×10^{-6} m. Fifty-one percent is in the near infrared region between 0.7 and 3.5×10^{-6}. About 9 percent is in the ultraviolet below 0.4×10^{-6} m.

A part of the solar radiation entering the earth's atmosphere is scattered by gas and water vapor molecules and by cloud and dust particles. The blue color of the sky is a result of the scattering of some of the shorter wavelengths from the visible portion of the spectrum. The familiar red at sunset results from the scattering of longer wavelengths by dust or cloud particles near the earth. Some radiation (particularly ultraviolet) may be absorbed by ozone in the upper atmosphere, and other radiation is absorbed by water vapor near the earth's surface. That part of the radiation that is not scattered or absorbed and reaches the earth's surface is called *direct radiation*. It is accompanied by radiation that has been scattered or reemitted, called *diffuse radiation*. Radiation may also be reflected onto a surface from nearby surfaces. The total irradiation G_t on a surface normal to the sun's rays is thus made up of normal direct irradiation G_{ND}, diffuse irradiation G_d, and reflected irradiation G_R.

$$G_t = G_{ND} + G_d + G_R \tag{6-10}$$

The depletion of the sun's rays by the earth's atmosphere depends on the composition of the atmosphere (cloudiness, dust and pollutants present, atmospheric pressure, and humidity). With a given composition on a clear day the depletion is also strongly dependent on the length of the path of the rays through the atmosphere. In the morning or evening, for example, the sun's rays must travel along a much longer path through the atmosphere than they would at noontime. Likewise the sun's rays that hit the polar regions at midday have passed through a longer atmospheric path than those that hit the tropical regions at midday. This length is described in terms of the air mass m, the ratio of the mass of atmosphere in the actual sun–earth path to the mass that would exist if the sun were directly overhead at sea level. The air mass is for practical purposes equal to the cosecant of the solar altitude β multiplied by the ratio of actual atmospheric pressure to standard atmospheric pressure.

Figure 6-6 also shows the spectral distribution of direct solar radiation normally incident on a surface at sea level, with air masses equal to $1(\beta = 90 \text{ deg})$ and to $5 (\beta = 11.5 \text{ deg})$, for a specified value of water vapor and dust in the air denoted by w and d. The area under each of the curves is proportional to the total irradiation that would strike a surface under that particular condition. It can be easily seen that the total radiation is significantly depleted and the spectral distribution is altered by the atmosphere.

ASHRAE Clear Sky Model

It is important to recognize that the value of the solar constant is for a surface outside the earth's atmosphere and does not take into account the absorption and scattering of the earth's atmosphere, which can be significant even for clear days. The value of the solar irradiation* at the surface of the earth on a clear day is given by the ASHRAE Clear Sky Model (4):

$$G_{ND} = \frac{A}{\exp(B/\sin \beta)}$$ sun rays (6-11)

where

G_{ND} = the normal direct irradiation, Btu/(hr-ft^2) or W/m^2
A = apparent solar irradiation at air mass equal zero, Btu/(hr-ft^2) or W/m^2
B = atmospheric extinction coefficient
β = solar altitude

Values of A and B are given in Table 6-1 from reference 5 for the twenty-first day of each month and for an atmospheric clearness number C_N of unity. The data in Table 6-1, when used in Eq. 6-11, do not give the maximum value of G_{ND} that can occur in any given month, but are representative of conditions on average cloudless days. The values of C_N expressed as a percent are given in Fig. 6-7 for nonindustrial locations in the United States (4). The use of these values will be shown below.

The diffuse radiation on a horizontal surface is given by the use of the factor C from Table 6-1:

$$G_d = (C)(G_{ND})$$ (6-12)

where C is obviously the ratio of diffuse on a horizontal surface to direct normal irradiation. The parameter C is assumed to be a constant for an average clear day for a particular month. In reality the diffuse radiation varies directionally (Fig. 6-8) (6) and changes during the day in a fairly predictable way (see Fig. 6-14) (7). Also given in Table 6-1 are monthly values of the equation of time, and solar declination.

An evaluation of the ASHRAE Clear Sky Model was made by Powell (8), who suggested corrections for the effects of refraction at solar zenith angles in excess of 70 degrees and to account for the decrease in atmospheric path length with increased elevation. His corrections make the model applicable to situations where the clearness numbers may not be known.

A critical review of the ASHRAE Solar Radiation Model is given by Galanis and Chatigny (9) in which they show the contradiction of using the concept of a clearness number as a multiplying factor for both the normal direct irradiation and the diffuse radiation. They suggested dividing by the square of the clearness number to give the correct behavior for the diffuse component

horizontal
$$G_d = \frac{(C)(G_{ND})}{(C_N)^2}$$ (6-13)

---

* Some references refer to irradiation as "intensity"; however, most heat transfer texts reserve the term *intensity* for a different quantity.

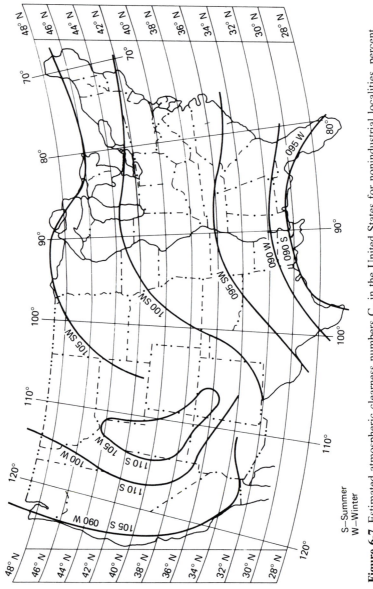

Figure 6-7 Estimated atmospheric clearness numbers C_n in the United States for nonindustrial localities, percent. (Reprinted by permission from *ASHRAE Handbook, Fundamentals Volume*, 1989.)

S—Summer
W—Winter

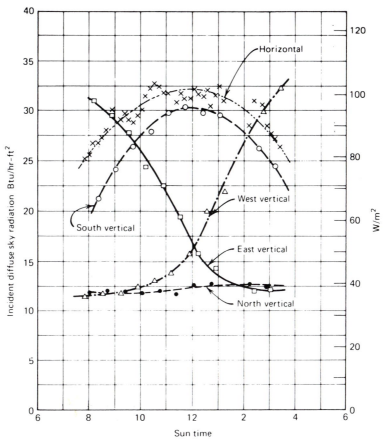

Figure 6-8 Variation of diffuse solar radiation from a clear sky. (Adapted by permission from *ASHRAE Transactions,* Vol. 69, p. 29.)

Their model would then combine Eq. 6-13 with an equation for the direct radiation G_D on a surface of arbitrary orientation, corrected for clearness

$$G_D = C_N G_{ND} \cos \theta \tag{6-14}$$

where θ is the angle of incidence between the sun's rays and the normal to the surface. The result is, for a horizontal surface where $\cos \theta = \sin \beta$,

$$G_t = G_D + G_d = \left[\cos \theta + \frac{C}{(C_N)^3} \right] C_N G_{ND} \tag{6-15}$$

Reference 9 also gives an expression for a cloudy sky model, using Eq. 6-15 as a starting point. This model involves the use of cloud cover information reported in meterological observations.

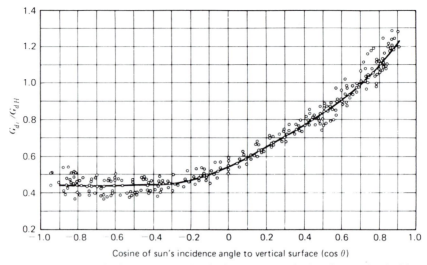

Figure 6-9 Ratio of diffuse sky radiation incident on a vertical surface to that incident on a horizontal surface during clear days. (Reprinted by permission from *ASHRAE Transactions*, Vol. 69, p. 29.)

To estimate the rate at which diffuse radiation $G_{d\theta}$ strikes a nonhorizontal surface on a clear day, the following equation is used:

$$G_{d\theta} = \frac{CG_{ND}F_{ws}}{C_N^2} \tag{6-16}$$

in which F_{ws} is the configuration factor or angle factor between the wall and the sky. The *configuration factor* is the fraction of the diffuse radiation leaving one surface that would fall directly on another surface. This factor is sometimes referred to in the literature as the *angle factor* or *the view, shape, interception,* or *geometrical factor*. For diffuse radiation this factor is a function only of the geometry of the surface or surfaces to which it is related. Because the configuration factor is useful for any type of diffuse radiation, information obtained in illumination, radio, or nuclear engineering studies is often useful to engineers interested in thermal radiation.

The symbol for configuration factor always has two subscripts designating the surface or surfaces that it describes. For example, the configuration factor F_{12} applies to the two surfaces numbered 1 and 2. F_{12} would be the fraction of the diffuse radiation leaving surface 1 that would fall directly on surface 2. F_{11} would be the fraction of the diffuse radiation leaving surface 1 that would fall on itself and obviously applies only to nonplane surfaces.

A very important and useful characteristic of configuration factors is the reciprocity relationship:

$$A_1F_{12} = A_2F_{21} \tag{6-17}$$

Its usefulness is in determining configuration factors when the reciprocal factor is known or when the reciprocal factor is more easily obtained than the desired factor. For example, the fraction of the diffuse radiation in the sky that strikes a given surface would be difficult to determine directly. The fraction of the energy that leaves the surface and ''strikes'' the sky directly, F_{ws}, however, can be easily determined from the geometry:

$$F_{ws} = \frac{(1 + \cos \Sigma)}{2}$$ (6-18)

Σ is the tilt angle of the surface from horizontal, $\Sigma = (90 - \alpha)$ degrees.

The rate at which diffuse radiation from the sky strikes a given surface of area A_w is, per unit area of surface,

$$\frac{\dot{q}}{A_w} = \frac{A_s G_d F_{sw}}{A_w}$$

By reciprocity

$$A_s F_{sw} = A_w F_{ws}$$

Therefore

$$\frac{\dot{q}}{A_w} = G_d F_{ws}$$

Thus, although the computation involves the irradiation of the sky on the surface or wall, the configuration factor most convenient to use is F_{ws}, the one describing the fraction of the surface radiation that strikes the sky.

The use of the configuration factor assumes that diffuse radiation comes uniformly from the sky in all directions. This is not true for either cloudy or clear skies. In both clear and cloudy skies the location of the sun can distort the uniformity of the diffuse radiation. In the case of cloudiness, the nonuniformity of the clouds in the sky may also cause the diffuse radiation to be nonuniform. Figure 6-8 shows the variation of diffuse solar radiation with sun position on a clear day (6).

EXAMPLE 6-3

Calculate the clear day direct, diffuse, and total solar radiation rate on a horizontal surface at 36 deg N latitude and 84 deg W longitude on June 21 at 12:00 noon CST.

SOLUTION

Equations 6-11 and 6-12 are used together with the parameters A, B, and C from Table 6-1. To calculate the angles β and θ used in these equations, the local solar time must be calculated. The declination is also obtained from Table 6-1.

$$LST = LCT + (\text{equation of time})$$

$$= 12:00 + \left(\frac{90 - 84}{15}\right)(60) + (-1.4) = 12:22.6$$

The hour angle h is given by

$$h = \frac{(22.6)(15)}{60} = 5.65 \text{ deg} \qquad \text{and} \qquad d = 23.45 \text{ min}$$

$$\sin \beta = \cos l \cos d \cos h + \sin l \sin d$$

$$\sin \beta = (0.809)(0.917)(0.995) + (0.588)(0.398)$$

$$\sin \beta = 0.972$$

For a horizontal surface $\cos \theta = \sin \beta$

$$G_{ND} = \frac{A}{\exp\left(\dfrac{B}{\sin \beta}\right)} = \frac{346.3}{\exp\left(\dfrac{0.185}{0.972}\right)} = 286 \text{ Btu/(hr-ft}^2)$$

The direct radiation is

$$G_D = G_{ND} \cos \theta = (286)(0.972) = 278 \text{ Btu/(hr-ft}^2)$$

The diffuse radiation is

$$G_d = CG_{ND} = (0.137)(286) = 39.2 \text{ Btu/(hr-ft}^2)$$

The total radiation is

$$G_t = G_D + G_d = 278 + 39.2 = 317 \text{ Btu/(hr-ft}^2)$$

A particularly useful curve (Fig. 6-9) gives the ratio of diffuse sky radiation on a vertical surface to that incident on a horizontal surface on a clear day (6). See Example 6-4.

In determining the total rate at which radiation strikes an arbitrarily oriented surface at any time, one must also consider the energy reflected onto the surface. The most common case is reflection of solar energy from the ground to a tilted surface or vertical wall. For such a case the rate at which energy is reflected to the wall is

$$G_R = G_{tH} \rho_g F_{wg} \tag{6-19}$$

where

G_R = rate at which energy is reflected onto the wall, Btu/(hr-ft^2) or W/m^2

G_{tH} = rate at which the total radiation (direct plus diffuse strikes the horizontal surface or ground in front of the wall, Btu/(hr-ft^2) or W/m^2

ρ_g = reflectance of ground or horizontal surface

F_{wg} = configuration or angle factor from wall to ground, defined as the fraction of the radiation leaving the wall of interest that strikes the horizontal surface or ground directly.

For a surface or wall at a tilt angle Σ to the horizontal ($\Sigma = 90 - \alpha$)

$$F_{wg} = \frac{(1 - \sin \Sigma)}{2} \tag{6-20}$$

EXAMPLE 6-4

Calculate the total incidence of solar radiation on a window facing south located 6 ft above the ground. In front of the window is a concrete parking area that extends 50 ft south and 50 ft to each side of the window. The window has no setback. The following parameters have been previously computed: $\beta = 69$ degrees 13 min, $\phi = 17$ degrees 18 min, $G_{ND} = 278$ Btu/(hr-ft²), $G_{tH} = 293$ Btu/(hr-ft²), $G_{dH} = 33$ Btu/(hr-ft²), $\rho_1 = 0.33$, and $F_{wg} = 0.433$.

SOLUTION

The angle of incidence for the window is first computed:

$$\gamma = \phi + \psi; \qquad \psi = 0$$
$$\gamma = \phi = 17 \text{ degrees } 18 \text{ min}$$
$$\cos \theta = \cos \beta \cos \gamma = 0.339$$
$$G_{DV} = G_{ND} \cos \theta = 287(0.339) = 94 \text{ Btu/(hr-ft}^2)$$

From Fig. 6-9

$$\frac{G_{dV}}{G_{dH}} = 0.75$$
$$G_{dV} = 0.75(33) = 25 \text{ Btu/(hr-ft}^2)$$

The reflected component is given by Eq. 6-19 where

$$G_R = 0.33(293)(0.433) = 42 \text{ Btu/(hr-ft}^2)$$

Then

$$G_{tV} = G_{DV} + G_{dV} + G_R = 94 + 25 + 42 = 161 \text{ Btu/(hr-ft}^2)$$

6-6 HEAT GAIN THROUGH FENESTRATIONS

The term *fenestration* refers to any glazed aperture in a building envelope. The components of fenestrations include:

- Glazing material, either glass or plastic
- Framing, mullions, muntins, and dividers
- External shading devices
- Internal shading devices
- Integral (between-glass) shading systems

Because of their glazing, fenestrations transmit solar radiation, and this permits heat gains into a building that are quite different from the heat gains of the nontransmitting parts of the building envelope. This behavior is best seen by referring to Fig. 6-10.

When solar radiation strikes an unshaded window (Fig. 6-10), about 8 percent of the radiant energy is typically reflected back outdoors, from 5 to 50 percent is absorbed within the glass, depending on the composition and thickness of the glass, and the remainder is transmitted directly indoors, to become part of the cooling load. The solar gain is the sum of the transmitted radiation and the portion of the absorbed radiation that flows inward. Because heat is also conducted through the glass whenever there is an outdoor–indoor temperature difference, the total rate of heat admission is

Total heat admission through glass = Radiation transmitted through glass
+ Inward flow of absorbed solar radiation + Conduction heat gain (6-21a)

The first two quantities are related to the amount of solar radiation falling on the glass, and the third quantity occurs whether or not the sun is shining. In winter the

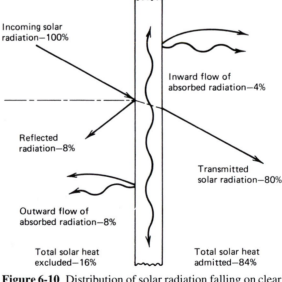

Figure 6-10 Distribution of solar radiation falling on clear plate glass.

conduction heat flow may well be outward rather than inward. We can rewrite Eq. 6-21a to read

$$\text{Total heat gain} = \text{Solar heat gain} + \text{Conduction heat gain} \qquad \textbf{(6-21b)}$$

The conduction heat gain per unit area is the product of the overall coefficient of heat transfer U for the existing fenestration and the outdoor–indoor temperature difference $(t_o - t_i)$. Values of U for a number of widely used glazing systems are given in Table 5-8.

The hourly solar heat gains that occur in a unit area of double-strength sheet glass (DSA) for a given orientation and time are called the *solar heat gain factors* (SHGF). The term accounts for the combined effects of both transmitted solar heat gain and absorbed solar heat gain conducted into the space. Because of refinements made in the method for calculating cooling loads, the transmitted and the absorbed solar heat gain components are now treated separately (10). The transmitted solar heat gain that occurs in a unit area of DSA glass for a given orientation and time is referred to as the *transmitted solar heat gain factor* (TSHGF). The absorbed solar heat gain that occurs in a unit area of DSA glass for a given orientation and time is referred to as the *absorbed solar heat gain factor* (ASHGF).

Both solar heat gain factors are calculated assuming that the direct solar irradiation G_D and the diffuse solar irradiation G_d have already been determined. The procedures for doing this will now be given.

The transmittance τ_D of DSA glass to direct (beam) radiation incident at an angle θ is

$$\tau_D = 0.83804$$

$$\tau_D = \sum_{j=0}^{5} t_j [\cos \theta]^j \qquad \textbf{(6-22)}$$

where t_j = Transmission coefficients for glass (Table 6-2) (4).

The transmittance τ_d of DSA glass to diffuse radiation is given by

$$\tau_d = 2(0.39950) = 0.79900$$

$$\tau_d = 2 \sum_{j=0}^{5} t_j / (j + 2) \qquad \textbf{(6-23)}$$

Note that both calculations use the transmission coefficients for glass found in Table 6-2. These coefficients give a normal transmittance for DSA glass of 0.88, which is slightly higher than values sometimes used. The transmitted solar heat gain factor is

$$\text{TSHGF} = G_D \sum_{j=0}^{5} t_j [\cos \theta]^j + 2G_d \sum_{j=0}^{5} t_j / (j + 2) \qquad \textbf{(6-24)}$$

The units of TSHGF will be consistent with the units of G_D and G_d.

Table 6-2 Coefficients for DSA
Glass for Calculation of
Transmittance and Absorptance

i	a_j	t_j
0	0.01154	− 0.00885
1	0.77674	2.71235
2	− 3.94657	− 0.62062
3	8.57811	− 7.07329
4	− 8.38135	9.75995
5	3.01188	− 3.89922

Source: Reprinted by permission from *ASHRAE Handbook, Fundamentals Volume,* 1989.

The fraction of direct (beam) solar radiation incident at an angle θ that is absorbed by DSA glass is

$$\alpha_D = \sum_{j=0}^{5} a_j [\cos \theta]^j \tag{6-25}$$

where a_j = Absorption coefficients for glass (Table 6-2) (4).

The fraction of diffuse solar radiation absorbed by DSA glass is given by

$$\alpha_d = 2 \sum_{j=0}^{5} a_j/(j + 2) \tag{6-26}$$

The absorbed solar heat gain factor is then given by

$$\text{ASHGF} = G_D \sum_{j=0}^{5} a_j [\cos \theta]^j + 2G_d \sum_{j=0}^{5} a_j/(j + 2) \tag{6-27}$$

Shading Coefficients (SC)

Procedures for estimating solar heat gain assume that a constant ratio exists between the solar heat gain through any given type of fenestration system and the solar heat gain (under exactly the same solar conditions) through DSA glass. This ratio, called the *shading coefficient* and abbreviated SC, is unique for each type of fenestration or each combination of glazing and internal shading device:

$$\text{SC} = \frac{\text{Solar heat gain of fenestration}}{\text{Solar heat gain of DSA glass}} \tag{6-28}$$

The shading coefficients are determined experimentally based on total solar heat gain, lumping the transmitted and absorbed components together. Because shading coefficients are not available for the individual solar gain components, the overall shading coefficient defined by Eq. 6-28 is used as an approximation for the individual components. Thus the transmitted solar heat gain TSHG is given by

$$TSHG = (SC)(TSHGF) \tag{6-29}$$

and the absorbed solar heat gain ASHG is given by

$$ASHG = (SC)(ASHGF)(N_i) \tag{6-30}$$

where N_i = inward flowing fraction of absorbed solar heat gain.

The inward flowing fraction of the absorbed solar heat gain is dependent on the relative magnitude of the indoor and outdoor heat transfer coefficients and is given approximately by

$$N_i = h_i/(h_i + h_o) \tag{6-31}$$

The values of SC in Tables 6-3 to 6-5 are based on natural convection conditions at the inner surface of the fenestration, and a 7.5 mph (3.35 m/s) wind at the outer surface

Table 6-3 Shading Coefficients for Single Glass and Insulating Glass

A. Single Glass

Type of Glass	Nominal Thickness in.	Nominal Thickness mm	Solar Trans.[a]	Shading Coefficient $h_o = 4.0$ Btu hr-ft²-F	Shading Coefficient $h_o = 3.0$ Btu hr-ft²-F
Clear	$\frac{1}{8}$	3.2	0.86	1.00	1.00
	$\frac{1}{4}$	6.4	0.78	0.94	0.95
	$\frac{3}{8}$	9.5	0.72	0.90	0.92
	$\frac{1}{2}$	12.7	0.67	0.87	0.88
Heat absorbing	$\frac{1}{8}$	3.2	0.64	0.83	0.85
	$\frac{1}{4}$	6.4	0.46	0.69	0.73
	$\frac{3}{8}$	9.5	0.33	0.60	0.64
	$\frac{1}{2}$	12.7	0.24	0.53	0.58

B. Insulating Glass[b]

Type of Glass	Nominal Thickness in.	Nominal Thickness mm	Solar Trans.[a]	Shading Coefficient $h_o = 4.0$ Btu hr-ft²-F	Shading Coefficient $h_o = 3.0$ Btu hr-ft²-F
Clear out, clear in	$\frac{1}{8}$[c]	3.2	0.71[e]	0.88	0.88
Clear out, clear in	$\frac{1}{4}$	6.4	0.61	0.81	0.82
Heat absorbing[d] out, clear in	$\frac{1}{4}$	6.4	0.36	0.55	0.58

[a]Refer to manufacturer's literature of values.

[b]Refers to factory-fabricated units with $\frac{3}{16}$-, $\frac{1}{4}$-, or $\frac{1}{2}$-in. air space or to prime windows plus storm sash.

[c]Thickness of each pane of glass, not thickness of assembled unit.

[d]Refers to gray-, bronze-, and green-tinted heat-absorbing float glass.

[e]Combined transmittance for assembled unit.

Source: Reprinted by permission from *ASHRAE Handbook, Fundamentals Volume,* 1989.

Table 6-4 Shading Coefficients for Single Glass with Indoor Shading by Venetian Blinds or Roller Shades

	Nominal Thickness,[a] inches	Solar Trans.[b]	Type of Shading				
			Venetian Blinds		Roller Shade		
					Opaque		Translucent
			Medium	Light	Dark	White	Light
Clear	$\frac{3}{32}$[c]	0.87–0.79	0.74[d] (0.63)[e]	0.67[d] (0.58)[e]	0.81	0.39	0.44
Clear	$\frac{1}{4}$–$\frac{1}{2}$	0.80–0.71	0.57	0.53	0.45	0.30	0.36
Clear pattern	$\frac{1}{8}$–$\frac{1}{2}$	0.87–0.79					
Heat-absorbing pattern	—	—					
Tinted	$\frac{3}{16}$, $\frac{7}{32}$	0.74, 0.71					
Heat-absorbing[f]	$\frac{3}{16}$, $\frac{1}{4}$	0.46	0.54	0.52	0.40	0.28	0.32
Heat-absorbing pattern	$\frac{3}{16}$, $\frac{1}{4}$	—					
Tinted	$\frac{1}{8}$, $\frac{7}{32}$	0.59, 0.45					
Heat-absorbing or pattern		0.44–0.30	0.42	0.40	0.36	0.28	0.31
Heat-absorbing[f]	$\frac{3}{8}$	0.34					
Heat-absorbing or pattern		0.29–0.15					
		0.24					
Reflective coated glass	SC[g] = 0.30		0.25	0.23			
	= 0.40		0.33	0.29			
	= 0.50		0.42	0.38			
	= 0.60		0.50	0.44			

[a] Refer to manufacturer's literature for values.

[b] For vertical blinds with opaque white and beige louvers in the tightly closed position, SC is 0.25 and 0.29 when used with glass of 0.71 to 0.80 transmittance.

[c] Typical residential glass thickness

[d] For 45° open venetian blinds, 35° solar incidence, and 35° profile angle.

[e] Values for closed venetian blinds. Use these values only when operation is automated for solar gain reduction (as opposed to daylight use).

[f] Refers to gray-, bronze-, and green-tinted heat-absorbing glass.

[g] SC for glass with no shading device.

Source: Reprinted by permission from *ASHRAE Handbook, Fundamentals Volume,* 1989.

Table 6-5 Shading Coefficients for Insulating Glass^a with Indoor Shading by Venetian Blinds or Roller Shades

Type of Glass	Nominal Thickness, Each Light	Solar Trans.^{b}		Type of Shading				
		Outer Pane	Inner Pane	Venetian Blinds^c		Roller Shade		
						Opaque		Translucent
				Medium	Light	Dark	White	Light
Clear out	$\frac{3}{32}, \frac{1}{8}$ in.	0.87	0.87	0.62^d	0.58^d	0.71	0.35	0.40
Clear in				(0.63)^e	(0.58)^e			
Clean in	$\frac{1}{4}$ in.	0.80	0.80					
Heat-absorbing^f out	$\frac{1}{4}$ in.	0.46	0.80	0.39	0.36	0.40	0.22	0.30
Clear in								
Reflective coated glass								
SC^g = 0.20				0.19	0.18			
= 0.30				0.27	0.26			
= 0.40				0.34	0.33			

^{a}Refers to factory-fabricated units with $\frac{3}{16}$-, $\frac{1}{4}$-, or $\frac{1}{2}$-in. air space, or to prime windows plus storm windows.

^{b}Refer to manufacturer's literature for exact values.

^{c}For vertical blinds with opaque white or beige louvers, tightly closed, SC is approximately the same as for opaque roller shades.

^{d}For 45° open venetian blinds, 35° solar incidence, and 35° profile angle

^{e}Values for closed venetian blinds. Use these values only when operation is automated for solar gain reduction (as opposed to daylight use).

^{f}Refers to bronze- or green-tinted, heat-absorbing glass.

^{g}SC for glass with no shading device.

Source: Reprinted by permission from *ASHRAE Handbook, Fundamentals Volume*, 1989.

(4). For these conditions h_i is 1.46 Btu/(hr-ft^2-F) [(0.257 W/(m^2-C)] and h_o is 4.0 Btu/(hr-ft^2-F) [(0.704 W/(m^2-C)], which yields a value of N_i equal to 0.267. For significantly different conditions N_i may be recalculated. Finally, the instantaneous solar heat gain is SHG = TSHG + ASHG.

EXAMPLE 6-5

Calculate SHGF for 2.30 P.M. local solar time, for a SW facing window at 36 deg N latitude, 97 deg W longitude. The direct normal solar irradiation G_D = 131 Btu/(hr-ft^2) and the diffuse solar irradiation G_d = 64.1 Btu/(hr-ft^2).

SOLUTION

The transmitted solar heat gain factor is found from Eq. 6-24:

$$\text{TSHGF} = G_D \sum_{j=0}^{5} t_j[\cos \theta]^j + 2G_d \sum_{j=0}^{5} t_j/(j + 2)$$

$$= (131 \times 0.7835) + (64.1 \times 2 \times 0.3995) = 154 \text{ Btu/(hr-ft}^2)$$

where $\cos \theta = \cos(61.7) = 0.4741$. The absorbed solar heat gain factor is found from Eq. 6-27:

$$\text{ASHGF} = G_D \sum_{j=0}^{5} a_j[\cos \theta]^j + 2G_d \sum_{j=0}^{5} a_j/(j + 2)$$

$$= (131 \times 0.0556) + (64.1 \times 2 \times 0.272) = 10.8 \text{ Btu/(hr-ft}^2)$$

The total solar heat gain factor is the sum of transmitted and the inward-flowing part of the absorbed energy:

$$\text{SHGF} = 154 + (10.8 \times 0.267) = 157 \text{ Btu/(hr-ft}^2)$$

Solar calorimeter tests are required for accurate determination of shading coefficients for combinations of glazing materials and internal shading devices. The shading coefficients for several types and combinations of glass are given in Table 6-3. The shading coefficient for any fenestration will rise above the tabulated values when the inner surface coefficient is increased and when the outer surface coefficient is decreased. The converse is also true. Table 6-3 shows the effect for outer film coefficients of 3 and 4 Btu/(hr-ft^2-F) or 17 and 23 W/(m^2-C), where the larger value is for 7.5 mph or 3.4 m/s wind, and the smaller value is for a lower wind velocity. The effect of the film coefficient on the shading coefficient is related to energy absorbed by the glass and then transferred away by convection.

Blinds, shades, and drapes or curtains that are often installed on the inside next to windows decrease the solar heat gain. The shading coefficient is used to express this effect. Tables 6-4 and 6-5 give representative data for single-sheet and insulating glass with indoor shading by blinds and shades. Note that the shading coefficient applies to the combination of glass and shading device.

Shading coefficients for draperies are a complex function of color and weave of the fabric. Although other variables also have an effect, reasonable correlation has been obtained using only color and openness of the weave. Figure 6-11 and Table 6-6 present a brief summary of information (4). Many manufacturers of drapery materials furnish data on the radiation properties of their products. This is usually in the form of the reflectance and transmittance, the abscissa and ordinate for Fig. 6-11. The shading coefficient index letter is read from Fig. 6-11 (A to J) and used in Table 6-6 to determine the shading coefficient for the combination glass and drapery system. The reflectance and transmittance of the drapery are often not known. In this case the index letter may be estimated from the openness of the weave and color of the material. Openness is classified as: open, I; semiopen, II; and closed, III. Color is classified as: dark, D; medium, M; and light, L. A light-colored, closed-weave material would then be classified III_L. From Fig. 6-11 the index letter varies from G to J for this classification, and judgment is required in making a final selection.

There are other data available for shading coefficients of such things as shade screen, domed skylights, and hollow glass blocks. The *ASHRAE Handbook* (4) or manufacturers' catalogs may be consulted for these cases.

External Shading

A fenestration may be shaded by roof overhangs, side fins or other parts of the building, trees, shrubbery, or another building. External shading of fenestrations is effective in reducing solar heat gain to a space and may produce reductions of up to 80 percent. The shading coefficient SC discussed above is *not* appropriate to use in determining the effect of external shade since its purpose is to account only for the effect of the fenestration and its internal shading devices. The SC will be used in heat gain calculations whether the fenestration is externally shaded or not. What is needed in considering heat gains affected by external shade are the areas of the fenestrations that are externally shaded. These areas on which external shade falls can be calculated from the geometry of the external objects creating the shade and from a knowledge of the sun angles for that particular time and location.

Figure 6-12 illustrates a window that is set back into the structure where shading may occur on the sides and top depending on the time of day and the direction the window faces. It can be shown that the dimensions x and y in Fig. 6-12 are given by

$$x = b \tan \gamma \qquad (6\text{-}32)$$

$$y = b \tan \delta \qquad (6\text{-}33)$$

where

$$\tan \delta = \frac{\tan \beta}{\cos \gamma} \qquad (6\text{-}34)$$

Table 6-6 Shading Coefficients for Single and Insulating Glass with Draperies

Glazing	Glass Transmission	Glass alone SC (no drapes)	A	B	C	D	E	F	G	H	I	J
Single Glass												
$\frac{1}{8}$-in. clear	0.86	1.00	0.87	0.82	0.74	0.69	0.64	0.59	0.53	0.48	0.42	0.37
$\frac{1}{4}$-in. clear	0.80	0.95	0.80	0.75	0.70	0.65	0.60	0.55	0.50	0.45	0.40	0.35
$\frac{1}{2}$-in. clear	0.71	0.88	0.74	0.70	0.66	0.61	0.56	0.52	0.48	0.43	0.39	0.35
$\frac{1}{4}$-in. heat abs.	0.46	0.67	0.57	0.54	0.52	0.49	0.46	0.44	0.41	0.38	0.36	0.33
$\frac{1}{2}$-in. heat abs.	0.24	0.50	0.43	0.42	0.40	0.39	0.38	0.36	0.34	0.33	0.32	0.30
Reflective coated[a]	—	0.60	0.57	0.54	0.51	0.49	0.46	0.43	0.41	0.38	0.36	0.33
	—	0.50	0.46	0.44	0.42	0.41	0.39	0.38	0.36	0.34	0.33	0.31
	—	0.40	0.36	0.35	0.34	0.33	0.32	0.30	0.29	0.28	0.27	0.26
	—	0.30	0.25	0.24	0.24	0.23	0.23	0.23	0.22	0.21	0.21	0.20
Insulating glass, $\frac{1}{4}$-in. air space ($\frac{1}{8}$ in. out and $\frac{1}{8}$ in. in)	0.76	0.89	0.75	0.71	0.65	0.63	0.57	0.53	0.48	0.45	0.38	0.36
Insulating glass, $\frac{1}{2}$-in. air space clear out and clear in	0.64	0.83	0.66	0.62	0.58	0.56	0.52	0.48	0.45	0.42	0.37	0.35
Heat abs. out and clear in	0.37	0.55	0.49	0.47	0.45	0.43	0.41	0.39	0.37	0.35	0.33	0.32
Reflective coated[a]	—	0.40	0.38	0.37	0.37	0.36	0.34	0.32	0.31	0.29	0.28	0.28
	—	0.30	0.29	0.28	0.27	0.27	0.26	0.26	0.25	0.25	0.24	0.24
	—	0.20	0.19	0.19	0.18	0.18	0.17	0.17	0.16	0.16	0.15	0.15

[a]See manufacturers' literature for exact values.

Source: Reprinted by permission from *ASHRAE Handbook, Fundamentals Volume*, 1989.

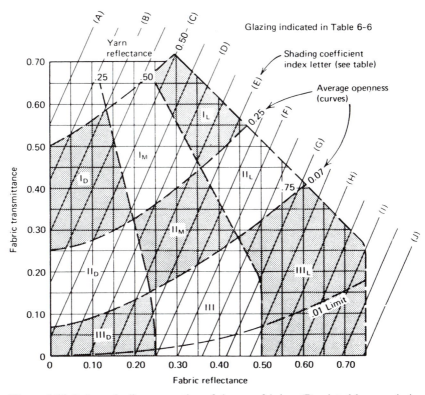

Figure 6-11 Indoor shading properties of drapery fabrics. (Reprinted by permission from *ASHRAE Handbook, Fundamentals Volume*, 1985.)

and

β = sun's altitude angle from Eq. 6-4
γ = wall solar azimuth angle ($\phi \pm \psi$)
ϕ = solar azimuth from Eq. 6-7
ψ = wall azimuth measured east or west from the south

The following rules aid in the computation of the wall solar azimuth angle γ.

For morning hours with walls facing east of south and afternoon hours with walls facing west of south:

$$\gamma = |(\phi - \psi)| \qquad \text{LST} \qquad \text{(6-35a)}$$

For afternoon hours with the walls facing east of south and morning hours with walls facing west of south:

$$\gamma = |(\phi + \psi)| \qquad \text{LST} \qquad \text{(6-35b)}$$

If γ is greater than 90 deg, the surface is in the shade. Equation 6-33 can be used for

and
$q_D = 0$

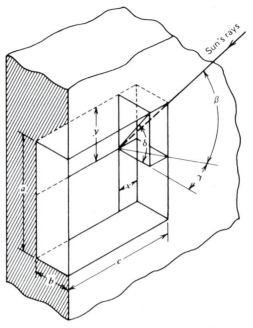

Figure 6-12 Shading of window set back from the plane of a building surface.

an overhang at the top and perpendicular to the window provided that the overhang is wide enough for the shadow to extend completely across the window.

EXAMPLE 6-6

A 4×5 ft, $\frac{1}{4}$-in. regular plate glass window faces southwest. The top of the window has a 2-ft overhang that extends a great distance on each side of the window. Compute the shaded area of the window on July 21 at 3:00 P.M. solar time at 40 deg N latitude.

SOLUTION

To find the area, the dimension y from Eq. 6-33 must be computed. From Eqs. 6-4 and 6-7, β and ϕ are 47.0 and 76.6 deg, respectively. The wall azimuth for a window facing southwest is 45 deg. Then for a wall facing west of south and for afternoon hours on July 21 at 3:00 P.M. solar time at 40 deg N latitude.

$$\gamma = |(\phi - \psi)| = |(76.6 - 45)| = 31.6 \deg$$

Then

$$y = \beta \tan \delta = \frac{b \tan \beta}{\cos \gamma}$$

$$y = \frac{2 \tan (47.0)}{\cos (31.6)} = 2.52 \text{ ft}$$

The shaded area is then

$$A_{sh} = 2.52 \times 4 = 10.1 \text{ ft}^2$$

and the sunlit portion has an area of

$$A_{sl} = A - A_{sh} = 20 - 10.1 = 9.9 \text{ ft}^2$$

The shaded portion of a window receives only indirect (diffuse) radiation.

After the locations of shadow lines on the glass have been found, the glass solar cooling load (SCL) is calculated separately for the externally unshaded and externally shaded portions as follows:

$$\dot{q}_{(unsh)} = A_{(unsh)} \, (\text{SC}) \, \text{SCL}_{(unsh)} \tag{6-36}$$

$$\dot{q}_{(sh)} = A_{(sh)} \, (\text{SC}) \, \text{SCL}_{(sh)} \tag{6-37}$$

Values of SCL are given in Chapter 8, where cooling load calculations are discussed. For latitudes greater than 24 deg, the solar cooling load for any shaded area of the glass may be approximated by using the SCL for a north orientation. This is not a good approximation for latitudes less than 24 deg.

The total glass solar cooling load is given by

$$\dot{q} = q_{(unsh)} + q_{(sh)} \tag{6-38}$$

If there is any doubt regarding the exterior shading or if the glass area is covered by exterior shade for only short periods during the day, the SCL for sunlit glass should be used.

6-7   ENERGY CALCULATIONS

Equations 6-11 through 6-20 are useful for design purposes where cooling loads are to be estimated because these equations are for clear days, when solar conditions are most severe.

In some cases it is desirable to be able to estimate the solar radiation for typical conditions, including both clear and cloudy days. In many of these cases one might want to know the total irradiation on a surface over an entire day or series of days or weeks. An example would be the computation of the irradiation on a solar collector panel.

Another would be in the determination of energy requirements of a building for a simulation study. In such cases the best information would be based on historical weather data for that particular location. If available, local weather data are usually best.

Hourly total (global) horizontal insolation data are available for 26 sites in the SOLMET database, available from the National Climatic Center, Asheville, North Carolina, 28801. These are rehabilitated data derived from data taken by the National Weather Bureau Service over a time period that generally covers 1953 through 1975.

This database was expanded to cover 222 additional sites by applying correlations to those locations having historical records of percentage sunshine and/or cloud cover observations. This database is known as the SOLMET/ERZATZ database and is also available from the National Climatic Center.

Typical Meteorological Year (TMY) Data were developed from the SOLMET and ERZATZ data for a collection of 12 typical months over a typical year. This is also available from NCC.

ASHRAE has made available four weather data tapes, based on the Weather Year for Energy Calculations (WYEC) for 51 cities in the United States and Canada (4).

In the most common situation the typical weather data that are available give the total (or global) solar insolation on a horizontal surface. To use these data for making predictions of insolation on nonhorizontal surfaces, there must be some procedure for determining the relative proportion of the total horizontal radiation that is direct and the proportion that is diffuse. Each part can then be used to determine the rate at which direct and diffuse radiation strikes the surface of interest. In addition, the energy reflected onto the surface must be determined.

Figure 6-13 illustrates the logic involved. The total radiation on a horizontal surface is first divided into the direct and diffuse components, step *a*. One way this can be accomplished is by using the methods described in references 11 and 12.

If hourly calculations are being made, information relating hourly diffuse to daily diffuse radiation and hourly total to daily total radiation must be used. These curves are given in Fig. 6-14 (7).

With the total radiation thus divided, one can use Eq. 6-14 to determine the direct radiation on the "tilted" surface (step *b*), and Eq. 6-16 to determine the diffuse radiation on the "tilted" surface (step *c*). Equation 6-19 is then used with the total horizontal insolation data to determine the radiation reflected (step *d*). The three components are then added to give the total insolation on the surface of interest.

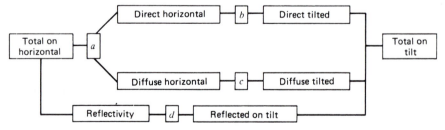

Figure 6-13 Conversion of horizontal insolation to insolation on tilted surface.

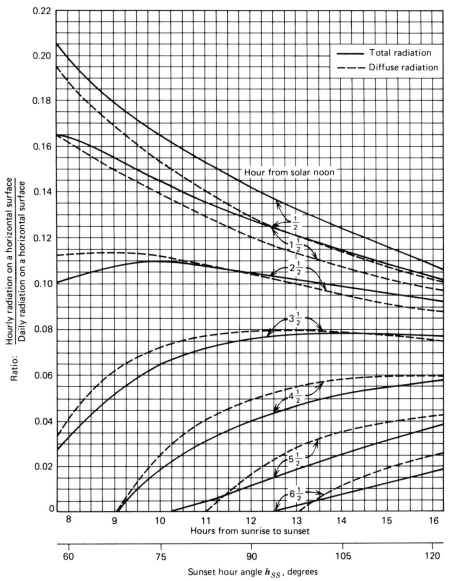

Figure 6-14 Relationships between daily radiation and hourly radiation on a horizontal surface. (Reprinted by permission from *ASHRAE Low Temperature Engineering Application of Solar Energy,* 1967.)

REFERENCES

1. U. S. Nautical Almanac Office, "The American Ephemeris and Nautical Almanac," U. S. Naval Observatory, Washington, DC (published annually).
2. C. Frohlich and R. W. Brusa, "Solar Radiation and Its Variation in Time," *Solar Physics,* Vol. 74, p. 209, 1981.

3. J. L. Threlkeld and R. C. Jordan, "Direct Solar Radiation Available on Clear Days," *ASHRAE Transactions,* Vol. 64, p. 50, American Society of Heating, Refrigerating and Air-Conditioning Engineers, Inc., Atlanta, GA.

4. *ASHRAE Handbook, Fundamentals Volume,* American Society of Heating, Refrigerating and Air-Conditioning Engineers, Inc., Atlanta, GA, 1989.

5. M. A. Machler and M. Iqbal, "A Modification of the ASHRAE Clear Sky Model," *ASHRAE Transactions,* Vol. 91, Pt. 1, 1985.

6. J. K. Threlkeld, "Solar Irradiation of Surfaces on Clear Days," *ASHRAE Transactions,* Vol. 69, p. 29, American Society of Heating, Refrigerating and Air-Conditioning Engineers, Atlanta, GA.

7. *ASHRAE Low Temperature Engineering Application of Solar Energy,* 1967, American Society of Heating, Refrigerating and Air-Conditioning Engineers, Inc., Atlanta, GA.

8. Gary L. Powell, "The ASHRAE Clear Sky Model—An Evaluation," *ASHRAE Journal,* pp. 32–34, November 1982.

9. N. Galanis and R. Charigny, "A Critical Review of the ASHRAE Solar Radiation Model," *ASHRAE Transactions,* Vol. 92, Pt. 1, 1986.

10. *Cooling and Heating Load Calculation Manual,* 2nd ed. American Society of Heating, Refrigerating and Air-Conditioning Engineers, Inc., Atlanta, GA, 1992.

11. *Applications of Solar Energy for Heating and Cooling of Buildings,* ASHRAE GRP 170, American Society of Heating, Refrigerating and Air-Conditioning Engineers, Inc., Atlanta, GA, 1977.

12. P. J. Lunde, *Solar Thermal Engineering,* John Wiley & Sons, Inc., New York, 1980.

PROBLEMS

6-1. Find the local solar time (LST) on August 21 for the following local times and locations:

(a) 9:00 A.M. EST and 70 deg W longitude

(c) 10:00 A.M. MST and 100 deg W longitude

(b) 1:00 P.M. CST and 95 deg W longitude

(d) 3:00 P.M. PST and 125 deg W longitude

6-2. What are the hour angles corresponding to the following local solar times: (a) 9:32 A.M., (b) 11:38 A.M., (c) 3:20 P.M., and (d) 12:38 P.M.?

6-3. Compute the time for sunrise and sunset on July 21 in (a) Stillwater, Oklahoma, (b) Chicago, Illinois, (c) Tallahassee, Florida, and (d) San Francisco, California.

6-4. Calculate the sun's altitude and azimuth angles at 9:00 A.M. solar time on September 21 at 40 deg N latitude.

6-5. Determine the solar time and azimuth angle for sunrise at 50 deg N latitude on (a) June 21 and (b) December 21.

6-6. On what month, day, and time does the maximum solar altitude angle β occur in (a) Stillwater, Oklahoma, 36 deg N latitude, (b) Chicago, Illinois, 42 deg N latitude, and (c) central Florida, 29 deg N latitude?

6-7. Compute the wall solar azimuth γ for a surface facing 12 deg west of south located at 40 deg N latitude and 90 deg W longitude on October 21 at 3:30 P.M. central daylight saving time (CDST).

6-8. Calculate the angle of incidence for the surface of Problem 6-7 for (a) a vertical orientation, and (b) a 20 degree tilt from the vertical.

6-9. For a location at 85 deg W longitude and 43 deg N latitude on July 21, determine (a) the incidence angle of the sun for a horizontal surface at 4:00 P.M. central daylight saving time and (b) the time of sunset in central daylight saving time.

6-10. Calculate the angle of incidence at 10:30 A.M. EST on July 21 for 36 deg N latitude and 78 deg W longitude for (a) a horizontal surface, (b) a surface facing southeast, and (c) a surface inclined 40 deg from the vertical and facing south.

6-11. Calculate the total clear day irradiation of a surface tilted at an angle of 30 deg from the vertical located at 36 deg latitude and 78 deg longitude on July 21 at 2:00 P.M. central daylight saving time. The surface faces the southwest. Neglect reflected radiation.

6-12. Compute the reflected irradiation of a window facing southwest over a large lake on a clear day. The location is 36 deg N latitude and 92 deg W longitude. The time is June 21 at 3:00 P.M. CST. Assume that the water has a diffuse reflectance of 0.06.

6-13. Determine magnitudes of direct, diffuse, and reflected clear day solar radiation incident on a small vertical surface facing south on March 21 at solar noon for a location at 36 deg N latitude having a clearness number of 0.95. The reflecting surface is snow-covered ground of infinite extent with a diffuse reflectance of 0.85.

6-14. Estimate the total clear day irradiation of a roof with a one-to-one slope that faces southwest at 36 deg N latitude. The date is August 21 and the time is 1:00 P.M. LST. Include reflected radiation from the ground with a reflectance of 0.3. The view factor from the roof to the ground is given approximately by $F_{rg} = (\frac{1}{2})(1 - \cos \Sigma)$, where Σ is the slope angle of the roof.

6-15. Compute the direct normal solar irradiation and the diffuse solar irradiation for a west-facing window at 36 deg N latitude, 90 deg W longitude for all 24 hours of a clear day on July 21. It is suggested that a short computer program or spreadsheet be used.

6-16. Use the results of Problem 6-15 to compute the total and absorbed heat gain factors (TSHGF and ASHGF) for DSA glass for all 24 hours of the day. Use a computer program or a spreadsheet.

6-17. Determine the amount of diffuse, direct, and total radiation that would strike a south-facing surface tilted at 45 deg on a clear December 21 in St. Louis, Missouri:
(a) At 12 noon solar time   (c) For all 24 hours
(b) At 3 P.M. solar time

6-18. Use the results of Problem 6-17 to compute the total and absorbed solar heat gain factors for DSA glass.

6-19. For the hours of 1 P.M., 2 P.M., 3 P.M., 4 P.M., 5 P.M., 6 P.M., and 7 P.M. CDST, estimate the rate at which solar energy will strike a west-facing window, 3 ft wide by 5 ft high, with no setback. Assume a clear July 21 day at 40 deg N latitude and 84 deg W longitude.

6-20. A south-facing window is 4 ft wide by 6 ft tall and is set back into the wall a distance of 8 in. For Atlanta, Georgia, estimate the percent of the window that is shaded for
(a) April 21, 9:00 A.M. solar time   (c) September 21, 5:00 P.M. solar
(b) July 21, 12:00 noon solar time   time

6-21. Work Problem 6-20 assuming a long 2-ft overhang located 2 ft above the top of the window.

6-22. Work Problem 6-19 assuming a 6-in. setback for the window.

6-23. Work Problem 6-19 for a clear day on December 21.

6-24. Work Problem 6-19 assuming a long overhang of 2 ft that is 3 ft above the top of the window.

6-25. Write a computer program to predict the altitude and zenith angles for the sun and the normal direct and diffuse irradiation. Make a table of these quantities for the twenty-first day of each month for each hour between sunrise and sunset.

6-26. Write a computer program to predict the total irradiation of any building surface by the sun. Include reflected radiation from the ground. (*Note:* This problem is an extension of Problem 6-25. Test the program with hand calculations.)

6-27. Write a computer program to predict the fraction of sunlit area of a vertical window that may face any arbitrary direction in the northern hemisphere. The window may have setback or an overhang extending to the left and right a great distance. Test the program at some critical conditions, such as a west window at solar noon.

6-28. Determine the shading coefficient for clear, $\frac{1}{4}$-in. single-sheet glass under summer design conditions for (a) no internal shade, (b) shading with medium-color venetian blinds, (c) shading with white, tightly woven drapes.

Chapter 7

Space Heat Load

Prior to the design of the heating system, an estimate must be made of the maximum probable heat loss of each room or space to be heated. There are two kinds of heat losses: (1) the heat transmitted through the walls, ceiling, floor, glass, or other surfaces and (2) the heat required to warm outdoor air entering the space.

The actual heat loss problem is transient because the outdoor temperature, wind velocity, and sunlight are constantly changing. The transfer function method, discussed in Chapter 8 in connection with the cooling load, may be used under winter conditions to account for changing solar radiation, outdoor temperature, and the energy storage capacity of the structure. During the coldest months, however, sustained periods of very cold, cloudy, and stormy weather with relatively small variation in outdoor temperature may occur. In this situation heat loss from the space will be relatively constant, and in the absence of internal heat gains will peak during the early morning hours. Therefore, for design purposes the heat loss is usually estimated for the early morning hours assuming steady-state heat transfer. Transient analyses are often used to study the actual energy requirements of a structure in simulation studies. In such cases solar effects and internal heat gains are taken into account.

The procedures for calculation of design heat losses of a structure are outlined in the following sections. Reference 1 may be consulted for further details related to the heating load.

7-1 OUTDOOR DESIGN CONDITIONS

The ideal heating system would provide just enough heat to match the heat loss from the structure. However, weather conditions vary considerably from year to year, and heating systems designed for the worst weather conditions on record would have a great excess of capacity most of the time. The failure of a system to maintain design conditions during brief periods of severe weather is usually not critical. However, close regulation of indoor temperature may be critical for some industrial processes.

Table C-1 contains outdoor temperatures that have been recorded for selected locations in the United States and Canada. The data are based on official weather station records of the U. S. Weather Bureau, U. S. Air Force, U. S. Navy, and the Canadian Department of Transport. At this point we shall consider only data pertaining to the winter months. The $97\frac{1}{2}$ percent value is the temperature equaled or exceeded $97\frac{1}{2}$

221

percent of the total hours (2160) in December, January, and February. During a normal winter there would be about 54 hours at or below the $97\frac{1}{2}$ percent value. For the Canadian stations the $97\frac{1}{2}$ percent value pertains only to hours in January.

The *ASHRAE Handbook* (2) has a more extensive tabulation of weather data; it also includes temperatures that are equaled or exceeded 99 percent of the total hours in December, January, and February.

The outdoor design temperature should generally be the $97\frac{1}{2}$ percent value as specified by ASHRAE Energy Standards. If the structure is of lightweight construction (low heat capacity), is poorly insulated, or has considerable glass or space temperature control is critical, however, the 99 percent values should be considered. The designer must remember that should the outdoor temperature fall below the design value for some extended period, the indoor temperature may do likewise. The performance expected by the owner is a very important factor, and the designer should make clear to the owner the various factors considered in the design.

Abnormal local conditions should be considered. It is good practice to seek local knowledge relative to design conditions.

7-2   INDOOR DESIGN CONDITIONS

One purpose of Chapter 4 was to define indoor conditions that make most of the occupants comfortable. Therefore, the theories and data presented there should serve as a guide to the selection of the indoor temperature and humidity for heat-loss calculation. It should be kept in mind, however, that the purpose of heat-loss calculations is to obtain data on which the heating system components are sized. In most cases, the system will never be set to operate at the design conditions. Indeed, it is practically impossible to control the environment to operate at the design point. Therefore, the use and occupancy of the space is a general consideration from the design temperature point of view. Later, when the energy requirements of the building are computed, the actual conditions in the space and outdoor environment, including internal heat gains, must be considered.

The indoor design temperature should be kept relatively low so that the heating equipment will not be oversized. ASHRAE Standard 90.1 does not specify specific design temperature and humidity conditions for load calculations but does specify that the conditions shall be in accordance with the comfort criteria established in ASHRAE Standard 55 (see Chapter 4). A design temperature of 70 F or 22 C is commonly used with relative humidity less than or equal to 30 percent. Although a relative humidity of 30 percent with a temperature of 70 F or 22 C is in the lower part of the comfort zone, maintaining a higher humidity must be given careful consideration because severe condensation may occur on windows and other surfaces depending on window and wall insulation and construction. Even properly sized equipment operates under partial load, at reduced efficiency, most of the time; therefore, any oversizing aggravates this condition and lowers the overall system efficiency. The indoor design value of relative humidity should be compatible with a healthful environment and the thermal and moisture integrity of the building envelope.

Frequently, unheated rooms or spaces exist in a structure. These spaces will be at temperatures between the indoor and outdoor design temperatures discussed earlier. The temperature in an unheated space is needed to compute the heat loss and may be estimated by assuming steady-state heat transfer and making an energy balance on the space. This can best be explained by an example.

EXAMPLE 7-1

Estimate the temperature in the crawl space of Fig. 7-1. The conductance for the floor is 0.20 Btu/(hr-ft²-F) including the air film on each side. The conductance for the foundation wall including the insulation and inside and outside air film resistances is 0.12 Btu/(hr-ft²-F). Assume an indoor temperature of 70 F and an outdoor temperature of 10 F at a location where the degree days are about 3000. The building dimensions are 50 × 75 ft.

SOLUTION

The first step is to make an energy balance on the crawl space. Heat will be gained through the floor and heat will be lost through the foundation wall and through the ground around the perimeter much like a floor slab. Infiltration of outdoor air will also represent a heat loss but that will be neglected in this example.

$$\dot{q}_{fl} = \dot{q}_{fo} + \dot{q}_{ground}$$

or

$$C_{fl}A_{fl}(t_i - t_c) = C_{fo}A_{fo}(t_c - t_o) + U'P(t_c - t_o)$$

$$t_c = \frac{t_o(CA)_{fo} + U'Pt_o + t_i(CA)_{fl}}{(CA)_{fl} + (CA)_{fo} + U'P}$$

Now the area of the floor is 50 × 75 = 3750 ft² and assuming that the foundation wall averages a height of 2 ft, the area of the foundation wall is 2(2 × 50) + (2 × 75) =

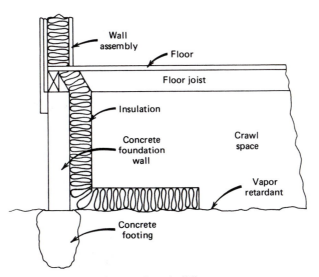

Figure 7-1 A crawl space for a building.

500 ft². The perimeter of the building is (2 × 50) + (2 × 75) = 250 ft. Referring to Table 5-12, the heat-loss coefficient is estimated to be 0.8 Btu/(hr-ft-F). Then

$$t_c = \frac{10[(0.12 \times 500) + (0.8 \times 250)] + 70(0.20 \times 3750)}{(0.2 \times 3750) + (0.12 \times 500) + (0.8 \times 250)} = 54.6\,\text{F}$$

If the infiltration had been considered, the crawl-space temperature would be lower. Many crawl spaces are ventilated to prevent moisture problems, and infiltration would be significant even when the vents are closed.

EXAMPLE 7-2

Estimate the temperature in the unheated room shown in Fig. 7-2. The structure is built on a slab. The exterior walls of the unheated room have an overall heat-transfer coefficient of 0.2 Btu/(hr-ft²-F), the ceiling–attic–roof combination has an overall coefficient of 0.07 Btu/(hr-ft²-F), and the interior walls have an overall coefficient of 0.06 Btu/(hr-ft²-F). Inside and outdoor design temperatures are 72 F and −5 F, respectively.

SOLUTION

The heat transferred from the heated to the unheated room has a single path, neglecting the door, and may be represented as

$$\dot{q} = U_i A_i (t_i - t_u) \tag{7-1}$$

The heat transferred from the unheated room to the outdoor air has parallel paths through the ceiling and walls (neglecting the door and floor):

$$\dot{q} = U_c A_c (t_u - t_o) + U_o A_o (t_u - t_o) \tag{7-2}$$

The heat loss to the floor can be neglected because of the anticipated low temperature in the room. Equation 7-2 may be written

$$\dot{q} = \frac{(t_u - t_o)}{R'_c} + \frac{(t_u - t_o)}{R'_o} = \frac{(t_u - t_o)}{R'} \tag{7-3}$$

where

$$\frac{1}{R'} = \frac{1}{R'_c} + \frac{1}{R'_o} = U_c A_c + U_o A_o \tag{7-4}$$

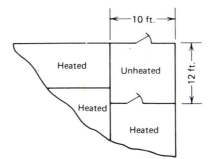

Figure 7-2 A structure with an attached unheated room.

By combining Eqs. 7-1 and 7-3, we obtain

$$\frac{(t_i - t_u)}{R'_i} = \frac{(t_u - t_o)}{R'} \qquad (7\text{-}5)$$

and solving for t_u,

$$t_u = \frac{t_i + (R'_i/R')t_o}{1 + (R'_i/R')} \qquad (7\text{-}6)$$

$$R'_i = \frac{1}{U_i A_i} = \frac{1}{0.06(8)(22)} = 0.095$$

$$\frac{1}{R'} = U_c A_c + U_o A_o = 0.07(120) + 0.2(22)8 = 43.6$$

$$R' = 0.023 \ (\text{hr-F})/\text{Btu}$$

Then from Eq. 7-6

$$t_u = \frac{72 + (0.095/0.023)(-5)}{1 + (0.095/0.023)} = 10 \, \text{F}$$

The assumption of negligible heat transfer through the slab is justified by the result.

The temperature of unheated basements is generally between the ground temperature (about 50 F or 10 C) and the inside design temperature unless there are many windows. Therefore a reasonable estimate of the basement temperature is not difficult. However, for a more precise value, the energy balance procedure may be used with data from Chapter 5.

7-3   TRANSMISSION HEAT LOSSES

The heat transferred through walls, ceiling, roof, window glass, floors, and doors is all sensible heat transfer, referred to as *transmission heat loss* and computed from

$$\dot{q} = UA(t_i - t_o) \tag{7-7}$$

The overall heat-transfer coefficient is determined as discussed in Chapter 5, where the area A is the net area for the given component for which U was calculated. A separate calculation is made for each different surface in each room of the structure. To ensure a thorough job in estimating the heat losses manually, a worksheet should be used. A worksheet provides a convenient and orderly way of recording all the coefficients and areas. Summations are conveniently made by room and for the complete structure. The spreadsheet computer programs available for all personal computers are quite useful in simple load calculation, or a particular program may be written to carry out the calculations and print the results. Many such programs are available.

7-4   INFILTRATION

All structures have some air leakage or infiltration. This means a heat loss because the cold dry outdoor air must be heated to the inside design temperature and moisture must be added to increase the humidity to the design value. The sensible heat required (to increase the temperature) is given by

$$\dot{q}_s = \dot{m}_o c_p (t_i - t_o) \tag{7-8a}$$

where

$\dot{m}_o$ = mass flow rate of the infiltrating air, lbm/hr or kg/s
c_p = specific heat capacity of the air, Btu/(lbm-F) or J/(kg-C)

Infiltration is usually estimated on the basis of volume flow rate at outdoor conditions. Equation 7-8 then becomes

$$\dot{q}_s = \frac{\dot{Q} c_p (t_i - t_o)}{v_o} \tag{7-8b}$$

where

$\dot{Q}$ = volume flow rate, ft³/hr or m³/s
v_o = specific volume, ft³/lbm or m³/kg

The latent heat required to humidify the air is given by

$$\dot{q}_l = \dot{m}_o (W_i - W_o) i_{fg} \tag{7-9a}$$

where

$(W_i - W_o)$ = difference in design humidity ratio, lbmv/lbma or kgv/kga
i_{fg} = latent heat of vaporization at indoor conditions, Btu/lbmv or J/kgv

In terms of volume flow rate of air, Eq. 7-9 becomes

$$\dot{q}_l = \frac{\dot{Q}}{v_o}(W_i - W_o)i_{fg} \tag{7-9b}$$

It is easy to show, using Eqs. 7-8a and 7-9a, that infiltration can account for a large portion of the heating load.

Various methods are used in estimating air infiltration in building structures (1). In this book two approaches to the problem will be discussed. In one method the estimate is based on the characteristics of the windows, walls, and doors and the pressure difference between inside and outside. This is known as the *crack method* because of the cracks around window sash and doors. The other approach is the *air-change method,* which is based on an assumed number of air changes per hour based on experience. The crack method is generally considered to be the most accurate when the crack and pressure characteristics can be properly evaluated. However, the accuracy of predicting air infiltration is restricted by the limited information on the air leakage characteristics of the many components that make up a structure (3). The pressure differences are also difficult to predict because of variable wind conditions and stack effect in tall buildings.

Air-change Method

Experience and judgment are required to obtain satisfactory results with this method. Experienced engineers will often simply make an assumption of the number of air changes per hour (ACH) that a building will experience based on their appraisal of the building type, construction, and use. The range will usually be from 0.5 ACH (very low) to 2.0 ACH (very high). Modern office buildings may experience infiltration rates as low as 0.1 ACH. This approach is usually satisfactory for design load calculation but not recommended for the beginner. The infiltration rate is related to ACH and space volume as follows:

$$\dot{Q} = \text{ACH}\,(V)/C \tag{7-10}$$

where

$\dot{Q}$ = infiltration rate, cfm or m^3/s
V = gross space volume, ft^3 or m^3
C = constant, 60 for English units and 3600 for SI

Crack Method

Outdoor air infiltrates the indoor space through cracks around doors, windows, lighting fixtures, and joints between walls and floor and even through the building material itself. The amount depends on the total area of the cracks, the type of crack, and the pressure difference across the crack. The volume flow rate of infiltration may be represented by

$$\dot{Q} = A\,C\,\Delta P^n \tag{7-11}$$

where

A = effective leakage area of the cracks

C = flow coefficient, which depends on the type of crack and the nature of the flow in the crack

ΔP = outside–inside pressure difference, $P_o - P_i$

n = exponent that depends on the nature of the flow in the crack, $0.4 < n < 1.0$.

Experimental data are required to use Eq. 7-11 directly; however, the relation is useful in understanding the problem. For example, Fig. 7-3 shows the leakage rate for some windows and doors as a function of the pressure difference and the type of crack. The curves clearly exhibit the behavior of Eq. 7-11.

The pressure difference of Eq. 7-11 results from three different effects:

$$\Delta P = \Delta P_w + \Delta P_s + \Delta P_p \qquad (7\text{-}12)$$

where

ΔP_w = pressure difference due to the wind

ΔP_s = pressure difference due to the stack effect

ΔP_p = pressure difference due to building pressurization

All of the pressure differences are positive when each causes flow of air to the inside of the building.

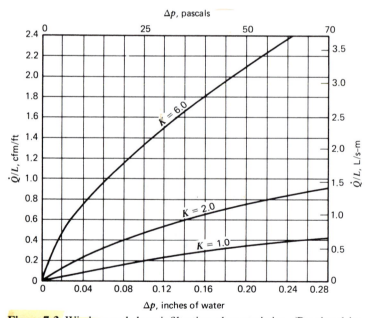

Figure 7-3 Window and door infiltration characteristics. (Reprinted by permission from *ASHRAE Cooling and Heating Load Calculation Manual*, 2nd ed., 1992.)

The pressure difference due to the wind results from an increase or decrease in air velocity and is described by

$$\Delta P_w = \frac{\rho}{2g_c}(\overline{V}_w^2 - \overline{V}_f^2) \tag{7-13a}$$

where ΔP_w has the unit of lbf/ft^2 or Pascals when consistent English or SI units are used. The velocity $\overline{V}_f$ is the final velocity of the wind at the building boundary. Note that ΔP_w is positive when $\overline{V}_w > \overline{V}_f$, which gives an increase in pressure. The velocity $\overline{V}_f$ is not known or easily predictable; therefore, it is assumed equal to zero in this application and a pressure coefficient, defined by

$$C_p = \Delta P_w / \Delta P_{wt} \tag{7-14}$$

is used to account for the fact that $\overline{V}_f$ is not zero. The pressure difference ΔP_{wt} is the computed pressure difference when $\overline{V}_f$ is zero. The pressure coefficient may be positive or negative. Finally, Eq. 7-13a may be written

$$\frac{\Delta P_w}{C_p} = \frac{\rho}{2g_c}\overline{V}_w^2 \tag{7-13b}$$

The pressure coefficient depends on the shape and orientation of the building with respect to the wind. To satisfy conditions of flow continuity, the air velocity must increase as it flows around or over a building; therefore, the pressure coefficient will change from a positive to a negative value in going from the windward to the leeward side. The pressure coefficients will also depend on whether the wind approaches normal to the side of the building or at an angle. Buildings are classifed as Low-Rise or High-Rise, where high-rise is defined as those with height greater than three times the crosswind width ($H > 3W$). Figure 7-4 gives average wall pressure coefficients for low-rise buildings. The average roof pressure coefficient for a low-rise building with roof inclined less than 20 degrees is approximately -0.5. Figures 7-5 and 7-6 give average pressure coefficients for high-rise buildings. There is an increase in pressure coefficient with height; however, the variation is well within the approximations of the data in general.

The stack effect occurs when the air density differs between the inside and outside of a building. On winter days, the lower outdoor temperature causes a higher pressure at ground level on the outside and consequent infiltration. Buoyancy of the warm inside air leads to upward flow, a higher inside pressure at the top of the building, and exfiltration of air. In the summer, the process reverses with infiltration in the upper portion of the building and exfiltration in the lower part.

Considering only the stack effect, there is a level in the building where no pressure difference exists. This is defined as the neutral pressure level. Theoretically, the neutral pressure level will be at the midheight of the building if the cracks and other openings are distributed uniformly in the vertical direction. When larger openings predominate in the lower portion of the building, the neutral pressure level will be lowered. Similarly, the neutral pressure level will be raised by larger openings in the upper portion of the

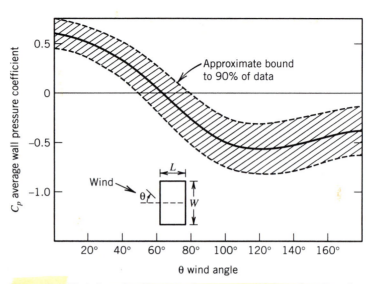

Figure 7-4 Variation of wall averaged pressure coefficients for a low-rise building. (Reprinted by permission from *ASHRAE Handbook, Fundamentals Volume,* 1989.)

building. Normally the larger openings will occur in the lower part of the building because of doors. The theoretical pressure difference with no internal separations is given by

$$\Delta P_{st} = \frac{P_o h}{R_a} \frac{g}{g_c} \left(\frac{1}{T_o} - \frac{1}{T_i} \right) \tag{7-15}$$

where

P_o = outside pressure, psia or Pascals
$\ h$ = vertical distance from neutral pressure level, ft or m
T_o = outside temperature, R or K
T_i = inside temperature, R or K
R_a = gas constant for air, (ft-lbf)/(lbm-R) or J/(kg-K)

The floors in a conventional building offer resistance to vertical air flow. Furthermore, this resistance varies depending on how stairwells and elevator shafts are sealed. When the resistance can be assumed equal for each floor, a single correction, called the *draft coefficient*, can be used to relate the actual pressure difference ΔP_s to the theoretical value ΔP_{st}:

$$C_d = \frac{\Delta P_s}{\Delta P_{st}} \tag{7-16}$$

The flow of air from floor to floor causes a decrease in pressure at each floor; therefore,

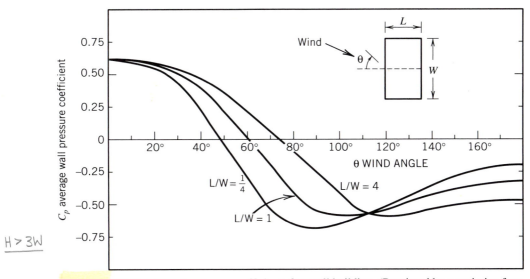

Figure 7-5 Wall averaged pressure coefficients for a tall building. (Reprinted by permission from *ASHRAE Handbook, Fundamentals Volume,* 1989.)

H > 3W

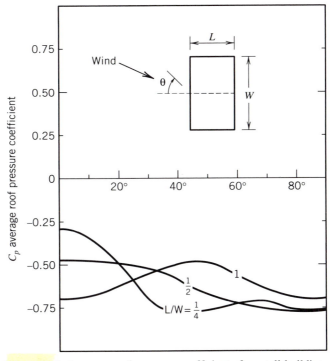

Figure 7-6 Average roof pressure coefficients for a tall building. (Reprinted by permission from *ASHRAE Handbook, Fundamentals Volume,* 1989.)

ΔP_s will be less than ΔP_{st} and C_d is less than one. Using the draft coefficient, Eq. 7-15 becomes

$$\Delta P_s = \frac{C_d P_o hg}{R_a g_c} \left(\frac{1}{T_o} - \frac{1}{T_i} \right) \qquad (7\text{-}17)$$

Figure 7-7 is a plot of Eq. 7-17 for an inside temperature of 75 F or 24 C, sea level outside pressure, and winter temperatures; however, Fig. 7-7 can be used for summer stack effect with little loss in accuracy.

The draft coefficient depends on the tightness of the doors in the stairwells and elevator shafts. Values of C_d range from 1.0 for buildings with no doors in the stairwells and varies from about 0.65 to 0.85 for modern office buildings.

Pressurization of the indoor space is accomplished by introducing more makeup air than exhaust air and depends on the design of the air distribution system rather than

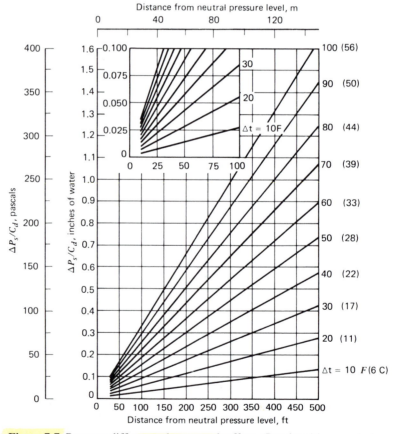

Figure 7-7 Pressure difference due to stack effect. (Reprinted by permission from ASHRAE GRP 158, *Cooling and Heating Load Calculation Manual,* 2nd ed., 1992.)

natural phenomena. The space may be depressurized by improper or maladjusted equipment, which is usually undesirable. For purposes of design, the designer must assume a value for ΔP_p and care must be taken to use a value that can actually be achieved in practice.

Calculation Aids

Figures 7-3, 7-8, and 7-9 and associated Tables 7-1, 7-2, and 7-3 give the infiltration rates, based on experimental evidence, for windows and doors, curtain walls, and commercial swinging doors. Note that the general procedure is the same in all cases, except that curtain wall infiltration is given per unit of wall area rather than crack length. The pressure differences are estimated by the methods discussed earlier and the values for the coefficient K are given in Tables 7-1, 7-2, and 7-3. The use of storm sash and storm doors is common. The addition of a storm sash with crack length and a K value equal to the prime window reduces infiltration by about 35 percent.

Commercial buildings often have a rather large number of people going and coming, which can increase infiltration significantly. Figures 7-10 and 7-11 have been developed to estimate this kind of infiltration for swinging doors. The infiltration rate per door is given in Fig. 7-10 as a function of the pressure difference and a traffic coefficient that depends on the traffic rate and the door arrangement. Figure 7-11 gives the traffic coefficients as a function of the traffic rate and two door types. Single-bank doors open directly into the space; however, there may be two or more doors at one location. Vestibule-type doors are best characterized as two doors in series so as to form an air

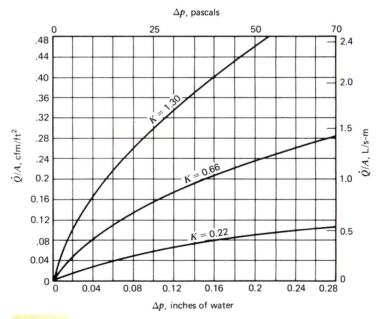

Figure 7-8 Curtain wall infiltration for one room or one floor. (Reprinted by permission from *ASHRAE Cooling and Heating Load Calculation Manual*, 2nd ed., 1992.)

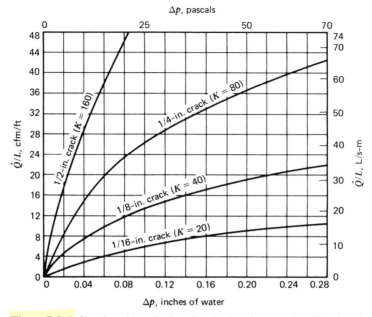

Figure 7-9 Infiltration through closed swinging door cracks. (Reprinted by permission from *ASHRAE Cooling and Heating Load Calculation Manual,* 2nd ed., 1992.)

lock between them. These doors often appear as two pairs of doors in series, which amounts to two vestibule-type doors.

The stack effect seems to be small in low-rise buildings, and wall infiltration is usually very low; therefore, only wind effects and crackage need be considered. In high-rise buildings the stack effect may be dominant, with a relatively large amount of leakage through the walls and around fixed window panels. All pressure effects as well as window, door, and wall leakage should be considered for high-rise buildings.

Theoretically, it is possible to predict which sides of a building will experience infiltration and which will experience exfiltration by use of the pressure coefficient. However, buildings usually do not have uniformly distributed openings on all sides. This will be particularly true for low-rise buildings. Because flow continuity must be satisfied, it is recommended that the infiltration for low-rise buildings be based on double the identifiable crack length for windows and doors to account for other obscure cracks. Assume that air infiltrates on all sides and leaves through openings and cracks in and near the ceiling. Base the pressure difference on wind alone for the windward side. There is room for innovation by the designer in making infiltration calculations. Each situation must be evaluated and a rational approach developed. The pressure coefficient approach is more feasible for high-rise buildings because the stack effect tends to cause infiltration at the lower levels and exfiltration at the higher levels in winter and the reverse in summer. Nonuniformity of the cracks and openings tends to be less important for flow continuity here. The following examples demonstrate the use of the data and methods described previously.

Table 7-1 Window Classification (For Fig. 7-3)

	Wood Double-hung (Locked)	Other Types
Tight-fitting window $K = 1.0$	Weatherstripped average gap ($\frac{1}{64}$-in. crack)	Wood casement and awning windows; weatherstripped Metal casement windows; weatherstripped
Average-fitting window $K = 2.0$	Non-weatherstripped average gap ($\frac{1}{64}$-in. crack) or Weatherstripped large gap ($\frac{3}{32}$-in. crack)	All types of vertical and horizontal sliding windows; weatherstripped. Note: If average gap ($\frac{1}{64}$-in. crack) this could be tight-fitting window Metal casement windows; Nonweatherstripped. Note: If large gap ($\frac{3}{32}$-in. crack) this could be a loose-fitting window
Loose-fitting window $K = 6.0$	Non-weatherstripped large gap ($\frac{3}{32}$-in. crack)	Vertical and horizontal sliding windows; nonweatherstripped

Source: Reprinted by permission from *ASHRAE Cooling and Heating Load Calculation Manual,* 2nd ed., 1992.

Table 7-2 Door Classification (For Fig. 7-3)

Tight-fitting door $K = 1.0$	Very small perimeter gap and perfect fit weatherstripping—often characteristic of new doors
Average-fitting door $K = 2.0$	Small perimeter gap having stop trim fitting properly around door and weatherstripped
Loose-fitting door $K = 6.0$	Larger perimeter gap having poor fitting stop trim and weatherstripped or Small perimeter gap with no weatherstripping

Source: Reprinted by permission from *ASHRAE Cooling and Heating Load Calculation Manual,* 2nd ed., 1992.

Table 7-3 Curtain Wall Classification (For Fig. 7-8)

Leakage Coefficient	Description	Curtain Wall Construction
$K = 0.22$	Tight-fitting wall	Constructed under close supervision of workmanship on wall joints. When joints seals appear inadequate, they must be redone
$K = 0.66$	Average-fitting wall	Conventional construction procedures are used
$K = 1.30$	Loose-fitting wall	Poor construction quality control or an older building having separated wall joints

Source: Reprinted by permission from *ASHRAE Cooling and Heating Load Calculation Manual,* 2nd ed., 1992.

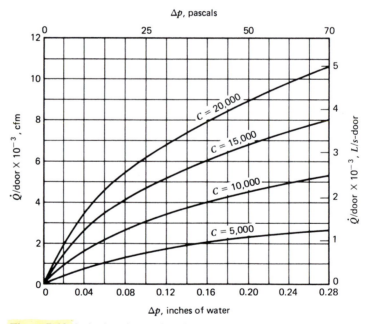

Figure 7-10 Swinging door infiltration characteristics with traffic. (Reprinted by permission from *ASHRAE Cooling and Heating Load Calculation Manual,* 2nd ed., 1992.)

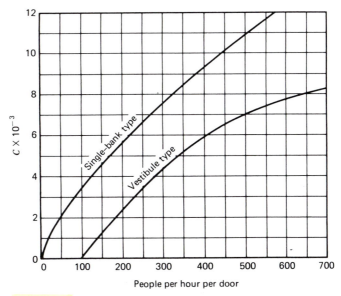

Figure 7-11 Flow coefficient dependence on traffic rate. (Reprinted by permission from *ASHRAE Cooling and Heating Load Calculation Manual,* 2nd ed., 1992.)

EXAMPLE 7-3

A 12-story office building is 120 ft tall with plan dimensions of 120 × 80 ft. The structure is of conventional curtain wall construction with all windows fixed in place. There are double vestibule-type doors on all four sides. Under winter design conditions, a wind of 15 mph blows normal to one of the long dimensions. Estimate the pressure differences for all walls for the first and twelfth floors. Consider only wind and stack effects. The indoor–outdoor temperature difference is 60 F.

SOLUTION

The pressure difference for each effect must first be computed and then combined to find the total. First consider the wind: Equation 7-13b expresses the wind pressure difference where the pressure coefficients may be obtained from Fig. 7-5 for a normal wind. Then using standard sea level density:

Windward Side: $C_p = 0.60$

$$\Delta P_w = \frac{0.60(0.0765)(15 \times 1.47)^2 (12)}{2(32.17)62.4} = 0.066 \text{ in. } wg$$

Leeward: $C_p = -0.30$

$$\Delta P_w = \frac{0.105}{0.60}(-0.30) = -0.033 \text{ in. } wg$$

Sides: $C_p = -0.60$

$$\Delta P_w = \frac{0.066(-0.60)}{0.60} = -0.066 \text{ in. } wg$$

The wind effect will be assumed constant with respect to height.

The pressure difference due to stack effect can be computed from Eq. 7-17 or more easily determined from Fig. 7-7. Because there are more openings in the lower part of the building, assume that the neutral pressure level is at the fifth floor instead of the sixth. Also assume that the draft coefficient is 0.8. Then for the first floor, $h = 50$ ft and from Fig. 7-7

$$\frac{\Delta P_s}{C_d} = 0.10$$

and

$$\Delta P_s = 0.10(0.8) = 0.08 \text{ in. } wg$$

For the twelfth floor, $h = 70$ ft and

$$\frac{\Delta P_s}{C_d} = -0.12$$

and

$$\Delta P_s = -0.12(0.8) = -0.096 \text{ in. } wg$$

The negative sign indicates that the pressure is greater inside the building than on the outside.

The pressure differences may now be summarized for each side where $\Delta P = (\Delta P_w + \Delta P_s)$ in. wg.

Orientation	1st Floor	12th Floor
Windward	0.146	−0.030
Sides	0.014	−0.162
Leeward	0.047	−0.129

These results show that air will tend to infiltrate on most floors on the windward wall. Infiltration will occur on about the lower four floors on the leeward wall. All other surfaces will have exfiltration.

EXAMPLE 7-4

Estimate the infiltration rate for the leeward doors of Example 7-3. The doors have $\frac{1}{8}$-in. cracks and the traffic rate is low except at 5:00 P.M., when the traffic rate is 350 people per hour per door for a short time.

SOLUTION

This problem is solved in two steps to account for crack leakage and infiltration due to traffic. For the design condition, the effect of traffic is negligible; however, it is of interest to compute this component for 5:00 P.M. Figure 7-9 pertains to crack leakage for commercial swinging doors. For a pressure difference of 0.047 in. wg and $\frac{1}{8}$-in. cracks, the leakage rate is 8 cfm/ft. The crack length for standard double swinging doors is

$$L = 3(6.75) + 2(6) = 32 \text{ ft}$$

Then

$$\dot{Q} = \left(\frac{\dot{Q}}{L}\right) L = 8(32) = 256 \text{ cfm}$$

Vestibule-type doors will tend to decrease the infiltration rate somewhat like storm sash or a storm door. Assume a 30 percent reduction; then

$$\dot{Q} = (1 - 0.3)256 = 179 \text{ cfm}$$

Figures 7-10 and 7-11 are used to estimate the infiltration due to traffic. The traffic coefficient C is read from Fig. 7-11 for 350 people per hour and for vestibule-type doors at 5000. Then, from Fig. 7-10 at a pressure difference of 0.047 in. wg,

$$\dot{Q}/\text{door} = 800 \text{ cfm/door}$$

and for two doors

$$\dot{Q} = 1600 \text{ cfm}$$

A part of the crack leakage should be added to this; however, this is somewhat academic. Care should be exercised in including the traffic infiltration in the design heat load. It will usually be a short-term effect.

EXAMPLE 7-5

Estimate the leakage rate for the twelfth floor of the building in Example 7-3. Neglect the roof.

SOLUTION

Referring to the pressure differences computed in Example 7-3, it is obvious that the leakage will be from the inside out on the twelfth floor. Therefore, a great deal of air must be entering the space from the stairwells and elevator shafts. Because the twelfth floor has no movable openings, except to the roof, all leakage is assumed to be through the walls. Figure 7-8 gives data for this case where $K = 0.66$ for conventional construction.

Windward wall: $\quad \Delta P = -0.030$ in. $wg;$ $\dot{Q}/A = -0.065$ cfm/ft^2

$$\dot{Q} = -0.065(120)10 = -78 \text{ cfm}$$

Side walls: $\quad \Delta P = -0.162$ in. $wg,$ $\dot{Q}/A = -0.210$ cfm/ft^2

$$\dot{Q} = -0.21(80)(10)2 = -336 \text{ cfm}$$

Leeward wall: $\quad \Delta P = -0.129$ in. $wg,$ $\dot{Q}/A = -0.18$ cfm/ft^2

$$\dot{Q} = -0.18(120)10 = -216 \text{ cfm}$$

The net leakage rate is then:

$$\dot{Q}_{net} = -78 - 336 - 216 = -630 \text{ cfm}$$

where the negative sign indicates that the flow is from the inside out. The net leakage flow of 630 cfm entered the building at other locations where the heat loss should be assigned.

EXAMPLE 7-6

A single-story building is oriented so that a 15 mph wind approaches 45 degrees to the windward sides. There are 120 ft of crack for the windows and 20 ft of crack for a door on the windward side. The sides have 130 ft of window cracks and 18 ft of door crack. All windows and doors are average fitting. Estimate the infiltration.

SOLUTION

The major portion of the infiltration for this kind of building will be through the cracks. It is approximately true that air will enter on the sides and flow out with most of the heat loss imposed on the rooms where the air enters. As suggested, we will use double the total crack length and assume that most of the air leaves through the ceiling area with a pressure difference computed for a quartering wind on the windward side. Using Eq. 7-13b, Fig. 7-4, and Table 7-1.

$$\Delta P_w = \frac{0.70(0.0765)(15 \times 1.47)^2 (12)}{2(32.17)62.4} = 0.078 \text{ in. } wg$$

where standard sea level air density has been used. From Tables 7-2 and 7-3, the K factor for the windows and doors is read as 2.0. Then from Fig. 7-3, the leakage per foot of crack is

$$\frac{\dot{Q}}{L} = 0.375 \text{ cfm/ft}$$

and the total infiltration for the space is

$$\dot{Q}_1 = 0.375(250 + 38) = 108 \text{ cfm}$$

Exhaust fans, chimneys, and flues can increase infiltration dramatically or necessitate the introduction of outdoor air. In either case the heat loss of the structure is increased.

Direct-fired warm-air furnaces are sometimes installed within the confines of the conditioned space. If combustion air is not brought in from outdoors, conditioned air from the space will be drawn in and exhausted through the flue. Infiltration or outdoor air must then enter the structure to make up the loss and contributes to a higher heat loss. Many codes require that combustion air be introduced directly to the furnace from outdoors. Indeed, this should always be the rule. For natural gas (methane) the ratio of air to gas on a volume basis is about 10. This is equivalent to 10 ft^3 or 0.28 m^3 of air per 1000 Btu or 1.06×10^6 J input to the furnace.

7-5   HEAT LOSSES FROM AIR DUCTS

The losses of a duct system can be considerable when the ducts are not in the conditioned space. Proper insulation will reduce these losses but cannot completely eliminate them. The loss may be estimated using the following relation:

$$\dot{q} = UA_s \, \Delta t_m \tag{7-18}$$

where

U = overall heat transfer coefficient, Btu/(hr-ft^2-F) or W/(m^2-C)
A_s = outside surface area of the duct, ft^2 or m^2
Δt_m = mean temperature difference between the air in the duct and the environment, F or C.

When the duct is covered with 1 or 2 in. of fibrous glass insulation with a reflective covering, the heat loss will usually be reduced sufficiently to assume that the mean temperature difference is equal to the difference in temperature between the supply air temperature and the environment temperature. Unusually long ducts should not be treated in this manner and a mean air temperature should be used instead.

EXAMPLE 7-7

Estimate the heat loss from 1000 cfm of air at 120 F flowing in a 16-in. round duct 25 ft in length. The duct has 1 in. of fibrous glass insulation and the overall heat-transfer coefficient is 0.2 Btu/(hr-ft^2-F). The environment temperature is 12 F.

SOLUTION

Equation 7-18 will be used to estimate the heat loss assuming that the mean temperature difference is given approximately by

$$\Delta t_m = t_s - t_a = 12 - 120 = -108 \, \text{F}$$

The surface area of the duct is

$$A_s = \frac{\pi(16 + 2)(25)}{12} = 117.8 \, \text{ft}^2$$

Then

$$\dot{q} = 0.2(117.8)(-108) = -2540 \, \text{Btu/hr}$$

The temperature of the air leaving the duct may be computed from

$$\dot{q} = \dot{m}c_p(t_2 - t_1) = \dot{Q}\rho c_p(t_2 - t_1)$$

or

$$t_2 = t_1 + \frac{\dot{q}}{\rho c_p}$$

$$t_2 = 120 + \frac{-2540}{1000(60)(0.067)(0.24)}$$

$$t_2 = 117 \, \text{F}$$

Although insulation drastically reduces the heat loss, the magnitude of the temperature difference and surface area must be considered in each case.

Minimum insulation of supply and return ducts is presently specified by ASHRAE Standard 90.1 as follows:

All duct systems shall be insulated to provide a thermal resistance, excluding film resistance, as shown in Table 7-4, where Δt is the design temperature differential between the air in the duct and the surrounding air in F or C. Heat losses from the supply ducts become part of the space heat load and should be summed with transmission and infiltration heat losses. Heat losses from the return air ducts are not part of the space heat loss but should be added to the heating equipment load.

7-6 AUXILIARY HEAT SOURCES

The heat energy supplied by people, lights, motors, and machinery may be estimated, but any actual allowance for these heat sources requires careful consideration. People may not occupy certain spaces in the evenings, or weekends, and during other periods, but these spaces must generally be heated to a reasonably comfortable temperature prior

Table 7-4 Duct Insulation Required

Δt		R	
F	C	(hr-ft²-F)/Btu	(m²-C)/W
<15	<8	None required	None required
>15	>8	3.3	0.58
>40	>22	5.0	0.88

to occupancy. In industrial plants heat sources, if available during occupancy, should be substituted for part of the heating requirement. In fact, there are situations where so much heat energy is available that outdoor air must be used to cool the space. However, sufficient heating equipment must still be provided to prevent freezing of water pipes during periods when a facility is shut down.

7-7 INTERMITTENTLY HEATED STRUCTURES

When a structure is not heated on a continuous basis, the heating equipment capacity may have to be enlarged to assure that the temperature can be raised to a comfortable level within a reasonable period of time. The heat capacity of the building and occupant comfort are important factors when considering the use of intermittent heating. Occupants may feel discomfort if the mean radiant temperature falls below the air temperature. To conserve energy it is a common practice to set back thermostats or to completely shut down equipment during the late evening, early morning, and weekend hours. This is effective and is accompanied by only small sacrifices in comfort when the periods of shutdown are adjusted to suit outdoor conditions and the mass of the structure.

7-8 SUPPLY AIR FOR SPACE HEATING

Computing the air required for heating was discussed in Chapter 3 and took into account sensible and latent effects as well as outdoor air. That procedure is always recommended. However, there are many cases when the air quantity is conveniently computed from

$$\dot{q} = \dot{m}c_p(t_s - t_r) \qquad \text{(7-19a)}$$

or

$$\dot{q} = \frac{\dot{Q}c_p}{v_s}(t_s - t_r) \qquad \text{(7-19b)}$$

where

v_s = specific volume of supplied air, ft^3/lbm or m^3/kg
t_s = temperature of supplied air, F or C
t_r = room temperature, F or C

The temperature difference $(t_s - t_r)$ is normally less than 100 F or 38 C. Light commercial equipment operates with a temperature rise of 60 to 80 F or 16 to 27 C, whereas commercial applications will allow higher temperatures. The temperature of the air to be supplied must not be high enough to cause discomfort to occupants before it becomes mixed with room air.

With unit-type equipment typically used for small commercial buildings, each size is able to circulate a relatively fixed quantity of air. Therefore the air quantity is fixed within a narrow range when the heating equipment is selected. These units have different capacities that change in increments of 10 to 20,000 Btu/hr or about 5 kW according to the model. A slightly oversized unit is usually selected with the capacity to circulate a larger quantity of air than theoretically needed. Another condition that leads to greater quantities of circulated air for heating than needed is the greater air quantity usually required for cooling and dehumidifying. The same fan is used throughout the year and must therefore be large enough for the maximum air quantity required. Some units have different fan speeds for heating and for cooling.

After the total air-flow rate required for the complete structure has been determined, the next step is to allocate the correct portion of the air to each room or space. This is necessary for design of the duct system. Obviously the air quantity for each room should be apportioned according to the heating load for that space. Then

$$\dot{Q}_{rn} = \dot{Q}(\dot{q}_{rn}/\dot{q}) \tag{7-20}$$

where

$\dot{Q}_{rn}$ = volume flow rate of air supplied to room n, ft³/min or m³/s
$\dot{q}_{rn}$ = total heat loss of room n, Btu/hr or W

REFERENCES

1. *ASHRAE Cooling and Heating Load Calculation Manual,* 2nd ed., American Society of Heating, Refrigerating and Air-Conditioning Engineers, Inc., Atlanta, GA, 1992.
2. *ASHRAE Handbook, Fundamentals Volume,* American Society of Heating, Refrigerating and Air-Conditioning Engineers, Inc., Atlanta, GA, 1989.
3. P. E. Janssen, et al., "Calculating Infiltration: An Examination of Handbook Models," *ASHRAE Transactions,* Vol. 86, Pt. 2, 1980.

PROBLEMS

7-1. Select an indoor design relative humidity for structures located in the cities given below. Assume an indoor design dry bulb temperature from Table C-1. Windows in the building are double glass, aluminum frame with thermal break. Other external surfaces are well insulated.
   (a) Sioux City, Iowa
   (b) Atlanta, Georgia
   (c) Syracuse, New York
   (d) Denver, Colorado
   (e) San Francisco, California
   (f) Bismarck, North Dakota
   (g) Rapid City, South Dakota

7-2. Estimate the temperature in the crawl space of a building with floor plan dimensions of 30 × 60 ft. The concrete foundation has an average height of 2 ft and the wall is 6-in. thick. Use winter design conditions for Oklahoma City, Oklahoma.

7-3. Estimate the temperature in the knee space shown in Fig. 7-12. The roof is equivalent to 0.5 in. (25 mm) of wood on 2 × 6-in. rafters on 24-in. (0.6-m) centers. The walls are 2 × 4-in. studs on 16-in. (0.4-m) centers with 3.5 in. (90 mm) of insulation. The joist spaces all have 6 in. (150 mm) of insulation. Inside and outside temperatures are 70 F (21 C) and 10 F (−12 C).

7-4. Consider the knee space shown in Fig. 7-12. The vertical dimension is 8 ft, the horizontal dimension is 3 ft, and the space is 20 ft long. The walls and roof surrounding the space all have an overall heat-transfer coefficient of about 0.09

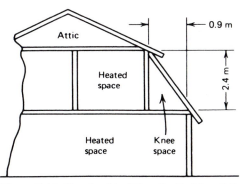

Figure 7-12 Sketch of building for Problem 7-3.

Btu/(hr-ft²-F). Assuming an outdoor temperature of 2 F and an indoor temperature of 68 F, make a recommendation concerning the placement of water pipes in the knee space.

7-5. Estimate the temperature in an unheated basement that is completely below ground level with heated space above. Assume no insulation and dimensions of 20 × 20 × 7 ft. The basement is located in Lincoln, Nebraska. Use standard design conditions.

7-6. Rework Example 7-1 assuming that the crawl space experiences an infiltration rate of 0.25 air changes per hour.

7-7. Rework Example 7-2 assuming an infiltration rate of 0.2 air changes per hour.

7-8. A large single-story business office is fitted with nine loose-fitting, double-hung woodsash windows 3 ft wide by 6 ft high. If the outside wind is 15 mph at a temperature of − 10 F, what is the percent reduction in sensible heat loss if the windows are weather stripped? Assume an inside temperature of 70 F. Base your solution on a quartering wind.

7-9. Using the crack method compute the infiltration for a tight-fitting swinging door that is used occasionally. The door has dimensions of 0.9 × 2.0 m and is on the windward side of a house exposed to a 9 m/s wind. Neglect internal pressurization and stack effect.

7-10. A room in a single-story building has two 3 × 5-ft wood, double-hung windows of average fit that are not weather stripped. The wind is 20 mph and normal to the wall with negligible pressurization of the room. Find the infiltration rate assuming that the entire crack is admitting air.

7-11. Compute the design infiltration rate and heat gain for the house described in Problem 7-16 assuming an orientation normal to a 15 mph wind. The windows and doors are tight fitting.

7-12. Refer to Example 7-3. (a) Estimate the total pressure difference for each wall for the third and ninth floors. (b) Using design conditions for Denver, Colorado, estimate the heat load due to infiltration for the third floors and ninth floors.

7-13. Refer to Examples 7-3 and 7-4. (a) Estimate the infiltration rates for the windward and side doors for a low traffic rate. (b) Estimate the curtain wall infiltration for the first floor. (c) Compute the heat load due to infiltration for the first floor if the building is located in Cleveland, Ohio.

7-14. A 20-story office building has plan dimensions of 100×60 ft and is oriented at 45 degrees to a 20 mph wind. All windows are fixed in place. There are double vestibule-type swinging doors on the 60-ft walls. The walls are average curtain wall construction and the doors have about $\frac{1}{8}$-in. cracks. (a) Compute the pressure differences for each wall due to wind and stack effect for the first, fifth, fifteenth, and twentieth floors. (b) Plot pressure difference versus height for each wall and estimate which surfaces have infiltration and exfiltration. (c) Compute the total infiltration rate for the first floor assuming a very low traffic rate. (d) Compute the infiltration rate for the fifteenth floor. (e) Compute the infiltration rate for the twentieth floor. Neglect any leakage through the roof.

7-15. Refer to Problem 7-14. (a) Compute the heat gain due to infiltration for the first floor with the building located in Dallas, Texas. (b) Compute the heat gain due to infiltration for the fifteenth floor. (c) What is the heat gain due to infiltration for the twentieth floor?

7-16. Compute the transmission heat loss for the structure described below. Use design conditions recommended by ASHRAE Standards.

Location:	Kansas City, Missouri
Walls:	Table 5-7a, case 2
Floor:	Concrete slab with 1-in. edge insulation, both horizontal and vertical, with heating ducts under the slab
Windows:	Double-insulating glass, $\frac{1}{4}$-in. air space, 3×3-ft, double-hung, aluminum frame with thermal break, three on each side.
Doors:	Wood, $1\frac{3}{4}$ in. with wood storm doors, three each, 3×6.75 ft
Roof–ceiling:	Same as Example 5-3, height of 8 ft
House plan:	Single story, 40×55 ft

7-17. Rework Problem 7-16 for Lincoln, Nebraska. Include infiltration in the analysis.

7-18. Consider a building located in Syracuse, New York. Using ASHRAE Standard recommendations, compute the transmission loss for (a) a wall like Table 5-7a, case 2, which is 3×4 m with a window of 1 m^2 area, (b) a single pane window with aluminum frame and no thermal break, (c) a roof–ceiling combination that is like Table 5-7b, case 1, with an area of 16 m^2.

7-19. Compute the heating load for the structure described by the plans and specifications furnished by the instructor.

7-20. A small commercial building has a computed heat load of 200,000 Btu/hr sensible and 25,000 Btu/hr latent. Assuming a 50 F temperature rise for the heating unit, compute the quantity of air to be supplied by the unit using the following methods: (a) Use a psychrometric chart with room conditions of 70 F and 30 percent relative humidity. (b) Calculate the air quantity based on the sensible heat transfer.

Chapter **8**

The Cooling Load

As explained in Chapter 7, heat loss is usually based on steady-state heat transfer, and the results obtained are usually quite adequate. In design for cooling, however, transient analysis must be used. The instantaneous heat gain into a conditioned space is quite variable with time, primarily because of the strong transient effect created by the hourly variation in solar radiation. There may be an appreciable difference between the heat gain of the structure and the heat removed by the cooling equipment at a particular time. This difference is caused by the storage and subsequent transfer of energy from the structure and contents to the circulated air. If this is not taken into account, the cooling and dehumidifying equipment will usually be grossly oversized.

8-1 HEAT GAIN, COOLING LOAD, AND HEAT EXTRACTION RATE

It is important to differentiate between *heat gain, cooling load,* and *heat extraction rate.* *Heat gain* is the rate at which energy is transferred to or generated within a space. It has two components, sensible heat and latent heat, which must be computed and tabulated separately. Heat gains usually occur in the following forms:

1. Solar radiation through openings.
2. Heat conduction through boundaries with convection and radiation from the inner surface into the space.
3. Sensible heat convection and radiation from internal objects.
4. Ventilation (outside air) and infiltration air.
5. Latent heat gains generated within the space.

The *cooling load* is the rate at which energy must be removed from a space to maintain the temperature and humidity at the design values. The cooling load will generally differ from the heat gain because the radiation from the inside surface of walls and interior objects as well as the solar radiation coming directly into the space through openings does not heat the air within the space directly. This radiant energy is mostly absorbed by floors, interior walls, and furniture, which are then cooled primarily by convection as they attain temperatures higher than that of the room air. Only when the room air receives the energy by convection does this energy become part of the cooling load. Figure 8-1 illustrates the phenomenon. The heat storage characteristics of the structure and interior objects determine the thermal lag and therefore the relationship

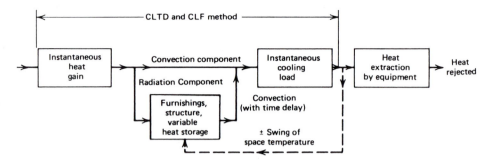

Figure 8-1 Schematic relation of heat gain to cooling load.

between heat gain and cooling load. For this reason the thermal mass (product of mass and specific heat) of the structure and its contents must be considered in such cases. The reduction in peak cooling load because of the thermal lag can be quite important in sizing the cooling equipment.

Figure 8-2 shows the relation between heat gain and cooling load and the effect of the mass of the structure. The attenuation and delay of the peak heat gain is very evident, especially for heavy construction. Figure 8-3, shows the cooling load for fluorescent lights that are used only part of the time. The sensible heat component from people and equipment acts in a similar way. The part of the energy produced by the lights,

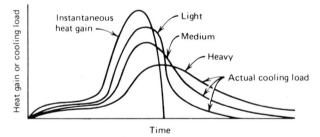

Figure 8-2 Actual cooling load and solar heat gain for light, medium, and heavy construction.

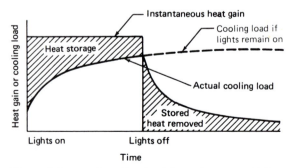

Figure 8-3 Actual cooling load from fluorescent lights.

equipment, or people that is radiant energy is momentarily stored in the surroundings. The energy convected directly to the air by the lights and people, and later by the surrounding objects, goes into cooling load. The area under the curves of Fig. 8-3 are all approximately equal. This means that about the same total amount of energy must be removed from the structure during the day; however, a larger portion is removed during the evening hours for heavier constructions.

The *heat extraction rate* is the rate at which energy is removed from the space by the cooling and dehumidifying equipment. This rate is equal to the cooling load when the space conditions are constant and the equipment is operating. However, this is rarely true because some fluctuation in room temperature is necessary for the control system to operate. Because the cooling load is also below the peak or design value most of the time, intermittent or variable operation of the cooling equipment is required.

To obtain some insight to the nature of the problem, consider the heat conduction through a wall or roof with a variable outdoor temperature and with a variable solar radiation input on the outside surface. Mathematical modeling leads to the heat conduction equation with nonlinear, time-dependent boundary conditions. Walls are usually a complex assembly of materials and may be two-dimensional. However, if the wall or roof is a single homogeneous slab, the governing differential equation is

$$\frac{\partial t}{\partial \theta} = \frac{k}{\rho c} \frac{\partial^2 t}{\partial x^2} \tag{8-1}$$

where

$$t = \text{local temperature at a point in the slab, F or C}$$
$$\theta = \text{time, hr}$$
$$k/\rho c = \text{thermal diffusivity of the slab, ft}^2/\text{hr or m}^2/\text{hr}$$
$$x = \text{length, ft or m}$$

A nonlinear, time-dependent boundary condition at the outside surface is a great obstacle in obtaining a solution to Eq. 8-1.

An equally difficult problem exists to determine the cooling load for a space after the heat gains are determined. The construction and furnishing of the space have a significant effect on the rate at which the heat gain is either convected directly to the air or absorbed and later convected to the air. Like the conduction problem discussed above, this problem must usually be solved using a computer.

To simplify and reduce the time required for computations, transform methods have been applied to these problems (1, 2, 3). A digital computer is necessary, but the speed of calculations is rapid even for small computers. Design cooling loads for one day as well as long-term energy calculations may be done using the transfer function approach. Details of this method are discussed in Section 8-3.

Although preferable, it is not always practical to compute the cooling load using the transfer function method; therefore, a hand calculation method has been developed from the transfer function procedure and is referred to as the *Cooling Load Temperature Difference, Solar Cooling Load, Cooling Load Factor* (CLTD/SCL/CLF) method (4, 5). The method involves extensive use of tables and charts and various factors to express the dynamic nature of the problem. Following a discussion of design conditions, which are the same for both calculation methods, the two procedures will be described.

8-2   OUTDOOR AND INDOOR DESIGN CONDITIONS

The problem of selecting outdoor design conditions for cooling is similar to that for heating. Again it is not reasonable to design for the worst conditions on record because a great excess of capacity will result. The heat storage capacity of the structure also plays an important role in this regard. A massive structure will reduce the effect of overload from short intervals of outdoor temperature above the design value. The *ASHRAE Handbook, Fundamentals Volume* (6) gives extensive outdoor design data. Tabulation of dry bulb and mean coincident wet bulb temperatures that are equaled or exceeded, 1, $2\frac{1}{2}$, and 5 percent of the total hours during June through September (2928 hours) are given. For example, a normal summer in Stillwater, Oklahoma, will have about 30 hours at 100 F dry bulb or greater, about 75 hours at 96 F or greater, and about 150 hours at 93 F or greater. Table C-1 gives the $2\frac{1}{2}$ percent values that are recommended for design purposes by ASHRAE Standards. The *daily range* of temperature given in Table C-1 is the difference between the average maximum and average minimum for the warmest month. The daily range is usually larger for the higher elevations, where temperatures may be quite low late at night and during the early morning hours. The daily range has an effect on the energy stored by the structure.

The *ASHRAE Handbook* gives prevailing wind data for some stations, which is the wind direction occurring most frequently with the $2\frac{1}{2}$ percent dry bulb temperature. The local wind velocity for summer conditions is usually taken to be about $7\frac{1}{2}$ mph or 3.4 m/s.

The indoor design conditions are governed by principles outlined in Chapter 4. For the average job in the United States and Canada, a condition of 75 F or 24 C dry bulb and relative humidity of 50 percent is typical when activity and dress of the occupants are light. ASHRAE Standard 90.1 sets the indoor design temperature and relative humidity within the comfort envelope defined in Figure 4-8. The designer should be alert for unusual circumstances that may lead to uncomfortable conditions. Certain activities may require occupants to engage in active work or require heavy protective clothing, both of which would require lower design temperatures.

Read 8-3 thru 8-6

8-3   THE TRANSFER FUNCTION METHOD

The general procedure referred to as the *transfer function method* (TFM) was developed by the ASHRAE Task Group on Energy Requirements (6). More recently, considerable work has been done to produce improved data for use with the transfer function method (2, 3). The transfer function method is based on a number of simplifications. Therefore, it is instructive to briefly consider the foundation for all calculation methods before presenting the TFM.

The Heat Balance Method

The heat balance method ensures that all energy flows in each zone are balanced and involves the solution of a set of energy balance equations for the zone air and the interior and exterior surfaces of each wall, roof, and floor. These energy balance equations are combined with equations for transient conduction heat transfer through walls and roofs and algorithms or data for weather conditions including outdoor air dry bulb temperature, wet bulb temperature, solar radiation, and so on. To illustrate the heat balance method, consider a simple zone with six surfaces: four walls, a roof, and a floor. The zone has solar energy coming through windows, heat conducted through exterior walls and roof, and internal heat gains due to lights, equipment, and occupants. The heat balance on each of the six surfaces is represented by

$$\dot{q}_{i,\theta} = \left[h_{ci}(t_{a,\theta} - t_{i,\theta}) + \sum_{j=1, J \neq i}^{m} g_{ij}(t_{j,\theta} - t_{i\theta}) \right] A_i + \dot{q}_{si,\theta} + \dot{q}_{li,\theta} + \dot{q}_{ei,\theta} \qquad \textbf{(8-2)}$$

where

$i = 1, 2, 3, 4, 5, 6$
m = number of surfaces in the room
$\dot{q}_{i,\theta}$ = rate of heat conducted into surface i at the inside surface at time θ
A_i = area of surface i
h_{ci} = convective heat transfer coefficient at interior surface i
g_{ij} = linearized radiation heat transfer factor between interior surface i and interior surface j
$t_{a,\theta}$ = inside air temperature at time θ
$t_{i,\theta}$ = average temperature of interior surface i at time θ
$t_{j,\theta}$ = average temperature of interior surface j at time θ
$\dot{q}_{si,\theta}$ = rate of solar heat coming through the windows and absorbed by surface i at time θ
$\dot{q}_{li,\theta}$ = rate of heat from the lights absorbed by surface i at time θ
$\dot{q}_{ei,\theta}$ = rate of heat from equipment and occupants absorbed by surface i at time θ

The equations governing conduction heat transfer must be solved simultaneously with Eq. 8-2. Typically these equations are formulated as conduction transfer functions (CTF) in the form

$$\dot{q}_{i,\theta} = \sum_{m=1}^{M} Y_{k,m} t_{o,\theta-m+1} - \sum_{m=1}^{M} Z_{k,m} t_{i,\theta-m+1} + \sum_{m=1}^{k} F_m \dot{q}_{i,\theta-m} \qquad \textbf{(8-3)}$$

where

i = inside surface subscript
k = order of CTF
m = time index variable
M = the number of nonzero CTF values
o = outside surface subscript
t = temperature
θ = time
Y = cross CTF values
Z = interior CTF values
F_m = flux history coefficients

Note that the interior surface temperature $t_{i,\theta}$ is found in both Eqs. 8-2 and 8-3, hence the need for simultaneous solution.

In addition, the equation representing an energy balance on the zone air must also be solved simultaneously; then the cooling load is given by

$$\dot{q}_\theta = \left[\sum_{i=1}^{6} h_{ci}(t_{i,\theta} - t_{a,\theta}) \right] A_i + \rho c_p \dot{Q}_{i,\theta}(t_{o,\theta} - t_{a,\theta})$$

$$+ \rho c_p \dot{Q}_{v,\theta}(t_{v,\theta} - t_{a,\theta}) + \dot{q}_{s,\theta} + \dot{q}_{l,\theta} + \dot{q}_{e,\theta} \quad \textbf{(8-4)}$$

where

ρ = air density
c_p = air specific heat
$\dot{Q}_{i,\theta}$ = volume flow rate of outdoor air infiltrating into the room at time θ
$t_{o,\theta}$ = outdoor air temperature at time θ
$\dot{Q}_{v,\theta}$ = volume rate of flow of ventilation air at time θ
$t_{v,\theta}$ = ventilation air temperature at time θ
$\dot{q}_{s,\theta}$ = rate of solar heat coming through the windows and convected into the room air at time θ
$\dot{q}_{l,\theta}$ = rate of heat from the lights convected into the room air at time θ
$\dot{q}_{e,\theta}$ = rate of heat from equipment and occupants convected into the room air at time θ

Note that the zone air temperature is allowed to float. By fixing the zone air temperature, the cooling load need not be determined simultaneously.

In the past, this approach (as well as the TFM) was generally considered to be too complicated for practical design calculations, though both were commonly used for energy analysis. But with desktop personal computers that outperform yesterday's mainframe computers, this is no longer true. In fact, there is no reason that either method cannot be used to perform design load calculations. However, the transfer function method has become commonly accepted.

Transfer Functions

The transfer function method (TFM) is based on two important concepts; conduction transfer functions (CTF) and room transfer functions (RTF). Both types of transfer functions are time series that relate a current variable to past values of itself and other variables, at discrete time intervals. The discrete time intervals are usually one-hour periods for transfer functions used in building analysis.

Transfer functions are commonly derived from response factors. Response factors are infinite series that relate a current variable to past values of other variables, at discrete time intervals. A transfer function converts the theoretically infinite set of response factors into a finite number of terms that multiply both past values of the variable of interest and past values of other variables. Figure 8-4 gives an overview of the transfer function method.

Room Transfer Functions

Room transfer functions relate the hourly cooling load due to individual types of heat gains to past values of that type of heat gain and previous values of the cooling load due

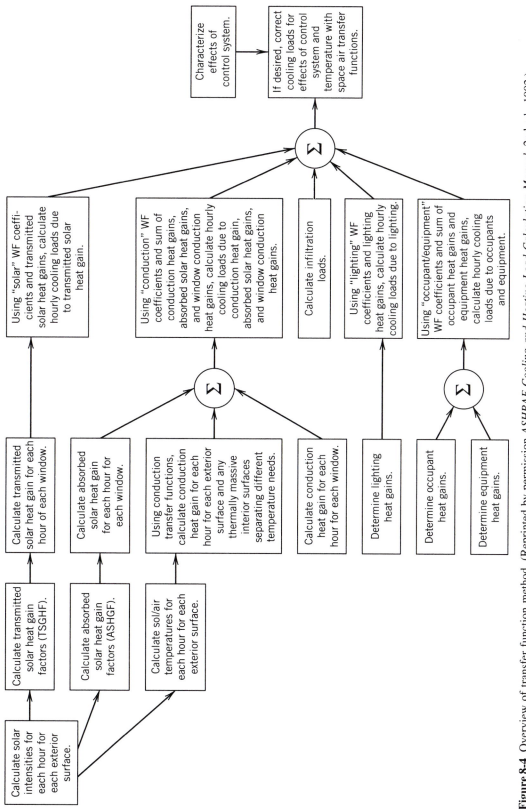

Figure 8-4 Overview of transfer function method. (Reprinted by permission *ASHRAE Cooling and Heating Load Calculation Manual*, 2nd ed., 1992.)

to that type of heat gain. Room transfer functions are sometimes referred to as weighting factors.

For example, consider a particular zone with equipment turned on for a certain fraction of the day. The equipment heat gain is partially radiant and partially convective. The convective portion immediately becomes space cooling load; the radiant portion is absorbed by thermal mass in the zone and convected to the room air later. If this zone is modeled with a set of heat balance equations, and pulsed with a unit heat gain, the space cooling load each hour will be a fractional number indicating how much of the original pulse will be convected into the room air each hour. This set of fractional numbers is a set of thermal response factors.

There are many methods that can be used to convert response factors to transfer functions. The room transfer functions used herein were determined using the DOE-2 program (7). The methodology is described by the *DOE-2 Engineers Manual* (8). Also, there are many possible combinations or number of heat gain terms and number of cooling load terms that may be used. For the purposes of room transfer functions, a standard form has been chosen:

$$\dot{q}_{c,\theta} = v_0 \dot{q}_{i,\theta} + v_1 \dot{q}_{i,\theta-\Delta} + v_2 \dot{q}_{i,\theta-2\Delta} - w_1 \dot{q}_{c,\theta-\Delta} - w_2 \dot{q}_{c,\theta-2\Delta} \qquad \textbf{(8-5)}$$

where Δ = time interval. The terms v_0, v_1, v_2, w_1, and w_2 are the coefficients of the room transfer function

$$K(z) = \frac{v_0 + v_1 z^{-1} + v_2 z^{-2}}{1 + w_1 z^{-1} + w_2 z^{-2}} \qquad \textbf{(8-6)}$$

which relates the transform of the corresponding parts of the cooling load and of the heat gain. These coefficients are dependent on the time interval Δ, the nature of the heat gain (fraction radiant and location in room), and the thermal characteristics of the room. The time interval for load calculations is always assumed to be one hour.

Extensive research (3) resulted in the calculation of room transfer function coefficients, the vs and Ws in Eq. 8-5, for over 200,000 zone types. The 14 parameters along with the variable levels are shown in Table 8-1. Since such a large database of weighting factors would be very unwieldy, the zones were grouped into sets of similar response, and a custom database and software to access the database were developed. This software is referred to collectively as the *RTF routines* and is part of the ASHRAE *Cooling and Heating Load Calculation Manual* (5).

The intended use of these weighting factors is that users will develop TFM computer programs that access the database to choose the sets of weighting factors that most closely match the actual zone. Some engineering judgment is still required to choose the set of parameters that most closely describe the actual zone.

Zone geometry, zone height, interior shade, furniture, and floor covering all primarily affect the thermal and solar radiation interchange. The number of exterior walls, partition type, zone location, midfloor type, slab type, ceiling type, roof type, and glass percent, when taken together indicate the construction of each of the six surfaces in the zone. For example, a bottom floor zone with one exterior wall has three partition walls defined by the partition type variable, one exterior wall defined by the exterior construction type variable and the percent glass variable, a slab on grade floor, and a

Table 8-1 Zone Parametric Level Definitions

No.	Parameter	Meaning	Levels (in normal order)
1	zg	Zone geometry	100 ft $\times$ 20 ft, 15 ft $\times$ 15 ft 100 ft $\times$ 100 ft
2	zh	Zone height	8 ft, 10 ft, 20 ft
3	nw	No. exterior walls	1, 2, 3, 4, 0
4	is	Interior shade	100, 50, 0%
5	fn	Furniture	With, without
6	ec	Exterior wall construction	1, 2, 3, 4 (Table 8-2)
7	pt	Partition type	$\frac{5}{8}$-in. gypsum board-air space $\frac{5}{8}$-in. gypsum board, 8-in. concrete block
8	zl	Zone location	Single-story, top floor, bottom floor, mid-floor
9	mf	Midfloor type	8-in. concrete, 2.5-in. concrete, 1-in. wood
10	st	Slab type	Mid-floor type, 4-in. slab on 12-in. soil
11	ct	Ceiling type	$\frac{3}{4}$-in. acoustic tile and air space, w/o ceiling
12	rt	Roof type	1, 2, 3, 4 (Table 8-4)
13	fc	Floor covering	Carpet with rubber pad, vinyl tile
14	gl	Glass percent	10, 50, 90

Source: Reprinted by permission from ASHRAE *Cooling and Heating Load Calculation Manual,* 2nd ed., 1992.

ceiling described by the midfloor type variable and the ceiling type variable. (The ceiling type variable indicates whether or not there is a suspended ceiling.)

A brief subjective discussion of each of the 14 parameters follows. The term *weak variable* refers to one that usually has little effect on the results as it is varied. The term *strong variable* refers to one that can significantly effect the results as it is varied. It is important to keep in mind that the following discussion is somewhat subjective and the database does have some counterintuitive features.

Zone Geometry (zg). This variable primarily affects the intersurface radiation exchange and the distribution of solar energy. It tends to be a weak variable.

Zone Height (zh). This variable also primarily affects the intersurface radiation exchange and the distribution of solar energy. It also tends to be a weak variable.

Number of Exterior Walls (nw). The number of exterior walls parameter specifies how many of the four walls are described by the exterior construction type and how many are described by the partition type. This can be a strong variable, depending on the difference between the partition type and the exterior construction type.

Interior Shade (is). The interior shade parameter primarily affects the distribution of solar radiation. Any solar radiation intercepted by an interior shade is immediately convected into the room air. Thus, increasing the level of interior shading usually increases the dynamic response of a zone with windows to solar radiation. However, in cases where the zone is lightweight (rapid dynamic response), say with wooden floors and furniture, varying the interior shade has little effect. Interior shades may also affect the distribution of thermal radiation and have a weak effect on nonsolar transfer functions.

Furniture (fn). Furniture acts in a manner similar to interior shading, in that it intercepts solar radiation and convects it into the room air. Likewise, it may have little effect on a zone that is lightweight.

Exterior Wall Construction (ec). Exterior wall construction specifies the construction of the exterior walls. However, there are only four choices, and the engineer must choose the construction that most closely matches the actual wall. Mass is a primary consideration. Further description of the four exterior wall construction types is found in Table 8-2.

Partition Type (pt). Partition type specifies the construction of the partitions, i.e., the walls that are not exterior walls. There are two choices: a lightweight (drywall) partition, and a more thermally massive partition made of 8 in. concrete block.

Zone Location (zl). The zone location parameter determines the floor and ceiling constructions, as shown in Table 8-3.

Midfloor Type (mf). This parameter determines the floor construction for midfloor zones and top-floor zones. It determines the ceiling construction for bottom-floor zones and midfloor zones.

Table 8-2 Exterior Wall Construction Types

Type	Description
1	Outside surface resistance, 1-in. stucco, 1-in. insulation, $\frac{3}{4}$-in. plaster or gypsum, inside surface resistance (A0, A1, B1, E1, E0)[a]
2	Outside surface resistance, 1-in. stucco, 8-in. HW concrete, $\frac{3}{4}$-in. plaster or gypsum, inside surface resistance (A0, A1, C10, E1, E0)
3	Outside surface resistance, steel siding, 3-in. insulation, steel siding, inside surface (A0, A3, B12, A3, E0)[a]
4	Outside surface resistance, 4-in. face brick, 2-in. insulation, 12-in. HW concrete, $\frac{3}{4}$-in. plaster or gypsum, inside surface resistance (A0, A2, B3, C11, E1, E0)[a]

[a]Code letters are defined in Table 8-21.

Source: Reprinted by permission from ASHRAE *Cooling and Heating Load Calculation Manual,* 2nd ed., 1992.

Table 8-3 Floor and Ceiling Types Specified by Zone Location Parameter

Zone Location	Floor	Ceiling
Single story	Slab-on-grade	Roof
Top floor	Midfloor	Roof
Bottom floor	Slab-on-grade	Midfloor
Midfloor	Midfloor	Midfloor

Source: Reprinted by permission from ASHRAE *Cooling and Heating Load Calculation Manual,* 2nd ed., 1992.

Slab Type (st). Originally, this parameter was intended to allow for bottom-floor zones to have floors other than a slab-on-grade. For practical reasons, a decision was made to allow slab-on-grade floors only for zones located on the bottom floor; therefore, this parameter has no effect. A *bottom-floor* zone that is not on a slab should be treated as a midfloor zone.

Ceiling Type (ct). The ceiling type parameter is used to indicate whether or not a suspended ceiling is present. Either the roof type parameter or the midfloor parameter, depending on the zone, indicate the construction of the ceiling.

Roof Type (rt). The roof type parameter indicates the type of roof for top-floor zones and single-story zones. Again, there are only four choices, and the engineer must choose the roof type that most closely matches the actual roof. Further description of the four roof types is found in Table 8-4.

Floor Covering (fc). The floor covering parameter indicates whether or not the floor is covered with carpeting or vinyl floor tile. This can be a strong parameter, particularly if the floor is thermally massive. Covering a thermally massive floor with carpet reduces the effect of the thermal mass and increases the dynamic response of the zone.

Glass Percent (gl). The glass percent parameter indicates the percentage of exterior wall area that is taken up by glazing. This does not affect the solar gain, which is accounted for separately. Rather, this parameter affects the dynamic response of the zone. Increasing the percentage of glass on an exterior wall decreases the effect of the wall's thermal mass and increases the dynamic response of the zone.

Tables 8-5*a*, 8-5*b*, and 8-5*c* give some room weighting factors taken from the ASHRAE database. These factors are limited to single-story buildings; therefore, the required number of parameters needed to define the zone is reduced. Tables 8-6*a*, 8-6*b*, and 8-6*c* may be used to select the zone type for use in Tables 8-5. Room weighting factors may also be obtained using the routines found in reference 5.

Table 8-4 Roof Construction Types

Type	Description
1	Outside surface resistance, $\frac{1}{2}$-in. slag or stone, $\frac{3}{8}$-in. felt membrane, 1-in. insulation, steel siding, inside surface resistance (A0, E2, E3, B4, A3, E0)[a]
2	Outside surface resistance, $\frac{1}{2}$-in. slag or stone, $\frac{3}{8}$-in. felt membrane, 6-in. LW concrete, inside surface resistance (A0, E2, E3, C15, E0)[a]
3	Outside surface resistance, $\frac{1}{2}$-in. slag or stone, $\frac{3}{8}$-in. felt membrane, 2-in. insulation, steel siding, ceiling air space, acoustic tile, inside surface resistance (A0, E2, E3, B6, A3, E4, E5, E0)[a]
4	Outside surface resistance, $\frac{1}{2}$-in. slag or stone, $\frac{3}{8}$-in. felt membrane, 8-in. LW concrete, ceiling air space, acoustic tile, inside surface resistance (A0, E2, E3, C16, E4, E5, E0)[a]

[a]Code letters are defined in Table 8-21.

Source: Reprinted by permission from ASHRAE *Cooling and Heating Load Calculation Manual,* 2nd ed., 1992.

Table 8-5a Room Weighting Factors, Single Story, Perimeter Rooms

Zone Type	Component	v_0	v_1	v_2	w_1	w_2
1	Solar	0.56355	−0.57965	0.09904	−1.18944	0.27238
	Conduction	0.69882	−0.76616	0.15198	−1.18077	0.26541
	Gen. light	0.69444	−0.74631	0.13983	−1.15478	0.24274
	People	0.64142	−0.69475	0.12997	−1.18318	0.25982
2	Solar	0.67373	−0.74892	0.11906	−1.34372	0.38759
	Conduction	0.79164	−0.97078	0.21064	−1.33098	0.36788
	Gen. light	0.77275	−0.92994	0.19525	−1.32181	0.35987
	People	0.76733	−0.92886	0.19843	−1.33098	0.36788
3	Solar	0.47617	−0.50514	0.08913	−1.26779	0.32795
	Conduction	0.63137	−0.67468	0.12056	−1.17015	0.24740
	Gen. light	0.61116	−0.67267	0.12987	−1.19718	0.26554
	People	0.61309	−0.65498	0.11386	−1.16547	0.23744
4	Solar	0.56355	−0.57965	0.09904	−1.18944	0.27238
	Conduction	0.73930	−0.87810	0.20074	−1.31608	0.37802
	Gen. light	0.69696	−0.78366	0.15367	−1.28606	0.35303
	People	0.70074	−0.81332	0.17452	−1.31608	0.37802
5	Solar	0.41780	−0.40122	0.06334	−1.18051	0.26043
	Conduction	0.68342	−0.77042	0.16406	−1.22530	0.30236
	Gen. light	0.64359	−0.71163	0.14143	−1.20988	0.28327
	People	0.62088	−0.66215	0.12035	−1.16903	0.24811
6	Solar	0.56355	−0.57965	0.09904	−1.18944	0.27238
	Conduction	0.73930	−0.87810	0.20074	−1.31608	0.37802
	Gen. light	0.72002	−0.82708	0.17276	−1.29108	0.35678
	People	0.70074	−0.81332	0.17452	−1.31608	0.37802
7	Solar	0.56355	−0.57965	0.09904	−1.18944	0.27238
	Conduction	0.74738	−0.84432	0.17882	−1.22032	0.30220
	Gen. light	0.69696	−0.78366	0.15367	−1.28606	0.35303
	People	0.67245	−0.77612	0.16628	−1.29367	0.35628
8	Solar	0.47617	−0.50514	0.08913	−1.26779	0.32795
	Conduction	0.68342	−0.77042	0.16406	−1.22530	0.30236
	Gen. light	0.64359	−0.71163	0.14143	−1.20988	0.28327
	People	0.62088	−0.66215	0.12035	−1.16903	0.24811
9	Solar	0.56355	−0.57965	0.09904	−1.18944	0.27238
	Conduction	0.73930	−0.87810	0.20074	−1.31608	0.37802
	Gen. light	0.76054	−0.88689	0.19225	−1.28204	0.34794
	People	0.70074	−0.81332	0.17452	−1.31608	0.37802
10	Solar	0.56355	−0.57965	0.09904	−1.18944	0.27238
	Conduction	0.74738	−0.84432	0.17882	−1.22032	0.30220
	Gen. light	0.69444	−0.74631	0.13983	−1.15478	0.24274
	People	0.67245	−0.77612	0.16628	−1.29367	0.35628

Table 8-5a Room Weighting Factors, Single Story, Perimeter Rooms (*Continued*)

Zone Type	Component	v_0	v_1	v_2	w_1	w_2
11	Solar	0.59094	−0.63490	0.09545	−1.30715	0.35864
	Conduction	0.73930	−0.87810	0.20074	−1.31608	0.37802
	Gen. light	0.76054	−0.88689	0.19225	−1.28204	0.34794
	People	0.70074	−0.81332	0.17452	−1.31608	0.37802
12	Solar	0.47617	−0.50514	0.08913	−1.26779	0.32795
	Conduction	0.69882	−0.76616	0.15198	−1.18077	0.26541
	Gen. light	0.69444	−0.74631	0.13983	−1.15478	0.24274
	People	0.64142	−0.69475	0.12997	−1.18318	0.25982
13	Solar	0.56355	−0.57965	0.09904	−1.18944	0.27238
	Conduction	0.68342	−0.77042	0.16406	−1.22530	0.30236
	Gen. light	0.64359	−0.71163	0.14143	−1.20988	0.28327
	People	0.62088	−0.66215	0.12035	−1.16903	0.24811
14	Solar	0.59094	−0.63490	0.09545	−1.30715	0.35864
	Conduction	0.73930	−0.87810	0.20074	−1.31608	0.37802
	Gen. light	0.72002	−0.82708	0.17276	−1.29108	0.35678
	People	0.70074	−0.81332	0.17452	−1.31608	0.37802

Source: Reprinted by permission from ASHRAE *Cooling and Heating Load Calculation Manual,* 2nd ed., 1992.

EXAMPLE 8-1

Select room parameters for the following structure. The zone is a 70 × 30-ft retail store in a strip shopping mall. The store is one end of the mall, so that the north 30-ft wall is an interior wall adjoining a store room; the west 70-ft wall is an interior wall adjoining another store; the east 70-ft wall is an exterior wall with no glazing; and the south wall is an exterior wall with a large (30 × 7 ft) display window. The exterior wall is constructed of 4-in. face brick, 8-in. concrete block, 0.75-in. polystyrene insulation with 1-in. furring strips (2-in. wide, spaced 16 in. o.c.), and 0.5-in. drywall. The roof is constructed from built-up roofing, rigid roof deck insulation, 2-in. concrete on a corrugated metal deck, and a suspended ceiling. The interior walls are constructed from 0.5-in. drywall and steel studs. The store is built on a 4-in. concrete slab-on-grade.

SOLUTION

Room parameters are as follows:

Zone Geometry. This zone is 30 × 70 ft in size. Of the three choices available for zone size, this zone most closely matches type 1, which is 100 × 20 ft in size.

Zone Height. The distance between the floor and ceiling is 12 ft, which most closely matches type 2, which is 10 ft.

Number of Exterior Walls. There are two exterior walls.

Interior Shade. There is no interior shading, which corresponds to type 3, 0 percent.

Furniture. There are a number of book shelves, which act the same as furniture, so type 1, with furniture is chosen.

Table 8-5b Room Weighting Factors, Single Story, Single Zone

Zone Type	Component	v_0	v_1	v_2	w_1	w_2
1	Solar	0.737220	−0.768950	0.129390	−1.134080	0.231740
	Conduction	0.818200	−0.940620	0.195140	−1.231910	0.304630
	Gen. light	0.803928	−0.922162	0.189654	−1.233200	0.304620
	People	0.792191	−0.859114	0.155903	−1.137340	0.226320
2	Solar	0.700910	−0.767740	0.140880	−1.199820	0.273870
	Conduction	0.818200	−0.940620	0.195140	−1.231910	0.304630
	Gen. light	0.803928	−0.922162	0.189654	−1.233200	0.304620
	People	0.829548	−0.991988	0.201130	−1.277850	0.316540
3	Solar	0.700910	−0.767740	0.140880	−1.199820	0.273870
	Conduction	0.818200	−0.940620	0.195140	−1.231910	0.304630
	Gen. light	0.803928	−0.922162	0.189654	−1.233200	0.304620
	People	0.792191	−0.859114	0.155903	−1.137340	0.226320
4	Solar	0.737220	−0.768950	0.129390	−1.134080	0.231740
	Conduction	0.818200	−0.940620	0.195140	−1.231910	0.304630
	Gen. light	0.803928	−0.922162	0.189654	−1.233200	0.304620
	People	0.829548	−0.991988	0.201130	−1.277850	0.316540
5	Solar	0.427892	−0.426402	0.072210	−1.192600	0.266300
	Conduction	0.746378	−0.855214	0.177946	−1.240070	0.309180
	Gen. light	0.723098	−0.788946	0.147838	−1.161520	0.243510
	People	0.703696	−0.762833	0.141127	−1.161520	0.243510
6	Solar	0.509380	−0.509750	0.072340	−1.234240	0.306210
	Conduction	0.746378	−0.855214	0.177946	−1.240070	0.309180
	Gen. light	0.762500	−0.830700	0.159570	−1.153170	0.244540
	People	0.744000	−0.804800	0.142600	−1.145010	0.226810
7	Solar	0.427892	−0.426402	0.072210	−1.192600	0.266300
	Conduction	0.746378	−0.855214	0.177946	−1.240070	0.309180
	Gen. light	0.762500	−0.830700	0.159570	−1.153170	0.244540
	People	0.744000	−0.804800	0.142600	−1.145010	0.226810
8	Solar	0.509380	−0.509750	0.072340	−1.234240	0.306210
	Conduction	0.746378	−0.855214	0.177946	−1.240070	0.309180
	Gen. light	0.762500	−0.830700	0.159570	−1.153170	0.244540
	People	0.743639	−0.805102	0.152833	−1.153170	0.244540
9	Solar	0.700910	−0.767740	0.140880	−1.199820	0.273870
	Conduction	0.818200	−0.940620	0.195140	−1.231910	0.304630
	Gen. light	0.803928	−0.922162	0.189654	−1.233200	0.304620
	People	0.829548	−0.991988	0.201130	−1.277850	0.316540
10	Solar	0.700910	−0.767740	0.140880	−1.199820	0.273870
	Conduction	0.818200	−0.940620	0.195140	−1.231910	0.304630
	Gen. light	0.803928	−0.922162	0.189654	−1.233200	0.304620
	People	0.829548	−0.991988	0.201130	−1.277850	0.316540

Table 8-5b Room Weighting Factors, Single Story, Single Zone (*Continued*)

Zone Type	Component	v_0	v_1	v_2	w_1	w_2
11	Solar	0.509380	−0.509750	0.072340	−1.234240	0.306210
	Conduction	0.746378	−0.855214	0.117946	−1.240070	0.309180
	Gen. light	0.762500	−0.830700	0.159570	−1.153170	0.244540
	People	0.743639	−0.805102	0.152833	−1.153170	0.244540
12	Solar	0.427892	−0.426402	0.072210	−1.192600	0.266300
	Conduction	0.746378	−0.855214	0.177946	−1.240070	0.309180
	Gen. light	0.762500	−0.830700	0.159570	−1.153170	0.244540
	People	0.743639	−0.805102	0.152833	−1.153170	0.244540

Source: Reprinted by permission from ASHRAE *Cooling and Heating Load Calculation Manual,* 2nd ed., 1992.

Table 8-5c Room Weighting Factors, Single Story, Interior Rooms

Zone Type	Component	v_0	v_1	v_2	w_1	w_2
1	Solar	0.654898	−0.768295	0.155077	−1.266910	0.308590
	Conduction	0.785821	−0.883188	0.134466	−1.211551	0.248650
	Gen. light	0.784411	−0.881027	0.133716	−1.211550	0.248650
	People	0.771348	−0.861002	0.126754	−1.211500	0.248650
2	Solar	0.396279	−0.394969	0.066340	−1.206960	0.274610
	Conduction	0.696060	−0.754340	0.128940	−1.173130	0.243790
	Gen. light	0.685780	−0.740180	0.125060	−1.173130	0.243790
	People	0.667935	−0.715608	0.118333	−1.173130	0.243790

Source: Reprinted by permission from ASHRAE *Cooling and Heating Load Calculation Manual,* 2nd ed., 1992.

Exterior Construction. There are four choices for exterior construction type in Table 8-2. They can be thought of as lightweight with little insulation, heavyweight with no insulation, lightweight with moderate insulation, and heavyweight with moderate insulation. The nearest choice for the wall here is type 4, which is a heavyweight wall with moderate insulation.

Partition Type. This store has lightweight frame construction partitions. They correspond to the type 1 partition, which is $\frac{5}{8}$-in. gypsum, an airspace, and $\frac{5}{8}$-in. gypsum.

Zone Location. The store is part of a single-story building. Therefore, the zone location type is type 1, corresponding to a single-story building.

Midfloor Type. This parameter has no meaning for single-story zones.

Slab Type. This building is built on a slab-on-grade; therefore the slab type is the 4-in. slab on 12-in. soil, which is type 2.

Ceiling Type. The ceiling type parameter indicates whether or not there is a suspended ceiling. This zone has a suspended ceiling, which corresponds to ceiling type 1, not a significant parameter in this case.

Roof Type. The roof type is chosen from Table 8-4 to be type 2, not a significant parameter in this case.

Floor Covering. The slab is covered with vinyl tile, which corresponds to floor covering type 2.

Glass Percent. This parameter indicates the percentage of exterior wall that is consumed by glazing. For this zone, the approximate total exterior wall area is

$$(12 \times 70 \, \text{ft}) + (12 \times 30 \, \text{ft}) = 1200 \, \text{ft}^2$$

The glazing area is

$$7 \times 30 \, \text{ft} = 210 \, \text{ft}^2$$

Table 8-6a Zone Types, Single Story, Perimeter Rooms

Zone Type	Zone Classification Parameters								
	zg	zh	nw	is	fn	ec	gl	pt	fc
1	2	2	2	2	1	2	1	2	1
	2	2	1	2	1	2	2	2	1
	2	2	1	2	1	2	1	2	1
2	2	2	1	1	1	3	1	1	1
	2	2	1	1	1	3	2	1	1
	2	2	2	3	1	3	3	1	1
	2	2	2	2	1	3	1	1	1
	2	2	1	2	1	3	1	1	1
	2	2	1	2	1	3	2	1	1
	2	2	2	2	1	3	2	1	1
3	2	2	1	2	1	2	1	2	2
	1	2	2	3	1	4	1	1	2
4	2	2	1	2	1	3	1	1	2
5	2	2	1	2	1	2	2	2	2
	2	2	2	2	1	2	1	2	1
6	2	2	1	2	1	3	2	1	2
	2	2	2	2	1	3	1	1	2
7	2	2	2	2	1	2	2	2	1
8	2	2	2	2	1	2	2	2	2
9	2	2	2	3	1	3	3	1	2
	2	2	2	2	1	3	2	1	2
10	2	2	1	2	1	2	3	2	1
11	2	2	2	2	1	2	3	2	1
12	2	2	2	3	1	2	3	2	2
	2	2	2	2	1	2	3	2	2
13	2	2	2	1	1	2	1	2	2
14	2	2	2	1	1	3	1	1	2

Table 8-6*b* Zone Types, Single Zone with Four Walls

Zone Type	Room Classification Parameters						
	zg	zh	is	fn	ec	gl	fc
1	1	2	2	1	2	1	1
2	1	2	2	1	3	1	1
3	1	2	2	1	2	2	1
4	1	2	2	1	3	2	1
5	1	2	2	1	2	1	2
6	1	2	2	1	3	1	2
7	1	2	2	1	2	2	2
8	1	2	2	1	3	2	2
9	1	2	1	1	2	3	1
10	1	2	3	1	2	3	1
11	1	2	1	1	2	3	2
12	1	2	3	1	2	3	2

Table 8-6*c* Zone Types, Single Story, Interior Rooms

Zone Type	Room Classification Parameters			
	fn	pt	fc	ct
1	1	1	1	1
	1	2	1	1
2	1	1	2	1
	1	2	2	1

The glass percentage is

$$\frac{210}{1200} = 17.5 \text{ percent}$$

Obviously, 10 percent or case 1 is the closest choice to the actual percentage.

With these parameters the weighting factors can be determined using the routines of reference 5 or the zone may be identified as type 3 in Table 8-6*a* and the weighting factors read from Table 8-5*a*. See Example 8-5 for application of the weighting factors for lights.

Wall and Roof Transfer Functions

Conduction transfer functions are used by the TFM to describe the heat flux at the inside of a wall, roof, partition, ceiling, or floor as a function of previous values of the heat flux and previous values of inside and outside temperatures.

The TFM uses fixed, combined convection and radiation coefficients on the inside and outside surfaces, so that the conduction transfer functions are driven by sol-air temperature (defined in Section 8-4) on the outside and room temperature on the inside. Furthermore, the inside air temperature is assumed to be constant. This allows the following formulation, since previous inside temperatures are constant:

$$\dot{q}_{i,\theta} = A\left[\sum_{n=0} b_n(t_{e,\theta-n\Delta}) - \sum_{n=1} d_n(\dot{q}_{i,\theta-n\Delta}/A) - t_i \sum_{n=0} c_n \right] \qquad (8\text{-}7)$$

where

$\dot{q}_{i\theta}$ = heat gain through wall, roof, partition, etc., Btu/hr, at calculation hour θ
A = indoor surface area of a wall or roof, ft^2
θ = time, hr
Δ = time interval, hr
n = summation index (each summation has as many terms as there are nonzero values of the coefficients)
$t_{e,\theta-n\Delta}$ = sol-air temperature at time $\theta - n\Delta$, F
t_i = constant indoor room temperature, F
b_n, c_n, d_n = conduction transfer function coefficients

Eq. 8-7 must be solved iteratively because the heat flux history terms on the right-hand side are not known beforehand when analyzing a 24-hour time period. Typically, the heat flux history terms are assumed to be zero, and Eq. 8-7 is calculated for successive 24-hour periods until convergence is reached. At that time the results are independent of the values assumed initially.

The method described herein uses a set of construction types with precalculated CTF coefficients. Software was developed to access a database of conduction transfer function coefficients for representative roof and wall assemblies. Conduction transfer function coefficients depend only on the physical properties of the wall or roof and not on the construction of the zone in general.

The CTF coefficients stored in the database are based on 41 wall types and 42 roof types. A procedure was developed (2) to categorize the thermal response of arbitrarily constructed walls such that CTFs from a representative wall or roof type can be selected. This procedure has been automated and subroutines are given in the ASHRAE *Cooling and Heating Load Calculation Manual* to select the representative assembly and return the CTF coefficients. The *b* and *c* coefficients must then be corrected by multiplying by the ratio of the *U*-factor for the actual wall or roof to the *U*-factor of the representative wall or roof.

Methods for determining the *U*-factor are described in Chapter 5. It is important that the actual *U*-factor is used for correcting the *b* coefficients. In cases where thermal bridging is important, its effect must be included in the *U*-factor calculation. Thermal bridging in walls and roofs does not significantly affect the dynamic response. However, the reduced resistance is very important and must be accounted for.

In order to determine the CTF coefficients for a wall, the following parameters must be specified in the CTF routines referenced above:

R-value Range. Chosen from 17 different ranges.
Primary Wall Material. Chosen from 25 different material categories.
Mass Location. Inside, outside, or integral. This is the location of thermal mass in the wall relative to the insulation.
Secondary Wall Material. Chosen from 6 different material categories.

In order to determine the CTF coefficients for a roof, the following parameters must be specified:

R-value Range. Chosen from 6 different ranges.
Roof Material. Chosen from 20 different material categories, including roof terrace systems.
Mass Location. Inside, outside, or integral. The mass location parameters for roofs are identical to those for walls.
Suspended Ceiling. With or without.

Users should also be aware that the CTFs are calculated with exterior and interior heat transfer coefficients. The outdoor heat transfer coefficient h_o is 3.0 Btu/(hr-ft^2-F) or 17 W/(m^2-C) and the interior heat transfer coefficient h_i is 1.46 Btu/(hr-ft^2-F) or 8.3 W/(m^2-C). The use of $h_o = 3.0$ Btu/(hr-ft^2-F) in the calculation of CTF coefficients presumes that the same value will be used in the calculation of the sol-air temperatures, described below.

Tables 8-7 and 8-8 are examples of transfer function coefficients taken from the ASHRAE database. Tables 8-20 and 8-22 in Section 8-7 may be used to determine the appropriate wall or roof type rather than using the ASHRAE software (5). Remember that the *b* coefficients must be corrected by multiplying by the ratio of the actual *U*-factor to the *U*-factor for the wall or roof type chosen.

EXAMPLE 8-2

Determine wall CTF coefficients for a wall composed of 4-in. face brick, 8-in. normal weight concrete block, $\frac{3}{4}$-in. polystyrene insulation with 1-in. furring strips (spaced 16 in.), and $\frac{1}{2}$-in. gypsum wall board.

SOLUTION

First, the wall must be categorized. To do this, choose materials from Table 8-21 to determine the *R*-value for the wall or use the data and procedures of Chapter 5. Considering the furring strips, the *R*-value is approximately 6.7 (hr-ft^2-F)/Btu and the *U*-factor is 0.15 Btu/(hr-ft^2-F). The mass location is mass out. The resistance range is No. 9, 6.5 to 7.75 (hr-ft^2-F)/Btu. The primary wall material is the 8-in. H. W. concrete block, C-8 Table 8-21. The secondary wall material is the 4-in. face brick (A-2 Table 8-21).

When these parameters are passed to the ASHRAE computer routine, a wall type of 17 is returned with the following CTF coefficients and a *U*-factor of 0.043 Btu/(hr-ft^2-F). This may also be done by using Tables 8-22 in Section 8-7 and Table 8-7.

Table 8-7 Transfer Function Coefficients for Walls and Partitions[a]

WALL TYPE 1: E0, A3, B1, B13, A3, A0
STEEL SIDING W/ 4" INS.

	$n=0$	$n=1$	$n=2$	$n=3$	$n=4$	$n=5$	$n=6$	
b_n	.76794E − 02	.34978E − 01	.71911E − 02	.55759E − 04	.94018E − 08	.13353E − 13	.15183E − 15	c_n = .049904
d_n	.10000E + 01	−.24072E + 00	.16764E − 02	−.51488E − 06	.25837E − 11	−.39151E − 19	.65086E − 32	U = .065581

WALL TYPE 2: E0, E1, B14, A1, A0
FRAME WALL W/ 5" INS.

	$n=0$	$n=1$	$n=2$	$n=3$	$n=4$	$n=5$	$n=6$	
b_n	.15618E − 03	.54544E − 02	.96078E − 02	.21547E − 02	.51664E − 04	.47736E − 07	.25103E − 12	c_n = .017425
d_n	.10000E + 01	−.93389E + 00	.27396E + 00	−.25609E − 01	.14239E − 03	−.19319E − 08	.60005E − 17	U = .055386

WALL TYPE 3: E0, C3, B5, A6, A0
4" H.W. CONC. BLK. W/ 1" INS.

	$n=0$	$n=1$	$n=2$	$n=3$	$n=4$	$n=5$	$n=6$	
b_n	.41066E − 02	.32305E − 01	.14744E − 01	.46550E − 03	.28937E − 06	.62274E − 13	−.43623E − 15	c_n = .051621
d_n	.10000E + 01	−.76963E + 00	.40143E − 01	−.41970E − 03	.27284E − 09	−.12997E − 21	.20665E − 36	U = .191122

WALL TYPE 4: E0, E1, B6, C12, A0
2" INS. W/ 2" H.W. CONC.

	$n=0$	$n=1$	$n=2$	$n=3$	$n=4$	$n=5$	$n=6$	
b_n	.10663E − 04	.10790E − 02	.38405E − 02	.18711E − 02	.13427E − 03	.87708E − 06	.19822E − 09	c_n = .006936
d_n	.10000E + 01	−.13758E + 01	.61544E + 00	−.93889E − 01	.22104E − 02	−.89008E − 06	.21209E − 11	U = .046874

WALL TYPE 5: E0, A6, B21, C7, A0
1.36" INS. W/ 8" L.W. CONC. BLK.

	$n=0$	$n=1$	$n=2$	$n=3$	$n=4$	$n=5$	$n=6$	
b_n	.80326E − 04	.44353E − 02	.10182E − 01	.29601E − 02	.10453E − 03	.23140E − 06	.97085E − 11	c_n = .017762
d_n	.10000E + 01	−.11604E + 01	.32547E + 00	−.27463E − 01	.20980E − 03	−.15254E − 07	.40784E − 14	U = .128916

WALL TYPE 6: E0, E1, B2, C5, A1, A0
1" INS. W/ 4" H.W. CONC.

	$n=0$	$n=1$	$n=2$	$n=3$	$n=4$	$n=5$	$n=6$	
b_n	.50539E − 03	.93807E − 02	.10572E − 01	.12730E − 02	.88344E − 05	.60571E − 09	.43398E − 16	c_n = .021740
d_n	.10000E + 01	−.11758E + 01	.30071E + 00	−.15606E − 01	.58619E − 05	−.12039E − 11	.56442E − 24	U = .198884

WALL TYPE 7: E0, A6, C5, B3, A3, A0
4" H.W. CONC. W/ 2" INS.

	$n=0$	$n=1$	$n=2$	$n=3$	$n=4$	$n=5$	$n=6$	
b_n	.98849E − 03	.83611E − 02	.36141E − 02	.69313E − 04	.10034E − 07	.18942E − 13	−.33976E − 16	$c_n = .013033$
d_n	.10000E + 01	−.93970E + 00	.46636E − 01	−.21210E − 05	−.17233E − 10	−.43539E − 19	.87073E − 30	$U = .121878$

WALL TYPE 8: E0, A2, C12, B5, A6, A0
FACE BRICK & 2" H.W. CONC. W/ 1" INS.

	$n=0$	$n=1$	$n=2$	$n=3$	$n=4$	$n=5$	$n=6$	
b_n	.14401E − 03	.45986E − 02	.73304E − 02	.13529E − 02	.24650E − 04	.18978E − 07	.97973E − 13	$c_n = .013451$
d_n	.10000E + 01	−.12001E + 01	.27937E + 00	−.10391E − 01	−.53403E − 04	−.868423 − 09	.47859E − 17	$U = .195190$

WALL TYPE 9: E0, A6, B15, B10, A0
6" INS. W/ 2" WOOD

	$n=0$	$n=1$	$n=2$	$n=3$	$n=4$	$n=5$	$n=6$	
b_n	.35081E − 07	.63821E − 02	.85866E − 03	.14648E − 02	.50642E − 03	.37673E − 04	.52063E − 06	$c_n = .002932$
d_n	.10000E + 01	−.16335E + 01	.86971E + 00	−.18121E + 00	.14453E − 01	−.30536E − 03	.99873E − 06	$U = .042413$

WALL TYPE 10: E0, E1, C2, B5, A2, A0
4" L.W. CONC. BLK. W/ 1" INS. & FACE BRICK

	$n=0$	$n=1$	$n=2$	$n=3$	$n=4$	$n=5$	$n=6$	
b_n	.72162E − 05	.10184E − 02	.44117E − 02	.25989E − 02	.23871E − 03	.24701E − 05	.13649E − 08	$c_n = .008277$
d_n	.10000E + 01	−.16636E + 01	.82440E + 00	−.11098E + 00	.35088E − 02	−.43699E − 05	.10989E − 09	$U = .155161$

WALL TYPE 11: E0, E1, C8, B6, A1, A0
8" H.W. CONC. BLK. W/ 2" INS.

	$n=0$	$n=1$	$n=2$	$n=3$	$n=4$	$n=5$	$n=6$	
b_n	.34085E − 05	.61426E − 05	.28902E − 02	.18273E − 02	.18442E − 03	.22725E − 05	.20763E − 08	$c_n = .005522$
d_n	.10000E + 01	−.15248E + 01	.67146E + 00	−.98438E − 01	.23897E − 02	−.23571E − 05	.39632E − 09	$U = .109090$

WALL TYPE 12: E0, E1, B1, C10, A1, A0
8" H.W. CONC.

	$n=0$	$n=1$	$n=2$	$n=3$	$n=4$	$n=5$	$n=6$	
b_n	.15708E − 04	.19814E − 02	.81619E − 02	.46745E − 02	.43793E − 03	.51955E − 05	.41275E − 08	$c_n = .015277$
d_n	.10000E + 01	−.15166E + 01	.64261E + 00	−.83816E − 01	.28873E − 02	−.72458E − 05	.49361E − 09	$U = .338747$

Table 8-7 Transfer Function Coefficients for Walls and Partitions[a] (*Continued*)

WALL TYPE 13: E0, A2, C5, B19, A6, A0
FACE BRICK & 4" H.W. CONC. W/ 0.61" INS.

	n = 0	n = 1	n = 2	n = 3	n = 4	n = 5	n = 6	
b_n	.26083E − 04	.20290E − 02	.60073E − 02	.23346E − 02	.12949E − 03	.75936E − 06	.20618E − 09	c_n = .010527
d_n	.10000E + 01	−.14135E + 01	.48697E − 00	−.32176E − 01	−.56718E − 03	−.80080E − 06	.81209E − 11	U = .251432

WALL TYPE 14: E0, A2, A2, B6, A6, A0
FACE BRICK & FACE BRICK W/ 2" INS.

	n = 0	n = 1	n = 2	n = 3	n = 4	n = 5	n = 6	
b_n	.11169E − 05	.29943E − 03	.16719E − 02	.12341E − 02	.15533E − 03	.29997E − 05	.63766E − 08	c_n = .003365
d_n	.10000E + 01	−.15299E + 01	.62059E + 00	−.63288E − 01	−.19603E − 02	−.63858E − 05	.18049E − 08	U = .114460

WALL TYPE 15: E0, A6, C17, B1, A7, A0
8" L.W. CONC, BLK. (FILLED) & FACE BRICK

	n = 0	n = 1	n = 2	n = 3	n = 4	n = 5	n = 6	
b_n	.85864E − 08	.30915E − 04	.60237E − 03	.14516E − 02	.73638E − 03	.87564E − 04	.22493E − 05	c_n = .002911
d_n	.10000E + 01	−.20000E + 01	.13680E + 01	−.37388E + 00	−.38846E − 01	−.14032E − 02	.14291E − 04	U = .091982

WALL TYPE 16: E0, A6, C18, B1, A7, A0
8" H.W. CONC. BLK. (FILLED) & FACE BRICK

	n = 0	n = 1	n = 2	n = 3	n = 4	n = 5	n = 6	
b_n	.12864E − 06	.14018E − 03	.16884E − 02	.27022E − 02	.86402E − 03	.56177E − 04	.61038E − 06	c_n = .005452
d_n	.10000E + 01	−.20026E + 01	.13289E + 01	−.32486E − 01	−.23607E − 01	−.52205E − 03	.14500E − 05	U = .222421

WALL TYPE 17: E0, A2, C2, B15, A0
FACE BRICK & 4" L.W. CONC. BLK. W/ 6" INS.

	n = 0	n = 1	n = 2	n = 3	n = 4	n = 5	n = 6	
b_n	.36855E − 09	.45258E − 05	.13363E − 03	.43665E − 03	.30010E − 03	.51204E − 04	.21448E − 05	c_n = .000928
d_n	.10000E + 01	−.20087E + 01	.13712E + 01	−.37897E + 00	.39616E − 01	−.16467E − 02	.22889E − 04	U = .043219

[a]U, bs, and cs are in Btu/(hr-ft² − F), and d is dimensionless. To convert U, bs and cs to W/(m²·C), multiply by 5.6783.

Source: Reprinted by permission from ASHRAE *Cooling and Heating Load Calculation Manual*, 2nd ed., 1992.

Table 8-8 Transfer Function Coefficients for Roofs, Floors, and Ceilings

ROOF TYPE 1: E0, A3, B25, E3, E2, A0
STEEL SHEET W/ 3.33" INS.

	$n = 0$	$n = 1$	$n = 2$	$n = 3$	$n = 4$	$n = 5$	$n = 6$	
b_n	.48703E − 02	.34736E − 01	.13652E − 01	.35823E − 03	.23339E − 06	.77659E − 12	−.19875E − 15	$c_n = .053616$
d_n	.10000E + 01	−.35451E + 00	.22670E − 01	−.46908E − 04	.30920E − 09	−.13754E − 16	.32374E − 28	$U = .080251$

ROOF TYPE 2: E0, A3, B14, E3, E2, A0
STEEL SHEET W/ 5" INS.

	$n = 0$	$n = 1$	$n = 2$	$n = 3$	$n = 4$	$n = 5$	$n = 6$	
b_n	.55584E − 03	.12022E − 01	.12817E − 01	.14330E − 02	.12926E − 04	.45185E − 08	.20582E − 13	$c_n = .026841$
d_n	.10000E + 01	−.60064E + 00	.86016E − 01	−.13527E − 02	.12858E − 05	−.10446E − 10	.84718E − 18	$U = .055454$

ROOF TYPE 3: E0, E5, E4, C12, E2, A0
2" H.W. CONC. W/ SUSP. CEIL.

	$n = 0$	$n = 1$	$n = 2$	$n = 3$	$n = 4$	$n = 5$	$n = 6$	
b_n	.61344E − 02	.39829E − 01	.13754E − 01	.24851E − 03	.90887E − 07	.25257E − 13	−.39498E − 15	$c_n = .059966$
d_n	.10000E + 01	−.75615E + 00	.14387E − 01	−.60003E − 04	.71457E − 10	−.73722E − 21	.00000E + 00	$U = .232261$

ROOF TYPE 4: E0, E1, B15, E4, B7, A0
ATTIC ROOF W/ 6" INS.

	$n = 0$	$n = 1$	$n = 2$	$n = 3$	$n = 4$	$n = 5$	$n = 6$	
b_n	.29438E − 05	.65306E − 05	.33867E − 02	.23970E − 02	.28879E − 03	.48569E − 05	.74294E − 08	$c_n = .006733$
d_n	.10000E + 01	−.13466E + 01	.59384E + 00	−.92953E − 01	.29637E − 02	−.57855E − 05	.18771E − 08	$U = .042814$

ROOF TYPE 5: E0, B14, C12, E3, E2, A0
5" INS. W/ 2" H.W. CONC.

	$n = 0$	$n = 1$	$n = 2$	$n = 3$	$n = 4$	$n = 5$	$n = 6$	
b_n	.55526E − 04	.25555E − 02	.47745E − 02	.10009E − 02	.21154E − 04	.28998E − 07	.94604E − 12	$c_n = .008408$
d_n	.10000E + 01	−.11040E + 01	.26169E + 00	−.47482E − 02	.21450E − 04	−.77618E − 09	.10491E − 14	$U = .054946$

ROOF TYPE 6: E0, C5, B17, E3, E2, A0
4" H.W. CONC. W/ 0.3" INS.

	$n = 0$	$n = 1$	$n = 2$	$n = 3$	$n = 4$	$n = 5$	$n = 6$	
b_n	.29012E − 02	.31434E − 01	.21138E − 01	.12012E − 02	.17968E − 05	.30135E − 11	.36508E − 15	$c_n = .056676$
d_n	.10000E + 01	−.97905E + 00	.13444E + 00	−.27189E − 02	.54600E − 08	−.10116E − 15	.23403E − 28	$U = .371225$

Table 8-8 Transfer Function Coefficients for Roofs, Floors, and Ceilings (*Continued*)

ROOF TYPE 7: E0, B22, C12, E3, E2, C12, A0
1.67″ INS. W/ 2″ H.W. CONC. RTS

	$n = 0$	$n = 1$	$n = 2$	$n = 3$	$n = 4$	$n = 5$	$n = 6$	
b_n	.59119E − 03	.86704E − 02	.68771E − 02	.37485E − 03	.38508E − 06	.96297E − 11	−.34560E − 15	c_n = .016518
d_n	.10000E + 01	−.11177E + 01	.23731E + 00	−.81562E − 04	.55144E − 08	−.78474E − 14	.92552E − 26	U = .138145

ROOF TYPE 8: E0, B16, C13, E3, E2, A0
0.15″ INS. W/ 6″ H.W. CONC.

	$n = 0$	$n = 1$	$n = 2$	$n = 3$	$n = 4$	$n = 5$	$n = 6$	
b_n	.98388E − 03	.19384E − 01	.20827E − 01	.21888E − 02	.14924E − 04	.24665E − 08	.72781E − 15	c_n = .043398
d_n	.10000E + 01	−.11023E + 01	.20750E + 00	−.28654E − 02	.24761E − 05	−.51142E − 11	.12905E − 20	U = .424307

ROOF TYPE 9: E0, E5, E4, B12, C14, E3, E2, A0
3″ INS. W/ 4″ L.W. CONC. & SUSP. CEIL.

	$n = 0$	$n = 1$	$n = 2$	$n = 3$	$n = 4$	$n = 5$	$n = 6$	
b_n	.31296E − 06	.24455E − 03	.21733E − 02	.25081E − 02	.55406E − 03	.23441E − 04	.15058E − 06	c_n = .005504
d_n	.10000E + 01	−.14060E + 01	.58814E + 00	−.90336E − 01	.44423E − 02	−.57037E − 04	.67593E − 07	U = .057246

ROOF TYPE 10: E0, E5, E4, C15, B16, E3, E2, A0
6″ L.W. ConC. W/ 0.15″ INS. & SUSP. CEIL.

	$n = 0$	$n = 1$	$n = 2$	$n = 3$	$n = 4$	$n = 5$	$n = 6$	
b_n	.29155E − 06	.24937E − 03	.24091E − 02	.30337E − 02	.73964E − 03	.35543E − .04	.28164E − 06	c_n = .006468
d_n	.10000E + 01	−.15570E + 01	.73120E + 00	−.11774E − 04	.59992E − 02	−.75519E − .04	.17266E − 06	U = .103708

ROOF TYPE 11: E0, C5, B15, E3, E2, A0
4″ H.W. CONC. W/ 6″ INS.

	$n = 0$	$n = 1$	$n = 2$	$n = 3$	$n = 4$	$n = 5$	$n = 6$	
b_n	.25044E − 06	.12506E − 03	.97422E − 03	.10173E − 02	.19801E − 03	.68017E − 05	.29359E − 07	c_n = .002322
d_n	.10000E + 01	−.16147E + 01	.79142E + 00	−.13242E + 00	.61096E − 02	−.81333E − 04	.36169E − 07	U = .046113

ROOF TYPE 12: E0, C13, B16, E3, E2, C12, A0
6″ H.W. CONC. −0.15″ INS. −2″ H.W. CONC. RTS

	$n = 0$	$n = 1$	$n = 2$	$n = 3$	$n = 4$	$n = 5$	$n = 6$	
b_n	.47450E − 04	.35559E − 02	.10578E − 01	.40431E − 02	.18888E − 03	.64267E − 06	.92034E − 10	c_n = .018414
d_n	.10000E + 01	−.15927E + 01	.72160E + 00	−.82751E − 01	.29414E − 03	−.16165E − 06	.37449E − 12	U = .396230

ROOF TYPE 13: E0, C13, B6, E3, E2, A0
6" H.W. CONC. W/ 2" INS.

	$n = 0$	$n = 1$	$n = 2$	$n = 3$	$n = 4$	$n = 5$	$n = 6$	
b_n	.18820E − 04	.13616E − 02	.37252E − 02	.12912E − 02	.55471E − 04	.17856E − 06	.24310E − 10	$c_n = $ .006453
d_n	.10000E + 01	−.13445E + 01	.44285E + 00	−.43441E − 01	.15834E − 03	−.68216E − 07	.16836E − 12	$U = $ .117195

ROOF TYPE 14: E0, E5, E4, C12, B13, E3, E2, A0
2" H.W. CONC. W/ 4" INS. & SUSP. CEIL.

	$n = 0$	$n = 1$	$n = 2$	$n = 3$	$n = 4$	$n = 5$	$n = 6$	
b_n	.48865E − 05	.46075E − 03	.14326E − 02	.56641E − 03	.30458E − 04	.15234E − 06	.35244E − 10	$c_n = $ .002495
d_n	.10000E + 01	−.13374E + 01	.41454E + 00	−.33463E − 01	.30859E − 03	−.14791E − 06	.43718E − 12	$U = $ .056736

ROOF TYPE 15: E0, E5, E4, C5, B6, E3, E2, A0
4" H.W. CONC. W/ 2" INS. & SUSP. CEIL.

	$n = 0$	$n = 1$	$n = 2$	$n = 3$	$n = 4$	$n = 5$	$n = 6$	
b_n	.98911E − 05	.66225E − 03	.16257E − 02	.48742E − 03	.18491E − 04	.63809E − 07	.95544E − 11	$c_n = $ .002806
d_n	.10000E + 01	−.12435E + 01	.28741E + 00	−.12738E − 01	.90481E − 04	−.46519E − 07	.11675E − 12	$U = $ .089674

ROOF TYPE 16: E0, E5, E4, C13, B20, E3, E2, A0
6" H.W. CONC. W/ 0.76" INS. & SUSP. CEIL.

	$n = 0$	$n = 1$	$n = 2$	$n = 3$	$n = 4$	$n = 5$	$n = 6$	
b_n	.61572E − 05	.59629E − 03	.19718E − 02	.86082E − 03	.54900E − 04	.37610E − 06	.12787E − 09	$c_n = $ .003490
d_n	.10000E + 01	−.13918E + 01	.46336E + 00	−.47139E − 01	.58021E − 03	−.12080E − 05	.35322E − 11	$U = $ .139616

U, bs, and cs are in Btu/(hr-ft²-F), and d is dimensionless. To convert U, bs, and cs to W/(m²-C), multiply by 5.6783.

Source: Reprinted by permission from ASHRAE *Cooling and Heating Load Calculation Manual*, 2nd ed., 1992.

n	b_n	d_n
0	0.000000	1.000000
1	0.000005	−2.008750
2	0.000134	1.371200
3	0.000436	−0.378967
4	0.000300	0.039616
5	0.000051	−0.001647
6	0.000002	0.000023

Now the CTF coefficients must be unnormalized. This requires the actual U-factor for the wall. Transfer function coefficients are unnormalized by multiplying the b coefficients by

$$\frac{U_{\text{actual}}}{U_{\text{from wall type}}} = \frac{0.15}{0.043} = 3.488$$

The unnormalized CTF coefficients are

n	b_n	d_n
0	0.000000	1.000000
1	0.000017	−2.008750
2	0.000465	1.371200
3	0.001517	−0.378967
4	0.001041	0.039616
5	0.000178	−0.001647
6	0.000007	0.000023

$$c_n = \Sigma b_n = 0.003237$$

8-4   DETERMINATION OF HEAT GAIN

This section discusses the calculation of heat gains due to conduction through walls, roofs, partitions, and so on, transmission of solar radiation through fenestration, and heat gains due to occupants, equipment, and lighting.

Solar Radiation

The calculation of solar irradiation is integral to calculation of sol-air temperatures and solar heat gain through windows. The calculations associated with the prediction of the total solar radiation incident on a surface, including the effect of shading, are covered in Chapter 6. The application of those relations to heat gain through walls, roofs, and fenestration will be covered here.

Exterior Walls and Roofs

A sampling of the transfer function coefficients is shown in Tables 8-7 and 8-8. The outside surface film coefficient used for the calculations was equal to 3.0 Btu/(hr-ft^2-F)

8.3 W/(m²-C). The time interval Δ was one hour. Sol-air temperatures should also be based on a film coefficient of 3.0 Btu/(hr-ft²-F) or 17 W/(m²-C).

The sol-air temperature as a function of time, based on a heat balance on an external surface, is required to solve Eq. 8-7 for the heat gain through a wall or roof. Equation 8-8 expresses the sol-air temperature as a function of the dry bulb temperature, emittance and absorptance of the surface, total incident solar radiation, and the convective heat transfer coefficient for the surface:

$$t_e = t_o + \alpha \frac{G_t}{h_o} - \frac{\epsilon \Delta R}{h_o} \qquad \text{(8-8)}$$

The term α/h_o varies from about 0.15 (hr-ft²-F)/Btu or 0.026 (m²-C)/W for a light-colored surface to a maximum of about 0.30 (hr-ft²-F)/Btu or 0.053 (m²-C)/W, whereas the term $\epsilon \Delta R/h_o$ varies from about zero for a vertical surface to about 7 F or 4 C for a horizontal surface. The total radiation incident on the surface is computed by the methods of Chapter 6.

The dry bulb temperature required in Eq. 8-8 is computed from the design dry bulb temperature and the daily range of temperature given in Table C-1 with the percentage of daily range given in Table 8-9 as

$$t_o = t_d - DR(X) \qquad \text{(8-9)}$$

where

t_d = design dry bulb temperature, F or C
DR = daily range, F or C
X = percentage of daily range divided by 100

Equation 8-7 can now be solved for a particular wall or roof using Eq. 8-8 and transfer coefficient data from Table 8-7 or 8-8. The last term in Eq. 8-7 is a constant because the inside design temperature is assumed to be constant. The only practical way of doing this is by digital computer; however, some of the equations will be set up to demonstrate the procedure. With a time interval of one hour, the value for which coefficients are available, Eq. 8-7 may be expanded to give 24 separate equations for θ

Table 8-9 Percentage of the Daily Range

Time, hr	Percent	Time, hr	Percent	Time, hr	Percent	Time, hr	Percent
1	87	7	93	13	11	19	34
2	92	8	84	14	3	20	47
3	96	9	71	15	0	21	58
4	99	10	56	16	3	22	68
5	100	11	39	17	10	23	76
6	98	12	23	18	21	24	82

Source: Reprinted by permission from ASHRAE *Cooling and Heating Load Calculation Manual*, 2nd ed., 1992.

ranging from 1 to 24. The first equation for 1:00 A.M. is

$$\frac{\dot{q}_{i,1}}{A} = \begin{bmatrix} b_0 t_{e,1} \\ + b_1 t_{e,24} \\ + b_2 t_{e,23} \\ + b_3 t_{e,22} \\ + \cdots \end{bmatrix} - \begin{bmatrix} + d_1 \dot{q}_{i,24} \\ + d_2 \dot{q}_{i,23} \\ + d_3 \dot{q}_{i,22} \\ + d_4 \dot{q}_{i,21} \\ + \cdots \end{bmatrix} - [t_i c_n] \qquad \textbf{(8-7a)}$$

The equation for 2:00 A.M. is

$$\frac{\dot{q}_{i,2}}{A} = \begin{bmatrix} + b_0 t_{e,2} \\ + b_1 t_{e,1} \\ + b_2 t_{e,24} \\ + b_3 t_{e,23} \\ + \cdots \end{bmatrix} - \begin{bmatrix} + d_1 \dot{q}_{i,1} \\ + d_2 \dot{q}_{i,24} \\ + d_3 \dot{q}_{i,23} \\ + d_4 \dot{q}_{i,22} \\ + \cdots \end{bmatrix} - [t_i c_n] \qquad \textbf{(8-7b)}$$

and the remaining 22 equations could be written in a similar way. The complete set of equations is solved sequentially starting with Eq. 8-7a and continuing until the solution to each set does not change more than a specified amount. Three or four calculation cycles are usually required for an acceptable solution. To start the calculation the unknown heat gains $\dot{q}_{i,\theta}$ are set equal to zero. The effect of this assumption is negligible after successive calculation cycles. The final result of these calculations is the heat gain to a space for a particular wall, roof, or partition as a function of time.

When Eq. 8-7 is used to compute the heat gain for a partition, floor, or ceiling, the sol-air temperation t_e must be changed to the temperature in the adjacent uncooled space t_a, and its variation with time must be known. If the temperature in the uncooled space is constant, Eq. 8-7 reduces to the steady-state situation given by

$$\dot{q}_i = UA(t_a - t_i) \qquad \textbf{(8-10)}$$

EXAMPLE 8-3

Compute the solar radiation and sol-air temperature for a southwest facing wall at 36 deg N latitude, 97 deg W longitude corresponding to the hour between 2:00 and 3:00 P.M. Local Solar Time on June 21. Assume an outdoor temperature of 97 F, surface absorptance of 0.9, and an air film coefficient of 4.0 Btu/(hr-ft²-F). Use solar radiation at 2:30 to represent the whole hour.

SOLUTION

Solar time = 2:30 P.M.

Hour angle, $h = 0.25$ deg/min $[(2 \times 60) + 30]$ min $= 37.5$ degrees

Declination, d $= +23.45$ deg, From Table 6-1

Solar altitude, $\beta = \sin^{-1} [\cos (36) \cos (23.45) \cos (37.5)$

$\qquad + \sin (36 \sin (+23.45)] = 55.4$ degrees

$$\text{Solar azimuth, } \phi = \cos^{-1}\left[\frac{(\sin \beta \sin l) - \sin d}{\cos \beta \cos l}\right] = \pm 79.3 \text{ degrees}$$

Angles to the west of south are positive; therefore, $\phi = 79.3$ deg

$$\text{Wall azimuth, } \psi = 45 \text{ deg (facing SW)}$$

$$\text{Wall solar azimuth, } \gamma = \phi - \psi = 79.3 - 45 = 34.3 \text{ deg}$$

$$\text{Angle of incidence, } \theta = \cos^{-1}[\cos \beta \cos \gamma] = \cos^{-1}[\cos 55.4 \cos 34.3]$$

$$= 61.7 \text{ deg}$$

$$\text{Direct normal irradiation, } G_{DN} = Ae^{(-B/\sin \beta)}$$

where

$A = 346.3$, Table 6-1
$B = 0.185$, Table 6-1
$C = 0.137$, Table 6-1

Then

$$G_{DN} = 346.3e^{-(.185/\sin 55.4)} = 276.6 \text{ Btu/(hr-ft}^2\text{-F)}$$

$$\text{Direct irradiation, } G_D = G_{DN}\cos \theta = 131.1 \text{ Btu/(hr-ft}^2\text{)}$$

The diffuse radation is computed as follows: For vertical surfaces the ratio between diffuse radiation incident on the surface and diffuse radiation incident on a horizontal surface is given by Figure 6-9 with $\cos \theta = 0.474$.

$$Y = G_{dv}/G_{dH} = 0.83$$

$$\text{Diffuse irradiation, } G_d = G_{ds} + G_{dg}$$
$$= CYG_{DN} + G_{DN}(C + \sin \beta) \times \rho_g \times 0.5$$

$$= (0.137 \times 0.83 \times 276.9) + 276.9\,(0.1237 + 0.823) \times 0.2 \times 0.5$$

$$G_d = 31.4 + 26.2 = 57.6 \text{ Btu/(hr-ft}^2\text{)}$$

$$G_t = G_D + G_d = 131.1 + 57.6 = 188.7 \text{ Btu/(hr-ft}^2\text{)}$$

$$\text{Sol-air temperature, } t_e = t_o + \frac{\alpha G_t}{h_o} + \frac{\epsilon \Delta R}{h_o}$$

$$= 97 + 0.9\frac{(188.7)}{4.0} - \frac{0.0}{4.0} = 139.5.\text{F}$$

where $\epsilon \Delta R$ is a typical value for a dark-colored vertical surface (6).

Once the sol-air temperatures have been determined, Eq. 8-7 is used to calculate the hourly heat gain through walls, roofs, floors, and partitions, as discussed above. Example 8-4 demonstrates the calculation procedure.

EXAMPLE 8-4

Compute the conduction heat gain for a wall composed of 4-in. face brick, 8-in. normal weight concrete block, $\frac{3}{4}$-in. polystyrene insulation with 1-in. furring strips (spaced 16 in.) and $\frac{1}{2}$-in. gypsum wall board under specified design conditions.

SOLUTION

As determined in Example 8-2, the appropriate CTF coefficients are

n	b_n	d_n
0	0.000000	1.000000
1	0.000017	−2.008750
2	0.000465	1.371200
3	0.001517	−0.378967
4	0.001041	0.039616
5	0.000178	−0.001647
6	0.000007	0.000023

$$c_n = \Sigma b_n = 0.003237$$

The design conditions (sol-air temperatures) are specified in Table 8-10 and the room temperature is 78 F. The heat flux history terms were initially assumed zero. Equation 8-7 was then calculated for successive 24-hour periods and convergence was essentially achieved by the end of the fourth 24-hour period. The conduction heat gains are shown in Table 8-10.

Fenestration

Solar heat gain factors (SHGF), the hourly solar heat gains that occur due to a unit area of double-strength sheet glass (DSA) for a given orientation and time, were developed in Chapter 6 with other discussion of solar radiation. The relations developed there are used to compute the solar component of heat gain for windows and other light-transmitting openings.

Internal Sources

These internal sources fall into the general categories of people, lights, and miscellaneous equipment.

People

The heat gain from human beings has two components, sensible and latent. The total and relative amounts of sensible and latent heat vary depending on the level of activity. Table 8-11 gives heat gain data from occupants in conditioned spaces. These data are based on metabolic heat generation data from the comfort and health chapter. Note that

Table 8-10 Design Conditions and Results for
Example 8-4

Time, hr	Sol-Air Temperature, F	Conduction Heat Gain, Btu/(hr-ft^2)
1	77.1	3.80
2	76.1	3.61
3	75.3	3.38
4	74.7	3.13
5	74.5	2.87
6	76.6	2.61
7	80.4	2.35
8	84.2	2.11
9	88.5	1.88
10	93.1	1.69
11	97.8	1.54
12	106.4	1.44
13	120.5	1.40
14	131.0	1.41
15	136.8	1.49
16	136.7	1.66
17	130.8	1.93
18	118.1	2.30
19	97.7	2.72
20	85.1	3.15
21	82.9	3.54
22	80.9	3.81
23	79.3	3.94
24	78.1	3.92

the data in the last three columns are adjusted, based on the normally expected percentages of men, women, and children for the listed application. These data are recommended for usual load calculations. Although the data of Table 8-11 are quite accurate, large errors are often made in the computation of heat gain from occupants because of poor estimates of the periods of occupancy or the number of occupants. Great care should be taken to be realistic about the allowance for the number of people in a structure. It should be kept in mind that rarely will a complete office staff be present or a classroom be full. On the other hand, a theater may often be completely occupied and sometimes may contain more occupants than it is designed for. Each design problem must be judged on its own merits. With the exception of theaters and other high-occupancy spaces, most spaces are designed with too large an allowance for its occupants. One should not allow for more than the equivalent full-time occupants.

The latent heat gain is computed from

$$\dot{q}_1 = NF_u \dot{q}_1'$$

(8-11)

Table 8-11 Rates of Heat Gain from Occupants of Conditioned Spaces[a]

Degree of Activity	Typical Application	Total Heat Adults, Male		Total Heat Adjusted,[b]		Sensible Heat		Latent Heat	
		Btu/hr	W	Btu/hr	W	Btu/hr	W	Btu/hr	W
Seated at theater	Theater—Matinee	390	114	330	97	225	66	105	31
Seated at theater	Theater—Evening	390	114	350	103	245	72	105	31
Seated, very light work	Offices, hotels, apartments	450	132	400	117	245	72	155	45
Moderately active office work	Offices, hotels, apartments	475	139	450	132	250	73	200	59
Standing, light work; walking	Department store, retail store	550	162	450	132	250	73	200	59
Walking; standing	Drugstore, bank	550	162	500	146	250	73	250	73
Sedentary work	Restaurant[c]	490	144	550	162	275	81	275	81
Light bench work	Factory	800	235	750	220	275	81	475	139
Moderate dancing	Dance hall	900	264	850	249	305	89	545	160
Walking 3 mph; light machine work	Factory	1000	293	1000	293	375	110	625	183
Bowling[d]	Bowling alley	1500	440	1450	425	580	170	870	255
Heavy work	Factory	1500	440	1450	425	580	170	870	255
Heavy machine work; lifting	Factory	1600	469	1600	469	635	186	965	283
Athletics	Gymnasium	2000	586	1800	528	710	208	1090	320

[a]Tabulated values are based on 75 F room dry-bulb temperature. For 80 F room dry-bulb, the total heat remains the same, but the sensible heat values should be decreased by approximately 20%, and the latent heat values increased accordingly.

[b]Adjusted heat gain is based on normal percentage of men, women, and children for the application listed, with the postulate that the gain from an adult female is 85% of that for an adult male, and that the gain from a child is 75% of that for an adult male.

[c]Adjusted total gain for Sedentary work, Restaurant, includes 60 Btu/hr for food per individual (30 Btu/hr sensible and 30 Btu/hr latent).

[d]For Bowling, figure one person per alley actually bowling, and all others as sitting (400 Btu/hr) or standing and walking slowly (550 Btu/hr).

Source: Reprinted by permission from ASHRAE Cooling and Heating Load Calculation Manual, 2nd ed., 1992.

and the sensible heat gain is given by

$$\dot{q}_s = N F_u \dot{q}'_s \qquad (8\text{-}12)$$

where

N = maximum number of people expected to occupy the space
F_u = use factor, unity for a room
$\dot{q}'$ = heat gain per person, Table 8-11, Btu/(hr-person) or W

The latent and sensible heat gain for occupants should be computed separately except in the case of estimating the building refrigeration load, where the two components may be combined. The latent heat gain is assumed to instantly become cooling load, whereas the sensible heat gain is partially delayed depending on the nature of the conditioned space. The sensible heat gain for people generally is assumed to be 30 percent convective (instant cooling load) and 70 percent radiative (the delayed portion).

Lights

Since lighting is often the major internal load component, an accurate estimate of the space heat gain it imposes is needed. The rate of heat gain at any given moment can be quite different from the heat equivalent of power supplied instantaneously to those lights.

Some of the energy emitted by the lights is in the form of radiation that is absorbed in the space. The absorbed energy is later transferred to the air by convection. The manner in which the lights are installed, the type of air distribution system, and the mass of the structure are important. Obviously a recessed light fixture will tend to transfer heat to the surrounding structure, whereas a hanging fixture will transfer more heat directly to the air. Some light fixtures are designed so that air returns through them, absorbing heat that would otherwise go into the space. Lights are often turned off to save energy, which makes accurate computations difficult. Lights left on 24 hours a day approach an equilibrium condition where the cooling load equals the power input.

The primary source of heat from lighting comes from the light-emitting elements, or lamps, although significant additional heat may be generated from associated components in the light fixtures housing such lamps. Generally, the instantaneous rate of heat gain from electric lighting may be calculated from

$$\dot{q} = 3.41 \, W F_u F_s \qquad (8\text{-}13)$$

where

W = total installed light wattage
F_u = use factor, ratio of wattage in use to total installed wattage
F_s = special allowance factor (ballast factor in the case of fluorescent and metal halide fixtures)

The *total light wattage* is obtained from the ratings of all lamps installed, both for general illumination and for display use.

The *use factor* is the ratio of the wattage in use, for the conditions under which the load estimate is being made, to the total installed wattage. For commercial applications such as stores, the use factor would generally be unity.

The *special allowance factor* is for fluorescent and metal halide fixtures or for fixtures that are ventilated or installed so that only part of their heat goes to the conditioned space. For *fluorescent fixtures,* the special allowance factor accounts primarily for ballast losses and can be as high as 2.19 for 32-W single-lamp high-output fixtures on 277-V circuits. Rapid-start, 40-W lamp fixtures have special allowance factors varying from a low of 1.18 for two lamps at 277 V to a high of 1.30 for one lamp at 118 V, with a recommended value of 1.20 for general applications. *Industrial fixtures* other than fluorescent, such as sodium lamps, may have special allowance factors varying from 1.04 to 1.37, depending on the manufacturer.

For *ventilated* or *recessed fixtures,* manufacturer's or other data must be sought to establish the fraction of the total wattage expected to enter the conditioned space directly (and subject to time lag effect) versus that which must be picked up by return air or in some other appropriate manner. For ordinary design load estimation, the heat gain for each component may simply be calculated as a fraction of the total lighting load, by using judgment to estimate *heat-to-space* and *heat-to-return* percentages. The heat from fixtures ranges from 40 to 60 percent heat-to-return for ventilated fixtures or 15 to 25 percent for unventilated fixtures.

The heat gain to the space from fluorescent fixtures is often assumed to be divided 59 percent radiative and 41 percent convective (3). The heat gain from incandescent fixtures is typically assumed to be divided 80 percent radiative and 20 percent convective (6).

Miscellaneous Equipment

Estimates of heat gain in this category tend to be even more subjective than for people and lights. However, considerable data are available, which, when used judiciously, will yield reliable results (5). Careful evaluation of the operating schedule and the load factor for each piece of equipment is essential.

When equipment is operated by electric motor within a conditioned space, the heat equivalent is calculated as

$$\dot{q}_m = C(P/E_m)F_1F_u \tag{8-14}$$

where

$\dot{q}_m$ = heat equivalent of equipment operation, Btu/hr or kW
P = motor power rating (shaft), horsepower or kW
E_m = motor efficiency, as decimal fraction <1.0
F_l = motor load factor
F_u = motor use factor
C = constant, 2545 or 1.0

The *motor use factor* may be applied when motor use is known to be intermittent with significant nonuse during all hours of operation (i.e., overhead door operator). For conventional applications, its value is 1.0.

The motor load factor is the fraction of the rated load delivered under the conditions of the cooling load estimate. In Eq. 8-14, both the motor and the driven equipment are assumed to be within the conditioned space. If the motor is outside the space or airstream with the driven equipment within the conditioned space,

$$\dot{q}_m = C(P)F_l F_u \tag{8-15}$$

When the motor is inside the conditioned space or airstream but the driven machine is outside,

$$\dot{q}_m = C(P)[(1.0 - E_m)/E_m]F_l F_u \tag{8-16}$$

Equation 8-16 also applies to a fan or pump in the conditioned space that exhausts air or pumps fluid outside that space.

Unless the manufacturer's technical literature indicates otherwise, the heat gain should be divided about 70 percent radiant and 30 percent convective for subsequent cooling load calculations (3).

In a cooling load estimate, heat gain from all appliances—electric, gas, or steam— should be taken into account. The tremendous variety of appliances, applications, usage schedules, and installations makes estimates very subjective.

To establish a heat gain value, actual input data values and various factors, efficiencies, or other judgmental modifiers are preferred. Where no data are available, the maximum hourly heat gain can be estimated as 50 percent of the total nameplate or catalog input ratings because of the diversity of appliance use and the effect of thermostatic controls, giving a usage factor of 0.50. Radiation contributes up to 32 percent of the heat gain for hooded appliances. The convective heat gain is assumed to be removed by the hood. Therefore, the heat gain may be estimated for hooded, steam, and electric appliances to be

$$\dot{q}_a = 0.5(0.32)\dot{q}_i = 0.16\dot{q}_i \tag{8-17}$$

where $\dot{q}_i$ is the catalog input rating.

Direct fuel-fired cooking appliances require more heat input than electric or steam equipment of the same type and size. In the case of gas fuel, the American Gas Association established an overall figure of approximately 60 percent more. Where appliances are installed under an effective hood, only radiant heat adds to the cooling load; convected and latent heat from the cooking process and combustion products are exhausted and do not enter the kitchen. It is therefore necessary to adjust Eq. 8-17 for use with hooded fuel-fired appliances, to compensate for the 60 percent higher input ratings, since the appliance surface temperatures are the same and the extra heat input from combustion products are exhausted to outdoors. This correction is made by the introduction of a flue loss factor of 1.60. Then, for hooded fuel-fired appliances,

$$\dot{q}_a = (0.16/1.6)\dot{q}_i = 0.10\dot{q}_i \tag{8-18}$$

Reference 5 gives recommended rates of heat gain for restaurant equipment, both hooded and unhooded. For unhooded appliances the sensible heat gain is often divided 70 percent radiant and 30 percent convective for cooling load estimates (3). In the case of hooded appliances, all the heat gain to the space is assumed to be radiant for cooling load calculation.

As with large kitchen installations, hospital and laboratory equipment is a major source of heat gain in conditioned spaces. Care must be taken in evaluating the probability and duration of simultaneous usage when many components are concentrated in one area, such as in a laboratory, operating room, and so on. The chapters related to health facilities and laboratories in the *ASHRAE Handbook, Applications Volume* (9) should be consulted for further information.

Reference 5 gives recommended rates of heat gain for hospital equipment. The sensible heat gain is approximately 70 percent radiant and 30 percent convective (3).

Offices in general with computer display terminals at most desks, and other typical equipment such as personal computers, printers, and copiers, have heat gains ranging up to 15 Btu/(hr-ft^2) or 0.05 kW/m^2.

Computer rooms housing mainframe or minicomputer equipment must be considered individually. Computer manufacturers have data pertaining to various individual components. In addition, computer usage schedules and the like should be considered. The chapter related to data processing systems of the *ASHRAE Handbook, Applications Volume* should be consulted for further information about design of large computer rooms and facilities. Reference 5 gives recommended rates of heat gain for office equipment.

The heat gain from office equipment is assumed to be split approximately 70 percent radiative and 30 percent convective for cooling load calculation.

Air Distribution System

The heat gains of the duct system must be considered when the ducts are not in the conditioned space. Proper insulation will reduce these losses but cannot completely eliminate them. The heat gain may be estimated using the following relation:

$$\dot{q} = UA_s\Delta t_m \tag{8-19}$$

where

U = overall heat transfer coefficient, Btu/(hr-ft^2-F) or W/(m^2-C)
A_s = outside surface area of the duct, ft^2 or m^2
Δt_m = mean temperature difference between the air in the duct and the environment, degree F or C

When the duct is covered with 1 or 2 in. of insulation with a reflective covering, the heat gain will usually be reduced sufficiently to assume that the mean temperature difference is equal to the difference in temperature between the supply air temperature and the environment temperature. Since the duct surface area is not known at this point, it is a common practice to simply assume that a small percentage of the sensible load is a loss or a gain. When the ducts are insulated, 1 to 3 percent is reasonable. These heat gains are often included in the psychrometric analysis rather than the space load analysis.

Heat Gain from Infiltration

The necessity of introducing outdoor air was discussed in Chapter 4 and current recommendations were given. Computation of the heat gain from outdoor-make-up air is properly a part of the air quantity calculation, as discussed in Chapter 3; however, it may be computed separately to obtain an estimate of the total load. In such a case the method given below for infiltration may be used.

The methods used to estimate the quantity of infiltration air were discussed in Chapter 7 when the heating load was considered. The same methods apply to cooling load calculations. Both a sensible and latent heat gain will result and are computed as follows:

$$\dot{q}_s = \dot{m}_a c_p (t_o - t_i) = \frac{\dot{Q}_o c_p}{\upsilon_o}(t_o - t_i) \tag{8-20}$$

$$\dot{q}_l = \dot{m}_a (W_o - W_i) i_{fg} = \frac{\dot{Q}_o}{\upsilon_o}(W_o - W_i) i_{fg} \tag{8-21}$$

These heat gains may also be computed using the psychrometric chart described in Chapter 3.

Wind velocities are lower in the summer, making an appreciable difference in the computed infiltration rate. The direction of the prevailing winds usually changes from winter to summer. This should be considered in making infiltration estimates because the load will be imposed mainly in the space where the air enters. During the summer, infiltration will enter the upper floors of high-rise buildings instead of the lower floors.

8-5   CONVERSION OF HEAT GAINS INTO COOLING LOAD

Where the heat gain is given at distinct time intervals, the corresponding cooling load can be related to the current value of heat gain and the preceding values of cooling load and heat gain by Eq. 8-5. The coefficients are chosen from the database using the routines described previously (5). The thermal characteristics of the zone are specified as zone parameters for the database.

The procedure for converting heat gains into cooling load is essentially the same for each type of heat gain. Once the hourly heat gains have been determined, and the zone parameters have been specified, four sets of weighting factors are required:

1. Solar—applied to transmitted solar heat gain.
2. Conduction—applied to conduction loads from exterior surfaces, interior partitions, and absorbed solar heat gains.
3. Lighting—applied to lighting loads.
4. Occupant or equipment—applied to either occupant or equipment sensible loads.

All that remains to be done is to select the correct weighting factors for each type of load and apply Eq. 8-5.

Solar and Conduction Heat Gains

The determination of transmitted and absorbed solar heat gains was described earlier. The transmitted solar heat gains are converted to cooling loads with the use of υ and w coefficients labeled *solar*. The absorbed solar heat gains are converted to cooling loads with the use of υ and w coefficients labeled *conduction*. (Absorbed solar heat gain in glass is conducted through the glass into the space, where it is convected and radiated in the same manner as heat conducted through walls and roofs. Therefore, the room response is very similar and the same coefficients are used.)

Occupant and Equipment Heat Gains

Quantifying heat gains due to people and equipment was described before. For any zone type, the coefficients apply to both people and equipment. This is due to the fact that the heat gain is assumed to be split 70 percent radiant and 30 percent convective for both people and equipment. If in fact the split is significantly different, the coefficients may be corrected as described later.

Lighting

Heat gains due to lighting was described before and are converted to cooling load using the v and w coefficients labeled *lighting*. The heat gain of the lighting is assumed to be split 59 percent radiant and 41 percent convective. This corresponds to a recessed fluorescent troffer. If in fact the split is significantly different, the coefficients may be corrected as described later.

EXAMPLE 8-5

Convert the lighting heat gain to cooling load for the store in the strip shopping mall described in Example 8-1. This store has 10,000 Btu/hr of lighting on from 10 A.M. to 10 P.M. solar time.

SOLUTION

Using the zone characterization parameters determined in the example, the following WF coefficients for lighting are given in the database:

$$v_0 = 0.61116$$
$$v_1 = -0.67267$$
$$v_2 = 0.12987$$
$$w_1 = -1.19718$$
$$w_2 = 0.26554$$

Using Eq. 8-5 and assuming previous heat gains and cooling loads to be zero, the cooling load profile can be calculated in an iterative manner until the results converge. Table 8-12 shows the results of the cooling load calculation using the room transfer function coefficients, Eq. 8-5, and a heat gain of 10,000 Btu/hr from hour 10 through hour 21 (10:00 A.M. to 10:00 P.M.). The cooling loads essentially converge by the fifth iteration.

Table 8-12 Results from WF Calculation, Example 8-5

Hour	Load	Hour	Load	Hour	Load	Hour	Load	Hour	Load
1	0	25	2066	49	2243	73	2258	97	2259
2	0	26	1848	50	2007	74	2021	98	2022
3	0	27	1663	51	1807	75	1820	99	1821
4	0	28	1501	52	1631	76	1642	100	1643
5	0	29	1355	53	1472	77	1483	101	1483
6	0	30	1223	54	1330	78	1339	102	1340
7	0	31	1105	55	1201	79	1209	103	1210
8	0	32	998	56	1085	80	1092	104	1093
9	0	33	901	57	980	81	986	105	987
10	0	34	814	58	885	82	891	106	891
11	6112	35	6847	59	6911	83	6916	107	6917
12	6702	36	7366	60	7423	84	7428	108	7429
13	7084	37	7683	61	7736	85	7740	109	7740
14	7385	38	7926	62	7973	86	7977	110	7978
15	7643	39	8132	63	8175	87	8179	111	8179
16	7873	40	8315	64	8353	88	8357	112	8357
17	8079	41	8479	65	8513	89	8516	113	8516
18	8266	42	8626	66	8657	90	8660	114	8660
19	8434	43	8759	67	8787	91	8790	115	8790
20	8585	44	8879	68	8905	92	8907	116	8907
21	8722	45	8988	69	9011	93	9013	117	9013
22	8846	46	9086	70	9107	94	9108	118	9109
23	2846	47	3063	71	3081	95	3083	119	3083
24	2357	48	2553	72	2570	96	2571	120	2571

Nonstandard Radiative/Convective Split

There may be certain conditions for which the heat gain due to people, lights, and equipment may be significantly different from the standard values. An example is a hooded appliance for which the convective heat is removed by the hood and the remaining heat gain is essentially 100 percent radiative. For these cases a correction may be made as follows:

1. Determine the actual composition of the heat gain (percent radiative/percent convective).
2. Let r = actual radiative fraction ($0 \leq r \leq 1$).
   Let r_1 = the radiative fraction on which the weighting factors are based.
3. Then modify the weighting factors as follows:

$$v'_0 = \left(\frac{r}{r_1} v_0 + \left(1 - \frac{r}{r_1}\right) \right) \tag{8-22a}$$

$$v'_1 = \left(\frac{r}{r_1} v_1 + \left(1 - \frac{r}{r_1}\right) \right) w_1 \tag{8-22b}$$

$$v_2' = \left(\frac{r}{r_1} v_2 + \left(1 - \frac{r}{r_1}\right) w_2\right) \tag{8-22c}$$

where v_0', v_1', v_2', represent the modified weighting factors. The values for w_1 and w_2 remain the same.

EXAMPLE 8-6

Correct the weighting factors for nonstandard radiative/convective split as described below. A zone that can be described with the zone parameters shown in Table 8-13 has weighting factors for people and equipment as shown in Table 8-14. One piece of equipment in the zone is a hooded range with four burners. The heat gain to the space is assumed to be 100 percent radiative. Then r, the actual radiative fraction, is 1.0.

SOLUTION

Applying the equations from step 3 above results in the new set of weighting factors given in Table 8-15. Assuming an actual heat gain to the space of 1000 Btu/hr for the

Table 8-13 Zone Parameters for Example Zone

zg	1	ec	2	st	1
zh	1	gl	1	fc	2
nw	1	pt	2	rt	2
is	1	zl	3	ct	2
fn	1	mf	2		

Table 8-14 Weighting Factors for Example Zone

v_0	0.57748
v_1	−0.63235
v_2	0.11875
w_1	−1.20074
w_2	0.26462

Table 8-15 Revised Weighting Factors for Example Zone

v_0	0.39640
v_1	−0.38875
v_2	−0.056234
w_1	−1.20074
w_2	0.26462

Table 8-16 Cooling Loads for Example Zone[a]

Hour	Heat Gain	Unmodified Cooling Load	Modified Cooling Load
1	0	143	205
2	0	131	186
3	0	119	170
4	0	108	154
5	0	98	140
6	0	89	128
7	1000	659	513
8	1000	713	589
9	1000	745	636
10	1000	770	672
11	1000	791	702
12	1000	810	729
13	1000	827	753
14	1000	843	776
15	1000	857	796
16	1000	870	814
17	1000	882	831
18	0	315	450
19	0	264	376
20	0	233	333
21	0	210	300
22	0	191	272
23	0	173	248
24	0	158	225

[a]Heat gain and cooling loads in Btu/hr.

seventh through the seventeenth hours of the day and performing the weighting factor calculation gives the results shown in Table 8-16. For comparison purposes, the third column gives the results if the unmodified weighting factors are used and the fourth column gives the results if the modified weighting factors are used.

Summary

At this point where hourly cooling loads have been determined for each type of heat gain, heat gain due to infiltration must also be accounted for. Heat gain due to infiltration is assumed to instantaneously become a cooling load, that is, the cooling load due to infiltration is equal to the heat gain. For many cases, all that remains to be done is to sum the cooling loads for each zone for each hour. However, if zone temperatures are varied significantly, it may be desirable to calculate the heat extraction rate, as described in the next section.

8-6 HEAT EXTRACTION RATE AND ROOM TEMPERATURE

The cooling load calculation procedure discussed in the previous section produces data that may be used to compute the heat extraction rate and the room temperature obtained with a particular terminal unit. This is accomplished by the use of a transfer function in a somewhat analogous manner to that used to find the cooling load. When the cooling equipment removes heat energy from the space air at a rate equal to the cooling load, the space air temperature will remain constant and the heat extraction rate $\dot{q}_x$ is equal to the cooling load $\dot{q}_c$. However, this is seldom true. A simple energy balance on the space air yields

$$\dot{q}_x - \dot{q}_c = (mc)_{air} \frac{dt_r}{d\theta} \tag{8-23}$$

Equation 8-23 shows the general behavior of the system; however, the actual process is complex because of the dependence of $\dot{q}_x$ and $\dot{q}_c$ on the air temperature and time. Therefore, a transfer function has been devised, which has an obvious similarity to Eq. 8-23, to describe the process, and a digital computer has been utilized to compute the transfer function coefficients. The room air transfer function is

$$\sum_{i=0}^{1} p_i(\dot{q}_{x,\theta-i\Delta} - \dot{q}_{c,\theta-i\Delta}) = \sum_{i=0}^{2} g_i(t_i - t_{r,\theta-i\Delta}) \tag{8-24}$$

where

p_i, g_i = transfer function coefficients
$\dot{q}_c$ = cooling load at the various times
t_i = room temperature used for cooling load calculations
t_r = actual room temperature at the various times

The room air transfer function coefficients are given in Table 8-17. Note that the g coefficients have been normalized to unit floor area. The coefficients g_0 and g_1 depend on the heat conductance to the surroundings UA and the infiltration and ventilation rate to the space:

Table 8-17 Normalized Coefficients of the Room Air Transfer Function[a]

Room Envelope Construction	Btu/(hr-ft²-F)[b]			Dimensionless	
	g_0^*	g_1^*	g_2^*	p_0	p_1
Light	+1.68	−1.73	+0.05	1.0	−0.82
Medium	+1.81	−1.89	+0.08	1.0	−0.87
Heavy	+1.85	−1.95	+0.10	1.0	−0.93

[a]For all cases, the room is assumed furnished.
[b]To convert gs to W/(m²-C) multiply by 5.6783.
Source: Reprinted by permission from ASHRAE *Cooling and Heating Load Calculation Manual,* 2nd ed., 1992.

$$g = g_0^* A_{fl} + [UA + \rho c_p(\dot{Q}_i + \dot{Q}_v)]p_0 \qquad \text{(8-25a)}$$

$$g_1 = g_1^* A_{fl} + [UA + \rho c_p(\dot{Q}_i + \dot{Q}_v)]P_1 \qquad \text{(8-25b)}$$

where

g_i^* = normalized g coefficients, Btu/(hr-ft^2-F) or W/(m^2-C)
A_{fl} = floor area, ft^2 or m^2
UA = conductance between the space and the surrounding (average), Btu/(hr-F) or W/C
$\dot{Q}_i$ = infiltration rate, ft^3/hr or m^3/s
$\dot{Q}_v$ = ventilation rate, ft^3/hr or m^3/s
p_i = coefficients from Table 8-17, dimensionless

The remaining g coefficient, g_2, is given by

$$g_2 = g_2^* A_{fl} \qquad \text{(8-26)}$$

The next step is to describe the characteristic of the terminal unit. This can often be done by a linear expression of the form

$$\dot{q}_{x,\theta} = W + S t_{r,\theta} \qquad \text{(8-27)}$$

where W and S are parameters that characterize the equipment at time θ. The equipment being modeled here is actually the cooling coil and the associated control system that matches the coil load to the space load. A face and bypass system would be an example.

The ability of the cooling unit to extract heat energy from the air varies from some minimum to some maximum value over the throttling range of the control system (thermostat). Equation 8-27 is valid only in the throttling range. The throttling range Δt_r may be described as the temperature at which the thermostat calls for maximum cooling minus the temperature at which the thermostat calls for minimum cooling. Figure 8-5

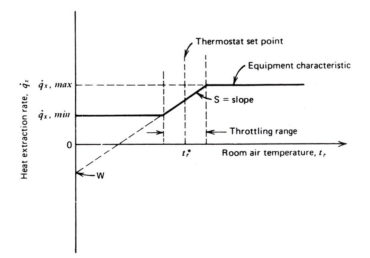

Figure 8-5 Cooling equipment characteristic.

shows the relationship of the heat extraction rate $\dot{q}_x$ and the room air temperature according to Eq. 8-27. When the room air temperature lies outside the throttling range, the heat extraction rate is either $\dot{q}_{x,\min}$ or $\dot{q}_{x,\max}$ depending on whether t_r is above or below the set-point temperature t_r^*. The value of S is the slope of the characteristic in the throttling range:

$$S = \frac{[\dot{q}_{x,\max} - \dot{q}_{x,\min}]}{\Delta t_r} \qquad (8\text{-}28)$$

The value of W is the intercept of Eq. 8-27 with the $\dot{q}_x$ axis and is defined by

$$W = \frac{[\dot{q}_{x,\max} + \dot{q}_{x,\min}]}{2} - St_{r,\theta}^* \qquad (8\text{-}29)$$

where $t_{r,\theta}^*$, the set-point temperature, is assumed to be in the center of the throttling range. The set-point temperature may be changed during the day.

Equations 8-24 and 8-27 may be combined and solved for $\dot{q}_{x,\theta}$:

$$\dot{q}_{x,\theta} = \frac{g_o W}{S + g_o} + \frac{S}{S + g_o} G_\theta \qquad (8\text{-}30)$$

where

$$G_\theta = t_i \sum_{i=0}^{2} g_i - \sum_{i=1}^{2} g_i(t_{y,\theta - i\Delta}) + \sum_{i=0}^{1} p_i(\dot{q}_{c,\theta - i\Delta}) - \sum_{i=1}^{1} p_i(\dot{q}_{x,\theta - i\Delta}) \qquad (8\text{-}31)$$

When the value of $\dot{q}_{x,\theta}$ computed by Eq. 8-30 is greater than $\dot{q}_{x,\max}$ it is made equal to $\dot{q}_{x,\max}$, and when it is less than $\dot{q}_{x,\min}$ it is made equal to $\dot{q}_{x,\min}$. Finally, Eqs. 8-27 and 8-30 can be combined and solved for $t_{r,\theta}$:

$$t_{r,\theta} = \frac{(G_\theta - \dot{q}_{x,\theta})}{g_o} \qquad (8\text{-}32)$$

The solution of Eq. 8-30 is carried out in the same manner as Eqs. 8-5 and 8-7 for the heat gain and cooling load discussed in Section 8-4. To summarize, Eq. 8-30 is written for each hour of the day and solved successively beginning with hour 1. Cyclic calculations are carried on until there is no significant change in $\dot{q}_{x,\theta}$ or $t_{r,\theta}$ for each hour. The transfer function coefficients are all constant for a given space unless the infiltration and ventilation rates change with time. In this case the values of g_0 and g_1 must be recalculated for each set of conditions.

Figure 8-6 illustrates results obtained by using the foregoing method. Both continuous and intermittent operation are shown using somewhat undersized cooling equipment. Notice that only a small sacrifice in space temperature results from intermittent operation, but the total heat energy removed seems to be less, which would reduce operation costs.

The transfer function method just described is useful in finding the space-cooling load and the coil load on an hour-to-hour basis for a design day. The same method may

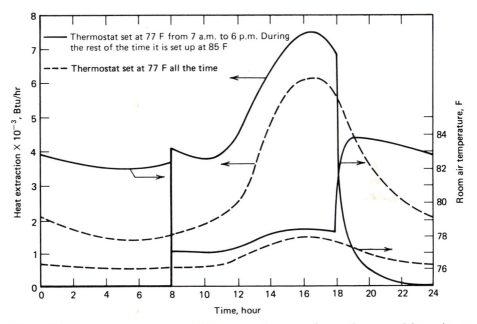

Figure 8-6 Room air temperature and heat-extraction rates for continuous and intermittent operation.

be used to find design heating loads. In such a case the method is identical, except that the solar heat gain and the internal heat gains should be set equal to zero.

8-7 THE CLTD/SCL/CLF METHOD

The methodology discussed in this section is a manual load calculation procedure that produces cooling load on an hourly basis if desired.

The CLTD/SCL/CLF method makes use of a temperature difference (CLTD) in the case of walls and roofs and solar cooling load factors (SCL) in the case of solar gain through windows and cooling load factors (CLF) for internal heat sources. The CLTD, SCL, and CLF vary with time and are a function of environmental conditions and building parameters. They have been derived from computer solutions using the transfer function procedure. A great deal of care has been taken to sample a wide variety of conditions in order to obtain reasonable accuracy. In general, calculations proceed as follows:

For walls and roofs

$$\dot{q}_\theta = UA(\text{CLTD})_\theta \tag{8-33}$$

where

U = overall heat transfer coefficient, Btu/(hr-ft^2-F) or W/(m^2-C)
A = area, ft^2 or m^2
CLTD = temperature difference which gives the cooling load at time θ, F or C

CLTD – cooling load temp. difference
SCL – solar cooling load factor
CLF – cooling load factor

The CLTD accounts for the thermal response (lag) in the heat transfer through the wall or roof, as well as the response (lag) due to radiation of part of the energy from the interior surface of the wall to objects within the space.

For solar gain through glass

$$\dot{q}_\theta = A(SC)SCL_\theta \qquad (8\text{-}34)$$

where

A = area, ft^2 or m^2
SC = shading coefficient (internal shade)
SCL_θ = solar cooling load factor, Btu/(hr-ft^2-F) or W/(m^2-C)

The SCL accounts for the variation of the solar heat gain with time, the massiveness of the structure, and the geographical location. Again the SCL accounts for the thermal response (lag) of the radiant part of the solar input:

For internal heat gains

$$\dot{q}_\theta = \dot{q}_i(CLF)_\theta \qquad (8\text{-}35)$$

where

$\dot{q}_i$ = instantaneous heat gain from lights, people, and equipment, Btu/hr or W
CLF_θ = cooling load factor for time θ

The CLF accounts for the thermal response of the space to the various internal heat gains and is slightly different for each.

If the CLTD/SCL/CLF procedure has been adapted to a computer or when the transfer function method is used directly, the cooling load for each hour of the day is computed and the hours of interest may be selected. However, when the calculations are made by hand, calculations for all 24 hours in the day are too time consuming. In such a case, the time of day when the peak cooling load will occur may be estimated. In fact, two different types of peaks may need to be determined. First, the time of the peak load for each room is needed in order to compute the air quantity for that room. Second, the time of the peak load for a zone served by a central unit is required to size the unit. It is at these peak times that cooling load calculations should be made. The estimated times when the peak load will occur are determined from the tables of CLTD, SCL, and CLF values together with the orientation and physical characteristics of the room or space. These tables are described in the following section. The times of the peak cooling load for walls, roofs, and windows are obvious in the tables, and the most dominant cooling load components will then determine the peak time for the entire room or zone. For example, rooms facing west with no exposed roof will experience a peak load in the late afternoon or early evening. East-facing rooms tend to peak during the morning hours. High internal loads may dominate the cooling load in some cases and cause an almost uniform load throughout the day.

The details of computing the various cooling load components will be discussed in the following sections. Most of the tables to follow have been abridged and many of the footnotes have been included in the narrative to conserve space. Reference 5 should be consulted for complete data and an original source of information.

Cooling Load—External Surfaces

The calculation procedure is similar for walls, roofs, and conduction through glass. A different procedure is used for the glass solar gain.

Walls and Roofs

The cooling load temperature difference (CLTD) is used to determine the cooling load for walls and roofs using Eq. 8-33. Tables 8-18 and 8-19 give CLTDs for roofs and walls for July 21, 36 deg N latitude. These CLTDs are based on conditions that apply to a wide range of construction types. Reference 5 has software designed to generate CLTD tables for other months, latitudes, and zone conditions.

The values in Tables 8-18 and 8-19 were generated for standard conditions listed below.

Outdoor conditions

- July 21, at 36 deg N latitude
- No exterior shading
- Ground reflectance of 0.20
- Clear sky with a clearness number of 1.0
- Outside surface of roofs and walls with a ratio of the absorptance to the film coefficient (α/h_0) of 0.30
- Outside air maximum dry bulb temperature of 95 F or 35 C with a daily range of 21 F or 12 C.

Inside conditions

- Room dry bulb temperature constant at 78 F or 26 C
- Inside film coefficient for still air

The tabulated CLTDs must be corrected for inside and outside temperature and daily range when conditions differ from the above as follows:

$$\text{CLTD}_{\text{cor}} = \text{CLTD} + (78 - t_i) + (t_{om} - 85) \qquad \text{(8-36)}$$

where

CLTD = tabular CLTD, F
t_i = actual inside design dry bulb temperature, F or C
t_{om} = t_o − (DR/2), mean outside design dry bulb temperature, F or C

where

t_o = outside design dry bulb temperature, F or C ⎫ Table C-1
DR = daily range, F or C ⎭

Roof CLTD Selection

To obtain CLTDs for a particular roof, four charcteristics of the roof must first be determined in order to find the roof number in Table 8-18. These are

Mass placement with respect to insulation
Overall *R*-value of the roof, $1/U$
Principal roof material
Presence or absence of a suspended ceiling

Table 8-18 Cooling Load Temperature Differences for Calculating Cooling Load from Flat Roofs—36° N Latitude, July 21

Roof No.	Solar time, hr																							
	1	2	3	4	5	6	7	8	9	10	11	12	13	14	15	16	17	18	19	20	21	22	23	24
1	0	−2	−4	−5	−6	−6	0	12	28	45	61	75	84	90	90	84	74	60	42	26	15	9	5	2
2	2	0	−2	−4	−5	−6	−4	4	16	32	48	63	76	84	88	87	81	70	55	39	25	15	9	5
3	12	8	5	−2	0	−2	−1	4	13	24	35	47	58	67	73	75	74	68	59	48	38	30	23	17
4	17	11	7	3	1	−1	−3	−3	0	7	17	29	42	55	66	74	78	79	75	67	56	45	34	24
5	21	16	12	8	5	3	1	2	5	12	21	31	42	52	61	67	70	70	66	59	51	42	34	27
8	29	25	21	17	14	12	10	9	11	15	21	28	35	42	49	54	57	58	56	53	48	43	38	33
9	32	26	21	16	13	9	6	4	4	7	12	19	27	37	46	54	60	64	65	63	59	52	45	38
10	37	32	27	23	19	15	12	10	9	9	12	17	23	30	38	45	51	56	58	58	56	52	47	42
13	34	31	28	25	23	20	18	16	16	17	20	24	28	33	38	43	47	49	50	49	47	44	41	37
14	35	32	30	28	25	23	21	20	19	20	22	25	28	32	36	40	43	45	46	46	44	42	40	37

Notes: 1. Direct application of data

• Dark surface

• Indoor temperature of 78 F

• Outdoor maximum temperature of 95 F with mean temperature of 85 F and daily range of 21 F

• Solar radiation typical of clear day on 21st day of month

• Outside surface film resistance of 0.333 (hr-ft^2-F)/Btu

• With or without suspended ceiling but no ceiling plenum air return systems

• Inside surface resistance of 0.685 (hr-ft^2-F)/Btu

2. Adjustments to table data

• Design temperatures:

Corr. CLTD = CLTD + $(78 - t_r) + (t_m - 85)$.

where

t_r = inside temperature

t_m = mean outdoor temperature

t_m = maximum outdoor temperature − (daily range)/2

• No adjustment recommended for color

• No adjustment recommended for ventilation of air space above a ceiling

• Latitudes other than 36 degrees N

A table for a specific latitude may be generated. See text.

• Months other than July

For design purposes, the data will suffice for about 2 weeks from the 21st day of given month.

Tables may be generated for a specific month. See text.

3. CLTDs may be converted to degree C by multiplying by 5/9.

Source: Reprinted by permission from ASHRAE *Cooling and Heating Load Calculation Manual,* 2nd ed., 1992.

Mass Placement. Table 8-20 has provisions for roofs with massive layer(s) placed inside of insulating layers; mass evenly distributed; and massive layers placed outside of the insulation. Inside refers to the location in the roof nearest the conditioned space.

A massive layer is defined to be any layer of building material composed of brick, concrete, concrete block, or clay tile with a thickness of 2 in. (50 mm) or more. Referring to Table 8-21, Thermal Properties and Code Numbers of Layers for Walls and Roofs, this includes materials A2, A7, and C1 through C20 only. The insulating materials include B2 through B6 and B12 through B27.

Overall Roof R-value. The R-value for a given roof is generally calculated by summing all of the individual thermal resistances of the roof components including the inside and outside film resistances E0 and A0 of Table 8-21. See Chapter 5 for complete discussion and information.

Principal Roof Material. The principal roof materials are considered to be the massive material previously mentioned, as well as the nonmassive wood layers B7, B8, and B9, the nonmassive steel deck A3 and the attic ceiling combination. Note that Table 8-20 contains only one massive and three nonmassive materials. These were judged to be adequate for normal calculations.

Suspended Ceiling. Table 8-20 provides for roofs with and without suspended ceilings. A suspended ceiling is defined as an air space and acoustic tile or similar material with or without insulation located below the roof assembly.

With these four roof characteristics, the roof type can be determined from Table 8-20, which contains 10 roof types, covering the range of roof characteristics most commonly used. The roof type then defines the CLTD values in Table 8-18. Note that the CLTD values may require correction for indoor and outdoor temperature and daily range.

EXAMPLE 8-7

A single-story building located in Las Vegas, Nevada, has a roof described as follows:

Layer	Unit Resistance, R (hr-ft²-F)/Btu
Outside surface, A0	0.33
Slag and stone, E2	0.05
Felt and membrane, E3	0.29
Heavyweight concrete, 2 in., C12	0.17
Steel deck, A3	0.00
Air space, E4	1.00
Insulation, 3 in., B4	10.00
Acoustic tile, E5	1.79
Inside surface, E0	0.69
Total	14.32

heaviest (Table 8-21) → Heavyweight concrete, 2 in., C12

outside insulation → Air space, E4

Table 8.21 p. 304

$U = \frac{1}{R} = 0.07$

Compute the cooling load per 100 ft² at 10:00 A.M. and 5:00 P.M. solar time using ASHRAE Standard 90.1 design conditions for July.

Table 8-19 Cooling Load Temperature Differences for Calculating Cooling Load from Sunlit Walls—36° N Latitude, July 21

Wall No. 1

Wall Facing	1	2	3	4	5	6	7	8	9	10	11	12	13	14	15	16	17	18	19	20	21	22	23	24
																			Solar Time, hr					
N	1	0	-1	-2	-3	-1	7	12	12	14	17	21	25	28	29	29	28	30	27	17	11	7	5	3
NE	1	0	-1	-2	-3	1	23	41	48	46	38	30	28	29	30	29	28	24	19	14	10	7	5	3
E	1	0	-1	-2	-3	1	26	49	62	64	59	48	36	31	31	30	28	24	20	14	10	7	5	3
SE	1	0	-1	-2	-3	-1	13	31	44	52	55	52	45	36	32	30	28	24	19	14	10	7	5	3
S	1	0	-1	-2	-3	-3	0	4	9	18	29	39	45	47	46	40	31	25	20	14	10	7	5	3
SW	2	0	-1	-2	-2	-3	0	4	8	13	17	23	36	50	61	67	67	59	43	23	13	8	5	3
W	2	1	-1	-2	-2	-2	0	4	8	13	17	21	27	42	59	73	81	78	60	31	15	9	6	3
NW	2	0	-1	-2	-2	-2	0	4	8	13	17	21	25	29	40	53	63	65	54	28	14	8	5	3

Wall No. 2

Wall Facing	1	2	3	4	5	6	7	8	9	10	11	12	13	14	15	16	17	18	19	20	21	22	23	24
																			Solar Time, hr					
N	5	3	2	0	-1	-2	-1	3	7	9	12	15	18	21	24	27	28	28	29	27	22	17	12	8
NE	5	3	2	0	-1	-2	-1	12	26	36	41	39	35	32	30	30	29	28	26	23	18	14	10	7
E	5	3	2	0	-1	-2	2	14	31	46	54	56	52	45	38	34	32	30	27	23	19	14	11	8
SE	5	3	2	0	-1	-2	0	7	19	31	41	48	50	47	42	37	34	31	27	23	19	14	11	8
S	5	3	2	0	-1	-2	-2	-1	3	6	12	21	30	37	42	44	41	37	31	25	20	15	11	8
SW	7	4	2	1	-1	-1	-2	-1	2	5	9	13	19	28	39	50	58	62	60	51	38	26	17	11
W	8	5	3	1	0	-1	-2	-1	2	5	9	13	17	23	33	46	59	69	73	65	49	33	21	13
NW	7	4	2	1	-1	-2	-2	-1	2	5	9	13	17	21	26	33	43	53	58	55	42	29	19	12

Wall No. 3

Wall Facing	1	2	3	4	5	6	7	8	9	10	11	12	13	14	15	16	17	18	19	20	21	22	23	24
																			Solar Time, hr					
N	7	5	3	2	1	0	2	5	7	9	11	14	17	20	23	25	26	27	27	24	20	16	13	10
NE	7	5	3	2	0	0	6	17	26	32	34	33	31	30	30	30	29	28	25	22	18	15	12	9
E	7	5	4	2	1	0	7	20	32	42	47	47	44	40	37	35	33	31	28	24	20	16	13	10
SE	7	5	4	2	1	0	4	11	21	30	37	42	43	41	38	36	34	31	28	24	20	16	13	10
S	8	5	4	2	1	0	0	1	3	7	14	21	28	34	37	38	36	33	29	25	20	16	13	10
SW	11	8	6	4	2	1	0	1	3	6	9	13	19	28	38	47	53	55	52	43	34	27	20	15
W	14	10	7	4	2	1	1	2	4	6	9	13	17	24	34	45	56	63	62	53	42	32	24	18
NW	12	8	6	4	2	1	0	1	3	6	9	13	16	20	26	33	42	49	51	44	35	27	21	16

Wall No. 4

Wall Facing	1	2	3	4	5	6	7	8	9	10	11	12	13	14	15	16	17	18	19	20	21	22	23	24
N	11	8	6	4	2	0	0	0	3	5	8	10	13	16	19	22	24	26	27	27	26	23	19	15
NE	10	7	5	3	2	0	0	4	12	21	29	33	34	33	32	31	31	30	29	27	24	20	16	13
E	10	8	5	4	2	1	0	4	14	26	37	45	48	47	44	40	37	34	32	29	25	21	17	14
SE	10	8	5	3	2	0	0	2	8	16	26	34	40	44	43	41	38	35	33	29	25	21	17	14
S	11	8	5	3	2	0	-1	-1	0	2	5	11	18	25	32	36	39	38	36	32	27	23	18	14
SW	17	12	8	5	3	1	0	0	0	2	4	8	11	17	25	34	43	51	55	54	49	40	31	24
W	21	15	10	7	4	2	0	0	0	2	5	8	11	15	21	30	41	52	60	63	59	50	39	29
NW	18	13	9	6	3	1	0	0	0	2	4	8	11	15	19	24	31	39	46	51	48	41	33	2

Wall No. 5

Wall Facing	1	2	3	4	5	6	7	8	9	10	11	12	13	14	15	16	17	18	19	20	21	22	23	24
N	13	10	8	6	5	3	2	3	5	6	8	10	12	15	17	20	21	23	24	25	23	21	18	15
NE	13	10	8	6	5	3	3	7	14	20	25	28	29	29	29	29	29	29	28	26	24	21	18	15
E	14	11	9	7	5	4	4	8	16	25	33	38	40	40	38	37	35	34	32	29	26	23	20	17
SE	14	11	9	7	5	4	3	5	10	17	24	30	35	37	37	36	35	33	32	29	26	23	20	17
S	14	11	9	7	5	3	2	2	2	4	7	11	17	22	27	31	33	33	31	29	26	23	19	16
SW	21	17	13	11	8	6	4	3	4	4	6	8	11	16	23	31	38	44	47	46	42	36	30	25
W	24	20	16	12	10	7	5	4	4	5	7	9	11	15	20	28	37	46	52	54	49	43	36	30
NW	21	17	13	10	8	6	4	3	3	4	6	8	11	14	17	22	28	35	41	43	41	36	30	25

Wall No. 6

Wall Facing	1	2	3	4	5	6	7	8	9	10	11	12	13	14	15	16	17	18	19	20	21	22	23	24
N	13	11	9	8	6	5	4	5	6	7	8	10	12	14	16	18	20	21	23	23	22	20	18	15
NE	14	12	10	8	7	5	6	9	15	20	24	26	26	26	27	27	28	27	27	25	23	21	19	16
E	16	13	11	9	7	6	6	11	17	25	31	35	36	36	35	34	34	33	31	29	26	24	21	18
SE	15	13	11	9	7	6	5	8	12	17	23	28	32	33	34	33	33	32	30	28	26	23	20	18
S	15	13	10	9	7	5	4	4	4	5	8	12	16	21	25	28	30	30	29	27	25	22	20	17
SW	22	19	16	13	11	8	7	6	6	6	8	9	12	17	23	29	35	40	42	41	38	34	30	26
W	25	22	18	15	12	10	8	7	7	7	8	10	12	15	20	27	35	42	47	47	44	39	34	30
NW	21	18	15	13	10	8	7	6	6	6	8	9	11	14	17	21	27	33	37	39	36	33	29	25

Table 8-19 Cooling Load Temperature Differences for Calculating Cooling Load from Sunlit Walls—36° N Latitude, July 21 *Continued*

Wall No. 7

Wall Facing	1	2	3	4	5	6	7	8	9	10	11	12	13	14	15	16	17	18	19	20	21	22	23	24
N	13	12	10	9	7	6	6	7	8	8	9	11	12	14	16	17	19	20	21	20	19	18	16	15
NE	15	13	12	10	9	8	9	13	17	21	23	23	24	24	25	25	26	26	25	24	22	20	18	17
E	17	15	13	12	10	9	11	15	21	26	30	32	32	32	32	31	31	30	29	27	25	23	21	19
SE	17	15	13	11	10	8	9	11	15	19	24	27	29	30	30	30	30	29	28	26	24	22	20	18
S	15	13	12	10	9	7	7	6	7	8	10	14	17	21	24	25	26	26	25	24	22	20	19	17
SW	22	19	17	15	13	11	10	9	9	10	10	12	14	19	24	29	33	36	37	35	32	30	27	24
W	25	22	19	17	15	13	11	11	10	11	11	13	14	17	22	28	34	39	42	40	37	34	31	28
NW	20	18	16	14	12	10	9	9	10	9	10	11	13	15	18	22	26	31	34	33	31	28	25	23

Solar Time, hr

Wall No. 9

Wall Facing	1	2	3	4	5	6	7	8	9	10	11	12	13	14	15	16	17	18	19	20	21	22	23	24
N	17	15	13	11	9	7	5	4	4	4	5	7	8	10	12	15	17	19	21	22	23	23	22	20
NE	18	15	13	11	9	7	5	5	6	10	15	20	24	26	27	28	28	28	28	28	27	25	23	20
E	20	17	14	12	10	8	6	5	7	12	19	26	32	35	37	37	37	36	34	33	31	29	26	23
SE	20	17	14	12	10	8	6	5	5	8	13	18	24	29	33	34	35	34	34	32	31	28	25	23
S	19	16	14	11	9	7	5	4	3	3	4	5	8	13	17	22	26	29	31	31	29	27	25	22
SW	30	25	21	17	14	11	9	7	5	5	5	6	7	10	13	18	25	31	37	41	43	42	39	34
W	35	30	25	20	17	13	11	8	7	6	6	6	8	10	13	17	22	30	37	44	48	48	45	40
NW	29	25	21	17	14	11	9	7	5	5	6	6	7	9	12	14	18	23	29	35	38	39	37	33

Solar Time, hr

Wall No. 10

Wall Facing	1	2	3	4	5	6	7	8	9	10	11	12	13	14	15	16	17	18	19	20	21	22	23	24
N	17	15	13	11	9	7	6	5	5	5	6	7	8	10	12	15	17	19	20	22	22	22	21	19
NE	18	16	13	11	9	7	6	6	8	12	16	20	23	25	26	27	27	28	28	27	26	25	23	20
E	20	17	15	12	10	8	7	7	9	14	20	26	30	33	35	35	35	35	34	33	31	28	26	23
SE	20	17	15	12	10	8	6	6	7	10	14	19	24	28	31	32	33	33	33	32	30	28	25	23
S	19	17	14	12	10	8	6	5	4	4	4	6	9	13	18	22	25	28	29	29	28	26	25	22
SW	29	25	22	18	15	12	10	8	6	6	6	7	8	10	14	19	25	31	36	39	40	39	37	33
W	34	30	25	21	18	14	12	9	8	7	7	7	8	10	13	17	23	30	37	42	45	45	42	38
NW	28	25	21	18	15	12	10	8	6	6	6	6	8	10	12	15	19	24	29	34	36	36	34	32

Solar Time, hr

Wall No. 11

Wall Facing	1	2	3	4	5	6	7	8	9	10	11	12	13	14	15	16	17	18	19	20	21	22	23	24
NE	18	17	15	13	12	10	9	9	11	14	17	20	22	23	23	24	25	25	25	25	24	23	22	20
E	21	19	17	16	14	12	11	11	13	17	21	25	28	30	31	31	31	31	31	30	29	27	25	23
SE	20	19	17	15	13	12	10	10	11	13	16	20	23	26	28	29	29	29	29	28	27	26	24	22
S	19	17	15	14	12	10	9	8	7	7	8	9	11	14	16	17	20	24	25	25	24	23	22	20
SW	27	24	22	19	17	15	13	12	11	10	10	11	12	13	16	19	24	28	32	34	35	33	31	29
W	30	27	25	22	20	17	15	14	12	12	11	11	12	13	15	18	23	28	33	37	39	38	36	33
NW	25	23	21	18	16	14	13	11	10	10	9	10	11	12	13	15	18	22	26	30	31	31	29	27

Wall No. 12

Wall Facing	1	2	3	4	5	6	7	8	9	10	11	12	13	14	15	16	17	18	19	20	21	22	23	24
N	16	15	13	12	11	10	9	8	8	8	8	9	10	11	12	14	15	16	18	19	19	19	18	17
NE	19	17	16	14	13	12	11	10	12	14	17	19	21	22	23	23	24	24	25	24	24	23	22	20
E	22	20	18	16	15	13	12	12	14	17	21	25	27	29	30	30	30	30	30	29	28	27	25	23
SE	21	19	17	16	14	13	11	11	12	13	16	20	23	25	27	28	28	28	28	28	27	26	24	22
S	19	17	16	14	13	11	9	9	8	8	8	9	11	14	17	20	22	23	24	24	23	23	21	20
SW	26	24	22	20	18	16	14	13	12	11	11	11	12	13	16	19	23	27	30	32	33	32	30	28
W	30	27	25	23	20	18	16	15	14	13	12	13	13	14	16	19	23	27	32	35	37	36	34	32
NW	25	23	21	19	17	15	13	12	11	11	10	11	11	12	14	15	18	22	25	28	30	30	28	27

Wall No. 13

Wall Facing	1	2	3	4	5	6	7	8	9	10	11	12	13	14	15	16	17	18	19	20	21	22	23	24
N	15	14	13	12	11	10	9	9	9	9	9	9	10	10	11	13	14	15	16	17	18	18	17	16
NE	19	17	16	15	14	12	11	12	13	16	18	19	21	21	22	22	23	23	24	23	23	22	21	20
E	22	20	19	17	16	15	14	14	16	19	22	25	27	28	28	29	29	29	29	28	27	26	25	23
SE	21	19	18	16	15	14	13	12	13	15	17	20	23	24	26	26	27	27	27	26	25	25	23	22
S	18	17	16	14	13	12	11	10	9	9	10	11	12	15	17	19	21	22	22	22	21	20	20	19
SW	25	23	22	20	18	17	15	14	13	13	13	13	13	15	17	20	23	26	29	31	31	30	28	27
W	28	26	24	23	21	19	17	16	15	14	14	15	15	15	17	20	23	27	31	33	34	33	12	30
NW	23	22	20	19	17	16	14	13	12	12	12	12	13	13	14	16	18	21	25	27	28	27	26	25

Table 8-19 Cooling Load Temperature Differences for Calculating Cooling Load from Sunlit Walls—36° N Latitude, July 21 *Continued*

Wall No. 14

Wall Facing	Solar Time, hr																							
	1	2	3	4	5	6	7	8	9	10	11	12	13	14	15	16	17	18	19	20	21	22	23	24
N	15	15	14	13	12	11	10	10	10	10	10	10	11	11	12	13	14	15	16	16	17	17	17	16
NE	19	18	17	16	15	14	13	13	14	15	17	18	19	20	21	21	22	22	23	23	22	22	21	20
E	23	22	20	19	18	17	16	15	16	18	20	23	25	26	27	27	27	28	28	27	27	26	25	24
SE	22	20	19	18	17	16	15	14	14	15	17	19	21	23	24	25	25	26	26	26	25	25	24	23
S	18	17	16	15	14	13	12	12	11	11	11	11	12	14	16	17	19	20	21	21	21	21	20	19
SW	25	24	23	21	20	19	17	16	15	15	14	14	14	15	16	18	21	24	26	28	28	28	28	27
W	28	27	25	24	22	21	19	18	17	16	16	16	16	16	17	19	21	24	27	28	28	28	28	27
NW	23	22	21	20	18	17	16	15	14	13	13	13	13	14	14	16	17	21	22	24	25	25	25	24

Wall No. 15

Wall Facing	Solar Time, hr																							
	1	2	3	4	5	6	7	8	9	10	11	12	13	14	15	16	17	18	19	20	21	22	23	24
N	19	18	16	14	12	11	9	7	6	6	6	6	7	8	10	11	13	15	17	19	20	21	21	20
NE	21	19	17	15	13	11	9	8	7	9	11	14	18	21	23	24	25	26	27	27	27	26	25	23
E	24	22	19	17	15	12	10	9	8	10	13	18	22	27	30	32	33	34	34	33	32	31	29	27
SE	24	22	19	17	14	12	10	8	8	8	10	13	17	21	25	28	30	31	32	32	31	30	28	26
S	23	20	18	16	14	12	10	8	7	5	5	5	7	9	12	15	19	22	25	27	27	27	26	25
SW	33	30	27	24	21	18	15	12	10	9	8	7	7	8	10	13	17	22	27	32	35	37	37	36
W	38	35	31	28	24	21	17	14	12	10	9	8	7	9	10	12	16	21	26	32	38	41	42	41
NW	31	29	26	23	20	17	14	12	10	8	7	7	7	8	9	11	14	17	21	26	30	33	34	33

Wall No. 16

Wall Facing	Solar Time, hr																							
	1	2	3	4	5	6	7	8	9	10	11	12	13	14	15	16	17	18	19	20	21	22	23	24
N	18	17	16	14	13	11	10	9	8	7	7	7	8	9	10	11	13	14	16	17	19	19	20	19
NE	21	20	18	16	14	13	11	10	10	10	12	15	17	19	21	22	24	24	25	25	25	25	24	23
E	25	23	21	19	17	15	13	11	11	12	15	18	22	25	28	30	31	31	32	31	31	30	28	27
SE	24	22	20	18	16	14	12	11	10	10	12	14	17	20	23	26	27	29	29	30	29	29	27	26
S	22	20	18	16	15	13	11	10	8	7	7	7	8	10	12	15	18	21	23	24	25	25	24	23
SW	31	29	27	24	22	19	17	15	13	11	10	10	10	10	11	12	14	17	21	25	29	33	34	33
W	35	33	30	28	25	22	19	17	15	13	12	11	11	11	12	14	16	20	25	30	34	37	37	37
NW	29	27	25	23	21	18	16	14	12	11	10	9	10	11	12	14	17	20	24	27	30	33	37	30

Notes: 1. Direct application of data

- Dark surface
- Indoor temperature of 78 F
- Outdoor maximum temperature of 95 F with mean temperature of 85 F and daily range of 21 F
- Solar radiation typical of clear day on 21st day of month
- Outside surface film resistance of 0.333 (hr-ft^2-F)/Btu
- Inside surface resistance of 0.685 (hr-ft^2-F)/Btu

2. Adjustments to table data

- Design temperatures:

Corr. CLTD = CLTD + $(78 - t_r) + (t_m - 85)$

where

t_r = inside temperature

t_m = mean outdoor temperature

t_m = maximum outdoor temperature − (daily range)/2

- No adjustment recommended for color
- Latitudes other than 36 degrees N

A table for a specific latitude may be generated. See text.

- Months other than July

For design purposes, the data will suffice for about 2 weeks from the 21st day of given month.

Tables may be generated for a specific month. See text.

3. CLTDs may be converted to degrees C by multiplying by 5/9.

Source: Reprinted by permission from ASHRAE *Cooling and Heating Load Calculation Manual,* 2nd ed., 1992.

Table 8-20 Roof Classification for Use with Table 8-18

Mass Location[a]	Suspended Ceiling	R-Factor, (ft^2-hr-F)/Btu	B7, Wood 1 in.	C12, HW Concrete 2 in.	A3, Steel Deck	Attic-Ceiling Combination
		0 to 5	b	2	b	b
		5 to 10	b	2	b	b
	Without	10 to 15	b	4	b	b
		15 to 20	b	4	b	b
		20 to 25	b	5	b	b
Mass inside		25 to 30	b	b	b	b
the insulation		0 to 5	b	5	b	b
		5 to 10	b	8	b	b
	With	10 to 15	b	13	b	b
		15 to 20	b	13	b	b
		20 to 25	b	14	b	b
		25 to 30	b	b	b	b
		0 to 5	1	2	1	1
		5 to 10	2	b	1	2
	Without	10 to 15	2	b	1	2
		15 to 20	4	b	2	2
		20 to 25	4	b	2	4
Mass evenly		25 to 30	b	b	b	b
placed		0 to 5	b	3	1	b
		5 to 10	4	b	1	b
	With	10 to 15	5	b	2	b
		15 to 20	9	b	2	b
		20 to 25	10	b	4	b
		25 to 30	10	b	b	b
		0 to 5	b	2	b	b
		5 to 10	b	3	b	b
	Without	10 to 15	b	4	b	b
		15 to 20	b	5	b	b
		20 to 25	b	5	b	b
Mass outside		25 to 30	b	b	b	b
the insulation		0 to 5	b	3	b	b
		5 to 10	b	3	b	b
	With	10 to 15	b	4	b	b
		15 to 20	b	5	b	b
		20 to 25	b	b	b	b
		25 to 30	b	b	b	

[a]The 2-in. concrete is considered massive and the others nonmassive.

[b]Denotes a roof that is not possible with the chosen parameters.

Source: Reprinted by permission from ASHRAE *Cooling and Heating Load Calculation Manual,* 2nd ed., 1992.

SOLUTION

Example Solution:

The cooling load is calculated using Eq. 8-33 with $U = 1/R = 1/14.32 = 0.07$ Btu/(hr-ft^2-F). The CLTD is determined using Tables 8-20 and 8-18 and corrected for local conditions. Outside design temperature and daily range are 106 F and 30 F respectively (Table C-1). Note that Las Vegas, Nevada, is located at 36 deg, 1 min N latitude. Assume an inside design temperature of 75 F. Note that the principal material in the roof, 2-in. heavyweight concrete, is located outside the insulation. There is a suspended ceiling and the R-factor range is 10 to 15. Then from Table 8-20 the roof type is No. 4 and the CLTDs at 10:00 A.M. or 1000 hours and 5:00 P.M. or 1700 hours are 7 F and 78 F, respectively. Equation 8-36 is used to correct these values:

$$t_{om} = 106 - (30/2) = 91\,\text{F}$$

$$10\!:\!00\ \text{AM}\!:\quad CLTD_{10} = 7 + (78 - 75) + (91 - 85) = 16\,\text{F}$$

$$5\!:\!00\ \text{pm}\!:\quad CLTD_{17} = 78 + (78 - 75) + (91 - 85) = 87\,\text{F}$$

The cooling load at 10:00 A.M. is

$$\dot{q}_c = UA(CLTD_{10}) = 0.07(100)(16) = 112\,\text{Btu/hr}$$

and at 5:00 P.M.

$$\dot{q}_c = UA(CLTD_{17}) = 0.07(100)(87) = 609\,\text{Btu/hr}$$

Wall CLTD Selection

To obtain CLTDs for a particular wall, three characteristics of the wall must first be determined in order to find the wall number in Table 8-19. These are

Mass placement with respect to insulation
Overall R-value of the wall, $1/U$
Principal wall material

Mass Placement. Table 8-22a is for walls with massive layer(s) placed inside of insulating layers. Table 8-22b is for walls with no massive layer, no insulation, or mass evenly distributed. Table 8-22c is for walls with massive layers placed outside of the insulation.

A massive layer is defined to be any layer of building material composed of brick, concrete, concrete block, or clay tile with a thickness of 2 in. (50 mm) or more. Referring to Table 8-21 this includes materials A2, A7, and C1 through C20 only. The insulating materials include B2 through B6 and B12 through B27.

Overall Wall R-value. The R-value for a given wall is calculated by summing all of the individual thermal resistances of the wall components including the inside and outside film resistances E0 and A0 from Table 8-21 or by procedures discussed in Chapter 5.

Principal Wall Material. The principal wall materials are considered to be the massive materials previously defined, as well as the nonmassive wood layers B7, B8, and B9, the nonmassive steel layer A3.

Table 8-21 Thermal Properties and Code Numbers of Layers for Walls and Roofs

Code Number	Description	L	k	ρ	c_p	R	Mass
		\multicolumn{6}{c}{Thickness and Thermal Properties[a]}					
A0	Outside surface resistance	0.0	0.0	0.0	0.0	0.33	0.0
A1	1-in. stucco	0.0833	0.4	116.0	0.20	0.21	9.7
A2	4-in. face brick	0.333	0.77	125.0	0.22	0.43	41.7
A3	Steel siding	0.005	26.0	480.0	0.10	0.00	2.4
A4	$\frac{1}{2}$-in. slag	0.0417	0.11	70.0	0.40	0.38	2.2
A5	Outside surface resistance	0.0	0.0	0.0	0.0	0.33	0.0
A6	Finish	0.0417	0.24	78.0	0.26	0.17	3.3
A7	4-in. face brick	0.333	0.77	125.0	0.22	0.43	41.7
B1	Airspace resistance	0.0	0.0	0.0	0.0	0.91	0.0
B2	1-in. insulation	0.083	0.025	2.0	0.2	3.33	0.2
B3	2-in. insulation	0.167	0.025	2.0	0.2	3.33	0.3
B4	3-in. insulation	0.25	0.025	2.0	0.2	10.00	0.5
B5	1-in. insulation	0.0833	0.025	5.7	0.2	3.33	0.5
B6	2-in. insulation	0.167	0.025	5.7	0.2	6.67	1.0
B7	1-in. wood	0.0833	0.07	37.0	0.6	10.00	3.1
B8	2.5-in. wood	0.2083	0.07	37.0	0.6	2.98	7.7
B9	4-in. wood	0.333	0.07	37.0	0.6	4.76	12.3
B10	2-in. wood	0.167	0.07	37.0	0.6	2.39	6.2
B11	3-in. wood	0.25	0.07	37.0	0.6	3.57	9.3
B12	3-in. insulation	0.25	0.025	5.7	0.2	10.00	1.4
B13	4-in. insulation	0.333	0.025	5.7	0.2	13.33	1.9
B14	5-in. insulation	0.417	0.025	5.7	0.2	16.67	2.4
B15	6-in. insulation	0.500	0.025	5.7	0.2	20.00	2.9
B16	0.15-in. insulation	0.0126	0.025	5.7	0.2	0.50	0.1
B17	0.3-in. insulation	0.0252	0.025	5.7	0.2	1.00	0.1
B18	0.45-in. insulation	0.0379	0.025	5.7	0.2	1.50	0.2
B19	0.61-in. insulation	0.0505	0.025	5.7	0.2	2.00	0.3
B20	0.76-in. insulation	0.0631	0.025	5.7	0.2	2.50	0.4
B21	1.36-in. insulation	0.1136	0.025	5.7	0.2	4.50	0.6
B22	1.67-in. insulation	0.1388	0.025	5.7	0.2	5.50	0.8
B23	2.42-in. insulation	0.2019	0.025	5.7	0.2	8.00	1.2
B24	2.73-in. insulation	0.2272	0.025	5.7	0.2	9.00	1.3
B25	3.33-in. insulation	0.2777	0.025	5.7	0.2	11.00	1.6
B26	3.64-in. insulation	0.3029	0.025	5.7	0.2	12.00	1.7
B27	4.54-in. insulation	0.3786	0.025	5.7	0.2	15.00	2.2
C1	4-in. clay tile	0.333	0.33	70.0	0.2	1.01	23.3
C2	4-in. lightweight concrete block	0.333	0.22	38.0	0.2	1.51	12.7
C3	4-in. heavyweight concrete block	0.333	0.47	61.0	0.2	0.71	20.3
C4	4-in. common brick	0.333	0.42	120.0	0.2	0.79	40.0
C5	4-in. heavyweight concrete	0.333	1.0	140.0	0.2	0.33	46.7
C6	8-in. clay tile	0.667	0.33	70.0	0.2	2.00	46.7
C7	8-in. lightweight concrete block	0.667	0.33	38.0	0.2	2.00	25.3
C8	8-in. heavyweight concrete block	0.667	0.6	61.0	0.2	1.11	40.7
C9	8-in. common brick	0.667	0.42	120.0	0.2	1.59	80.0
C10	8-in. heavyweight concrete	0.667	1.0	140.0	0.2	0.67	93.4
C11	12-in. heavyweight concrete	1.0	1.0	140.0	0.2	1.00	140.0
C12	2-in. heavyweight concrete	0.167	1.0	140.0	0.2	0.17	23.3
C13	6-in. heavyweight concrete	0.5	1.0	140.0	0.2	0.50	70.0
C14	4-in. lightweight concrete	0.333	0.1	40.0	0.2	3.33	13.3
C15	6-in. lightweight concrete	0.5	0.1	40.0	0.2	5.00	20.0
C16	8-in. lightweight concrete	0.667	0.1	40.0	0.2	6.67	26.7

Table 8-21 Thermal Properties and Code Numbers of Layers for Walls and Roofs (*Continued*)

Code Number	Description	Thickness and Thermal Properties[a]					
		L	k	ρ	c_p	R	Mass
C17	8-in. lightweight conc. block (filled)	0.667	0.08	18.0	0.2	8.34	12.0
C18	8-in. heavyweight conc. block (filled)	0.667	0.34	53.0	0.2	1.96	35.4
C19	12-in. lightweight conc. block (filled)	1.000	0.08	19.0	0.2	12.50	19.0
C20	12-in. heavyweight conc. block (filled)	1.000	0.39	56.0	0.2	2.56	56.4
E0	Inside surface resistance	0.0	0.0	0.0	0.0	0.69	0.0
E1	$\frac{3}{4}$-in. plaster or gypsum	0.0625	0.42	100.0	0.2	0.15	6.3
E2	$\frac{1}{2}$-in. slag or stone	0.0417	0.83	11.0	0.40	0.05	2.3
E3	$\frac{3}{8}$-in. felt and membrane	0.0313	0.11	70.0	0.40	0.29	2.2,
E4	Ceiling air space	0.0	0.0	0.0	0.0	1.00	0.0
E5	Acoustic tile	0.0625	0.035	30.0	0.2	1.79	1.9

[a]L = thickness, ft; k = thermal conductivity, Btu/(hr-ft²-F); ρ = density; lb/ft²; c_p = specific heat, Btu/(lb-F); R = thermal resistance, (ft²-hr-F)/Btu; Mass = unit mass, lb/ft². See inside front cover for conversion factors to SI units.

Source: Reprinted by permission from ASHRAE *Cooling and Heating Load Calculation Manual*, 2nd ed., 1992.

With these three wall characteristics and the secondary wall material, the wall type can be determined in Tables 8-22, which contain 15 wall types covering the range of wall characteristics most commonly used. The wall type, latitude, and direction define the CLDT values in Tables 8-19. Note that the CLTD values may require correction for inside and outdoor temperature and daily range.

EXAMPLE 8-8

The building of Example 8-7 has walls consisting of 4-in. face brick plus 8-in. H.W. concrete block with 1-in. insulation between. Compute the cooling load per 100 ft² at 10:00 A.M. and 5:00 P.M. solar time for the south wall using ASHRAE Standard 90.1 design conditions for July.

SOLUTION

The cooling load is computed using Eq. 8-33 with the thermal resistances for each layer from Table 8-21 summed to obtain U. The CLTDS are obtained using Table 8-22 and 8-19 and corrected using Eq. 8-36.

Layer	Unit Resistance, R (hr-ft²-F)/Btu	mass (Table 8-21)
Outside surface, A0	0.33	—
Face brick, 4 in., A7	0.43	41.7
Insulation, 1 in., B5	3.33	—
Concrete block, 8 in., C8	1.11	40.7
Inside surface, E0	0.69	—
Total	5.60	

Then $U = 1/R = 1/5.60 = 0.18$ Btu/(hr-ft²-F).

Table 8-22a Wall Types, Mass Located Inside Insulation, for Use with Table 8-19[a]

Secondary Material	R-Factor, (ft²-F)/Btu	Principal Wall Material[a]														
		A1	A2	B7	B10	B9	C1	C2	C3	C4	C5	C6	C7	C8	C17	C18
	0.0–2.0	b	b	b	b	b	b	b	b	b	b	b	b	b	b	b
	2.0–2.5	b	5	b	b	b	b	b	b	b	5	b	b	b	b	b
	2.5–3.0	b	5	b	b	b	3	b	2	5	6	b	b	5	b	b
	3.0–3.5	b	5	b	b	b	4	2	2	5	6	b	b	6	b	b
	3.5–4.0	b	5	b	b	b	4	2	3	6	6	10	4	6	b	5
	4.0–4.75	b	6	b	b	b	5	2	4	6	6	11	5	10	b	10
	4.75–5.5	b	6	b	b	b	5	2	4	6	6	11	5	10	b	10
	5.5–6.5	b	6	b	b	b	5	2	5	10	7	12	5	11	b	10
	6.5–7.75	b	6	b	b	b	5	4	5	11	7	16	10	11	b	11
	7.75–9.0	b	6	b	b	b	5	4	5	11	7	b	10	11	4	11
	9.0–10.75	b	6	b	b	b	5	4	5	11	7	b	10	11	4	11
	10.75–12.75	b	6	b	b	b	10	4	5	11	11	b	10	11	9	11
	12.75–15.0	b	10	b	b	b	10	5	5	11	11	b	11	12	10	12
	15.0–17.5	b	10	b	b	b	10	5	9	11	11	b	15	16	10	16
	17.5–20.0	b	11	b	b	b	10	9	9	16	11	b	15	16	10	16
	20.0–23.0	b	11	b	b	b	b	b	b	b	b	b	16	16	15	16
	23.0–27.0	b	b	b	b	b	b	b	b	b	b	b	b	b	b	b
Stucco and/or plaster	0.0–2.0	b	b	b	b	b	b	b	b	b	b	b	b	b	b	b
	2.0–2.5	b	3	b	b	b	b	b	2	3	5	b	b	b	b	b
	2.5–3.0	b	5	b	b	b	2	b	2	5	3	b	b	5	b	b
	3.0–3.5	b	5	b	b	b	3	1	2	5	5	b	b	5	b	b
	3.5–4.0	b	5	b	b	b	3	2	2	5	5	6	3	5	b	5
	4.0–4.75	b	6	b	b	b	4	2	2	5	5	10	4	6	b	5

306

Wall facing	Thickness range																
Steel or other lightweight siding	4.75–5.5	b	b	b	b	b	6	6	2	2	5	11	6	6	6	b	6
	5.5–6.5	3	b	b	b	b	6	6	2	2	5	11	6	6	6	b	6
	6.5–7.75	5	11	b	5	b	6	6	2	2	5	11	6	6	6	b	6
	7.75–9.0	5	12	b	6	b	6	6	2	2	5	12	6	6	6	b	6
	9.0–10.75	5	12	10	6	b	6	6	2	2	5	12	6	6	5	4	6
	10.75–12.75	6	13	11	6	b	6	6	3	2	5	12	7	6	6	4	11
	12.75–15.0	6	13	11	6	b	6	6	3	2	5	b	7	10	10	5	11
	15.0–17.5	6	13	11	6	b	10	10	3	4	6	b	7	10	10	9	11
	17.5–20.0	6	13	16	10	b	10	10	4	4	10	b	11	11	10	10	11
	20.0–23.0	10	14	16	10	b	11	11	4	4	10	b	11	11	10	10	16
	23.0–27.0	10	14	16	10	b	b	b	5	5	b	b	b	b	10	11	16
Face brick	0.0–2.0	6	b	b	b	b	b	b	b	b	b	b	b	b	b	b	b
	2.0–2.5	b	b	b	b	b	b	b	b	b	b	b	b	b	b	b	b
	2.5–3.0	11	b	b	b	b	11	11	6	6	b	b	b	b	b	b	b
	3.0–3.5	12	5	b	5	b	12	12	11	11	b	b	b	b	b	12	b
	3.5–4.0	12	6	b	6	b	12	12	12	11	b	b	b	b	b	12	b
	4.0–4.75	13	6	10	6	b	13	12	12	11	b	13	10	6	11	b	16
	4.75–5.5	13	6	11	6	b	13	13	12	11	b	13	11	6	16	b	b
	5.5–6.5	13	6	11	6	b	13	b	12	12	b	13	11	6	b	b	b
	6.5–7.75	13	6	11	6	b	13	b	13	12	b	13	11	6	b	b	b
	7.75–9.0	13	10	16	10	b	14	b	13	13	b	16	16	10	b	b	b
	9.0–10.75	14	10	16	10	b	14	b	13	13	16	16	16	10	b	16	b
	10.75–12.75	14	10	16	10	b	14	b	13	16	16	16	16	10	b	16	b
	12.75–15.0	b	11	16	11	b	b	b	13	16	b	b	b	11	b	b	b
	15.0–17.5	10	11	b	11	b	b	b	b	16	b	b	b	11	b	b	b
	17.5–20.0	10	11	b	11	b	b	b	b	b	b	b	b	11	b	b	b
	20.0–23.0	11	15	b	15	b	b	b	b	b	b	b	b	15	b	b	b
	23.0–27.0	b	b	b	b	b	b	b	b	16	b	b	b	b	b	b	b

[a] See Table for definition of code letters.

[b] Denotes a wall that is not possible with the chosen set of parameters

Source: Reprinted by permission from ASHRAE *Cooling and Heating Load Calculation Manual*, 2nd ed., 1992.

Table 8-22b Wall Types, Mass Evenly Distributed, for Use with Table 8-19

Secondary Material	R-Factor, (ft²-F)/Btu	Principal Wall Material[a]														
		A1	A2	B7	B10	B9	C1	C2	C3	C4	C5	C6	C7	C8	C17	C18
	0.0–2.0	1	3	b	b	b	b	b	1	3	3	b	b	b	b	b
	2.0–2.5	1	3	1	b	b	2	b	2	4	4	b	b	5	b	b
	2.5–3.0	1	4	1	b	b	2	2	2	4	4	b	b	5	b	b
	3.0–3.5	1	b	1	b	b	2	2	b	b	b	10	4	5	b	4
	3.5–4.0	1	b	1	2	b	b	4	b	b	b	10	4	b	b	4
	4.0–4.75	1	b	1	2	b	b	b	b	b	b	10	4	b	b	4
	4.75–5.5	1	b	1	2	b	b	b	b	b	b	b	b	b	b	b
	5.5–6.5	1	b	2	4	10	b	b	b	b	b	b	b	b	b	b
	6.5–7.75	1	b	2	4	11	b	b	b	b	b	b	b	b	b	b
	7.75–9.0	1	b	2	4	16	b	b	b	b	b	b	b	b	b	b
	9.0–10.75	1	b	2	4	16	b	b	b	b	b	b	b	b	4	b
	10.75–12.75	1	b	2	5	b	b	b	b	b	b	b	b	b	4	b
	12.75–15.0	2	b	2	5	b	b	b	b	b	b	b	b	b	b	b
	15.0–17.5	2	b	2	5	b	b	b	b	b	b	b	b	b	b	b
	17.5–20.0	2	b	2	9	b	b	b	b	b	b	b	b	b	b	b
	20.0–23.0	2	b	4	9	b	b	b	b	b	b	b	b	b	b	b
	23.0–27.0	b	b	b	9	b	b	b	b	b	b	b	b	b	b	b
Stucco and/or plaster	0.0–2.0	1	3	b	b	b	b	b	1	3	2	b	b	b	b	b
	2.0–2.5	1	3	1	b	b	2	b	1	3	2	b	b	3	b	b
	2.5–3.0	1	4	1	b	b	2	1	2	4	4	b	b	3	b	b
	3.0–3.5	1	b	1	b	b	4	1	b	b	b	5	2	4	b	4
	3.5–4.0	1	b	1	2	b	b	2	b	b	b	5	2	b	b	4
	4.0–4.75	1	b	1	2	b	b	b	3	b	b	10	4	b	b	4

308

This page presents a rotated, dense data table (ASHRAE wall cooling-load data). The row labels are wall R-value (or thickness) ranges grouped under two wall types: "Steel or other lightweight siding" and "Face brick". Most cells are marked "b" (denoting a wall not possible with the chosen parameters). The following is a best-effort reading.

Steel or other lightweight siding

Range	Values (best reading)
4.75–5.5	1
5.5–6.5	1, 2, 10
6.5–7.75	1, 4, 11, 5
7.75–9.0	1, 4, 16, 5, 15
9.0–10.75	1, 4, 16, 5, 10, 16, 2
10.75–12.75	1, 4, 5, 10, 16, 4
12.75–15.0	1, 5, 5, 10, 16
15.0–17.5	1, 5, 5, 11, 16
17.5–20.0	1, 5, 5, 15, 16
20.0–23.0	2, 9, 9, 15
23.0–27.0	b, 9, 9, 16

Face brick

Range	Values (best reading)
0.0–2.0	3, 6
2.0–2.5	3, 10, 5, 5, 10, 10
2.5–3.0	4, 10, 5, 5, 5, 10, 10, 10
3.0–3.5	11, 5, 5, 5, 11, 10, 10, 15
3.5–4.0	11, 5, 5, 10, 11, 11, 16, 16
4.0–4.75	11, 10, 5, 11, 10, 16, 16, 16
4.75–5.5	11, 10, 5, 10, 10, 11, 16, 16
5.5–6.5	16, 10, 9, 10, 11, 16, 16, 16
6.5–7.75	16, 11, 9, 11, 16, 16, 16, 16
7.75–9.0	16, 15, 9, 15, 15, 16, 16
9.0–10.75	16, 15, 10, 16, 16, 16, 10, 10, 10
10.75–12.75	16, 16, 10, 16, 16, 16, 15, 15
12.75–15.0	16, 16, 10, 16, 16, 10, 15, 15
15.0–17.5	16, 16, 10, 11, 10, 16, 16
17.5–20.0	16, 16, 15, 16, 10, 16, 16
20.0–23.0	16, 15, 15
23.0–27.0	16, 15, 16

[a] See Table 8-21 for definition of code letters.

[b] Denotes a wall that is not possible with the chosen parameters.

Source: Reprinted by permission from ASHRAE *Cooling and Heating Load Calculation Manual*, 2nd ed., 1992.

Table 8-22c Wall Types, Mass Located Outside Insulation, for Use with Table 8-19

Secondary Material	R-Factor, (ft²-°F)/Btu	Principal Wall Material[a]														
		A1	A2	B7	B10	B9	C1	C2	C3	C4	C5	C6	C7	C8	C17	C18
	0.0–2.0	b	b	b	b	b	b	b	b	b	b	b	b	b	b	b
	2.0–2.5	b	3	b	b	b	b	b	2	3	5	b	b	b	b	b
	2.5–3.0	b	3	b	b	b	2	2	2	4	5	b	b	5	b	b
	3.0–3.5	b	3	b	b	b	2	2	2	5	5	b	b	5	b	b
	3.5–4.0	b	3	b	b	b	2	2	2	5	5	10	4	6	b	5
	4.0–4.75	b	4	b	b	b	4	2	2	5	5	10	4	6	b	9
	4.75–5.5	b	4	b	b	b	4	2	2	5	6	11	5	10	b	10
Stucco	5.5–6.5	b	5	b	b	b	4	2	2	5	6	11	5	10	b	10
and/or	6.5–7.75	b	5	b	b	b	4	2	2	5	6	11	5	10	b	10
plaster	7.75–9.0	b	5	b	b	b	5	2	4	5	6	16	10	10	b	10
	9.0–10.75	b	5	b	b	b	5	4	4	5	6	16	10	10	4	11
	10.75–12.75	b	5	b	b	b	5	4	4	10	6	16	10	10	9	11
	12.75–15.0	b	5	b	b	b	5	4	4	10	10	b	10	11	9	11
	15.0–17.5	b	5	b	b	b	5	4	4	10	10	b	10	11	10	16
	17.5–20.0	b	5	b	b	b	9	4	4	10	10	b	10	15	10	16
	20.0–23.0	b	9	b	b	b	9	9	9	15	10	b	10	15	15	16
	23.0–27.0	b	b	b	b	b	b	b	b	b	b	b	15	b	15	16
	0.0–2.0	b	b	b	b	b	b	b	b	b	b	b	b	b	b	b
	2.0–2.5	b	3	b	b	b	b	b	2	3	2	b	b	b	b	b
	2.5–3.0	b	3	b	b	b	2	b	2	3	2	b	b	b	b	b
	3.0–3.5	b	3	b	b	b	2	1	2	4	3	b	b	4	b	b
	3.5–4.0	b	3	b	b	b	2	2	2	4	3	5	2	5	b	4
	4.0–4.75	b	3	b	b	b	2	2	2	4	3	10	3	5	b	5

310

Wall material	Weight (lb/ft²)														
Steel or other light-weight siding	4.75–5.5	5	b	5	4	10	3	5	2	2	2	b	b	3	b
	5.5–6.5	5	b	5	4	10	3	5	2	2	2	b	b	4	b
	6.5–7.75	6	b	5	5	11	4	5	2	2	2	b	b	4	b
	7.75–9.0	6	b	5	5	11	4	5	2	2	2	b	b	5	b
	9.0–10.75	10	4	5	5	11	4	5	2	2	2	b	b	5	b
	10.75–12.75	10	4	5	5	11	5	5	2	2	4	b	b	5	b
	12.75–15.0	10	5	10	5	11	5	5	2	2	4	b	b	5	b
	15.0–17.5	10	9	10	9	16	5	5	4	4	4	b	b	5	b
	17.5–20.0	10	10	10	9	16	5	9	4	4	4	b	b	5	b
	20.0–23.0	11	10	10	10	16	9	9	4	4	4	b	b	9	b
	23.0–27.0	15	10	b	10	16	b	b	b	4	b	b	b	b	b
Face brick	0.0–2.0	b	b	b	b	b	b	b	b	b	b	b	b	b	b
	2.0–2.5	b	b	b	b	b	11	b	b	b	b	b	5	3	b
	2.5–3.0	b	b	11	b	b	11	10	5	b	5	b	5	3	b
	3.0–3.5	b	b	11	b	b	11	11	5	5	5	b	5	3	b
	3.5–4.0	b	b	11	10	b	11	11	6	5	10	b	5	3	b
	4.0–4.75	16	b	16	11	b	12	11	10	9	10	b	10	3	b
	4.75–5.5	16	b	16	15	b	12	11	10	10	10	b	10	3	b
	5.5–6.5	16	15	17	16	b	12	12	10	10	10	b	10	4	b
	6.5–7.75	16	15	b	16	b	12	12	10	10	11	b	10	4	b
	7.75–9.0	b	15	b	16	b	12	16	10	10	11	b	15	5	b
	9.0–10.75	b	b	b	16	b	b	16	10	10	11	b	15	5	b
	10.75–12.75	b	b	b	16	b	b	b	11	10	11	b	16	5	b
	12.75–15.0	b	b	b	b	b	b	b	11	11	15	b	16	5	b
	15.0–17.5	b	b	b	b	b	b	b	15	11	16	b	16	5	b
	17.5–20.0	b	b	b	b	b	b	b	15	15	16	b	16	5	b
	20.0–23.0	b	b	b	b	b	b	b	15	15	b	b	16	9	b
	23.0–27.0	b	b	b	b	b	b	b	15	15	b	b	b	b	b

[a]See Table 8-21 for definition of code letters.

[b]Denotes a wall that is not possible with the chosen parameters.

Source: Reprinted by permission from ASHRAE *Cooling and Heating Load Calculation Manual*, 2nd ed., 1992.

Table 8-23 Cooling Load Temperature Difference for

Solar time, hr											
1	2	3	4	5	6	7	8	9	10	11	12
CLTD, F											
1	0	−1	−2	−2	−2	−2	0	2	4	7	9

"The CLTD may be converted to C by multiplying by 5/9.

[b]*Corrections:* The values in the table were calculated for an inside temperature of 78 F and an outdoor Eq. 8-36.

Source: Reprinted by permission from ASHRAE *Cooling and Heating Load Calculation Manual,* 2nd ed.,

The mass is evenly distributed with the concrete blocks being the principal material and the face brick the secondary material. Using Table 8-22*b,* the *R*-value range is 5.5 to 6.5 and the wall type is No. 16. From Table 8-19 for a south wall, the CLTDs are 7 F and 18 F for 10:00 A.M. or 1000 hours and 5:00 P.M. or 1700 hours, respectively. Equation 8-36 is used to correct these values using the design conditions from Example 8-7.

$$\text{CLTD}_{10} = 7 + (78 - 75) + (91 - 85) = 16\,\text{F}$$

$$\text{CLTD}_{17} = 18 + (78 - 75) + (91 - 85) = 27\,\text{F}$$

The cooling load at 10:00 A.M. is

$$\dot{q}_c = UA(\text{CLTD}_{10}) = 0.18(100)(16) = 288\,\text{Btu/hr}$$

and at 5:00 P.M.

$$\dot{q}_c = UA(\text{CLTD}_{17}) = 0.18(100)(27) = 486\,\text{Btu/hr}$$

Fenestration

The cooling load due to fenestration is divided into radiant and conductive loads. The cooling load due to conduction is calculated using the CLTD method and Eq. 8-33. Due to the generally light weight and the small magnitude of this component, the effect of mass and latitude are neglected. CLTDs are given in Table 8-23 and can also be used for doors with reasonable accuracy. Corrections for indoor and outdoor design temperatures and the daily range are the same as those for roofs and walls.

Solar Cooling Load (SCL). The cooling load per square foot of unshaded fenestration due to solar radiation transmitted through and absorbed by the glass is determined by Eq. 8-34.

The subject of solar radiation is discussed in Chapter 6, where it is shown that the Solar Cooling Load (SCL) for a particular zone is dependent on latitude, direction, and internal zone parameters, which affect the absorption and release of radiant energy. SCLs for July 21, 36 deg N latitude and four zone types are given in Table 8-24. To determine the correct zone type, refer to Tables 8-25 where zone types (A, B, C, or D) are given as a function of the various zone parameters.

Shading Coefficients (SC). The ASHRAE procedure for estimating solar heat gain assumes that a constant ratio exists between the solar heat gain through any given type

Conduction Through Glass and Conduction Through Doors[a]

13	14	15	16	17	18	19	20	21	22	23	24
12	13	14	14	13	12	10	8	6	4	3	2

maximum temperature of 95 F with an outdoor daily range of 21 degrees F. Correct for other conditions using

1992.

of fenestration system and the solar heat gain (under exactly the same solar conditions) through unshaded clear sheet glass (i.e., the reference glass, see Chapter 6). This ratio, called the shading coefficient, is unique for each type of fenestration or each combination of glazing and internal shading device. Tables of shading coefficients for various combinations of glass and internal shade are given in Chapter 6.

EXAMPLE 8-9

The building of Example 8-7 has a 4 × 5-ft single glass window with no thermal break in the south wall. The window has light-colored venetian blinds. Compute the cooling load due to the window at 10:00 A.M. and 5:00 P.M. solar time for July using ASHRAE Standard 90.1 design conditions.

SOLUTION

There are two components of cooling load for the window. One is due to heat conduction, the other to solar radiation. Equation 8-33 gives the cooling load due to conduction where the overall coefficient is obtained from Chapter 5 as 1.23 Btu/(hr-ft^2-F). The CLTD values are given in Table 8-23 as 4 F and 13 F for 10:00 A.M. and 5:00 P.M., respectively. Because the design temperature is 106 F and the daily average temperature is 91 F, the table values should be increased by 9 F. The cooling load due to conduction at 10:00 A.M. is

$$\dot{q}_c = 20(1.23)(4 + 9) = 320 \text{ Btu/hr}$$

and for 5:00 P.M.

$$\dot{q}_c = 20(1.23)(13 + 9) = 541 \text{ Btu/hr}$$

The radiation component is computed using Eq. 8-34 and Tables 8-24 and 8-25a. Assume that the zone has 1 or 2 walls, the floor is carpeted, and the partitions are concrete block. Then from Table 8-25a, the zone type is B and from Table 8-24 for south-facing glass the SCLs are 47 and 33 Btu/(hr-ft^2) for 10:00 A.M. and 5:00 P.M., respectively. The shading coefficient (SC) is 0.58 from Table 6-4, Chapter 6. The radiation cooling load component at 10:00 A.M. is

$$\dot{q}_c = 20(0.58)47 = 545 \text{ Btu/hr}$$

and at 5:00 P.M.

$$\dot{q}_c = 20(0.58)33 = 383 \text{ Btu/hr}$$

The total cooling load due to the window at 10:00 A.M. is

$$\dot{q}_c = 320 + 545 = 865 \text{ Btu/hr}$$

and at 5:00 P.M.

$$\dot{q}_c = 541 + 383 = 924 \text{ Btu/hr}$$

The shaded portion of a window is easily estimated using the methods and data presented in Chapter 6. Separate calculations are then made for the sunlit and shaded portions of the glass to obtain the cooling load; however, the cooling load due to conduction is the same for both parts. Common situations are overhangs, side projections, or setback.

The shaded portion of the glass is treated as a north-facing window with the SCL read from Table 8-24 for the north orientation. The shading coefficient is the same for the sunlit and shaded parts.

EXAMPLE 8-10

Compute the cooling load for a 4 × 6 ft-clear plate glass window in an aluminum frame with thermal break facing southwest at 36 deg N latitude. The solar time is 3:00 P.M. on 21 July. A drape of semiopen weave and medium color covers the window on the inside. There is a 2-ft overhang at the top of the glass, which is much wider than the window. Assume ASHRAE Standard 90.1 design conditions. Assume a zone type C.

SOLUTION

The cooling load due to heat conduction is not affected by any external shade that may exist. Therefore, using an overall coefficient of 1.10 for single glass with internal shade and a CLTD of 14 F from Table 8-23

$$\dot{q}_c = 24(1.10)14 = 370 \text{ Btu/hr}$$

The cooling load due to solar heat gain must be done separately for the sunlight and shaded portions; however, the shading coefficient will be the same for each calculation. The drapery material has a II_M classification. Using Fig. 6-10, the index letter ranges from about D to F. To be conservative, D will be used. Then from Table 6-6 the shading coefficient is about 0.65.

To find the shaded portion of the glass, the distance the shadow extends downward is computed using Eqs. 6-32 and 6-33. The solar altitude and azimuth angles are 47.2 and 76.7 deg, respectively, and the wall azimuth is 45 deg. Then

$$\gamma = |\phi - \psi| = |76.7 - 45| = 31.7 \text{ deg}$$

and

$$y = b \tan \beta / \cos \gamma$$

$$y = 2 \tan (47.2) / \cos (31.7) = 2.53 \text{ ft}$$

The shaded area is then

$$A_{sh} = 4(2.53) = 10.12 \text{ ft}^2$$

and the sunlit area is

$$A_{sl} = A - A_{sh} = 24 - 10.12 = 13.88 \text{ ft}^2$$

The cooling load for each part may now be computed. For the sunlit portion, the SCL is 116 Btu/(hr-ft^2) from Table 8-24. Then

$$\dot{q}_{sl} = 13.88(0.65)116 = 1047 \text{ Btu/hr}$$

For the shaded portion, the SCL is 32 Btu/(hr-ft^2) from Table 8-24. Then

$$\dot{q}_{sh} = 10.12(0.65)32 = 211 \text{ Btu/hr}$$

The total cooling load for the window is

$$\dot{q}_t = q_c + q_{sl} + q_{sh}$$

$$\dot{q}_t = 370 + 1047 + 211 = 1628 \text{ Btu/hr}$$

Cooling Load—Internal Sources

Internal sources of heat energy may contribute significantly to the total cooling load of a structure, and poor judgment in the estimation of their magnitude can lead to unsatisfactory operation and/or high costs when part of the capacity is unneeded. These internal sources fall into the general categories of people, lights, and miscellaneous equipment.

People

Chapter 4 contains detailed information containing the rates at which heat and moisture are given up by occupants engaged in different levels of activity, and Table 8-11 summarizes the data needed for heat gain calculations. Refer to Section 8-4 for additional discussion relative to heat gain from occupants.

The heat gain from people has two components: sensible and latent. Latent heat gain goes directly into the air in the space; therefore, this component immediately becomes

Table 8-24 Solar Cooling Load for Sunlit Glass, Btu/(hr-ft²)—36° N Latitude, July

Zone Type A

Glass Facing												Solar Time, hr												
	1	2	3	4	5	6	7	8	9	10	11	12	13	14	15	16	17	18	19	20	21	22	23	24
N	0	0	0	0	0	25	29	28	32	36	39	40	41	39	36	32	33	36	12	6	3	1	1	0
NE	0	0	0	0	0	79	129	139	120	84	58	50	45	41	37	32	26	17	7	3	2	1	0	0
E	0	0	0	0	0	86	153	184	182	155	107	67	54	45	39	33	26	17	7	3	2	1	0	0
SE	0	0	0	0	0	42	90	125	142	140	119	86	58	48	40	34	27	17	7	3	2	1	0	0
S	0	0	0	0	0	8	17	24	36	53	70	80	79	68	52	38	29	18	7	3	2	1	0	0
SW	0	0	0	0	0	8	17	24	30	35	38	57	90	122	141	144	127	85	32	15	8	4	2	1
W	0	0	0	0	0	8	17	24	30	35	38	40	66	115	159	188	191	149	53	25	12	6	3	2
NW	0	0	0	0	0	8	17	24	30	35	38	40	40	56	93	129	148	127	43	21	10	5	2	1
hor	0	0	0	0	0	20	66	120	171	215	246	263	265	251	221	178	124	66	28	13	7	3	2	1

Zone Type B

Glass Facing												Solar Time, hr												
	1	2	3	4	5	6	7	8	9	10	11	12	13	14	15	16	17	18	19	20	21	22	23	24
N	2	2	1	1	1	21	25	25	29	33	36	38	38	38	35	32	33	35	15	10	7	5	4	3
NE	2	1	1	1	1	68	109	120	108	81	61	54	50	46	42	37	30	22	12	9	6	5	3	3
E	2	2	1	1	1	73	130	158	161	143	106	75	63	55	48	41	34	25	14	10	7	5	4	3
SE	2	2	1	1	1	36	77	107	124	125	111	85	64	55	48	41	33	24	14	10	7	5	4	3
S	2	2	1	1	1	7	14	21	31	47	61	71	72	65	52	41	33	24	13	9	7	5	4	3
SW	6	4	3	3	2	8	15	21	27	31	35	51	80	108	126	131	119	86	43	29	20	14	11	8
W	8	6	5	4	3	9	16	22	27	32	35	37	60	101	140	166	172	141	63	42	29	20	15	11
NW	6	5	4	3	2	8	15	21	27	31	35	37	38	52	84	115	132	117	49	32	22	16	11	8
hor	8	6	5	4	3	19	57	103	148	188	218	237	244	237	215	182	137	88	53	37	26	19	14	11

Zone Type C

Glass Facing												Solar Time, hr												
	1	2	3	4	5	6	7	8	9	10	11	12	13	14	15	16	17	18	19	20	21	22	23	24
N	5	5	4	4	3	24	25	24	27	31	33	35	35	34	32	29	31	34	14	10	8	7	6	6
NE	7	6	6	5	5	71	106	111	95	68	51	48	46	44	41	37	32	24	16	13	11	10	9	8
E	9	8	8	7	6	77	128	148	145	124	89	62	56	52	47	43	37	29	20	17	15	13	12	11
SE	8	8	7	6	5	40	77	102	114	112	97	73	55	49	45	40	35	27	18	15	13	12	11	9
S	6	6	5	4	4	10	17	22	31	45	58	65	65	57	45	35	30	22	14	11	10	9	8	7

Solar Time, hr

Glass Facing	1	2	3	4	5	6	7	8	9	10	11	12	13	14	15	16	17	18	19	20	21	22	23	24
SW	13	12	10	9	8	14	20	25	29	32	35	50	77	102	116	118	105	74	34	26	21	18	16	14
W	16	15	13	12	11	16	22	27	31	34	36	37	59	98	132	154	155	122	48	34	28	24	21	18
NW	12	11	10	9	8	14	20	25	29	32	35	36	36	50	80	108	122	104	37	26	21	18	15	14
hor	24	22	19	17	16	31	66	107	145	178	203	217	220	212	192	161	122	81	53	44	38	34	30	27

Zone Type D

Solar Time, hr

Glass Facing	1	2	3	4	5	6	7	8	9	10	11	12	13	14	15	16	17	18	19	20	21	22	23	24
N	8	7	6	6	5	21	22	21	25	27	30	31	32	32	31	29	30	32	17	14	12	11	10	9
NE	11	10	9	8	7	59	87	93	82	63	51	49	47	46	43	40	35	29	22	19	17	15	14	12
E	15	13	12	11	10	65	105	123	124	110	84	65	60	57	53	48	43	36	28	25	22	20	18	16
SE	13	12	11	10	9	36	65	85	96	97	87	70	56	52	49	45	40	33	25	22	20	18	16	15
S	9	9	8	7	6	11	16	20	27	39	49	56	57	52	43	36	31	26	19	16	15	13	12	11
SW	20	18	16	14	13	17	21	25	28	31	33	45	67	87	100	103	95	72	41	35	30	27	24	22
W	25	22	20	18	16	20	24	27	30	33	34	36	53	84	112	131	134	111	55	45	39	34	31	28
NW	18	17	15	14	12	16	20	24	27	30	32	34	34	45	69	92	104	92	42	34	29	26	23	21
hor	37	33	30	27	24	35	62	94	125	153	174	189	195	191	179	157	128	95	72	63	56	51	46	41

Notes: 1. Direct application of data
- Standard double-strength glass with no inside shade
- Clear sky, 21st day of month

2. Adjustments to table data
- Latitudes other than 36 degrees N

  A table for a specific latitude may be generated. See text.
- Months other than July

  For design purposes, the data will suffice for about 2 weeks from the 21st day of given month.

  Tables may be generated for a specific month. See text.
- Other types of glass and internal shade

  Use shading coefficients as multiplier. See text.
- Externally shaded glass

  Use north orientation. See text.

3. To convert SCL to W/m^2 multiply by 3.154.

Source: Reprinted by permission from ASHRAE *Cooling and Heating Load Calculation Manual*, 2nd ed., 1992.

cooling load with no delay. However, the sensible component from a person is delayed due to storage of a part of this energy in the room and furnishings. The cooling load factor is used to express this delayed effect in Eq. 8-35. The CLF depends on the total hours the occupants are in the space and varies from the time of entry. Table 8-26 gives CLF values for people and Tables 8-25a through 8-25e give the zone types. The data are for continuously operating cooling equipment. When the cooling equipment is turned on and off when the occupants arrive and leave, the CLF is 1.0. This is because any energy stored in the building at closing time is still essentially there the next morning. When the cooling equipment is operated for several hours after occupancy, however, the CLF values are about the same as for continuous equipment operation. When the density of people is large, such as in a theater, a CLF of 1.0 should be used.

EXAMPLE 8-11

An office suite is designed with 10 private offices, a secretarial area with space for four secretaries, a reception area and waiting room, and an executive office with a connecting office for a secretary. Estimate the cooling load from the occupants at 3:00 P.M. solar time. Assume zone type B.

SOLUTION

In the absence of additional data the following approach seems reasonable. For the private offices assume that an average of 7 out of the 10 will be occupied between 8:00 A.M. and 5:00 P.M. Assume that 3 of the 4 secretaries are always present and a receptionist is always there. The waiting room will have a transient occupancy, assume 2 people. The executive office will probably experience variable occupancy. However, an average of one occupant is about right. Assume that the secretary is always present. The total number of people for which the heat gain is to be based is then 15. We will assume sedentary, very light work and use data from Table 8-11. Assuming an adjusted group of males and females and very light work, the sensible and latent heat gains per person are 245 and 155 Btu/hr, respectively. The latent cooling load due to people is

$$\dot{q}_l = 15(155) = 2325 \text{ Btu/hr}$$

There is a question about where the occupants go for lunch. Assuming that they stay in the space, then the total hours in the space are 9, and 3:00 P.M. represents the seventh hour after entry. From Table 8-26, CLF is 0.93. The cooling load due to sensible heat is given by Eq. 8-35 as

$$\dot{q}_s = 15(245)0.93 = 3418 \text{ Btu/hr}$$

and the total cooling load at 3:00 P.M. is

$$\dot{q} = \dot{q}_l + \dot{q}_s = 2325 + 3418 = 5743 \text{ Btu/hr}$$

Note that the occupants could enter and leave the space on different schedules. In such a case, make a separate calculation for each group.

Lights

The cooling load due to lighting is often the major component of the space load and an accurate estimate is essential. See Section 8-4 for complete discussion of light heat gain.

The designer should be careful to use the correct wattage in making calculations. It is not good practice to assume nominal wattage per unit area. Additionally, all the installed lights may not be used all the time. The instantaneous heat gain for lights may be expressed by Eq. 8-13 and cooling load is then given by Eq. 8-35.

The cooling load factor is a function of the building mass, air-circulation rate, type of fixture, and time. Table 8-27 gives cooling load factors as a function of time for lights that are on for 8 through 16 hours. The data in Table 8-27 are for typical recessed fluorescent fixtures. Tables 8-25a through 8-25e give the zone types for use in Table 8-27. The CLF values of Table 8-27 are for the case of continuously operating cooling equipment. If the cooling equipment is turned on and off on the same schedule as the lights, the CLF is equal to 1.0 because this condition is similar to the case when lights and cooling equipment are on continuously. When the cooling equipment is operated for several hours after the lights are used, however, the system behaves as if the cooling equipment were operating continuously.

In the case of light fixtures that are cooled by the return air (vented fixtures or

Table 8-25a Zone Types for Use with SCL and CLF Tables, Single-Story Building

No. Walls	Floor Covering	Partition Type	Inside Shade	Glass Solar	People and Equipment	Lights
1 or 2	Carpet	Gypsum	[b]	A	B	B
1 or 2	Carpet	Concrete block	[b]	B	C	C
1 or 2	Vinyl	Gypsum	Full	B	C	C
1 or 2	Vinyl	Gypsum	Half to None	C	C	C
1 or 2	Vinyl	Concrete block	Full	C	D	D
1 or 2	Vinyl	Concrete block	Half to None	D	D	D
3	Carpet	Gypsum	[b]	A	B	B
3	Carpet	Concrete block	Full	A	B	B
3	Carpet	Concrete block	Half to None	B	B	B
3	Vinyl	Gypsum	Full	B	C	C
3	Vinyl	Gypsum	Half to None	C	C	C
3	Vinyl	Concrete block	Full	B	C	C
3	Vinyl	Concrete block	Half to None	C	C	C
4	Carpet	Gypsum	[b]	A	B	B
4	Vinyl	Gypsum	Full	B	C	C
4	Vinyl	Gypsum	Half to None	C	C	C

[a]A total of 14 zone parameters is fully defined in Table 8-1. Those not shown in this table were selected to achieve minimum error band.

[b]The effect of inside shade is negligible in this case.

Source: Reprinted by permission from ASHRAE *Cooling and Heating Load Calculation Manual,* 2nd ed., 1992.

Table 8-25b Zone Types for Use with SCL and CLF Tables, Top Floor of Multistory Building

No. Walls	Mid-Floor Type	Ceiling Type	Floor Covering	Partition Type	Inside Shade	Glass Solar	People and Equipment	Lights
	Zone Parameters[a]					Zone Type		
1 or 2	2.5-in. concrete	With	Carpet	Gypsum	Full	A	A	B
	2.5-in. concrete	With	Carpet	Gypsum	Half to None	A	A	B
	2.5-in. concrete	With	Carpet	Concrete block	Full	B	A	C
	2.5-in. concrete	With	Carpet	Concrete block	Half to None	B	A	C
	2.5-in. concrete	With	Vinyl	Gypsum	Full	B	A	C
	2.5-in. concrete	With	Vinyl	Gypsum	Half to None	C	A	C
	2.5-in. concrete	With	Vinyl	Concrete block	Full	C	B	D
	2.5-in. concrete	With	Vinyl	Concrete block	Half to None	C	B	D
	2.5-in. concrete	Without	Carpet	Gypsum	b	A	A	B
	2.5-in. concrete	Without	Carpet	Concrete block	b	A	A	C
	2.5-in. concrete	Without	Vinyl	Gypsum	b	A	A	B
	2.5-in. concrete	Without	Vinyl	Concrete block	Full	B	A	C
	2.5-in. concrete	Without	Vinyl	Concrete block	Half to None	B	A	C
	1-in. wood	b	b	Gypsum	b	A	A	B
	1-in. wood	b	b	Concrete block	b	A	B	C
3	2.5-in. concrete	With	Carpet	Gypsum	Full	A	B	B
	2.5-in. concrete	With	Carpet	Gypsum	Half to None	A	A	B
	2.5-in. concrete	With	Carpet	Concrete block	Full	A	A	C

Zone	Mid-floor	Partition	Floor covering	Interior wall	Furniture			
	2.5-in. concrete	With	Carpet	Concrete block	Half to None	B	A	C
	2.5-in. concrete	With	Vinyl	Gypsum	Full	B	A	C
	2.5-in. concrete	With	Vinyl	Gypsum	Half to None	C	A	C
	2.5-in. concrete	With	Vinyl	Concrete block	Full	B	A	C
	2.5-in. concrete	With	Vinyl	Concrete block	Half to None	C	A	C
	2.5-in. concrete	Without	Carpet	Gypsum	*b*	A	A	B
	2.5-in. concrete	Without	Vinyl	Gypsum	Full	A	A	B
	2.5-in. concrete	Without	Vinyl	Concrete block	Half to None	A	A	B
	2.5-in. concrete	Without	Vinyl	Concrete block	Full	A	A	B
	2.5-in. concrete	Without	Vinyl	Gypsum	Half to None	B	A	B
	1-in. wood	*b*	*b*	Gypsum	*b*	A	A	B
	1-in. wood	*b*	*b*	Concrete block	*b*	A	A	B
4	2.5-in. concrete	With	Carpet	Gypsum	Full	A	A	B
	2.5-in. concrete	With	Carpet	Gypsum	Half to None	B	B	B
	2.5-in. concrete	With	Vinyl	Gypsum	Full	B	C	C
	2.5-in. concrete	With	Vinyl	Gypsum	Half to None	B	B	C
	2.5-in. concrete	Without	Carpet	Gypsum	Full	A	A	B
	2.5-in. concrete	Without	Carpet	Gypsum	Half to None	A	A	B
	2.5-in. concrete	Without	Vinyl	Gypsum	Full	A	A	B
	2.5-in. concrete	Without	Vinyl	Gypsum	Half to None	A	A	B
	1-in. wood	*b*	*b*	*b*	*b*	B	B	B

[a] A total of 14 zone parameters is fully defined in Table 8-1. Those not shown in this table were selected to achieve minimum error band.

[b] The effect of this parameter is negligible in this case.

Source: Reprinted by permission from ASHRAE *Cooling and Heating Load Calculation Manual,* 2nd ed., 1992.

Table 8-25c Zone Types for Use with SCL and CLF Tables, First Floor of Multistory Building

No. Walls	Zone Parameters[a]					Zone Type		
	Mid-Floor	Ceiling Type	Floor Covering	Partition Type	Inside Shade	Glass Solar	People and Equipment	Lights
1 or 2	2.5-in. concrete	With	Carpet	Gypsum	Full	A	C	B
	2.5-in. concrete	With	Carpet	Gypsum	Half to None	B	C	B
	2.5-in. concrete	With	Carpet	Concrete block	Full	B	D	C
	2.5-in. concrete	With	Carpet	Concrete block	Half to None	C	D	C
	2.5-in. concrete	With	Vinyl	Gypsum	Full	C	D	D
	2.5-in. concrete	With	Vinyl	Gypsum	Half to None	C	D	D
	2.5-in. concrete	With	Vinyl	Concrete block	Full	D	D	D
	2.5-in. concrete	With	Vinyl	Concrete block	Half to None	D	D	D
	2.5-in. concrete	Without	Carpet	Gypsum	b	B	C	B
	2.5-in. concrete	Without	Carpet	Concrete block	Full	C	D	C
	2.5-in. concrete	Without	Carpet	Concrete block	Half to None	C	D	C
	2.5-in. concrete	Without	Vinyl	Gypsum	Full	C	D	D
	2.5-in. concrete	Without	Vinyl	Gypsum	Half to None	D	D	D
	2.5-in. concrete	Without	Vinyl	Concrete block	Full	C	D	D
	2.5-in. concrete	Without	Vinyl	Concrete block	Half to None	D	D	D
	1-in. wood	b	Carpet	Gypsum	Full	A	A	B
	1-in. wood	b	Carpet	Gypsum	Half to None	B	A	B
	1-in. wood	b	Carpet	Concrete block	Full	B	B	C
	1-in. wood	b	Carpet	Concrete block	Half to None	C	B	C
	1-in. wood	b	Vinyl	Gypsum	Full	B	B	B
	1-in. wood	b	Vinyl	Gypsum	Half to None	C	B	B
	1-in. wood	b	Vinyl	Concrete block	Full	C	C	D
	1-in. wood	b	Vinyl	Concrete block	Half to None	D	C	D
3	2.5-in. concrete	With	Carpet	Gypsum	Full	A	C	B

2.5-in. concrete	With	Carpet	Gypsum	Half to None	B	C	B
2.5-in. concrete	With	Carpet	Concrete block	b	B	C	B
2.5-in. concrete	With	Vinyl	Gypsum	Full	C	D	C
2.5-in. concrete	With	Vinyl	b	Half to None	C	D	C
2.5-in. concrete	With	Vinyl	Concrete block	Full	C	D	C
2.5-in. concrete	Without	Carpet	Gypsum	b	B	C	B
2.5-in. concrete	Without	Carpet	Concrete block	Full	B	C	B
2.5-in. concrete	Without	Carpet	Concrete block	Half to None	C	C	B
2.5-in. concrete	Without	Vinyl	Gypsum	b	C	D	C
2.5-in. concrete	Without	Vinyl	Gypsum	Half to None	C	D	C
2.5-in. concrete	Without	Vinyl	Concrete block	Full	C	D	C
1-in. wood	b	Carpet	Gypsum	Full	A	A	B
1-in. wood	b	Carpet	Gypsum	Half to None	B	A	B
1-in. wood	b	Carpet	Concrete block	b	B	B	B
1-in. wood	b	Vinyl	Gypsum	Full	B	B	C
1-in. wood	b	Vinyl	Gypsum	Half to None	C	B	C
1-in. wood	b	Vinyl	Concrete block	Full	C	B	C
1-in. wood	b	Vinyl	Concrete block	Half to None	C	B	C
4	2.5-in. concrete	With	Carpet	Gypsum	Full	A	B	B
2.5-in. concrete	With	Carpet	Gypsum	Half to None	B	B	B
2.5-in. concrete	With	Vinyl	Gypsum	b	C	C	C
2.5-in. concrete	Without	Carpet	Gypsum	b	B	B	C
2.5-in. concrete	Without	Vinyl	Gypsum	b	B	B	C
1-in. wood	b	Carpet	Gypsum	Full	A	B	A
1-in. wood	b	Carpet	Gypsum	Half to None	A	B	A
1-in. wood	b	Vinyl	Gypsum	Full	B	B	B
1-in. wood	b	Vinyl	Gypsum	Half to None	C	B	B

[a] A total of 14 zone parameters is fully defined in Table 8-1. Those not shown in this table were selected to achieve minimum error band.

[b] The effect of this parameter is negligible in this case.

Source: Reprinted by permission from ASHRAE *Cooling and Heating Load Calculation Manual*, 2nd ed., 1992.

Table 8-25d Zone Types for Use with SCL and CLF Tables, Middle Floor of Multistory Building

No. Walls	Zone Parameters[a]					Zone Type		
	Mid-Floor Type	Ceiling Type	Floor Covering	Partition Type	Inside Shade	Glass Solar	People and Equipment	Lights
1 or 2	2.5-in. concrete	With	Carpet	Gypsum	b	B	B	C
	2.5-in. concrete	With	Carpet	Concrete block	Full	C	C	C
	2.5-in. concrete	With	Carpet	Concrete block	Half to None	C	C	C
	2.5-in. concrete	With	Vinyl	Gypsum	Full	C	D	D
	2.5-in. concrete	With	Vinyl	Gypsum	Half to None	D	D	D
	2.5-in. concrete	With	Vinyl	Concrete block	b	D	D	D
	2.5-in. concrete	Without	Carpet	Gypsum	b	B	B	D
	2.5-in. concrete	Without	Carpet	Concrete block	b	C	C	D
	2.5-in. concrete	Without	Vinyl	Gypsum	Full	B	C	C
	2.5-in. concrete	Without	Vinyl	Gypsum	Half to None	C	C	C
	2.5-in. concrete	Without	Vinyl	Concrete block	b	C	C	D
	1-in. wood	b	b	Gypsum	b	A	A	A
	1-in. wood	b	b	Concrete block	b	B	A	B
3	2.5-in. concrete	With	Carpet	Gypsum	b	B	B	C
	2.5-in. concrete	With	Carpet	Concrete block	Full	B	C	C
	2.5-in. concrete	With	Carpet	Concrete block	Half to none	C	C	C

Group	Middle floor	Interior shade	Floor covering	Partition	Inside shade				
	2.5-in. concrete	With	Vinyl	Gypsum	Full	C	C	D	D
	2.5-in. concrete	With	Vinyl	Gypsum	Half to None	D	D	D	D
	2.5-in. concrete	With	Vinyl	Concrete block	Full	C	C	D	D
	2.5-in. concrete	With	Vinyl	Concrete block	Half to None	D	D	D	D
	2.5-in. concrete	Without	Carpet	Gypsum	b	B	B	B	C
	2.5-in. concrete	Without	Carpet	Concrete block	Full	C	C	C	C
	2.5-in. concrete	Without	Carpet	Concrete block	Half to None	B	B	C	C
	2.5-in. concrete	Without	Vinyl	Gypsum	Full	B	C	C	C
	2.5-in. concrete	Without	Vinyl	Gypsum	Half to None	C	B	C	C
	2.5-in. concrete	Without	Vinyl	Concrete block	Full	B	C	C	C
	1-in. wood	b	b	Gypsum	b	A	A	A	A
	1-in. wood	b	b	Concrete block	b	A	A	A	B
4	2.5-in. concrete	With	Carpet	Gypsum	b	B	B	C	C
	2.5-in. concrete	With	Vinyl	Gypsum	Full	C	C	C	C
	2.5-in. concrete	With	Vinyl	Gypsum	Half to None	C	C	C	C
	2.5-in. concrete	Without	Carpet	Gypsum	b	B	A	B	C
	2.5-in. concrete	Without	Vinyl	Gypsum	Full	B	A	A	C
	2.5-in. concrete	Without	Vinyl	Gypsum	Half to None	B	A	A	C
	1-in. wood	b	b	b	b	A	A	A	B

[a]A total of 14 zone parameters is fully defined in Table 8-1. Those not shown in this table were selected to achieve minimum error band.

[b]The effect of this parameter is negligible in this case.

Source: Reprinted by permission from ASHRAE *Cooling and Heating Load Calculation Manual*, 2nd ed., 1992.

Table 8-25e Zone Types for Use with CLF Tables, Interior Rooms

	Zone Parameters[a]			Zone Type	
Room Location	Middle Floor	Ceiling Type	Floor Covering	People and Equipment	Lights
Single story	N/A	N/A	Carpet	C	B
	N/A	N/A	Vinyl	D	C
Top floor	2.5-in. concrete	With	Carpet	D	C
	2.5-in. concrete	With	Vinyl	D	D
	2.5-in. concrete	Without	[b]	D	B
	1-in. wood	[b]	[b]	D	B
Bottom floor	2.5-in. concrete	With	Carpet	D	C
	2.5-in. concrete	[b]	Vinyl	D	D
	2.5-in. concrete	Without	Carpet	D	D
	1-in. wood	[b]	Carpet	D	C
	1-in. wood	[b]	Vinyl	D	D
Mid-floor	2.5-in. concrete	N/A	Carpet	D	C
	2.5-in. concrete	N/A	Vinyl	D	D
	1-in. wood	N/A	[b]	C	B

A total of 14 zone parameters is fully defined in Table 8-1. Those not shown in this table were selected to achieve a minimum error band.

[b]The effect of this parameter is negligible in this case.

Source: Reprinted by permission from ASHRAE *Cooling and Heating Load Calculation Manual,* 2nd ed., 1992.

recessed fixtures in a ceiling return plenum), the cooling load given by Eq. 8-35 is not all imposed on the space. The return air carries part of the load directly back to the coil. The amount of load absorbed by the return air is a function of many variables but can be up to 50 percent of the light power.

EXAMPLE 8-12

The office suite of Example 8-11 has total installed light wattage of 8400 W. The fluorescent light fixtures are recessed with 40-W lamps. Supply air is through the ceiling with the air returning through the ceiling plenum. The lights are turned on at 8:00 A.M. and turned off at 6:00 P.M. Estimate the cooling load at 10:00 A.M. and 4:00 P.M. Assume a zone type B.

SOLUTION

Assuming that about 15 percent of the lights are off all the time due to unoccupied offices, the use Factor F_u is 0.85. The ballast factor F_s is 1.2. Then from Eq. 8-13

$$\dot{q}_i = 3.41(8400)0.85(1.2) = 29{,}217 \, \text{Btu/hr}$$

From Table 8-27, CLF_{10} is 0.86 and CLF_{16} is 0.96. Then

$$\dot{q}_{10} = 29,217(0.86) = 25,127 \text{ Btu/hr}$$

$$\dot{q}_{15} = 29,217(0.96) = 28,048 \text{ Btu/hr}$$

Since the return air is flowing through the ceiling space over the light fixtures, some of the cooling load due to the lights never reaches the space. Let us assume a 20 percent reduction in the load to the space. Then

$$\dot{q}_{10} = (1 - 0.2)(25,127) = 20,102 \text{ Btu/hr}$$

$$\dot{q}_{15} = (1 - 0.2)(28,048) = 22,438 \text{ Btu/hr}$$

One must remember, however, that the 20 percent reduction in space load must be added to the coil load.

Appliances and Equipment

Many appliances have both a sensible and a latent component. The latent component of heat gain immediately becomes cooling load. The sensible component of heat gain from unhooded appliances and equipment is assumed to have the same radiant and convective split as people, allowing use of the same data, from Table 8-26, to obtain the cooling load factor (CLF). See Section 8-4 for additional discussion of heat gain for appliances and equipment.

The convective portion of the heat gain from hooded appliances is assumed to be completely exhausted from the space, leaving only the radiant portion to influence the cooling load. Separate cooling load factors are given for hooded appliances in Table 8-28.

When details about the appliance or equipment are known (size and input), the sensible cooling load due to the appliance is calculated using Eq. 8-35.

The CLF is dependent on the zone type. The CLF = 1.0 when the cooling system does not operate 24 hours per day.

EXAMPLE 8-13

Suppose the office suite previously described has a collection of typewriters, duplicating machines, electric coffee pot, and the like, with a total nameplate rating of 3.7 kW. It is estimated that only about one-half of the equipment is in use on a continuous basis. Estimate the cooling load at 5:00 P.M. The zone type is B.

SOLUTION

Assume that the equipment is turned on at 8:00 A.M. and operates until 4:00 P.M. From Table 8-26 the CLF is 0.31 because 5:00 P.M. is 9 hours after startup time and the equipment operates only 8 hours. The cooling load is

$$\dot{q} = 3.7(3412)(0.31)/2 = 1950 \text{ Btu/hr}$$

Table 8-26 Cooling Load Factors for People and Unhooded Equipment[a]

Hours in Space	Number of Hours after Entry into Space or Equipment Turned On																							
	1	2	3	4	5	6	7	8	9	10	11	12	13	14	15	16	17	18	19	20	21	22	23	24
Zone Type A																								
2	0.75	0.88	0.18	0.08	0.04	0.02	0.01	0.01	0.01	0.01	0.00	0.00	0.00	0.00	0.00	0.00	0.00	0.00	0.00	0.00	0.00	0.00	0.00	0.00
4	0.75	0.88	0.93	0.95	0.22	0.10	0.05	0.03	0.02	0.02	0.01	0.01	0.01	0.01	0.00	0.00	0.00	0.00	0.00	0.00	0.00	0.00	0.00	0.00
6	0.75	0.88	0.93	0.95	0.97	0.97	0.33	0.11	0.06	0.04	0.03	0.02	0.02	0.01	0.01	0.01	0.01	0.00	0.00	0.00	0.00	0.00	0.00	0.00
8	0.75	0.88	0.93	0.95	0.97	0.97	0.98	0.98	0.24	0.11	0.06	0.04	0.03	0.02	0.02	0.01	0.01	0.01	0.01	0.01	0.00	0.00	0.00	0.00
10	0.75	0.88	0.93	0.95	0.97	0.97	0.98	0.98	0.99	0.99	0.24	0.12	0.07	0.04	0.03	0.02	0.02	0.01	0.01	0.01	0.01	0.01	0.00	0.00
12	0.75	0.88	0.93	0.96	0.97	0.98	0.98	0.98	0.99	0.99	0.99	0.99	0.25	0.12	0.07	0.04	0.03	0.02	0.02	0.02	0.01	0.01	0.01	0.01
14	0.76	0.88	0.93	0.96	0.97	0.98	0.98	0.99	0.99	0.99	0.99	0.99	1.00	1.00	0.25	0.12	0.07	0.05	0.03	0.03	0.02	0.02	0.01	0.01
16	0.76	0.89	0.94	0.96	0.97	0.98	0.98	0.99	0.99	0.99	0.99	0.99	1.00	1.00	1.00	1.00	0.25	0.12	0.07	0.05	0.03	0.03	0.02	0.02
18	0.77	0.89	0.94	0.96	0.97	0.98	0.98	0.99	0.99	0.99	0.99	1.00	1.00	1.00	1.00	1.00	1.00	1.00	0.25	0.12	0.07	0.05	0.03	0.03
Zone Type B																								
2	0.65	0.74	0.16	0.11	0.08	0.06	0.05	0.04	0.03	0.02	0.02	0.01	0.01	0.01	0.01	0.00	0.00	0.00	0.00	0.00	0.00	0.00	0.00	0.00
4	0.65	0.75	0.81	0.85	0.24	0.17	0.13	0.10	0.07	0.06	0.04	0.03	0.03	0.02	0.02	0.01	0.01	0.01	0.01	0.00	0.00	0.00	0.00	0.00
6	0.65	0.75	0.81	0.85	0.89	0.91	0.29	0.20	0.15	0.12	0.09	0.07	0.05	0.04	0.03	0.02	0.02	0.01	0.01	0.01	0.01	0.01	0.00	0.00
8	0.65	0.75	0.81	0.85	0.89	0.91	0.93	0.95	0.31	0.22	0.17	0.13	0.10	0.08	0.06	0.05	0.04	0.03	0.02	0.02	0.01	0.01	0.01	0.01
10	0.65	0.75	0.81	0.85	0.89	0.91	0.93	0.95	0.96	0.97	0.33	0.24	0.18	0.14	0.11	0.08	0.06	0.05	0.04	0.03	0.02	0.02	0.01	0.01
12	0.66	0.76	0.81	0.86	0.89	0.92	0.94	0.95	0.96	0.97	0.98	0.98	0.34	0.24	0.19	0.14	0.11	0.08	0.06	0.05	0.04	0.03	0.02	0.02
14	0.67	0.76	0.82	0.86	0.89	0.92	0.94	0.95	0.96	0.97	0.98	0.98	0.99	0.99	0.35	0.25	0.19	0.15	0.11	0.09	0.07	0.05	0.04	0.03
16	0.69	0.78	0.83	0.87	0.90	0.92	0.94	0.95	0.96	0.97	0.98	0.98	0.99	0.99	0.99	0.99	0.35	0.25	0.19	0.15	0.11	0.09	0.07	0.05
18	0.71	0.80	0.85	0.88	0.91	0.93	0.95	0.96	0.97	0.98	0.98	0.99	0.99	0.99	0.99	0.99	1.00	1.00	0.35	0.25	0.19	0.15	0.11	0.09

Zone Type C

2	0.60	0.68	0.14	0.11	0.09	0.07	0.06	0.05	0.04	0.03	0.03	0.02	0.02	0.01	0.01	0.01	0.01	0.01	0.01	0.00	0.00	0.00	0.00	0.00
4	0.60	0.68	0.74	0.79	0.23	0.18	0.14	0.12	0.10	0.08	0.06	0.05	0.04	0.04	0.03	0.02	0.02	0.02	0.01	0.01	0.01	0.01	0.01	0.01
6	0.61	0.69	0.74	0.79	0.83	0.86	0.28	0.22	0.18	0.15	0.12	0.10	0.08	0.07	0.06	0.05	0.04	0.03	0.03	0.02	0.02	0.01	0.01	0.01
8	0.61	0.69	0.75	0.79	0.83	0.86	0.89	0.91	0.32	0.26	0.21	0.17	0.14	0.11	0.09	0.08	0.06	0.05	0.04	0.04	0.03	0.02	0.02	0.02
10	0.62	0.70	0.75	0.80	0.83	0.86	0.89	0.91	0.92	0.94	0.35	0.28	0.23	0.18	0.15	0.12	0.10	0.08	0.07	0.06	0.05	0.04	0.03	0.03
12	0.63	0.71	0.76	0.81	0.84	0.87	0.89	0.91	0.93	0.94	0.95	0.96	0.37	0.29	0.24	0.19	0.16	0.13	0.11	0.09	0.07	0.06	0.05	0.04
14	0.65	0.72	0.77	0.82	0.85	0.88	0.90	0.92	0.93	0.94	0.95	0.96	0.97	0.97	0.38	0.30	0.25	0.20	0.17	0.14	0.11	0.09	0.08	0.06
16	0.68	0.74	0.79	0.83	0.86	0.89	0.91	0.92	0.94	0.95	0.96	0.96	0.97	0.98	0.98	0.98	0.39	0.31	0.25	0.21	0.17	0.14	0.11	0.09
18	0.72	0.78	0.82	0.85	0.88	0.90	0.92	0.93	0.94	0.95	0.96	0.97	0.97	0.98	0.98	0.99	0.99	0.99	0.39	0.31	0.26	0.21	0.17	0.14

Zone Type D

2	0.59	0.67	0.13	0.09	0.08	0.06	0.05	0.05	0.04	0.04	0.03	0.03	0.02	0.02	0.02	0.01	0.01	0.01	0.01	0.01	0.01	0.01	0.01	0.00
4	0.60	0.67	0.72	0.76	0.20	0.16	0.13	0.11	0.10	0.08	0.07	0.06	0.05	0.05	0.04	0.03	0.03	0.03	0.02	0.02	0.02	0.01	0.01	0.01
6	0.61	0.68	0.73	0.77	0.80	0.83	0.26	0.20	0.17	0.15	0.13	0.11	0.09	0.08	0.07	0.06	0.05	0.05	0.04	0.03	0.03	0.03	0.02	0.02
8	0.62	0.69	0.74	0.77	0.80	0.83	0.85	0.87	0.30	0.24	0.20	0.17	0.15	0.13	0.11	0.10	0.08	0.07	0.06	0.05	0.05	0.04	0.04	0.03
10	0.63	0.70	0.75	0.78	0.81	0.84	0.86	0.88	0.89	0.91	0.33	0.27	0.22	0.19	0.17	0.14	0.12	0.11	0.09	0.08	0.07	0.06	0.05	0.05
12	0.65	0.71	0.76	0.79	0.82	0.84	0.87	0.88	0.90	0.91	0.92	0.93	0.35	0.29	0.24	0.21	0.18	0.16	0.13	0.12	0.10	0.09	0.08	0.07
14	0.67	0.73	0.78	0.81	0.83	0.86	0.88	0.89	0.91	0.92	0.93	0.94	0.95	0.95	0.37	0.30	0.25	0.22	0.19	0.16	0.14	0.12	0.11	0.09
16	0.70	0.76	0.80	0.83	0.85	0.87	0.89	0.90	0.92	0.93	0.94	0.95	0.95	0.96	0.96	0.97	0.38	0.31	0.26	0.23	0.20	0.17	0.15	0.13
18	0.74	0.80	0.83	0.85	0.87	0.89	0.91	0.92	0.93	0.94	0.95	0.95	0.96	0.97	0.97	0.97	0.98	0.98	0.39	0.32	0.27	0.23	0.20	0.17

[a]See Table 8-25 for zone type. Data based on a radiative/convective fraction of 0.70/0.30.

Adjustment for other radiative/convective fractions is discussed in Sec. 8-5.

Source: Reprinted by permission from ASHRAE *Cooling and Heating Load Calculation Manual*, 2nd ed., 1992.

Table 8-27 Cooling Load Factors for Lights[a]

Lights On For	\multicolumn{24}{c}{Number of Hours after Lights Turned On}

Lights On For	1	2	3	4	5	6	7	8	9	10	11	12	13	14	15	16	17	18	19	20	21	22	23	24
Zone Type A																								
8	0.85	0.92	0.95	0.96	0.97	0.97	0.97	0.98	0.13	0.06	0.04	0.03	0.02	0.02	0.02	0.01	0.01	0.01	0.01	0.01	0.01	0.01	0.01	0.01
10	0.85	0.93	0.95	0.97	0.97	0.97	0.98	0.98	0.98	0.98	0.14	0.07	0.04	0.03	0.02	0.02	0.02	0.02	0.02	0.02	0.01	0.01	0.01	0.01
12	0.86	0.93	0.96	0.97	0.97	0.98	0.98	0.98	0.98	0.98	0.98	0.98	0.14	0.07	0.04	0.03	0.03	0.02	0.02	0.02	0.02	0.02	0.02	0.02
14	0.86	0.93	0.96	0.97	0.98	0.98	0.98	0.98	0.98	0.98	0.99	0.99	0.99	0.99	0.15	0.07	0.05	0.03	0.03	0.03	0.02	0.02	0.02	0.02
16	0.87	0.94	0.96	0.97	0.98	0.98	0.98	0.99	0.99	0.99	0.99	0.99	0.99	0.99	0.99	0.99	0.15	0.08	0.05	0.04	0.03	0.03	0.03	0.02
Zone Type B																								
8	0.75	0.85	0.90	0.93	0.94	0.95	0.95	0.96	0.23	0.12	0.08	0.05	0.04	0.04	0.03	0.03	0.03	0.02	0.02	0.02	0.02	0.02	0.02	0.01
10	0.75	0.86	0.91	0.93	0.94	0.95	0.95	0.96	0.96	0.97	0.24	0.13	0.08	0.06	0.05	0.04	0.04	0.03	0.03	0.03	0.03	0.02	0.02	0.02
12	0.76	0.86	0.91	0.93	0.95	0.95	0.96	0.96	0.97	0.97	0.97	0.97	0.24	0.14	0.09	0.07	0.05	0.05	0.04	0.04	0.03	0.03	0.03	0.03
14	0.76	0.87	0.92	0.94	0.95	0.96	0.96	0.97	0.97	0.97	0.97	0.98	0.98	0.98	0.25	0.14	0.09	0.07	0.06	0.05	0.05	0.04	0.04	0.03
16	0.77	0.88	0.92	0.95	0.96	0.96	0.97	0.97	0.97	0.98	0.98	0.98	0.98	0.98	0.98	0.99	0.25	0.15	0.10	0.07	0.06	0.05	0.05	0.04
Zone Type C																								
8	0.72	0.80	0.84	0.87	0.88	0.89	0.90	0.91	0.23	0.15	0.11	0.09	0.08	0.07	0.07	0.06	0.05	0.05	0.05	0.04	0.04	0.03	0.03	0.03
10	0.73	0.81	0.85	0.87	0.89	0.90	0.91	0.92	0.92	0.93	0.25	0.16	0.13	0.11	0.09	0.08	0.08	0.07	0.06	0.06	0.05	0.05	0.04	0.04
12	0.74	0.82	0.86	0.88	0.90	0.91	0.92	0.92	0.93	0.94	0.94	0.95	0.26	0.18	0.14	0.12	0.10	0.09	0.08	0.08	0.07	0.06	0.06	0.05
14	0.75	0.84	0.87	0.89	0.91	0.92	0.92	0.93	0.94	0.94	0.95	0.95	0.96	0.96	0.27	0.19	0.15	0.13	0.11	0.10	0.09	0.08	0.08	0.07
16	0.77	0.85	0.89	0.91	0.92	0.93	0.93	0.94	0.95	0.95	0.95	0.96	0.96	0.97	0.97	0.97	0.28	0.20	0.16	0.13	0.12	0.11	0.10	0.09
Zone Type D																								
8	0.66	0.72	0.76	0.79	0.81	0.83	0.85	0.86	0.25	0.20	0.17	0.15	0.13	0.12	0.11	0.10	0.09	0.08	0.07	0.06	0.06	0.05	0.04	0.04
10	0.68	0.74	0.77	0.80	0.82	0.84	0.86	0.87	0.88	0.90	0.28	0.23	0.19	0.17	0.15	0.14	0.12	0.11	0.10	0.09	0.08	0.07	0.06	0.06
12	0.70	0.75	0.79	0.81	0.83	0.85	0.87	0.88	0.89	0.90	0.91	0.92	0.30	0.25	0.21	0.19	0.17	0.15	0.13	0.12	0.11	0.10	0.09	0.08
14	0.72	0.77	0.81	0.83	0.85	0.86	0.88	0.89	0.90	0.91	0.92	0.93	0.94	0.94	0.32	0.26	0.23	0.20	0.18	0.16	0.14	0.13	0.12	0.10
16	0.75	0.80	0.83	0.85	0.87	0.88	0.89	0.90	0.91	0.92	0.93	0.94	0.94	0.95	0.96	0.96	0.34	0.28	0.24	0.21	0.19	0.17	0.15	0.14

[a]See Table 8-25 for zone types. Data based on a radiative/convective fraction of 0.59/0.41.
Adjustments for other radiative/convective fractions is discussed in Sec. 8-5.

Source: Reprinted by permission from ASHRAE *Cooling and Heating Load Calculation Manual,* 2nd ed., 1992.

It should be noted that all the equipment in a space does not have to be lumped together to make a calculation. Equipment can be started and stopped at different times and separate calculations made for each.

Air Distribution System

Strictly speaking heat gain to the air moving through the duct system is not an internal load component; however, any heat transfer to the air in the supply system is a load on the space. Heat transfer to the air in the return system creates a load on the central system. The components are estimated as discussed in Section 8-4.

Infiltration

Methods to estimate the quantity of infiltration air are covered in Chapter 7 and the procedure to compute the heat gain is covered in Section 8-4. Recall that two components of heat gain exist for infiltration air, sensible and latent. Both are convective in nature and immediately become cooling load.

Cooling Load Summary

It is important to adopt a systematic approach to the calculation and summation of the cooling load for a structure. If this is not done, some important parts of the problem may be overlooked. The computer is exceptionally good for this purpose. However, many jobs are too small to justify the use of a computer. In these situations a calculation form or worksheet is a necessity. Figure 8-7 is an example of such a worksheet designed to use the CLTD method. The worksheet has a number of advantages over a page-by-page analysis of the problem. A worksheet provides a concise record of the load estimate that can be easily checked.

The calculations discussed so far pertain to the load estimates needed for a large commercial building that is divided into many zones with separate temperature control in each. Each zone usually has different outside exposures and internal load patterns. Therefore, as explained earlier, the peak load for each zone and the peak coincident load for all zones served by a central unit are required and hour-by-hour loads are helpful to have. A variable-volume system is an example of a system requiring such load estimates.

At the other end of the spectrum of building and system types are small detached buildings with one central, constant-flow system with one thermostat to control the temperature in all spaces. This is similar to a single-family detached house. Recent studies (10) have shown that buildings and systems of this type require design loads different from the peak values mentioned earlier. The reason for this is related to the design of the air-distribution system. For a small detached building with one central system, the calculation procedure is modifed as follows:

1. Exterior walls. Use a daily average CLTD from Table 8-19.
2. Roof. Use an average CLTD for the sunlit hours of the day from Table 8-18.
3. Glass. Use an average SCL for the sunlit hours of the day from Table 8-24.
4. People, lights, and equipment. Compute as usual, assuming they peak in the late afternoon or when turned off.

Table 8-28 Cooling Load Factors for Hooded Equipment[a]

Hours in Operation	Number of Hours after Equipment Turned On																						
	1	2	3	4	5	6	7	8	9	10	11	12	13	14	15	16	17	18	19	20	21	23	24
Zone Type A																							
2	0.64	0.83	0.26	0.11	0.06	0.03	0.01	0.01	0.01	0.01	0.00	0.00	0.00	0.00	0.00	0.00	0.00	0.00	0.00	0.00	0.00	0.00	0.00
4	0.64	0.83	0.90	0.93	0.31	0.14	0.07	0.04	0.03	0.03	0.01	0.01	0.01	0.01	0.00	0.00	0.00	0.00	0.00	0.00	0.00	0.00	0.00
6	0.64	0.83	0.90	0.93	0.96	0.96	0.33	0.16	0.09	0.06	0.04	0.03	0.03	0.01	0.01	0.01	0.01	0.00	0.00	0.00	0.00	0.00	0.00
8	0.64	0.83	0.90	0.93	0.96	0.96	0.97	0.97	0.34	0.16	0.09	0.06	0.04	0.03	0.03	0.01	0.01	0.01	0.01	0.01	0.01	0.01	0.00
10	0.64	0.83	0.90	0.93	0.96	0.96	0.97	0.97	0.99	0.96	0.34	0.17	0.10	0.06	0.04	0.03	0.03	0.01	0.01	0.01	0.01	0.01	0.00
12	0.64	0.83	0.90	0.94	0.96	0.97	0.97	0.99	0.99	0.99	0.99	0.99	0.36	0.17	0.10	0.06	0.04	0.03	0.03	0.03	0.03	0.03	0.01
14	0.66	0.83	0.90	0.94	0.96	0.97	0.97	0.99	0.99	0.99	0.99	0.99	1.00	1.00	0.36	0.17	0.10	0.07	0.04	0.04	0.03	0.03	0.01
16	0.66	0.84	0.91	0.94	0.96	0.97	0.97	0.99	0.99	0.99	0.99	0.99	1.00	1.00	1.00	1.00	0.36	0.17	0.10	0.07	0.04	0.04	0.03
18	0.67	0.84	0.91	0.94	0.96	0.97	0.97	0.99	0.99	0.99	0.99	1.00	1.00	1.00	1.00	1.00	1.00	1.00	0.36	0.17	0.10	0.07	0.04
Zone Type B																							
2	0.50	0.63	0.23	0.16	0.11	0.09	0.07	0.06	0.04	0.03	0.03	0.01	0.01	0.01	0.01	0.00	0.00	0.00	0.00	0.00	0.00	0.00	0.00
4	0.50	0.64	0.73	0.79	0.34	0.24	0.19	0.14	0.10	0.09	0.06	0.04	0.04	0.03	0.03	0.01	0.01	0.01	0.01	0.00	0.00	0.00	0.00
6	0.50	0.64	0.73	0.79	0.84	0.87	0.41	0.29	0.21	0.17	0.13	0.10	0.07	0.06	0.04	0.03	0.03	0.01	0.01	0.01	0.01	0.01	0.00
8	0.50	0.64	0.73	0.79	0.84	0.87	0.90	0.93	0.44	0.31	0.24	0.19	0.14	0.11	0.09	0.07	0.06	0.04	0.03	0.03	0.01	0.01	0.01
10	0.50	0.64	0.73	0.79	0.84	0.87	0.90	0.93	0.94	0.96	0.47	0.34	0.26	0.20	0.16	0.11	0.09	0.07	0.06	0.04	0.03	0.03	0.01
12	0.51	0.66	0.73	0.80	0.84	0.89	0.91	0.93	0.94	0.96	0.97	0.97	0.49	0.34	0.27	0.20	0.16	0.11	0.09	0.07	0.06	0.04	0.03
14	0.53	0.66	0.74	0.80	0.84	0.89	0.91	0.93	0.94	0.96	0.97	0.97	0.99	0.99	0.50	0.36	0.27	0.21	0.16	0.13	0.10	0.07	0.06
16	0.56	0.69	0.76	0.81	0.86	0.89	0.91	0.93	0.94	0.96	0.97	0.97	0.99	0.99	0.99	0.99	0.50	0.36	0.27	0.21	0.16	0.13	0.10
18	0.59	0.71	0.79	0.83	0.87	0.90	0.93	0.94	0.96	0.97	0.97	0.99	0.99	0.99	0.99	0.99	1.00	1.00	0.50	0.36	0.27	0.21	0.16

Zone Type C

Hr																							
2	0.43	0.54	0.20	0.16	0.13	0.10	0.09	0.07	0.06	0.04	0.04	0.03	0.03	0.01	0.01	0.01	0.01	0.01	0.01	0.00	0.00	0.00	0.00
4	0.43	0.54	0.63	0.70	0.33	0.26	0.20	0.17	0.14	0.11	0.09	0.07	0.06	0.06	0.04	0.03	0.03	0.03	0.01	0.01	0.01	0.01	0.01
6	0.44	0.56	0.63	0.70	0.76	0.80	0.40	0.31	0.26	0.21	0.17	0.14	0.11	0.10	0.09	0.07	0.06	0.04	0.04	0.03	0.03	0.01	0.01
8	0.44	0.56	0.64	0.70	0.76	0.80	0.84	0.87	0.46	0.37	0.30	0.24	0.20	0.16	0.13	0.11	0.09	0.07	0.06	0.06	0.04	0.03	0.03
10	0.46	0.57	0.64	0.71	0.76	0.80	0.84	0.87	0.89	0.91	0.50	0.40	0.33	0.26	0.21	0.17	0.14	0.11	0.10	0.09	0.07	0.06	0.04
12	0.47	0.59	0.66	0.73	0.77	0.81	0.84	0.87	0.90	0.91	0.93	0.94	0.53	0.41	0.34	0.27	0.23	0.19	0.16	0.13	0.10	0.09	0.07
14	0.50	0.60	0.67	0.74	0.79	0.83	0.86	0.89	0.90	0.91	0.93	0.94	0.96	0.96	0.54	0.43	0.36	0.29	0.24	0.20	0.16	0.13	0.11
16	0.54	0.63	0.70	0.76	0.80	0.84	0.87	0.89	0.91	0.93	0.94	0.94	0.96	0.97	0.97	0.97	0.56	0.44	0.36	0.30	0.24	0.20	0.16
18	0.60	0.69	0.74	0.79	0.83	0.86	0.89	0.90	0.91	0.93	0.94	0.96	0.96	0.97	0.97	0.99	0.99	0.99	0.56	0.44	0.37	0.30	0.24

Zone Type D

Hr																						
2	0.41	0.53	0.19	0.13	0.11	0.09	0.07	0.07	0.06	0.06	0.04	0.03	0.03	0.03	0.01	0.01	0.01	0.01	0.01	0.01	0.01	0.01
4	0.43	0.53	0.60	0.29	0.23	0.19	0.16	0.14	0.11	0.10	0.09	0.07	0.07	0.06	0.04	0.04	0.04	0.03	0.03	0.03	0.01	0.01
6	0.44	0.54	0.61	0.71	0.76	0.37	0.29	0.24	0.21	0.19	0.16	0.13	0.11	0.10	0.09	0.07	0.07	0.06	0.06	0.04	0.04	0.03
8	0.46	0.56	0.63	0.71	0.76	0.79	0.81	0.43	0.34	0.29	0.24	0.21	0.19	0.16	0.14	0.11	0.10	0.09	0.07	0.07	0.06	0.06
10	0.47	0.57	0.64	0.73	0.77	0.80	0.83	0.84	0.87	0.47	0.39	0.31	0.27	0.24	0.20	0.17	0.16	0.13	0.11	0.10	0.09	0.07
12	0.50	0.59	0.66	0.74	0.77	0.81	0.83	0.86	0.87	0.89	0.90	0.50	0.41	0.34	0.30	0.26	0.23	0.19	0.17	0.14	0.13	0.11
14	0.53	0.61	0.69	0.76	0.80	0.83	0.84	0.87	0.89	0.90	0.91	0.93	0.93	0.53	0.43	0.36	0.31	0.27	0.23	0.20	0.17	0.16
16	0.57	0.66	0.71	0.79	0.81	0.84	0.86	0.89	0.90	0.91	0.93	0.93	0.94	0.94	0.96	0.54	0.44	0.37	0.33	0.29	0.24	0.21
18	0.63	0.71	0.76	0.81	0.84	0.87	0.87	0.89	0.90	0.91	0.93	0.94	0.96	0.96	0.96	0.97	0.97	0.56	0.46	0.39	0.33	0.29

[a]See Table 8-25 for zone type. Data based on a radiative/convective fraction of 100/0.

Adjustment for other radiative/convective fractions are discussed in Sec. 8-5.

Source: Reprinted by permission from ASHRAE *Cooling and Heating Load Calculation Manual*, 2nd ed., 1992.

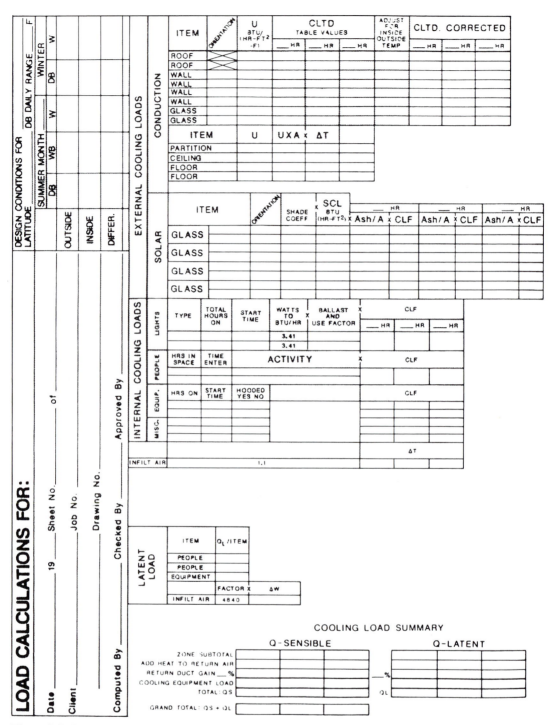

Figure 8-7 Cooling load calculation sheet. (Reprinted by permission *ASHRAE Cooling and Heating Load Calculation Manual,* 2nd ed., 1992.)

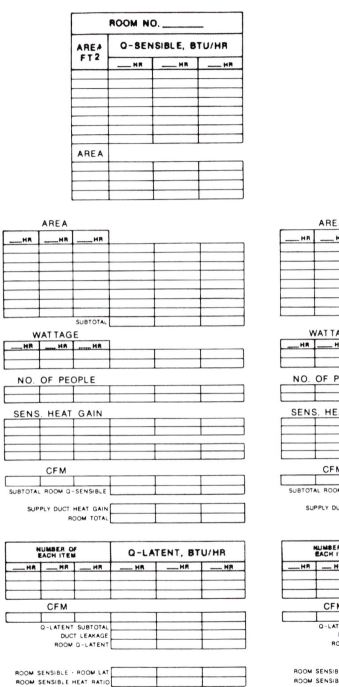

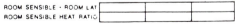

Figure 8-7 Cooling load calculation sheet. (*Continued*)

Room-by-room calculations will then lead to a much improved proportioning of the supply air to each space, while the block load for the building will be about the same as the peak coincident load calculated on an hour-by-hour basis.

There are of course many buildings and systems that fall between the two cases discussed. Design engineers may need to be somewhat artful in using the results of their load calculations.

The next steps are to determine the air quantities and to select the equipment. These steps may be reversed depending on the type of equipment to be utilized.

8-8 SUPPLY AIR QUANTITIES

The preferred method of computing air quantity for cooling and dehumidification has been described in Section 3-6 of Chapter 3. That method should always be used when the conditions and the size of the cooling load warrant specification of special equipment. This means that the cooling and dehumidifying coil is designed to match the sensible and latent heat requirements of a particular job and that the fan is sized to handle the required volume of air. The fan, cooling coil, control dampers, and the enclosure for these components are referred to as an *air handler*. These units are assembled at the factory in a wide variety of coil and fan models to suit almost any requirement. The design engineer usually specifies the entering and leaving moist-air conditions, the volume flow rate of the air, and the total pressure the fan must produce.

Specially constructed equipment cannot be justified for small commercial applications. Furthermore, these applications generally have a higher sensible heat factor, and dehumidification is not as critical as it is in large commercial buildings. Therefore, the equipment is manufactured to operate at or near one particular set of conditions. For example, typical light commercial unitary cooling equipment operates with a coil SHF of 0.75 to 0.8 with the air entering the coil at about 80 F or 27 C dry bulb and 67 F or 19 C wet bulb temperature. This equipment usually has a capacity of less than about 20 tons or 70 kW. When the peak cooling load and latent heat requirements are appropriate, this less expensive type of equipment may be used. In this case the air quantity is determined in a different way. The unit is first selected on the basis of the block sensible cooling load using the nearest available size but not less than the sensible cooling load. Next, the latent capacity of the unit must be equal to or greater than the computed latent cooling load. This procedure assures that the unit will handle both the sensible and latent load even though an exact match does not exist. The air quantity is specified by the manufacturer for each unit and is 350 to 400 cfm/ton or about 0.0537 $\text{m}^3/(\text{s-kW})$. The total air quantity is then divided among the various rooms according to the cooling load of each room.

At the conclusion of the load calculation phase, the designer is ready to proceed with other aspects of the system design discussed in the following chapters.

REFERENCES

1. ASHRAE, *Procedure for Determining Heating and Cooling Loads for Computerizing Energy Calculations, Algorithms for Building Heat Transfer Subroutines,* American Society of Heating, Refrigerating and Air-Conditioning Engineers, Atlanta, GA, 1975.

2. Harris, S. M., and F. C. McQuiston, "A Study to Categorize Walls and Roofs on the Basis of Thermal Response," *ASHRAE Transactions,* Vol. 94, Part 2, 1988.
3. Sowell, E. F., "Classification of 200,640 Parametric Zones for Cooling Load Calculations," *ASHRAE Transactions,* Vol. 94, Part 2, 1988.
4. Spitler, J. D., K. Lindsey, F. C. McQuiston, "The CLTD/SCL/CLF Cooling Load Calculation Method," ASHRAE Transactions, Vol. X, Part X, 1993.
5. *Cooling and Heating Load Calculation Manual,* 2nd ed., American Society of Heating, Refrigerating and Air-Conditioning Engineers, Inc., Atlanta, GA, 1992.
6. *ASHRAE Handbook, Fundamentals Volume,* American Society of Heating, Refrigerating and Air-Conditioning Engineers, Inc., Atlanta, GA 1989.
7. Lawrence Berkeley Laboratory, *DOE-2 Reference Manual* (Vol. 2.1A), LBL Report LBL-8706, 1981.
8. York, D. A., C. C. Cappiello, eds., *DOE-2 Engineers Manual,* U. S. Department of Energy, 1981.
9. *ASHRAE Handbook, Applications Volume,* American Society of Heating, Refrigerating and Air-Conditioning Engineers, Inc., Atlanta, GA, 1991.
10. F. C. McQuiston, "A Study and Review of Existing Data to Develop a Standard Methodology for Residential Heating and Cooling Load Calculations," *ASHRAE Transactions,* Vol. 90, Part 2, 1984.

PROBLEMS

8-1. Describe a situation where the heat gain to a space (a) is greater than the cooling load at a given time; (b) less than the cooling load at a given time; (c) equal to the cooling load at a given time.

8-2. Southern coastal regions of the United States experience periods of very high humidity. Explain how this might influence selection of design conditions.

8-3. Determine the ASHRAE Standard 90.1 design conditions for the following locations. Include the maximum outdoor temperature, the outdoor mean coincident wet bulb temperature, indoor dry bulb temperature, relative humidity, the elevation and latitude. (a) Washington, D.C., (b) San Francisco, California, (c) Denver, Colorado, (d) Dallas, Texas.

8-4. Determine the wall conduction transfer function coefficients (CTF) for a wall composed of 4-in. face brick, $\frac{1}{2}$-in. sheathing (R-1.32), $3\frac{1}{2}$-in. mineral fiber insulation (R-11), and $\frac{1}{2}$-in. gypsum board.

8-5. Change the insulation in Problem 8-4 to R-19 and determine the CTF.

8-6. A roof is composed of $\frac{1}{2}$-in. slag, $\frac{3}{8}$-in. felt membrane, $2\frac{1}{2}$-in. high-density insulation, 2-in. concrete, and steel deck exposed to the inside space. Determine the conduction transfer coefficients (CTF).

8-7. The roof of Problem 8-6 is changed to have a suspended ceiling with a 12-in. air space above it. Determine the CTF.

8-8. Compute the solar irradiation and sol-air temperatures for a west-facing wall at 36 deg N latitude 90 deg W longitude for each hour of the day on July 21. Assume an outdoor design temperature of 96 F with daily range of 20 F, surface absorptance of 0.9, and an air film coefficient of 4.0 Btu(hr-ft^2-F).

8-9. Compute the solar irradiation and sol-air temperatures for a flat roof for the conditions of Problem 8-8.

8-10. Compute the conduction heat gain per square foot for the wall of Problem 8-4 using the sol-air temperatures of Problem 8-8. Assume a space temperature of 75 F.

8-11. Compute the conduction heat gain per square foot for the wall of Problem 8-5 using the sol-air temperatures of Problem 8-8. Assume a room temperature of 75 F.

8-12. Compute the conduction heat gain per square foot for the roof of Problem 8-6 using the sol-air temperatures of Problem 8-9. Room temperature is 78 F.

8-13. Compute the conduction heat gain per square foot for the roof of Problem 8-7 using the sol-air temperatures of Problem 8-9. Room temperature is 78 F.

8-14. A large office space has an average occupancy of 20 people from 8:00 A.M. to 5:00 P.M. LST. Lighting is 2.5 W/ft^2 of recessed, unvented fluorescent fixtures on from 8:00 A.M. to 6:00 P.M. Miscellaneous equipment operates intermittently and amounts to an average of 3 hp. Compute the heat gain at 4:00 P.M. for the space assuming a floor area of 4000 ft^2.

8-15. A space has an occupancy of 40 people engaged in sedentary activity from 8:00 A.M. to 5:00 P.M. LST. The average light level is 20 W/m^2 of vented fluorescent fixtures on for 10 hours with a ceiling plenum return. Equipment amounts to 6 kW. Estimate the heat gain for a floor area of 750 m^2 at 4:00 P.M. LST.

8-16. A large room has 5000 W of vented fluorescent light fixtures on from 6:00 A.M. to 6:00 P.M. LST. The air flows from the lights through a ducted return. Compute the heat gain at 5:00 P.M. LST assuming that 20 percent of heat from the lights is convected to the return air.

8-17. A large office complex has a variable occupancy pattern. Twenty people arrive at 8:00 A.M. and leave at 4:00 P.M. Forty people arrive at 10:00 A.M. and leave at 4:00 P.M. Ten people arrive at 1:00 P.M. and leave at 5:00 P.M. Assume seated, light activity and compute the heat gain at 4:00 P.M. and 6:00 P.M. Assume LST.

8-18. A space contains 90 people at 8:00 A.M. The people start leaving at 10:00 A.M. at the rate of 10 people per hour. What is the heat gain at 6:00 P.M. assuming seated, very light work? Assume LST.

8-19. The wall of Problem 8-4 is in a small perimeter room in a single-story building with only one wall, 15 × 10 ft. The room has furniture and 50 percent interior shade at the windows, which make up 10 percent of the wall area. Partitions are gypsum board with air space. The floor is covered with vinyl tile. (a) Compute the cooling load for the heat gains of Problem 8-10. (b) Compute the cooling load for the heat gains of Problem 8-11.

8-20. An interior room in a single-story building has floor plan dimensions of 20 × 20 ft. The room is furnished with carpeted floors. Partitions are of 8-in. concrete block and there is a suspended ceiling. Compute the hourly cooling loads using the heat gains from Problem 8-13.

8-21. Refer to Problem 6-16 and compute the cooling loads for the absorbed and transmitted heat gains. The windows, with area of 40 ft^2, are in a single-story, perimeter room, zone type 11.

8-22. Refer to Problem 8-14 and compute the hourly cooling loads for (a) people and equipment and (b) lighting. Assume that people, equipment, and lights all operate on the same schedule. The space is a single zone in a single-story building, zone type 9.

8-23. Refer to Problem 8-17 and compute hourly cooling loads for the occupants. The space is a single zone in a single-story building, zone type 5.

8-24. For a window with shading coefficient of 0.8, an area of 300 ft^2, and the following transmitted and absorbed solar heat gain factors, determine the hourly transmitted and absorbed solar heat gains. [ASHGF and TSHFG are 0 for all hours not shown; 0600 represents the 6th hour of the day (5:00 A.M.–6:00 A.M.)]

Time	TSHGF	ASHGF	Time	TSHGF	ASHGF
0600	5	0	1200	62	5
0700	14	1	1300	130	9
0800	21	2	1400	187	11
0900	27	2	1500	219	15
1000	31	3	1600	223	15
1100	35	3	1700	186	11
			1800	76	6

Using the following v and w coefficients, determine the hourly cooling loads due to the transmitted solar heat gains calculated above.

$$v_0 = 0.59095 \qquad v_1 = -0.63490 \qquad v_2 = 0.09545$$
$$w_1 = -1.30715 \qquad w_2 = 0.35864$$

8-25. A zone in a building of heavy construction has a design cooling load profile as shown in Table 8-29, assuming an interior temperature of 75 F. The zone is occupied from 8:00 A.M. to 5:00 P.M.; the cooling system operates from 6:00 A.M. to 9:00 P.M. and is turned off the remainder of the time. The infiltration rate has been estimated to be 300 cfm. The floor area is 2000 ft^2 and the UA product is 450 Btu/(hr-F). The cooling system has a capacity of 42,000 Btu/hr at 76 F and 0 Btu/hr at 70 F. The thermostat has a throttling range of 2 F with a set-point of 74 F. Calculate the heat extraction rate and zone temperature for each hour of the day.

Table 8-29 Cooling Loads for Problem 8-25, Btu/hr

Hour	Cooling Load	Hour	Cooling Load	Hour	Cooling Load
1	16,810	9	13,575	17	39,125
2	14,870	10	16,575	18	37,860
3	13,100	11	25,155	19	35,610
4	11,530	12	29,315	20	33,110
5	10,260	13	33,010	21	30,690
6	9,655	14	36,090	22	28,250
7	10,025	15	38,260	23	21,120
8	11,330	16	39,225	24	18,875

8-26. Compute the total cooling load for the structure described by the plans and specifications furnished by the instructor using the transfer function method with a computer program.

8-27. Determine the cooling load temperature difference (CLTD) correction for each of the locations in Problem 8-3 assuming that the inside dry bulb temperature is 75 F and the month is July.

8-28. Compute the cooling load for the south wall of a building at 36 deg N latitude in July. The solar time is 4:00 P.M. The wall is brick veneer and frame with an overall heat-transfer coefficient of 0.08 Btu/(hr-ft^2-F). The inside design temperature is 75 F and the outside temperature is 101 F. The daily range of temperature is 22 deg F. The wall is 8 × 16 ft with a 4 × 5-ft window. (Do not make a calculation for the window.)

8-29. The building of Problem 8-28 has a roof–attic–ceiling combination with $U =$ 0.12 Btu/(hr-ft^2-F). Compute the cooling load for a 16 × 20-ft room using data from Problem 8-28 (roof only).

8-30. Compute the peak cooling load per square meter for a west wall like that of Table 5-7*a*. Assume that the wall is located at 36 deg N latitude. The date is July 21. At what time of day does the peak occur? The outdoor and indoor design temperatures are 38 C and 25 C, respectively, and the daily range is 12 deg C.

8-31. Compute the cooling load for a 8 × 40 m roof–ceiling combination like that of Table 5-7*b* at 6:00 P.M. LST in July. Assume 36 deg N latitude, 39 C and 24 C outdoor and indoor temperatures, and a daily range of 13 deg C.

8-32. A west wall for a building in Albuquerque, New Mexico, is made of 4-in. concrete with 2 in. of high-density insulation on the outside. The outside is finished with stucco and the inside is finished with gypsum board. Compute the cooling load per square foot of wall assuming standard design conditions and LST equal to 1700 hours.

8-33. A roof–ceiling combination has a metal deck with 2 in. of heavy concrete and 3-in. of rigid insulation on top. The outside layer is typical built-up membrane and stone, whereas the ceiling is acoustical tile below a ceiling air space. The building is located in Albuquerque, New Mexico. Assume standard ASHRAE design conditions and find the cooling load per square foot of roof at 1700 hours LST.

8-34. Solve Problem 8-32 for a building located in (a) Norfolk, Virginia and (b) Las Vegas, Nevada.

8-35. Solve Problem 8-33 for a building in (a) Norfolk, Virginia and (b) Las Vegas, Nevada.

8-36. Determine the solar cooling load factors at 0800 and 1700 hours LST in July for the major directions for vertical and horizontal glass for Las Vegas, Nevada. Assume a zone in a single-story building with 2 walls, vinyl floor covering, concrete block partitions, and full inside shade.

8-37. Compute the cooling load for a window facing southeast at 36 deg N latitude, 90 deg W longitude at 10:00 A.M. Central Daylight Time in July. The window is regular insulating glass with $\frac{1}{2}$-in. air space in an aluminum frame with thermal break. Inside drapes have a transmittance of 0.35 and reflectance of 0.35. The indoor design temperature is 75 F and the outdoor temperature is 100 F with a 24 deg F daily range. Window dimensions are 2 ft wide by 8 ft high. Assume a zone type of D.

8-38. The window in Problem 8-37 has an 18-in. overhang at the top. Compute the cooling load for the window.

8-39. Compute the cooling load for a window facing south at 36 deg N latitude at 1:00 P.M. sun time in July. The glass is 5-mm sheet in an aluminum frame with no thermal break with venetian blinds of medium color. The inside temperature is 25 C while the outdoor temperature is 35 C with a 13 deg daily range. The window is 1 × 2 m. Assume a zone type B.

8-40. The window in Problem 8-39 has an overhang at the top of 0.6 m. What percentage of the window is shaded, assuming that the overhang is much wider than the window? Compute the cooling load for the window.

8-41. Compute the cooling load for the south windows on a middle floor of an office building that has no external shading. The windows are the insulating type with regular plate glass inside and out. Drapes with a reflectance of 0.2 and transmittance of 0.5 are fully closed. The total window area is 400 ft². Assume that the indoor temperature is 78 F and the outdoor temperatures are 100 F in July with a 22 deg F daily range. Make the calculation for 12:00 noon at 36 deg N latitude. To determine the zone type, make assumptions for a modern office building.

8-42. A 6 × 6-ft regular plate glass window faces west. There is a 2-ft overhang at the top. Compute the cooling load at 3:00 P.M. solar time in July at 36 deg N latitude. Assume the outdoor–indoor temperature differential is 25 deg F and the outdoor temperature is 100 F. Assume carpeted floor, medium-weight construction, and no inside shade.

8-43. A 1-m² regular plate glass window has venetian blinds of medium color. The window is located at 36 deg N latitude. Calculate the cooling load for the window facing west at 2:00 P.M. in July with a mean outdoor temperature of 32 C. Assume an indoor temperature of 25 C, and use solar time. Assume zone type B.

8-44. Refer to Problem 8-14 and compute the cooling load assuming a zone type D.

8-45. Refer to Problem 8-15 and compute the cooling load assuming a zone type D.

8-46. Refer to Problem 8-16. The zone is located in the corner on a top floor of a multistory building. The construction is medium to heavy with carpet, suspended ceiling, and partial inside shade. Compute the cooling load at 5:00 P.M. LST.

8-47. Refer to Problem 8-17. Compute the cooling load at 4:00 and 6:00 P.M. The zone is described in Problem 8-46.

8-48. Refer to Problem 8-18. Compute the cooling load at 6:00 P.M. LST. Assume a zone type C.

8-49. A space contains a die-casting machine, which produces a heat gain of 20,000 Btu/hr (5.9 kW). The machine is hooded with about 40 percent of the heat gain exhausted to the outdoors. The remaining heat gain is all radiative. The zone is a single room in a single-story building with some inside shade. Compute the cooling load at 4:00 P.M. LST assuming that the machine was started at 6:00 P.M. LST.

8-50. Compute the total cooling load for the building described by the plans and specifications furnished by your instructor using the CLTD/SCL/CLF method.

Chapter 9

Energy Calculations

Following the calculation of the design heat load and selection of the heating system, it is often desirable to estimate the quantity of energy necessary to heat and cool the structure under typical weather conditions and with typical inputs from internal heat sources. This is a distinctly different procedure from design load calculations, which are usually made to determine size or capacity for one set of design conditions. The many factors such as solar effects and internal heat sources that influence energy requirements vary throughout the prediction period, making calculations that take all variations into account very complex. The calculations usually involve simulation of the system, and this requires a digital computer.

Many computer codes are available that model building systems quite well, and they may be used with all types of structures ranging from residential to heavy commercial. The building simulation is usually carried out for a whole year considering heating, cooling, and other energy requirements.

There are cases, however, where computer simulation cannot be justified. Residential and light commercial buildings fall into this category. Reasonable results can be obtained in this case using simple methods such as the degree day or bin method. Reference 1 discusses energy estimating in detail.

9-1 THE DEGREE DAY PROCEDURE

The traditional degree day procedure (1) for computing fuel requirements is based on the assumption that on a long-term basis, solar and internal gains for a residential structure will offset heat loss when the mean daily outdoor temperature is 65 F (18 C). It is further assumed that fuel consumption will be proportional to the difference between the mean daily temperature and 65 F or 18 C.

For cities in the United States and Canada, Table C-2 lists the average number of degree days that have occurred over a period of many years; they are listed by month, and the yearly totals of these averages are given. Degree days are defined by the relationship

$$DD = \frac{(t - t_a)N}{24} \tag{9-1}$$

where N is the number of hours for which the average temperature t_a is computed and t is 65 F or 18 C. Residential insulation and construction practices have improved dramatically over the last 40 years, however, and internal heat gains have increased. These changes indicate that a temperature less than 65 F should be used for the base; nevertheless, the data now available are based on 65 F. Another factor, which is not accounted for in traditional methods, is the decrease in efficiency of fuel-fired furnaces and heat pumps under partial load. A modified degree day procedure that accounts for these factors in an approximate way is in current use. The general relation for fuel calculations using this procedure is

$$F = \frac{24(\text{DD})\dot{q}C_D}{\eta(t_i - t_o)H} \qquad (9\text{-}2)$$

where

F = the quantity of fuel required for the period desired; the units depend on H
DD = the degree days for period desired, F-day or C-day
$\dot{q}$ = the total calculated heat loss based on design conditions, t_i and t_o, Btu/hr or W
η = an efficiency factor that includes the effects of rated full-load efficiency, part-load performance, oversizing, and energy conservation devices
H = the heating value of fuel, Btu or kW-hr per unit volume or mass
C_D = the interim correction factor for degree days based on 65 F or 18 C (Fig. 9-1)

Figure 9-1 gives values for the correction factor C_D as a function of yearly degree days. These values were calculated using typical modern single-family construction, and generally agree with electric utility experience (2).

The efficiency factor η of Eq. 9-2 is empirical and will vary from about 0.6 for older heating equipment to about 0.9 for new high-efficiency equipment. For electric resistance heat, η has a value of 1.0. This method is not recommended for cooling-energy calculations at all.

It is recommended that more sophisticated methods of energy estimating be consid-

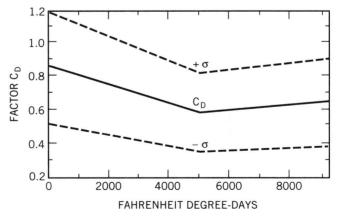

Figure 9-1 Correction factor for use in Eq. 9-2. (Reprinted by permission from *ASHRAE Handbook, Fundamentals Volume*, 1989.)

ered even for residential structures. The availability and simplicity of personal computers makes more refined methods practical. A serious shortcoming of the degree day method is inability to model equipment whose performance depends on outdoor ambient conditions. A heat pump is an example. Degree days are useful in comparing the heating requirements from one location to another. Sometimes degree days are used as a parameter in studying energy data such as utility costs. In Chapter 5 degree days were used in presenting data for floor slab heat losses.

EXAMPLE 9-1

Estimate the amount of natural gas required to heat a residence in Stillwater, Oklahoma, using the modified degree day method. The heating value of the fuel is 1000 Btu/std ft^3. The calculated heat loss from the house is 80,000 Btu/hr with indoor and outdoor design temperatures of 70 F and 0 F, respectively. The furnace efficiency factor is approximately 0.55.

SOLUTION

The average winter temperature and the degree days for Stillwater are estimated to be 48 F and 3725 from Table C-2. Equation 9-2 will give an estimate of the fuel required by the prescribed method.

The correction factor C_D is 0.66 from Fig. 9-1 for 3725 degree days:

should be 24 (Fnotc)

$$F = \frac{\cancel{3}(3725)80,000(0.66)}{0.55(70 - 0)1000} = 122,790 \text{ std ft}^3$$

or $F = 123$ mcf of natural gas.

Variable-Base Degree Day Procedure

The variable-base degree day method is a generalization of the degree day method. The concept is unchanged but counts degree days based on the balance point temperature, defined as the temperature where the building requires neither heating nor cooling. This method recognizes that internal heat gains that offset heating requirements may vary from one building to another. Therefore, the procedure accounts for only the energy required to offset the heat losses due to transmission and infiltration. Reference 1 gives details of this method. Again, this method is not recommended for heat pump or coolig applications.

9-2   BIN METHOD

The energy-estimating method discussed previously is based on average conditions and does not take into account actual day-to-day weather variations and the effect of temperature on equipment performance. The *bin method* is a hand-calculation procedure where energy requirements are determined at many outdoor temperature conditions. Reference 1 describes this method in detail.

Weather data are required in the form of 5 F bins with the hours of occurrence for each bin. The data should be divided into several shifts and the mean coincident wet bulb temperature for each bin should also be given so that latent load due to infiltration can be computed if desired. Table C-3 is an example of annual bin data for Oklahoma City, Oklahoma, taken from reference 3.

The bin method is based on the concept that all the hours during a month, season, or year when a particular temperature (bin) occurs can be grouped together and an energy calculation made for those hours with the equipment operating under those particular conditions. The bin method can be as simplified or complex as the situation may require and applies to both heating and cooling energy calculations. A somewhat simplified approach will first be used to introduce the method. Refinements will then be introduced and discussed.

The Classical Bin Method

The bin method requires a load profile for the building, that is, the heating or cooling required to maintain the conditioned space at the desired conditions as a function of outdoor temperature. Figure 9-2 shows a simplified profile. In some cases more than one profile may be required to accommodate different uses of the building, such as occupied and unoccupied periods. The load profiles may be determined in a number of ways. A comprehensive approach will be described later; however, more simplified profiles are often satisfactory when only heating is considered and will be used here. The design heating load represents an estimate of one point on the unoccupied load profile, since the design load does not include internal loads or solar effects and occurs in the early morning hours when the building is not occupied. This is point d in Fig. 9-2. There is some outdoor temperature where the heating load will be zero, such as point 0 in Fig. 9-2. Solar effects influence the location of point 0. The occupied load profile $d' - 0'$ is influenced by the internal loads due to people and equipment as well as solar effects. A method will be shown later to estimate the balance points where neither heating nor

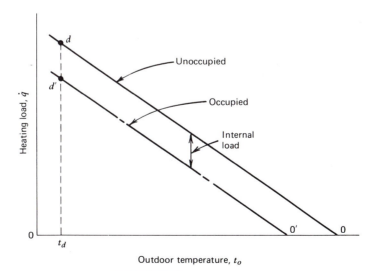

Figure 9-2 Simplified load profiles.

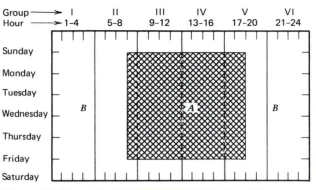

Figure 9-3 Converting bin hours into shifts A and B.

cooling is required for both profiles. For the present let us rely on experience. For a residence the balance point is approximately 60 F (16 C). The balance point for a commercial building will be lower, depending on occupancy and other internal loads. Assuming that points d and 0 on the load profile have been determined, a straight line may be drawn and a linear equation determined to express the load as a function of outdoor temperature.

The hours of each day in a typical week are divided into six four-hour groups. Assuming that two loads (occupied and unoccupied) are to be used, it is then necessary to reduce the bin data in the six time groups (Table C-3) to two time groups or shifts. This is most easily done as shown in Fig. 9-3, where the occupied and unoccupied hours are shown schematically as A and B, respectively. Table 9-1 shows computation of the fraction of the bin hours in each time group that fall in each shift.

Table 9-2a shows the calculation of the bin hours in each time group for each bin and the summations for each shift. For convenience, Table 9-2b summarizes the annual bin data for Oklahoma City, which was used to develop Table 9-2a. To summarize, the shift A bin hours are used with the occupied load profile and shift B bin hours are used with the unoccupied load profile.

The operating characteristics of the heating equipment as a function of the outdoor

Table 9-1 Computation of Fraction of Bin Hours in Each Shift

Time Group	Hours in Shift A in Each Group	Days in Shift A in Each Group	Total Occupied Hours in Each Group	Total Hours in Each Group	Shift A Fraction in Each Group	Shift B Fraction in Each Group
I	0	0	0	28	0.0	1.0
II	1	5	5	28	0.18	0.82
III	4	5	20	28	0.71	0.29
IV	4	5	20	28	0.71	0.29
V	2	5	10	28	0.36	0.64
VI	0	0	0	28	0.0	1.0

temperature are required. This information is supplied by the equipment manufacturer. The efficiency of fossil-fueled equipment such as gas- or oil-fired boilers and furnaces is relatively independent of outdoor temperature; however, the coefficient of performance (COP) of a heat pump is greatly dependent on outdoor conditions, and this must be taken into account. Another factor that should be considered for all equipment is the effect of operating at a partial load. Practically all manufacturers' performance data assume full-load steady-state operation when in fact the equipment operates at partial load most of the time. Figure 9-4 shows the operating characteristics for an air-to-air heat pump with fixed conditions for the heating coil. Table 9-3 is an example of air-to-air heat pump performance data from a manufacturer's catalog from which the curves of Fig. 9-4 may be plotted. Note that the performance depends on indoor temperature and air flow rate as well as outdoor temperature.

Detailed part-load performance of large equipment is often available from the manufacturer; however, for smaller unit-type equipment a method developed at the National Institute for Standards and Testing (NIST) is normally used. A parital load factor is defined as

$$\text{PLF} = \frac{\text{Theoretical energy required at part load}}{\text{Actual energy required at part load}} \quad (9\text{-}3)$$

Table 9-2a Calculation of Bin Hours for Each Shift

| Bin Temperature | Shift A Hours Each Time Group | | | | | | Shift A Hours | Shift B Hours |
	I 0.00[a]	II 0.18[a]	III 0.71[a]	IV 0.71[a]	V 0.36[a]	VI 0.00[a]		
102	0	0	0	1	0	0	1	1
97	0	0	4	50	10	0	64	40
92	0	0	39	109	32	0	179	117
87	0	0	82	103	43	0	229	178
82	0	6	105	109	60	0	280	338
77	0	17	94	82	52	0	244	532
72	0	40	98	84	42	0	264	745
67	0	29	70	70	33	0	202	545
62	0	18	67	96	35	0	216	426
57	0	19	75	77	42	0	212	389
52	0	19	97	67	48	0	230	454
47	0	17	71	47	31	0	166	403
42	0	22	70	48	33	0	172	495
37	0	28	63	38	27	0	156	465
32	0	25	53	28	18	0	124	380
27	0	9	19	17	9	0	54	175
22	0	7	16	12	8	0	44	127
17	0	7	12	1	2	0	21	74
12	0	1	2	0	0	0	3	15
						TOTAL	2861	5899

[a]Shift A fraction.

Table 9-2b Annual Bin Hours for Oklahoma City, Oklahoma

Bin Temperature, F	Time Group 1–4 I	5–8 II	9–12 III	13–16 IV	17–20 V	21–24 VI	Total Hours
102	0	0	0	2	0	0	2
97	0	0	5	70	29	0	104
92	0	0	55	153	88	0	296
87	2	0	116	145	120	24	407
82	20	33	148	153	168	96	618
77	121	93	132	115	144	171	776
72	229	221	138	118	117	186	1009
67	161	161	98	98	93	136	747
62	120	99	95	135	96	97	642
57	87	104	105	108	116	81	601
52	96	103	137	94	133	121	684
47	98	96	100	66	87	122	569
42	150	121	98	67	91	140	667
37	144	153	89	54	76	105	621
32	107	140	74	40	50	93	504
27	63	51	27	24	24	40	229
22	36	41	23	17	23	31	171
17	19	37	17	1	5	16	95
12	7	7	3	0	0	1	18

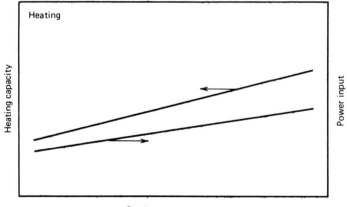

Figure 9-4 Heat-pump operating characteristics.

Table 9-3 Heat Pump Heating Capacities at 6000 CFM

Outdoor Temperature, F	Heating Capacity, Btu/hr × 1000 at Indoor Dry Bulb Temperature, F				Total Power Input, kW at Indoor Dry Bulb Temperature, F			
	60	70	75	80	60	70	75	80
−3	70.5	68.8	68.0	67.2	12.9	13.3	13.5	13.7
2	78.7	76.9	75.9	75.0	13.4	13.8	14.0	14.2
7	87.0	84.9	83.9	82.9	13.8	14.2	14.5	14.7
12	95.2	93.0	91.8	90.7	14.3	14.7	14.9	15.2
17	103.0	101.0	99.8	98.6	14.7	15.2	15.4	15.7
22	111.0	109.0	108.0	106.0	15.0	15.5	15.7	16.0
27	120.0	117.0	115.0	114.0	15.3	15.8	16.0	16.3
32	128.0	125.0	123.0	121.0	15.5	16.0	16.3	16.6
37	140.0	136.0	135.0	133.0	16.0	16.6	16.8	17.1
42	158.0	154.0	152.0	150.0	16.9	17.4	17.7	18.0
47	176.0	172.0	170.0	168.0	17.7	18.3	18.6	18.9
52	188.0	184.0	182.0	179.0	18.2	18.8	19.1	19.4
57	201.0	196.0	193.0	191.0	18.7	19.3	19.7	20.0
62	213.0	208.0	205.0	202.0	19.2	19.9	20.2	20.5
67	225.0	219.0	217.0	214.0	19.7	20.4	20.7	21.0

Note: Correction factor—Value at other air flow = (Value at 6000 CFM × (Cor. Fac.)

Air flow	5250	6750
Heating cap.	0.99	1.01
Power input	1.01	0.99

The theoretical energy required at part load is based on the steady-state operating efficiency, the COP in case of a heat pump. The actual part-load energy required takes into account the loss in efficiency due to startup and shutdown, or other part-load operation. To quantify Eq. 9-3

$$\text{PLF} = 1 - D_c \left(1 - \frac{\text{Building load}}{\text{Unit capacity}} \right) \tag{9-4}$$

where D_c is the degradation coefficient, which may be specified by the manufacturer or taken as 0.25 as a default value. For unitary equipment that is controlled by starting and stopping the unit, the part load factor may also be expressed as

$$\text{PLF} = \frac{\text{Theoretical run time}}{\text{Actual run time}} \tag{9-5}$$

The Bin Calculation Method can now be outlined for a typical bin:

1. Determine the building load from the profile shown in Fig. 9-2.
2. Determine the unit capacity from Fig. 9-4.

3. Compute the theoretical run-time fraction, as the ratio of building load to unit capacity.
4. Compute the partial-load fraction from Eq. 9-4.
5. Compute the actual run-time fraction, Eq. 9-5.
6. Compute actual run time as (bin hours) × (actual run-time fraction).
7. Determine the rate of unit input from Fig. 9-4.
8. Compute the energy use, (unit input) × (actual run time).
9. Determine energy cost per unit of energy from local utility rate schedule.
10. Compute energy cost for this bin as ($/kWh) × (energy use).
11. Repeat steps 1 through 10 for all bins.

Certain refinements may be required. For example, a heat pump may not be able to supply enough heat when the outdoor temperature is very low and supplemental electrical resistance heat may be required. Often the fan energy associated with the heat pump may not be accounted for in the performance data and must be added to the calculation. Also when the building load exceeds the unit capacity, the PLF is assumed to be 1.0 because the unit will run continuously. The bin calculation procedure may be carried out by hand; however, a personal computer is especially useful. An example for an air-to-air heat pump is presented next.

EXAMPLE 9-2

Consider a building in Oklahoma City, Oklahoma, which is operated on two shifts as shown in Fig. 9-3. The load profiles as shown in Fig. 9-2 are given by

Shift A, Occupied

$$\dot{q}_o = 267{,}000 - 4860t_o \text{ Btu/hr} \tag{9-6}$$

Shift B, Unoccupied

$$\dot{q}_{uo} = 316{,}000 - 4860t_o \text{ Btu/hr} \tag{9-7}$$

The heat pump performance is shown in Table 9-3 with a degradation coefficient of 0.25. Compute the energy required to heat the building assuming all the applicable bins in Table 9-2 occur during the heating season and the building is maintained at 70 F during both shifts.

SOLUTION

The load profiles are given in a convenient form for use with the bin method. The balance temperature for each shift may be found by setting $\dot{q}_o$ and $\dot{q}_{uo}$ equal to zero.

Occupied:

$$t_o = 267{,}000/4860 = 55 \text{ F}$$

and

Unoccupied:

$$t_{uo} = 316{,}000/4860 = 65 \text{ F}$$

Therefore, bin temperatures greater than 65 F do not have to be considered.

Equations to express the steady-state heat pump performance can be derived from Table 9-3 as follows. Assuming linear dependence on the outdoor temperature, select two operating capacities and temperatures such as 101,000 Btu/hr at 17 F and 172,000 Btu/hr at 47 F, and fit the points to a linear equation of the form $Y = ax + b$. For this case the heating capacity is

$$C = 2367t_o + 60{,}767 \text{ Btu/hr} \tag{9-8}$$

Using the same approach for the power input

$$P = 0.103t_o + 13.4 \text{ kW} \tag{9-9}$$

Table 9-4 shows the calculation procedure in tabular form. The calculations and source of data are explained for each of the numbered columns. Reading across for a single bin makes the procedure evident. Note that there is a duplicate calculation for each shift. Also note that supplemental heat in the form of electrical resistance is required at about 32 F for each shift.

It should be noted that annual bin data have been used in the preceding example. This was done for brevity and clarity. It would be more accurate to assemble the bin data for all the months during the heating season, say October through April, for the heating-energy calculation and to use data for the months of May through September for a cooling-energy calculation. The reason for this is that a few hours where the bin temperature is below the balance point for heating occur during the summer months, when heat will not actually be supplied. The same is true for cooling. A few hours occur in the winter when cooling may be indicated but the air-conditioning system is off. Reference 3 gives bin data on a monthly basis.

The Modified Bin Method

The classical bin method of computing energy requirements was introduced above. That method is equally valid for computing the energy to cool a building. However, because of solar effects a more refined load profile is desirable. A procedure will now be outlined to obtain a more realistic load profile. Reference 1 may be consulted for full details.

It is important that a building be separated into zones when part of the building requires heat while the remainder requires cooling. If the zones are combined, the heating and cooling requirements will cancel, a solution that may not be representative of the actual situation. The components of the load profile are

Solar gain through walls and roof
Solar gain through glass
Conduction gain through walls and roof
Conduction gain through glass
Lights
People
Equipment
Infiltration

Table 9-4 Bin Energy Calculation for Example 9-2

Bin Temperature	Occupied Hours	Unoccupied Hours	Occupied Load Btu/hr	Unoccupied Load Btu/hr	Equipment Capacity Btu/hr	Occupied PLF
1	2	3	4	5	6	7
			Given	Given	Given	Eq. 9-4
Table 9-2	Table 9-2	Table 9-2	Eq. 9-6	Eq. 9-7	Eq. 9-8	$D_c = 0.25$
62	216	426	0	14,680	207,521	0.75
57	212	389	0	38,980	195,686	0.75
52	230	454	14,280	63,280	183,851	0.77
47	166	403	38,580	87,580	172,016	0.81
42	172	495	62,880	111,880	160,181	0.85
37	156	465	87,180	136,180	148,346	0.90
32	124	380	111,480	160,480	136,511	0.95
27	54	175	135,780	184,780	124,676	1.00
22	54	127	160,080	209,080	112,841	1.00
17	21	74	184,380	233,380	101,006	1.00
12	3	15	208,680	257,680	89,171	1.00

Development of each load component as a linear function of outdoor temperature permits combination of the components into a total load profile for winter and summer conditions and determination of the balance points.

The CLTD data used for a design load calculations are useful in developing the load components for walls and roofs. In order to take into account the fact that the sky is not clear all the time, it is necessary to separate the solar component from the conduction component in the CLTD. The corrected CLTD may be expressed as

$$CLTD_c = (t_o - t_i) + ESTD \tag{9-10}$$

where

$$ESTD = \text{equivalent solar temperature difference, F or C}$$
$$t_o = \text{outdoor temperature, F or C}$$
$$t_i = \text{indoor temperature, F or C}$$

Now

$$CLTD_c = CLTD + (78 - t_i) + t_d - \left(\frac{DR}{2}\right) - 85 \tag{8-36}$$

where the CLTD is from Table 8-18 or 8-19 and t_d is the outdoor design temperature. Solving for ESTD from Eq. 9-10 and (8-36)

$$ESTD = CLTD - 7 - \left(\frac{DR}{2}\right) + t_d - t_o \tag{9-11a}$$

Table 9-4 (*Continued*)

Unoccupied PLF	Occupied Run Time Hr	Unoccupied Run Time Hr	Power Input kW	Occupied Electrical Resistance Input kW	Unoccupied Electrical Resistance Input kW	Total Energy kWh
8	9	10	11	12	13	14
Eq. 9-4			Given			$9 \times (11 + 12) +$
$D_c = 0.29$	$4 \times 2/(6 \times 7)$	$5 \times 3/(6 \times 8)$	Eq. 9-9	$4 - 6$	$5 - 6$	$10 \times (11 + 13)$
0.77	0.0	39.3	19.8			776.7
0.80	0.0	96.9	19.3			1,867.0
0.84	23.2	186.9	18.8			3,941.1
0.88	46.2	233.9	18.2			5,108.8
0.92	79.6	373.9	17.7			8,039.4
0.98	102.2	435.8	17.2			9,259.8
1.00	106.1	446.7	16.7		7.0	12,368.6
1.00	58.8	259.4	16.2	3.3	17.6	9,908.6
1.00	62.4	235.3	15.7	13.8	28.2	12,165.8
1.00	38.3	171.0	15.2	24.4	38.8	10,741.5
1.00	7.0	43.3	14.6	35.0	49.4	3,123.8
					TOTAL	77,301.1

It may be shown that

$$t_d = \bar{t}_o + \frac{DR}{2} \tag{9-12}$$

where

$$\bar{t}_o = \sum_{i=1}^{24} \frac{t_{oi}}{24} \tag{9-13}$$

then

$$\text{ESTD} = \text{CLTD} - 7 + \bar{t}_o - t_o \tag{9-11b}$$

and the average ESTD for a day is

$$\overline{\text{ESTD}} = \left[\sum_{i=1}^{24} \frac{\text{CLTD}_i}{24} \right] - 7 \tag{9-14}$$

The average solar transmission component for a wall or roof is then given by

$$\dot{q}_{sn} = U_n A_n \times \overline{\text{ESTD}_n} \times \text{FPS} \tag{9-15}$$

where *n* refers to the particular wall or roof being considered and FPS is the fraction of possible sunshine from Table 9-5, which accounts for cloudiness. The conduction

Table 9-5 Percentage of Possible Sunshine for Selected Locations

City & State	January	July	Annual
Chicago, IL	44	73	59
Denver, CO	67	68	67
Oklahoma City, OK	57	76	68
Washington, DC	46	64	58

Source: From U.S. Weather Bureau publication. *Local Climatological Data.*

component is given by

$$\dot{q}_c = U_n A_n (t_o - t_i) \tag{9-16}$$

In this case orientation of the surface is not a factor but the various walls and roofs may vary in composition. Equation 9-16 also gives the conduction component for glass and doors. When the $CLTD_i$ in Eq. 9-14 are averaged, the same value results for all walls and roofs. Therefore, a simplified table may be constructed (Table 9-6).

Table 9-6 The 24-Hour Average Solar Component, ESTD

	JULY			
	Latitude			
Direction	24	32	40	48
North	7	7	6	5
Northeast	13	12	11	10
East	15	16	16	15
Southeast	11	13	15	15
South	5	8	11	12
Southwest	11	13	15	15
West	15	15	16	14
Northwest	13	12	11	9
Roof	24	25	24	24
	JANUARY			
Direction	Latitude			
North	2	1	1	0
Northeast	3	2	1	0
East	10	8	7	4
Southeast	17	16	15	12
South	23	22	21	17
Southwest	17	16	15	12
West	10	8	6	4
Northwest	3	2	1	0
Roof	14	9	5	1

The average solar load component for glass is given by

$$\dot{q}_{sg} = \frac{\sum_{i=1}^{n} (\text{TSCL}_i \times A_i \times \text{SC}_i \times \text{FPS})}{\theta} \tag{9-17}$$

where

TSCL_i = total solar cooling load, Btu/ft^2
A_i = area of the glass, ft^2
FPS = fraction of the possible sunshine for the location, Table 9-5
SC = classical shading coefficient, Tables 6-3 through 6-6
θ = run time of the system, hr

and

$$\text{TSCL}_i = \sum_{j=1}^{24} (\text{SCL}_j) \tag{9-18}$$

where

SCL_j = solar cooling load factors from Table 8-24, Btu/(hr-ft^2)

In summer the actual run time for the system should be used. In winter, however, a significant load occurs at night and a run time of 24 hours is appropriate. Outside shading may have a significant effect on the energy requirements of a building. Procedures are available to estimate the average shaded area for windows with overhangs and side projections (1); however, for brevity this information is not presented here.

The internal loads may constitute a significant contribution to the total load and are important in determining the balance point. In all cases of lights, people, and equipment, it is necessary to determine the average heat gain during the period under consideration. It may be necessary to plot the occupancy or lighting schedule to find a realistic average value. The internal loads are not a function of the outdoor temperature; therefore, these load components are constants. Note that the occupants have a latent component that should be computed separately.

The load component due to infiltration and outdoor ventilation air is a function of outdoor dry bulb and wet bulb temperatures given by

$$\dot{q}_{vs} = \dot{Q}\rho(60)(t_o - t_i) \tag{9-19}$$

and

$$\dot{q}_{vl} = \dot{Q}\rho(60)i_{fg}(W_o - W_i) \tag{9-20}$$

where $\dot{Q}$ is in cfm, the average amount of air entering the space during the period of interest. The mean coincident wet bulb temperature should be used in evaluating W_o in Eq. 9-20, whereas W_i depends on the average indoor conditions.

The general procedure has been outlined to obtain the various components of the total load profile. Recall that most buildings operate on at least two shifts, occupied and unoccupied, and some buildings may require more than one zone. July and January usually represent the warmest and coldest months of the year in the northern hemi-

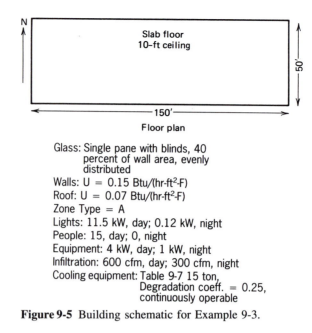

Figure 9-5 Building schematic for Example 9-3.

sphere. Therefore, the usual practice is to find points on the load profile at the design conditions in January and July. An equation is then found between the two points in the form

$$\dot{q} = at_o + bt_i + c \qquad (9\text{-}21)$$

Using Eq. 9-21 with $\dot{q} = 0$ and t_i known, the outdoor temperature at which the system balances may be computed. The following example will demonstrate the complete procedure.

EXAMPLE 9-3

A small single-story building located in Oklahoma City, Oklahoma, is described in Fig. 9-5. The indoor operating temperatures are 75 F and 72 F for summer and winter, respectively, and there are occupied (8:00 A.M.–7:00 P.M.) and unoccupied (7:00 P.M.–8:00 A.M. and weekends) shifts. Compute the energy to operate the system, assuming the times given are solar times.

SOLUTION

Table C-1 gives the summer and winter design temperatures as 96 F and 13 F, respectively. These temperatures will be used in determining the load profile. Let us first consider the solar component through the glass using Eq. 9-17. The total solar cooling load (TSCL) must first be computed using Eq. 9-18 and Table 8-24, assuming 36 deg N latitude.

Solar Through Glass, July

Surface	TSCL	A	SC	FPS	$\dot{q}_n$
N	469	600	0.55	0.78	120,721
E	1161	200	0.55	0.78	99,614
S	585	600	0.55	0.78	150,579
W	1162	200	0.55	0.78	99,700
				Total:	470,614 Btu/day

Since the cooling equipment is operable all the time, $\theta = 24$ hr. It is also convenient to reference each component to one square foot of floor area. Then

$$(\dot{q}_{sg})_{\text{July}} = 470,614/(24)(7500) = 2.615 \text{ Btu/(hr-ft}^2)$$

The same procedure is used for January.

Solar Through Glass, January

Surface	TSCL	A	SC	FPS	$\dot{q}_n$
N	194	600	0.55	0.57	36,491
E	602	200	0.55	0.57	37,745
S	1639	600	0.55	0.57	308,296
W	606	200	0.55	0.57	37,996
				Total:	420,528 Btu/day

$$(\dot{q}_{sg})_{\text{Jan}} = 420,528/(24)(7500) = 2.336 \text{ Btu/(hr-ft}^2)$$

Now to obtain the profile for this component the design temperatures are used to derive a linear equation as follows:

$$(\dot{q}_{sg}) = at_o + b$$

or

$$2.615 = 96a + b$$

and

$$2.336 = 13a + b$$

from which

$$a = 0.0034 \quad \text{and} \quad b = 2.29$$

and

$$(\dot{q}_{sg}) = (0.0034t_o + 2.29) \text{ Btu/(hr-ft}^2) \tag{9-22}$$

The same general procedure is used for the solar component through the walls and roof using Eq. 9-15 and Table 9-6.

Solar Through Walls and Roof, July

Surface	A	U	ESTD	FPS	$\dot{q}_{sw}$
N	900	0.15	6	0.78	632
E	300	0.15	16	0.78	562
S	900	0.15	10	0.78	1053
W	300	0.15	16	0.78	562
Roof	7500	0.07	24	0.78	9828
				Total:	12,636 Btu/hr

$$(\dot{q}_{sw})_{\text{July}} = 12{,}636/7500 = 1.685 \text{ Btu/(hr-ft}^2)$$

Solar Through Walls and Roof, January

Surface	A	U	ESTD	FPS	$\dot{q}_{sw}$
N	900	0.15	1	0.57	77
E	300	0.15	7	0.57	180
S	900	0.15	21	0.57	1616
W	300	0.15	7	0.57	180
Roof	7500	0.07	7	0.57	2095
				Total:	4,148 Btu/hr

$$(\dot{q}_{sw})_{\text{Jan}} = 4{,}148/7500 = 0.553 \text{ Btu/(hr-ft}^2)$$

Using the same procedure as before, the profile for this component becomes

$$(\dot{q}_{sw}) = 0.014t_o + 0.341 \text{ Btu/(hr-ft}^2) \tag{9-23}$$

Equation 9-16 will be used to find the load component for conduction.

Conduction for Walls, Roof, Glass

Surface	A	U	$A \times U$
Walls	2400	0.15	360
Roof	7500	0.07	525
Glass	1600	1.00	1600
		Total:	2,485 Btu/(hr-F)

$$(\dot{q}_{cw} = 2{,}485(t_o - t_i)/7500$$

or

$$\dot{q}_{cw} = (0.331t_o - 0.331t_i) \text{ Btu/(hr-ft}^2) \tag{9-24}$$

The load components considered so far are independent of the shift. The internal loads and infiltration to be considered next do depend on the shift.

Internal Loads, Assumed Constant the Year Round

$$\text{Lights: } \dot{q}_{ld} = 11.5(3412)/7500 = 5.23 \text{ Btu/(hr-ft}^2) \tag{9-25}$$

$$\dot{q}_{ln} = 0.12(3412)/7500 = 0.055 \text{ Btu/(hr-ft}^2) \tag{9-26}$$

$$\text{Equipment: } \dot{q}_{ed} = 4(3412)/7500 = 1.82 \text{ Btu/(hr-ft}^2) \tag{9-27}$$

$$\dot{q}_{en} = 1(3412)/7500 = 0.45 \text{ Btu/(hr-ft}^2) \tag{9-28}$$

$$\text{Occupants, sensible: } \dot{q}_{od} = 15(230)/7500 = 0.46 \text{ Btu/(hr-ft}^2) \tag{9-29}$$

$$\dot{q}_{on} = 0.0$$

$$\text{Occupants, latent: } \dot{q}_{od} = 15(190)/7500 = 0.38 \text{ Btu/(hr-ft}^2) \tag{9-30}$$

$$\dot{q}_{on} = 0.0$$

Infiltration (assume standard air)

$$\text{Day, sensible: } \dot{q}_{vs} = 1.1(600)(t_o - t_i)/7500 \tag{9-31}$$

$$= 0.088t_o - 0.088t_i$$

$$\text{Day, latent: } \dot{q}_{vl} = 4840(600)(0.015 - 0.009)/7500 \tag{9-32}$$

$$= 2.32 \text{ Btu/(hr-ft}^2)$$

$$\text{Night, sensible } \dot{q}_{vs} = 1.1(300)(t_o - t_i)/7500 \tag{9-33}$$

$$= 0.044t_o - 0.044t_i \text{ Btu/(hr-ft}^2)$$

$$\text{Night, latent } \dot{q}_{vl} = (4840)(300)(0.015 - 0.009)/7500 \tag{9-34}$$

$$= 1.16 \text{ Btu/(hr-ft}^2)$$

Infiltration would probably be lower during the summer months than in the winter; however, it will be assumed constant in this case. The internal and infiltration loads may be summarized as follows: For the day shift sum Eqs. 9-25, 9-27, 9-29, and 9-31 to obtain the sensible component

$$\dot{q}_{is,d} = 5.23 + 1.82 + 0.46 + 0.088t_o - 0.088t_i$$

$$\dot{q}_{is,d} = (0.088t_o - 0.088t_i + 7.51) \text{ Btu/(hr-ft}^2) \tag{9-35}$$

The latent component for the day shift is the sum of Eqs. 9-30 and 9-32:

$$\dot{q}_{il,d} = 0.38 + 2.32 = 2.70 \text{ Btu/(hr-ft}^2) \tag{9-36}$$

The sensible component for the night shift is the sum of Eqs. 9-26, 9-28, and 9-33:

$$\dot{q}_{is,n} = 0.055 + 0.45 - 0.44t_o - 0.044t_i$$

$$\dot{q}_{is,n} = (0.44t_o - 0.044t_i + 0.505)\ \text{Btu/(hr-ft}^2) \tag{9-37}$$

The latent component for the night shift is given by Eq. 9-34:

$$\dot{q}_{il,n} = 1.16\ \text{Btu/(hr-ft}^2) \tag{9-34}$$

The total sensible load profiles for each shift may now be determined. The latent components will be included after the balance points have been determined.

For the Day Shift

$$\dot{q}_{sg} = 0.0034t_o \qquad\qquad +2.29 \quad (9\text{-}22)$$

$$\dot{q}_{sw} = 0.014t_o \qquad\qquad +0.341\ (9\text{-}23)$$

$$\dot{q}_{cw} = 0.331t_o \quad -0.331t_i \qquad\qquad (9\text{-}24)$$

$$\dot{q}_{is} = 0.088t_o \quad -0.088t_i \quad +7.51 \quad (9\text{-}35)$$

$$\overline{\dot{q}_{sd} = 0.4364t_o - 0.419t_i + 10.14} \tag{9-38a}$$

For the Night Shift

$$\dot{q}_{sg} = 0.0034t_o \qquad\qquad +2.29 \quad (9\text{-}22)$$

$$\dot{q}_{sw} = 0.014t_o \qquad\qquad +0.341\ (9\text{-}23)$$

$$\dot{q}_{cw} = 0.331t_o \quad -0.331t_i \qquad\qquad (9\text{-}24)$$

$$\dot{q}_{is} = 0.044t_o \quad -0.044t_i \quad +0.505\ (9\text{-}37)$$

$$\overline{\dot{q}_{sn} = 0.3924t_o - 0.375t_i + 3.096} \tag{9-39a}$$

The balance point temperatures may now be computed from Eqs. 9-38a and 9-39a. For the cooling season and the occupied (day) shift the indoor temperature is 75 F and $\dot{q}_{sd} = 0$

$$t_o = [0.419(75) - 10.14]/0.4364 = 48.8\ \text{F}$$

and for the unoccupied (night) shift

$$t_o = [0.375(75) - 3.096]/0.3924 = 63.8\ \text{F}$$

These appear to be quite reasonable values. During the heating season the indoor temperature is 72 F. Using the same procedure we obtain the balance temperatures for the day and night shifts as 46 F and 60 F, respectively. The total sensible load profiles

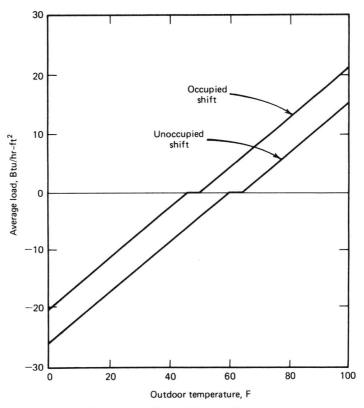

Figure 9-6 Sensible-load profiles for Example 9-3.

then appear as shown in Fig. 9-6.. Note that a span of about 3 F occurs where neither heating nor cooling is required. Before proceeding with the bin calculation, the latent component should be added to Eqs. 9-38a and 9-39a for the cooling energy calculation. This latent component is usually omitted from the analysis for the heating energy. Equations 9-38a and 9-39a, with $t_i = 75$ F, then become

$$\dot{q}_d = (0.44t_o - 18.59) \text{ Btu/(hr-ft}^2) \qquad \textbf{(9-38b)}$$

and

$$\dot{q}_n = (0.39t_o - 23.87) \text{ Btu/(hr-ft}^2) \qquad \textbf{(9-39b)}$$

These load profiles are superior to the simplified versions used in Example 9-2.

The bin analysis may now be carried out to determine the energy required. The procedure is identical to that of Example 9-2 and the occupied and unoccupied shifts are the same. Therefore, the bin hours for each shift may be taken from Table 9-2 for the bin temperatures of interest. The only required additional data are the air-conditioning equipment capacity and input as a function of outdoor temperature, which can be obtained from Table 9-7 for the 15-ton unit. A linear curve fit yields:

$$\text{Capacity} = (299{,}200 - 1220t_o)\ \text{Btu/hr} \tag{9-40}$$

$$\text{Input} = (9.1 + 0.14t_o)\ \text{kW} \tag{9-41}$$

A calculation sheet, by hand or computer, similar to Table 9-8, can now be constructed. To make a complete analysis, it will be assumed that a gas furnace with a steady-state output of 200,000 Btu/hr and efficiency of 75 percent is used to heat the building. The heating value of the natural gas is assumed to be 1000 Btu per standard cubic foot. Table 9-8 shows the results. Note that three bin temperatures occur when cooling is required during the day, whereas heating is required during the night hours. The negative loads are heating loads and the power shown in the last column includes fan energy when heating or cooling.

9-3 COMPREHENSIVE SIMULATION METHODS

Following design of the environmental control system for a building, it is often desirable to make a more detailed analysis of the anticipated energy requirements of the structure for heating, cooling, lighting, and other powered equipment. This same information is often required in energy conservation studies involving existing buildings. Simulation implies that the complete system configuration is already determined; therefore, this type of analysis is distinctly different from design, where sizing of components is the objective. Simulation might sometimes be a useful tool in design optimization, where alternatives are considered on the basis of energy use or operating cost.

Frequent evaluation is needed to obtain reasonably accurate estimates of annual energy consumption because of the constantly changing internal and external factors that determine thermal loads on commercial and industrial buildings. Many commercial buildings have high peak internal loads that necessitate a review of load patterns hourly and for different day types during the year. To use simulation methods, the mathematical model of the building and its systems must represent the thermal behavior of the structure (the loads model), the thermodynamic process of the air-conditioning system (the secondary systems model), and a mathematical relationship for load versus energy requirements of the primary energy conversion equipment (the plant model). Each model is usually formulated so that input quantities allow calculation of output quantities. Weather and internal load information are inputs to the loads model, allowing calculation of sensible loads on the spaces in the building, for example. The building must usually be divided into several thermally different zones for modeling rather than one zone for the building as a whole.

The secondary system, made up of coils, mixing boxes and fans, is modeled using the conservation of energy and mass equations. These relations are arranged so that the required rate of energy usage can be calculated for a given space load, calculated from the loads model. Several secondary system models may be required in a simulation, depending on the number and types of zones.

The required input energy rate, natural gas or electrical power for example, is given by the plant model using the required energy rates from the secondary system models. The plant model includes models for boilers, chillers, cooling towers, and so on.

Table 9-7 Performance Data for Some Unitary Air-Conditioning Equipment[a]

Outdoor Dry Bulb Temp., F	7½ Ton, 3000 cfm			10 Ton, 4000 cfm			15 Ton, 6000 cfm			20 Ton, 8000 cfm		
	Total Cap. Btu/hr × 1000	Sensible Cap. Btu/hr × 1000	EER[b] Btu/(W-hr)	Total Cap. Btu/hr × 1000	Sensible Cap. Btu/hr × 1000	EER[b] Btu/(W-hr)	Total Cap. Btu/hr × 1000	Sensible Cap. Btu/hr × 1000	EER[b] Btu/(W-hr)	Total Cap. Btu/hr × 1000	Sensible Cap. Btu/hr × 1000	EER[b] Btu/(W-hr)
75	101.2	68.4	11.0	133.4	92.4	106	206.8	131.5	10.4	272.5	193.0	10.4
85	96.2	66.8	9.8	126.7	89.8	9.5	195.5	127.5	9.3	257.7	187.3	9.2
95	91.0	64.8	8.7	120.0	87.4	8.5	184.0	123.4	8.3	244.0	182.0	8.2
105	85.2	62.7	7.6	112.8	84.7	7.5	171.1	118.9	7.2	229.0	176.3	7.3
115	79.5	60.7	6.7	105.5	82.1	6.6	158.2	114.5	6.2	214.1	170.7	6.4

[a]Indoor coil operating at 75 F db and 62 F wb.

[b]EER includes all fan power.

Table 9-8 Bin Energy Calculation for Example 9-3

Bin Temperature	Occupied hours	Unoccupied Hours	Occupied Load Btu/hr	Unoccupied Load Btu/hr	Cooling Equipment Capacity Btu/hr	Heating Equipment Capacity Btu/hr	Occupied PLF	Unoccupied PLF	Occupied Run Time Hr	Unoccupied Run Time Hr	Cooling Equipment Input kW	Heating Energy mcf	Cooling Energy kW-hr
Table 9-2	Table 9-2	Table 9-2	Given Eq. 9-38b	Given Eq. 9-39b	Given Eq. 9-40	Given	Eq. 9-4 $Dc = 0.25$	Eq. 9-4 $Dc = 0.25$	$4 \times 2/(6 \times 7)$	$5 \times 3/(6 \times 8)$	Given Eq. 9-41	$0.267 \times (9 + 10)$	$11 \times (9 + 10)$
1	2	3	4	5	6a	6b	7	8	9	10	11	12	13
102	1	1	197,175	119,325	174,760		1.00	0.92	1.1	0.7	23.38		43.7
97	64	40	180,675	104,700	180,860		1.00	0.89	64.0	25.9	22.68		2037.4
92	179	117	164,175	90,075	186,960		0.97	0.87	162.1	64.8	21.98		4986.9
87	229	178	147,675	75,450	193,060		0.94	0.85	186.1	82.1	21.28		5706.6
82	280	338	131,175	60,825	199,160		0.91	0.83	201.6	124.9	20.58		6723.3
77	244	532	114,675	46,200	205,260		0.89	0.81	153.2	148.5	19.88		5998.5
72	264	745	98,175	31,575	211,360		0.87	0.79	141.6	141.4	19.18		5426.7
67	202	545	81,675	16,950	217,460		0.84	0.77	89.9	55.2	18.48		2681.6
62	216	426	65,175	2,325	223,560		0.82	0.75	76.5	5.9	17.78		1465.3
57	212	389	48,675	−12,300	229,660		0.80	0.77	56.0	31.3	17.08	23.3	1489.6
52	230	454	32,175	−26,925	235,760		0.78	0.78	40.0	78.0	16.38	31.5	1933.2
47	166	403	15,675	−41,550	241,860	200,000	0.77	0.80	14.0	104.4	15.68	31.6	1857.2
42	172	495	−825	−56,175		200,000	0.75	0.82	0.9	169.5		45.5	
37	156	465	−17,325	−70,800		200,000	0.77	0.84	17.5	196.3		57.1	
32	124	380	−33,825	−85,425		200,000	0.79	0.86	26.5	189.4		57.6	
27	54	175	−50,325	−100,050		200,000	0.81	0.88	16.7	100.0		31.2	
22	44	127	−66,825	−114,675		200,000	0.83	0.89	17.6	81.5		26.5	
17	21	74	−83,325	−129,300		200,000	0.85	0.91	10.2	52.0		16.7	
12	3	15	−99,825	−143,925		200,000	0.87	0.93	1.7	11.6		3.6	
										TOTALS		324.6	40347.0

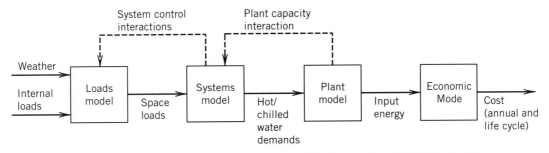

Figure 9-7 Flow diagram for calculating hourly heating cooling loads using ASHRAE algorithms.

Figure 9-7 shows how the various models are related. Dashed lines show the control interaction paths. Capacity limits and control characteristics of the system can affect the space load. Any variation in space temperature allowed by the system causes the space sensible load to vary. Also, capacity limits and control characteristics of the central plant can cause variation in secondary system performance, which in turn affect the loads.

An economic model is shown in Fig. 9-7 that calculates energy costs based on the computed input energy. Thus, the simulation model produces hourly energy and cost estimates based on input of weather and internal loads. Running sums of these quantities yields monthly or annual energy usage and costs.

Simulation models can only compare alternatives and predict trends. Unknown factors usually prevent accurate prediction of utility costs.

Space Sensible Load

The key to accurate building energy simulation is calculation of the space sensible load. The two most widely used methods for these calculations are the *heat balance method* and the *transfer function method*.

The more fundamental of the two methods is the heat balance method, which relies on the first law of thermodynamics and the principles of matrix algebra. It relies on fewer assumptions than the transfer function method and is therefore more flexible. The heat balance method does require more calculations and more computer time, however. This approach is discussed briefly in Chapter 8, The Cooling Load.

The transfer function method is a compromise between simpler methods, such as the bin method, and more complex methods, such as complete energy balance methods. The methodology is quite similar to the procedure outlined in Chapter 8 to calculate cooling load using the transfer function method. The main differences as compared with the design computation are as follows:

1. Hourly historical or assumed temperature data are used instead of the design maximum temperature and daily range.
2. Historical or assumed hourly solar radiation data are used instead of computed clear sky data so that the effect of cloudiness, haze, and so on, are taken into account.
3. More realistic occupancy and lighting schedules may be used.

4. Central equipment such as chillers and boilers is modeled to compute the energy required to meet the heating and cooling demand (heat addition or extraction rate).
5. The simulation is carried out on an hour-by-hour basis for periods of time greater than one day and generally up to a year.

Various weather data are available on magnetic media. One type, known as the Weather Year for Energy Calculations (WYEC), was developed by ASHRAE and is in popular use. This weather tape contains data that are typical of the weather patterns for many cities in the United States for a complete year on an hourly basis.

Central Plant Modeling

Modeling of the central cooling and heating plant can become quite complex; however, this doesn't have to be true. The model should take into account the effect of environmental conditions and load on the operating efficiency. For example, the coefficient of performance of a water chiller depends on the chilled water temperature and the condensing water temperature. The chilled water temperature may be relatively constant, but the condensing water temperature may depend on the outdoor wet bulb temperature and the load on the chiller. Boiler performance does not depend as much on environmental conditions, but the efficiency does drop rapidly with decreasing load. Reference 1 outlines various modeling approaches.

A useful and simple way of modeling all types of heating and cooling equipment is to normalize the energy input and the capacity with the rated full-load input and capacity. Then the normalized input is

$$Y = E/E_{\max} \qquad\qquad\qquad (9\text{-}42)$$

and the normalized capacity is

$$X = \dot{q}_x/\dot{q}_{x,\max} \qquad\qquad\qquad (9\text{-}43)$$

These quantities may then be plotted and a curve fitted that comprises a model. Figure 9-8 is an example of such a model for a hot water boiler and Fig. 9-9 is for a centrifugal chiller. To construct the curves it is necessary to have performance data for partial-load conditions. Most manufacturers can furnish such data. The plots of Y versus X tend to be straight lines. When this is true, it is possible to develop a model with a limited number of performance data.

The model for the central plant must also include pumps, fans, cooling towers, and any auxiliary equipment that uses energy. The energy consumed by the lights is also often included in the overall equipment model. An estimate of the total energy consumption of the building is the overall objective. For existing buildings, the total predicted energy usage may be compared with the actual utility data.

Many computer programs are available to carry out both the design and simulation analysis previously described. Among these are DOE-2, developed by the federal government, and BLAST, developed by the Army Corp of Engineers.

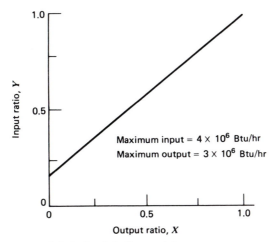

Figure 9-8 A simple boiler model.

Overall Modeling Strategies

The sequence and procedures used to solve the various equations is referred to as the *overall modeling strategy*. The accuracy of the results and the required computer resources are greatly dependent on this.

Most programs today use the sequential approach. With this strategy the loads are first computed for every hour of the period, followed by execution of the systems models, one at a time, for every hour of the period. Last, the plant simulation model is executed for the entire period. Each sequence processes the fixed output of the preceding step.

Certain phenomena cannot be modeled precisely by this loads–systems–plant sequence. For example, the systems model may not be able to meet the load and can only report the current load to the plant model. If the transfer function method is used, this

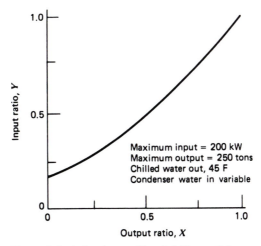

Figure 9-9 A simple centrifugal chiller model.

problem can be partially overcome provided the space temperature does not depart widely from the value used to compute the load initially. A similar problem can occur in plant simulation.

A conceivable alternative modeling strategy in which all calculations are carried out at each time step is possible. In this case all the equations making up the loads, systems, and plant models are solved simultaneously, and the unmet loads and imbalances discussed above cannot occur. The results should be improved over the sequential approach; however, the extra time and effort and computer resources may be difficult to justify.

REFERENCES

1. *ASHRAE Handbook, Fundamentals Volume,* American Society of Heating, Refrigerating and Air-Conditioning Engineers, Inc., Atlanta, GA, 1989.
2. W. J. Kelnhofer, *Evaluation of the ASHRAE Modified Degree Day Procedure for Predicting Energy Usage by Residential Gas Heating Systems,* American Gas Association, 1979.
3. *Bin and Degree Hour Weather Data for Simplified Energy Calculations,* American Society of Heating, Refrigerating and Air-Conditioning Engineers, Inc., Atlanta, GA, 1986.

PROBLEMS

9-1. Using the degree day method, estimate the quantity of natural gas required to heat a building located in Stillwater, Oklahoma. Design conditions are 70 F indoor and 12 F outdoor temperatures. The computed heat load is 225,000 Btu/hr. Assume an efficiency factor of 80 percent. The heating value of the fuel is 1000 Btu/std ft^3.

9-2. If electric resistance heat were used to heat the building mentioned in Problem 9-1, how much energy would be required in kW-hr assuming a 100 percent efficiency factor? If the electrical energy costs ten cents per kW-hr and natural gas costs $4.5 per mcf, what are the relative heating costs? Assuming a power plant efficiency of 33 percent, compare the total amounts of energy in terms of mcf of gas required to heat the building using a gas furnace and an electric furnace.

9-3. A light commercial building, located in Sioux City, Iowa, has construction and use characteristics much like a residence and a design heat load of 120,000 Btu/hr (35 kW). The structure is heated with a natural gas warm-air furnace and is considered energy efficient. Assuming standard design conditions, estimate the yearly heating fuel requirements.

9-4. Refer to Problem 9-3 and determine the simplified unoccupied load profile assuming a balance point temperature of 60 F (16 C).

9-5. Refer to Problems 9-3 and 9-4. The building has an average internal load of 20,000 Btu/hr (6 kW) due to lights, equipment, and people. Determine the simplified occupied load profile.

9-6. Consider a building that operates on two shifts. The first shift begins at 10:00 A.M. and ends at midnight and the second shift includes all the remaining hours.

Assume a 5-day work week. Compute the bin hours in each shift for Oklahoma City, Oklahoma. Consider bin temperatures of 62 F and less.

9-7. Solve Problem 9-6 for (a) Denver, Colorado, (b) Washington, D.C., (c) Chicago, Illinois.

9-8. Solve Example 9-2 using the shifts of Problem 9-6.

9-9. Solve Example 9-2 for (a) Denver, Colorado, (b) Washington, D.C., (c) Chicago, Illinois.

9-10. Solve Example 9-2 using the shifts of Problem 9-6 for (a) Denver, Colorado, (b) Washington, D.C., (c) Chicago, Illinois.

9-11. Estimate the energy requirements for the structure described by the plans and specifications furnished by the instructor using a computer program.

9-12. Solve Example 9-3 with the building located in (a) Denver, Colorado, (b) Chicago, Illinois, (c) Washington, D.C.

9-13. Solve Example 9-3 with two shifts where the day shift is 7:00 A.M. LST to 4:00 P.M. LST for (a) Oklahoma City, Oklahoma, (b) Denver, Colorado, (c) Chicago, Illinois, (d) Washington, D.C.

9-14. Use the bin method to compute the cooling energy for Problem 9-11.

9-15. Solve Example 9-3 assuming that the building is rotated 90 degrees clockwise.

Chapter **10**

Flow, Pumps, and Piping Design

The distribution of fluids by pipes, ducts, and conduits is essential to all heating and cooling systems. The fluids encountered are gases, vapors, liquids, and mixtures of liquid and vapor (two-phase flow). From the standpoint of overall design of the building system, water and air are of greatest importance. This chapter deals with the fundamentals of incompressible flow of fluids such as air and water in conduits, considers the basics of centrifugal pumps, and develops simple design procedures for water piping systems. Basic principles of the control of fluid-circulating systems including variable flow, secondary pumping, and the relationship between thermal and hydraulic performance of the system are covered.

10-1 FLUID FLOW BASICS

The adiabatic, steady flow of a fluid in a pipe or conduit is governed by the first law of thermodynamics, which may be written

$$\frac{P_1}{\rho_1} + \frac{\overline{V}_1^2}{2g_c} + \frac{gz_1}{g_c} = \frac{P_2}{\rho_2} + \frac{\overline{V}_2^2}{2g_c} + \frac{gz_2}{g_c} + w + \frac{g}{g_c}l_f \qquad \textbf{(10-1a)}$$

where

P = static pressure, lbf/ft^2 or N/m^2
ρ = mass density at a cross section, lbm/ft^3 or kg/m^3
$\overline{V}$ = average velocity at a cross section, ft/sec or m/s
g = local acceleration of gravity, ft/sec^2 or m/s^2
g_c = constant, 32.17 $(\text{lbm-ft})/(\text{lbf-sec}^2)$ or 1.0 $(\text{kg-m})/(\text{N-s}^2)$
z = elevation, ft or m
w = work, (ft-lbf)/lbm or J/kg
l_f = lost head, ft or m

Each term of Eq. 10-1a has the units of energy per unit mass or specific energy. The last term on the right in Eq. 10-1a is the internal conversion of energy due to friction. The first three terms on each side of the equality are the pressure energy, kinetic energy, and potential energy, respectively. A sign convention has been selected such that work done on the fluid is negative.

Another governing relation for steady flow in a conduit is the conservation of mass. For flow along a single conduit the mass rate of flow at any two cross sections 1 and 2 is given by

$$\dot{m} = \rho_1 \overline{V}_1 A_1 = \rho_2 \overline{V}_2 A_2 \tag{10-2}$$

where

$\dot{m}$ = mass flow rate, lbm/sec or kg/s
A = cross-sectional area normal to the flow, ft^2 or m^2

When the fluid is incompressible, Eq. 10-2 becomes

$$\dot{Q} = \overline{V}_1 A_1 = \overline{V}_2 A_2 \tag{10-3}$$

where

$$\dot{Q} = \text{volume flow rate, ft}^3\text{/sec or m}^3\text{/s}$$

Equation 10-1a has other useful forms. If it is multiplied by the mass density, assumed constant, an equation is obtained where each term has the units of pressure:

$$P_1 + \frac{\rho_1 \overline{V}_1^2}{2g_c} + \frac{\rho_1 g z_1}{g_c} = P_2 + \frac{\rho_2 \overline{V}_2^2}{2g_c} + \frac{\rho_2 g z_2}{g_c} + \rho w + \frac{\rho g l_f}{g_c} \tag{10-1b}$$

In this form the first three terms on each side of the equality are the static pressure, the velocity pressure, and the elevation pressure, respectively. The work term now has units of pressure, and the last term on the right is the pressure lost due to friction.

Finally, if Eq. 10-1a is multiplied by g_c/g, an equation results where each term has the units of length, commonly referred to as *head:*

$$\frac{g_c}{g} \frac{P_1}{\rho_1} + \frac{\overline{V}_1^2}{2g} + z_1 = \frac{g_c}{g} \frac{P_2}{\rho_2} + \frac{\overline{V}_2^2}{2g} + z_2 + \frac{g_c w}{g} + l_f \tag{10-1c}$$

The first three terms on each side of the equality are the static head, velocity head, and elevation head, respectively. The work term is now in terms of head, and the last term is the lost head due to friction.

Equations 10-1a and 10-2 are complementary because they have the common variables of velocity and density. When Eq. 10-1a is multiplied by the mass flow rate $\dot{m}$, another useful form of the energy equation results, assuming ρ = constant:

$$\dot{W} = \dot{m} \left[\frac{P_1 - P_2}{\rho} + \frac{\overline{V}_1^2 - \overline{V}_2^2}{2g_c} + \frac{g(z_1 - z_2)}{g_c} - \frac{g}{g_c} l_f \right] \tag{10-4}$$

where

$$\dot{W} = \text{power, work per unit time,} \frac{\text{ft-lbf}}{\text{sec}} \text{ or W}$$

All terms on the right-hand side of the equality may be positive or negative except the lost energy, which must always be positive.

Some of the terms in Eqs. 10-1a and 10-4 may be zero or negligibly small. When the fluid flowing is a liquid, such as water, the velocity terms are usually rather small and can be neglected. In the case of flowing gases, such as air, the potential energy terms are usually very small and can be neglected; however, the kinetic energy terms may be quite important. Obviously the work term will be zero when no pump, turbine, or fan is present.

The *total pressure*, a very important concept, is the sum of the static pressure and the velocity pressure:

$$P_0 = P + \frac{\rho \overline{V}^2}{2g_c} \tag{10-5a}$$

In terms of head Eq. 10-5a is written

$$\frac{g_c P_0}{g\rho} = \frac{g_c P}{g\rho} + \frac{\overline{V}^2}{2g} \tag{10-5b}$$

Equations 10-1c and 10-4 may be written in terms of total head and with rearrangement of terms become

$$\frac{g_c}{g} \frac{(P_{01} - P_{02})}{\rho} + (z_1 - z_2) = \frac{g_c w}{g} + l_f \tag{10-1d}$$

This form of the equation is much simpler to use with gases because the term $(z_1 - z_2)$ is negligible, and when no fan is in the system, the lost head equals the loss in total pressure head.

Lost Head

For incompressible flow in pipes and ducts the lost head is expressed as

$$l_f = f \frac{L}{D} \frac{\overline{V}^2}{2g} \tag{10-6}$$

where

f = Moody friction factor
L = length of the pipe or duct, ft or m
D = diameter of the pipe or duct, ft or m
$\overline{V}$ = average velocity in the conduit, ft/sec or m/s
g = acceleration due to gravity, ft/sec^2 or m/s^2

The lost head has the units of feet or meters of the fluid flowing. For conduits of noncircular cross section, the diameter may be replaced by the hydraulic diameter D_h given by

$$D_h = \frac{4(\text{cross-sectional area})}{\text{wetted perimeter}} \tag{10-7}$$

Use of the hydraulic diameter is restricted to turbulent flow and cross-sectional geometries without extremely sharp corners.

Figure 10-1 shows friction data correlated by Moody (1), which is commonly referred to as the Moody diagram. Figure 10-2 is an example of relative roughness data for various types of pipe and tubing. The friction factor is a function of the Reynolds number (Re) and the relative roughness e/D of the conduit in the transition zone; is a function of only the Reynolds number for laminar flow; and is a function of only relative roughness in the complete turbulence zone. Notice that the friction factor can be read directly from Fig. 10-2 if the Reynolds number and relative roughness are sufficiently large for the flow to be considered as in complete turbulence.

The Reynolds number is defined as

$$\text{Re} = \frac{\rho \overline{V} D}{\mu} = \frac{\overline{V} D}{\nu} \tag{10-8}$$

where

ρ = mass density of the flowing fluid, lbm/ft^3 or kg/m^3
μ = dynamic viscosity, lbm/(ft-sec) or (N-s)/m^2
ν = kinematic viscosity, ft^2/sec or m^2/s

The hydraulic diameter is used to calculate Re when the conduit is noncircular. Appendix B contains viscosity data for water, air, and refrigerants. The *ASHRAE Handbook* (2) has data on a wide variety of fluids.

To prevent freezing it is often necessary to use a brine solution, usually a mixture of ethylene glycol and water. Figure 10-3 gives specific gravity and viscosity data for water and various solutions of ethylene glycol and water. Note that the viscosity is given in units of centipoise [1 lbm/(ft-sec) = 1490 centipoises and 10^3 centipoise = 1 (N-s)/m^2]. The following example demonstrates calculation of lost head for pipe flow.

EXAMPLE 10-1

Compare the lost head for water and a 30 percent ethylene glycol solution flowing at the rate of 110 gallons per minute (gpm) in a 3-in. standard (Schedule 40) commercial steel pipe, 200 ft in length. The temperature of the water is 50 F.

SOLUTION

Equation 10-6 will be used. From Table D-1 the inside diameter of 3-in. nominal-diameter Schedule 40 pipe is 3.068 in. and the inside cross-sectional area for flow is

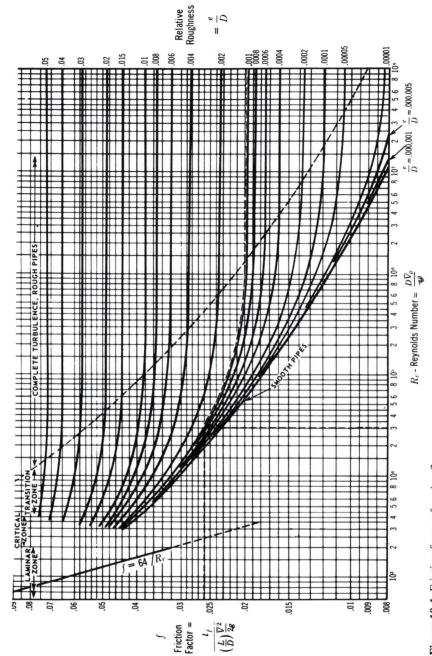

Figure 10-1 Friction factors for pipe flow.

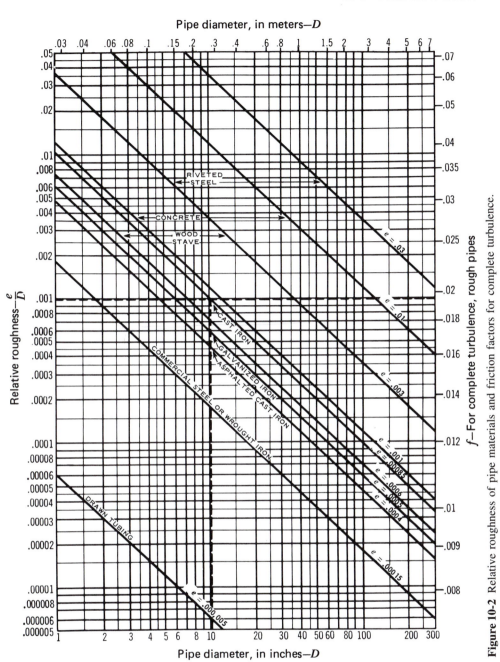

Figure 10-2 Relative roughness of pipe materials and friction factors for complete turbulence.

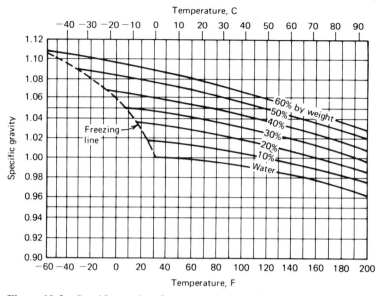

Figure 10-3a Specific gravity of aqueous ethylene glycol solutions. (Adapted by permission from *ASHRAE Handbook, Fundamentals Volume, 1989.*)

0.0513 ft^2. The Reynolds number is given by Eq. 10-8 and the average velocity in the pipe is

$$\overline{V} = \frac{\dot{Q}}{A} = \frac{110 \text{ gal/min}}{(7.48 \text{ gal/ft}^3)(0.0513 \text{ ft}^2)} = 287 \text{ ft/min} = 4.78 \text{ ft/sec}$$

The absolute viscosity of pure water at 50 F is 1.4 centipoises or 9.4×10^{-4} lbm/(ft-sec) from Fig. 10-3b. Then

$$\text{Re} = \frac{62.4(4.78)(3.068/12)}{9.4 \times 10^{-4}} = 8.1 \times 10^4$$

From Fig. 10-2 the relative roughness e/D is 0.00058 for commercial steel pipe. Then the flow is in the transition zone and the friction factor f is 0.021 from Fig. 10-1. The lost head for pure water is then computed using Eq. 10-6.

$$l_{fw} = 0.021 \times \frac{200}{(3.068/12)} \frac{(4.78)^2}{2(32.2)} = 5.83 \text{ ft of water}$$

The absolute viscosity of the 30 percent ethylene glycol solution is 3.1 centipoises from Fig. 10-3b, and its specific gravity is 1.042 from Fig. 10-3a. The Reynolds number for this case is

$$Re = \frac{1.042(62.4)(4.78)(3.068/12)}{(3.1)/1490} = 3.8 \times 10^4$$

and the friction factor is 0.024 from Fig. 10-1. Then

$$l_{fe} = 0.024 \times \frac{200}{(3.068/12)} \frac{(4.78)^2}{2(32.2)} = 6.66 \text{ ft of E.G.S.}$$

$$= 6.94 \text{ ft of water}$$

The increase in lost head with the brine solution is

$$\text{Percent increase} = \frac{100(6.94 - 5.83)}{5.83} = 19 \text{ percent}$$

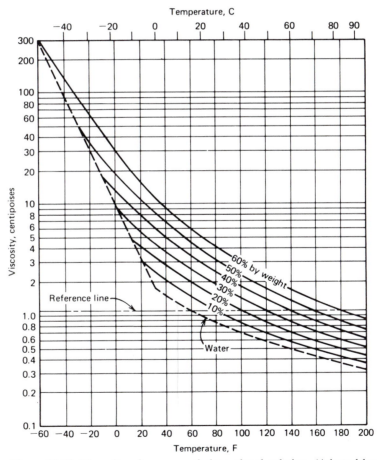

Figure 10-3b Viscosity of aqueous ethylene glycol solution. (Adapted by permission from *ASHRAE Handbook, Fundamentals Volume,* 1989.)

System Characteristic

The behavior of a piping system may be conveniently represented by plotting total head versus volume flow rate. Equation (10-1d) becomes

$$H_p = \frac{g_c(P_{01} - P_{02})}{g\rho} + (z_1 - z_2) - l_f \qquad \textbf{(10-1e)}$$

where H_p represents the total head required to produce the change in static, velocity, and elevation head and to offset the lost head. If a pump is present in the system, H_p is the total head it must produce for a given volume flow rate. Since the lost head and velocity head are proportional to the square of the velocity, the plot of total head versus flow rate is approximately parabolic, as shown in Fig. 10-4. Note that the elevation head is the same regardless of the flow rate. System characteristics are useful in analyzing complex circuits such as the parallel arrangement of Fig. 10-5. Circuit 1a2 and 1b2 each have a characteristic as shown in Fig. 10-6. The total flow rate is equal to the sum of $\dot{Q}_a$ and $\dot{Q}_b$ and the total head is the same for both circuits; therefore the characteristics are summed for various values of H_p to obtain the curve for the complete system shown as $(a + b)$. Series circuits have a common flow rate and the total heads are additive (Fig. 10-7). More discussion of system characteristics will follow introduction of pumps in Section 10-2.

Flow Measurement

It is important to make provisions for the measurement of flow rate in piping and duct systems. The most common devices for making these measurements are the pitot tube, orifice meter, or other restrictive device, and anemometers, which are generally direct

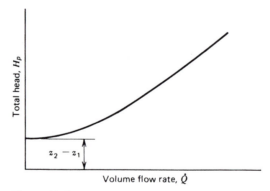

Figure 10-4 Typical system characteristic.

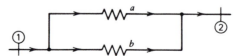

Figure 10-5 Arbitrary parallel flow circuit.

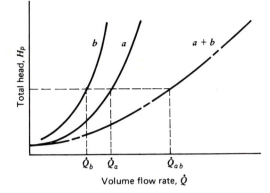

Figure 10-6 System characteristic for parallel circuits.

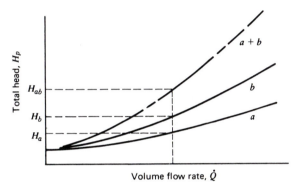

Figure 10-7 System characteristic for series circuits.

velocity-measuring instruments. The pitot tube and orifice meter will be discussed here because they have greatest application in duct and piping systems. Figure 10-8 shows a pitot tube installed in a duct. The pitot tube senses both total and static pressure. The difference, the velocity pressure, is measured with a manometer. If the pitot tube is very small relative to the duct size, then local measurements and traverses may be made. When Eq. 10-1a is applied to a streamline between the tip of the pitot tube and a point a short distance upstream, the following equation results (the lost energy is assumed to be negligibly small and the mass density constant):

$$\frac{P_1}{\rho} + \frac{V_1^2}{2g_c} = \frac{P_2}{\rho} = \frac{P_{02}}{\rho} \tag{10-9a}$$

or

$$\frac{P_{02} - P_1}{\rho} = \frac{\overline{V}_1^2}{2g_c} = Pv \tag{10-9b}$$

Solving for V_1,

$$\overline{V}_1 = \left[2g_c \frac{(P_{02} - P_1)}{\rho} \right]^{1/2} \tag{10-10}$$

Equation 10-10 yields the velocity upstream of the pitot tube. The velocity may be different at other transverse locations in the pipe. For laminar flow the velocity distribution is parabolic, whereas turbulent flow exhibits a more nearly uniform velocity distribution except near the wall.

Figure 10-9 shows a profile that might be typical of turbulent flow. It is generally necessary to traverse the pipe or duct and to integrate either graphically or numerically to find the average velocity in the duct. Equations 10-2 and 10-3 are then used to find the mass or volume flow rate.

The average velocity for incompressible steady flow is expressed by

$$\overline{V}A = \int_A V(x)\, dA \tag{10-11}$$

which is simply an adaptation of Eq. 10-3. For a circular duct the cross-sectional area A is given by

$$A = \pi x^2 \quad \text{and} \quad dA = 2\pi x\, dx \tag{10-12}$$

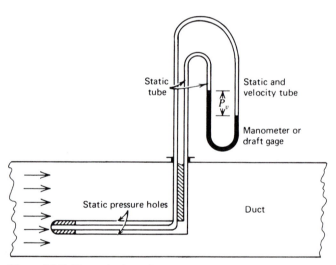

Figure 10-8 Pitot tube in a duct.

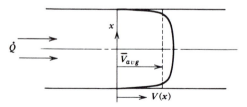

Figure 10-9 Velocity profile in a pipe or duct.

Then, letting x equal r at the wall, Eq. 10-11 becomes

$$\overline{V} = \frac{2}{r^2} \int_0^r V(x)x \, dx \qquad (10\text{-}13a)$$

In fully developed turbulent flow, the most common situation, the velocity profile can be well represented by

$$\frac{V}{V_{max}} = \left(1 - \frac{x}{r}\right)^{1/n} \qquad (10\text{-}14)$$

where n varies from 6 to 8 over the Reynolds number range of 4×10^3 to 5×10^5 (3). For practical cases of air flow in ducts, $\overline{V}/V_{max}$ is about 0.82 for fully developed flow. In practice it is usually necessary to determine the velocity profile using the pitot tube and to then integrate graphically or numerically. In Fig. 10-10 and Eq. 10-13a, the average velocity in the duct is given by

$$\overline{V} = \frac{2}{r^2} \sum_{i=1}^m x_i V_i(x) \, \Delta x_i \qquad (10\text{-}13b)$$

The number of elements m required in Eq. 10-13b depends on how $V(x)$ varies with x. There is usually a central core over which the velocity is relatively constant. There are cases when the velocity profile is not symmetrical, that is, where the velocity is not constant throughout the circular element for a particular radius. In this case velocity profiles should be measured along at least two diameters, which are oriented at 90 degrees. Then for each circular element in the summation of Eq. 10-13b, the velocity $V_i(x)$ is the average of the four measured values at the radius x_i.

Pitot tubes are more frequently used to measure the velocity of gases than liquids. With gases such as air, the available velocity head is often small, making precise measurements quite important. Micromanometers and inclined or Hook gages are used to measure $(P_{02} - P_1)$ in these cases. A form of the pitot tube is often used in liquid piping circuits to sense the flow rate for control purposes.

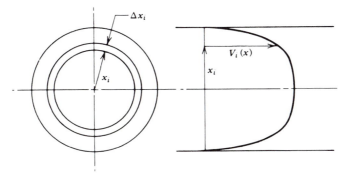

Figure 10-10 Integrating the velocity profile over the pipe cross-sectional area.

EXAMPLE 10-2

A pitot tube is installed in an air duct on the center line. The velocity pressure as indicated by an inclined gage is 0.32 in. of water, the static pressure is 0.25 in. of water, the air temperature is 60 F, and barometric pressure is 29.92 in. of mercury. Assuming that fully developed turbulent flow exists where the average velocity is approximately 82 percent of the center-line value, compute the volume and mass flow rates for a 10-in. diameter duct.

SOLUTION

The mass and volume flow rates are obtained from the average velocity, using Eqs. 10-2 and 10-3. The average velocity is fixed by the center-line velocity in this case, which is computed by using Eq. 10-10. Since the fluid flowing is air, the density term in Eq. 10-10 is that for air, ρ_a. The pressure difference $P_{02} - P_1$ in Eq. 10-10 is the measured pressure indicated by the inclined gage as 0.32 in. of water. The pressure equivalent of this column of water is given by

$$P_{02} - P_1 = (\text{ft. wg})\frac{g}{g_c}\rho_w$$

$$P_{02} - P_1 = \left(\frac{0.32}{12}\right)\text{ft}\left(\frac{32.2}{32.2}\right)\frac{\text{lb}_f}{\text{lbmw}}(62.4)\frac{\text{lbmw}}{\text{ft}^3} = 1.664\frac{\text{lb}_f}{\text{ft}^2}$$

To get the density of the air we assume an ideal gas

$$\rho_a = \frac{P_a}{R_a T_a} = \frac{(29.92)(0.491)(144)}{(53.35)(60 + 460)} = 0.076\frac{\text{lbma}}{\text{ft}^3}$$

which neglects the slight pressurization of the air in the duct. The center-line velocity is given by Eq. 10-10

$$V_{ce} = \left[\frac{(2)(32.2)(1.664)}{(0.076)}\right]^{1/2} = 37.6\,\text{ft/sec}$$

and the average velocity is

$$\overline{V} = 0.82\,V_{ce} = (0.82)(37.6) = 30.8\,\text{ft/sec}$$

The mass flow rate is given by Eq. 10-2 with the area given by

$$A = \frac{\pi}{4}\left(\frac{10}{12}\right)^2 = 0.545 \text{ ft}^2$$

$$\dot{m} = \rho_a \overline{V}A = 0.076(30.8)0.545 = 1.28 \text{ lbm/sec}$$

The volume flow rate is

$$\dot{Q} = \overline{V}A = 30.8(0.545)60 = 1007 \text{ ft}^3/\text{min}$$

using Eq. 10-3.

EXAMPLE 10-3

Air flows in a 0.3-m diameter circular duct. At the end of a long straight run the center-line velocity is measured to be 12 m/s. Estimate the average velocity and mass flow rate of the air assuming standard conditions for the air.

SOLUTION

Because the measurement follows a long straight run of pipe, we shall assume that the flow is fully developed and that the velocity profile is symmetrical. It will be necessary to integrate Eq. 10-13a using Eq. 10-14 to describe the velocity profile. The exponent n should be about 7 for this case and can be checked later. Substitution of Eq. 10-14 in Eq. 10-13a and integration with $n = 7$ yields

$$\overline{V} = \frac{2V_{max}}{r^{15/7}} \int_0^r x^{8/7}\, dx = \frac{14}{15} V_{max}$$

and

$$\overline{V} = \frac{14}{15}(12) = 11.2 \text{ m/s}$$

The mass flow rate is given by Eq. 10-2 as

$$\dot{m} = \rho \overline{V}A = 1.2(11.2)\frac{\pi}{4}(0.3)^2 = 0.95 \text{ kg/s}$$

where the mass density of standard air, 1.2 kg/m^3, has been used. To check the value of the exponent n, it is necessary to compute the Reynolds number.

$$\text{Re} = \frac{\rho \overline{V}D}{\mu} = \frac{1.2(11.2)0.3}{20.4 \times 10^{-6}} = 197,650$$

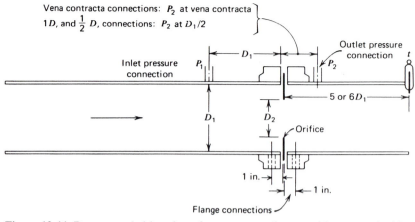

Figure 10-11 Recommended location of pressure taps for use with concentric thin-plate and square-edged orifices according to Reference 3.

where the dynamic viscosity is from Table B-4b. Reference 3 indicates that n equal to 7 is valid for the Reynolds number just computed.

Flow-measuring devices of the restrictive type use the pressure drop across an orifice, nozzle, or venturi to predict flow rate. The square-edged orifice is widely used because of its simplicity. Figure 10-11 shows such a meter with the location of the pressure taps (4). The flange-type pressure taps are widely used in HVAC piping systems and are standard fittings available commercially. The orifice plate may be fabricated locally or may be purchased. Reference 4 outlines the manufacturing procedure in detail.

The orifice meter is far from being an ideal flow device and introduces an appreciable loss in total pressure. An empirical *discharge coefficient* is

$$C = \frac{\dot{Q}_{\text{actual}}}{\dot{Q}_{\text{ideal}}} \tag{10-15}$$

The ideal flow rate may be derived from Eq. 10-1a with the lost energy equal to zero. Applying Eq. 10-1a between the cross sections defined by the pressure taps gives

$$\frac{P_1}{\rho} + \frac{\overline{V}_1^2}{2g_c} = \frac{P_2}{\rho} + \frac{\overline{V}_2^2}{2g_c} \tag{10-16}$$

To eliminate the velocity $\overline{V}_1$ from Eq. 10-16, Eq. 10-3 is recalled and

$$\overline{V}_1 = \overline{V}_2 \frac{A_2}{A_1} \tag{10-3a}$$

Substitution of Eq. 10-3a into Eq. 10-16 and rearrangement yields

$$\overline{V}_2 = \frac{1}{[1 - (A_2/A_1)^2]^{1/2}} \left[2g_c \frac{(P_1 - P_2)}{\rho} \right]^{1/2} \tag{10-17}$$

Then by using Eqs. 10-15 and 10-17 we get

$$= C_D A_z \left[\frac{2g_c (P_1 - P_2)}{\rho} \right]^{-1/2}$$

$$\dot{Q}_{actual} = \frac{CA_2}{[1 - (A_2/A_1)^2]^{1/2}} \left[2g_c \frac{(P_1 - P_2)}{\rho} \right]^{1/2} \tag{10-18}$$

The term $[1 - (A_2/A_1)^2]^{1/2}$ is referred to as the *velocity of approach factor*. In practice the discharge coefficient and velocity of approach factor are often combined and called the *flow coefficient C_d*:

$$C_d = \frac{C}{[1 - (A_2/A_1)^2]^{1/2}} \tag{10-19}$$

This is merely a convenience. For precise measurements other corrections and factors may be applied, especially for compressible fluids (4). Figure 10-12 shows representative values of the flow coefficient C_d. The data apply to pipe diameters over a wide range (1 to 8 in.) and to flange or radius taps within about 5 percent. When precise flow measurement is required, reference 4 should be consulted for more accurate flow coefficients.

In complex piping systems it is necessary to install meters at strategic locations so that the system can be balanced when it is put into operation. The designer must specify an orifice that, when the desired flow rate is attained, will have a reasonable pressure

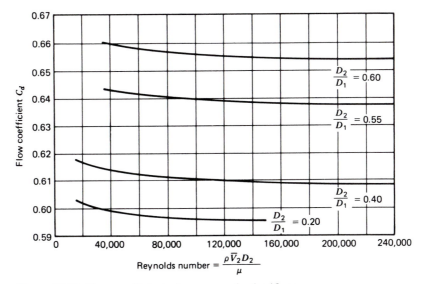

Figure 10-12 Flow coefficients for square-edged orifices.

drop but one not large enough to be wasteful of pumping power. The following example illustrates an orifice design procedure.

EXAMPLE 10-4

Design an orifice to be used in a 4-in. standard (schedule 40) commercial steel pipe that has a design flow rate of 200 gal per minute (gpm). The pipe will supply chilled water in summer at 45 F and in winter hot water at 150 F. A change in pressure, $P_1 - P_2$, of 5 to 10 in. of mercury is acceptable.

SOLUTION

Referring to Fig. 10-12, we note that the flow coefficient curves become independent of Reynolds number at high Reynolds numbers. It would therefore be desirable to operate in this region. It is also desirable to operate with a diameter ratio of 0.5 to 0.6 because the flow coefficient data seem to be most accurate in this range. Therefore, it is assumed that $D_2/D_1 = 0.55$ and $C_d = 0.637$. The change in pressure ΔP is calculated from Eq. 10-18.

$$P_1 - P_2 = \left(\frac{\rho \dot{Q}}{C_d A_2} \right)^2 \frac{1}{2g_c}$$

$$D_2 = 0.55 D_1 = 0.55(4.026) = 2.21 \text{ in.}$$

where the inside diameter of 4-in. pipe is 4.026 in.

$$A_2 = \frac{\pi}{4} \left(\frac{2.21}{12} \right)^2 = 0.027 \text{ ft}^2$$

$$\dot{Q} = \frac{200}{60(7.48)} = 0.446 \text{ ft}^3/\text{sec}$$

where 7.48 gal equals 1 ft^3

$$P_1 - P_2 = \left[\frac{0.446}{(0.637)0.027} \right]^2 \frac{(62.4)}{2(32.2)} = 652 \text{ lbf/ft}^2$$

or

$$P_{12} = 4.52 \text{ psi} = 9.22 \text{ in. of mercury}$$

where 1 in. of mercury is equal to 0.491 psi. This pressure loss is satisfactory. The Reynolds number will now be computed to check the assumed flow coefficient. The largest expected value of the viscosity is used to compute the Reynolds number because this yields the lowest Re number. If we use Eq. 10-8 and data from Fig. 10-3

$$\mathrm{Re} = \frac{\rho \overline{V}_2 D_2}{\mu} = \frac{62.4(0.446/0.027)(2.21/12)}{(1.6/1490)} = 1.8 \times 10^5$$

From Fig. 10-12 note that the flow coefficient C_d is nearly constant at Reynolds numbers greater than about 2×10^5. For this particular case the Reynolds number is sufficiently large to be acceptable.

10-2   CENTRIFUGAL PUMPS

The centrifugal pump is by far the most frequently used type of pump in HVAC systems. The essential parts of a centrifugal pump are the rotating member or impeller and the surrounding case. The impeller is usually driven by an electric motor that may be close-coupled (on the same shaft as the impeller) or flexible coupled (Fig. 10-13a). The fluid enters the center of the rotating impeller, is thrown into the volute, and flows outward through the diffuser (Fig. 10-13b). The fluid leaving the impeller has high kinetic energy that is converted to static pressure in the volute and diffuser as efficiently as possible. Although there are various types of impellers and casings (5), the principle of operation is the same. The pump shown in Fig. 10-13 is a single suction pump because the fluid enters the impeller from only one side. The double-suction type has fluid entering from both sides. A pump may be staged with more than one impeller on the same shaft with one casing. The fluid leaves the first stage and enters the impeller of the second stage before leaving the casing.

Figure 10-13a A single-inlet, flexible-coupled centrifugal pump. (Courtesy of ITT Bell and Gossett, Skokie, IL.)

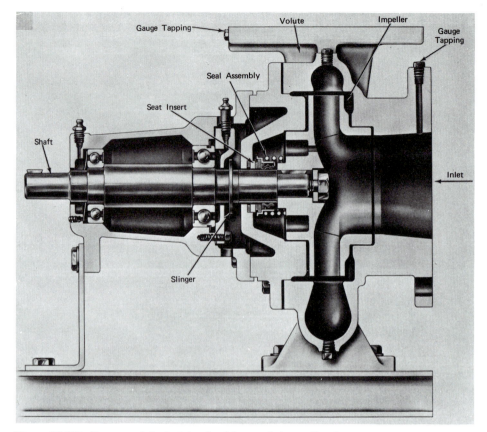

Figure 10-13b Cutaway of single-inlet, flexible-coupled centrifugal pump. (Courtesy of ITT Bell and Gossett, Skokie, IL.)

Pump performance is most commonly given in the form of curves. Figure 10-14 is an example of such data for a pump that may be operated at two different speeds with several different impellers. For each speed a different curve is given for each impeller. These curves give the total dynamic head, efficiency, shaft power, and the net positive suction head as a function of capacity.

The *total dynamic head* furnished by a pump can be understood by applying Eq. 10-1c to the fluid entering and leaving the pump:

$$H_p = \frac{w g_c}{g} = \frac{g_c(P_1 - P_2)}{g\rho} + \frac{\overline{V}_1^2 - \overline{V}_2^2}{2g} + (z_1 - z_2) \tag{10-20}$$

The lost head is unavailable as useful energy and is omitted from the equation. Losses are typically accounted for by the *efficiency,* defined as the ratio of the useful power actually imparted to the fluid to the shaft power input:

$$\eta_p = \frac{\dot{W}}{\dot{W}_s} = \frac{\dot{m}w}{\dot{W}_s} = \frac{\rho \dot{Q}w}{\dot{W}_s} \tag{10-21}$$

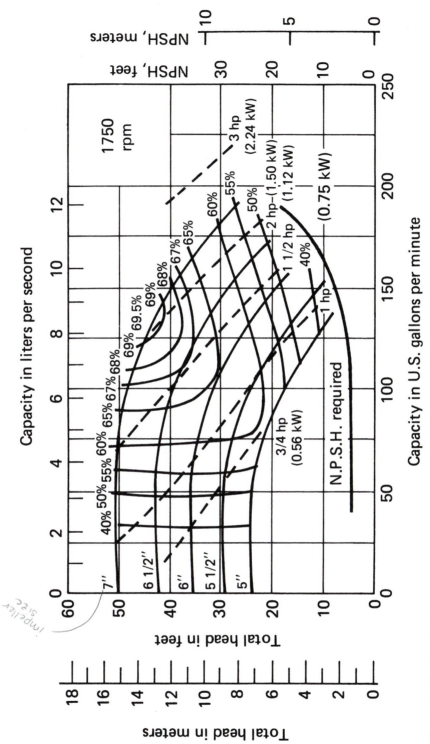

Figure 10-14a Centrifugal pump performance data.

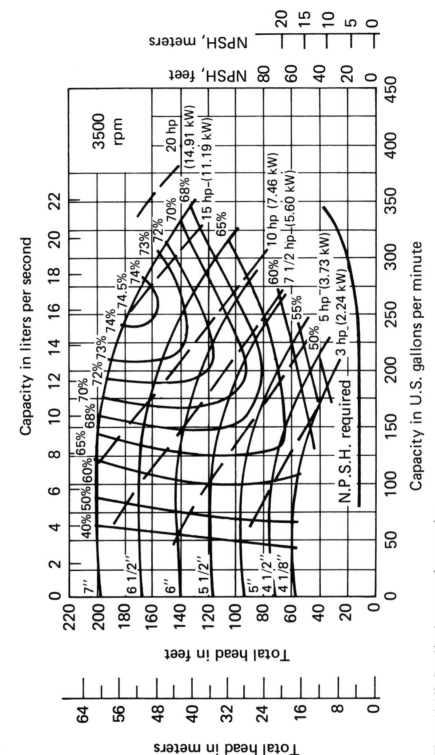

Figure 10-14b Centrifugal pump performance data.

The shaft power may be obtained from Eq. 10-21

$$\dot{W}_s = \frac{\dot{m}w}{\eta_p} = \frac{\rho\dot{Q}w}{\eta_p} = \frac{\rho\dot{Q}H_p g}{\eta_p g_c} \qquad \textbf{(10-22)}$$

Therefore a definite relationship exists between the curves for total head, efficiency, and shaft power in Fig. 10-14.

If the static pressure of the fluid entering a pump approaches the vapor pressure of the liquid too closely, vapor bubbles will form in the impeller passages. This condition is detrimental to pump performance and the collapse of the bubbles is noisy and may damage the pump. This phenomenon is known as *cavitation*. The amount of pressure in excess of the vapor pressure required to prevent cavitation (expressed as head) is known as the *required net positive suction head* (NPSHR). This is a characteristic of a given pump and varies considerably with speed and capacity. NPSHR is determined by the actual testing of each model.

Whereas each pump has its own NPSHR, each system has its own *available net positive suction head* (NPSHA).

$$\text{NPSHA} = \frac{P_s g_c}{\rho g} + \frac{\overline{V}_s^2}{2g} - \frac{P_v g_c}{\rho g} \qquad \textbf{(10-23a)}$$

where

$g_c P_s/\rho g$ = static head at the pump inlet, ft or m, absolute
$\overline{V}_s^2/2g$ = velocity head at the pump inlet, ft or m
$g_c P_v/\rho g$ = static vapor pressure head of the liquid at the pumping temperature, ft or m, absolute

The net positive suction head available must always be greater than the NPSHR or noise and cavitation will result.

EXAMPLE 10-5

Suppose the pump of Fig. 10-14 is installed in a system as shown in Fig. 10-15. The pump is operating at 3500 rpm with the 6-in. impeller and delivering 200 gpm. The

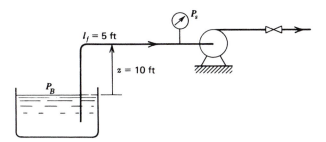

Figure 10-15 Open system with suction lift.

suction line is standard 4-in. pipe that has an inside diameter of 4.026 in. Compute the NPSHA and compare it with the NPSHR. The water temperature is 60 F.

SOLUTION

From Fig. 10-14 the NPSHR is 10 ft of head. The available net positive suction head is computed from Eq. 10-23; however, the form will be changed slightly through the application of Eq. 10-1c between the water surface and the pump inlet:

$$\frac{P_B g_c}{\rho g} = \frac{P_s g_c}{\rho g} + \frac{\overline{V}_s^2}{2g} + z_s + l_f$$

or

$$\frac{P_s g_c}{\rho g} + \frac{\overline{V}_s^2}{2g} = \frac{P_B g_c}{\rho g} - z_s - l_f$$

Then Eq. 10-23a becomes

$$\text{NPSHA} = \frac{P_B g_c}{\rho g} - z_s - l_f - \frac{P_v g_c}{\rho\ g} \qquad \textbf{(10-23b)}$$

Assuming standard barometric pressure

$$\frac{P_B g_c}{\rho g} = \frac{29.92}{12} \times 13.55 = 33.78 \text{ ft of water}$$

$$\frac{P_v g_c}{\rho g} = \frac{0.2562 \times 144}{62.4} = 0.59 \text{ ft of water}$$

where P_v is read from Table A-1a. Then from Eq. 10-23b

$$\text{NPSHA} = 33.78 - 10 - 5 - 0.59 = 18.19 \text{ ft of water}$$

which is almost twice as large as the NPSHR. If the water temperature is increased to 160 F and other factors remain constant, the NPSHA becomes

$$\text{NPSHA} = 33.78 - 10 - 5 - \left(\frac{4.74 \times 144}{61}\right) = 7.6 \text{ ft}$$

which is less than the NSPHR of 10 ft. Cavitation will undoubtedly result.

In an open system such as a cooling tower, the pump suction (inlet) should be flooded; that is, the inlet must be lower than the free water surface. This is particularly important if the fluid temperature is high such as in the case of condensate from a steam condenser. In an open reservoir or sump the end of the suction pipe should be adequately covered with fluid to prevent entrainment of air from the vortex formed at the pipe entrance. An inlet velocity of less than 3 ft/sec or 1 m/s will minimize vortex formation. Long runs of suction piping should be eliminated whenever possible, and care should be taken to eliminate air traps on the suction side of the pump. Care must be taken to locate the pump in a space where freezing will not occur and where maintenance may be easily performed.

The pump foundation, usually concrete, should be sufficiently rigid to support the pump base plate. This is particularly important for flexible-coupled pumps to maintain alignment between the pump and motor. The pump foundation should weigh from $1\frac{1}{2}$ to 3 times the total pump and motor weight for vibration and sound control.

It is extremely important to install the suction and discharge piping such that forces are not transmitted to the pump housing. Expansion joints are required on both the suction and discharge sides of the pump to isolate expansion and contraction forces, and the piping must be supported independent of the pump housing. In-line pumps, as in Fig. 2-12b, are an exception to this rule since they are designed to be supported by the piping. However, the piping must still be supported on both sides of the pump.

10-3   COMBINED SYSTEM AND PUMP CHARACTERISTICS

The combination of the system and pump characteristics (head versus capacity) is very useful in the analysis and design of piping systems. Figure 10-16 is an example of how a system with parallel circuits behaves with a pump installed. Recall that the total head H_p furnished by the pump is given by Eq. 10-20. Note that the combination operates at point t, where the characteristics cross. The pump and system must both operate on their characteristics; therefore, the point where they cross is the only possible operating condition. This concept is very important in understanding more complex systems. The

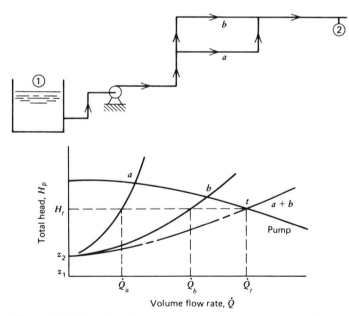

Figure 10-16 Combination of system and pump charcteristics for parallel circuits.

flow rate for each of the parallel circuits in Fig. 10-16 is quite obvious because the required change in total head from 1 to 2 is equal for both circuits.

Figure 10-17 illustrates a series-type circuit. When the valve is open, the operating point is at point a with flow rate $\dot{Q}_a$ and total head H_a. Partial closing of the valve introduces additional flow resistance (head loss) and is similar to adding series resistance in an electrical circuit. The new system characteristic crosses the pump curve at point c and the flow rate is $\dot{Q}_c$ with total head H_c. All piping systems should contain valves for control and adjustment purposes.

A typical design problem is one of pump selection. The following example illustrates the procedure.

EXAMPLE 10-6

A water piping system has been designed to distribute 150 gpm and the total head requirement is 36 ft. Select a pump using the data of Fig. 10-14 and specify the power rating for the electric motor.

SOLUTION

Figure 10-18 shows the characteristic for the piping system as it was designed. Point 0 denotes the operating capacity desired. Examination of Fig. 10-14 indicates that the low-speed version of the given pump covers the desired range. The desired operating point lies between the curves for the $6\frac{1}{2}$- and 7-in. impellers. The curves are sketched on Fig. 10-18. Obviously, the pump with the 7-in. impeller must be selected, but the flow

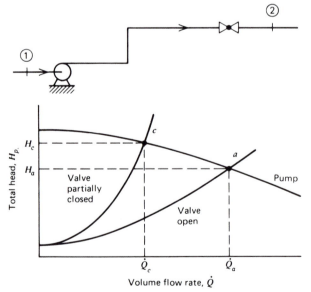

Figure 10-17 Combination of system and pump characteristics for series circuits.

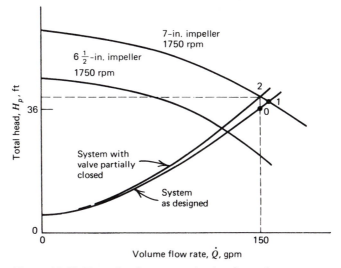

Figure 10-18 Example of a pump selection for a given system.

rate will be about 160 gpm as indicated by point 1. Therefore a valve must be adjusted (closed slightly) to modify the system characteristic as shown, to obtain 150 gpm at point 2. Again referring to Fig. 10-14, we read the shaft power requirement as about 2.3 hp. Electric motors usually have an efficiency of 85 to 90 percent, and a 3-hp motor should be specified.

Variable Pump Speed

It is common to vary the impellor rotational speed in controlling a water-distribution system. The flow rate, head, and shaft power are related to the new and old speeds. The laws governing this relationship are known as the *Affinity Laws for Pumps.* They may be stated as

$$\dot{Q}_n = \dot{Q}_o \frac{\text{rpm}_n}{\text{rpm}_o} \qquad (10\text{-}24)$$

$$H_{pn} = H_{po} \left[\frac{\text{rpm}_n}{\text{rpm}_o} \right]^2 \qquad (10\text{-}25)$$

$$\dot{W}_{sn} = \dot{W}_{so} \left[\frac{\text{rpm}_n}{\text{rpm}_o} \right]^3 \qquad (10\text{-}26)$$

The affinity laws may be used in conjunction with the system characteristic to generate a new pump head characteristic. The total system will operate where the new pump characteristic and old system characteristic cross.

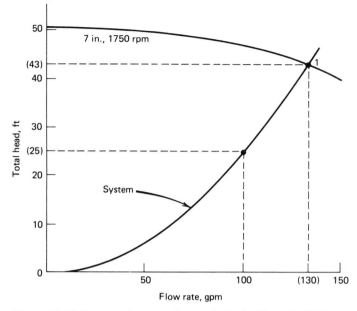

Figure 10-19 Pump and system characteristics for Example 10-7.

EXAMPLE 10-7

The 1750-rpm pump with 7-in. impeller of Fig. 10-14 is operating in a system as shown as point 1 of Fig. 10-19. It is desired to reduce the pump speed until the flow rate is 100 gpm. Find the new pump head, shaft power, and efficiency.

SOLUTION

From the system characteristic it may be observed that the pump must produce 25 ft of head at a flow rate of 100 gpm. This is one point on the new pump characteristic. The new pump speed can be found from either Eq. 10-24 or 10-25. Using Eq. 10-24

$$\text{rpm}_n = \text{rpm}_o \, (\dot{Q}_n/\dot{Q}_o)$$
$$= 1750 \, (100/130) = 1346$$

The new shaft power is given by Eq. 10-26 with $\dot{W}_{so} = 2.1$ hp from Fig. 10-14:

$$\dot{W}_{sn} = 2.1 \, (1346/1750)^3 = 0.96 \, \text{hp}$$

The pump efficiency could be recalculated using Eq. 10-21. However, it may be deduced from the affinity laws that the efficiency will remain constant at about 69.4 percent.

$$\frac{\eta_{pn}}{\eta_{po}} = \frac{\dot{Q}_n H_{pn}/\dot{W}_{sn}}{\dot{Q}_o H_{po}/\dot{W}_{so}} = 1$$

10-4 PIPING SYSTEM DESIGN

There are many different types of piping systems used with HVAC components, and there are many specialty items and refinements that make up these systems. Chapter 12 of reference 6 gives a detailed description of various arrangements of the components making up the complete system. Chapter 33 of reference 2 pertains to the sizing of pipe. The main thrust of the discussion to follow is to develop methods for the design of basic piping systems used to distribute hot and chilled water. The basic concepts will first be covered and applied to simple open- and closed-loop systems. The principles involved in designing larger variable-flow systems using secondary pumping will then be discussed in Section 10-5.

Open-loop System

A typical open-loop piping system is shown in Fig. 10-20. Characteristically an open-loop system will have some part of the circuit open to the atmosphere. The cooling tower of Fig. 10-20 shows the usual valves, filters, and fittings installed in this type of circuit. It is important to protect the pump with a filter. The isolation valves provide for maintenance without complete drainage of the system, whereas a ball or plug valve should be provided at the pump outlet for adjustment of the flow rate. Expansion joints and a rigid base support are shown to isolate the pump as previously discussed.

Closed-loop System

Two closed-loop systems are shown in Fig. 10-21 and are known as two-pipe and four-pipe systems. Although it is not necessary to install a filter in a closed-loop system, an expansion tank and air separator are required. The expansion tank protects the system from being damaged by a change in volume due to temperature variations; provides a collecting space for air removed by the separator; and creates a point of constant pressure in the system. The expansion tank is covered fully later. Closed-loop systems

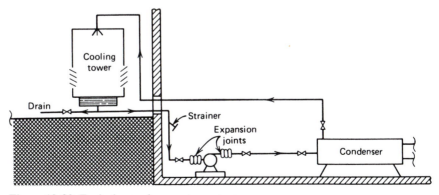

Figure 10-20 Typical open-loop water system.

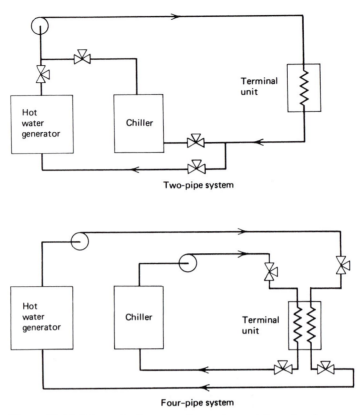

Figure 10-21 Schematic diagrams of two-pipe and four-pipe flow systems.

are normally pressurized so that the complete loop will be above atmospheric pressure. A pressure regulator is installed in the makeup water line, which is connected to the system at the same place as the expansion tank. Isolation and control values are required, and instrumentation to measure flow rates and temperatures at strategic locations is essential. This is especially important when parallel circuits are employed.

Pipe Sizing Criteria

Piping systems often pass through or near occupied spaces where noise generated by the flowing fluid may be objectionable. Therefore, a velocity limit of 4 ft/sec or 1.2 m/s for 2-in. pipe and smaller is generally imposed. For larger sizes a limit on the head loss of 4 ft per 100 ft of pipe is imposed. This corresponds to about 0.4 kPa/m in SI units. These criteria should not be treated as hard rules but rather as guides. Noise is caused by entrained air, abrupt pressure drops, and turbulence in general. If these factors can be minimized, the given criteria can be relaxed. Open systems such as cooling tower circuits are usually remote from occupied spaces. Therefore, somewhat higher velocities may be used in such a case. A reasonable effort to design a balanced system will prevent drastic valve adjustments and will contribute to a quieter system.

The piping layout for a heating and air-conditioning system depends on the location of the central and terminal equipment and the type of system to be used. All of the piping may be located in the central equipment room or may run throughout the building to terminal units in every room. In this case the available space may be a controlling factor. It must be kept in mind that piping for domestic hot and cold water, sewage, and other services must be provided in addition to the heating and air-conditioning requirements. The designer must constantly check to make sure the piping will fit into the allowed space.

Pipe Sizing

After the piping layout has been completed, the problem of sizing the pipe consists mostly of applying the design criteria discussed before. Where possible the pipes should be sized so that drastic adjustments are not required. Often an ingenious layout helps in this respect. The system and pump characteristics are also useful in the design process.

To facilitate the actual pipe sizing and computation of head loss, charts such as those shown in Figs. 10-22 and 10-23 for pipe and copper tubing have been developed. Figures 10-22 and 10-23 are based on 60 F or 16 C water and give head losses that are about 10 percent high for hot water. Examination of Figs. 10-22 and 10-23 shows that head loss may be obtained directly from the flow rate and nominal pipe size or from flow rate and water velocity. When the head loss and flow rate are known, a pipe size and velocity may be obtained.

Pipe fittings and valves also introduce losses in head. These losses are usually accounted for by use of a *resistance coefficient K,* which is the number of velocity heads lost because of the valve or fitting. Thus,

$$l_f = K\frac{\overline{V}^2}{2g}$$

(10-27a)

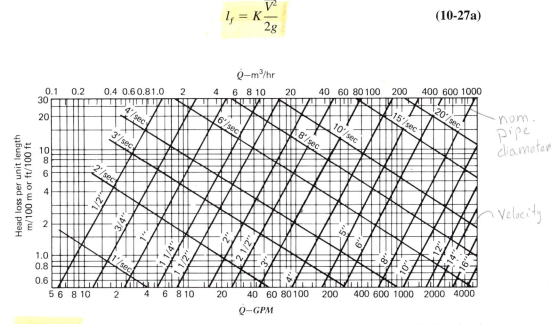

Figure 10-22 Friction loss due to flow of water in commercial steel pipe (schedule 40). (Reprinted by permission from *ASHRAE Handbook, Fundamentals Volume,* 1989.)

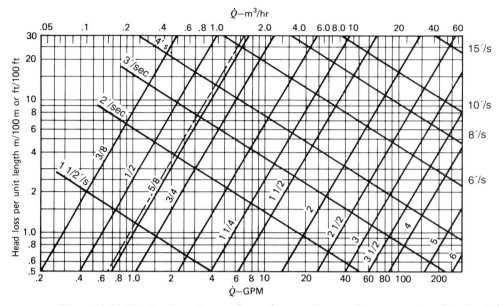

Figure 10-23 Friction loss due to flow of water in type L copper tube. (Reprinted by permission from *ASHRAE Handbook, Fundamentals Volume,* 1989.)

Comparing this definition with Eq. 10-6 it can be seen that

$$K = f\frac{L}{D} \tag{10-27b}$$

The ratio L/D is the equivalent length in pipe diameters of straight pipe that will cause the same pressure drop as the valve or fitting under the same flow conditions. This is a convenient concept to use when one is computing head loss in a piping system. Representative values of resistance coefficients for some common valves and fittings are given in Fig. 10-24a. (7) Conversions between K, L/D, and L can be obtained for various pipe sizes by the use of Fig. 10-24b. When using SI units it is suggested that the L/D ratio be determined from Fig. 10-24b, using the nominal pipe size. The equivalent length in meters may then be determined using the inside diameter D in meters. The lost head for a given length of pipe of constant diameter and containing fittings is computed as the product of the lost head per foot from Figs. 10-22 and 10-23 and the total equivalent length of the pipe and fittings.

EXAMPLE 10-8

Compute the lost head for a 150-ft run of standard pipe, having a diameter of 3 in. The pipe run has 3 standard 90 degree elbows, a globe valve, and a gate valve. One hundred gpm of water flows in the pipe.

SOLUTION

The equivalent length of the various fittings will first be determined by using Figs. 10-24a and 10-24b.

$$\text{Globe valve: } K_1 = 340 \, f_t, \, f_t = 0.018; \text{ Fig. 10-24}a \text{ and Table 10-1}$$

$$K_1 = 340(0.018) = 6.1$$

$$L = 86 \text{ ft; Fig. 10-24}b$$

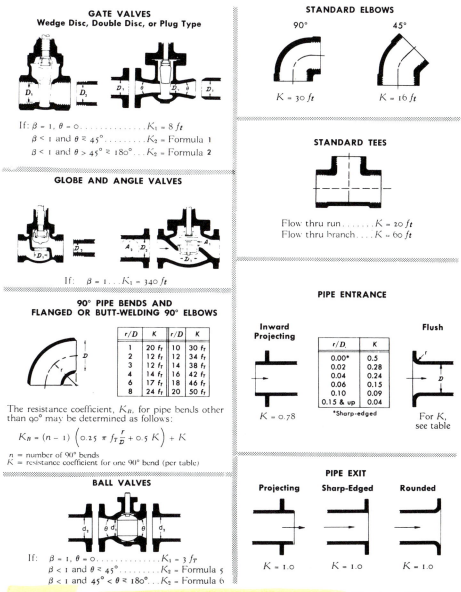

Figure 10-24a Resistance coefficients *K* for various valves and fittings. (Courtesy of the Crane Company, Technical Paper No. 410.)

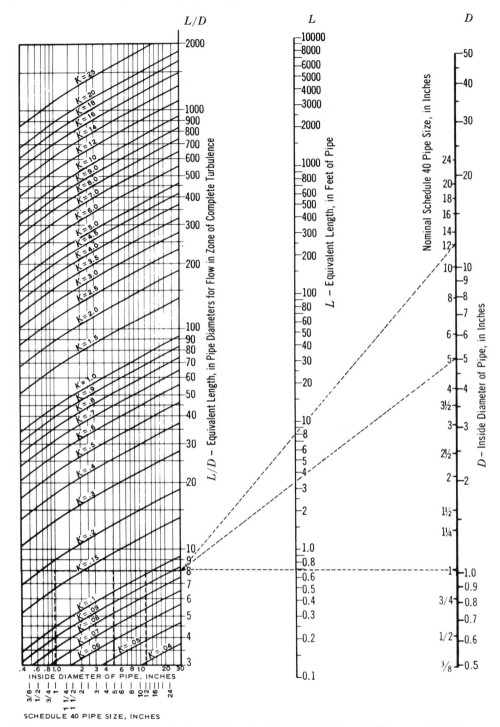

Figure 10-24b Equivalent lengths L and L/D and resistance coefficient K. (Courtesy of the Crane Company, Technical Paper No. 410.)

Table 10-1 Formulas, Definition of Terms, and Values of f_t for Figure 10-24

Formula 1 $K_2 = \dfrac{K_1 + \left(\sin \dfrac{\theta}{2}\right)\,0.8(1 - \beta^2) + 2.6(1 - \beta^2)^2}{\beta^4}$

Formula 2 $K_2 = \dfrac{K_1 + 0.5\left(\sin \dfrac{\theta}{2}\right)(1 - \beta^2) + (1 - \beta^2)^2}{\beta^4}$

$\beta = \dfrac{D_1}{D_2}; \qquad \beta^2 = \left(\dfrac{D_1}{D_2}\right)^2 = \dfrac{A_1}{A_2}; \qquad$ D_1 = smaller diameter
A_1 = smaller area

Nominal size	$\frac{1}{2}$ in.	$\frac{3}{4}$ in.	1 in.	$1\frac{1}{4}$ in.	$1\frac{1}{2}$ in.	2 in.
Friction factor (f_t)	0.027	0.025	0.023	0.022	0.021	0.019

$2\frac{1}{2}$, 3 in.	4 in.	5 in.	6 in.	8–10 in.	12–16 in.	18–24 in.
0.018	0.017	0.016	0.015	0.014	0.013	0.012

Elbow: $K = 30\,f_t, f_t = 0.018$

$K = 30(0.018) = 0.54$

$L = 8\,\text{ft}$

Gate valve: $K_1 = 8\,f_t, f_t = 0.018$

$K_1 = 8(0.018) = 0.14$

$L = 2\,\text{ft}$

The total equivalent length is then

Actual length of pipe	150 ft
One globe valve	86 ft
Three elbows	24 ft
One gate valve	2 ft
Total	262 ft

From Fig. 10-22 the lost head l_f' is 2.3 ft per 100 ft of length or

$$l_f' = 2.3 \times 10^{-2}\ \text{ft/ft of length}$$

The lost head for the complete pipe run is then given by

$$l_f = L_e l_f' = (262)2.3 \times 10^{-2} = 6.0\,\text{ft}$$

The lost head for control valves, check valves, strainers, and other such devices is often given in terms of a coefficient C_v. The coefficient is numerically equal to the flow rate of water at 60 F in gpm, which will give a pressure loss of one lbf/in.^{2} (2.31 ft of water). Because the head loss is proportional to the square of the velocity, the pressure loss or lost head may be computed at other flow rates:

$$\frac{l_{f1}}{l_{f2}} = \left(\frac{\dot{Q}_1}{\dot{Q}_2}\right)^2 \tag{10-28}$$

In terms of the coefficient C_v:

$$l_f = 2.31 \left(\frac{\dot{Q}}{C_v}\right)^2 \tag{10-29}$$

where $\dot{Q}$ and C_v are both in gpm and l_f is in feet of water.

It may be shown that the flow rate of any fluid is given by

$$\dot{Q} = C_v \left[\frac{\Delta P(62.4)}{\rho}\right]^{1/2} \tag{10-30}$$

where ΔP is in lbf/in.^{2} and ρ is in lbm/ft^3.

There is a relationship between C_v and the resistance coefficient K. By using Eqs. 10-3 and 10-6, we can show that

$$C_v = \frac{0.208 \, D^2}{\sqrt{K}}$$

where D is in feet. In SI units a flow coefficient C_{vs} is defined as the flow rate of water at 15 C in m^3/s with a pressure loss of 1 kPa given by

$$C_{vs} = 1.11 \frac{D^2}{\sqrt{K}} \tag{10-31}$$

where D is in meters.

EXAMPLE 10-9

A strainer has a C_v rating of 60. It is to be used in a system to filter 50 gpm of water. What pressure loss and head loss can be expected?

SOLUTION

Equation 10-29 will yield the desired result:

$$l_f = 2.31 \left(\frac{50}{60}\right)^2 = 1.6 \text{ ft of water}$$

The pressure loss ΔP is given by

$$\Delta P = \frac{1.6(62.4)}{144} \frac{g}{g_c} = 0.7 \text{ lbf/in.}^2$$

Heating and cooling units and terminal devices usually have head-loss information furnished by the manufacturer. The head loss is often used to indicate the flow rate for the adjustment of the system. Equation 10-28 may be used to estimate head loss at other than specified conditions.

There is no one set procedure for pipe sizing. The following example will demonstrate some approaches to the problem.

EXAMPLE 10-10

Figure 10-25 shows a two-pipe water system such as might be found in a central equipment room. The terminal units a, b, and c are air-handling units that contain air-to-water finned tube heat exchangers. An actual system would probably contain both a heater and a chiller; only one or the other is to be considered here. Size the piping and specify the pumping requirements.

SOLUTION

The first step is to select criteria for sizing of the pipe. Because the complete system is confined to a central equipment room, the velocity and head loss criteria may be relaxed somewhat. Let the maximum velocity be 5 ft/sec and the maximum head loss be about 7 ft per 100 ft in the main run. Somewhat higher values may be used in the parallel

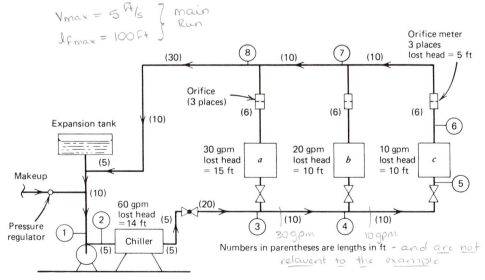

Figure 10-25 Two-pipe system design example.

circuits. By using Fig. 10-22 we select pipe sizes as follows:

$$D_{23} = D_{81} = 2\tfrac{1}{2}\,\text{in.}; \qquad \overline{V} = 4\,\text{ft/sec}; \qquad l_f' = 2.6\,\text{ft/100 ft}$$

$$D_{34} = D_{78} = 1\tfrac{1}{2}\,\text{in.}; \qquad \overline{V} = 4\tfrac{3}{4}\,\text{ft/sec}; \qquad l_f' = 6.5\,\text{ft/100 ft}$$

$$D_{45} = D_{67} = 1\,\text{in.}; \qquad \overline{V} = 3\tfrac{3}{4}\,\text{ft/sec}; \qquad l_f' = 6.5\,\text{ft/100 ft}$$

The lost head for the various sections is now computed. The equivalent lengths used for the various fittings are approximate in the following analysis and are summed with the actual pipe section lengths. In actual practice the fittings should be identified and the equivalent lengths found as shown in Examples 10-8 and 10-9. A three-dimensional piping diagram would also be used to describe the piping layout in detail.

Section 8 to 3 (exclusive of the heater or chiller)

$$L_{83} = 140\,\text{ft}$$
$$l_{f83} = 2.6(140)/100 = 3.6\,\text{ft} \;\; = \; (l_f')(L_{83})$$

Sections 3 to 4 and 7 to 8

$$L_e = L_{43} + L_{78} = 2(15) = 30\,\text{ft}$$
$$l_{f34} + l_{f78} = 30(6.5)/100 = 2.0\,\text{ft}$$

Sections 4 to 5 and 6 to 7

$$L_e = L_{45} + L_{67} = 2(30) = 60\,\text{ft}$$
$$l_{f45} + l_{f67} = 60(6.5)/100 = 3.9\,\text{ft}$$

Chiller, Unit c and Orifice

$$l_f = 14 + 10 + 5 = 29\,\text{ft}$$

The total lost energy for this pipe run including fittings and equipment is

(eq. 10.16)
$$\frac{(P_{02} - P_{01})}{\rho} = (l_f)_t \frac{g}{g_c} = 3.6 + 2.0 + 3.9 + 29 = 39.0\,\text{(ft-lbf)/lbm}$$

Applying Eq. 10-1d to just the pump, we get

$$w = \frac{(P_{01} - P_{02})}{\rho}$$

Therefore

$$w = -39.0\,\text{(ft-lbf)/lbm}$$

which is equivalent to stating that the pump must develop 39 ft of head.

The two remaining parallel circuits will now be sized to balance the system within reason.

Section 4 to 7

$$(l_{f47})_t = l_{f4567} = 3.9 + 10 + 5 = 18.9 \text{ ft}$$
$$L_{47} = 45 \text{ ft}$$
$$l_{f47} = 18.9 - 15 = 3.9 \text{ ft} \quad \Leftarrow \quad 18.9 - 10 - 5$$

$$(l'_f)_{47} = \frac{3.9(100)}{45} = 8.67 \text{ ft/100 ft}$$

From Fig. 10-22

$$D_{47} = 1\frac{1}{4} \text{ in.}; \qquad \overline{V} = 4.8 \text{ ft/sec}$$

Section 3 to 8

$$(l_{f38})_t = l_{f3478} = (l_{f47})_t + l_{f34} + l_{f78} = 18.9 + 2.0 = 20.9 \text{ ft}$$
$$L_{38} = 40 \text{ ft}$$
$$l_{f38} = 20.9 - 20 = 0.9 \text{ foot}$$

$$(l'_f)_{38} = \frac{0.9(100)}{40} = 2.25 \text{ ft/100 ft}$$

From Fig. 10-22

$$D_{38} = 2 \text{ in.}; \qquad \overline{V} = 3 \text{ ft/sec}$$

Only minor adjustments should be required when the system is put into operation.

The sizing of pipe and calculation of lost head follows the same procedure for larger and more complex systems. In the sections to follow this will become evident.

The Expansion Tank

The expansion tank is a much more important element of a piping system than generally assumed. It was mentioned earlier that the expansion tank provides for changes in volume, may be part of the air-elimination system, and establishes a point of constant pressure in the system. This last purpose is very important. A point of constant pressure is necessary to establish the pressure at other points of the closed-loop system; otherwise the system would be like an electrical circuit without a *ground*. The location of the expansion tank then becomes an important design consideration. One rule can be stated that has no exceptions. *A system, no matter how large or complex, must have only one expansion tank.* Consider the piping system shown in Fig. 10-25. The pressure regulator in the makeup water line establishes the pressure in the expansion tank and the pipe at point 1, except for a small amount of elevation head. The pressure at any other point in the system may then be computed relative to point 1 using Eq. 10-1a. Note that the arrangement shown in Fig. 10-25 will produce positive pressures throughout the system, assuming that the pressure at point 1 is positive. The tank pressure is usually between about 10 and 50 psig. If the expansion tank were located at point 2 in Fig. 10-

25, it would be possible to have negative pressures in the system depending on the lost head for the system. The pressure would be lowest at point 1. It is not possible to state one fixed rule for location of the expansion tank; however, it is usually best to locate the tank and pump as shown in Fig. 10-25 in a chilled water loop so that the pump is discharging into the system. A hot water boiler requires a different approach because it must be equipped with a safety relief valve, and improper location of the expansion tank and pump may cause unnecessary opening of the relief valve. Therefore, the expansion tank should be located at the boiler outlet with the pump located either just upstream or just downstream of the boiler. Again the pressures in the system should be analyzed to ensure that positive pressures occur throughout.

Sizing of the expansion tank is important and depends on the total volume of the system, the maximum and minimum system pressures and temperatures, the piping material, the type of tank, and how it is installed. Expansion tanks are of two types. The first type is simply a tank where air is compressed above the free liquid–air interface by system pressure. The second type has a balloonlike bladder within the tank that contains the air. The bladder does not fill the complete tank and is inflated, prior to filling the system, to the pressure setting of the makeup water pressure regulator. Either tank can be used in hot or chilled water systems; however, the first type is usually used in hot water systems because it provides a convenient place for air to collect when released from the heated water in the boiler. The second, bladder type is usually applied with chilled water systems because cold water tends to absorb the air in the free surface type of tank and release it elsewhere in the system where it is removed. This process may eventually lead to a *water-logged* system where no compressible volume exists. Drastic structural damage can occur with a water-logged system.

Relations may be derived for sizing of the expansion tanks by assuming that the air behaves as an ideal gas. The type of tank and the way it is employed in the system then influence the results. Consider the free liquid–air interface type where the water in the tank always remains at its initial temperature (uninsulated and connected by a small pipe); the expansion and compression of the air in the tank is isothermal; and the air in the tank is initially at atmospheric pressure. The resulting relation for the tank volume is

$$V_T = \frac{V_w\left[\left(\dfrac{v_2}{v_1} - 1\right) - 3\alpha\,\Delta t\right]}{\dfrac{P_a}{P_1} - \dfrac{P_a}{P_2}} \tag{10-32}$$

where

V_T = expansion tank volume, ft^3 or m^3
V_w = volume of water in the system, ft^3 or m^3
P_a = local barometric pressure, psia or kPa
P_1 = pressure at lower temperature, t_1 (regulated system pressure), psia or kPa
P_2 = pressure at higher temperature, t_2 (some maximum acceptable pressure), psia or kPa
t_1 = lower temperature (initial fill temperature for hot water system or operating temperature for chilled water system), F or C
t_2 = higher temperature (some maximum temperature for both hot and chilled water systems), F or C

v_1 = specific volume of water at t_1, ft³/lbm or m³/kgm
v_2 = specific volume of water at t_2, ft³/lbm or m³/kgm
α = linear coefficient of thermal expansion for the piping, F^{-1} or C^{-1}, 6.5×10^{-6}
 F^{-1} or $11.7 \times 10^{-6} C^{-1}$ for steel pipe, and $9.3 \times 10^{-5} F^{-1}$ or 16.74×10^{-6}
 C^{-1} for copper pipe
Δt = higher temperature minus the lower temperature, F or C

If the initial air charge in the tank is not compressed from atmospheric pressure but rather is forced into the tank at the design operating pressure, as with a bladder-type tank, and then expands or compresses isothermally, the following relation results:

$$V_T = \frac{V_w \left[\left(\dfrac{v_2}{v_1} - 1 \right) - 3\alpha \, \Delta t \right]}{1 - \dfrac{P_1}{P_2}}$$

(10-33)

where the variables are defined as for Eq. 10-32.

The expansion tank must be installed so that the assumptions made in deriving Eqs. 10-32 and 10-33, are valid. This generally means that the expansion tank is not insulated and is connected to the main system by a relative long, small-diameter pipe so that water from the system does not circulate into the expansion tank. The following example demonstrates the expansion tank problem.

EXAMPLE 10-11

Compute the expansion tank volume for a chilled water system that contains 2000 gal of water. The system is regulated to 10 psig at the tank with an operating temperature of 45 F. It is estimated that the maximum water temperature during extended shutdown would be 100 F and a safety relief valve in the system is set for 35 psia. Assume standard barometric pressure and steel pipe.

SOLUTION

A bladder type would be the best choice; however, calculations will be made for both types. Equation 10-32 will give the volume of the free liquid–air interface type tank where $v_2 = 0.01613$ ft³/lbm and $v_1 = 0.01602$ ft³/lbm from Table A-1a.

$$V_{TF} = \frac{2000 \left(\dfrac{0.01613}{0.01602} - 1 \right) - 3(6.5 \times 10^{-6})(55)}{\left(\dfrac{14.696}{24.696} - \dfrac{14.696}{49.696} \right)}$$

$V_{TF} = 38.7$ gal or 5.2 ft³

Equation 10-33 will give the volume of the bladder-type tank

$$V_{TB} = \frac{2000 \left[\left(\dfrac{0.01613}{0.01602} - 1 \right) - 3(6.5 \times 10^{-6})(55) \right]}{1 - \dfrac{24.696}{49.696}}$$

$$V_{TB} = 23.0 \text{ gal or } 3.1 \text{ ft}^3$$

Note that the volume of the bladder-type tank is less than the free-surface type. This is an advantage in large systems.

10-5 CONTROL OF HYDRONIC SYSTEMS

The small system shown in Fig. 10-25 has no provisions for control other than the hand-operated balancing valves and the expansion tank. As shown, the flow rate of water is constant everywhere in the system, the chiller cycles off at some selected minimum water temperature, and the air side of the system is controlled by a space thermostat. The cooling coils in such a system are said to *run wild*. The need to control the flow of water in the coils in response to the load so that the partial load characteristics of the space can be met was discussed in Chapter 3. The most feasible way of matching the water-side to the air-side load is to regulate the amount of water flowing through the coil. Two ways to do this are shown in Fig. 10-26. (*a*) A two-way valve may be used to throttle the flow to maintain a relatively fixed water temperature leaving the coil or (*b*) a

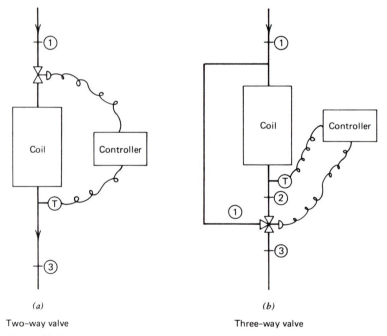

(*a*)

Two–way valve

(*b*)

Three–way valve

Figure 10-26 Basic control valve types.

three-way mixing valve may be used to bypass some of the flow with remixing downstream of the coil. In both cases (a) and (b) the coil receives the same flow of water and the temperature leaving the coil at T is the same; however, the overall effect of the two different control methods on the system is different. The two-way valve produces a variable flow rate with a fixed water temperature differential, whereas the three-way valve produces a fixed flow rate (downstream of the valve) and a variable water temperature differential. We will see later that the two-way valve control method is preferred although the three-way valve method has been the most popular. High energy costs have caused designers to reevaluate their design methods and as a consequence many advances in hydronic system design have evolved. One reason for the use of three-way valves is the requirement to maintain a constant flow rate of water through the source unit, the chiller or boiler. Note in Fig. 10-25 that if two-way valve control is used on each coil, the flow rate through the chiller will decrease as the load becomes lower and lower. This is intolerable and may cause severe damage to the chiller. Therefore, the three-way valve control method, which maintains a constant flow rate, is a better choice for a small system with a single chiller or boiler. As systems become large in capacity and have extensive piping systems that may serve one or more buildings from one central plant, it is not feasible to use one pump to serve the complete system. In such a case it is desirable to interconnect many subsystems into one integrated system containing several pumps. This type of system can use the two-way valve control, which results in water flow rates proportional to the load and more economical operation.

EXAMPLE 10-12

The coils in Fig. 10-26 are identical and require 20 gpm of water at full load. The water enters at 45 F and the flow controllers are set for 55 F. A partial-load condition exists where the flow rate through the coil is reduced to 12 gpm. Find the temperature of the water being returned to the chiller for each type of control valve.

SOLUTION

The two-way valve system will return the water at 55 F assuming the controller can maintain the water temperature leaving the coil at exactly 55 F. In the case of the three-way valve system, water at 45 F is mixed with 55 F water leaving the coil. An energy balance on the valve yields

$$\dot{Q}_1 T_1 + \dot{Q}_2 T_2 = \dot{Q}_3 T_3$$

$$T_3 = \frac{\dot{Q}_1 T_1 + \dot{Q}_2 T_2}{\dot{Q}_3} = \frac{8(45) + 12(55)}{20} = 51 \text{ F}$$

It is clear that the two-way valve leads to variable flow and a fixed temperature differential, whereas the three-way valve gives a constant flow rate with a variable temperature differential.

10-6   LARGE SYSTEM DESIGN

A piping system operating with an incompressible fluid like water is sensitive to disturbances such as valves opening or closing or pumps cycling on and off. Therefore, with several pumps all in the same system, ranging from large to small, care must be taken that valve action or pump cycling in one part of the system does not cause a problem in another section, such as overloading a pump or an idle pump being driven like a turbine. This leads to the conclusion that interconnected circuits each containing a pump must be dynamically and hydraulically isolated. This is accomplished by using *the principle of the common pipe.*

Consider Fig. 10-27a, which shows a secondary load circuit connected to a primary circuit. Under full design load the three-way valve is in its normally open position and there is no flow in Section C. This clearly places the secondary pump in series with the primary pump and sets up a potential for trouble. To overcome this situation a pipe common to both circuits is installed as shown in Fig. 10-27b. The common pipe is always designed to have lost head approaching zero by making it very short in length and of ample size. There is now no possibility of the primary circuit affecting the secondary pump because the secondary circuit is dynamically isolated from the primary circuit. To understand this concept, remember that the secondary pump must operate at the point where its characteristic crosses the characteristic of the secondary circuit and continuity must exist at every tee. Consider junction A with the three-way valve in its normally open position. The secondary circuit is designed for a given flow rate, say 100 gpm, using three-way valve control of the coils. The system is balanced so that the flow rate from the primary circuit is 100 gpm. Therefore, a mass balance at junction A shows that the common pipe has no flow and the 100 gpm leaves the secondary circuit at point B through the normally open (NO) port of the three-way valve. At a partial load, some of the 100 gpm flow in the secondary circuit, say 40 gpm, will recirculate in the common pipe and mix with 60 gpm of water from the primary supply circuit while the remaining 40 gpm of the 100 gpm from the supply circuit will flow through section C and the three-way valve, mixing with 60 gpm from the secondary circuit. Remember that the three-way valve will let water out of the secondary circuit only when it reaches a set temperature, say 55 F.

If the coils in the secondary circuit of Fig. 10-27b were controlled using two-way valves, it would be a variable-flow secondary circuit. In this case there would be no flow in the common pipe and the excess flow from the primary circuit would flow through section C. The primary supply circuit is still a constant flow system. The arrangement of Fig. 10-27b is not intended to be a model system but only a means of presenting the common pipe principle.

The systems of Fig. 10-27 using coils with either three-way or two-way valve control returns water to the source at a variable temperature depending on the total system load. This is generally not desirable in large chilled water systems because the chillers cannot operate at their full capacity unless the water returns at some reasonably high temperature compatible with their design. Because two-way valve control produces a fixed temperature rise in the water and a variable flow rate, the primary circuit should be of this type.

Variable Volume Flow

Variable volume flow has been studied extensively by Coad (8). Hansen (9) has also published on this subject.

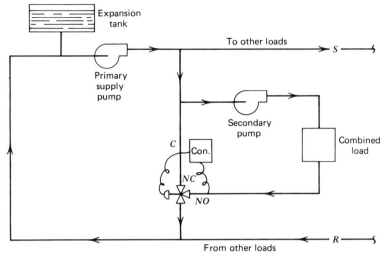

Figure 10-27a Potentially troublesome secondary piping circuit.

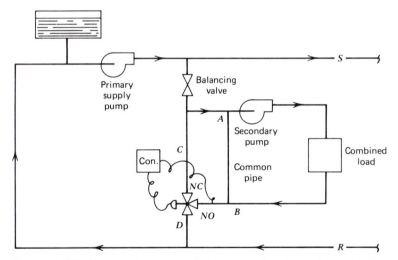

Figure 10-27b Common pipe used to isolate secondary from primary circuit.

Let us consider a secondary load circuit using two-way valves as shown in Fig. 10-28a with the primary circuit three-way valve replaced with a two-way valve. Suppose the secondary circuit is designed for 100 gpm and for the present assume $t_s = t_p$. Then when balanced at full load, the design flow will enter the tee at A, the secondary pump will take it all, and there will be no flow in the common pipe. Likewise at B, the design flow will exit the secondary circuit and flow through the two-way valve into the primary return. Now consider a partial load condition with a 60 gpm flow to the coils, caused by the response of the two-way coil valves to decreased space load. The secondary pump will respond as shown in Fig. 10-28b, backing up on its characteristic. Then, 60 gpm

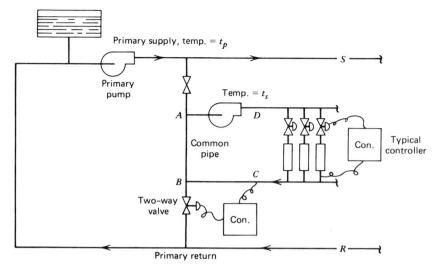

Figure 10-28a Variable-flow primary and secondary circuits.

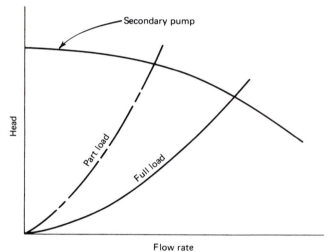

Figure 10-28b Secondary system and pump characteristics.

will enter the tee at A and flow to the pump. The same amount will exit at B with no flow in the common pipe. Note that this arrangement will produce a fixed temperature differential across the primary circuit and a variable flow rate to the primary supply circuit.

It is often the case that the primary supply water temperature t_p is lower than the secondary circuit requires, t_s. For this case the temperature sensor is moved from point C to point D in Fig. 10-28a. This arrangement will cause some secondary water to flow in the common pipe and mix at point A to obtain water at point D at a temperature t_s. We must now provide a suitable primary circuit to accommodate the chillers or boilers.

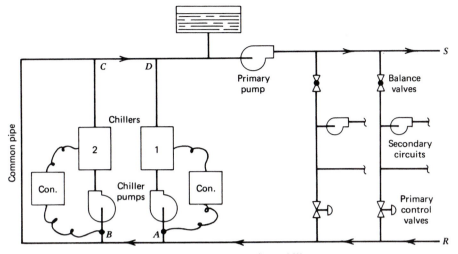

Figure 10-29 Variable-flow system with constant flow chillers.

The primary distribution system shown in Fig. 10-28a is simply another piping and pumping system. As the coil control valves throttle back the flow in the secondary circuits, the primary pump will back up on its characteristic curve in the same way as the secondary pump, as shown in Fig. 10-28b. Methods of controlling the pumps to achieve additional energy savings will be covered later. It is interesting to note the sequence of control events. The load on a coil dictates its flow requirement, and its two-way valve is the primary control. The secondary pump is the next response followed by the response of the primary pump. Note that only one simple control system is required, the two-way valves.

The next problem is to add the source system apparatus to the primary supply circuit. One method of configuration is shown in Fig. 10-29, where two chillers are shown, each with its own pump. There is no limit on the number of source elements (chillers or boilers), but there should be at least two and they do not have to be of the same capacity. Each source pump should be designed to match the circuit from B to C and A to D, whereas the primary pump is sized to match the primary distribution circuit, including the two-way valves but excluding the secondary circuits and the source (chiller or boiler) circuits. Note that the common pipe shown in the primary distribution circuit (Fig. 10-29) decouples the source pumps from the primary pump.

To understand the flow dynamics of the system shown in Fig. 10-29, assume a total design flow of 500 gpm, equal-sized chillers, and supply and return water temperatures of 42 F and 55 F, respectively. Fig. 10-28a shows the type of secondary circuits. Assume that the secondary circuits use 42 F water and that the two-way valves control the flow so that 55 F water leaves the coils. Then at full design load the total flow rate of 500 gpm is divided evenly between the two chillers; the primary pump is operating at 500 gpm and the total flow is divided among the various secondary circuits, which are operating at their full design load. The primary circuit common pipe has no flow in either direction. Now suppose that the various loads have decreased and the coil two-way valves have reduced the flow in the secondary circuits so that the total required flow is 400 gpm. The primary pump will back up on its characteristic to 400 gpm; however, the chiller pumps are unaffected by the actions of the secondary and primary pumps because of the common pipes and continue to move 250 gpm each. Consider the tee at *A*. Continuity requires that 250 gpm go to chiller 1 and 150 gpm flow to the tee at *B*. Chiller 2 requires 250 gpm; therefore, continuity requires that 100 gpm flow from the common pipe to the tee at *B*. In other words, 100 gpm of the 250 gpm entering the tee at *C* must be returned through the common pipe to the tee at *B*. The chillers are controlled by thermostats at *A* and *B*. At this partial-load condition note that chiller 1 receives water at 55 F and remains fully loaded. Chiller 2, however, now receives water at a lower temperature (about 50 F) and is not fully loaded. Chiller 2 is said to *unload,* which means it will operate at less than its full capacity using less power input. As the coil loads continue to decrease, the primary circuit flow will continue to decrease. When the total flow reaches 250 gpm, chiller 2 and its pump will cycle off because all of the flow through it will be diverted to the common pipe and reenter it at *B*. Further reduction in flow below 250 gpm will cause chiller 1 to unload and eventually cycle off when the total coil load reaches zero. As the coil loads increase from zero, the primary flow will increase, causing warm water to flow toward points *A* and *B*. Thermostat *A* will activate chiller 1 and its pump, which operates until the total flow exceeds 250 gpm, when thermostat *B* starts chiller 2 and its pump. Note that no matter how many chillers are used, they will unload from left to right and load from right to left. Also note that all the chillers that are operating are fully loaded except one that may be partially loaded. This permits maximum operating efficiency. Further, this type system provides the minimum flow of water to meet the space load, which leads to low pumping costs. Finally notice that only a bare minimum of controls are needed for a rather extensive distribution system. The common pipes dramatically reduce the complexity because any one circuit can be designed and analyzed independent of the others.

Figure 10-30 shows a variation on the location of the common pipe in the primary distribution circuit. Analysis of this arrangement shows that the chillers will load and unload equally, which means that most of the time none are fully loaded unless some extra controls are used to cycle one or more chillers off and on. This type of setup is

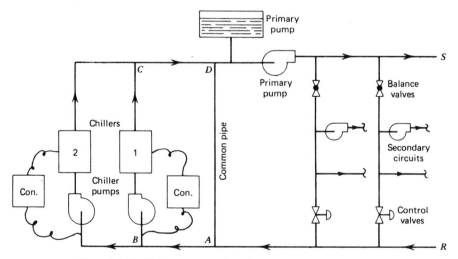

Figure 10-30 Chillers arranged to share the load equally.

sometimes used where the load is relatively constant and operating personnel are present to monitor the system and start and stop one or more units if necessary.

Figure 10-31 shows how the system of Fig. 10-29 can utilize thermal storage. Note that under partial load the extra chiller capacity cools the water in the storage tank and chiller 2 will not unload until water leaving the tank is at a temperature less than the system return water temperature. At some other time when the total system load exceeds the total chiller capacity, water flows through both chillers and through the storage tank out into the distribution system. Note that the primary pump has a capacity greater than the total capacity of the chiller pumps.

The design and sizing of the piping and pumps for the large variable-flow system follow the same general procedures given for constant-flow systems in Section 10-4. Each part of the variable-flow system is designed for full load. Partial-load operation is then controlled as described previously.

Pump Control

The primary and secondary pumps in a variable-flow system may be controlled in some way rather than allowing them to back up on their characteristics. This is desirable for operating economy and also prevents overpressuring of the two-way valves, which may vibrate or leak if too large a pressure differential exists. The control method most frequently used is to reduce pump speed in response to a critical pressure differential some place in the circuit. For example, the path to and from one particular coil in a secondary circuit will require the greatest pressure differential of all the coils in that circuit. Therefore, the differential pressure sensor for the pump motor speed control should be located across that coil and control valve and set so that the pump will always produce enough head for that coil. Frequently the critical coil is the one located farthest from the pump. The primary pump will be controlled in the same general way. In this case, the critical secondary circuit must be identified and the pressure sensor located accordingly.

The foregoing discussions of water system control and design have generally referred to chilled water systems. However, the concepts apply to all kinds of source elements for both heating and cooling.

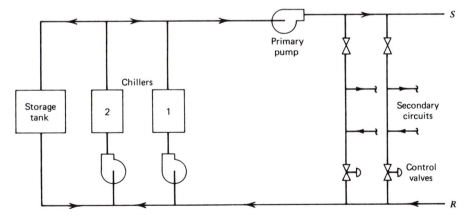

Figure 10-31 Variable-flow system adapted to a storage system.

Control Valve Characteristics and Selection

The selection of control valves is an important step in hydronic system design. There are two main considerations. These are the size or head loss at design flow and the relation of flow to valve plug lift, the *valve characteristic.*

In sizing control valves one must take care that the valve is not so large that its control range is very small. That is, it is undesirable for a large change in flow to result from a small lift of the valve plug. To prevent this the valve should be selected to have about the same head loss when fully open as the element being controlled. For example, a two-way valve for a coil with a head loss of 10 ft of water should have a head loss of about 10 ft of water. The valve head loss is determined from its C_v coefficient, as discussed earlier in the chapter.

The design of the valve plug depends on the liquid medium for the application such as hot water, chilled water, or steam. A valve plug designed as shown in Fig. 10-32b is said to be linear, as shown by curve A in Fig. 10-32a, whereas a plug shaped as shown in Fig. 10-32c is for an equal percentage valve, as shown by curve B in Fig. 10-32a. The requirement for different characteristics relates to the temperature changes for the fluids at decreased loads.

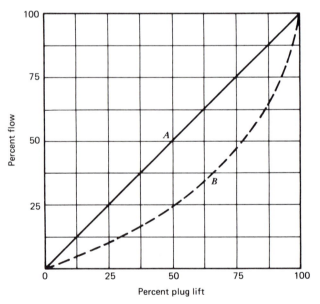

Figure 10-32a Relative flow versus plug lift for linear and equal percentage valves.

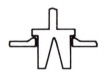

Figure 10-32b Linear or V-port valve.

Figure 10-32c Equal percentage valve.

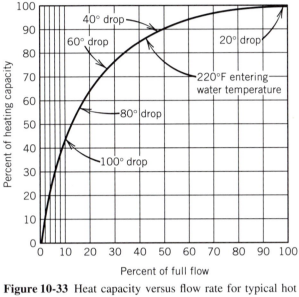

Figure 10-33 Heat capacity versus flow rate for typical hot water heating coil.

For steam the heat exchanger load is directly proportional to the flow rate because the condensing vapor is at about the same temperature for all flow rates and a linear valve is quite satisfactory (curve *A* Fig. 10-32*a*).

Hot water presents a different problem because a decrease in flow rate is accompanied by an increase in the temperature change of the water (Fig. 10-33). The net result may be only a small reduction in heat exchange for a large reduction in flow. To obtain a better relation between lift and output for this case, an equal percentage valve should be used (curve *B* of Fig. 10-32*a*). The net result is a nearly linear response in heating capacity.

Chilled water coils have a limited water temperature range of 10 to 15 F and can be adequately controlled by a properly sized linear valve.

REFERENCES

1. L. F. Moody, ''Friction Factors for Pipe Flow,'' *Transactions of ASME,* Vol. 66, 1944.
2. *ASHRAE Handbook, Fundamentals Volume,* American Society of Heating, Refrigerating and Air-Conditioning Engineers, Inc., Atlanta, GA, 1989.
3. H. Schlichting, *Boundary Layer Theory,* 4th ed., McGraw-Hill, New York, 1960.

4. *Fluid Meters, Their Theory and Application,* American Society of Mechanical Engineers, New York, 1959.

5. *ASHRAE Handbook, Applications Volume,* American Society of Heating, Refrigerating and Air-Conditioning Engineers, Atlanta, Inc., GA, 1991.

6. *ASHRAE Handbook, HVAC Systems and Equipment,* American Society of Heating, Refrigerating and Air-Conditioning Engineers, Inc., Atlanta, GA, 1992.

7. "Flow of Fluids Through Valves, Fittings, and Pipes," Technical Paper No. 410, The Crane Co., Chicago, IL, 1976.

8. W. J. Coad, "Variable Flow in Hydronic Systems for Improved Stability, Simplicity and Energy Economics," *ASHRAE Transactions,* Vol. 91, Part 1, 1985.

9. E. G. Hansen, *Hydronic System Design and Operation,* McGraw-Hill, New York, 1985.

PROBLEMS

10-1. Consider the system shown in Fig. 10-34, which has a nozzle at point 2. Compute (a) the work done on the water in (ft-lbf)/lbm, ft of head and lbf/in.^{2}, (b) the power delivered to the water, and (c) sketch the system characteristic.

10-2. The system shown in Fig. 10-35 transfers water to the tank at a rate of 0.015 m^3/s through standard commercial steel 4-in. pipe (I.D. = 103 mm). The total equivalent length of the pipe is 100 m. The increase in elevation is 40 m. Compute (a) the work done on the water in kJ/kg, (b) the power delivered to the water in kW, and (c) sketch the system characteristic.

10-3. The piping of Fig. 10-36 is part of a larger water distribution system. The pressure at point 1 is 10 psig (70 kPa), and the pipe is all the same size. (a) Compute the pressure at points 2, 3, and 4. (b) Sketch the system characteristic for the complete run of pipe. Assume a flow rate of 100 gpm (6.3 l/s).

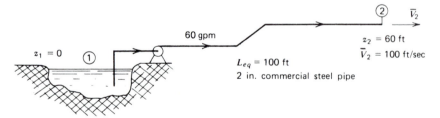

Figure 10-34 Sketch for Problem 10-1.

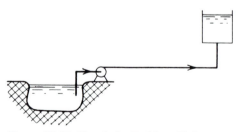

Figure 10-35 Sketch for Problem 10-2.

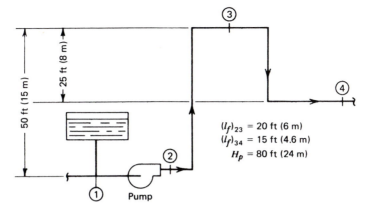

Figure 10-36 Sketch for Problem 10-3.

10-4. A pump located at ground level circulates water in the chilled water system for a 20-story building. Consider a vertical riser from the pump to an equipment room on the twentieth floor. The lost head in the riser is 25 ft (8 m) of water and the pump produces 120 ft (37 m) of head. What must the pressure be on the suction side of the pump for a pressure of 5 psig (35 kPa) to exist in the riser on the twentieth floor? Assume 12 ft (3.7 m) of elevation per floor.

10-5. For the building of Problem 10-4 it is required that the domestic service water pressure be the same on the twentieth floor as supplied by the city water main. Assuming a lost head of 20 ft (6 m) in the distribution riser to the twentieth floor, how much head must a booster pump produce?

10-6. Consider the piping system shown in Fig. 10-37. Sketch the characteristics for each separate part of the system and combine them to obtain the characteristic for the complete system.

10-7. The characteristic for a section of pipe may be represented by a function of the form $H = a\dot{Q}^2 + z$ where a is a constant, H is head, $\dot{Q}$ is flow rate, and z is elevation change. Derive an expression to represent the characteristic for two different pipe sections connected in (a) series and (b) parallel.

10-8. Refer to Problem 10-7 and derive an expression for the characteristic of a system comprised of the series elements (a) and the parallel elements (b) connected in series.

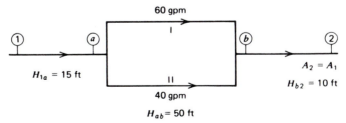

Figure 10-37 Schematic for Problem 10-6.

10-9. Compute the lost head for 100 gpm (0.006 m³/s) of 20 percent ethylene glycol solution flowing through 300 ft (100 m) of $2\frac{1}{2}$-in. (63-mm) commercial steel pipe. The temperature of the solution is 60 F (16 C).

10-10. A piping system has three parallel circuits. Circuit *A* requires 20 ft (6 m) of head with a flow rate of 50 gpm (3.2 L/s); circuit *B* requires 25 ft (7.5 m) of head with a flow rate of 30 gpm (1.9 L/s); and circuit *C* requires 30 ft (9 m) of head with a flow rate of 45 gpm (2.8 L/s). (a) Construct the characteristic for each circuit and find the characteristic for the combination of *A, B,* and *C*. (b) What is the flow rate in each circuit when the total flow rate is 100 gpm (6.3 L/s)? (c) How much head is required to produce a total flow rate of 125 gpm (7.9 L/s)? (d) What is the flow rate in each circuit of part (c)?

10-11. Solve Problem 10-10 assuming that the characteristic of each circuit can be represented by $H = a\dot{Q}^2$ where *a* is a constant for each circuit.

10-12. A pitot tube is being used to measure the flow rate of air in a 6-in. diameter duct. The pressure indicated by the inclined gage (velocity head) is 0.25 in. wg. The static pressure is essentially standard atmospheric and the air temperature is 120 F. Assume that the average velocity is 80 percent of the center-line velocity. Compute the volume flow rate in cubic feet per minute.

10-13. Assume the measurement made in Problem 10-12 is on the center line and the velocity profile can be described by Eq. 10-14 with $n = 7.5$. Find (a) the average velocity in the duct, (b) the volume flow rate, (c) the mass flow rate, and (d) the ratio of the average to the maximum velocity.

10-14. Saturated water vapor at 101.35 kPa flows in a standard 8-in. pipe (203 mm). A pitot tube located at the center of the pipe shows a velocity head of 10 mm of mercury. Find (a) the velocity of the water vapor at this location and (b) the mass flow rate assuming that the average velocity is 82 percent of the maximum velocity.

10-15. Integrate the velocity profile of Problem 10-14, using Eq. 10-14 and assuming $n = 8$. Find the average velocity of the steam. Compute the Reynolds number. Does this agree with the assumed value of *n*?

10-16. Design an orifice to be used in 2-in. standard commercial steel pipe to measure 60 gpm of 30 percent ethylene glycol solution at 50 F. A pressure drop of 5 to 10 in. of mercury is required.

10-17. A square-edged orifice is installed in standard 4-in. water pipe (103 mm I.D.). The orifice diameter is 50 mm and a head differential across the orifice of 98 mm of mercury is observed. Compute the volume flow rate of the water assuming a temperature of 10 C. What is the Reynolds number based on the orifice diameter? Does the Reynolds number agree with the flow coefficient?

10-18. The piping system of Problem 10-9 has an increase in elevation of 18 ft from inlet to outlet. (a) Select a pump using Fig. 10-14 and sketch both the system and pump characteristics. (b) How much power is delivered to the fluid in horsepower? (c) How much shaft power is required? (d) If the electric motor has an efficiency of 90 percent, what is the size of the motor required?

10-19. Two $6\frac{1}{2}$-in. 1750 rpm pumps as shown in Fig. 10-14 are used in parallel to deliver 240 gpm at 35 ft of head. (a) Sketch the system and pump characteristics. (b) What is the shaft power requirement of each pump? (c) If one pump fails, what are the flow rate and shaft power requirement of the pump still in operation? (d) Could this type of failure cause a problem in general?

10-20. A 7-in. 3500-rpm pump, shown in Fig. 10-14, is to be used to transfer lake water to a water-treatment plant. The flow rate is to be 300 gpm. What is the maximum height that the pump can be located above the lake surface to prevent cavitation? Assume that the water has a maximum temperature of 80 F, the lost head in the suction line is 2 ft of water, and barometric pressure is 28 in. of mercury.

10-21. Select a pump using Fig. 10-14 for a system that requires a flow rate of 250 gpm (15.8 L/s) at a head of 150 ft (45 m) of water. (a) Sketch the pump and system characteristics and show the operating flow rate, efficiency, and power assuming no adjustments. (b) Assume that the system has been adjusted to 250 gpm (15.8 L/s) and find the efficiency and power.

10-22. A system requires a flow rate of 210 gpm (13.25 L/s) and a head of 120 ft (37 m). (a) Select a pump from Fig. 10-14 that most closely matches the required flow rate and head and list its shaft power and efficiency. (b) Suppose a 7-in., 3500-rpm pump was selected for the system. At what flow rate and head would the system operate assuming no adjustments? Find the efficiency and power. (c) The system of part (b) is adjusted to a flow rate of 210 gpm (13.25 L/s). What is the efficiency and power? (d) Show the pump and system characteristics of (a), (b), and (c) above on the same graph.

10-23. Refer to Problem 10-22b. Suppose that the pump speed were reduced to obtain 210 gpm (13.25 L/s) and find the rpm, head, efficiency, and shaft power.

10-24. Use the data of Fig. 10-14 and the pump affinity laws and plot a characteristic for a pump with a 7-in. impeller rotating at 2500 rpm.

10-25. Size schedule 40 commercial steel pipe for the given flow rates. Comment on your selections. (a) 40 gpm (2.5 L/s), (b) 30 gpm (1.9 L/s), (c) 15 gpm (0.95 L/s), (d) 50 gpm (3.2 L/s), (e) 100 gpm (6.3 L/s), (f) 1500 gpm (95 L/s).

10-26. Find the lost head for each of the following fittings. (a) 2-in. standard elbow with flow rate of 50 gpm (3.2 L/s), (b) 4-in. globe valve with flow rate of 300 gpm (19 L/s), (c) branch of 3-in. standard tee with 150 gpm (9.5 L/s).

10-27. A control valve has a C_v of 80. It has been selected to control the flow in a coil that requires 120 gpm. What head loss can be expected for the valve?

10-28. Size the piping for the cooling tower circuit shown in Fig. 10-38. The water flow rate is 0.03 m³/s and the total equivalent length of the pipe and fittings is 200 m. Pressure loss for the condenser coil is 35 kPa and the strainer has a C_{vs} of 1.14×10^{-2} m³/s per kPa pressure loss. What is the head requirement for the pump?

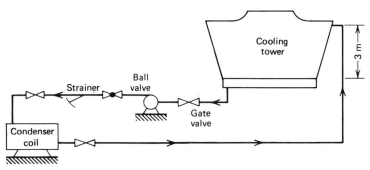

Figure 10-38 Sketch for Problem 10-28.

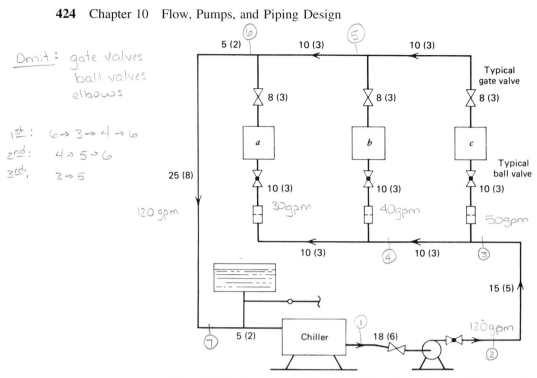

Omit: gate valves
      ball valves
      elbows

1st: 6 → 3 → 4 → 6
2nd: 4 → 5 → 6
3rd: 3 → 5

Figure 10-39 Sketch for Problem 10-29. Lengths are in feet with meters in parentheses.

10-29. Size the piping for the layout shown in Fig. 10-39 and specify the pump requirements. Assume that all the turns and fittings are as shown on the diagram. The pipe is commercial steel. Table 10-2 gives the required data.

10-30. Size the piping and specify pump requirements for a cooling tower installation similar to that shown in Fig. 10-20. The volume flow rate of the water is 300 gpm. The piping is commercial steel. Assume that fittings are as shown. The head loss in the condenser is 20 ft of water. C_v for the strainer is 200. The horizontal distance from the condenser to the cooling tower is 80 ft. The vertical distance from the pump to the top of the tower is 30 ft. The tower sump is 12 ft above the pump.

Table 10-2 Data for Problem 10-29

		Head Loss in Feet	
Unit	gpm	Coil	Orifice
a	30	15	6
b	40	12	6
c	50	10	6
Chiller	120	20 15	—

Table 10-3 Data for Problem 10-31

Unit	m³/s	Head Loss in Meters	
		Coil	Orifice
a	0.002	5	2
b	0.003	4	2
c	0.0033	4	2
Chiller	0.0083	10	—

10-31. Size the piping for the layout shown in Fig. 10-39. Assume that fittings are as shown and the pipe is standard commercial steel. Table 10-3 gives pertinent data. Specify pump requirements.

10-32. Determine the volume of a free surface expansion tank for a system similar to that shown in Fig. 10-39. The system volume is 500 gal (1.9 m³). Assume a system gage pressure of 12 psi (83 kPa) and an operating temperature of 42 F (6 C). A maximum temperature and pressure of 110 F (43 C) and 50 psig (345 kPa) are specified. Assume steel pipe.

10-33. Rework Problem 10-32 for a bladder-type expansion tank.

10-34. Find the volume of a free surface expansion tank for a hot water system with a volume of 1200 gal (4.6 m³). The system gage pressure is regulated to 20 psi (140 kPa) at the tank and is initially filled with water at 60 F (16 C). The pressure relief valve on the boiler is set for a gage pressure of 50 psi (345 kPa) and the maximum water temperature is expected to be 220 F (104 C). The system is predominantly copper tubing.

10-35. Consider the system shown in Fig. 10-27b operating at one-third load. Water is supplied at 42 F (6 C) and the control valve is set for 57 F (14 C). The secondary circuit uses three-way valves on its coils and the design full-load water flow rate is 150 gpm (9.5 L/s). (a) How much water is flowing into the tee at A and how much water is flowing in section C? (b) What is the temperature of the water entering the secondary pump? (c) What is the temperature of the water leaving the three-way valve at point D? (d) What size pipe should be used for the main sections of the secondary circuit and the connections to the primary circuit?

10-36. Consider the system shown in Fig. 10-28a. The primary supply water temperature is 40 F (4.5 C) and the controller, with sensor located at D, is set for 47 F (8.3 C). The controllers on the coil valves are set for 57 F (14 C). (a) If the full-load secondary circuit flow rate is 100 gpm (6.3 L/s), how much water must recirculate in the common pipe? (b) How much water is supplied and returned to the primary circuit? (c) Size the main sections of the secondary circuit (C and D), the common pipe, and the connections to the primary circuit.

10-37. Consider the system shown in Fig. 10-29, where the chillers are of equal size. Assume the system is designed to circulate 1000 gpm (63 L/s) under full-load design conditions at 45 F (7 C) and the secondary circuits utilize water at the same temperature. Water is returned from the secondary circuits at 60 F (15.6 C). At a part-load condition, 650 gpm (41 L/s) of water flows to the secondary circuits. (a) What is the flow rate of the water in the common pipe? (b) What is the temperature of the water at point B? (c) What is the load ratio (load/capacity) for chiller 2? (d) Size the pipe, based on full-load design conditions (except the secondary circuits). (e) The primary pump operates at 3500 rpm when fully loaded. Approximately what speed is required at the part-load condition? (f) What is the power reduction at part load?

10-38. Consider the system in Fig. 10-30, where the chillers are of equal size. Design and part-load operating conditions are the same as Problem 10-37. (a) What is the flow rate of the water in the common pipe? (b) What is the temperature of the water entering both chillers? (c) What is the load ratio for each chiller?

10-39. Consider the preliminary system layout in Fig. 10-40. The final design is to be a variable flow system similar to Fig. 10-29 with two-way valve control throughout. The primary system has three chillers of equal size and each secondary system has two or three air handlers. (a) Sketch the system in more detail showing pumps, common pipes, control valves, and so on. (b) The primary supply water is at 42 F (5.6 C); however, half the buildings require water at about 45 F (7 C). Show how the primary controller sensors are located in all

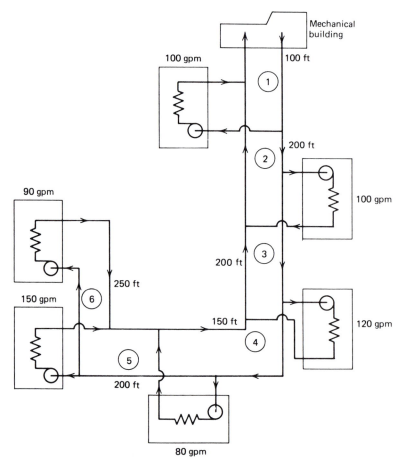

Figure 10-40 Preliminary sketch of a central chilled water system.

cases. (c) The coil valve controllers are set for 58 F (14 C). Show the flow rates and water temperatures at strategic points in the system at full load. (d) Variable-speed pumps are to be used in the primary and secondary circuits. Show where you would install the differential pressure sensors for the pump speed controllers.

10-40. Size the pipe for the secondary water circuit shown in Fig. 10-41. The pipe is type L copper. Notice that the lengths given are the *total* equivalent lengths excluding the coil and control valves. Select a pump from Fig. 10-14 and sketch the system and pump characteristics.

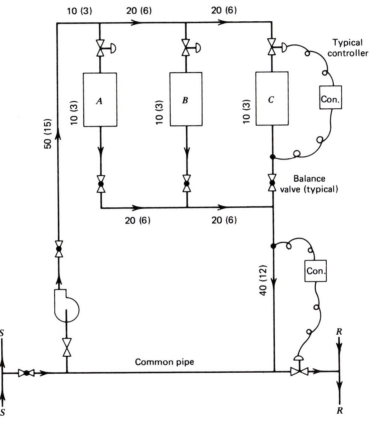

Note: Piping is type L copper
      All lengths are total equivalent lengths

Coil	Flow rate gpm (L/s)	Lost head ft (m) Coil	Con. valve
A	40 (2.5)	12 (3.7)	10 (3)
B	40 (2.5)	15 (4.6)	12 (3.7)
C	50 (3.2)	18 (5.5)	15 (4.6)

Figure 10-41 Schematic of secondary circuit for a variable flow system for problem 10-40.

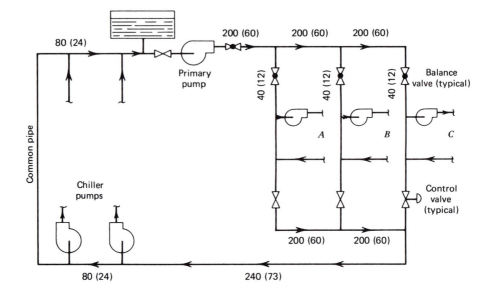

Notes: Pipe is schedule 40, commercial steel
All lengths are total equivalent lengths excluding control valves

Circuit	Flow rate gpm (L/s)	Control valve head loss ft (m)
A	60 (3.8)	40 (12)
B	70 (4.4)	50 (15)
C	70 (4.4)	50 (15)

Figure 10-42 Schematic of a primary water circuit for a variable flow system for Problem 10-41.

10-41. Size the pipe for the primary circuit shown in Fig. 10-42. The lengths shown are the *total equivalent lengths* for the section exclusive of the control valve. Specify the primary pump performance requirement.

Chapter 11

Space Air Diffusion

The major objective of an HVAC system is to provide comfort and suitable indoor air quality within the occupied zones of a building. An important step in the process is to furnish air to each space in such a way that any natural air currents or radiative effects within the space are counteracted, and to assure that temperatures, humidities, and air velocities within the occupied spaces are held at acceptable conditions. This is usually accomplished by introducing air into the spaces at optimum locations and with sufficient velocity so that entrainment of air already within the space will occur. The resulting mixing will permit energy stored in the warm air to be carried into the occupied spaces in the case of heating or the introduction of cool air and the carrying away of energy from the occupied spaces in the case of cooling. Additionally the mixing of the jet and the room air permits the carrying away of contaminants that may be generated within the spaces. The challenge is to provide good mixing without creating uncomfortable drafts and to assure that there is reasonable uniformity of temperature throughout the occupied spaces. This must be done without unacceptable changes in room conditions as the load requirements of the rooms change. The design also involves selection of suitable diffusing equipment so that noise and pressure drop requirements are met.

11-1 BEHAVIOR OF JETS

Conditioned air is normally supplied to air outlets at velocities much higher than would be acceptable in the occupied space. The conditioned air temperature may be above, below, or equal to the temperature of the air in the occupied space. Proper air distribution causes entrainment of room air by the primary air stream and the resultant mixing reduces the temperature differences to acceptable limits before the air enters the occupied space. It also counteracts the natural convection and radiation effects within the room.

The air projection from free round openings, grilles, perforated panels, ceiling diffusers, and other outlets is related to the average velocity at the face of the air supply opening. A free jet has four zones of expansion, and the center-line velocity in any of the zones is related to the initial velocity as shown in Fig. 11-1. Regardless of the type of opening, the jet will tend to assume a circular shape. The effect of nearby surfaces will be considered later. In zone III, the most important zone from the point of view of room

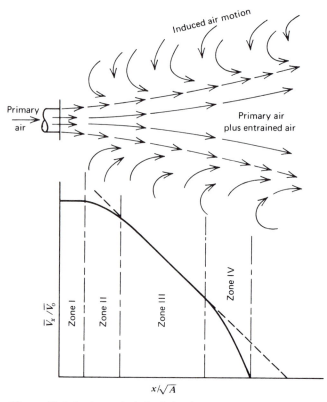

Figure 11-1 Isothermal air-jet behavior.

air distribution, the relation between the jet center-line velocity and the initial velocity is given by

$$\frac{\overline{V}_x}{\overline{V}_0} = K \frac{\sqrt{A_0}}{x}$$

(11-1a)

or

$$\overline{V}_x = \frac{K\dot{Q}_o}{(\sqrt{A_0}x)}$$

(11-1b)

where

$\overline{V}_x$ = center-line velocity at any x, ft/min or m/s
$\overline{V}_0$ = initial velocity, ft/min or m/s
A_0 = area corresponding to initial velocity, ft² or m²
x = distance from outlet to point of measurement of $\overline{V}_x$, ft or m
$\dot{Q}_o$ = air flow rate at outlet, cfm or m³/s
K = constant of proportionality, dimensionless

Equations 11-1a and 11-1b strictly pertain to isothermal free jets, but with the proper A and K the equations define the throw for any type of outlet. The *throw* is the distance from the outlet to where the maximum velocity in the jet has decreased to some specified value such as 50, 100, or 150 ft/min. The constant K varies from about 6 for free jets to about 1 for ceiling diffusers.

The jet expands because of entrainment of room air; the air beyond zone II is a mixture of primary and induced air. The ratio of the total volume of the jet to the initial volume of the jet at a given distance from the origin depends mainly on the ratio of initial velocity $\overline{V}_0$ to the terminal velocity $\overline{V}_x$. The *induction ratio* is

$$\frac{\dot{Q}_x}{\dot{Q}_o} = C \frac{\overline{V}_0}{\overline{V}_x} \tag{11-2}$$

where

$\dot{Q}_x$ = total air mixture at distance x from the outlet, cfm or m³/s
C = entrainment coefficient, 2 for round, free jet, dimensionless

In zone IV where the terminal velocity is low, Eq. 11-2 will give values about 20 percent high.

When a jet is projected parallel to and within a few inches of a surface, the induction or entrainment is limited on the surface side of the jet. A low-pressure region is created between the surface and the jet, and the jet attaches itself to the surface. This phenomenon results if the angle of discharge between the jet and the surface is less than about 40 degrees and if the jet is within about one foot of the surface. The jet from a floor outlet is drawn to the wall and the jet from a ceiling outlet is drawn to the ceiling. This surface effect increases the throw for all types of outlets and decreases the drop for horizontal jets. The drop is illustrated in Fig. 11-2.

In nonisothermal jets, buoyant forces cause the jet to rise when the air is warm and drop when cool. These conditions result in shorter throws for jet velocities less than 150 ft/min or 0.76 m/s. The lower sketch of Fig. 11-2 shows the drop for a cool jet of air.

The following general statements may be made concerning the characteristics of air jets.

1. Surface effect increases the throw and decreases the drop compared to free space conditions.
2. Increased surface effect may be obtained by moving the outlet away from the surface somewhat so that the jet spreads over the surface after impact.
3. Increased surface effect may be obtained by spreading the jet when it is discharged.
4. Spreading the air stream reduces the throw and drop.
5. Drop primarily depends on the quantity of air and only partially on the outlet size or velocity. Thus the use of more outlets with less air per outlet reduces drop.

Room Air Motion

Room air near the jet is entrained and must then be replaced by other room air. The room air always moves toward the supply and thus sets all the room air into motion. Whenever the average room air velocity is less than about 50 ft/min or 0.25 m/s, buoyancy effects may be significant. In general, about 8 to 10 air changes per hour are required to prevent stagnant regions (velocity less than 15 ft/min or 0.08 m/s). However,

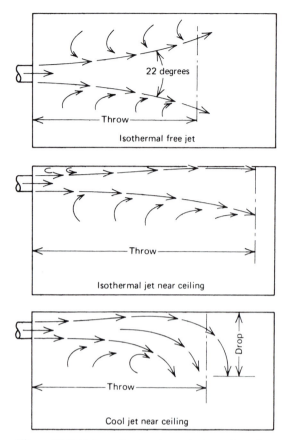

Figure 11-2 Schematic showing entrainment, surface effect, and drop.

stagnant regions are not necessarily a serious condition. The general approach is to supply air in such a way that the high-velocity air from the outlet does not enter the occupied space. The region within 1 ft of the wall and above about 6 ft from the floor is out of the occupied space for practical purposes.

Figure 11-3 shows velocity envelopes for a high sidewall outlet. Equation 11-1a has been used to estimate the throw for the terminal velocities shown. In order to interpret the air motion shown in terms of comfort, it is necessary to estimate the local air temperatures corresponding to the terminal velocities. The relationship between the center-line velocities and the temperature differences is given approximately (1) by

$$\Delta t_x = 0.8 \, \Delta t_o \frac{\overline{V}_x}{\overline{V}_0} \tag{11-3}$$

where Δt_x and Δt_o are the differences in temperature between the local stream temperature and the room ($t_x - t_r$) and the outlet air and the room ($t_o - t_r$). Temperatures calculated using Eq. 11-3 are shown in Fig. 11-3. On the opposite wall where the terminal velocity is 100 ft/min, the air temperature is 1.6 F below the room temperature.

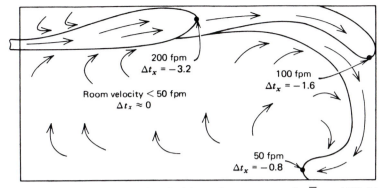

Figure 11-3 Jet and room air velocities and temperatures for $\overline{V}_o = 1000$ ft/min and $\Delta t_o = -20$ F.

The temperature difference for the 50 ft/min envelope shows that within nearly the entire occupied space the temperature is less than about 0.8 F below the room temperature and the room air motion is under 50 ft/min.

Basic Flow Patterns

The basic flow patterns for the most often used types of outlets are shown in Figs. 11-4 to 11-7 (2). The high-velocity primary air is shown by the dark shading, and the total air is represented by the light shading. These areas represent the high-momentum regions of the room air motion. Natural convection (buoyancy) effects are evident in all cases. Note that stagnant zones always have a large temperature gradient. When this occurs in the occupied space, air needs to be projected into the stagnant region to enhance mixing. An ideal condition would be uniform room temperature from the floor

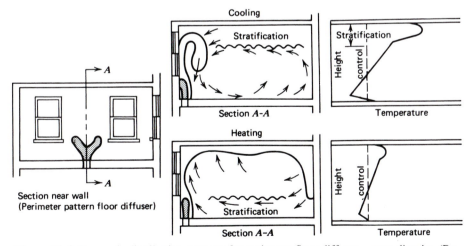

Figure 11-4 Room air distribution patterns for perimeter floor diffuser—spreading jet. (Reprinted by permission from *ASHRAE Handbook, Fundamentals Volume*, 1989.)

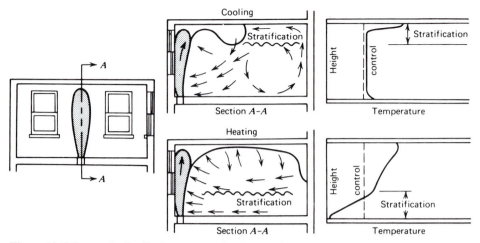

Figure 11-5 Room air distribution patterns for floor register—nonspreading jet. (Reprinted by permission from *ASHRAE Handbook, Fundamentals Volume, 1989.*)

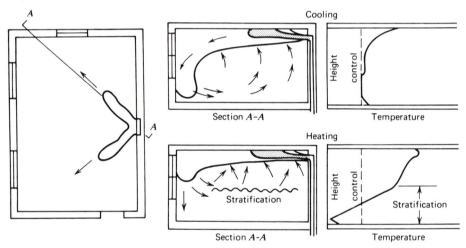

Figure 11-6 Room air distribution patterns for high sidewall register. (Reprinted by permission from *ASHRAE Handbook, Fundamentals Volume, 1989.*)

to about 6 ft above the floor. However, a gradient of about 4 F or 2 C should be acceptable to about 85 percent of the occupants.

The perimeter-type outlets shown in Fig. 11-4, ASHRAE Group C (2), are generally regarded as superior for heating applications. This is particularly true when the floor is over an unheated space or a slab and where considerable glass area exists in the wall. Diffusers with a wide spread are usually best for heating because buoyancy tends to increase the throw. For the same reason the spreading jet is not as good for cooling applications because the throw may not be adequate to mix the room air thoroughly. However, the perimeter outlet with a nonspreading jet, ASHRAE Group B (2), is quite satisfactory for cooling. Figure 11-5 shows a typical cooling application of the

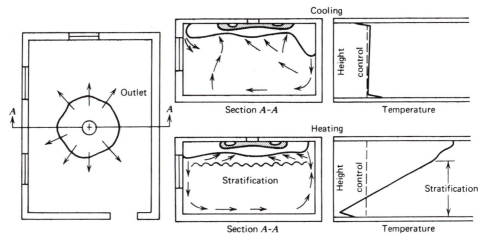

Figure 11-7 Room air distribution patterns for ceiling diffuser. (Reprinted by permission from *ASHRAE Handbook, Fundamentals Volume,* 1989.)

nonspreading perimeter diffuser. It can be seen that the nonspreading jet is less desirable for heating because a larger stratified zone will usually result. Diffusers are available that may be changed from the spreading to nonspreading type according to the season.

The high sidewall type of register, ASHRAE Group A (2), shown in Fig. 11-6 is often used in mild climates and on the second and succeeding floors of multistory buildings. This type of outlet is not recommended for cold climates or with unheated floors. Figure 11-6 shows that a considerable temperature gradient may exist between floor and ceiling when heating; however, this type outlet gives good air motion and uniform temperatures in the occupied zone for cooling application.

The ceiling diffuser, ASHRAE Group A (2), shown in Fig. 11-7 is very popular in commercial applications, and many variations of it are available. The air patterns shown in Fig. 11-7 are typical. Because the primary air is projected radially in all directions, the rate of entrainment is large, causing the high-momentum jet to diffuse quickly. This feature enables the ceiling diffuser to handle larger quantities of air at higher velocities than most other types. Figure 11-7 shows that the ceiling diffuser is quite effective for cooling applications but generally poor for heating. However, satisfactory results may be obtained in commercial structures when the floor is above a heated space.

The return air intake generally has very little effect on the room air motion, but the location may have a significant effect on the performance of the heating and cooling equipment. Because it is desirable to return the coolest air to the heating coil and the warmest air to the cooling coil, the return air intake should be located in a stagnant region.

Noise

Noise produced by the air diffuser can be annoying to the occupants of the conditioned space. Noise associated with air motion usually does not have distinguishable frequency characteristics and the level or loudness is basically a statistically representative sample of human reactions. Loudness contours or curves of equal loudness versus frequency can be established from such reactions.

A method of providing information on the spectrum content of noise is the use of the *noise criteria* (NC) curves and numbers. The NC curves are shown in Fig. 11-8. These are a series of curves constructed using loudness contours and the speech-interfering properties of noise and are used as a simple means of specifying sound-level limits for an environment by a simple, single-number rating. They have been found to be quite generally applicable for conditions of comfort. In general, levels below a NC of 30 are considered to be quiet, whereas levels above a NC of 50 or 55 are considered noisy. The activity within the space is a major consideration in determining an acceptable level.

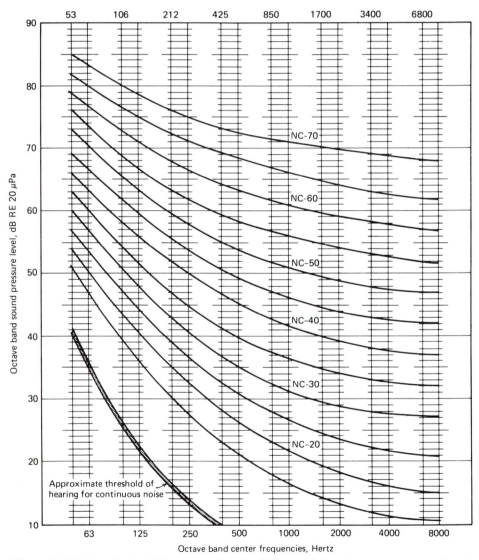

Figure 11-8 Noise criteria (NC) curves. (Reprinted by permission from *ASHRAE Handbook, Fundamentals Volume*, 1989.)

Table 11-1 Acceptable HVAC Noise Levels in
Unoccupied Rooms

Occupancy	Noise Criteria
Private residences	NC 25-30
Apartments	NC 30-35
Hotels/Motels	
Individual rooms or suites	NC 30-35
Meeting/banquet rooms	NC 30-35
Halls, corridors, lobbies	NC 35-40
Service/support areas	NC 40-45
Offices	
Executive	NC 25-30
Conference rooms	NC 25-30
Private	NC 30-35
Open-plan areas	NC 35-40
Business machines/computers	NC 40-45
Public circulation	NC 40-45
Hospitals and clinics	
Private rooms	NC 25-30
Wards	NC 30-35
Operating rooms	NC 25-30
Laboratories	NC 35-40
Corridors	NC 30-35
Public areas	NC 35-40
Churches	NC 30-35
Schools	
Lecture and classrooms	NC 25-30
Open-plan classrooms	NC 35-40
Libraries	NC 35-40
Courtrooms	NC 35-40
Legitimate theaters	NC 20-35
Movie theaters	NC 30-35
Restaurants	NC 40-45
Concert and recital halls	NC 15-20
Recording studios	NC 15-20
TV studios	NC 20-25

Source: Reprinted by permission from *ASHRAE Handbook, HVAC Applications Volume,* 1991.

Table 11-1 gives recommended noise criteria numbers for various applications (3). To determine the acceptability of a given space for a given specification, sound pressure level must be measured at several octave band center frequencies and compared with the specified NC curve of Fig. 11-8. To meet the particular NC rating, the actual octave band reading should lie on or below the NC curve.

Some manufacturers of air-diffusing equipment rate their products using the noise criteria concept. An example of these data is presented in the next section.

11-2 AIR-DISTRIBUTION SYSTEM DESIGN

This section discusses the selection and placement of the air outlets. If this is done purely on the basis of comfort, the preceding discussions on room air motion dictate the type of system and the location of the air inlets. However, the architectural design and the functional requirements of the building often override comfort.

When the designer is free to select the type of air-distribution system based on comfort, the perimeter type of system with vertical discharge of the supply air is to be preferred for exterior spaces when the heating requirements exceed 2000 degree (F) days. This type of system is excellent for heating and satisfactory for cooling when adequate throw is provided. When the floors are warmed and the degree (F) day requirement is between about 3500 and 2000, the high sidewall outlet with horizontal discharge toward the exterior wall is acceptable for heating and quite effective for cooling. When the heating requirement falls below about 2000 degree (F) days, the overhead ceiling outlet or high sidewall diffuser is recommended because cooling is the predominant mode. Interior spaces in commercial structures are usually provided with overhead systems because cooling is required most of the time.

Commercial structures often are constructed in such a way that ducts cannot be installed to serve the desired air-distribution system. Floor space is very valuable and the floor area required for outlets may be covered by shelving or other fixtures, making a perimeter system impractical. In this case an overhead system must be used. In some cases the system may be a mixture of the perimeter and overhead type.

Renovation of commercial structures may represent a large portion of a design engineer's work. Compromises are almost always required in this case, and the air-distribution system is often dictated by the nature of the existing structure.

In all cases where an ideal system cannot be used it is particularly important that the air-diffusing equipment be carefully selected and located. Although most manufacturers of air diffusers and grilles furnish extensive data on the performance of their products, there is no substitute for experience and good judgment in designing the air-distribution system.

Table 11-2 gives performance data for a type of diffuser that may be used for perimeter systems having a vertical discharge from floor outlets or as a linear diffuser in the ceiling or sidewall. Note that the data pertain to the capacity, throw, total pressure loss, noise criteria, and free area as a function of the size. It is important to read the notes given with these catalog data. Notice that throw values for three different terminal velocities are given. The diffuser may be almost any length, but its capacity is based on a length of 1 ft, whereas the throw is based on a 4-ft active length, and the NC is based on 10 ft of length. Some corrections are also required when the diffuser is used as a return intake.

Performance data for one type of round ceiling diffuser are shown in Table 11-3, and Table 11-4 shows data for an adjustable diffuser that would generally be used for high sidewall applications. The same general data are given. Note that the diffuser of Table 11-4 has adjustable vanes and throw data are given for three different settings, 0, $22\frac{1}{2}$, and 45 degrees. Figure 11-9 shows a T-bar type diffuser, which is used extensively with modular ceilings. These diffusers are often associated with variable air volume systems and sometimes have automatic flow control built into the diffuser itself. The diffuser shown produces horizontal throw parallel to the ceiling in opposing directions. Table 11-5 gives performance data for the T-bar diffuser.

Return grilles are quite varied in design. The construction of the grille has very little to do with the overall performance of the system except to introduce some loss in pressure and noise if not properly sized. The appearance of a return grille is important, and the louver design is usually selected on this basis. Table 11-6 gives data for one style of grille. Note that the capacity, pressure loss, and noise criteria are the main performance data given.

The Air Distribution Performance Index (ADPI)

A measure of the effective temperature difference between any point in the occupied space and the control conditions is called the *effective draft temperature*. It is defined by the equation proposed by Rydberg and Norback (4):

$$EDT = (t_x - t_r) - M(\overline{V}_x - \overline{V}_r) \tag{11-4}$$

where

t_r = average room dry bulb temperature, F or C
$\overline{V}_r$ = 30 ft/min or 0.15 m/s
t_x = local air stream dry bulb temperature, F or C
$\overline{V}_x$ = local air stream velocity, ft/min or m/s
M = 0.07 (F-min)/ft or 7.0 (C-s)/m

Equation 11-4 takes into account the feeling of coolness produced by air motion. It also shows that the effect of a one degree F temperature change is equivalent to a 15 ft/min velocity change. In summer the local air stream temperature t_x is usually below the control temperature. Hence both temperature and velocity terms are negative when the velocity $\overline{V}_x$ is greater than $\overline{V}_r$ and both of them add to the feeling of coolness. If in winter $\overline{V}_x$ is above $\overline{V}_r$, it will reduce the feeling of warmth produced by t_x. Therefore, it is usually possible to have zero difference in effective temperature between location x and the control point in winter but not in summer. Research indicates that a high percentage of people in sedentary occupations are comfortable where the effective draft temperature is between -3 F (-1.7 C) and $+2$ F (1.1 C) and the air velocity is less than 70 ft per minute (0.36 m/s). These conditions are used as criteria for developing the Air Distribution Performance Index (ADPI).

The ADPI is defined as the percentage of measurements taken at many locations in the occupied zone of a space that meet the -3 F to 2 F effective draft temperature criteria. The objective is to select and place the air diffusers so that an ADPI approaching 100 percent is achieved. Note that ADPI is based only on air velocity and effective draft temperature, a local temperature difference from the room average, and is not directly related to the level of dry bulb temperature or relative humidity. These effects and other factors such as mean radiant temperature must be accounted for as discussed in Chapter 4. The ADPI provides a means of selecting air diffusers in a rational way. There are no specific criteria for selection of a particular type of diffuser except as discussed before, but within a given type the ADPI is the basis for selecting the throw. The space cooling load per unit area is an important consideration. Heavy loading tends to lower the ADPI. However, loading does not influence design of the diffuser system significantly. Each type of diffuser has a characteristic room length, as shown in Table 11-7. Table 11-8 is the ADPI selection guide. Table 11-8 gives the

Table 11-2 Performance Data for a Typical Linear Diffuser

Size/Area	Total Pressure	0.009	0.020	0.036	0.057	0.080	0.109	0.143	0.182	0.225
2 in.	Flow, cfm/ft	22	33	44	55	66	77	88	99	110
	NC	—	—	12	18	23	27	31	34	37
0.055	Throw, ft—sill or floor	1-1-1	4-4-4	7-7-7	9-9-10	11-11-12	13-14-16	14-16-18	15-17-20	17-19-21
3 in.	Flow, cfm/ft	38	58	77	96	115	134	154	173	192
	NC	—	—	11	17	22	26	30	33	36
0.096	Throw, ft—sill or floor	2-2-2	7-7-7	10-10-11	12-13-14	15-16-17	18-19-20	20-21-23	23-24-25	25-25-26
4 in.	Flow, cfm/ft	56	83	111	139	167	195	222	250	278
	NC	—	—	12	18	23	27	31	34	37
0.139	Throw, ft—sill or floor	3-3-3	9-9-9	13-13-13	16-16-17	20-20-21	22-23-24	24-25-26	27-27-27	30-30-30
5 in.	Flow, cfm/ft	72	107	143	179	215	250	286	322	358
	NC	—	—	12	18	23	27	31	34	37
0.179	Throw, ft—sill or floor	4-4-4	10-10-10	14-14-14	18-18-18	22-22-23	24-24-24	27-27-28	30-30-31	32-32-32

6 in.									
Flow, cfm/ft	88	133	177	221	265	310	354	398	442
NC	—	—	13	19	24	28	32	35	38
0.221	Throw, ft— sill or floor	5-5-5	10-10-10	15-15-15	18-18-18	23-23-23	25-25-25	28-28-28	31-31-31	32-32-32

Area is given in square feet per foot of length.

1. All pressures are in in. wg.
2. Minimum throw values refer to a terminal velocity of 150 ft/min, middle to 100 ft/min and maximum to 50 ft/min, for a 4-ft active section with a cooling temperature differential of 20 F. The multiplier factors listed in the table are applicable for other lengths.

Terminal Velocity

Active Length	150 ft/min	100 ft/min	50 ft/min
1 ft	0.5	0.6	0.7
10 ft or continuous	1.6	1.4	1.2

3. The NC values are based on a room absorption of 80 Db, Re 10^{-12} watts and a 10-ft active section.

NC Correction for Length

Active length, ft	1	2	4	6	8	10	15	20	25	30
Correction	-10	-7	-4	-2	-1	0	+2	+3	+4	+5

Source: Reprinted by permission of Environmental Elements Corporation, Dallas, Texas.

Table 11-3 Performance Data for a Typical Round Ceiling Diffuser

Size	Neck Velocity, ft/min Velocity Pressure	400 0.010	500 0.016	600 0.023	700 0.031	800 0.040	900 0.051	1000 0.063	1200 0.090
6 in.	Total pressure	0.026	0.041	0.059	0.079	0.102	0.130	0.161	0.230
	Flow rate, cfm	80	100	120	140	160	180	200	235
	Radius of diffusion, ft	2-2-4	2-3-5	2-4-6	3-4-7	3-5-8	4-5-9	4-6-10	5-7-11
	NC	—	—	14	19	23	26	30	35
8 in.	Total pressure	0.033	0.052	0.075	0.101	0.130	0.166	0.205	0.292
	Flow rate, cfm	140	175	210	245	280	315	350	420
	Radius of diffusion, ft	2-4-6	3-4-7	4-5-9	4-6-10	5-7-11	5-8-13	6-9-14	7-11-17
	NC	—	15	21	26	31	34	37	44
10 in.	Total pressure	0.027	0.043	0.062	0.084	0.108	0.138	0.170	0.243
	Flow rate, cfm	220	270	330	380	435	490	545	655
	Radius of diffusion, ft	3-4-7	3-5-8	4-6-10	5-7-11	5-8-13	6-9-15	7-10-16	8-12-20
	NC	—	11	17	21	26	30	33	39
12 in.	Total pressure	0.026	0.042	0.060	0.081	0.105	0.134	0.166	0.236
	Flow rate, cfm	315	390	470	550	630	705	785	940
	Radius of diffusion, ft	3-5-8	4-6-10	5-7-12	6-8-13	6-10-15	7-11-17	8-12-19	10-14-23
	NC	—	11	17	22	26	30	33	39
18 in.	Total pressure	0.030	0.048	0.069	0.093	0.120	0.153	0.189	0.270
	Flow rate, cfm	710	885	1060	1240	1420	1590	1770	2120
	Radius of diffusion, ft	5-7-12	6-9-15	7-11-18	9-13-21	10-15-24	11-17-27	12-19-30	15-22-36
	NC	—	15	21	26	30	34	37	43

24 in.	Total pressure	0.024	0.038	0.054	0.073	0.094	0.120	0.148	0.211
	Flow rate, cfm	1260	1570	1880	2200	2510	2820	3140	3770
	Radius of diffusion, ft	6-9-15	8-12-19	9-14-22	11-16-26	12-19-30	14-21-34	16-23-37	19-28-45
	NC	—	13	19	24	28	32	35	41

1. All pressures are in in. wg.

2. Minimum radii of diffusion are to a terminal velocity of 150 ft/min, middle to 100 ft/min, and maximum to 50 ft/min.

3. The NC values are based on a room absorption of 18 Db, Re 10^{-13} watts or 8 Db, Re 10^{-12} watts.

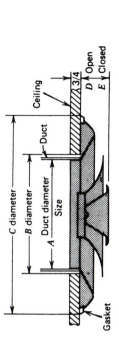

Dimensions

	Size			
A	B	C	D	E
6	$6\frac{1}{2}$	$11\frac{1}{8}$	$1\frac{3}{4}$	$1\frac{1}{8}$
8	$8\frac{1}{2}$	$14\frac{3}{4}$	$2\frac{1}{8}$	$1\frac{1}{2}$
10	$10\frac{1}{2}$	$18\frac{1}{4}$	$2\frac{7}{8}$	$2\frac{1}{8}$
12	$12\frac{1}{2}$	22	$3\frac{1}{8}$	$2\frac{3}{8}$
24	$24\frac{1}{2}$	$43\frac{3}{4}$	$7\frac{3}{4}$	$6\frac{5}{8}$

Source: Reprinted by permission of Environmental Elements Corporation, Dallas, Texas.

Table 11-4 Performance Data for an Adjustable Type, High Sidewall Diffuser

Size, in.	Velocity, ft/min	300	400	500	600	700	800	1000	1200
	Velocity Pressure	0.006	0.010	0.016	0.022	0.030	0.040	0.062	0.090
	Total Pressure 0	0.010	0.017	0.028	0.038	0.052	0.069	0.107	0.156
	$22\frac{1}{2}$	0.011	0.019	0.031	0.043	0.058	0.078	0.120	0.175
	45	0.016	0.029	0.047	0.064	0.088	0.117	0.181	0.263
8 × 4 7 × 5	cfm	55	70	90	110	125	145	180	215
	NC				10	15	19	25	31
6 × 6 $A_c = 0.18\ \text{ft}^2$	Throw, ft 0	4-7-13	6-8-15	7-11-17	9-13-19	10-15-20	11-16-22	14-17-24	15-19-26
	$22\frac{1}{2}$	3-6-10	5-6-12	6-9-14	7-10-15	8-12-16	9-13-18	11-14-19	12-15-21
	45	2-3-7	3-4-8	4-5-9	4-7-10	5-7-10	6-8-11	7-9-12	8-10-13
10 × 4 8 × 5	cfm	65	90	110	130	155	175	220	265
	NC				10	15	19	25	31
7 × 6 $A_c = 0.22\ \text{ft}^2$	Throw, ft 0	4-7-14	7-10-17	8-12-19	9-15-21	11-16-23	13-17-24	16-19-27	17-21-29
	$22\frac{1}{2}$	3-6-11	6-8-14	6-10-15	7-12-17	9-13-18	10-14-19	13-15-22	14-17-23
	45	2-4-7	3-5-9	4-6-10	5-7-10	6-8-11	6-9-12	8-10-13	9-11-15
12 × 4 10 × 5	cfm	80	105	130	155	180	210	260	310
	NC				11	16	20	26	32
8 × 6 $A_c = 0.26\ \text{ft}^2$	Throw, ft 0	5-8-16	7-11-19	9-13-21	10-16-23	12-17-24	14-19-26	17-21-29	19-23-32
	$22\frac{1}{2}$	4-6-13	6-9-15	7-10-17	8-13-18	10-14-19	11-15-21	14-17-23	15-18-26
	45	3-4-8	4-5-9	4-7-10	5-8-11	6-9-12	7-9-13	8-11-15	9-12-16

16 × 4 12 × 5	cfm NC	100	135 12	170	205 12	240 17	270 21	340 27	410 33
10 × 6 $A_c = 0.34\ \text{ft}^2$	Throw, ft	0 $22\frac{1}{2}$ 45							
	0	5-9-18	8-12-21	10-15-24	12-19-26	14-20-28	16-22-30	20-24-33	22-26-37
	$22\frac{1}{2}$	4-7-14	6-10-17	8-12-19	10-15-21	11-16-22	13-18-24	16-19-26	18-21-30
	45	3-4-9	4-6-11	5-8-11	6-9-13	7-10-14	8-11-15	10-12-17	11-13-18
18 × 4 14 × 5	cfm NC	115	155	195	235 13	275 18	310 22	390 28	470 34
12 × 6 8 × 8 $A_c = 0.39\ \text{ft}^2$	Throw, ft 0	6-9-19	9-13-23	11-16-25	13-19-28	15-22-30	17-23-32	21-26-36	23-27-40
	$22\frac{1}{2}$	5-7-15	7-10-18	9-13-20	10-15-22	12-18-24	14-18-26	17-21-29	18-22-32
	45	3-5-10	4-6-11	5-18-13	7-10-14	8-11-15	9-12-16	11-13-18	12-14-20

0 deg.
Deflection

$22\frac{1}{2}$ deg.
Deflection

45 deg.
Deflection

Source: Reprinted by permission of Environmental Elements Corporation, Dallas, Texas.

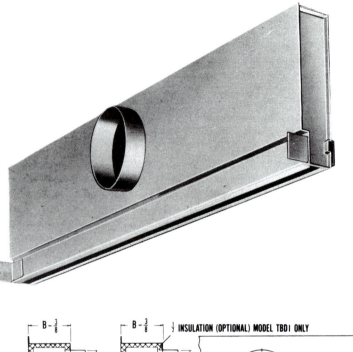

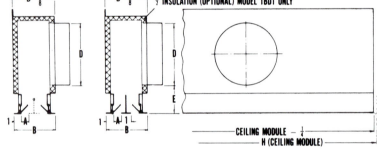

Model	A	H	B	C	D	E
27	½	24	4	12	5	5⅞
		48			7	3⅞
28	¾	24	4½	12	6	4⅞
		48			8	2⅞

Figure 11-9 A typical T-bar type diffuser assembly. (Courtesy of Environmental Corporation, Dallas, TX.)

Table 11-5 Performance Data for the T-Bar Diffusers of Figure 11-9

Model 27

H-24									
Cfm	55	62	68	80	95	110	120	135	150
Horiz. proj.	2-3-4	2-3-4	2-3-5	2-4-6	3-5-7	3-6-8	3-6-9	4-7-10	4-8-11
Total press.	0.04	0.06	0.07	0.10	0.14	0.18	0.22	0.28	0.34
NC	—	11	14	19	]24	28	32	35	38

H-48									
Cfm	104	120	135	160	185	215	240	270	295
Horiz. proj.	2-4-5	2-4-6	3-5-7	3-6-8	4-6-9	4-7-10	5-8-12	5-9-13	6-10-14
Total press.	0.04	0.05	0.07	0.10	0.13	0.18	0.22	0.28	0.34
NC		14	17	22	27	31	35	38	41

Model 28

H-24									
Cfm	80	90	100	120	140	160	180	200	215
Horiz. proj.	2-3-5	2-3-5	2-4-6	3-5-7	3-6-8	4-7-9	4-6-10	5-8-12	5-8-12
Total press.	0.05	0.06	0.08	0.11	0.15	0.20	0.25	0.31	0.36
NC	17	21	24	29	34	38	42	45	48

H-48									
Cfm	140	155	175	210	245	280	315	350	385
Horiz. proj.	2-4-6	3-4-6	3-5-7	4-6-8	5-7-10	5-8-11	5-9-13	6-10-14	7-12-16
Total press.	0.04	0.05	0.06	0.08	0.11	0.15	0.19	0.23	0.28
NC	15	19	22	27	32	36	40	43	46

1. All pressures are in in. wg.
2. Minimum projection is to a terminal velocity of 150 fpm, middle to 100 fpm, and maximum to 50 fpm.
3. NC values are based on room absorption of 10 dB, Re 10^{-12} watts.

Source: Reprinted by permission of Environmental Elements Corporation, Dallas, Texas.

Table 11-6 Performance Data for One Type of Return Grille

A_c	Size, in.		Core Velocity fpm	200	300	400	500	600	700	800
			Velocity Pressure	0.002	0.006	0.010	0.016	0.023	0.031	0.040
			Negative Static Pressure	0.011	0.033	0.055	0.088	0.126	0.170	0.220
0.34 ft²	16 × 4, 12 × 5	10 × 6	cfm	70	100	135	170	205	240	270
			NC			13	20	25	30	33
0.39 ft²	18 × 4, 14 × 5	12 × 6, 8 × 8	cfm	80	115	155	195	235	275	310
			NC			14	21	26	31	34
0.46 ft²	20 × 4, 16 × 5	14 × 6, 10 × 8	cfm	90	140	185	230	275	320	370
			NC			15	22	27	32	35
0.52 ft²	24 × 4, 18 × 5	16 × 6	cfm	105	155	210	260	310	365	415
			NC			16	23	28	33	36
0.60 ft²	28 × 4, 20 × 5	18 × 6, 12 × 8	10 × 10	cfm	120	180	240	300	360	420	480
			NC			17	24	29	34	37
0.69 ft²	30 × 4, 24 × 5	20 × 6, 14 × 8	12 × 10	cfm	140	205	275	345	415	485	550
			NC			17	24	29	34	37
0.81 ft²	36 × 4, 28 × 5	22 × 6, 16 × 8	14 × 10	cfm	160	245	325	405	485	565	650
			NC		10	18	25	30	35	38
0.90 ft²	40 × 4, 30 × 5	26 × 6, 18 × 8	16 × 10, 12 × 12	cfm	180	270	360	450	540	630	720
			NC		11	19	26	31	36	39
1.07 ft²	48 × 4, 36 × 5	30 × 6, 18 × 10	14 × 12	cfm	215	320	430	535	640	750	855
			NC		12	20	27	32	37	40

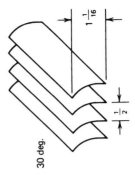

30 deg.

ft²	Dimensions						cfm / NC							
1.18 ft²	34 × 6	20 × 10	14 × 14			cfm	235	355	470	590	710	825	945
	24 × 8	16 × 12				NC	—	13	21	28	33	38	41
1.34 ft²	60 × 4	36 × 6	16 × 14			cfm	270	400	535	670	805	940	1070
	48 × 5	18 × 12				NC	—	13	21	28	33	38	41
1.60 ft²	72 × 2	24 × 10	18 × 14			cfm	320	480	640	800	960	1120	1280
	30 × 8	22 × 12	16 × 16			NC	—	14	22	29	34	39	42
1.80 ft²	60 × 5	36 × 12	24 × 12	18 × 16		cfm	360	540	720	900	1080	1260	1440
	48 × 6	30 × 10	20 × 14			NC	—	15	23	30	35	40	43
2.08 ft²	72 × 5	40 × 8	30 × 12	20 × 16		cfm	415	625	830	1040	1250	1460	1660
	60 × 6	36 × 10	24 × 14	18 × 18		NC	—	16	24	31	36	41	44
2.45 ft²	72 × 6	32 × 12	24 × 16			cfm	490	735	980	1220	1470	1720	1960
	48 × 8	26 × 14	20 × 18			NC	—	17	25	32	37	42	45
2.78 ft²	36 × 12	26 × 16	22 × 20			cfm	555	835	1110	1390	1670	1950	2220
	30 × 14	24 × 18				NC	—	18	26	33	38	43	46
3.11 ft²	60 × 8	40 × 12	30 × 16	24 × 20		cfm	620	935	1240	1560	1870	2180	2490
	40 × 10	36 × 14	26 × 18			NC	—	19	27	34	39	44	47
3.61 ft²	72 × 8	48 × 12	30 × 18			cfm	720	1080	1440	1800	2170	2530	2890
	60 × 10	36 × 16	24 × 24			NC	—	20	28	35	40	45	48

1. All pressures are in in. wg.

2. The NC values are based on a room absorption of 8 dB, Re 10^{-12} watts, and one return.

Source: Reprinted by permission of Environmental Elements Corporation, Dallas, Texas.

Table 11-7 Characteristic Room Length for Several Diffuser Types

Diffuser Type	Characteristic Length, L
High sidewall grille	Distance to wall perpendicular to jet
Circular ceiling diffuser	Distance to closest wall or intersecting jet
Sill grille	Length of room in the direction of the jet
Ceiling slot diffuser	Distance to wall or midplane between outlets

Source: Reprinted by permission from *ASHRAE Handbook, Fundamentals Volume,* 1989.

recommended ratio of throw to characteristic length that should maximize the ADPI. A range of throw-to-length ratios are also shown that should give a minimum ADPI. Note that the throw is based on a terminal velocity of 50 ft/min for all diffusers except the ceiling slot type. The general procedure for use of Table 11-8 is as follows:

1. Determine the air flow requirements and the room size.
2. Select the type of diffuser to be used.
3. Determine the room characteristic length.
4. Select the recommended throw-to-length ratio from Table 11-8.
5. Calculate the throw.
6. Select the appropriate diffuser from catalog data such as that in Tables 11-2, 11-3, 11-4, or 11-5.
7. Make sure any other specifications are met (noise, total pressure, etc.).

To illustrate the use of diffuser and grille performance data to design air distribution systems, let us consider a few examples.

EXAMPLE 11-1

The room shown in Fig. 11-10 is part of a single-story office building located in the central United States. A perimeter type of air-distribution system is used since heating will be important. The air quantity required for the room is 250 cfm. Select diffusers for the room.

SOLUTION

Diffusers of the type shown in Table 11-2 should be used for this application. Because the room has two exposed walls, an air outlet should be placed under each window in the floor near the wall (Fig. 11-10c). This will help to counteract the cold air moving downward from the window as a result of natural convection. The total air quantity is divided equally between the two diffusers. According to Table 11-1 the NC should be about 30 to 40. If we assume that the room has an 8-ft ceiling, the room characteristic length is 8 ft. Table 11-8 gives a throw-to-length ratio ranging from 1.7 to 0.9 for a straight vane diffuser. We will assume a room load of 40 Btu/(hr-ft^2), then

$$x_{50}/L = 1.3$$

Table 11-8 ADPI Selection Guide

Terminal Device	Room Load Btu/(hr-ft²)[a]	x_{50}/L for Max. ADPI	Maximum ADPI	For ADPI Greater Than	Range of x_{50}/L
High	80	1.8	68	—	—
sidewall	60	1.8	72	70	1.5–2.2
grilles	40	1.6	78	70	1.2–2.3
	20	1.5	85	80	1.0–1.9
Circular	80	0.8	76	70	0.7–1.3
ceiling	60	0.8	83	80	0.7–1.2
diffusers	40	0.8	88	80	0.5–1.5
	20	0.8	93	90	0.7–1.3
Sill grille	80	1.7	61	60	1.5–1.7
straight	60	1.7	72	70	1.4–1.7
vanes	40	1.3	86	80	1.2–1.8
	20	0.9	95	90	0.8–1.3
Sill grille	80	0.7	94	90	0.8–1.5
spread	60	0.7	94	80	0.6–1.7
vanes	40	0.7	94	—	—
	20	0.7	94	—	—
Ceiling	80	0.3[b]	85	80	0.3–0.7
slot	60	0.3[b]	88	80	0.3–0.8
diffusers	40	0.3[b]	91	80	0.3–1.1
	20	0.3[b]	92	80	0.3–1.5

[a]To convert to W/m² multiply by 3.155.

[b]x_{100}/L.

Source: Reprinted by permission from *ASHRAE Handbook, Fundamentals Volume,* 1989.

and

$$x_{50} = 1.3(8) = 10.4 \, \text{ft}$$

From Table 11-2 a 4 × 12-in. diffuser with 125 cfm has a throw, corrected for length, between

$$x_{50} = 13(0.7) = 9.1 \, \text{ft}$$

and

$$x_{50} = 17(0.7) = 11.9 \, \text{ft}$$

because 125 cfm lies between 111 cfm and 139 cfm. The NC is quite acceptable and is between 12 and 18 uncorrected for length. The total pressure required by the diffuser is between 0.036 and 0.057 in. wg and is about

$$\Delta P = (125/111)^2(0.036) = 0.046 \, \text{in. wg}$$

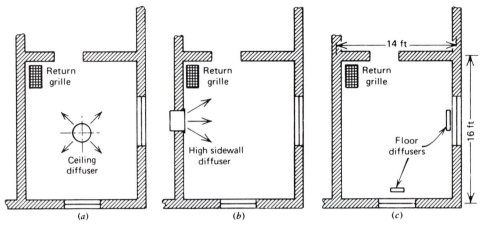

Figure 11-10   Plan view of a room showing location of different types of outlets.

An acceptable solution is listed as follows:

Size, in.	Capacity, cfm	Throw, ft	NC	ΔP_0, in. wg
4 × 12	125	10.5	5	0.046

The loss in total pressure for the diffuser is an important consideration. The value shown above would be acceptable for a light commercial system.

EXAMPLE 11-2

Suppose the room of Fig. 11-10 is located in the southern latitudes where overhead systems are recommended. Select a round ceiling diffuser system and a high sidewall system. Also select a return grille.

SOLUTION

The data of Table 11-3 with information from Tables 11-7 and 11-8 will be used to select a ceiling diffuser. The characteristic length is 7 or 8 ft and the throw-to-length ratio is 0.8; then

$$x_{50} = 0.8(7.0) = 5.6 \, \text{ft}$$

The best choice would be

Size, in.	Throw, ft	NC	ΔP_0, in. of wg
10	$7\frac{1}{2}$	10	0.035

The throw is larger than desired but the throw-to-length ratio is within the range to give a minimum ADPI of 80 percent. Figure 11-10 shows this application. A high sidewall diffuser may be selected from Table 11-4. In this case the throw-to-length ratio should be about 1.6 and the characteristic length is 14 ft; then

$$x_{50} = 1.6(14) = 22.4 \text{ ft}$$

The following units using the $22\frac{1}{2}$ degree spread would be acceptable:

Size, in.	Throw, ft	NC	ΔP_0, in. wg
16 × 4			
12 × 5	$22\frac{1}{2}$	18	0.063
10 × 6			

Figure 11-10b shows the diffuser location. It would be desirable to locate the return air intake near the floor for heating purposes and in the ceiling for cooling. However, two different intakes are not generally used except in extreme cases, and the return will be located to favor the cooling case or to accommodate the building structure. For the room shown in Figure 11-10 it will be assumed that the building design prevents practical location of the return near the floor and the return is located in the ceiling as shown. We may select the following grilles from Table 11-6:

Size, in.	NC	ΔP_0, in. wg
24 × 4		
16 × 6		
18 × 5	23	0.067
12 × 6		
8 × 8	27	0.12

EXAMPLE 11-3

Figure 11-11 shows a sketch of a recreational facility with pertinent data on ceiling height, air quantity, and building dimensions. The elevated seating rises 6 ft from the floor. The floor area and walls are not available for air outlets in the locker rooms. The structure is located in Topeka, Kansas. Select an air diffuser system for the complete structure.

SOLUTION

It would be desirable to use a perimeter type of system throughout the structure; floor area is not available in all of the spaces, however, and air motion will be enhanced in the central part of the gymnasium by an overhead system.

The entry area is subject to large infiltration loads and has a great deal of glass area. Therefore, outlets should be located in the floor around the perimeter. There is 50 ft of perimeter wall with 12 ft taken up by doors. Then about 38 ft of linear diffuser could be used if required. Noise is not a limiting factor and the throw should be about 12 ft based on the ADPI (Table 11-8). If we refer to Table 11-2, the 2-in. size has a throw of 12 ft,

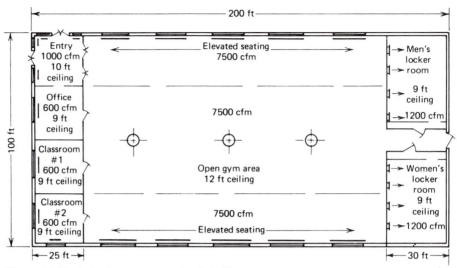

Figure 11-11 A single-story recreational facility.

total pressure loss of 0.08 in. wg, an NC of 23, and a capacity of 66 cfm/ft. The total length of the required diffusers would then be

$$L_d = \frac{1000}{66} = 15 \text{ or } 16 \text{ ft}$$

This total length should be divided into four equal sections and located as shown in Figure 11-11.

The office and classrooms should also be equipped with perimeter air inlets. The throw should be about 12 to 15 ft and a NC of about 30 would be acceptable. Referring to Table 11-2, a 3-in. size may be used with a capacity of 115 cfm/ft. The NC is 22 and the throw is 15 ft with a loss in total pressure of 0.08 in. wg. The total length of diffuser is then computed as

$$L_d = \frac{600}{115} = 5.2 \text{ or } 5 \text{ ft}$$

The total length may be divided into two sections, or a single 5-ft length will function adequately as shown in Fig. 11-11. The corner classroom should have two outlets.

The elevated seating on each side of the gym should also be equipped with perimeter upflow air outlets because of the exposed walls and glass. A throw of 10 ft would be acceptable because the seating is elevated about 6 ft. Noise is not a major factor. There is about 145 ft of exposed wall on each side and 7500 cfm is required. Therefore, a capacity of at least 52 cfm/ft is required. From Table 11-2, a 2-in. size with a capacity of 55 cfm/ft will give a throw of 10 ft with a loss in total pressure of 0.057 in. wg. The total length of diffuser is computed as

$$L_d = \frac{7500}{55} = 136.4 \text{ or } 136 \text{ ft}$$

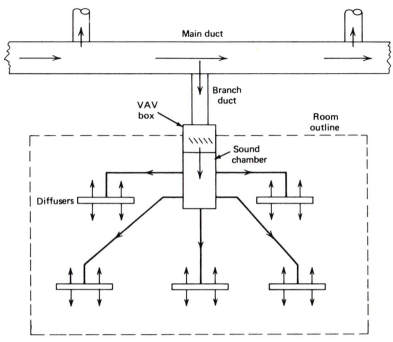

Figure 11-12 Schematic of VAV air-distribution system for a room.

The total length should be divided into at least five sections and located beneath each window as shown in Fig. 11-11.

The central portion of the gymnasium should be equipped with round ceiling diffusers. Table 11-3 has data for this type of outlet. The total floor area is divided into imaginary squares and a diffuser selected with a capacity to serve that area with a throw just sufficient to reach the boundary of the area. If the total area is divided into 12 equal squares of about 25 × 25 ft, a 12-in. diffuser in each area with a capacity of about 630 cfm would be in the acceptable range, although the throw of 15 ft is slightly large. Since this is a gym area, some sacrifice of comfort is acceptable and a more economical system will result if large diffusers are used. Imagine that the area is divided into three equal squares of about 50 × 50 ft. Then each diffuser should provide about 2500 cfm and have a throw of about 25 ft. A 24-in. size, which has a capacity of 2510 cfm and a throw of 30 ft, would be acceptable. The loss in total pressure is about 0.094 in. wg. The throw is slightly high, but is within the range given in Table 11-8. Three diffusers should be located as shown in Fig. 11-11.

The locker room areas may be equipped either with ceiling-type diffusers or with high sidewall outlets because the floor area is all covered near the walls. Since moisture will tend to collect near the ceiling and condense on ceiling diffusers, a high sidewall system will be used. If four 16 × 4 in. diffusers with capacity of 300 cfm are selected from Table 11-4, a throw of 31 ft (zero deflection) will result in a loss in total pressure of 0.085 in. with an NC of 24. This throw is less than that required by the ADPI of Table 11-8. A different diffuser type should be used in a critical situation. The diffusers should be equally spaced about 6 to 12 in. below the ceiling as shown in Fig. 11-11.

The air return grilles should all be placed in the ceiling unless the structure has a basement, which would make placement of grilles near the floor feasible if desired.

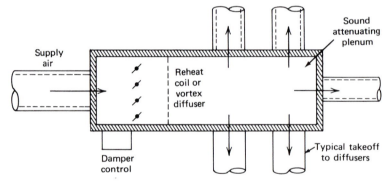

Figure 11-13 A single-duct VAV terminal box.

Because cooling and ventilation will be important factors in the gym and locker room area, a ceiling type of return air system will be utilized. The locker rooms should have a separate exhaust system to remove a total of 2400 cfm. Return grilles may be selected from Table 11-6 as follows:

No.	Size, in.	Capacity, cfm	ΔP_0, in. wg	NC	Location
1	24 × 12	900	0.07	30	Entry
1	24 × 8	590	0.07	28	Office
1	24 × 8	590	0.07	28	Classroom 1
1	24 × 8	590	0.07	28	Classroom 2
10	30 × 24	2320	0.07	36	Gym
1	24 × 16	1220	0.07	32	Men's L.R.
1	24 × 16	1220	0.07	32	Women's L.R.

It has been assumed that all of the air, except for the locker rooms, will flow back through the air return before any of it is exhausted.

Variable air volume air-distribution systems usually involve the use of linear or T-bar diffusers and a thermostat-controlled metering device, referred to as a VAV terminal box. Figure 11-12 shows how such a device is used in relation to the main air supply and the diffusers. Figure 11-13 is a photo and schematic of a typical terminal box. There are almost infinite variations in these devices depending on the manufacturer. Some are self-powered, using energy from the flowing air, whereas others use power from an external source. Many of the self-powered boxes require a relatively high static pressure and therefore are adaptable only to high-velocity systems. However, there are models available that operate with pressures compatible with low-velocity systems. The layout and selection of the diffusers follow the principles and methods previously discussed.

REFERENCES

1. Alfred Koestel, "Computing Temperature and Velocities in Vertical Jets of Hot or Cold Air," *ASHVE Transactions,* Vol. 60, 1954.
2. *ASHRAE Handbook, Fundamentals Volume,* American Society of Heating, Refrigerating and Air-Conditioning Engineers, Inc., Atlanta, GA, 1989.
3. *ASHRAE Handbook, HVAC Applications,* American Society of Heating, Refrigerating and Air-Conditioning Engineers, Inc., Atlanta, GA 1991.
4. J. Rydberg and P. Norback, "ASHVE Research Report No. 1362—Air Distribution and Draft," *ASHVE Transactions,* Vol. 65, 1949.

PROBLEMS

11-1. A free isothermal jet is discharged horizontally from a circular opening. There is no nearby surface. The initial velocity and volume flow rate are 700 ft/min and 200 cfm, respectively. Compute (a) the throw for terminal velocities of 50, 100, and 150 ft/min and (b) the total volume flow rate of the jet for each terminal velocity in part (a).

11-2. Air is discharged from a circular pipe 100 mm in diameter to an open space. The discharge velocity is 5 m/s. Assuming standard air, compute (a) the throw for terminal velocities of 0.25 and 0.5 m/s and (b) the total volume flow rate of the air at each terminal velocity in part (a).

11-3. A free jet is discharged horizontally below a ceiling. The initial velocity and volume flow rate are 1000 ft/min and 300 cfm. The initial jet temperature is 100 F, whereas the room is to be maintained at 75 F. If the coefficient K in Eq. 11-1a has a value of 9, compute the throw and the difference in temperature between the center line of the jet and the room at terminal velocities of 50, 100, and 150 ft/min.

11-4. Air at 49 C is discharged into a room at 20 C with a velocity of 4 m/s. Estimate the stream temperature where the jet velocities are 0.25 and 0.5 m/s.

11-5. A free isothermal jet of 100 cfm, 6-in. diameter is discharged vertically from the floor toward a ventilation hood 10 ft above the floor. Approximately what capacity must the hood exhaust fan have to capture all of the air stream at the entrance to the hood?

11-6. To ventilate a space, it is desired to discharge free isothermal jets vertically downward from a ceiling 30 ft above the floor. The terminal velocity of the jets should be no more than 50 ft/min, 6 ft above the floor. Determine a reasonable diameter and initial volume flow rate for each jet (D_o < 12 in.).

11-7. Suppose a given space requires a very large quantity of circulated air for cooling purposes. What type of diffuser system would be best? Why?

11-8. A space has a low but essentially constant occupancy with a moderate cooling load. What type of air diffuser system would be best for heating and cooling? Explain. Assume that the space is on the ground floor.

11-9. Consider a single-story structure with many windows. What would be the best all-around air-distribution system for (a) the northern part of the United States and (b) the southern states? Explain.

11-10. Consider a relatively large open space with a small cooling load and low occupancy located in the southern part of the United States. What type of air-distribution system would be best? Explain.

11-11. A linear floor diffuser is required for a space with an air supply rate of 480 cfm. The room has a 12-ft ceiling and cooling load of 40 Btu/(hr-ft^2). (a) Select a diffuser from Table 11-2 for this application. (b) Determine the total pressure and NC for your selection.

11-12. Suppose a round ceiling diffuser is to be used in the situation described in Problem 11-11. The room has plan dimensions of 16 × 18 ft. (a) Select a diffuser from Table 11-3 for this application. (b) Determine the total pressure and NC for the diffuser.

11-13. Assume that two high sidewall diffusers are to be used for the room described in Problems 11-11 and 11-12, and they are to be installed in the wall with the longest dimension. (a) Select suitable diffusers from Table 11-4. (b) Determine the total pressure and NC for the diffusers.

11-14. Select a suitable return grille from Table 11-6 for the room described in Problem 11-11. Total pressure for the grille should be less than 0.10 in. wg and one dimension should be 12. in.

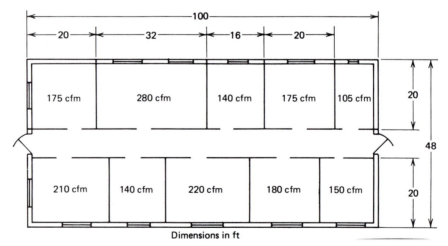

Figure 11-14 Floor plan for Problem 11-15.

11-15. Select a perimeter-type diffuser system for the building shown in Fig. 11-14. It is general office space. Dimensions given are in feet.

11-16. Select a round ceiling diffuser system for the building in Problem 11-15.

11-17. Select a high sidewall diffuser system for the building in Problem 11-15.

11-18. Select return air grilles for the building in Problem 11-15. Assume that the return system must be placed in the attic and each room must have a return.

11-19. Select an air-distribution system and diffusers and grilles for the structure described by plans and specifications furnished by the instructor.

11-20. Consider a room with a 20-ft exposed wall that has two windows. The other dimension is 30 ft. The room is part of a variable air volume system. (a) Lay out and select T-bar diffusers from Table 11-5 if the room requires a total air quantity of 800 cfm and the maximum total pressure available is 0.10 in. wg. (b) Note the total pressure, the throw to where the maximum velocity has decreased to 100 ft/min, and the NC for each diffuser.

11-21. Consider a 18 × 30 ft room in the southwest corner of a zone. There are windows on both exterior walls and the peak air quantity for the room is 1000 cfm. (a) Lay out and select T-bar diffusers from Table 11-5 using a maximum total pressure of 0.15 in. wg. (b) Note the total pressure, the throw to where the maximum velocity has decreased to 100 ft/min, and the NC for each diffuser.

11-22. Select perimeter-type diffusers for the room shown in Fig. 11-15. The perimeter distribution is for the heating system that is secondary to the VAV cooling system. The perimeter system requires 1800 cfm (850 L/s) evenly distributed along the exterior walls. Locate the diffusers on the floor plan. Limit the total pressure to 0.10 in. wg.

11-23. Select round ceiling diffusers for the room shown in Fig. 11-15. The room has a cooling load of 80,000 Btu/hr (23.5 kW) and a design air supply rate of 3000

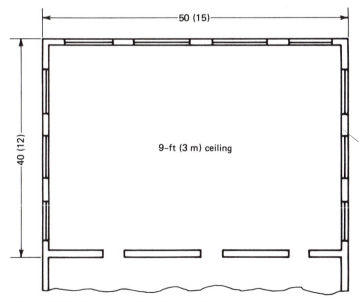

Figure 11-15 Floor plan for a large office space.

cfm (1416 L/s). Locate the diffusers on the floor plan. A maximum total pressure of 0.12 in. wg is allowed.

11-24. Select T-bar (slot) type diffusers for the room of Fig. 11-15 and locate them on the floor plan. The cooling load is 90,000 Btu/hr (26.4 kW) and the design air supply rate is 3200 cfm (1500 L/s). The maximum allowable total pressure is 0.10 in. wg.

11-25. Select and locate a return grille(s) for the room of Problem 11-23. A quiet system is desirable.

11-26. Select and locate return grilles for the room of Problem 11-24. Limit the NC to less than 30.

Chapter 12

Fans and Building Air Distribution

The previous chapter considered the distribution and movement of the air within the conditioned space. It assumed optimum location of diffusers and that the proper amount of air at the required total pressure was delivered to each diffuser. This chapter discusses the details of distributing the air to the various spaces in the structure. Proper design of the duct system and the selection of appropriate fans and accessories are essential. A poorly designed system may be noisy or inefficient or lead to discomfort of occupants. Correction of a poorly designed duct system is expensive and sometimes practically impossible. In some HVAC systems the energy costs for moving the air approaches that of the chiller system.

12-1 FANS

The fan is an essential component of almost all heating and air-conditioning systems. Except in those cases where free convection creates air motion, a fan is used to move air through ducts and to induce air motion in the space. An understanding of the fan and its performance is necessary if one is to design a satisfactory duct system (1, 2).

The *centrifugal fan* is the most widely used because it can efficiently move large or small quantities of air over a wide range of pressures. The principle of operation is similar to the centrifugal pump in that a rotating impeller mounted inside a scroll type of housing imparts energy to the air or gas being moved. Figure 12-1 shows the various components of a centrifugal fan. The impeller blades may be forward-curved, backward-curved, or radial. The blade design influences the fan characteristics and will be considered later.

The *vaneaxial fan* is mounted on the center line of the duct and produces an axial flow of the air. Guide vanes are provided before and after the wheel to reduce rotation of the air stream. The *tubeaxial fan* is quite similar to the vaneaxial fan but does not have the guide vanes. Figure 12-2 illustrates both types.

Axial flow fans are not capable of producing pressures as high as those of the centrifugal fan but can move large quantities of air at low pressure. Axial flow fans generally produce higher noise levels than centrifugal fans.

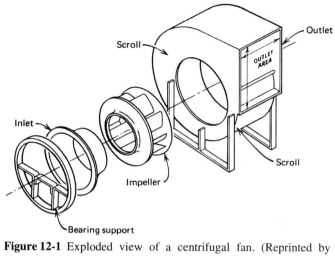

Figure 12-1 Exploded view of a centrifugal fan. (Reprinted by permission from *ASHRAE Handbook, Systems and Equipment Volume,* 1992.)

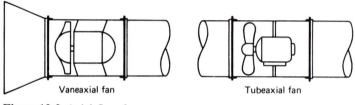

Figure 12-2 Axial flow fans.

12-2 FAN PERFORMANCE

The performance of fans is generally given in the form of a graph showing pressure, efficiency, and power as a function of capacity. The energy transferred to the air by the impeller results in an increase in static and velocity pressure; the sum of the two pressures gives the total pressure. Although these quantities are pressure, they are often expressed in inches or millimeters of water. When Eq. 10-1c is applied to a fan with elevation effects neglected and constant density assumed, the following result is obtained:

$$\frac{g_c w}{g} = \frac{g_c}{g}\left(\frac{P_1}{\rho} - \frac{P_2}{\rho}\right) + \frac{1}{2g}(\overline{V}_1^2 - \overline{V}_2^2) = \frac{g_c}{g}\frac{(P_{01} - P_{02})}{\rho} \qquad \textbf{(12-1a)}$$

In this form the equation expresses the increase in total head of the air. Multiplying Eq. 12-1a by g/g_c gives

$$w = \frac{P_{01} - P_{02}}{\rho} \qquad \textbf{(12-1b)}$$

which is an expression for the energy imparted to the air per unit mass. Multiplication of Eq. 12-1b by the mass flow rate of the air produces an expression for the *total power* imparted to the air:

$$\dot{W}_t = \frac{\dot{m}(P_{01} - P_{02})}{\rho} \tag{12-2}$$

The *static power* is the part of the total power that is used to produce the change in static pressure:

$$\dot{W}_s = \frac{\dot{m}(P_1 - P_2)}{\rho} = \dot{Q}(P_1 - P_2) \tag{12-3}$$

where

$\dot{Q}$ = volume flow rate, ft^3/min or m^3/s

Fan efficiency may be expressed in two ways. The total fan efficiency is the ratio of total air power $\dot{W}_1$ to the shaft power input $\dot{W}_{sh}$:

$$\eta_t = \frac{\dot{W}_t}{\dot{W}_{sh}} = \frac{\dot{m}(P_{01} - P_{02})}{\rho\dot{W}_{sh}} = \dot{Q}\frac{(P_{01} - P_{02})}{\dot{W}_{sh}} \tag{12-4a}$$

It has been common practice in the United States for $\dot{Q}$ to be in ft^3/min, $(P_{01} - P_{02})$ to be in in. wg, and $\dot{W}_{sh}$ to be in horsepower. In this special case

$$\eta_t = \frac{\dot{Q}(P_{01} - P_{02})}{6350\,\dot{W}_{sh}} \tag{12-4b}$$

The *static fan efficiency* is the ratio of the static air power to the shaft power input:

$$\eta_s = \frac{\dot{W}_s}{\dot{W}_{sh}} = \frac{\dot{m}(P_1 - P_2)}{\rho\dot{W}_{sh}} = \frac{\dot{Q}(P_1 - P_2)}{\dot{W}_{sh}} \tag{12-5a}$$

Using the units of Eq. 12-4b, we get

$$\eta_s = \frac{\dot{Q}(P_1 - P_2)}{6350\dot{W}_{sh}} \tag{12-5a}$$

Figures 12-3, 12-4, 12-5, and 12-6 illustrate typical performance curves for centrifugal and vaneaxial fans. Note the difference in the pressure characteristics for the different types of blade. Also note the point of maximum efficiency with respect to the point of maximum pressure. The power characteristics of vaneaxial fans are distinctly different from centrifugal fans. Note that the power increases as the flow rate approaches zero for a vaneaxial fan, which is opposite to that of a centrifugal fan. Otherwise the general behavior is similar. Emphasis will be given to the centrifugal fan in discussion to follow with comments related to vaneaxial fans when appropriate. Fan characteristics are discussed in greater detail later.

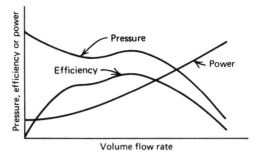

Figure 12-3 Forward-tip fan characteristics.

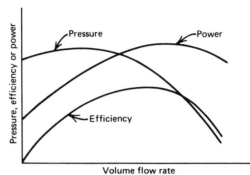

Figure 12-4 Backward-tip fan characteristics.

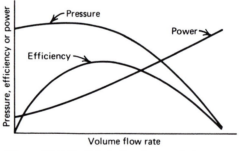

Figure 12-5 Radial-tip fan characteristics.

A conventional representation of fan performance is shown in Fig. 12-7 for a specific backward-curved blade fan. In this case total pressure and total efficiency are also given. Note that the zone for desired application is marked. When data from this zone are plotted on a logarithmic scale, the curves appear as shown in Fig. 12-8. This plot has some advantages over the conventional representation (3). Many different fan speeds can be conveniently shown, and the system characteristic is a straight line parallel to the efficiency lines.

Table 12-1 compares some of the more important characteristics of centrifugal fans. The noise emitted by a fan is of great importance in many applications. For a given

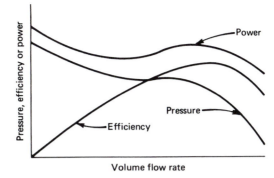

Figure 12-6 Vaneaxial fan characteristics.

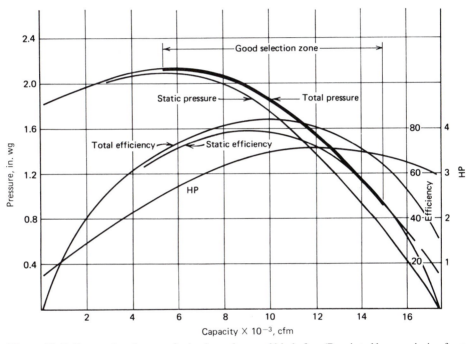

Figure 12-7 Conventional curves for backward-curved blade fan. (Reprinted by permission from
ASHRAE Journal, Vol. 14, Part I, No. 1, 1972.)

pressure the noise level is proportional to the tip speed of the impeller and to the air
velocity leaving the wheel. Furthermore, fan noise is roughly proportional to the
pressure developed regardless of the blade type. However, backward-curved fan blades
are generally considered to have the better (lower) noise characteristics.

The pressure developed by a fan is limited by the maximum allowable speed. If noise
is not a factor, the straight radial blade is superior. Fans may be operated in series to
develop higher pressures, and multistage fans are also constructed. However, difficul-
ties may arise when fans are used in parallel. Surging back and forth between fans may

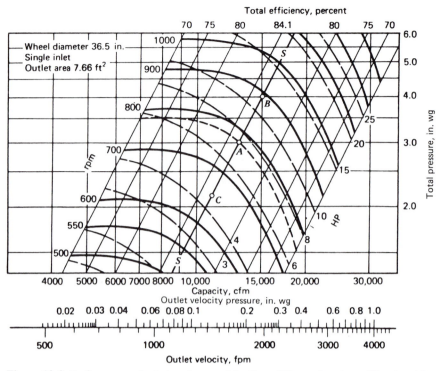

Figure 12-8 Performance chart showing combination of fan and system. (Reprinted by permission from *ASHRAE Journal,* Vol. 14, Part I, No. 1, 1972.)

Table 12-1 Comparison of Centrifugal Fan Types

Item	Forward-curved Blades	Radial Blades	Backward-curved Blades
Efficiency	Medium	Medium	High
Space required	Small	Medium	Medium
Speed for given pressure rise	Low	Medium	High
Noise	Fair	Poor	Good

develop, particularly if the system demand is changing. Forward-curved blades are particularly unstable when operated at the point of maximum efficiency.

Combining both the system and fan characteristics on one plot is very useful in matching a fan to a system and ensuring fan operation at the desired conditions. Figure 12-9 illustrates the desired operating range for a forward-curved blade fan. The range is to the right of the point of maximum efficiency. The backward-curved blade fan has a selection range that brackets the range of maximum efficiency and is not so critical to the point of operation; however, this type should always be operated to the right of the point of maximum pressure. Figure 12-8 shows the system characteristic using the logarithmic plot. The system characteristic is line *S–S.* For a given system the efficiency

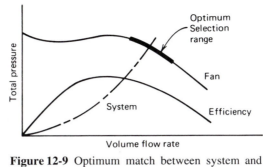

Figure 12-9 Optimum match between system and forward-curved blade fan.

does not change with speed; however, capacity, total pressure, and power all depend on the speed. Changing the fan speed will not change the relative point of intersection between the system and fan characteristic. This can be done only by changing fans.

Figure 12-10 shows fan characteristics for a forward-curved blade fan using SI units, except that capacity is in m^3/min instead of m^3/s.

There are several simple relationships between fan capacity, pressure, speed, and power, which are referred to as the *fan laws*. The first three fan laws are the most useful and are stated as follows:

1. The capacity is directly proportional to the fan speed.
2. The pressure (static, total, or velocity) is proportional to the square of the fan speed.
3. The power required is proportional to the cube of the fan speed.

The last three fan laws are of less utility but are nevertheless useful.

4. The pressure and power are proportional to the density of the air at constant speed and capacity.
5. Speed, capacity, and power are inversely proportional to the square root of the density at constant pressure.
6. The capacity, speed, and pressure are inversely proportional to the density, and the power is inversely proportional to the square of the density at a constant mass flow rate.

EXAMPLE 12-1

A centrifugal fan is operating as shown in Fig. 12-11 at point 1. Estimate the capacity, total pressure, and power requirement when the speed is increased to 1050 rpm. The initial power requirement is 2 hp.

SOLUTION

The first three fan laws may be used to estimate the new capacity, total pressure, and power.

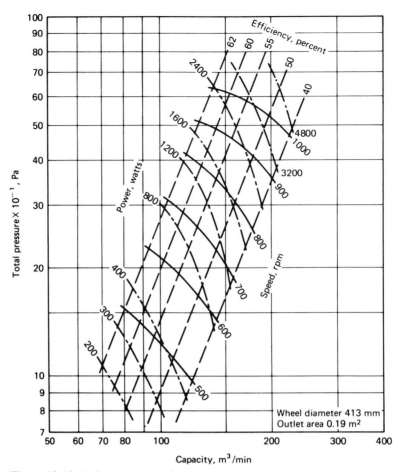

Figure 12-10 Performance data for a forward-curved blade fan.

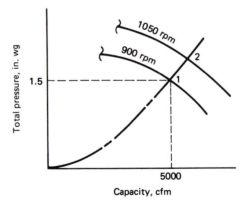

Figure 12-11 Fan and system characteristics for Example 12-1.

Capacity:

$$\frac{\dot{Q}_1}{\dot{Q}_2} = \frac{\text{rpm}_1}{\text{rpm}_2}$$

or

$$\dot{Q}_2 = \dot{Q}_1 \frac{\text{rpm}_2}{\text{rpm}_1} = 5000 \left(\frac{1050}{900}\right) = 5833 \text{ ft}^3/\text{min.} \quad \text{or} \quad \text{cfm}$$

Total pressure:

$$\frac{P_{01}}{P_{02}} = \left(\frac{\text{rpm}_1}{\text{rpm}_2}\right)^2$$

$$P_{02} = P_{01} \left(\frac{\text{rpm}_2}{\text{rpm}_1}\right)^2 = 1.5 \left(\frac{1050}{900}\right)^2 = 2.04 \text{ in. wg}$$

Power:

$$\frac{\dot{W}_1}{\dot{W}_2} = \left(\frac{\text{rpm}_1}{\text{rpm}_2}\right)^3$$

$$\dot{W}_2 = \dot{W}_1 \left(\frac{\text{rpm}_2}{\text{rpm}_1}\right)^3 = 2 \left(\frac{1050}{900}\right)^3 = 3.2 \text{ hp}$$

12-3   FAN SELECTION

It is not practical to satisfy all of the many different fan applications with one fan design. Therefore, the engineer is faced with selecting the right fan for an application. The following discussion explains the specific characteristics of different types of fans as they relate to their selection. Radial-bladed fans are not usually used in heating, ventilating, and air-conditioning systems and are not discussed.

Backward-curved Blade Fans

This type is used for general heating, ventilating, and air-conditioning systems, especially where system size permits significant horsepower savings. Such fans can be used in low-, medium-, and high-pressure HVAC systems. These are the highest efficiency designs of all centrifugal fan types. For a given duty, these fans will operate at the highest speed of the different centrifugal fans.

The performance curve is stable, and this type of fan has a load-limiting horsepower characteristic (Fig. 12-7). The horsepower curve reaches a maximum near the peak efficiency area and becomes lower for free delivery. Since there is a maximum horsepower requirement, if the fan is equipped with a motor of such size that this maximum requirement is met, there will be no danger of overloading the motor.

These fans are also used in industrial applications where power savings will be significant. The airfoil type blade must be used only in those applications where the air is clean and the airfoil blade is not subject to erosion or corrosion.

Forward-curved Blade Fan

This type of fan is usually used in low-pressure heating, ventilation, and air-conditioning applications, such as domestic furnaces, central station units, and packaged air-conditioning equipment. This design tends to have the lowest efficiency of the centrifugal fans and will operate at the lowest speed of the various centrifugal fans.

The pressure curve is less steep than that of the other designs. There is a dip in the pressure curve to the left of peak pressure and the highest efficiency occurs just to the right of peak pressure. The fan should be rated well to the right of the peak pressure point. The horsepower curve rises continually toward free delivery, and this must be taken into account when the fan is applied and the motor is selected.

Vaneaxial Fan

This type of fan is becoming more commonly used in heating, ventilating, and air-conditioning systems in low-, medium-, and high-pressure applications and is particularly advantageous where straight-through flow is required.

Vaneaxial fans usually have blades of airfoil design, which permits medium- to high-pressure capability at relatively high efficiency.

The performance curve (Fig. 12-6) shows the highest pressure characteristics of the axial designs at medium volume flow rate. The performance curve includes a break to the left of peak pressure, which is caused by dynamic stall. Application on this part of the curve should be avoided.

Some fans of this design have the capability of changing the pitch of the blade to meet different application requirements. In some cases this is accomplished by shutting the fan down and changing the blade angle to a new position and restarting the fan. In other cases, the pitch of the fan blade can be changed with the fan in operation. This latter method provides good control characteristics for the fan in VAV systems.

Performance Data

To select a fan for a given system it is necessary to know the capacity and total pressure requirement of the system. To assist in the actual fan selection, manufacturers furnish graphs such as those shown in Figs. 12-7 and 12-8 with the areas of preferred operation shown. The static pressure is often given but not the total pressure. The total pressure may be computed from the capacity and the fan outlet dimensions. Data pertaining to noise are also available from most manufacturers and are generally similar to those discussed in Chapter 11.

In many cases manufacturers present their fan performance data in the form of tables. Tables 12-2*a* and 12-2*b* are examples of such data for two forward-curved blade fans. Note that the static pressure is given instead of the total pressure; however, the outlet velocity is given, which makes it convenient to calculate the velocity pressure to find the total pressure.

It is important that the fan be quiet. Generally a fan will generate the least noise when operated near the peak efficiency. Operation considerably beyond the point of maximum efficiency will be noisy. Forward-curved blades operated at high speeds will be noisy and straight blades are generally noisy, especially at high speed. Backward-

Table 12-2a Pressure-Capacity Table for a Forward-Curved Blade Fan

Volume Flow Rate cfm	Outlet Velocity ft/min	$\frac{1}{4}$ in. wg[a]		$\frac{5}{8}$ in. wg		$\frac{3}{4}$ in. wg		1 in. wg		$1\frac{1}{4}$ in. wg		$1\frac{1}{2}$ in. wg	
		rpm	bhp[b]	rpm	bhp	rpm	bhp	rpm	bhp	rpm	bhp	rpm	bhp
851	1200	848	0.13	933	0.16	1018	0.19	—	—	—	—	—	—
922	1300	866	0.15	945	0.18	1019	0.21	—	—	—	—	—	—
993	1400	884	0.17	957	0.20	1030	0.23	1175	0.30	—	—	—	—
1064	1500	901	0.19	973	0.22	1039	0.26	1182	0.32	—	—	—	—
1134	1600	926	0.22	997	0.24	1057	0.29	1190	0.35	1320	0.43	1436	0.55
1205	1700	954	0.25	1020	0.27	1078	0.31	1200	0.38	1325	0.46	1440	0.59
1276	1800	983	0.28	1044	0.31	1100	0.34	1210	0.42	1330	0.50	1447	0.63
1347	1900	1011	0.31	1068	0.35	1126	0.38	1230	0.46	1341	0.54	1458	0.66
1418	2000	1039	0.35	1092	0.39	1152	0.42	1250	0.50	1352	0.59	1470	0.72
1489	2100	1068	0.39	1115	0.43	1178	0.47	1275	0.54	1370	0.62	1482	0.77
1560	2200	1096	0.44	1147	0.47	1204	0.51	1300	0.59	1390	0.67	1500	0.83
1631	2300	1124	0.48	1179	0.52	1230	0.56	1325	0.64	1420	0.73	1525	0.88
1702	2400	1152	0.53	1210	0.58	1256	0.62	1350	0.70	1448	0.78		

[a]Static pressure.

[b]bhp = shaft power in horsepower.

Note. Data are for a 9-in. wheel diameter and an outlet of 0.71 ft.^{2}

Table 12-2b Pressure-Capacity Table for a Forward-Curved Blade Fan

Volume Flow Rate m³/s	Outlet Velocity m/s	0.7 kPa[a]		0.8 kPa		0.9 kPa		1.0 kPa		1.1 kPa		1.2 kPa	
		rpm	kW	rpm	kW	rpm	kW	rpm	kW	rpm	kW	rpm	kW
3.35	7	692	4.33										
3.83	8	688	4.79										
4.32	9	679	5.20	737	5.44	778	6.90	825	7.68				
4.78	10	664	5.48	732	6.06	770	7.46	819	8.43				
5.27	11	654	5.82	721	6.48	755	7.98	808	9.02	864	9.47	900	11.2
5.75	12	656	6.38	704	6.82	743	8.43	790	9.47	855	10.1	887	11.7
6.23	13	663	7.12	699	7.31	741	8.87	781	9.84	840	10.5	871	12.3
6.72	14	674	7.90	702	7.98	747	9.62	781	10.6	825	11.0	855	12.7
7.18	15	686	8.95	710	8.72	755	10.7	787	11.6	817	11.6	853	13.5
7.67	16	702	10.1	720	9.77	765	11.6	797	12.6	820	12.5	860	14.5
8.13	17			733	10.8	778	12.9	808	13.9	828	13.6	869	15.8
8.62	18			748	12.0	793	14.3	822	15.4	839	14.8	880	17.3
9.10	19									851	16.3	891	18.9

[a]Static pressure.

[b]Outlet area = 0.479 m². Wheel diameter = 660 mm. Tip speed = rpm × 2.07 m/s.

curved blades may be operated on both sides of the peak efficiency at relatively high speeds with less noise than the other types of fans.

EXAMPLE 12-2

Comment on the suitability of using the fan described by Fig. 12-8 to move 15,000 cfm at 3. 5 in. wg total pressure. Estimate the speed and power requirement.

SOLUTION

Examination of Fig. 12-8 shows that the fan would be quite suitable. The operating point would be just to the right of the point of maximum efficiency and the fan speed is between 800 and 900 rpm. Therefore, the fan will operate in a relatively quiet manner. The speed and power required may be estimated directly from the graph as 830 rpm and 9.5 hp, respectively.

EXAMPLE 12-3

Determine whether the fan given in Table 12-2a is suitable for use with a system requiring 1250 cfm at 1.8 in. wg total pressure.

SOLUTION

There is a possibility that the fan could be used. At 1250 cfm the outlet velocity is about 1750 ft/min. To utilize velocity in ft/min and to give the velocity pressure in in. wg

$$P_v = \frac{\overline{V}^2}{2g} = \frac{\overline{V}}{2g} \frac{\rho_a}{\rho_w} \frac{12}{3600} = \left(\frac{\overline{V}}{4005}\right)^2$$

where

$\overline{V}$ = average velocity, ft/min
P_v = velocity pressure, in. wg

Then

$$P_v = \left(\frac{1750}{4005}\right)^2 = 0.19 \text{ in. wg}$$

The static pressure at the fan outlet is then computed as

$$P_s = P_t - P_v = 1.8 - 0.19 = 1.61 \text{ in. wg}$$

From Table 12-2a the maximum static pressure recommended for the fan is 1.5 in. wg and the speed is relatively high even at that condition. If the fan speed were to be further increased, fan noise would probably be unacceptable. A larger fan should be selected. The fan shown in Table 12-2a would be suitable for use with 1250 cfm and total pressures of 0.75 to 1 in. wg. The speed would range from about 975 to 1100 rpm.

EXAMPLE 12-4

A duct system requires a fan that will deliver 6 m³/s of air at 1.2 kPa total pressure. Is the fan of Table 12-2b suitable? If so, determine the speed, shaft power, and total efficiency.

SOLUTION

The required volume flow rate falls between 5.75 and 6.23 m³/s in the left-hand column of Table 12-2b. The corresponding outlet velocities are 12 and 13 m/s and the velocity pressure for each case is

$$(P_v)_{5.75} = \rho_a \frac{\overline{V}^2}{2} = 1.2 \frac{(12)^2}{2} = 86.4 \, \text{Pa}$$

$$(P_v)_{6.23} = \frac{1.2(13)^2}{2} = 101.4 \, \text{Pa}$$

Moving across the table to the 1.1 kPa static pressure column, the total pressure at 5.75 m³/s is

$$(P_0)_{5.75} = 1100 + 86.4 = 1186.4 \, \text{Pa}$$

and at 6.23 m³/s

$$(P_0)_{6.23} = 1100 + 101.4 = 1201.4 \, \text{Pa}$$

By interpolation the total pressure at 6 m³/s is

$$(P_0)_{6.0} = 1186.4 + \frac{6 - 5.75}{6.23 - 5.75}(1201.4 - 1186.4)$$

$$= 1190 \, \text{Pa} \quad \text{or} \quad 1.19 \, \text{kPa}$$

Although the total pressure at 6 m³/s is barely adequate, the fan speed can be increased to obtain total pressures up to almost 1.3 kPa at a capacity of 5.75 to 6.23 m³/s.
The fan speed may be determined by interpolation to be

$$\text{rpm} = 840 - \frac{6 - 5.75}{6.23 - 5.75}(840 - 825) = 832$$

and the shaft power is likewise found to be

$$\dot{W}_{sh} = 10.5 + \frac{6 - 5.75}{6.23 - 5.75}(0.5) = 10.76 \, \text{kW}$$

The total power imparted to the air is given by Eq. 12-2:

$$\dot{W}_t = \frac{\dot{m}}{\rho}(P_{01} - P_{02}) = \dot{Q}(P_{01} - P_{02}) \qquad \text{(12-2)}$$

where $\dot{Q}$ is in m^3/s, $(P_{01} - P_{02})$ is in N/m^2 or Pa, and W_t is in watts. Then

$$\dot{W}_t = (6)(1.2)(1000)/(1000) = 7.2\,kW$$

The total efficiency is then given by

$$\eta_t = \frac{\dot{W}_t}{\dot{W}_{sh}} = 7.2/10.76 = 0.67$$

12-4   FAN INSTALLATION

The performance of a fan can be reduced drastically by improper connection to the duct system. In general, the duct connections should be such that the air may enter and leave the fan as uniformly as possible with no abrupt changes in direction or velocity. The designer must rely on good judgment and ingenuity in laying out the system. Space is often limited for the fan installation, and a less than optimum connection may have to be used. In this case the designer must be aware of the penalties (loss in total pressure and efficiency).

If a fan and system combination does not seem to be operating at the volume flow rate and pressure specified, the difficulty may be that the system is not the same as specified in the design. The point of operation will not be at the design point of rating on the fan curve. In Fig. 12-12 point B is the specified point of operation, but tests may show that the actual point of operation is point A. The important thing to notice in this case is that the difference between the specified point of rating is due to a change in the system characteristic curve and not the fan. The fan curve is in its original position and the problem is simply to get the system characteristic curve to cross the fan curve at the appropriate point.

An entirely different change in the point of operation between the fan and the system can come about by an actual change in the fan performance curve. For example, fans are designed for uniform and straight flow into the inlet of the fan. If this flow is upset in any way, the fan will not perform on the original performance curve but will produce a new performance curve. This change in fan performance is discussed next.

The Air Movement and Control Association, Inc. (AMCA), has published system effect factors in their Fan Application Manual (4), which express the effect of various fan connections on system performance. These factors are in the form of total pressure loss that is added to the computed system total pressure loss.

System Effect Factors

The total pressure requirements of a fan are calculated by methods discussed in Section 12-7 and are the result of pressure losses in ductwork, fittings, heating and cooling coils, dampers, filters, process equipment, and similar sources. All of these sources of pressure loss are based on uniform velocity profiles. The velocity profile at the fan outlet

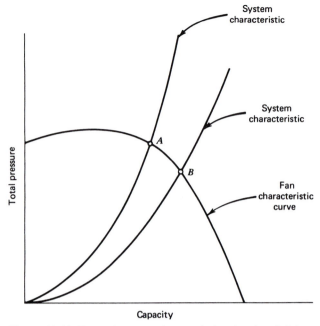

Figure 12-12 Fan and system characteristics showing deficient operation.

is not uniform, and fittings at or near the fan outlet will develop pressure losses greater than the rated value.

Fans are normally tested with open inlets and the flow to the wheel is uniform. In actual installations many other inlet configurations are encountered, and these will adversely affect the performance. This effect on fan performance is in addition to the usual pressure loss of ductwork, fittings, and equipment normally computed.

In order to properly rate the fan, the inlet and outlet effects must be taken into account and the pressure requirements of the fan as normally calculated must be increased. These effects, identified as *system effect factors,* may be estimated by using the procedure outlined next.

Fan Outlet Condition

As shown in Fig. 12-13, the outlet velocity profiles of fans are not uniformly distributed across the outlet duct until the air has traveled through a certain length of the duct. This length is identified as *one effective duct length*. To make best use of energy developed by the fan, this length of duct should be provided at the fan outlet. Preferably, the outlet duct should be the same size as the fan outlet, but good flow can be obtained if the duct is not greater in area than about 110 percent nor less in area than about 85 percent of the fan outlet. The slope of transition elements should not be greater than 15 degrees for the converging elements nor greater than 7 degrees for the diverging elements.

One effective duct length is a function of fan outlet velocity as shown in Table 12-3. If the duct is rectangular, the equivalent duct diameter is given by

$$D = (4 \times H \times W/\pi)^{1/2} \tag{12-6}$$

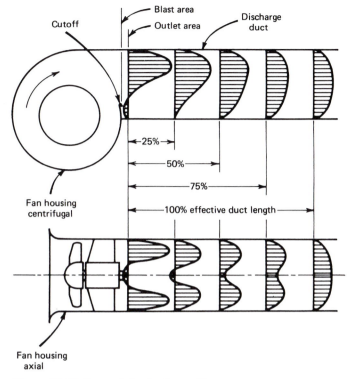

Figure 12-13 Fan-outlet velocity profiles.

Table 12-3 Effective Duct Length

Duct Velocity		One Effective Duct Length
fpm	m/s	Duct Diameters
2500	12.5	2.5
3000	15.0	3.0
4000	20.0	4.0
5000	25.0	5.0
6000	30.0	6.0
7000	35.0	7.0
8000	40.0	8.0

where

D = equivalent duct diameter, ft or m
H = rectangular duct height, ft or m
W = rectangular duct width, ft or m

In those cases where a shorter discharge duct length is used, an additional pressure loss will occur, and this additional pressure must be added to the fan total pressure requirements. The additional pressure loss is calculated by

$$\Delta P_0 = C_0 \times P_v \tag{12-7}$$

and

$$P_v = \rho(\overline{V}/k)^2 \tag{12-8}$$

where

$\Delta P_0 =$ pressure loss, in. wg or Pa
$P_v =$ velocity pressure, in. wg or Pa
$\rho =$ air density, lbm/ft³ or kg/m³
$\overline{V} =$ velocity at outlet plane, ft/min or m/s
$k =$ constant, 1097 for English units or 1.414 for SI
$C_0 =$ loss coefficient based on discharge duct area

The *blast area,* shown in Fig. 12-13, is smaller than the outlet area due to the *cutoff.* The blast area ratio used in determining loss coefficients is defined as

Blast area ratio = Blast area/outlet area

The blast area should be obtained from the fan manufacturer for the particular fan being considered. For estimating purposes values of blast area ratio are given in Table 12-4.

Table 12-5 gives loss coefficients for the case of a fan discharging into a plenum. Note that at least 50 percent effective duct length is required for best fan performance.

To obtain the rated performance from the fan, the first elbow fitting should be at least one effective duct length from the fan outlet (Fig. 12-14). If this length cannot be provided, an additional pressure loss will result and must be added to the fan total pressure requirements. The additional pressure loss may be determined from Eq. 12-7 with a loss coefficient from Table 12-6.

The coefficients in Table 12-6 are for single-wheel-single-inlet (SWSI) fans. For double-wheel-double-inlet (DWDI) fans, apply multipliers of 1.25 for position B, 0.85 for position *D*, and 1.0 for positions *A* and *C*.

There are other types of fittings that have an effect similar to the outlet elbow that are

Table 12-4 Blast Area Ratios

Fan Type		Blast Area Ratio
Centrifugal		
Backward-curved		0.70
Radial		0.80
Forward-curved		0.90
Axial		
Hub Ratio	0.3	0.90
	0.4	0.85
	0.5	0.75
	0.6	0.65
	0.7	0.50

Table 12-5 Loss Coefficients, Centrifugal Fan Discharging Into a Plenum

$L/L_e \rightarrow$ A_b/A_o	C_o				
	0.00	0.12	0.25	0.50	1.00
0.4	2.00	1.00	0.40	0.18	0.00
0.5	2.00	1.00	0.40	0.18	0.00
0.6	1.00	0.67	0.33	0.14	0.00
0.7	0.80	0.40	0.14	0.00	0.00
0.8	0.47	0.22	0.10	0.00	0.00
0.9	0.22	0.14	0.00	0.00	0.00
1.0	0.00	0.00	0.00	0.00	0.00

Source: Reprinted by permission from *ASHRAE Duct Fitting Database,* 1992.

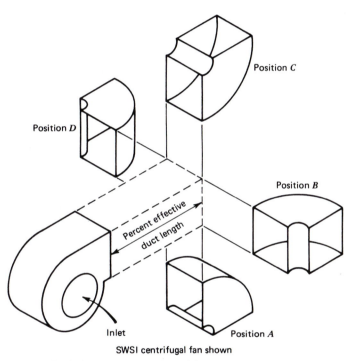

Figure 12-14 Outlet-duct elbow positions.

not covered here. Loss coefficients for axial fans are also available. Reference 4 should be consulted for full details.

Inlet Conditions

If it is necessary to install an elbow on the fan inlet, a straight run of duct is recommended between the elbow and the fan, and a long-radius elbow should be used.

Table 12-6 Outlet Duct Elbow Loss Coefficients

Blast Area Ratio	Outlet Elbow Position	Effective Duct Length, Percent			
		0	12	25	50
0.4	A	3.20	2.50	1.80	0.80
	B	3.80	3.20	2.20	1.00
	C & D	5.50	4.50	3.20	1.60
0.5	A	2.20	1.80	1.20	0.53
	B	2.90	2.20	1.60	0.67
	C & D	3.80	3.20	2.20	1.00
0.6	A	1.60	1.40	0.80	0.40
	B	2.00	1.60	1.20	0.53
	C & D	2.90	2.50	1.60	0.80
0.7	A	1.00	0.80	0.53	0.26
	B	1.40	1.00	0.67	0.33
	C & D	2.00	1.60	1.00	0.53
0.8	A	0.80	0.67	0.47	0.18
	B	1.00	0.80	0.53	0.26
	C & D	1.40	1.20	0.80	0.33
0.9	A	0.53	0.47	0.33	0.18
	B	0.80	0.67	0.47	0.18
	C & D	1.20	0.80	0.67	0.26
1.0	A	0.53	0.47	0.33	0.18
	B	0.67	0.53	0.40	0.18
	C & D	1.00	0.80	0.53	0.26

Source: Reprinted by permission from *ASHRAE Duct Fitting Database,* 1992.

Inlet elbows create an additional loss, which must be added to the fan total pressure requirements. Table 12-7 shows loss coefficients for both vaned and unvaned elbows. The additional loss may be calculated from Eq. 12-7. Loss factors for inlet elbows with axial fans are also available (4).

Enclosure Restrictions

In those cases where a fan (or several fans) is built into a fan cabinet construction or is installed in a plenum, it is recommended that the walls be at least one inlet diameter from the fan housing and that a space of at least two inlet diameters be provided between fan inlets. If these recommendations cannot be met, additional pressure losses will result and must be added to the fan total pressure requirements. Every effort must be made to keep the inlet of the fan free of obstructions. The fan inlet should be located so that the inlet is not obstructed by other equipment, walls, pipes, beams, columns, and so on, since such obstructions will degrade the performance of the fan (4).

Table 12-7 Inlet Duct Elbow Loss Coefficients

Figure No.	Duct Radius Ratio R/D	Duct Length Ratio, L/D or L/H		
		0.0	2.0	5.0
	0.50	1.80	1.00	0.53
	0.75	1.40	0.80	0.40
12-15a	1.00	1.20	0.67	0.33
	1.50	1.10	0.60	0.33
	2.00	1.00	0.53	0.33
	3.00	0.67	0.40	0.22
12-15b		3.20	2.00	1.00
	0.50	2.50	1.60	0.80
	0.75	1.60	1.00	0.47
12-15c	1.00	1.20	0.67	0.33
	1.50	1.10	0.60	0.33
	2.00	1.00	0.53	0.33
	3.00	0.80	0.47	0.26
	R/H			
	0.50	2.50	1.60	0.80
	0.75	2.00	1.20	0.67
12-15d	1.00	1.20	0.67	0.33
	1.50	1.00	0.57	0.30
	2.00	0.80	0.47	0.26
	0.50	0.80	0.47	0.26
12-15e	0.75	0.53	0.33	0.18
	1.50	0.40	0.28	0.16
	2.00	0.26	0.22	0.14

Source: Reprinted by permission from *ASHRAE Duct Fitting Database,* 1992.

EXAMPLE 12-5

A single-wheel-single-inlet (SWSI) backward-curved blade fan is operating with both inlet and outlet duct elbows. The outlet duct elbow is in position *C*, Fig. 12-14, and is located one duct diameter from the fan outlet. The average velocity in the duct is 4000 ft/min. The fan inlet is configured as shown in Fig. 12-15d, with a duct length ratio of two and *R/H* of 0.75.

SOLUTION

The first consideration is the effective duct length for the outlet. From Table 12-3, one effective duct length is four duct diameters for a duct velocity of 4000 ft/min. However, the elbow is located at one duct diameter; therefore, an additional pressure loss will

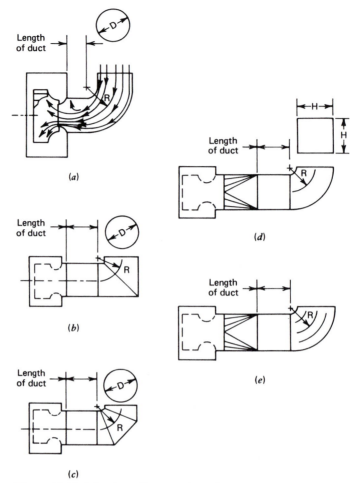

Figure 12-15 Inlet-duct elbow configurations.

result for both the outlet duct and the elbow. The percent effective duct length is $\frac{1}{4}$, or 25 percent. The blast area ratio is 0.7 from Table 12-4. The discharge duct loss coefficient is then 1.0 from Table 12-6, and the additional lost pressure for the duct using Eqs. 12-7 and 12-8 and assuming standard atmospheric pressure is

$$\Delta P_{Od} = 1.0 \times 0.075(4000/1097)^2 = 1.00 \text{ in. wg}$$

The inlet duct elbow loss coefficient is given as 1.2 in Table 12-7 for the Fig. 12-15d configuration with a duct length ratio of 2 and R/H of 0.75. Then using Eq. 12-7 and assuming the fan inlet velocity is one half the outlet velocity

$$\Delta P_{0i} = 1.2(0.075)(2000/1097)^2 = 0.30 \text{ in. wg}$$

Finally the total lost pressure for inlet and outlet system effects is

$$\Delta P_0 = \Delta P_{Od} + \Delta P_{0i}$$
$$= 1.00 + 0.30 = 1.30 \text{ in. wg}$$

This must be added to the computed system total pressure to obtain the actual total pressure that the fan must produce. This is illustrated in Fig. 12-16. Notice that a fan selected on the basis of zero system effect would operate at point *C* instead of point *B*. The fan selected, taking into account the system effect, operates at point *A*, producing the desired flow rate.

12-5   FIELD PERFORMANCE TESTING

The design engineer is often responsible for checking the fan installation when it is put into operation. In cases of malfunction or a drastic change in performance, the engineer must find and recommend corrective action. The logarithmic plot of fan performance is again quite convenient. From the original system design the specified capacity and total pressure are known, and the fan model number and description establishes the fan characteristics as shown in Fig. 12-17. The system characteristic is shown as line *S–D–A–S*. The system shown was designed to operate at about 13,000 cfm and 3 in. wg total pressure. To check the system, measurements of capacity and total pressure are made in the field using a pitot tube and an inclined manometer. The use of these instruments was discussed in Chapter 10.

Several different conditions may be indicated by capacity and pressure measurements. First, if the measurements indicate operation at point *A* in Fig. 12-17, the system and fan are performing as designed. Operation at points *B* or *C* indicates that the fan is performing satisfactorily, but that the system is not operating as designed. At point *B* the system has more pressure loss than anticipated, and at point *C* the system has less pressure loss than desired. To obtain the desired capacity the fan speed must be increased to about 900 rpm for point *B* and reduced to about 650 rpm for point *C*. Operation at point *D* indicates that the fan is not performing as it should. This may be because an incorrect belt drive or belt slippage has caused the fan to operate at the incorrect speed. Poorly designed inlet and outlet connections to the fan may have altered fan performance as previously discussed. The correct capacity may be obtained by correcting the fan speed or by eliminating the undesirable inlet and outlet connections. Operation at point *E* indicates that neither the system nor the fan is operating as designed, which is the usual case found in the field. Although in this situation corrective action could be made by decreasing fan speed to about 700 rpm, any undesirable features of the fan inlet or outlet should first be eliminated to maintain a high fan efficiency. After any increase in fan speed, the change in power requirements should be carefully ascertained, because fan power is proportional to the cube of the fan speed.

EXAMPLE 12-6

A duct system was designed to handle 2.5 m^3/s of air with a total pressure requirement of 465 Pa. The fan of Fig. 12-10 was selected for the system. Field measurements indicate that the system is operating at 2.4 m^3/s at 490 Pa total pressure. Recommend corrective action to bring the system up to the design capacity.

SOLUTION

The system characteristics for the design condition and the actual condition may be sketched on Fig. 12-10 and are parallel to the efficiency lines. The fan is performing as

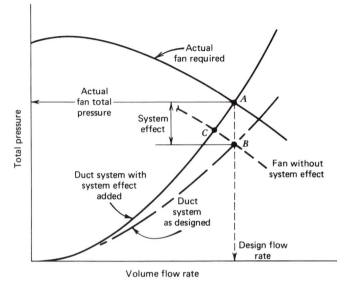

Figure 12-16 Illustration of system effect for Example 12-5.

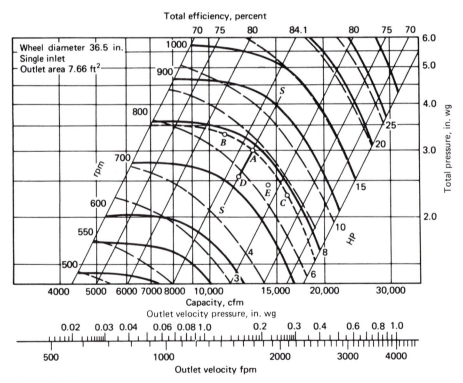

Figure 12-17 Performance curves showing field test combinations. (Reprinted by permission from *ASHRAE Journal,* Vol. 14, Part I, No. 1, 1972.)

specified; however, the system has more flow resistance than it was designed for. To obtain the desired volume flow rate, we must reduce the system flow resistance or increase the fan speed. A check should first be made for unnecessary flow restrictions or closed dampers and, where practical, adjustments should be made to lower the flow resistance. As a last resort the fan speed must be increased, keeping in mind that the power requirements and noise level will increase.

Assuming that the duct system cannot be altered, the fan speed for this example must be increased from 900 rpm to 975 rpm to obtain 2.5 m^3/s of flow. The total pressure produced by the fan will increase from 490 to 564 Pa. The shaft power requirement at the design condition was 2100 W and will be increased to 2400 W at the higher speed. In this case the required increase in fan speed is moderate and the increase in noise level should be minimal. The motor must be checked, however, to be sure that an overload will not occur at the higher speed.

12-6   FANS AND VARIABLE AIR VOLUME SYSTEMS

The variable volume air distribution system is usually designed to supply air to a large number of spaces with the total amount of circulated air varying between some minimum and the full-load design quantity. Normally the minimum is about 20 to 25 percent of the maximum. The volume flow rate of the air is controlled independently of the fan by the terminal boxes, and the fan must respond to the system. Because the fan capacity is directly proportional to fan speed, and power is proportional to the cube of the speed, the fan speed should be decreased as volume flow rate decreases. There are practical and economic considerations involved, however. A variable-speed electric motor is ideal. Fan drives that make use of magnetic couplings have been developed and are referred to as *eddy current drives*. These devices have almost infinite adjustment of fan speed. Their disadvantage is high cost. Another approach is to change the fan speed by changing the diameter of the V-belt drive pulley by adjusting the pulley shives. This requires a mechanism that will operate while the drive is turning. The disadvantage of this approach is maintenance. The motor speed system that has emerged as the best from the standpoint of cost, reliability, and efficiency is the *adjustable frequency control system*. This type of controller will operate with most alternating current motors, although motors of high quality are desired.

Another approach to control of the fan is to throttle and introduce a swirling component to the air entering the fan, which alters the fan characteristic in such a way that somewhat less power is required at the lower flow rates. This is done with variable inlet vanes, which are a radial damper system located at the inlet to the fan. Figure 12-18 illustrates the use of fan inlet vanes. Gradual closing of the vanes reduces the volume flow rate of air and changes the fan characteristic as shown in Fig. 12-19. This approach is not as effective in reducing fan power as fan speed reduction, but the cost and maintenance are low.

The fan speed or inlet vane position is normally controlled to maintain a constant static pressure at some location in the duct system. Static pressure could be sensed at the fan outlet; however, this will lead to abnormally low pressure at some terminal boxes located a large distance from the fan. Therefore, the static pressure sensor should be located downstream at a location such that the most distant terminal box will have an acceptable level of static pressure. This will result in a minimum pressure level throughout the system and the best operating economy.

Figure 12-18 Centrifugal fan inlet vanes. (Reproduced by permission of the Trane Company, LaCrosse, WI.)

Consider the response of a fan in a VAV system with static pressure control as discussed earlier. The minimum and maximum flow rates are shown in Figs. 12-19 and 12-20 for inlet vanes and variable-speed, respectively. Without any fan control, the operating point must move along the constant-speed and the full-open vane characteristics. This results in high system static pressure, low efficiency, and wasted fan power.

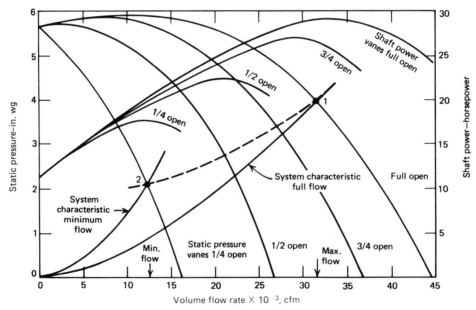

Figure 12-19 Variable-inlet vane fan in a variable-volume system.

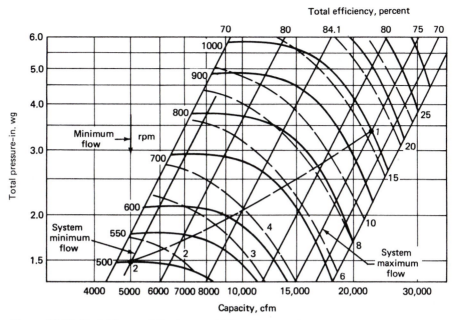

Figure 12-20 Variable-speed fan in a variable-volume system.

Further, it may not be possible to have stable operation of the fan at the minimum flow rate. When the fan speed is reduced or inlet vanes are closed to maintain a fixed static pressure at some downstream location in the duct system, the fan static pressure actually decreases to that shown at point 2 in Fig. 12-19 and 12-20. This occurs because the lost pressure between the fan and the sensing point decreases as the flow rate decreases. This is predictable because duct pressure loss is proportional to the air velocity squared. However, the complete analysis of a variable volume system is difficult because there are infinite variations of the terminal box dampers, fan speed, and inlet vanes. It is possible to locate point 2 in Figs. 12-29 and 12-20 and system operation will then be between points 1 and 2.

12-7   AIR FLOW IN DUCTS

The general subject of fluid flow in ducts and pipes was discussed in Chapter 10. The special topic of air flow is treated in this section. Although the basic theory is the same, certain simplifications and computational procedures will be adopted to aid in the design of air ducts.

Equation 10-1c applies to the adiabatic flow of air in a duct. Neglecting the elevation head terms, assuming that no fan is present, Eq. 10-1c becomes

$$\frac{g_c}{g}\frac{P_1}{\rho_1} + \frac{\overline{V}_1^2}{2g} = \frac{g_c}{g}\frac{P_2}{\rho_2} + \frac{\overline{V}_2^2}{2g} + l_f \tag{12-9a}$$

and in terms of the total head with ρ constant

$$\frac{g_c}{g}\frac{P_{01}}{\rho} = \frac{g_c}{g}\frac{P_{02}}{\rho} + l_f \tag{12-9b}$$

Equations 12-9 provide insight into the duct flow problem. The only important terms remaining in the energy equation are the static head, the velocity head, and the lost head. The static and velocity heads are interchangeable and may increase or decrease in the direction of flow depending on the duct cross-sectional area. Because the lost head l_f must be positive, the total pressure always decreases in the direction of flow. Figure 12-21 illustrates these principles.

For duct flow units of pressure are desired for each term in Eq. 12-9a. Equation 12-9a then takes the following form

$$P_1 + \frac{\rho\overline{V_1^2}}{2g_c} = P_2 + \frac{\rho\overline{V_2^2}}{2g_c} + \frac{\rho g l_f}{g_c} \tag{12-9c}$$

where l_f has the units of feet or meters as defined in Eq. 10-6. To simplify the notation, the equations may be written

$$P_{s1} + P_{v1} = P_{s2} + P_{v2} + \Delta P_f \tag{12-9d}$$

where

$$\Delta P_f = \frac{\rho g l_f}{g_c}$$

and

$$P_{01} = P_{02} + \Delta P_f \tag{12-9e}$$

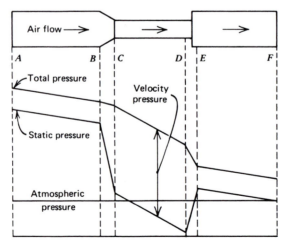

Figure 12-21 Pressure changes during flow in ducts.

In this form each term has the units of pressure in any system of units. For air at standard conditions and English units, pressure is usually in in. wg.

$$P_v = \rho(\overline{V}/1097)^2 = \left(\frac{\overline{V}}{4005}\right)^2 \text{ in. wg} \tag{12-10}$$

where $\overline{V}$ is in ft/min and ρ is in lbm/ft^3.
In SI units

$$P_v = \rho(\overline{V}/1.414)^2 = \left(\frac{\overline{V}}{1.29}\right)^2 \text{ Pa} \tag{12-11}$$

where $\overline{V}$ is in m/s and ρ is in kg/m^3.

The lost head due to friction in a straight, constant area duct is given by Eq. 10-6 and the computational procedure is the same as discussed in Section 10-2. Because this approach becomes tedious when designing ducts, special charts have been prepared. Figures 12-22, and 12-23 are examples of such charts for air flowing in galvanized steel ducts with approximately 40 joints per 100 ft or 30 m. The charts are based on standard air and fully developed flow. For the temperature range of 50 F or 10 C to about 100 F or 38 C there is no need to correct for viscosity and density changes. Above 100 F or 38 C, however, a correction should be made. The density correction is also small for moderate pressure changes. For elevations below about 2000 ft or 610 m the correction is small. The correction for density and viscosity will normally be less than one. For example, dry air at 100 F at an elevation of 2000 ft would exhibit a pressure loss about 10 percent less than given in Fig. 12-22. For average to rough ducts, a correction factor for density and viscosity may be expressed as

$$C = \left(\frac{\rho_a}{\rho_s}\right)^{0.9} \left(\frac{\mu_a}{\mu_s}\right)^{0.10} \tag{12-12}$$

where

ρ = air density
μ = air viscosity

and subscripts a and s refer to actual and standard conditions, respectively. The actual lost pressure is then given by

$$\Delta P_{0a} = C\Delta P_{0s} \tag{12-13}$$

where ΔP_{0s} is from Figures 12-22 and 12-23.

The effect of roughness is the most important consideration. A common problem to designers is determination of the roughness effect of fibrous glass duct liners and fibrous ducts. This material is manufactured in several grades with various degrees of absolute roughness. Further, the joints and fasteners necessary to install the material affect the overall pressure loss. Smooth galvanized ducts typically have a friction factor of about 0.02, whereas fibrous liners and duct materials will have friction factors varying from about 0.03 to 0.06 depending on the quality of the material and joints, and the duct

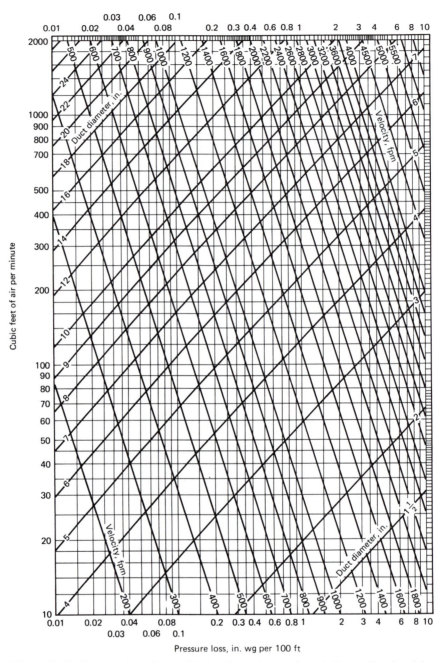

Cubic feet of air per minute

Pressure loss, in. wg per 100 ft

Figure 12-22 Pressure loss due to friction for galvanized steel ducts. (Reprinted by permission from *ASHRAE Handbook, Fundamentals Volume*, 1977.)

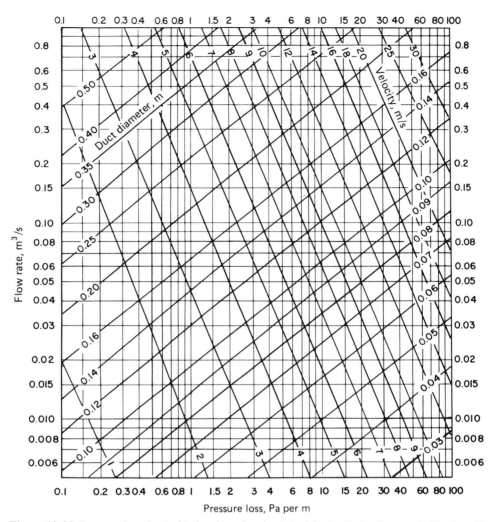

Figure 12-23 Pressure loss due to friction for galvanized steel ducts. (*Using SI Units in Heating, Air Conditioning and Refrigeration,* by W. F. Stoecker, 1975, Business News Publishing Company.)

diameter. The usual approach to account for this roughness effect is to use a correction factor that is applied to the pressure loss obtained for galvanized metal duct such as Fig. 12-22. Figure 12-24 shows a range of data for commercially available fibrous duct liner materials. These correction factors probably do not allow for typical joints and fasteners.

A more refined approach to the prediction of pressure loss in rough or lined ducts is to generate a chart, such as Fig. 12-22, using Eq. 10-6 and the Colebrook function (2)

$$\frac{1}{\sqrt{f}} = -2\log_{10}\left[\frac{e}{3.7D} + \frac{2.51}{\mathrm{Re}_D\sqrt{f}}\right] \qquad \textbf{(12-14)}$$

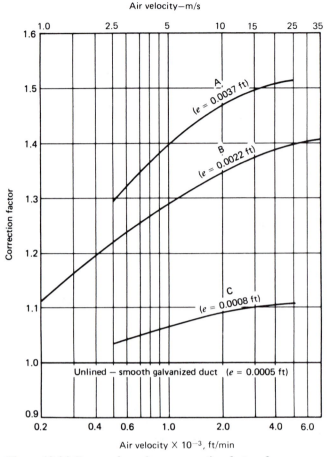

Figure 12-24 Range of roughness correction factors for commercially available duct liners.

to express the friction factor. Equation 12-14 is valid in the transition region where f depends on the absolute roughness e, the duct diameter D, and the Reynolds number Re_D. Equations 10-6 and 12-14 and the ideal gas property relation may be easily programmed for a small computer to calculate the lost pressure for a wide range of temperatures, pressures, and roughness. This general approach eliminates the need for corrections of any kind.

The pressure loss due to friction is greater for a rectangular duct than for a circular duct of the same cross-sectional area and capacity. For most practical purposes ducts of aspect ratio not exceeding 8:1 will have the same lost head for equal length and mean velocity of flow as a circular duct of the same hydraulic diameter. When the duct sizes are expressed in terms of hydraulic diameter D_h and when the equations for friction loss in round and rectangular ducts are equated for equal length and capacity, Eq. 12-15 for the circular equivalent of a rectangular duct is obtained:

$$D_e = 1.3 \frac{(ab)^{5/8}}{(a + b)^{1/4}} \tag{12-15}$$

where a and b are the rectangular duct dimensions in any consistent units. Table 12-8 has been compiled using Eq. 12-15. A more complete table is given in the *ASHRAE Handbook* (2).

Oval ducts are sometimes used in commercial duct systems. The frictional pressure loss may be treated in the same manner as rectangular ducts using the circular equivalent of the oval duct as defined by

$$D_e = \frac{1.55 A^{0.625}}{P^{0.25}}$$

(12-16a)

with

$$A = \frac{b^2}{4} + b(a - b)$$

(12-16b)

and

$$P = b + 2(a - b)$$

(12-16c)

where

a = major diameter of oval duct, in. or m
b = minor diameter of oval duct, in. or m

Equations 12-16 are valid for aspect ratios ranging from 2 to 4 (2).

12-8 AIR FLOW IN FITTINGS

Whenever a change in area or direction occurs in a duct or when the flow is divided and diverted into a branch, substantial losses in total pressure may occur. These losses are usually of greater magnitude than the losses in the straight pipe and are referred to as *dynamic losses.*

Dynamic losses vary as the square of the velocity and are conveniently represented by

$$\Delta P_0 = C_0 (P_v)$$

(12-17a)

where the loss coefficient C_0 is a constant and Eqs. 12-10 or 12-11 express P_v. When different upstream and downstream areas are involved as in an expansion or contraction, either the upstream or downstream value of P_v may be used in Eq. 12-17a and C will be different in each case.

Consider a transition such as that shown in Table 12-10. The loss coefficients are referenced to section zero. However, the coefficient referenced to section 1 is obtained as follows:

$$\Delta P_0 = C_0 P_{v0} = C_1 P_{v1}$$

(12-17b)

or

$$C_1 = C_0 (P_{v0}/P_{v1}) = C_0 (\overline{V}_0 / \overline{V}_1)^2$$

(12-18)

Notation for the loss coefficients is as follows:

Table 12-8 Circular Equivalents of Rectangular Ducts for Equal Friction and Capacity—Dimensions in Inches, Feet, or Meters

Side Rectangular Duct	6	7	8	9	10	11	12	13	14	15	16	17	18	19	20	22	24
6	6.6																
7	7.1	7.7															
8	7.5	8.2	8.8														
9	8.0	8.6	9.3	9.9													
10	8.4	9.1	9.8	10.4	10.9												
11	8.8	9.5	10.2	10.8	11.4	12.0											
12	9.1	9.9	10.7	11.3	11.9	12.5	13.1										
13	9.5	10.3	11.1	11.8	12.4	13.0	13.6	14.2									
14	9.8	10.7	11.5	12.2	12.9	13.5	14.2	14.7	15.3								
15	10.1	11.0	11.8	12.6	13.3	14.0	14.6	15.3	15.8	16.4							
16	10.4	11.4	12.2	13.0	13.7	14.4	15.1	15.7	16.3	16.9	17.5						
17	10.7	11.7	12.5	13.4	14.1	14.9	15.5	16.1	16.8	17.4	18.0	18.6					
18	11.0	11.9	12.9	13.7	14.5	15.3	16.0	16.6	17.3	17.9	18.5	19.1	19.7				
19	11.2	12.2	13.2	14.1	14.9	15.6	16.4	17.1	17.8	18.4	19.0	19.6	20.2	20.8			
20	11.5	12.5	13.5	14.4	15.2	15.9	16.8	17.5	18.2	18.8	19.5	20.1	20.7	21.3	21.9		
22	12.0	13.1	14.1	15.0	15.9	16.7	17.6	18.3	19.1	19.7	20.4	21.0	21.7	22.3	22.9	24.1	
24	12.4	13.6	14.6	15.6	16.6	17.5	18.3	19.1	19.8	20.6	21.3	21.9	22.6	23.2	23.9	25.1	26.2
26	12.8	14.1	15.2	16.2	17.2	18.1	19.0	19.8	20.6	21.4	22.1	22.8	23.5	24.1	24.8	26.1	27.2
28	13.2	14.5	15.6	16.7	17.7	18.7	19.6	20.5	21.3	22.1	22.9	23.6	24.4	25.0	25.7	27.1	28.2
30	13.6	14.9	16.1	17.2	18.3	19.3	20.2	21.1	22.0	22.9	23.7	24.4	25.2	25.9	26.7	28.0	29.3
32	14.0	15.3	16.5	17.7	18.8	19.8	20.8	21.8	22.7	23.6	24.4	25.2	26.0	26.7	27.5	28.9	30.1
34	14.4	15.7	17.0	18.2	19.3	20.4	21.4	22.4	23.3	24.2	25.1	25.9	26.7	27.5	28.3	29.7	31.0
36	14.7	16.1	17.4	18.6	19.8	20.9	21.9	23.0	23.9	24.8	25.8	26.6	27.4	28.3	29.0	30.5	32.0
38	15.0	16.4	17.8	19.0	20.3	21.4	22.5	23.5	24.5	25.4	26.4	27.3	28.1	29.0	29.8	31.4	32.8
40	15.3	16.8	18.2	19.4	20.7	21.9	23.0	24.0	25.1	26.0	27.0	27.9	28.8	29.7	30.5	32.1	33.6

Source: Reprinted by permission from *ASHRAE Handbook, Fundamentals Volume,* 1989.

C_n—used for constant flow fittings; C is based on the velocity at section n.

C_{ij}—used for converging or diverging fittings. Subscript i refers to the section, c, s, or b and subscript j refers to the path. If the path and section are the same, only one subscript is used.

Fittings are classified as either *constant flow,* such as an elbow or transition, or as *divided flow,* such as a wye or tee. Tables 12-9 through 12-11, give loss coefficients for different types of constant flow fittings. It should be kept in mind that the quality and type of construction may vary considerably for a particular type of fitting. Some manufacturers provide data for their own products.

An extensive database of duct fitting coefficients for over 200 fittings has been developed by ASHRAE (5) and is available on magnetic media. Individual fittings may be accessed or the database may be used with a computer program.

EXAMPLE 12-7

Compute the lost pressure in a 6-in., 90 degree pleated elbow that has 150 cfm of air flowing through it. The ratio of turning radius to diameter is 1.5. Assume standard air.

SOLUTION

The lost pressure will be computed from Eq. 12-17. From Table 12-9 the loss coefficient is read as 0.43 and the average velocity in the elbow is computed as

$$\overline{V} = \frac{\dot{Q}}{A} = \frac{\dot{Q}}{(\pi/4)D^2} = \frac{(150)4(144)}{\pi(36)} = 764 \text{ ft/min}$$

then P_v is given by Eq. 12-10 and

$$\Delta P_0 = C_0 \left(\frac{\overline{V}}{4005}\right)^2 = 0.43 \left(\frac{764}{4005}\right)^2 = 0.016 \text{ in. wg}$$

In SI units the elbow diameter is 15.24 cm and the flow rate is 4.25 m^3/min. The average velocity is then

$$\overline{V} = \frac{\dot{Q}}{A} = \frac{4.25}{(\pi/4)(0.1524)^2(60)} = 3.88 \text{ m/s}$$

The loss coefficient C_0 is dimensionless and is therefore unchanged. Using Eq. 12-11, we get

$$\Delta P_0 = C_0 \left(\frac{\overline{V}}{1.29}\right)^2 = 0.43 \left(\frac{3.88}{1.29}\right)^2 = 3.89 \text{ Pa}$$

This result compares favorably with the result obtained in English units since 0.016 in. wg may be converted directly to 3.89 Pa.

Elbows are generally efficient fittings in that their losses are small when the turn is gradual. When an abrupt turn is used without turning vanes, the lost pressure will be four or five times larger.

When considering the lost pressure in divided flow fittings, the loss in the straight-through section as well as the loss through the branch outlet must be considered. Tables 12-12 and 12-13 give data for branch-type fittings. The angle of the branch takeoff has a great influence on the loss coefficient. Converging flow (Table 12-13) differs from diverging flow (Table 12-12). It has been customary to base the loss coefficient on the velocity in the common section of the fitting; however, to accommodate modern duct design methods, the convention has been changed to base the coefficients on the branch and main sections. Equation 12-18 may be used to change the base for the coefficients when desired.

Tables 12-12 and 12-13 also give representative data for the straight-through flow for the common types of divided flow fittings. The velocity may increase, decrease, or remain constant through the fitting. In every case there will be some loss in total pressure. At velocities below about 1000 ft/min or 5 m/s, the loss may be neglected; however, it is good practice to allow for a small loss.

EXAMPLE 12-8

Compute the loss in total pressure for a round 90 degree branch and straight-through section, a tee. The main section is 12×10 in. with a flow rate of 1100 cfm. The branch flow rate is 250 cfm through a 6-in. duct.

SOLUTION

It is first necessary to compute the average velocity in each section of the fitting.

$$\overline{V}_c = \frac{\dot{Q}_c}{A_c} = \frac{1100}{(\pi/4)(12/12)^2} = 1400 \text{ ft/min}$$

$$\overline{V}_b = \frac{\dot{Q}_b}{A_b} = \frac{850}{(\pi/4)(6/12)^2} = 1273 \text{ ft/min}$$

$$\overline{V}_s = \frac{\dot{Q}_s}{A_s} = \frac{850}{(\pi/4)(10/12)^2} = 1558 \text{ ft/min}$$

The ratio of the branch to common flow rate is

$$\frac{\dot{Q}_b}{\dot{Q}_c} = \frac{250}{1100} = 0.23$$

and the ratio of the main to common flow rate is

$$\frac{\dot{Q}_s}{\dot{Q}_c} = \frac{850}{1100} = 0.77$$

The ratio of the branch to common areas is

$$\frac{A_b}{A_c} = \left(\frac{6}{12}\right)^2 = 0.25$$

The ratio of the main to common area is

$$\frac{A_s}{A_c} = \left(\frac{10}{12}\right)^2 = 0.69$$

The loss coefficient for the branch is then read from Table 12-12B as 1.55 and

$$\Delta P_{0b} = C_b \left[\frac{V_b}{4005}\right]^2 = 1.55 \left(\frac{1273}{4005}\right)^2 = 0.16 \text{ in. wg}$$

The lost pressure for the straight-through section is also determined from Table 12-12B. The loss coefficient is about 0.14 and

$$\Delta P_{0s} = 0.14 \left(\frac{1558}{4005}\right)^2 = 0.021 \text{ in. wg}$$

Some of the data just discussed for branch-type fittings are for pipes of circular cross section. When rectangular duct is used, the pressure losses will vary depending on the design of the fitting. Extensive data for fittings of all kinds are given in reference 2.

It is often convenient to express the effect of fittings in terms of equivalent length as was done in Chapter 10 for pipes. The equivalent length is a function of the air velocity and the size (diameter) of the fitting. In the case of pipes with water flowing, the Reynolds number is sufficiently large to assume that the fully turbulent friction factor is always valid. This is not generally true for air flowing in ducts unless the duct is large. However, for low-velocity systems a friction factor can be assumed, based on expected flow conditions, to derive equivalent lengths from the loss coefficient data of Tables 12-8 through 12-13. Recall that

$$l_f = f\frac{L}{D}\frac{\overline{V}^2}{2g} = C\frac{\overline{V}^2}{2g} \tag{12-19}$$

and

$$\frac{L}{D} = \frac{C}{f} \tag{12-20}$$

Table 12-14 gives friction factors that will generally be valid for various duct diameters.

Figures 12-25 and 12-26 give some equivalent lengths commonly used for residential and light commercial system design. The lengths shown have an unknown amount of conservatism built in. This approach simplifies calculations and is helpful in the design of simple low-velocity systems.

Table 12-9 Total Pressure Loss Coefficients for Elbows

A. Elbow, Pleated, $r/D = 1.5$

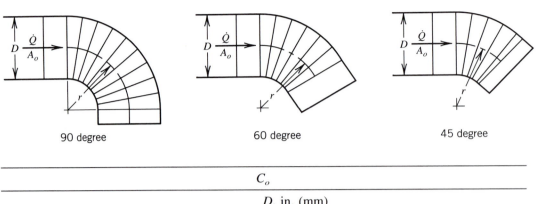

				C_o			
				D, in. (mm)			
Angle	4(100)	6(150)	8(200)	10(250)	12(300)	14(350)	16(400)
90	0.57	0.43	0.34	0.28	0.26	0.25	0.25
60	0.45	0.34	0.27	0.23	0.20	0.19	0.19
45	0.34	0.26	0.21	0.17	0.16	0.15	0.15

B. Elbow, Mitered, with Single-Thickness Vanes, Rectangular

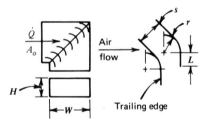

		C_o		
Design		Dimensions, in. (mm)		
No.	r	s	L	C_o
1	2.0 (50)	1.5 (40)	0.0	0.11
2	2.0 (50)	1.5 (40)	0.75 (20)	0.12
3	4.5 (110)	2.25 (60)	0.0	0.15
4	4.5 (110)	3.25 (80)	0.0	0.33

Table 12-9 Total Pressure Loss Coefficients for Elbows (*Continued*)

C. Elbow, Mitered, Rectangular

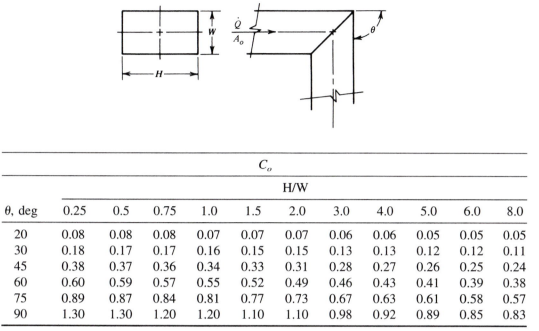

	C_o										
	H/W										
θ, deg	0.25	0.5	0.75	1.0	1.5	2.0	3.0	4.0	5.0	6.0	8.0
20	0.08	0.08	0.08	0.07	0.07	0.07	0.06	0.06	0.05	0.05	0.05
30	0.18	0.17	0.17	0.16	0.15	0.15	0.13	0.13	0.12	0.12	0.11
45	0.38	0.37	0.36	0.34	0.33	0.31	0.28	0.27	0.26	0.25	0.24
60	0.60	0.59	0.57	0.55	0.52	0.49	0.46	0.43	0.41	0.39	0.38
75	0.89	0.87	0.84	0.81	0.77	0.73	0.67	0.63	0.61	0.58	0.57
90	1.30	1.30	1.20	1.20	1.10	1.10	0.98	0.92	0.89	0.85	0.83

Source: Reprinted by permission from *ASHRAE Duct Fitting Database,* 1992.

EXAMPLE 12-9

Compute the equivalent lengths for the fittings in the duct system of Fig. 12-27. The fittings are an entrance, a 45 degree wye, the straight-through section of the wye fitting, a 45 degree elbow, and a 90 degree elbow.

SOLUTION

Table 12-11A gives the loss coefficient for an entrance. In this case, θ is either 0 or 180 degrees and C_0 is 0.5. Then using Eq. 12-20, Table 12-14 for f, and a 10-in. diameter

$$\frac{L_i}{D} = \frac{0.5}{0.022} = 22.7$$

and

$$L_i = 22.7 \left(\frac{10}{12}\right) = 19 \, \text{ft}$$

Table 12-12A gives the loss coefficients for the branch of a wye. For this case

$$\frac{\dot{Q}_b}{\dot{Q}_c} = \frac{120}{400} = 0.3$$

and

$$\frac{A_b}{A_c} = \left(\frac{6}{10}\right)^2 = 0.36$$

Then for θ of 45 degrees, $C_b = 0.60$. Since C_b is based on the branch velocity, use the diameter of the branch of 6 in. Then

$$\frac{L_b}{D} = \frac{0.60}{0.028} = 21.4$$

and

$$L_b = 21.4 \left(\frac{6}{12}\right) = 11 \text{ ft}$$

Table 12-10 Total Pressure Loss Coefficients for Transitions

A. Transition, Round to Round

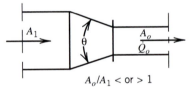

$$A_o/A_1 < \text{or} > 1$$

			C_o				
				θ, deg			
A_o/A_1	10	20	45	90	120	150	180
0.10	0.05	0.05	0.07	0.19	0.29	0.37	0.43
0.17	0.05	0.04	0.06	0.18	0.28	0.36	0.42
0.25	0.05	0.04	0.06	0.17	0.27	0.35	0.41
0.50	0.05	0.05	0.06	0.12	0.18	0.24	0.26
1.00	0.00	0.00	0.00	0.00	0.00	0.00	0.00
2.00	0.44	0.76	1.32	1.28	1.24	1.20	1.20
4.00	2.56	4.80	9.76	10.24	10.08	9.92	9.92
10.00	21.00	38.00	76.00	83.00	84.00	83.00	83.00
16.00	53.76	97.28	215.04	225.28	225.28	225.28	225.28

B. Transition, Rectangular, Two Sides Parallel

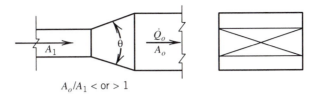

$A_o/A_1 < $ or $ > 1$

			C_o				
				θ, deg			
A_o/A_1	10	20	45	90	120	150	180
0.10	0.05	0.05	0.07	0.19	0.29	0.37	0.43
0.17	0.05	0.04	0.05	0.18	0.28	0.36	0.42
0.25	0.05	0.04	0.06	0.17	0.27	0.35	0.41
0.50	0.06	0.05	0.06	0.14	0.20	0.26	0.27
1.00	0.00	0.00	0.00	0.00	0.00	0.00	1.00
2.00	0.56	0.60	1.40	1.52	1.48	1.44	1.40
4.00	2.72	3.52	9.60	11.20	11.04	10.72	10.56
10.00	24.00	36.00	69.00	93.00	93.00	92.00	91.00
16.00	66.56	102.40	181.76	256.00	253.44	250.88	258.88

Source: Reprinted by permission from *ASHRAE Dust Fitting Database,* 1992.

Data for the straight-through section of the branch fitting is given in Table 12-12A:

$$\frac{\dot{Q}_s}{\dot{Q}_c} = \frac{280}{400} = 0.7$$

$$\frac{A_s}{A_c} = \left(\frac{9}{10}\right)^2 = 0.81$$

Then $C_s = 0.13$ and

$$\frac{L_s}{D} = \frac{0.13}{0.0225} = 5.8$$

and

$$L_s = 5.8 \left(\frac{9}{12}\right) = 4.4 \text{ ft}$$

Table 12-11 Total Pressure Loss Coefficients for Duct Entrances

A. Conical Converging Bellmouth with End Wall, Round and Rectangular

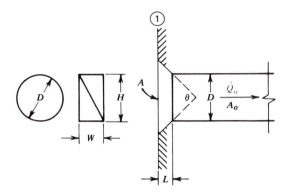

Rectangular: $D = 2\,HW/(H + W)$

				C_o					
					θ, deg				
$\dfrac{L}{D}$	0	10	20	30	40	60	100	140	180
0.025	0.50	0.47	0.45	0.43	0.41	0.40	0.42	0.45	0.50
0.05	0.50	0.45	0.41	0.36	0.33	0.30	0.35	0.42	0.50
0.075	0.50	0.42	0.35	0.30	0.26	0.23	0.30	0.40	0.50
0.10	0.50	0.39	0.32	0.25	0.22	0.18	0.27	0.38	0.50
0.15	0.50	0.37	0.27	0.20	0.16	0.15	0.25	0.37	0.50
0.60	0.50	0.27	0.18	0.13	0.11	0.12	0.23	0.36	0.50

B. Smooth Converging Bellmouth with End Wall

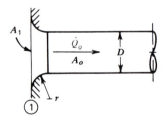

r/D	0	0.01	0.02	0.03	0.04	0.05
C_o	0.50	0.43	0.36	0.31	0.26	0.22
r/D	0.06	0.08	0.10	0.12	0.16	≥ 0.20
C_o	0.20	0.15	0.12	0.09	0.06	0.03

Source: Reprinted by permission from *ASHRAE Duct Fitting Database*, 1992.

Table 12-12 Total Pressure Loss Coefficients for Diverging Flow Fittings

A. Diverging Wye, Round, 45 deg

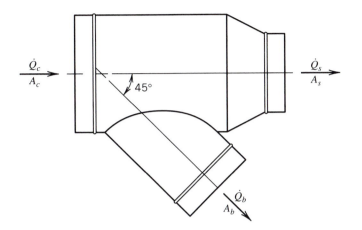

	Branch, C_b							
	$\dot{Q}_b/\dot{Q}_c$							
A_b/A_c	0.1	0.2	0.3	0.4	0.5	0.6	0.7	0.8
0.1	0.38	0.39	0.48					
0.2	2.25	0.38	0.31	0.39	0.46	0.48	0.45	
0.3	6.29	1.02	0.38	0.30	0.33	0.39	0.44	0.48
0.4	12.41	2.25	0.74	0.38	0.30	0.31	0.35	0.39
0.5	20.58	4.01	1.37	0.62	0.38	0.30	0.30	0.32
0.6	30.78	6.29	2.25	1.02	0.56	0.38	0.31	0.30
0.7	43.02	9.10	3.36	1.57	0.85	0.52	0.38	0.31
0.8	57.29	12.41	4.71	2.25	1.22	0.74	0.50	0.38
0.9	73.59	16.24	6.29	3.06	1.69	1.02	0.67	0.48

	Main, C_s							
	$\dot{Q}_s/\dot{Q}_c$							
A_s/A_c	0.1	0.2	0.3	0.4	0.5	0.6	0.7	0.8
0.1	0.13	0.16						
0.2	0.20	0.13	0.15	0.16	0.28			
0.3	0.90	0.13	0.13	0.14	0.15	0.16	0.20	
0.4	2.88	0.20	0.14	0.13	0.14	0.15	0.15	0.16
0.5	6.25	0.37	0.17	0.14	0.13	0.14	0.14	0.15
0.6	11.88	0.90	0.20	0.13	0.14	0.13	0.14	0.14
0.7	18.62	1.71	0.33	0.18	0.16	0.14	0.13	0.15
0.8	26.88	2.88	0.50	0.20	0.15	0.14	0.13	0.13
0.9	36.45	4.46	0.90	0.30	0.19	0.16	0.15	0.14

Table 12-12 Total Pressure Loss Coefficients for Diverging Flow Fittings (*Continued*)

B. Diverging Tee, Round

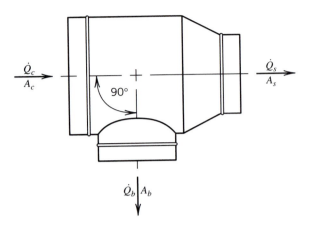

	Branch, C_b								
	$\dot{Q}_b/\dot{Q}_c$								
A_b/A_c	0.1	0.2	0.3	0.4	0.5	0.6	0.7	0.8	0.9
0.1	1.20	0.62	0.80	1.28	1.99	2.92	4.07	5.44	7.02
0.2	4.10	1.20	0.72	0.62	0.66	0.80	1.01	1.28	1.60
0.3	8.99	2.40	1.20	0.81	0.66	0.62	0.64	0.70	0.80
0.4	15.89	4.10	1.94	1.20	0.88	0.72	0.64	0.62	0.63
0.5	24.80	6.29	2.91	1.74	1.20	0.92	0.77	0.68	0.63
0.6	35.73	8.99	4.10	2.40	1.62	1.20	0.96	0.81	0.72
0.7	48.67	12.19	5.51	3.19	2.12	1.55	1.20	0.99	0.85
0.8	63.63	15.89	7.14	4.10	2.70	1.94	1.49	1.20	1.01
0.9	80.60	20.10	8.99	5.13	3.36	2.40	1.83	1.46	1.20

	Main, C_s								
	$\dot{Q}_s/\dot{Q}_c$								
A_s/A_c	0.1	0.2	0.3	0.4	0.5	0.6	0.7	0.8	0.9
0.1	0.13	0.16							
0.2	0.20	0.13	0.15	0.16	0.28				
0.3	0.90	0.13	0.13	0.14	0.15	0.16	0.20		
0.4	2.88	0.20	0.14	0.13	0.14	0.15	0.15	0.16	0.34
0.5	6.25	0.37	0.17	0.14	0.13	0.14	0.14	0.15	0.15
0.6	11.88	0.90	0.20	0.13	0.14	0.13	0.14	0.14	0.15
0.7	18.62	1.71	0.33	0.18	0.16	0.14	0.13	0.15	0.14
0.8	26.88	2.88	0.50	0.20	0.15	0.14	0.13	0.13	0.14
0.9	36.45	4.46	0.90	0.30	0.19	0.16	0.15	0.14	0.13

Source: Reprinted by permission from *ASHRAE Duct Fitting Database,* 1992.

Table 12-13 Total Pressure Loss Coefficients for Converging Flow Fittings

A. Converging Wye (45 deg), Round

Branch, C_b

A_s/A_c	A_b/A_c	0.1	0.2	0.3	0.4	0.5	0.6	0.7	0.8	0.9
					$\dot{Q}_b/\dot{Q}_c$					
0.4	0.3	−21.41	−2.85	−0.10	0.63	0.87	0.96	1.00	1.06	1.26
	0.4	−39.30	−6.02	−1.05	0.28	0.72	0.87	0.91	0.92	1.00
	0.5	−62.10	−9.96	−2.16	−0.06	0.63	0.85	0.90	0.88	0.86
	0.6	−89.77	−14.65	−3.42	−0.38	0.61	0.93	0.99	0.95	0.90
	0.7	−122.46	−20.19	−4.88	−0.74	0.61	1.04	1.12	1.06	0.95
	0.8	−160.18	−26.56	−6.55	−1.15	0.62	1.18	1.29	1.19	1.01

Table 12-13 Total Pressure Loss Coefficients for Converging Flow Fittings (*Continued*)

Branch, C_b

A_s/A_c	A_b/A_c	$\dot{Q}_b/\dot{Q}_c$								
		0.1	0.2	0.3	0.4	0.5	0.6	0.7	0.8	0.9
0.5	0.3	−14.10	−1.39	0.40	0.84	0.97	1.00	1.02	1.07	1.28
	0.4	−26.48	−3.53	−0.24	0.59	0.83	0.89	0.88	0.85	0.86
	0.5	−41.84	−5.96	−0.80	0.51	0.88	0.97	0.95	0.90	0.87
	0.6	−60.61	−8.90	−1.46	0.43	0.97	1.09	1.06	0.97	0.90
	0.7	−82.80	−12.36	−2.22	0.35	1.09	1.25	1.20	1.08	0.93
	0.8	−108.39	−16.35	−3.09	0.27	1.24	1.45	1.38	1.20	0.96

Main, C_s

A_s/A_c	A_b/A_c	$\dot{Q}_s/\dot{Q}_c$								
		0.1	0.2	0.3	0.4	0.5	0.6	0.7	0.8	0.9
0.4	0.3	−33.68	−6.60	−1.98	−0.53	0.05	0.31	0.41	0.45	0.45
	0.4	−25.24	−4.51	−1.13	−0.13	0.25	0.40	0.46	0.47	0.45
	0.5	−18.83	−3.04	−0.57	0.13	0.37	0.46	0.48	0.48	0.46
	0.6	−13.99	−1.97	−0.17	0.31	0.46	0.50	0.50	0.48	0.46
	0.7	−10.27	−1.17	0.12	0.44	0.52	0.53	0.51	0.49	0.46
	0.8	−7.32	−0.54	0.35	0.54	0.57	0.55	0.52	0.49	0.46
0.5	0.3	−53.80	−10.77	−3.45	−1.17	−0.27	0.11	0.26	0.30	0.29
	0.4	−40.66	−7.54	−2.16	−0.57	0.02	0.25	0.32	0.33	0.30
	0.5	−30.68	−5.27	−1.30	−0.18	0.21	0.33	0.36	0.34	0.30
	0.6	−23.15	−3.62	−0.69	0.09	0.33	0.39	0.38	0.35	0.30
	0.7	−17.34	−2.38	−0.24	0.29	0.42	0.43	0.40	0.35	0.30
	0.8	−12.75	−1.41	0.11	0.44	0.49	0.47	0.41	0.36	0.30

B. Converging Tee, Round

Branch, C_b

A_s/A_c	A_b/A_c	$\dot{Q}_b/\dot{Q}_c$								
		0.1	0.2	0.3	0.4	0.5	0.6	0.7	0.8	0.9
0.4	0.3	−23.96	−3.65	−0.48	0.43	0.75	0.88	0.91	0.91	0.86
	0.4	−42.98	−7.03	−1.46	0.11	0.67	0.87	0.93	0.91	0.84
	0.5	−67.44	−11.35	−2.69	−0.26	0.59	0.90	0.97	0.94	0.84
	0.6	−97.39	−16.60	−4.17	−0.69	0.52	0.95	1.06	1.01	0.87

Table 12-13 Total Pressure Loss Coefficients for Converging Flow Fittings (*Continued*)

Branch, C_b

A_s/A_c	A_b/A_c									

A_s/A_c	A_b/A_c	0.1	0.2	0.3	0.4	0.5	0.6	0.7	0.8	0.9
		\multicolumn{9}{c}{$\dot{Q}_b/\dot{Q}_c$}								
	0.7	−132.88	−22.81	−5.91	−1.17	0.46	1.03	1.17	1.11	0.92
	0.8	−173.96	−29.99	−7.90	−1.73	0.40	1.15	1.33	1.24	1.00
0.5	0.3	−16.99	−2.35	−0.07	0.57	0.80	0.89	0.91	0.90	0.87
	0.4	−30.49	−4.67	−0.72	0.38	0.76	0.89	0.92	0.90	0.85
	0.5	−47.82	−7.61	−1.50	0.19	0.75	0.93	0.97	0.93	0.85
	0.6	−69.03	−11.17	−2.42	−0.03	0.76	1.01	1.05	0.98	0.88
	0.7	−94.17	−15.37	−3.49	−0.26	0.80	1.13	1.17	1.07	0.93
	0.8	−123.30	−20.22	−4.71	−0.50	0.87	1.29	1.33	1.20	1.02

Main, C_s

A_s/A_c	A_b/A_c	0.1	0.2	0.3	0.4	0.5	0.6	0.7	0.8	0.9
		\multicolumn{9}{c}{$\dot{Q}_s/\dot{Q}_c$}								
0.4	0.3	49.68	7.59	2.74	1.42	0.90	0.65	0.51	0.43	0.37
	0.4	30.96	5.51	2.21	1.23	0.82	0.61	0.49	0.42	0.37
	0.5	21.00	4.40	1.92	1.13	0.78	0.59	0.48	0.42	0.37
	0.6	15.43	3.78	1.76	1.07	0.75	0.58	0.48	0.41	0.37
	0.7	12.36	3.44	1.67	1.04	0.74	0.57	0.48	0.41	0.37
	0.8	10.86	3.27	1.63	1.02	0.73	0.57	0.47	0.41	0.37
0.5	0.3	65.94	10.28	3.65	1.79	1.05	0.68	0.48	0.35	0.27
	0.4	38.84	7.27	2.87	1.51	0.93	0.63	0.45	0.34	0.27
	0.5	25.07	5.74	2.47	1.37	0.87	0.60	0.44	0.33	0.26
	0.6	17.98	4.95	2.27	1.29	0.84	0.58	0.43	0.33	0.26
	0.7	14.69	4.58	2.17	1.26	0.82	0.58	0.43	0.33	0.26
	0.8	13.78	4.48	2.15	1.25	0.82	0.57	0.43	0.33	0.26

Source: Reprinted by permission from *ASHRAE Duct Fitting Database*, 1992.

Table 12-14 Friction Factors for Various Galvanized Steel Ducts

| Diameter | | Friction |
in.	mm	Factor
4	10	0.035
6	15	0.028
8	20	0.023
10	25	0.022
12	30	0.019
14	36	0.017
16	40	0.016
20	50	0.015
24	60	0.014

which is small as expected.

Assume that the 90 degree elbow is the pleated type with a r/D ratio of 1.5. Then the loss coefficient is 0.43 from Table 12-9A and

$$\frac{L_e}{D} = \frac{0.43}{0.028} = 15.4$$

and

$$L_e = 15.4 \left(\frac{6}{12} \right) = 7.7 \text{ ft}$$

A 45 degree elbow will have about one-half the equivalent length of a 90 degree elbow.

To compute the lost pressure, we add the equivalent length of the fitting to the adjacent pipe length. The following example shows the procedure.

EXAMPLE 12-10

Compute the lost pressure for each branch of the simple duct system shown in Fig. 12-27 using the equivalent length approach with data from Figs. 12-25 and 12-26 and SI units.

SOLUTION

The lost pressure will first be computed from 1 to a; then sections a to 2 and a to 3 will be handled separately.

The equivalent length of section 1 to a consists of the actual length of the 25-cm duct

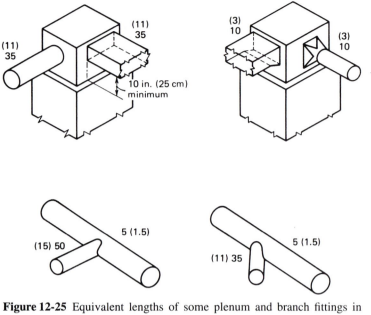

Figure 12-25 Equivalent lengths of some plenum and branch fittings in feet, with meters in parentheses. (Reprinted by permission from *ASHRAE Handbook, Systems and Equipment Volume,* 1992.)

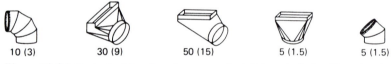

Figure 12-26 Equivalent lengths of common duct fittings in feet, with meters in parentheses. (Reprinted by permission from *ASHRAE Handbook, Systems and Equipment Volume,* 1992.)

plus the equivalent length for the entrance from the plenum of 11 m given in Fig. 12-25. Then

$$L_{1a} = 15 + 11 = 26 \text{ m}$$

From Fig. 12-23 at 0.19 m³/s and for a pipe diameter of 25 cm, the lost pressure is 0.85 Pa/m of pipe. Then

$$\Delta P_{01a} = (0.85)(26) = 22.1 \text{ Pa}$$

Section *a* to 3 has an equivalent length equal to the sum of the actual length, and the equivalent length for the 45 degree branch takeoff, one 45 degree elbow, and one 90 degree elbow:

$$L_{a3} = 12 + 11 + 1.5 + 3 = 27.5 \text{ m}$$

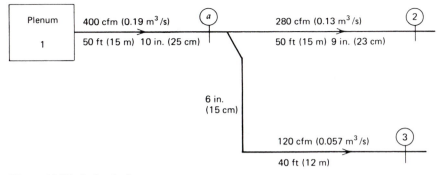

Figure 12-27 A simple duct system.

From Fig. 12-23, at 0.057 m^3/s and for a 15-cm diameter pipe, the lost pressure is 1.0 Pa/m of pipe. Then

$$\Delta P_{0a3} = (1.0)(27.5) = 27.5\,\text{Pa}$$

Section a to 2 has an equivalent length equal to the sum of the actual length and the equivalent length for the straight-through section of the branch fitting. A length of 2 m is assumed for this case:

$$L_{a2} = 15 + 2 = 17\,\text{m}$$

From Fig. 12-23, the loss per meter of length is 0.6 Pa and

$$\Delta P_{0a2} = (0.6)(17) = 10.2\,\text{Pa}$$

Then the lost pressure for section 1 to 2 is

$$\Delta P_{012} = \Delta P_{01a} + \Delta P_{0a2} = 22.1 + 10.2 = 32.3\,\text{Pa}$$

and for section 1 to 3

$$\Delta P_{013} = \Delta P_{01a} + \Delta P_{0a3} = 22.1 + 27.5 = 49.6\,\text{Pa}$$

12-9   TURNING VANES AND DAMPERS

The two main accessory devices used in duct systems are vanes and dampers. It is the responsibility of the design engineer to specify the location and use of the devices.

Turning vanes have the purpose of preventing turbulence and consequent high loss in total pressure where turns are necessary in rectangular ducts. Although large-radius turns may be used for the same purpose, this requires more space. When turning vanes are used, an abrupt 90 degree turn is made by the duct, but the air is turned smoothly by the vanes. Turning vanes are of two basic designs. The airfoil type is the more efficient of the two, but it is more expensive to fabricate than the single-piece flat vane.

Dampers are necessary to balance a system and to control makeup and exhaust air.

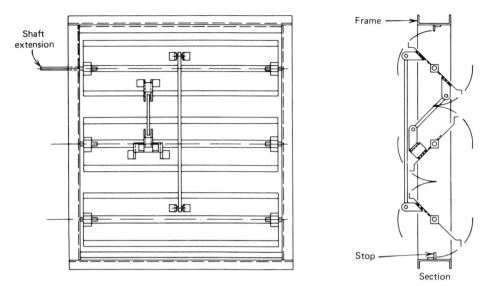

Figure 12-28 Typical opposed blade damper assembly.

The dampers may be hand operated and locked in position after adjustment or may be motor operated and controlled by temperature sensors or by other remote signals. The damper may be a single blade on a shaft or a multiblade arrangement as shown in Fig. 12-28. The blades may also be connected to operate in parallel. The damper causes a pressure loss even when full open. The loss coefficient C_0 is about 0.52 with the blades in the full open position.

12-10 DUCT DESIGN—GENERAL CONSIDERATIONS

The purpose of the duct system is to deliver a specified amount of air to each diffuser in the conditioned space at a specified total pressure. This is to ensure that the space load will be absorbed and the proper air motion within the space will be realized. The method used to lay out and size the duct system must result in a reasonably quiet system and must not require unusual adjustments to achieve the proper distribution of air to each space. A low noise level is achieved by limiting the air velocity, by using sound-absorbing duct materials or liners, and by avoiding drastic restrictions in the duct such as nearly closed dampers. Figure 12-29 gives recommended velocities and pressure losses for duct systems. A low-velocity duct system will generally have a pressure loss of less than 0.15 in. wg per 100 ft (1.23 Pa/m), whereas high-velocity systems may have pressure losses up to about 0.7 in. wg per 100 ft (5.7 Pa/m).

The use of fibrous glass duct materials has gained acceptance in recent times because they are very effective for noise control. These ducts are also attractive from the fabrication point of view because the duct, insulation, and reflective vapor barrier are all the same piece of material. Metal ducts are usually lined with fibrous glass material in

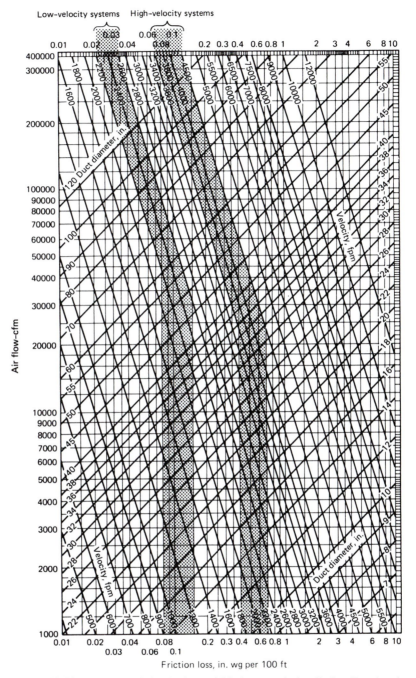

Figure 12-29 Recommended velocity and friction rate design limits. (Reprinted by permission from *ASHRAE Handbook, Fundamentals Volume,* 1985.)

the vicinity of the air distribution equipment and for some distance away from the equipment. The remainder of the metal duct is then wrapped or covered with insulation and a vapor barrier. Insulation on the outside of the duct also reduces noise.

The duct system should be relatively free of leaks, especially when the ducts are outside the conditioned space. Air leaks from the system to the outdoors result in a direct loss that is proportional to the amount of leakage and the difference in enthalpy between the outdoor air and the air leaving the conditioner. Presently 1 percent of the total volume flow of the duct system is an accepted maximum leakage rate in high-velocity systems. No generally accepted level has been set up for low-velocity systems. However, care should be taken to tape or otherwise seal all joints to minimize leakage in all duct systems. The sealing material should have a projected life of 20 to 30 years.

The layout of the duct system is very important to the final design of the system. Generally the location of the air diffusers and air-moving equipment is first selected with some attention given to how a duct system may be installed. The ducts are then laid out with attention given to space and ease of construction. It is very important to design a duct system that can be constructed and installed in the allocated space. If this is not done, the installer may make changes in the field that lead to unsatisfactory operation.

The total pressure requirements of a duct system are an important consideration. From the standpoint of first cost, the ducts should be small; however, small ducts tend to give high air velocities, high noise levels, and large losses in total pressure. Therefore, a reasonable compromise between first cost, operating cost, and practice must be reached. The cost of owning and operating an air-distribution system can be expressed in terms of system parameters, energy cost, life of the system, and interest rates such that an optimum velocity or friction rate can be established (2). A number of computer programs are available for this purpose.

The total pressure requirements of a duct system are determined in two main ways. For unit-type equipment, all of the heating, cooling, and air-moving equipment is determined by the heating and/or cooling load. Therefore, the fan characteristics are known before the duct design is begun. Furthermore, the pressure losses in all other elements of the system except the supply and return ducts are known. The total pressure available for the ducts is then the difference between the total pressure characteristic of the fan and the sum of the total pressure losses of all the other elements in the system excluding the ducts. Figure 12-30 shows a typical total pressure profile for a unitary-type system. In this case the fan is capable of developing 0.6 in. wg at the rated capacity. The return grille, filter, coils, and diffusers have a combined loss in total pressure of 0.38 in. wg. Therefore, the available total pressure for which the ducts must be designed is 0.22 in. wg. This is usually divided for low-velocity systems so that the supply-duct system has about twice the total pressure loss of the return ducts.

Large commercial and industrial duct systems are usually designed using velocity or pressure loss as a limiting criterion, with the fan requirements determined after the design is complete. For these larger systems the fan characteristics are specified and the correct fan is installed in the air handler at the factory or on the job.

It has been the practice of some designers to neglect velocity pressure. This assumption does not simplify the duct design procedure and is unnecessary. When the air velocities are high, the velocity pressure must be considered to achieve reasonable accuracy. If static and velocity pressure are computed separately as required by some design methods, the problem becomes very complex. The trend is to use total pressure exclusively in duct design procedures because it is simpler and accounts for all of the flow energy.

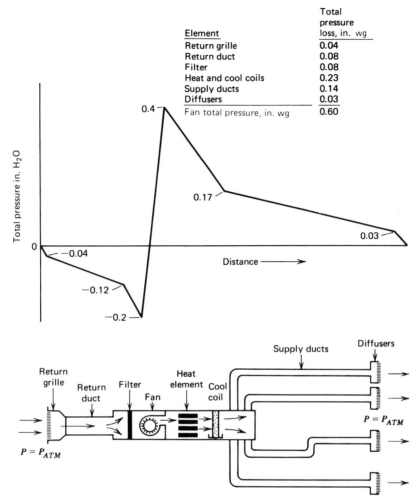

Element	Total pressure loss, in. wg
Return grille	0.04
Return duct	0.08
Filter	0.08
Heat and cool coils	0.23
Supply ducts	0.14
Diffusers	0.03
Fan total pressure, in. wg	0.60

Figure 12-30 Total pressure profile for a typical unitary system.

It is possible to lay out and size a duct system that cannot possibly function properly because the designer did not understand or failed to recognize the system pressure dynamics. In this connection, the concept of pressure and energy grade lines is useful in understanding a complex air distribution system.

Pressure Gradient Diagrams

As stated earlier, the total pressure in a duct system at any location is the sum of the static pressure and the velocity pressure. As frictional and dynamic effects occur in the airstream in the course of flow, an energy loss occurs that appears as a reduction in total pressure. Therefore, in any real duct system, except where energy is added with a fan, the total pressure will decrease in the direction of flow. This, however, is not necessarily true of the static and velocity pressures since the configuration of the air passages may cause trade-offs between the two.

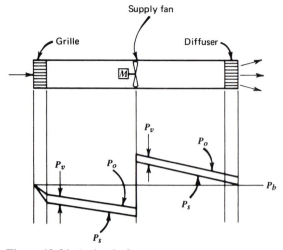

Figure 12-31 A simple fan system.

Figure 12-31 is a pressure gradient diagram for a simple fan system, where

P_0 = total pressure (in. wg or Pa)
P_s = static pressure (in. wg or Pa)
P_v = velocity pressure (in. wg or Pa)
P_b = barometric pressure

The line connecting all the respective total pressure points along the duct system is called the *energy grade line* (EGL). The line connecting all the static pressure points along the duct is called the *hydraulic grade line* (HGL). The vertical distance between these lines at any section is the velocity pressure. The reference pressure may be any value, although for simplicity local atmospheric pressure is normally used. For fans, fittings, entrances, exits, and the like, the loss is generally considered to be concentrated at the location of these devices, although the effects of these elements extend to some distance down- and upstream. Several points should be noted regarding the relationships depicted on the diagram. First, the only location at which there is a total pressure increase is at the point of external energy input, namely the fan. At all other locations, total pressure will decrease, in straight equal-area ducts at a rate equal to the pressure drop per unit length, and in fittings at a rate determined by the applicable dynamic losses. Second, the total pressure always equals or exceeds the static pressure by a quantity equal to the velocity pressure. Third, though the total pressure may not increase except at the fan, the static pressure may increase or decrease in relation to the arrangement of the system, at times in unusual manners.

Though this application may seem simple, similar applications occur frequently and are misunderstood because the pressure gradient diagram of the proposed system was not considered. Consider the simple industrial exhaust system shown in Figure 12-32. In this instance, the total pressure loss in section AB is 0.12 in. wg. However, the conversion of static pressure to velocity pressure at section B has caused another apparent loss in static pressure of one velocity head. A manometer placed at point B would read a negative static pressure of -0.37 in. wg relative to the room pressure. This

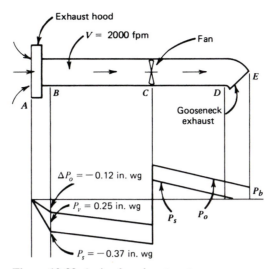

Figure 12-32 A simple exhaust system.

effect, in addition to the free discharge loss of an additional velocity head, has contributed significantly to the losses in this system in a manner difficult to visualize without the use of the pressure gradient diagram.

The pressure in any building served by an air system is dictated by the location of the fan and the duct system arrangement. The problem of determining, understanding, or controlling the relationship of the pressure in a building to the ambient or surrounding pressure is best understood through the use of the pressure gradient diagram. Figure 12-33 will begin to develop this concept more clearly. In this system, one additional element has been added: an outdoor-air intake. This system would be operable in this

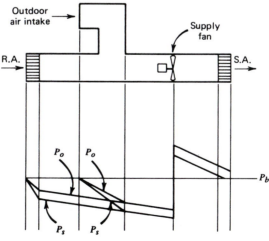

Figure 12-33 A simple duct system with outdoor air intake.

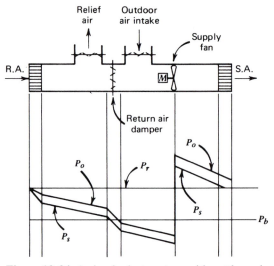

Figure 12-34 A simple duct system with outdoor air
intake and relief.

manner only if an exhaust system were operating to exhaust the outdoor air. Otherwise, the space pressure would have to be positive to permit the outdoor air to be exfiltrated from the building. There is no guarantee that this system will operate satisfactorily.

In many cases the return-air duct system is used as the route for the excess exhaust or relief air. Figure 12-34 displays such a pressure gradient diagram. In this case the pressure in the duct system must be above barometric pressure upstream of the return damper and below barometric pressure downstream of the return damper. In addition, and more importantly, the room pressure in this condition must exceed the barometric pressure by the static pressure losses in the return duct system plus the relief duct system. This condition will cause the familiar scene of exterior doors standing open with no possible means of correcting the problem short of modifying the return/relief duct system. In most cases the problem cannot be corrected without the use of an additional fan.

The use of a second fan can hold the building pressure closer to the barometric pressure. This system is shown in Fig. 12-35. In this case, the total energy required is still the same as in the case of Fig. 12-34 except that it is divided between the two fans. The return fan acts to negate the losses in the return/relief air system, furnishing the positive pressure necessary at the entrance to the relief point such that the outdoor air that has been introduced may be effectively exhausted. The pressure gradient diagram is invaluable as a tool for studying such systems to evaluate the methods and components for achieving proper system operation.

12-11 DESIGN OF LOW-VELOCITY DUCT SYSTEMS

The methods described in this section pertain to low-velocity systems as defined in Fig. 12-29. These methods can be used for high-velocity system design, but the results will not be satisfactory in some cases.

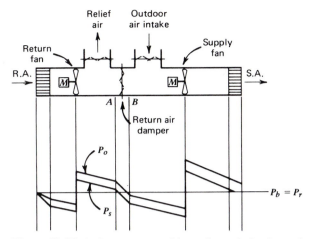

Figure 12-35 A duct system with outdoor air intake and relief with two fans.

Equal-friction Method

The principle of this method is to make the pressure loss per foot of duct length the same for the entire system. If the layout is symmetrical with all runs from fan to diffuser about the same length, this method will produce a good balanced design. However, most duct systems have a variety of duct runs ranging from long to short. The short runs will have to be dampered, which can cause considerable noise.

The usual procedure is to select the velocity in the main duct adjacent to the fan in accordance with Fig. 12-29. The known flow rate then establishes the duct size and the lost pressure per unit of length using Fig. 12-22. This same pressure loss per unit length is then used throughout the system. A desirable feature of this method is the gradual reduction of air velocity from fan to outlet, thereby reducing noise problems. After sizing the system, the designer must compute the total pressure loss of the longest run (largest flow resistance), taking care to include all fittings and transitions. When the total pressure available for the system is known in advance, the design loss value may be established by estimating the equivalent length of the longest run and computing the lost pressure per unit length.

EXAMPLE 12-11

Select duct sizes for the simple duct system of Fig. 12-36 using the equal friction method and SI units. The total pressure available for the duct system is 0.12 in. wg or 30 Pa and the loss in total pressure for each diffuser at the specified flow rate is 0.02 in. wg or 5 Pa.

SOLUTION

Because the system is small, the velocity in the main supply duct should not exceed about 1000 ft/min or 5 m/s and the branch duct velocities should not exceed about 600 ft/min or 3 m/s. The total pressure available for the ducts, excluding the diffusers, is 25

Pa. The total equivalent length of all three runs of duct are approximately the same, therefore let the longest run be 1-2-3. The equivalent-length method will be used to account for losses in the fittings. Then for simplicity if we use Figs. 12-25 and 12-26

$$L_{123} = (L_1 + L_{ent}) + (L_2 + L_{st}) + (L_3 + L_{wye} + L_{el} + L_{boot})$$

$$L_{123} = (6 + 11) + (4.6 + 1.5) + (3 + 11 + 3 + 9) = 49.1 \text{ m}$$

and

$$\Delta P_0' = \frac{\Delta P_0}{L_{123}} = \frac{25}{49.1} = 0.509 \text{ Pa/m}$$

This value will be used to size the complete system using Fig. 12-23. Table 12-15 summarizes the results showing the duct sizes, velocity in each section, and the loss in total pressure in each section. The rectangular sizes selected are rather arbitrary.

It is of interest to check the actual loss in total pressure from the plenum to each outlet.

$$(\Delta P_0)_{123} = 8.67 + 3.11 + 13.26 = 25.04 \text{ Pa}$$

$$(\Delta P_0)_{124} = 8.67 + 3.11 + 14.54 = 26.32 \text{ Pa}$$

$$(\Delta P_0)_{15} = 8.67 + 18.87 = 27.54 \text{ Pa}$$

The losses in total pressure for the three different runs are unequal when it is assumed that the proper amount of air is flowing in each. However, the actual physical situation is such that the loss in total pressure from the plenum to the conditioned space is equal for all runs of duct. That is, if the actual energy grade lines for each duct run were constructed, they would all begin at the plenum pressure and end at the room pressure. Therefore, the total flow rate from the plenum will divide itself among the three branches in order to satisfy the lost pressure requirement. If no adjustments are made to increase the lost pressure in sections 3 and 4, the flow rates in these sections will increase relative to section 5 and the total flow rate from the plenum will increase slightly because of the decreased system resistance. However, dampers in sections 3 and 4 could be adjusted to balance the system. Nevertheless this duct sizing method is used extensively. It is not always necessary that the system be designed to balance without adjustments.

Table 12-15 Solution to Example 12-11

Section Number	$\dot{Q}$ m³/s	D cm	$w \times h$ cm × cm	Velocity m/s	$\Delta P_0'$ Pa/m	L_e m	ΔP_0 Pa
1	0.237	30	40 × 18	3.5	0.51	17	8.67
2	0.166	26.5	38 × 16	3.2	0.51	6.1	3.11
3	0.071	19	28 × 11	2.6	0.51	26	13.26
4	0.095	21.5	32 × 13	2.8	0.51	28.5	14.54
5	0.071	19	28 × 11	2.6	0.51	37	18.87

Balanced-capacity Method

This method of duct design has been referred to as the "balanced pressure loss method" (6). However, it is the flow rate or capacity of each outlet that is balanced and not the pressure (7). As discussed earlier, the loss in total pressure automatically balances regardless of the duct sizes. The basic principle of this method of design is to make the loss in total pressure equal for all duct runs from fan to outlet when the required amount of air is flowing in each. In general all runs will have a different equivalent length and the pressure loss per unit length for each run will be different. It is theoretically possible to design every duct system to be balanced. This may be shown by combining Eqs. 10-6 and 12-9e to obtain

$$P_{01} - P_{02} = \Delta P_f = f \frac{L_e}{D} P_v \qquad (12\text{-}21)$$

Then for a given duct and fluid flowing

$$P_{01} - P_{02} = \Delta P_f(L_e, D, \overline{V}) \qquad (12\text{-}22\text{a})$$

Because the volume flow rate $\dot{Q}$ is a function of the velocity $\overline{V}$ and the diameter D, Eq. 12-22a may be written as

$$P_{01} - P_{02} = \Delta P_f(\dot{Q}, L_e) \qquad (12\text{-}22\text{b})$$

For a given equivalent length the diameter can always be adjusted to obtain the necessary velocity that will produce the required loss in total pressure. There may be cases, however, when the required velocity may be too high to satisfy noise limitations and a damper or other means of increasing the equivalent length will be required.

The design procedure for the balanced capacity method is the same as the equal friction method in that the design pressure loss per unit length for the run of longest equivalent length is determined in the same way depending on whether the fan characteristics are known in advance. The procedure then changes to one of determining the required total pressure loss per unit length in the remaining sections to balance the flow as required. The method shows where dampers may be needed and provides a record of the total pressure requirements of each part of the duct system. Example 12-12 demonstrates the main features of the procedure.

EXAMPLE 12-12

Design the duct system in Fig. 12-36 and Example 12-11 by using the balanced-capacity method. Use SI units.

SOLUTION

The total pressure available and the equivalent lengths will be the same as those in Example 12-11. In addition, the procedure for the design of the longest run L_{123} is exactly the same. Sections 4 and 5 must then be sized to balance the system. It is

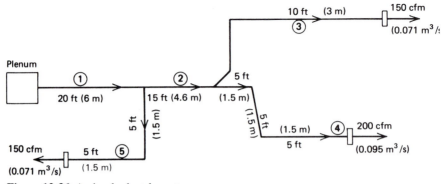

Figure 12-36 A simple duct layout.

obvious from Fig. 12-36 that the lost pressure in section 4 must equal that of section 3. Then from Table 12-15

$$\Delta P_{04} = \Delta P_{03} = 13.26 \, \text{Pa}$$

The equivalent length of section 4 is 28.5 m; therefore,

$$\Delta P'_{04} = \frac{\Delta P_{04}}{L_4} = \frac{13.26}{28.5} = 0.465 \, \text{Pa/m}$$

From Fig. 12-23, at $\Delta P'_{04} = 0.465$ and $\dot{Q} = 0.095 \, \text{m}^3/\text{s}$

$$D_4 = 22 \, \text{cm} \quad (26 \times 15 \, \text{cm rectangle})$$
$$\overline{V}_4 = 2.7 \, \text{m/s}$$

The loss in total pressure for section 5 is

$$\Delta P_{05} = \Delta P_{02} + \Delta P_{03} = 3.11 + 13.26 = 16.37 \, \text{Pa}$$

The equivalent length of section 5 is 37 m; therefore

$$\Delta P'_{05} = \frac{\Delta P_{05}}{L_5} = \frac{16.37}{37} = 0.442 \, \text{Pa/m}$$

From Fig. 12-23

$$D_5 = 19.5 \, \text{cm} \quad (22 \times 15 \, \text{cm rectangle})$$
$$\overline{V}_5 = 2.4 \, \text{m/s}$$

Comparison of the diameters or rectangular sizes obtained for sections 4 and 5 using the two different methods discussed thus far shows the sizes to be different, as suggested in the discussion following Example 12-11. For these simple examples the differences were not dramatic. In real systems the equivalent lengths of the various runs vary considerably and the balanced-capacity method is superior to the equal-friction method when system balance is critical.

The only limitation of the balanced-capacity method is the use of equivalent lengths for the fittings. Experience has shown this to be a minor error for low-velocity systems. Further, when the equivalent lengths are derived as illustrated earlier, the method is quite accurate within one's ability to describe the system.

EXAMPLE 12-13

Design the duct system of Fig. 12-37 by using the balanced-capacity method. The velocity in the duct attached to the plenum must not exceed 900 ft/min and the overall loss in total pressure should not exceed about 0.32 in. wg. Total pressure losses for the diffusers are all equal to 0.04 in. wg. Rectangular ducts are required. The lengths shown are the *total* equivalent lengths of each section. Use English units.

SOLUTION

In order to hold the duct sizes to a minimum, the maximum velocity criteria will be used to establish the design pressure loss for the longest run of duct. From Fig. 12-22 the pressure loss per 100 ft and equivalent diameter of section 1 is

$$\Delta P'_{01} = 0.096 \text{ in. wg}/100 \text{ ft}$$

$$D_1 = 12.9 \text{ in.}$$

If this pressure loss per 100 ft is used to design sections 1–2–3–4–5, the loss in total pressure will be

$$\Delta P_{015} = 0.096(55 + 25 + 40 + 45 + 95)/100 = 0.250 \text{ in. wg}$$

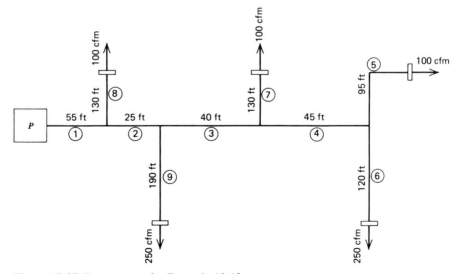

Figure 12-37 Duct system for Example 12-13.

When the diffuser loss is added, the maximum lost pressure criterion is still satisfied ($\Delta P_0 = 0.29$ in. wg); therefore, run 1–2–3–4–5 will be sized using $\Delta P'_{01} = 0.096$ in. wg/100 ft. Table 12-16 summarizes this procedure. Note that the main run of duct selected in this example is not the longest run. The total eqivalent length of section 6 is actually larger than section 5. However, the difference in small and the final results will be nearly identical.

Branches 6, 7, 8, and 9 must now be sized to balance the system. Because all of the diffusers have the same lost pressure, they need not be considered in the remainder of the analysis.

$$\Delta P_{06} = \Delta P_{05} = 0.091 \text{ in. wg}$$
$$L_6 = 120 \text{ ft}$$
$$\Delta P'_{06} = 0.091(100)/120 = 0.076 \text{ in. wg/100 ft}$$
$$D_6 = 8.7 \text{ in.} \quad (11 \times 6 \text{ rectangle})$$
$$\overline{V}_6 = 610 \text{ ft/min}$$

$$\Delta P_{07} = \Delta P_{04} + \Delta P_{05} = 0.044 + 0.091 = 0.135 \text{ in. wg}$$
$$L_7 = 130 \text{ ft}$$
$$\Delta P'_{07} = \frac{0.135(100)}{130} = 0.100 \text{ in. wg/100 ft}$$
$$D_7 = 5.8 \text{ in.} \quad (6 \times 5 \text{ rectangle})$$
$$\overline{V}_7 = 540 \text{ ft/min}$$

$$\Delta P_{09} = \Delta P_{03} + \Delta P_{07} = 0.038 + 0.135 = 0.173 \text{ in. wg}$$
$$L_9 = 190 \text{ ft}$$
$$\Delta P'_{09} = \frac{0.173(100)}{190} = 0.091 \text{ in. wg/100 ft}$$
$$D_9 = 8.4 \text{ in.} \quad (10 \times 6 \text{ rectangle})$$
$$\overline{V}_9 = 650 \text{ ft/min}$$

$$\Delta P_{08} = \Delta P_{09} + \Delta P_{02} = 0.173 + 0.024 = 0.197 \text{ in. wg}$$
$$L_8 = 130 \text{ ft}$$
$$\Delta P'_{08} = \frac{0.197(100)}{130} = 0.152 \text{ in. wg/100 ft}$$
$$D_8 = 5.4 \text{ in.} \quad (6 \times 4 \text{ rectangle})$$
$$\overline{V}_8 = 630 \text{ ft/min}$$

The resulting velocities for sections 8 and 9 are slightly high but are probably acceptable.

The previous examples have shown the balanced-capacity method to be straightforward and to produce much detailed information about the duct system, particularly with respect to the duct velocities and the placement of dampers. However, the calculation procedure is cumbersome and time-consuming when carried out for a complex system as shown. To alleviate these undesirable features, a worksheet has been devised. An example has been solved using the worksheet (Fig. 12-38). A full-sized worksheet is

Table 12-16 Data for Example 12-13

Section Number	$\dot{Q}$ cfm	D_e in.	$w \times h$ in. $\times$ in.	Velocity fpm	L_e ft	ΔP_0 in. wg
1	800	12.9	18 × 8	900	55	0.053
2	700	12.1	16 × 8	880	25	0.024
3	450	10.2	15 × 6	780	40	0.038
4	350	9.4	13 × 6	740	45	0.044
5	100	5.9	6 × 5	540	95	0.091
Diffuser 5	100	—	—	—	—	0.040
				TOTAL		0.290

enclosed in the packet at the end of the text, Chart 3. Note that the worksheet has provisions for recording information concerning the various elements of the air-distribution system including the fan, filters, cooling and heating coils, and the diffusers. One added feature of the worksheet is the provision for subbranches, which sometimes occur in complex systems. The following procedure is used with the duct design worksheet.

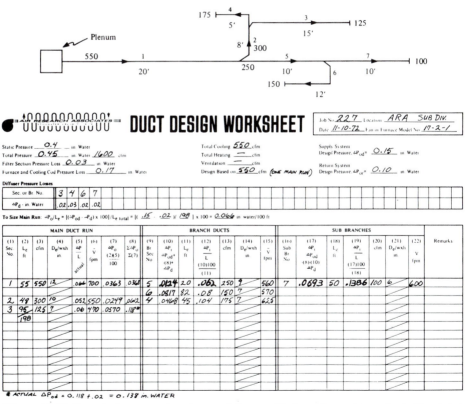

Figure 12-38 Duct design example using worksheet and balanced capacity method. (Reprinted with permission of Air Research Associates, Inc.)

1. Fill in the heading on the worksheet based on the equipment selected for the job.
2. Select diffusers for each outlet and record the total pressure requirements as shown.
3. Determine the main duct run (longest equivalent length). This is run 1 to 2 to 3 in the example; then section 4 is a branch, section 5 to 6 is a branch, and section 7 is a subbranch. Section 5 to 6 is selected as the branch because it has a larger equivalent length than section 5 to 7.
4. Enter the section numbers, equivalent lengths, and volume flow rate for each section as shown. Notice the arrangement of the section numbers with respect to the duct system. Reading across the first line, section 5 to 6 branches off section 1 and section 7 branches off section 5. Similarly, section 4 branches off section 2.
5. Calculate the loss in total pressure per 100 ft for the main duct run in the space provided and use the value and Fig. 12-22 to size the various sections. If rectangular ducts are to be used, record the exact duct diameter from the chart. If circular ducts are to be used, round off to the nearest standard diameter as shown for the example.
6. Record the actual loss per 100 ft in column 5, the velocity in column 6, and compute the actual loss in total pressure for each section; record in column 7. The accumulated loss in total pressure is recorded in column 8. The last number in column 8 is the actual loss in total pressure for the main duct run exclusive of the diffuser and should be less than or equal to the design value.
7. Size the branch ducts by computing the loss in total pressure for each branch as shown in column 10. Find $\Delta P/L$ in column 12 and select the duct diameter using Fig. 12-22. When a branch has two or more sections as shown, the branch is sized for the loss in total pressure and equivalent length for the complete branch. For this example, sections 5 and 6 must have a loss of 0.0817 in. wg and the total equivalent length is 102 ft. Then $\Delta P/L = 0.08$. From the friction chart $D_6 = 7$ in. and $D_5 = 8.5$ in. D_5 must be rounded up to 9 in. and then $\Delta P/L = 0.062$. The loss in total pressure for section 5 is finally calculated as 0.0124 in. wg.
8. The subbranches are sized in the same way as the branches. Only a few systems have more than one or two subbranches. Most systems have none.
9. If the ducts are to be rectangular, Table 12-8 may be used to convert the diameters to rectangular dimensions.

The completed worksheet makes a good record and provides a check on the velocities in each section. A check for computational errors may be made by summing the pressure losses horizontally in columns 8, 10, and 17 plus the lost pressure for the diffuser.

Return Air Systems

The design of the return system may be carried out using the methods described earlier. In this case the air flows through the branches into the main duct and back to the fan. Although the losses in constant-flow fittings are the same regardless of the flow direction, divided-flow fittings behave differently and different equivalent lengths or loss coefficients must be used. Table 12-13 gives loss coefficients for typical divided-flow fittings used in return systems. Reference 2 gives considerable data for converging-type fittings of both circular and rectangular cross section. It should be noted that for low velocity ratios the loss coefficient can become negative with converging flow streams. This seems contrary to natural laws; however, this behavior is a result of a high-velocity stream mixing with a low-velocity stream. Kinetic energy is transferred from the higher to the lower velocity air, which results in an increase in energy or total pressure of the slower stream. Low-velocity return systems are usually designed using the equal-friction method. The total pressure loss for the system is then estimated as discussed for supply duct systems. Dampers may be required just as with supply systems. In large commercial systems a separate fan for the return air may be required.

12-12 HIGH-VELOCITY DUCT DESIGN

Because space allocated for ducts in large commercial structures is very limited due to the high cost of building construction, alternatives to the low-velocity central system are often sought. One way this can be done is to move the air at much higher velocities. Figure 12-29 indicates that high-velocity systems may use velocities as high as 4500 ft/min (23 m/s) and lost pressures as high as 0.7 in. wg per 100 ft (5.7 Pa/m). The use of high velocities reduces the duct sizes dramatically, but introduces some new problems.

Noise is a serious consequence of high-velocity air movement. Special attention must be given to the design and installation of sound-attenuating equipment in the system. Generally, a sound absorber is installed just downstream of the fan and is similar to an automobile's muffler. Because the air cannot be introduced to the conditioned space at a high velocity, a device called a *terminal box* is used to throttle the air to a low velocity, control the flow rate, and attenuate the noise. The terminal box is located in the general vicinity of the space it serves and may distribute air to several outlets.

The energy required to move the air through the duct system at high velocity is also an important consideration. The total pressure requirement is typically on the order of several inches of water and is due to the large losses associated with the high velocities. To partially offset the high fan power requirements, variable-speed fans are often used to take advantage of low volume flow requirements of variable-volume systems at off-design conditions. Careful selection of duct fittings and the duct layout can also reduce the power requirement.

The higher static and total pressures required by high-velocity systems aggravate the duct leakage problem. Ordinary duct materials such as light-gauge steel and ductboard do not have suitable joints to prevent leakage and are also too weak to withstand the forces arising from the high pressure differentials. Therefore, an improved duct fabrication system has been developed for high-velocity systems. The duct is generally referred to as *spiral duct* and has either a round or oval cross section. The fittings are machine formed and are especially designed to have low pressure losses and close fitting joints to prevent leakage. The pressure losses in spiral pipe and fittings are generally different from those in ordinary galvanized duct and fittings. There is also a much wider variety of fittings available for high-velocity systems. For brevity, the data on pressure losses presented in Sections 12-7 and 12-8 will be used in this section. In actual practice, however, engineering data should be obtained from the manufacturer.

The criterion for designing high-velocity duct systems is somewhat different from that used for low-velocity systems. Emphasis is shifted from a self-balancing system to one that has minimum losses in total pressure. The high-velocity system achieves the proper flow rate at each outlet through the use of the terminal box. The terminal box is designed to operate with a minimum pressure loss of about 0.25 to 1.0 in. wg or 60 to

245 Pa. Some of these devices maintain a constant volume flow rate, whereas others vary the flow rate in proportion to the space heating or cooling requirement. A double-duct type may also be used where warm and cool air are blended according to the space requirement. The run of duct that has the largest potential flow resistance should be designed to have a total pressure loss as low as possible. This run will then determine the system total pressure. The remaining runs will presumably result in a lower lost pressure but will be balanced by the terminal box.

Total pressure losses must be estimated carefully using the loss-coefficient method (Eq. 12-17), because the equivalent-length method is not reliable at the velocities used with high-velocity systems. The most efficient fittings from the standpoint of lost pressure should be selected. For example, an ordinary square entrance to a duct from a plenum may be used in a low-velocity system, but when used with high-velocity air, the lost pressure becomes prohibitive. Table 12-11 indicates that the loss coefficient for a formed entrance is about one-tenth that of a square entrance. Similar savings in total pressure may be achieved in the selection of branch fittings. Manufacturer's data should be consulted and the best fitting selected. The designer should become acquainted with the standard practices for duct construction (8) before attempting an actual design problem.

There are a number of duct design procedures that may be used for high-velocity systems. The design procedures are relatively independent of the type of air-distribution system involved. For example, a variable-air-volume duct system should be designed for the full-load condition when air flow is at a maximum. At full load the VAV system operates the same as a constant-volume system. However, a VAV system may be very extensive, serving zones that peak at much different times. In such a case, each section of the duct system must be sized to handle the maximum amount of air required.

Static Regain Method

This method systematically reduces the air velocity in the direction of flow in such a way that the increase (regain) in static pressure in each transition just balances the pressure losses in the following section. This method is suitable for high-velocity, constant-volume systems having long runs of duct with many takeoffs. With this procedure approximately the same static pressure exists at the entrance to each branch, which simplifies terminal unit selection and system balancing. The main disadvantages of the method are (1) the very low velocities and large duct sizes that may result at the end of long runs; (2) the tedious bookkeeping and trial-and-error aspects of the method; and (3) the total pressure requirements of each part of the duct system are not readily apparent.

The general procedure for use of the static regain method is to first select a velocity for the duct attached to the fan or supply plenum. With the capacity this establishes the size of this main duct. The run of duct that appears to have the largest flow resistance is then designed first, using the most efficient fittings and layout possible. A velocity is assumed for the next section in the run and the static pressure regain is compared to the lost pressure for that section. Usually about two velocities must be checked to find a reasonable balance between the static pressure regain and the losses of a section. It must also be kept in mind that standard duct sizes must be used, which usually prevents an exact balance between regain and losses. The spiral duct is available in diameters of 3 to 24 in. using increments of 1 in. and from 24 to 50 in. using increments of 2 in. Standard metric pipe ranges from about 8 to 60 cm in 1-cm increments and from 60 to 120 cm in 2-cm increments. The following example demonstrates the procedure for a simple duct system.

EXAMPLE 12-14

Design the duct system shown in Fig. 12-39 using the static regain method. Each outlet has a terminal box that requires a minimum of 0.5 in. wg total pressure. Other pertinent data are shown on the sketch.

SOLUTION

The first step in the solution is to select a velocity for the section of duct connected to the plenum. Figure 12-29 indicates that a maximum velocity of 2500 fpm should be used when the duct carries 2000 cfm. The section may then be sized using Fig. 12-22. A 12-in. duct will result in a velocity of about 2550 fpm, which is acceptable. Then

$$D_1 = 12 \text{ in.} \qquad \overline{V}_1 = 2550 \text{ fpm} \qquad P_{v1} = 0.405 \text{ in. wg}$$

From Table 12-11B assume a loss coefficient for the formed entrance of 0.03 and from Fig. 12-22 a pressure loss per 100 ft of duct of $\Delta P'_{01} = 0.75$ in. wg/100 ft. Then

$$(\Delta P_0)_p = 0.03 \, P_{v1} = 0.03(0.405) = 0.012 \text{ in. wg}$$

$$(\Delta P_0)_{pa} = \frac{\Delta P'_{01} L}{100} = \frac{0.75(50)}{100} = 0.375 \text{ in. wg}$$

and

$$\Delta P_{01} = 0.387 \text{ in. wg}$$

The total pressure required at the plenum to meet the needs of section 1 is then given by

$$P_{0p} = P_{01} + P_{v1} = 0.387 + 0.405 = 0.792 \text{ in. wg}$$

If the static-region method could be followed exactly, this would be the system total pressure, exclusive of the terminal box requirement, because velocity pressure is to be converted to static pressure to offset the lost pressure.

The run of duct with the largest flow resistance appears to be sections 1–2–3–4, but this must be checked at the conclusion of the design. To size section 2 it is required that the increase in static pressure (decrease in velocity pressure) from point *a* to point *b*

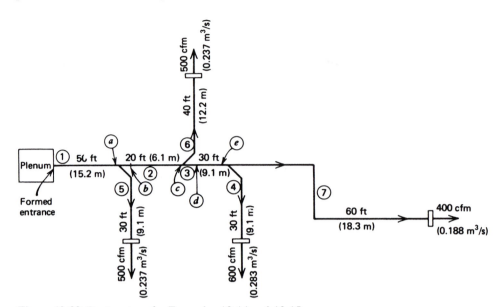

Figure 12-39 Duct system for Examples 12-14 and 12-15.

must equal the lost pressure from point b to point c. This is equivalent to

$$P_a - P_c = P_{vc} - P_{va} + (\Delta P_0)_{ab} + (\Delta P_0)_{bc} = 0 \qquad \text{(12-23a)}$$

or

$$P_a - P_c = P_{v2} - P_{v1} + \Sigma(\Delta P_0)_{ac} = 0 \qquad \text{(12-23b)}$$

and this may be generalized as

$$P_u - P_d = P_{vd} - P_{vu} + \Sigma(\Delta P_0)_{ac} = 0 \qquad \text{(12-23c)}$$

where the subscripts u and d refer to upstream and downstream, respectively. To size section 2 a diameter of 11 in. will be assumed and calculations made to check against Eq. 12-23c.

$$D_2 = 11 \text{ in.} \qquad \overline{V}_2 = 2000 \text{ fpm}; \qquad \Delta P'_{02} = 0.53 \text{ in. wg/100 ft}$$

$$P_{v2} = 0.25 \text{ in. wg} \qquad \dot{Q}_2 = 1500 \text{ cfm}$$

$$(\Delta P_0)_{ab} = 0.03 \text{ in. wg} \quad \text{(assumed, see Table 12-12B)}$$

$$(\Delta P_0)_{bc} = 0.53(20)/100 = 0.106 \text{ in. wg}$$

$$\Sigma(\Delta P_0)_{ac} = 0.136 \text{ in. wg}$$

Then using Eq. 12-23b

$$P_a - P_c = 0.25 - 0.405 + 0.136 = -0.019 \text{ in. wg}$$

This indicates that a larger increase in static pressure than required was achieved in the transition. Although the next smaller pipe size could be tried, the 11-in. duct would prove to be the best solution.

The same general procedure is used to size section 3 assuming a 10-in. diameter duct.

$$D_3 = 10 \text{ in.} \qquad \bar{V}_3 = 1850 \text{ fpm} \qquad \Delta P'_{03} = 0.5 \text{ in. wg/100 ft}$$

$$P_{v3} = 0.213 \text{ in. wg} \qquad \dot{Q}_3 = 1000 \text{ cfm}$$
$$(\Delta P_0)_{cd} = 0.015 \text{ in. wg} \quad \text{(assumed, see Table 12-12B)}$$

$$(\Delta P_0)_{de} = \frac{0.5(30)}{100} = 0.15 \text{ in. wg}$$

$$\Sigma(\Delta P_0)_{ce} = 0.165 \text{ in. wg}$$

$$P_c - P_e = 0.2313 - 0.25 + 0.165 = 0.128 \text{ in. wg}$$

Assume another diameter for section 3.

$$D_3 = 11 \text{ in.} \qquad \bar{V}_3 = 1520 \text{ cfm} \qquad \Delta P'_{03} = 0.31 \text{ in. wg/100 ft}$$

$$P_{v3} = 0.144 \text{ in. wg} \qquad \dot{Q}_3 = 1000 \text{ cfm}$$

$$(\Delta P_0)_{cd} = 0.017 \text{ in. wg} \quad \text{(assumed, see Table 12-12B)}$$

$$(\Delta P_0)_{de} = 0.31(30)/100 = 0.093 \text{ in. wg}$$

$$\Sigma(\Delta P_0)_{ce} = 0.11 \text{ in. wg}$$

$$P_c - P_e = 0.004 \text{ in. wg}$$

This is a very good solution for section 3. It is obvious that the design of a system using the foregoing procedure can become quite tedious. A tabular form reduces the work considerably and will be used to conclude this design. Table 12-17 contains all the design calculations for this example.

The total pressure requirement for the system is determined by the run with the maximum cumulative static pressure loss plus the velocity pressure in section 1.

$$\Delta P_0 = (\Delta P)_{1237} + P_{v1} = 0.392 + 0.405 = 0.797 \text{ in. wg}$$

where $\Delta P_{1237} = \Sigma(P_u - P_d)$ from Table 12-17. However, this does not include the pressure loss for the terminal box. In allowing for this pressure, remember that the total pressure just upstream of the box is equal to the velocity pressure. Section 7 has a velocity pressure of 0.052 in. wg. Because the terminal box requires at least 0.5 in. wg, the system total pressure requirement should be increased to

$$\Delta P_0 = 0.797 + (0.5 - 0.052) = 1.245 \text{ in. wg}$$

The required total pressure for this example is relatively low because of its small size. As high-velocity systems become larger, the maximum velocity may be increased according to Fig. 12-29 with a corresponding increase in required total pressure.

Table 12-17 Static-Regain Design Method[b]

Section Number	Item	D in.	$\dot{Q}$ cfm	$\bar{V}$ fpm	P_v in. wg	L ft	$\Delta P_0'$ or C in. wg per 100 ft	$\Sigma\Delta P_0$ in. wg	ΔP_0 in. wg	$(P_u - P_d)$[a] in. wg	$\Sigma(P_u - P_d)$ in. wg
1	Duct	12	2000	2550	0.405	50	0.75	0.375	0.387	0.387	0.387
	Entrance						0.03	0.012			
2	Duct	11	1500	2000	0.250	20	0.53	0.106	0.136	−0.019	0.368
	Fitting							0.030			
3	Duct	10	1000	1850	0.213	30	0.50	0.150	0.165	0.128	0.372
	Fitting							0.015			
4	Duct	11	1000	1520	0.144	30	0.31	0.093	0.110	0.004	0.390
	Fitting							0.017			
	Duct	9	600	1350	0.114	30	0.32	0.096	0.147	0.117	
	Fitting						0.45	0.051			
	Duct	10	600	1100	0.075	30	0.19	0.057	0.087	0.018	
	Fitting						0.40	0.030			
7	*Branch 7 off Section 3*										
	Duct	9	400	910	0.052	60	0.15	0.090	0.112	0.020	0.392
	Fitting							0.012			
	Fitting						0.200	0.010			
6	*Branch 6 off Section*										
	Duct	8	500	1450	0.131	40	0.41	0.164	0.216	0.097	0.332
	Fitting						0.40	0.052			
	Duct	9	500	1140	0.081	40	0.23	0.092	0.133	−0.036	
	Fitting						0.50	0.041			
5	*Branch 5 off Section 1*										
	Duct	8	500	1450	0.131	30	0.41	0.123	0.189	−0.085	0.302
	Fitting						0.50	0.066			

[a]Eq. 12-23c.

[b]Loss coefficients have been estimated for this example.

EXAMPLE 12-15

Suppose the duct system of Example 12-14 handles a total volume flow of 23 m^3/s or 49,000 cfm. Design section 1 for the maximum velocity allowable from Fig. 12-29 and estimate the total pressure required for the system assuming static regain design. Assume a minimum total pressure of 245 Pa for the terminal boxes.

SOLUTION

Figure 12-29 shows a maximum velocity of 4500 ft/min or 23 m/s for this case. Then using Fig. 12-23 the duct (section 1) is sized as

$$D_1 = 1.2 \text{ m} \qquad \overline{V}_1 = 22.5 \text{ m/s} \qquad \Delta P'_{01} = 3.2 \text{ Pa/m}$$

Assume the loss coefficient for the formed entrance is 0.03 and the velocity pressure in section 1 is

$$P_{v1} = \left(\frac{22.5}{1.29}\right)^2 = 304.2 \text{ Pa}$$

Then

$$(\Delta P_0)_p = 0.03(304.2) = 9.1 \text{ Pa}$$
$$(\Delta P_0)_{pa} = 3.2(15.2) = 48.6 \text{ Pa}$$

The total pressure loss for section 1 is then 58 Pa. A conservative estimate of the system total pressure is given by

$$P_0 = P_{01} + P_{v1} + (P_0)_{\text{box}}$$
$$P_0 = 58 + 304 + 245 = 607 \text{ Pa}$$

Optimization Procedures

The availability of duct fitting data bases (5) and sophisticated computers and software has made it possible to develop duct design programs which speed the process and allow more extensive analysis. As mentioned above, it is desirable to achieve peak performance in a cost effective way. Said in another way, the duct system should be optimized with respect to cost of construction and operation.

One optimization procedure is referred to as the T-Method (2, 9) which is comprised of the following major procedures.

System condensing. This procedure condenses a branched system into a single imaginary duct section with identical hydraulic characteristics and the same owning cost as the actual system. The entire supply and return system can be condensed into one section.

Fan selection. From the condensed system, the ideal fan total pressure is calculated and used to select a fan.

System expansion. The expansion process distributes the available fan pressure throughout the system. This procedure starts at the root section and proceeds toward the terminals.

Optimization. An objective function is developed based on present worth owning and operating basis. Initial cost, energy cost, hours of operation, annual escalation and interest rates, and the amortization period are required. The following constraints are necessary. Continuity must be maintained at all junctions. The loss in total pressure for each path must equal the fan total pressure. Nominal duct sizes are required; therefore, size rounding must be considered. Maximum allowable velocity is an acoustic limitation. Finally, architectural limitations on duct size must be considered.

Reference 9 describes the T-Method in detail. Computer programs are available to carry out the analysis.

Accessories

The main accessories used with high-velocity duct systems are sound absorbers and dampers. Sound absorbers were mentioned earlier and may take many forms. Although the design of these elements is beyond the scope of this book, manufacturer's data or other references on acoustics and noise should be consulted.

Balancing dampers may be used in high-velocity systems; however, this function is partially handled by the terminal box. Fire dampers are of great importance in this system. The function of the fire damper is to divert the flow of air to or from a particular area if a fire should occur. Large commercial structures have partitions, floors, and ceilings that confine a fire to a given area for some specified time. When an air duct passes through one of these fire barriers, a fire damper is generally required. Some of these dampers are held open by fusible links, whereas others are controlled by smoke detectors or other such devices.

The location and control of the fire dampers is the responsibility of the design engineer, who must become acquainted with the governing codes for the particular application and city and state in which he or she is practicing.

REFERENCES

1. *ASHRAE Handbook, Systems and Equipment Volume,* American Society of Heating, Refrigerating and Air-Conditioning Engineers, Inc., Atlanta, GA, 1992.
2. *ASHRAE Handbook, Fundamentals Volume,* American Society of Heating, Refrigerating and Air-Conditioning Engineers, Inc., Atlanta, GA, 1989.
3. J. B. Graham, "Methods of Selecting and Rating Fans," *ASHRAE Journal,* Vol. 14, No. 1, 1972.
4. *AMCA Fan Application Manual,* Parts 1, 2, and 3, Air Movement and Control Association, Inc., 30 West University Drive, Arlington Heights, IL, 1990.
5. *ASHRAE Duct Fitting Database,* American Society of Heating, Refrigerating and Air-Conditioning Engineers, Inc., Atlanta, GA, 1992.
6. Burgess H. Jennings, *Environmental Engineering Analysis and Practice,* International Textbook Company, Scranton, PA, 1970.

7. F. C. McQuiston, "Duct Design for Balanced Air Distribution in Low Velocity Systems," *Proceedings,* Conference on Improved Efficiency in HVAC Components, Purdue University, Lafayette, IN, 1974.
8. *High Pressure Duct Construction Standards,* 3rd ed., Sheet Metal and Air-Conditioning Contractors National Association, Washington, DC, 1975.
9. R. J. Tsal, H. F. Behls, and R. Mangel, "T-Method Duct Design: Part I and II," *ASHRAE Trans.,* Vol. 94, Part 2, 1988.

PROBLEMS

12-1. A centrifugal fan is delivering 1700 cfm (0.8 m^3/s) of air at a static pressure differential (across the fan) of 1.0 in. wg (250 Pa). The fan has an outlet area of 0.71 ft^2 (0.07 m^2) and requires 0.7 hp (0.52 kW) shaft input. Compute (a) the static power, (b) the static efficiency, (c) the total efficiency, and (d) the fan total pressure.

12-2. The fan of Problem 12-1 is operating at 1350 rpm. The fan speed is reduced to 1200 rpm. Compute the capacity, static and total pressure, and the shaft power at the lower speed.

12-3. Consider the fan curves shown in Fig. 12-7 and assume the data given are for a speed of 900 rpm. (a) Construct a new total pressure characteristic for a speed of 750 rpm for the good selection zone. (b) Construct a new shaft power curve for 750 rpm. (c) Sketch the total pressure and power characteristics for both speeds on the same graph with a system characteristic for 10,000 cfm (900 rpm). (d) What is the total pressure, volume flow rate, and shaft power at 750 rpm?

12-4. The fan of Fig. 12-7 is rated at standard sea level density. Suppose it is to be used in Denver, Colorado, where the elevation is 5280 ft. (a) Construct total pressure and shaft power characteristics for the selection zone for the new condition. (b) Compute the percent change in power between sea level and 5280 ft elevation for a volume flow rate of 10,000 cfm.

12-5. The fan of Fig. 12-10 is rated at sea level. Suppose it is to be used in Albuquerque, New Mexico, where the elevation is 1618 m. (a) Construct the total pressure characteristic for 800 rpm. (b) Compute the percent change in power from sea level to 1618 m for 800 rpm and a volume flow rate of 155 m^3/min.

12-6. Comment on the desirability of using the fan described in Fig. 12-7 to circulate (a) 10,000 cfm at about 1.8 in. wg total pressure, (b) 15,000 cfm at about 0.9 in. wg total pressure, (c) 2500 cfm at 2.0 in. wg total pressure.

12-7. Would the fan shown in Fig. 12-9 be suitable for use in a system requiring (a) 5000 cfm at 1.5 in. wg total pressure? (b) 30,000 cfm at 5.0 in. wg total pressure? (c) 15,000 cfm at 4.0 in. wg total pressure? Explain.

12-8. A duct system has been designed for 150 m^3/min at 500 Pa total pressure. Would the fan shown in Fig. 12-10 be suitable for this application? Explain and estimate the total efficiency, fan speed, and shaft power.

12-9. A small system requires 0.88 in. wg total pressure at a flow rate of 1420 cfm. Select a suitable fan using the data of Table 12-2a. (a) Sketch the system and fan characteristics showing the operating point. (b) What are the fan speed and power?

12-10. A system requires 1.0 kPa total pressure at a flow rate of 6.25 m³/s. (a) Select a fan using Table 12-2*b*. (b) Sketch the system and fan characteristics showing the operating point. (c) What are the fan speed and power?

12-11. A duct system has been designed to have 2.5 in. wg total pressure loss. It is necessary to use an elbow at the fan outlet that turns the air 90 degrees to the side. A system effect factor of 0.2 in. wg is given for this situation. The fan inlet also has an elbow with a system effect factor of 0.15 in. wg. (a) What total pressure should the fan for this system produce to circulate the desired amount of air? (b) Sketch the system and fan characteristics with and without the system effect factors. (c) The design volume flow rate is 15,000 cfm. Using Fig. 12-9 estimate the resulting flow rate when the system effect factors are not used.

12-12. The fan shown in Fig. 12-10 is operating in a system at 900 rpm with a flow rate of 150 m³/min. The system was designed to circulate 170 m³/min. Assuming that the duct pressure losses were accurately calculated, estimate the system effect factor for the fan.

12-13. A SWSI, backward-curved blade fan discharges air into a 12 × 16-in. rectangular duct at the rate of 6700 cfm. An elbow located 20 in. from the fan outlet turns to the right. Estimate the system effect for the elbow.

12-14. The fan of Problem 12-13 also has an inlet duct and elbow as shown in Fig. 12-15*b*. The diameter is 16 in. and the duct length is 32 in. Estimate the system effect.

12-15. A centrifugal fan discharges 10,000 cfm (4.7 m³/s) through a 20 × 20-in. (0.5 × 0.5-m) duct into a plenum. The duct length is 10 in. (0.25 m) and the blast area ratio is 0.8. Find the lost pressure (system effect) for this case.

12-16. A forward-curved blade fan with a 12 × 12-in. outlet is to be installed in a system to deliver 2500 cfm. What length of outlet duct is required to prevent any system effect if an elbow in position *A* is to be used?

12-17. The fan of Problem 12-16 has an inlet duct configured as shown in Fig. 12-15*e* with an *R/H* ratio of 1.0 and *H* of 12 in. About how long must the duct be so that the system effect will not exceed 0.16 in. wg?

12-18. The fan shown in Fig. 12-9 was selected for a constant-volume system requiring 15,000 cfm and a total pressure of 4.5 in. wg. The operating speed was expected to be about 900 rpm. (a) When the system was started, measurements were made and a flow rate of 9000 cfm and total pressure of 2.0 in. wg were observed. Explain the probable cause of the discrepancy. (b) At another time, after the problem of part (a) was corrected, a flow rate of 10,000 cfm and a total pressure of 4.8 in. wg was observed. Explain the possible problem in this case. (c) Several years later, after the problem of part (b) was solved, measurements show a flow rate of 13,000 cfm and a total pressure of 4.3 in. wg. Explain the probable cause of the decreased flow rate.

12-19. Refer to Fig. 12-20 and compute the percent difference in shaft power between flow conditions 1 and 2.

12-20. A variable-volume system using a fan as shown in Fig. 12-20 will operate at about 15,000 cfm a majority of the time. Estimate the power saved, in kW-hr for one day, as compared with a constant-volume system operating at point 1.

12-21. Consider the fan of Fig. 12-19, operating as shown in a VAV system. Compute the percent decrease in shaft power between flow conditions 1 and 2.

12-22. Assume that the fan and VAV system shown in Fig. 12-19 operate at an average capacity of 25,000 cfm over a given 24-hr period. (a) Estimate the power savings as compared with a constant-volume system operating at point 1. (b) Estimate the power saving as compared with a VAV system with no fan control, that is, where point 1 will move along the full open characteristic.

12-23. Estimate the lost pressure in 50 ft of 12 × 10-in. metal duct with an air flow rate of 2000 cfm. The duct is lined with 1 in. of the type B liner shown in Fig. 12-24.

12-24. Assuming the duct of Problem 12-23 is operating with standard sea level air, estimate the lost pressure for air at the same temperature but at an altitude of 5000 ft.

12-25. Compute the lost pressure for a 16-m length of 0.3 × 0.25-m duct lined with 25 mm of the type C liner shown in Fig. 12-24. The volume flow rate is 0.8 m³/s.

12-26. Compute the lost pressure for the duct of Problem 12-25 for an altitude of 1500 m.

12-27. A circular metal duct 20 ft in length has an abrupt contraction at the inlet and an abrupt expansion at the exit. Both have an area ratio of 0.6. The duct has a diameter of 10 in. with a flow rate of 600 cfm. Estimate the loss in total pressure for the duct including the contraction and expansion.

12-28. Compare the lost pressure for a bellmouth entrance, $r/D = 0.06$, Table 12-11B, with an abrupt entrance, $\theta = 180$ degrees, Table 12-11A, for (a) duct velocity of 1000 ft/min (5 m/s) and (b) 4000 ft/min (20 m/s).

12-29. Compute the lost pressure for a 14-in. (350-mm) pleated, 90 degree elbow with a volume flow rate of 1200 cfm (0.6 m³/s) of standard air.

12-30. Compute the lost pressure for a 16 × 16-in. (400 × 400-mm) 90 degree mitered elbow with a volume flow rate of 2500 cfm (1.2 m³/s) of standard air with (a) single-thickness vanes, design 3, and (b) without vanes.

12-31. Compute the lost pressure for a diverging wye fitting with a 45 degree branch. The flow rate in the 12-in. (30-cm) upstream section is 800 cfm (0.38 m³/s) and the flow rate in the 6-in. (15-cm) branch is 250 cfm (0.12 m³/s). The downstream section has a diameter of 10 in. (25 cm).

12-32. Use the data of Problem 12-31 and determine the loss coefficients based on the common (upstream) section.

12-33. Compute the lost pressure for a diverging tee fitting. Use the data of Problem 12-31.

12-34. Refer to Problem 12-33 and determine the loss coefficients based on the common (upstream) section.

12-35. Compute the lost pressure for a converging type fitting. The flow rate in the 8-in. (20-cm) upstream section is 500 cfm (0.24 m³/s) and the flow rate in the 8-in. (20-cm) branch is 500 cfm (0.24 m³/s). The downstream section has a diameter of 12 in. (25 cm). Assume (a) a 45 degree wye and (b) a tee.

12-36. The computed lost pressure for a fitting is 0.2 in. wg (50 Pa) based on sea level conditions. What is the lost pressure based on conditions in Denver, Colorado?

12-37. Compute the loss in total pressure for each run of the duct system shown in Fig. 12-40. The ducts are of round cross section. Turns and fittings are as shown. Use the loss coefficient and the equivalent length approaches (Fig. 12-25 and 12-26) and compare the answers. Use English units.

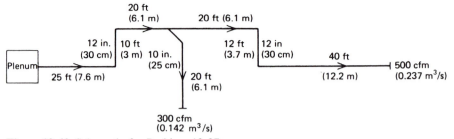

Figure 12-40 Schematic for Problem 12-37.

12-38. Refer to Problem 12-28 and compute the equivalent lengths for the two different entrances assuming a duct diameter of 12 in. (30 cm). Compute the lost pressure using the equivalent lengths.

12-39. Refer to Problem 12-29 and compute the equivalent lengths for the elbow. Compute the lost pressure using the equivalent lengths.

12-40. The duct system shown in Fig. 12-41 is one branch of a complete air-distribution system. The system is a perimeter type located below the floor. The diffuser boots are shown in Fig. 12-26. Size the various sections of the system using the equal friction method and round pipe. A total pressure of 0.13 in. wg is available at the plenum. Compute the actual loss in total pressure for each run assuming that the proper amount of air is flowing.

12-41. Consider the duct layout shown in Fig. 12-42. The system is supplied air by a rooftop unit that develops 0.25 in. wg total pressure external to the unit. The return air system requires 0.10 in. wg. The ducts are to be of round cross section and the maximum velocity in the main run is 850 ft/min, whereas the branch velocities must not exceed 650 ft/min. (a) Size the ducts using the equal-friction method. Show the location of any required dampers. Compute the total pressure loss for the system. (b) Size the ducts using the balanced-capacity method and the duct design worksheet.

12-42. Design the duct system shown in Fig. 12-43 for circular ducts. The fan produces a total pressure of 0.70 in. wg at 1000 cfm. The lost pressure in the filter, furnace, and evaporator is 0.35 in. wg. The remaining total pressure should be

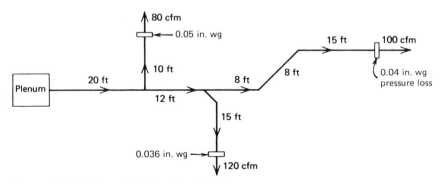

Figure 12-41 Schematic for Problem 12-40.

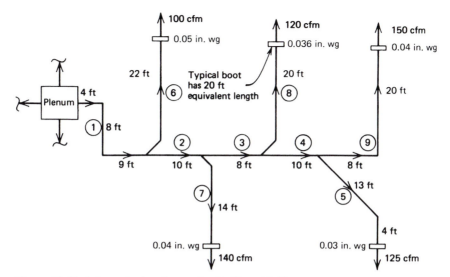

Figure 12-42 Schematic duct layout for Problem 12-41.

divided between the supply and return with 65 percent used for the supply system. Diffuser and grille losses are shown on the diagram. Use equivalent lengths to account for fitting losses. (a) Use the equal-friction method to size the ducts. (b) Use the balanced-capacity method and the duct design worksheet to size the ducts.

12-43. Design the duct system shown in Fig. 12-44 using the balanced-capacity method and the duct design worksheet. Circular ducts are to be used and installed below a concrete slab. The total pressure available at the plenum is 0.18 in. wg, and each diffuser has a loss in total pressure of 0.025 in. wg. Use the equivalent-length method of accounting for losses in the fittings.

12-44. A high-velocity duct layout is shown in Fig. 12-45. (a) Size the system using the static regain design method. Use maximum allowable velocities from Fig. 12-29. (b) Specify the required total pressure in the plenum. (c) Specify the fan shaft horsepower, efficiency, speed, and total pressure using Fig. 12-8.

12-45. Solve Problem 12-44 using SI units. Specify fan characteristics.

12-46. The schematic of a large office complex is shown in Fig. 12-46. Halls, closets, and so on, have been omitted for clarity. The air quantities shown are the maximum that each space will require, but the spaces will not all peak at the same time. (a) Lay out a duct system starting from the equipment room running due west and turning into two trunk ducts running north and south between the three rows of spaces. Each space requires a terminal box connected to the trunk duct with circular duct. (b) Estimate the maximum flow for the system and each section of the duct using the following guidelines: The east rooms will peak about 10:00 A.M. and remain constant until about 3:00 P.M. The interior rooms peak at 9:00 P.M., and remain constant until 4:00 P.M. The north and south rooms in the center peak at noon and demand gradually decreases after that. The west rooms reach one-half capacity at noon and peak at 5:00 P.M. The minimum overall load occurs at 5:00 A.M. and is about 20 percent of the maximum. (c)

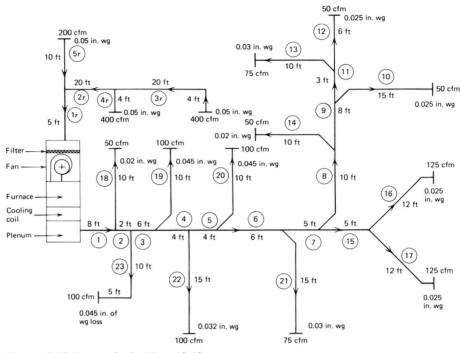

Figure 12-43 Layout for Problem 12-42.

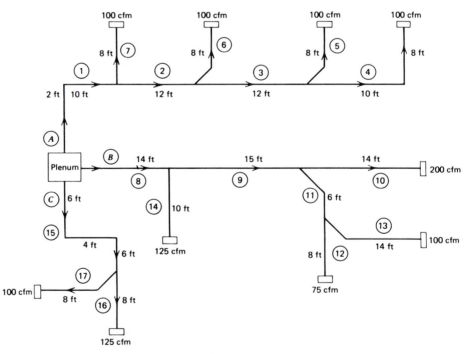

Figure 12-44 Schematic for Problem 12-43.

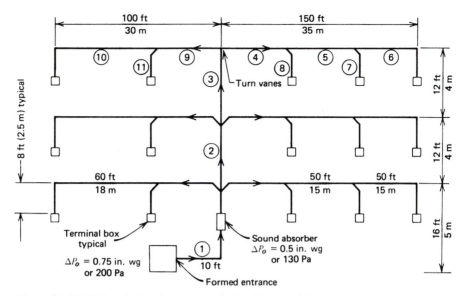

Figure 12-45 High-velocity duct system for Problem 12-44.

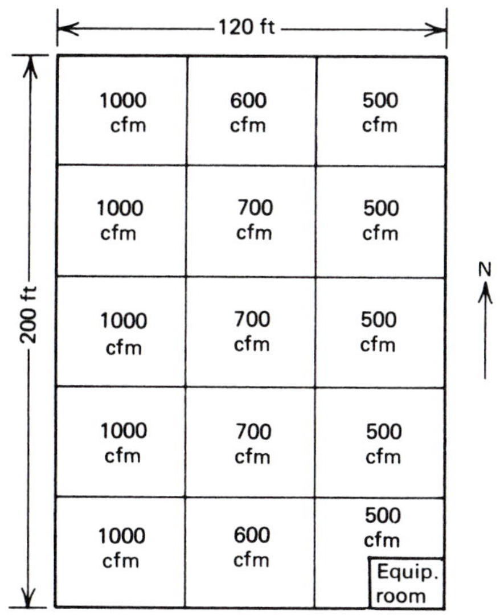

Figure 12-46 Schematic for Problem 12-46.

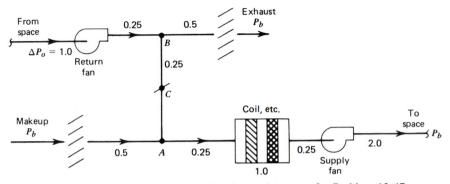

Figure 12-47 Schematic of a makeup and exhaust air system for Problem 12-47.

Size the ducts using a maximum velocity of about 2500 ft/min and the equal-friction method. (d) Estimate the total pressure requirements of the system at maximum flow. (e) Locate the static pressure sensor for the fan speed control. (f) Refer to Fig. 12-20 and locate operating points 1 and 2. Scale the size of the fan up or down as required by changing the capacity values along the abscissa. (g) How much power does the system require at the minimum and maximum operating points?

12-47. Refer to Fig. 12-47 and construct the energy grade line (total pressure versus length) for the system shown. The change in total pressure in in. wg is shown for each part of the system. What total pressure must each fan produce?

12-48. Refer to Problem 12-47 and assume that the return fan does not exist. Construct the energy grade lines and give the fan total pressure.

12-49. Refer to Fig. 12-47 and assume that the supply fan is moved just upstream of the coil section. Construct the energy grade lines and give the total pressure for the fans.

12-50. Refer to Problem 12-49 and assume that the return fan does not exist. Construct the energy grade lines and give the fan total pressure.

Chapter **13**

Direct Contact Heat and Mass Transfer

The process of air humidification was discussed in Chapter 3 on the basis of thermodynamics. In that case the condition line on the psychrometric chart was found to be solely a function of the water enthalpy, and all the water mixed with the air stream was assumed to evaporate and become a part of the air–water vapor mixture. In this chapter the physical aspects of mass transfer are briefly reviewed as they relate to the typical processes encountered in HVAC systems, such as air washers and cooling coils. We will then consider problems in which the quantity of water in contact with the air is much larger than the quantity added or withdrawn from the air stream. A variety of results are possible, depending on the moist air state and the water temperature. The air may be cooled and humidified or dehumidified, or heated and humidified by direct contact with water.

Only heat and mass transfer will be considered. As discussed in Chapter 4, the air may be cleansed of dust and water-soluble vapors by contact with water.

13-1 COMBINED HEAT AND MASS TRANSFER

Many problems in engineering are concerned with the simultaneous transfer of mass and heat. In this book these problems deal mainly with heated and cooled air–water vapor mixtures that result in the evaporation or condensation of the water.

It is well known that a link exists between the transport of momentum, heat, and mass (1). Consider the two-dimensional boundary layer equations for an incompressible, constant property fluid with a zero pressure gradient.

$$u \frac{\partial C}{\partial x} + v \frac{\partial C}{\partial y} = D \frac{\partial^2 C}{\partial y^2} \tag{13-1}$$

$$u \frac{\partial u}{\partial x} + v \frac{\partial u}{\partial y} = \nu \frac{\partial^2 u}{\partial y^2} \tag{13-2}$$

$$u \frac{\partial t}{\partial x} + v \frac{\partial t}{\partial y} = \alpha \frac{\partial^2 t}{\partial y^2} \tag{13-3}$$

543

The boundary conditions are

$$y = 0; \qquad u = 0, \qquad v = v_w, \qquad t = t_w, \qquad C = C_w \qquad \text{(13-4)}$$

$$y = \infty; \qquad u = u_\infty \qquad\qquad t = t_\infty, \qquad C = C_\infty \qquad \text{(13-5)}$$

The quantities D, v, and α are the diffusivities for mass, momentum, and heat, respectively. When these quantities are equal and the wall velocity v_w is zero, the solutions to Eqs. 13-1 through 13-3 are identical and the concentration C, velocity u, and temperature profiles t are similar. In this case the Prandtl number Pr and Schmidt number Sc are both equal to one, where

$$\text{Pr} = \frac{v}{\alpha} \qquad \text{(13-6)}$$

and

$$\text{Sc} = \frac{v}{D} \qquad \text{(13-7)}$$

The ratio of the Schmidt number to the Prandtl number is the Lewis number

$$\text{Le} = \frac{\text{Sc}}{\text{Pr}} = \frac{\alpha}{D} \qquad \text{(13-8)}$$

which is also equal to one in this case.

To clearly show the connection between heat, mass, and momentum transfer, consider an energy balance on a surface

$$h(t_\infty - t_w) = k \left(\frac{\partial t}{\partial y} \right)_{y=0} \qquad \text{(13-9)}$$

where h and k are the convective heat-transfer coefficient and the fluid thermal conductivity, respectively. When the solution to Eq. 13-3 for t and the distance y are nondimensionalized

$$t' = \frac{t - t_w}{t_\infty - t_w} \qquad y' = \frac{y}{L} \qquad \text{(13-10)}$$

and inserted in Eq. 13-9, the following result is obtained:

$$\text{Nu} = \frac{hL}{k} = \left(\frac{\partial t'}{\partial y'} \right)_{y=0} \qquad \text{(13-11)}$$

where Nu is the Nusselt number, a dimensionless heat-transfer coefficient.

Fick's law may be written as

$$\dot{m}_w = -DA \left(\frac{\partial C}{\partial y} \right)_{y=0} \qquad \text{(13-12a)}$$

where $\dot{m}_w$ is the mass-transfer rate, mass per unit time. A mass-transfer coefficient is introduced:

$$\dot{m}_w = h_m A (C_w - C_\infty) \tag{13-12b}$$

where h_m has dimensions of length per unit time and C is the concentration, mass per unit volume. Combination of Eqs. 13-12a and 13-12b and the use of the dimensionless concentration and length from the solution to Eq. 13-1

$$C' = \frac{(C_w - C)}{(C_w - C_\infty)} \qquad y' = \frac{y}{L} \tag{13-13}$$

yields

$$Sh = \frac{h_m L}{D} = \left(\frac{\partial C'}{\partial y'} \right)_{y=0} \tag{13-14}$$

where Sh is the Sherwood number, a dimensionless mass-transfer coefficient.

When the temperature and concentration profiles are similar, the dimensionless profiles C' and t' are identical. As a result the dimensionless heat- and mass-transfer coefficients given by Eqs. 13-11 and 13-14 are also identical. The link to momentum transfer or friction can be shown in a similar way. A similar development may be carried out to show the same general relationship between heat, mass, and momentum transfer in turbulent flow.

It is well known that expressions for Nusselt and Sherwood numbers have the form

$$Nu = f(Re, Pr) \tag{13-15a}$$

$$Sh = f(Re, Sc)$$

where the functional relations are given by

$$Nu = C_1 Re^a Pr^b \tag{13-15b}$$

$$Sh = C_1 Re^a Sc^b$$

Reynolds analogy was first used to show the connection between heat and momentum transfer and appears as

$$\frac{h}{\rho c_p \overline{V}} = \frac{f}{2} \tag{13-16a}$$

where f is the Fanning friction factor. The analogy was later extended to mass transfer:

$$\frac{h_m}{\overline{V}} = \frac{h}{\rho c_p \overline{V}} = \frac{f}{2} \tag{13-16b}$$

The Reynolds analogy has long been recognized as giving reasonable results when $Pr \approx Sc \approx 1$ and the temperature potential is moderate. Many other analogies have been proposed to account for the effect of Prandtl number. Chilton and Colburn (2) have proposed one of the most widely accepted analogies, called the *j*-factor analogy:

$$j = j_m = \frac{f}{2} \tag{13-17}$$

where

$$j = \frac{h}{\rho_a c_{pa} \overline{V}} Pr^{2/3} = \frac{h}{G c_{pa}} Pr^{2/3} \tag{13-18}$$

and

$$j_m = \frac{h_m}{\overline{V}} Sc^{2/3} \tag{13-19a}$$

From Eqs. 13-18 and 13-19

$$\frac{h}{\rho_a c_{pa} h_m} = \left(\frac{Sc}{Pr}\right)^{2/3} = Le^{2/3} \tag{13-20a}$$

where $Le^{2/3}$ is approximately 1.0 for moist air at usual conditions. In air-conditioning calculations it is generally more convenient to use the concentration in the form of the humidity ratio W rather than C in mass of water per unit volume. The relation between the two is

$$C = W\rho_a \tag{13-21}$$

where ρ_a is the mass density of the dry air, mass per unit volume. Equation 13-12b then becomes

$$\dot{m}_w = h_m A \rho_a (W_w - W_\infty) \tag{13-12c}$$

or

$$\dot{m}_w = h_d A (W_w - W_\infty) \tag{13-12d}$$

where

$$h_d = \rho_a h_m \tag{13-22}$$

The dimension of h_d is mass of dry air per unit area and time. The analogy of Eq. 13-20a then becomes

$$\frac{h}{c_{pa} h_d} = Le^{2/3} \tag{13-20b}$$

and the *j*-factor of Eq. 13-19a becomes

$$j_m = \frac{h_d}{\rho_a \overline{V}} Sc^{2/3} \tag{13-19b}$$

The use of the foregoing analogy in moist-air problems requires caution. Over the range of temperatures of 50 to 140 F or 10 to 60 C and from completely dry to saturated air, the Lewis number ranges from about 0.81 to 0.86 (3), whereas the Schmidt and Prandtl numbers have values of about 0.6 and 0.7. Therefore, the theoretical basis for the analogy is not entirely satisfied. The analogy is also based on ideal surface and flow field conditions. For example, there is undisputed evidence that the water deposited on a surface during dehumidification roughens the surface and upsets the analogy because h is for a smooth dry surface (4). There is also evidence that disturbances in the flow field may influence the transfer phenomena and sometimes render the analogy invalid (5). The analogy seems to be most valid when there is direct contact between the air and water. In situations such as a dehumidifying heat exchanger, the condensate that collects on the surface upsets the fundamental basis for the analogy. Research related to this phenomenon will be discussed in Chapter 14. We will use the *j*-factor analogy most extensively for direct contact processes.

13-2 THE AIR WASHER

The basic relations for all types of direct contact equipment are quite similar. The air washer is considered first; we shall introduce modifications later for spray dehumidifiers and cooling towers.

The primary reason for treating direct contact equipment as a separate group arises from the difficulty in evaluating the heat- and mass-transfer areas. For the air washer or any spray-type device that does not have packing materials, the heat- and mass-transfer areas are approximately equal. For a cooling tower the difference may be considerable.

The basic relations may be written as follows; see Fig. 13-1. The mass transfer is given by

$$-dG_l = G_a \, dW = h_d a_m (W_i - W) \, dL \qquad \textbf{(13-23)}$$

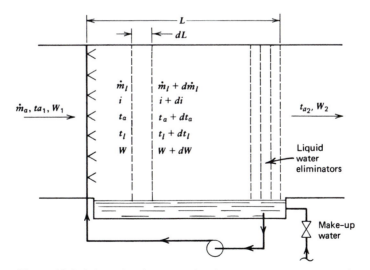

Figure 13-1 Schematic of a spray chamber.

where W_i is the humidity ratio at the interface between the water and moist air. The quantity a_m is the mass-transfer area per unit volume of the chamber. G_l and G_a are the water and air mass velocities $\dot{m}_l/A_c$ and $\dot{m}_a/A_c$ in mass per unit time and area where A_c is the cross-sectional area of the chamber. The water evaporated equals the increase in moisture of the air, which must equal the mass-transfer rate. The sensible heat transfer to the air is given by

$$G_a c_{pa}\, dt_a = h_a a_h (t_i - t_a)\, dL \tag{13-24}$$

where t_i is the interface temperature and a_h is the heat-transfer area per unit volume. The total energy transfer to the air is

$$G_a(c_{pa}\, dt_a + i_{fg}\, dW) = [h_d a_m (W_i - W)i_{fg} + h_a a_h (t_i - t_a)]\, dL \tag{13-25a}$$

The concept of enthalpy potential, which is discussed in detail in Chapter 14, may be used to simplify Eq. 13-25a. The term in brackets on the left-hand side of Eq. 13-25a is di, where i is the enthalpy of the moist air in Btu/lbma. If it is assumed that $a_h = a_m$ and the analogy of Eq. 13-20b is used to relate the heat- and mass-transfer coefficients h and h_d with $L_e = 1$, Eq. 13-25a becomes

$$G_a di = h_d a_m (i_i - i)\, dL \tag{13-25b}$$

where $(i_i - i)$ is the enthalpy potential. An energy balance yields

$$G_a di = \pm G_l c_l\, dt_l + c_l t_l dG_l \tag{13-26}$$

The negative sign refers to parallel flow of air and water, whereas the positive sign refers to counterflow. The last term in Eq. 13-26 is very small and will be neglected in the following development. The heat transfer to the water may be expressed as

$$G_l c_l\, dt_l = h_l a_h (t_l - t_i)\, dL \tag{13-27}$$

Equations 13-23 through 13-27 are the basic relations for direct-contact equipment with the possible exception of chambers that contain packing, such as cooling towers.

Various relations useful in equipment design may be derived from the basic relations. Combining Eqs. 13-25b, 13-26, and 13-27 gives

$$\frac{i - i_i}{t_l - t_i} = -\frac{h_l a_h}{h_d a_m} = -\frac{h_l}{h_d} \tag{13-28}$$

This shows that the ratio of driving potentials for total heat transfer through the air and liquid films is equal to the ratio of film resistances for the gas and liquid film. Combining Eq. 13-24 and 13-25b yields

$$\frac{di}{dt_a} = \frac{i - i_i}{t_a - t_i} \tag{13-29}$$

which use of Eqs. 13-23 and 13-24 gives

$$\frac{dW}{dt_a} = \frac{W - W}{t_a - t_i}$$ (13-30)

Equation 13-30 indicates that at any cross section in the spray chamber the instantaneous slope of the process path on the psychrometric chart is a straight line. This is illustrated in Fig. 13-2, where state 1 represents the state of the moist air entering the chamber and point 1_i represents the interface saturation state. The initial path of the process is then in the direction of the line connecting points 1 and 1_i. As the air is heated and humidified, the water is cooled and the interface state changes. In the case shown the interface state gradually moves downward along the saturation curve. Points 2 and 3 represent other states along the process curve. Note that the path is directed toward the interface in each case. The interface states are defined by Eqs. 13-26 and 13-28. Equation 13-26 relates the air enthalpy change to the water temperature change, whereas Eq. 13-28 describes the way in which the interface state changes to accommodate the transport coefficients and the air state. Solution of Eqs. 13-26 and 13-28 for the interface state is rather complex but can be done by trial and error or by the use of a complex graphical procedure. A simpler method utilizes a psychrometric chart with enthalpy and temperature as coordinates (6). The use of these coordinates makes it possible to plot Eqs. 13-26 and 13-28 for easy graphical solution. The following example illustrates the procedure.

EXAMPLE 13-1

A parallel-flow air washer is to be designed as shown in Fig. 13-1. The design conditions are as follows.

Water temperature at the inlet,	$t_{l1} = 90$ F
Water temperature at the outlet,	$t_{l2} = 75$ F
Air dry bulb temperature at the inlet,	$t_{a1} = 60$ F
Air wet bulb temperature at the inlet,	$t_{wb1} = 42$ F
Air mass flow rate per unit area,	$G_a = 1250$ lbm/(hr-ft^2)
Spray ratio,	$G_l/G_a = 0.75$
Air heat-transfer coefficient per unit volume,	$h_a a_h = 50$ Btu/(hr-F-ft^3)
Liquid heat-transfer coefficient per unit volume,	$h_l a_h = 666$ Btu/(hr-F-ft^3)
Air volume flow rate,	$\dot{Q} = 7000$ cfm

SOLUTION

The mass flow rate of the dry air is given by

$$\dot{m}_a = \frac{\dot{Q}}{v_1} = \frac{7000}{13.1} = 534 \text{ lb/min}$$

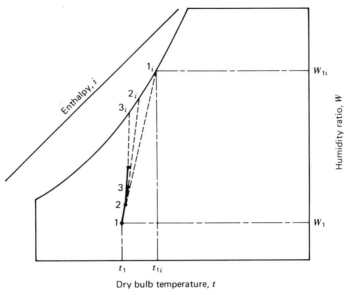

Figure 13-2 Air washer humidification process.

Then the spray chamber must have a cross-sectional area of

$$A_c = \frac{\dot{m}_a}{G_a} = \frac{534(60)}{1250} = 25.6 \text{ ft}^2$$

The Colburn analogy of Eq. 13-20b with Le = 1 will be used to obtain the mass-transfer coefficient (assuming $a_m = a_h$).

$$h_d a_m = \frac{h_a a_h}{c_{pa}} = \frac{50}{0.24} = 208 \text{ lbm/(hr-ft}^3)$$

In parallel flow the air entering at 60 F is in contact with the water at 90 F, and the air leaving at state 2 is in contact with the water at 75 F. This helps to understand the construction of Fig. 13-3, which shows the graphical solution for the interface states and the process path for the air passing through the air washer. The solution is carried out as follows:

1. Locate state 1 as shown at the intersection of t_{a1} and i_1. Point A is a construction point defined by the entering water temperature t_{l1} and i_1. Note that the temperature scale is used for both the air and water.
2. The energy balance line is constructed from point A to point B and is defined by Eq. 13-26:

$$\frac{d_i}{dt_1} = -\frac{G_l c_l}{G_a} = -\frac{G_l}{G_a} = -0.75$$

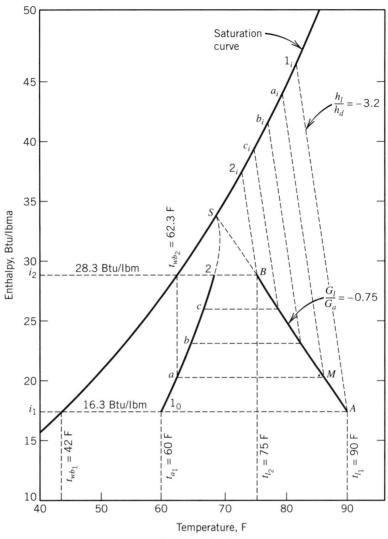

Figure 13-3 Graphical solution for Example 13-1.

since $c_l = 1$ Btu/(lbmw-F). Point B is determined by the temperature of leaving water $t_{l2} = 75$ F. The negative slope is a consequence of parallel flow. The line AB has no physical significance, except that it depends on the energy balance between the water and air.

3. The line $A1_i$ is called a *tie line* and is defined by Eq. 13-28.

$$\frac{i - i_i}{t_l - t_i} = \frac{-h_l a_h}{h_d a_m} = \frac{-666}{208} = -3.2$$

The intersection of the tie line having slope of -3.2 with the saturation curve defines the interface state 1_i. The combination of the line AB and line $A1_i$ represents graphical solutions of Eqs. 13-26 and 13-28.

4. The initial slope of the air process path is then given by a line from state 1 to state 1_i. The length of the line $1a$ depends on the required accuracy of the solution and the rate at which the curvature of the path is changing.

5. The procedure is repeated by constructing the line aM and the tie line Ma_i, which has the same slope as $A1_i$. The path segment ab is on a line from a to a_i. Continue in the same manner until the final state of the air at point 2 is reached. State point 2 is on a horizontal line passing through point B. The final state of the air is defined by $t_{a2} = 68$ F, $i_2 = 28.3$ Btu/lbma, and $t_{wb2} = 62.3$ F.

6. To complete the solution, it is necessary to determine the length of the air washer. Equation 13-25b gives

$$dL = \frac{G_a}{h_d a_m} \frac{di}{(i_i - i)} \tag{13-25c}$$

or

$$L = \frac{G_a}{h_d a_m} \int_1^2 \frac{di}{(i_i - i)} \tag{13-25d}$$

Equation 13-25d can be evaluated graphically or numerically. A plot of $1/(i_i - i)$ versus i is shown in Fig. 13-4 and the area under the curve represents the value of the integral. Using Simpson's rule with four equal increments yields

$$y = \int_1^2 \frac{di}{(i_i - i)} \approx \frac{\Delta i}{3} (y_1 + 4y_2 + 2y_3 + 4y_4 + y_5)$$

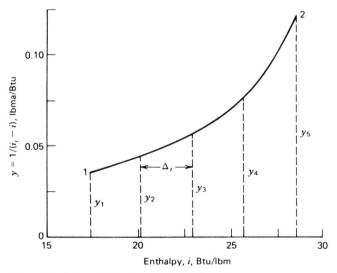

Figure 13-4 Graphical solution of $\int di/(i_i - i)$.

with $\Delta i = 2.75$

$$y = \left(\frac{2.75}{3}\right)(0.036 + 4(0.044) + 2(0.057) + 4(0.076) + 0.12) = 0.688$$

The design length is then

$$L = \frac{1250}{208}(0.688) = 6.0\,\text{ft}$$

The graphical procedure outline can be programmed rather easily for a small digital computer. If more than occasional calculations are needed, this would be worthwhile. Use of a computer would also permit refinements and variation of design parameters.

13-3 THE SPRAY DEHUMIDIFIER

The spray dehumidifier can often be used to advantage when a source of cold water is available. To be effective the spray dehumidifier must be used in counterflow, which can best be achieved as shown in Fig. 13-5. The following example illustrates the design procedure for this case.

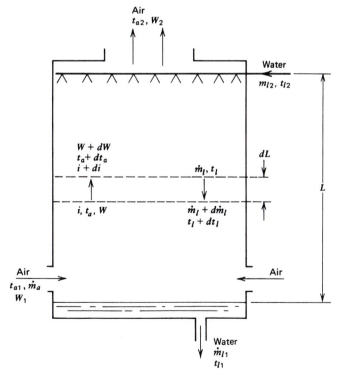

Figure 13-5 Schematic of counterflow spray chamber.

EXAMPLE 13-2

A counterflow spray dehumidifier is to be designed as shown schematically in Fig. 13-5. These are the design conditions:

Water temperature at the inlet,	$t_{l2} = 7$ C
Water temperature at the outlet,	$t_{l1} = 15$ C
Air dry bulb temperature at the inlet,	$t_{a1} = 28$ C
Air wet bulb temperature at the inlet,	$t_{wb1} = 22$ C
Air mass flow rate per unit area,	$G_a = 1.36$ kg/(s-m²)
Spray ratio,	$\dfrac{G_l c_l}{G_a} = 3.25$
Air heat-transfer coefficient per unit volume,	$h_a a_h = 1210$ W/(C-m³)
Liquid heat-transfer coefficient per unit volume,	$h_l a_h = 14700$ W/(C-m³)
Air volume flow rate,	$\dot{Q} = 2.83$ m³/s

Find the cross-sectional area and final state of the air.

SOLUTION

The mass flow rate of the dry air is

$$\dot{m}_a = \frac{\dot{Q}}{v_1} = \frac{2.83}{0.875} = 3.24 \text{ kg/s}$$

The chamber cross-sectional area is then

$$A_c = \frac{\dot{m}_a}{G_a} = \frac{3.24}{1.36} = 2.38 \text{ m}^2$$

Using the Colburn analogy and assuming that $a_m = a_h$

$$h_d a_m = \frac{h_a a_m}{c_{pa}} = \frac{1210}{1000} = 1.21 \text{ kg/(s-m}^3)$$

The graphical solution for the interface states and the process path is shown in Fig. 13-6. In the case of counterflow, the air entering at 28 C is in contact with the water at 15 C, and the air leaving is in contact with the water at 7 C. The procedure is quite similar to that given in Example 13-1. Note, however, that the energy balance line AB has a positive slope because of the counterflow arrangement. Notice also that because of counterflow, point A corresponds to t_{l1} and point B corresponds to t_{l2}. The final state of the air is defined by $t_{a2} = 14$ C and $t_{wb2} = 13.5$ C.

The height of the spray chamber is determined by using Eq. 13-25b with the procedure given in Example 13-1.

The use of spray dehumidifiers is less frequent than extended surface heat exchangers, which are discussed in Chapter 14. However, the procedure described above may be

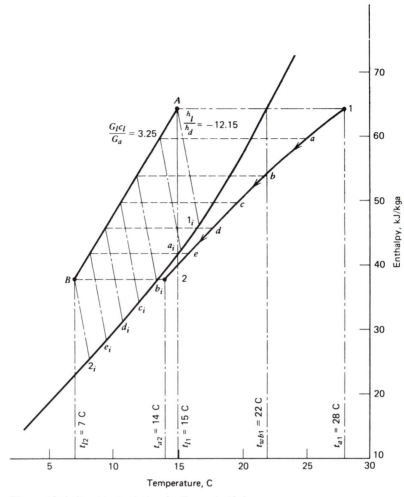

Figure 13-6 Graphical solution for Example 13-2.

applied to a dehumidifying heat exchanger by modifying Eq. 13-28 by replacing the liquid film coefficient h_l with an overall coefficient defined by

$$\frac{1}{UA} = \frac{1}{h_c A_c} + \frac{\Delta x}{k A_m} \qquad \text{(14-12a)}$$

The first term on the right is the refrigerant-side film resistance and the second term is the wall thermal resistance.

13-4 COOLING TOWERS

A typical cooling tower used in HVAC applications is shown in Fig. 2-11. The particular model shown is a packaged mechanical draft unit.

The function of the cooling tower is to reject heat to the atmosphere by reducing the temperature of water circulated through condensers or other heat-rejection equipment.

For this reason the state of the air and its path on the psychrometer chart is of little interest. The cooling range for the water (decrease in temperature) and the approach of the leaving water temperature to the ambient wet bulb temperature are most important in determining the size and cost of a cooling tower. This will be evident in the analysis to follow.

The general equations derived in Section 13-2 are still valid. To facilitate calculations, however, the procedure is somewhat different in this case. Equation 13-25b for the total energy transfer to the air is recalled.

$$G_a \, di = G_l c_l \, dt_l = h_d a_m (i_i - i) \, dL \qquad \textbf{(13-25b)}$$

To avoid consideration of the interfacial conditions, an overall coefficient U_i is adopted that relates the driving potential to the enthalpy i_l at the bulk water temperature t_l. Equation 13-25b then becomes

$$G_a \, di = G_l c_l \, dt_l = U_i a_m (i_l - i) \, dL \qquad \textbf{(13-29a)}$$

and, when Eq. 13-29a is integrated,

$$\frac{U_i a_m L}{G_l c_l} = \int \frac{dt_l}{(i_l - i)} \qquad \textbf{(13-29b)}$$

Now

$$\dot{m}_l = G_l A_c \qquad \text{and} \qquad V = A_c L$$

then

$$N = \frac{U_i a_m V}{\dot{m}_l c_l} = \int \frac{dt_l}{(i_l - i)} \qquad \textbf{(13-30)}$$

To review:

N = number of transfer units, NTU
U_i = overall mass-transfer coefficient between the water and air, lbm/(hr-ft²) or kg/(s-m²)
a_m = mass-transfer surface area per unit volume associated with U_i, ft²/ft³ or m²/m³
V = total cooling tower volume, ft³ or m³
$\dot{m}_l$ = mass flow rate of water through the tower, lbm/hr or kg/s
c_l = specific heat of the water, Btu/(lbm-F) or kJ/(kg-C)
t_l = water temperature at a particular location in the tower, F or C
i_l = enthalpy of saturated moist air at t_l, Btu/lbm or kJ/kg
i = enthalpy of the moist air at temperature t, Btu/lbm or kJ/kg

The left-hand side of Eq. 13-30 is a measure of the cooling tower size and has the familiar form of the NTU parameter used in heat-exchanger design.

Equation 13-30 cannot be integrated in a straightforward mathematical way; however, a step-by-step approach can be used. The following example illustrates the procedure. The remainder of the design procedure is considered after the example.

EXAMPLE 13-3

Water is to be cooled from 100 to 85 F in a counterflow cooling tower when the outside air has a 75 F wet bulb temperature. The water-to-air flow ratio $(\dot{m}_l/\dot{m}_a)$ is 1.0. Calculate the transfer units as defined by Eq. 13-30.

SOLUTION

Figure 13-7 is the cooling diagram for the given conditions. As the water is cooled from t_{l1} to t_{l2}, the enthalpy of the saturated air i_l follows the saturation curve from A to B. The air entering at wet bulb temperature t_{wb1} has enthalpy i_1. (This assumes that the air enthalpy is only a function of wet bulb temperature.) The leaving water temperature t_{l2} and the enthalpy i_1 define point C, and the initial driving potential is represented by the distance BC. The enthalpy increase of the air is a straight-line function with respect to the water temperature as defined by Eq. 13-29a. The slope of the air operating line CD is therefore $(c_l \dot{m}_l/\dot{m}_a)$.

Point C represents the air conditions at the inlet and point D represents the air

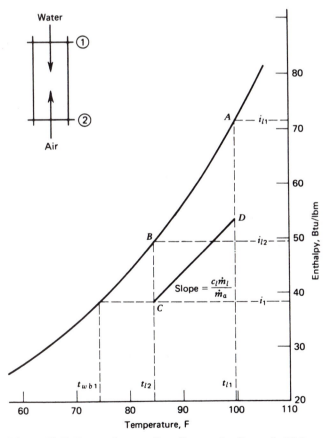

Figure 13-7 Counterflow cooling diagram for Example 13-3.

conditions leaving the tower. Note that the driving potential gradually increases from the bottom to the top of the tower. Counterflow integration calculations start at the bottom of the tower where the air conditions are known. Evaluation of the integral of Eq. 13-30 may be carried out in a manner similar to that described in Example 13-1 by plotting t_l versus $1/(i_l - i)$; however, another method will be used here (2). The step-by-step procedure is shown in Table 13-1. Water temperatures are listed in column 1 in increments of one to two degrees. Although smaller increments will give greater accuracy, this must be balanced against an increase in calculation time. The film enthalpies shown in column 2 are obtained from Table A-2a as the enthalpy of saturated air at the water temperatures. Column 3 shows the air enthalpy which is determined from Eq. 13-29a.

$$\Delta i = \frac{c_l \dot{m}_l}{\dot{m}_a} \Delta t_l \qquad\qquad \textbf{(13-29c)}$$

where the initial air enthalpy i_1 is 38.5 Btu/lbma, $c_l = 1.0$ Btu/(lbmw-F), $\dot{m}_l/\dot{m}_a = 1.0$, and Δt is read in column 1 of Table 13-1. The data of columns 4 and 5 are obtained from columns 2 and 3. Column 6 is the average of two steps from column 5 multiplied by the water temperature increment (column 1) for the same step. The number of transfer units are then given in column 7 as the summation of column 6. Column 8 gives the temperature range over which the water has been cooled. The last entry in column 7 is the number of transfer units required for this problem. It is evident from Table 13-1 that either an increase in the cooling range or a decrease in the leaving water temperature will increase the number of transfer units. As mentioned earlier, these two factors are quite important in cooling-tower design. The heat exchangers with which the cooling tower is connected should be designed with the cooling tower in mind. It may be more economical to enlarge the heat exchangers and/or increase the flow rate of the water than increase the size of the cooling tower.

Obviously the solution to this example could be easily programmed for a computer.

To continue the problem of tower design, we need information on the overall mass-transfer coefficient per unit volume, $U_i a_m$. There is little theory to predict this coefficient; therefore, we must rely on experiments. The tower characteristic may generally be represented as a log-log plot of $U_i a_m V/\dot{m}_l c_l$ for a constant air mass flow rate and water inlet temperature t_{l1} (Fig. 13-8). Although different air flow rates and water temperatures cause the characteristic to move up or down, variations in air flow rate of about ± 20 percent and water temperature variations of ± 10 F do not move the characteristic significantly. The slope of the characteristic varies from about -0.4 to -1.0. The required transfer units for a given set of water and air conditions are also shown in Fig. 13-8. The intersection of the two curves defines the operating point for the tower.

Assuming that the tower characteristics are known, we may determine the mass-transfer coefficient per unit volume $U_i a_m$. After many tests have been made on towers of a similar type, it is possible to predict $U_i a_m$ with reasonable accuracy. Then the volume of the tower required for a given set of conditions is given by

$$V = \frac{N \dot{m}_l c_l}{U_i a_m} \qquad\qquad \textbf{(13-31)}$$

Table 13-1 Counterflow Cooling Tower Integration Calculations

1	2	3	4	5	6	7	8
Water Temperature t_l degrees F	Enthalpy of Film i_l Btu/lbma	Enthalpy of Air i Btu/lbma	Enthalpy Difference $(i_l - i)$ Btu/lbma	Reciprocal of Enthalpy Difference $\dfrac{1}{(i_l - i)}$ lbma/Btu	Average $\dfrac{\Delta t_l}{(i_l - i)}$ F-lbma/Btu	Summation $\dfrac{\Delta t_l}{(i_l - i)}$ F-lbma/Btu	Cooling Range degrees F
85	49.4	38.5	10.9	0.0917	0.0905	0.0905	1
86	50.7	39.5	11.2	0.0893	0.1748	0.2653	3
88	53.2	41.5	11.7	0.0855	0.1661	0.4314	5
90	55.9	43.5	12.4	0.0806	0.1558	0.5872	7
92	58.8	45.5	13.3	0.0752	0.1451	0.7323	9
94	61.8	47.5	14.3	0.0699	0.1348	0.8671	11
96	64.9	49.5	15.4	0.0649	0.1248	0.9919	13
98	68.2	51.5	16.7	0.0599	0.1148	1.1067	15
100	71.7	53.5	18.2	0.0549			

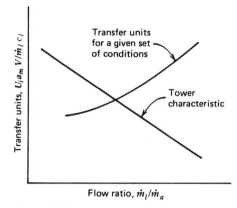

Figure 13-8 Tower characteristic.

where N is the number of transfer units given by Eq. 13-30. The cross-sectional area of the tower is defined by

$$A_c = \frac{\dot{m}_a}{G_a} = \frac{\dot{m}_l}{G_l} \qquad (13\text{-}32)$$

and the height of the tower is given by

$$L = \frac{V}{A_c} \qquad (13\text{-}33)$$

EXAMPLE 13-4

Suppose the cooling tower of Example 13-3 must handle 1000 gpm of water. It has been determined that an air mass velocity of 1500 lbma/(hr-ft^2) is acceptable without excessive water carry-over (drift). The overall mass-transfer coefficient per unit volume $U_i a_m$ is estimated to be 120 lbm/(hr-ft^3) for the type of tower to be used. Estimate the tower dimensions for the required duty.

SOLUTION

The transfer units N required for the tower were found to be 1.1067 in Example 13-3. Then the total volume of the tower is given by Eq. 13-31 as

$$V = \frac{1.1067(1000)8.33(60)(1.0)}{120} = 4609 \text{ ft}^3$$

The cross-sectional area of the tower may be determined from Eq. 13-32 using the mass velocity of the air and the water-to-air ratio.

$$A_c = \frac{\dot{m}_a}{G_a} = \left(\frac{\dot{m}_l}{\dot{m}_a}\right)\frac{\dot{m}_a}{G_a} = \frac{\dot{m}_l}{G_a}$$

$$A_c = \frac{1000(8.33)(60)}{1500} = 333 \text{ ft}^2$$

which is equivalent to an 18×18 ft cross section. Then from Eq. 13-33

$$L = \frac{V}{A_c} = \frac{4609}{333} = 13.8 \text{ ft}$$

Caution must be exercised in using mass-transfer data from the literature for tower design. There are many variations in construction that affect the transport coefficients dramatically. The scale of the tower also seems to be important because of the ratio of wall surface area to total volume.

For many HVAC applications, relatively small factory-assembled cooling towers are used. Performance data are usually presented in a form such that a certain standard size may be selected. Figure 13-9 and Table 13-2 are an example of what might be furnished by a manufacturer for a line of towers. The entering water temperature, air wet bulb temperature, and the water flow rate determine the model to be selected for a fixed leaving water temperature of 85 F. The example shown as a dotted line illustrates use of the chart. This procedure usually causes the tower to be slightly oversized. The cooling range may be computed from the entering water temperature and flow rate, plus the tower capacity.

In Table 13-2 the nominal rating in tons refers to a typical refrigeration system with

Table 13-2 Performance Data for Some Factory-Assembled Cooling Towers

Model	Nominal Rating Tons	gpm	No. of Cells	No. of Fans	cfm	Motor hp
A	50	120	1	1	10,500	5
B	100	240	1	2	21,000	10
C	100	240	1	2	21,000	2–5
D	150	360	1	3	31,500	15
E	200	480	1	4	42,000	20
F	200	480	1	4	42,000	2–10
G	250	600	1	5	52,500	25
H	300	720	1	6	63,000	30
I	300	720	1	6	63,000	2–15
J	350	840	1	8	84,840	2–20
K	400	960	1	8	84,000	2–20
L	500	1200	1	10	105,000	2–25
M	600	1440	1	12	126,000	2–30

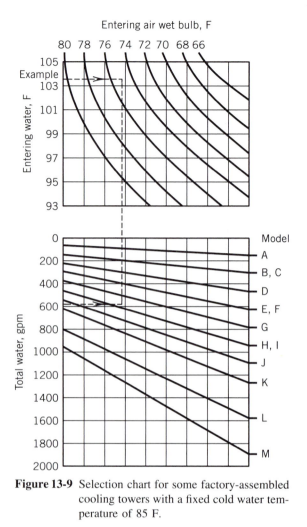

Figure 13-9 Selection chart for some factory-assembled cooling towers with a fixed cold water temperature of 85 F.

which the cooling tower may be used and includes only the heat transferred to the evaporator.

EXAMPLE 13-5

Select a cooling tower using Fig. 13-9 and Table 13-2 for the conditions of Examples 13-3 and 13-4. Compute the cooling range, approach, and heat-transfer rate.

SOLUTION

The entering water temperature and air wet bulb temperature are 100 F and 75 F, respectively, with a water flow rate of 1000 gpm. Referring to Fig. 13-9, a model L would be the obvious choice; however, it is somewhat oversized. Suppose that the heat exchangers (condensers) in the circuit could be changed so that the water would enter the tower at 103 F with a flow rate of 835 gpm. This is still the same duty of about 7.5×10^6 Btu/hr. Referring back to Fig. 13-9, a model K tower fits this situation. The cooling range is

$$t_{l1} - t_{l2} = 103 - 85 = 18 \, \text{F}$$

and the approach is

$$t_{l2} - t_{wb1} = 85 - 75 = 10 \, \text{F}$$

the heat-transfer rate from the water is

$$\dot{q} = 835(60)8.33(103 - 85) = 7.5 \times 10^6 \, \text{Btu/hr}$$

The reader is referred to the *HVAC Systems and Equipment Volume* of the *ASHRAE Handbook* (7), which has a great deal of information on cooling tower performance and selection.

REFERENCES

1. J. D. Parker and F. C. McQuiston, *Introduction to Fluid Mechanics and Heat Transfer,* Kendall/Hunt, Dubuque, Iowa, 1988.
2. T. H. Chilton and A. P. Colburn, ''Mass Transfer (Absorption) Coefficients—Prediction from Data on Heat Transfer and Fluid Friction,'' *Industrial and Engineering Chemistry,* November, 1934.
3. James L. Threlkeld, *Thermal Environmental Engineering,* 2nd ed., Prentice-Hall, Englewood Cliffs, NJ, 1970.
4. J. L. Guillory and F. C. McQuiston, ''An Experimental Investigation of Air Dehumidification in a Parallel Plate Exchanger,'' *ASHRAE Transactions,* Vol. 29, June 1973.
5. Wayne A. Helmer, ''Condensing Water Vapor—Air Flow in a Parallel Plate Heat Exchanger,'' Ph.D. Thesis, Purdue University, May 1974.
6. *ASHRAE Handbook, Fundamentals Volume,* American Society of Heating, Refrigerating and Air-Conditioning Engineers, Inc., Atlanta, GA, 1989.
7. *ASHRAE Handbook, HVAC Systems and Equipment Volume,* American Society of Heating, Refrigerating and Air-Conditioning Engineers, Inc., Atlanta, GA, 1992.

PROBLEMS

13-1. Compute the heat-transfer coefficient for air flowing over a surface at a velocity of 1200 ft/min with a mean temperature of 100 F. The *j*-factor is known to be 0.008.

13-2. Compute the heat-transfer coefficient for air flowing at a temperature of 40 C and a velocity of 6 m/s. The *j*-factor is 0.01.

13-3. It is necessary to estimate the rate at which water is removed from an air stream by a cooling coil. The mass-transfer coefficient is not known but the coil was tested at the same Reynolds number under sensible heat-transfer conditions, and the heat-transfer coefficient was 10 Btu/(hr-ft^2-F) [56.8 W/(m^2-C)]. Estimate the mass-transfer coefficient assuming a mean air temperature of 50 F (10 C).

13-4. Use Eqs. 13-21 and 13-12b to show that the units of the mass-transfer coefficient h_d are mass of dry air per unit area and time. Further show that c_{pa} in Eq. 13-20a must be on the basis of unit mass of dry air.

13-5. The Nusselt number for flow in a tube is given by Eq. 13-15b, where the constants C_1, a, and b are 0.023, 0.8, and 0.3, respectively. Estimate a mass-transfer coefficient for moist air flowing in a 12-in. (30-cm) tube at the rate of 600 cfm (0.3 m^3/s). Assume standard air.

13-6. Estimate the mass-transfer coefficient for moist air flowing normal to a 1-in. (25-mm) diameter tube (Nu $= 0.615$ Re$^{0.47}$). The air velocity is 100 ft/min (0.5 m/s.) Assume standard air.

13-7. Estimate the rate at which water is evaporated from a 1000-acre lake on an August day when the dry bulb and wet bulb temperatures are 100 and 75 F, respectively (43,560 ft^2 = 1 acre). Assume a heat-transfer coefficient h of 5 Btu/(hr-ft^2-F) between the moist air and the lake surface and a water surface temperature of 80 F.

13-8. To determine the latent cooling load produced by an indoor swimming pool, estimate the rate at which water is evaporated. The pool area is maintained at 75 F (24 C) dry bulb and 63 F (17 C) wet bulb while the pool water has a temperature of 80 F (27 C). The pool has dimensions of 300 × 150 ft (100 × 50 m). Assume a natural convection condition between the pool water and air of 1.5 Btu/(hr-ft^2-F) [8.5 W/(m^2-C)].

13-9. The sensible heat-transfer coefficient for a dry surface has been determined to be 9 Btu/(hr-ft^2-F) [50 W/(m^2-C)] at a certain Reynolds number. Estimate the total heat transfer to the surface per unit area at a location where the wall temperature is 50 F (10 C) and the state of the moist air flowing over the surface is given by 75 F (24 C) dry bulb and 65 F (18 C) wet bulb. Assume that the Reynolds number does not change and the Lewis number is 0.82. Assume that the condensate present on the surface will increase the transfer coefficients by about 15 percent.

13-10. A housekeeper hangs a wet blanket out to dry in a high wind. The blanket weighs 4 lb dry and 16 lb wet and has dimensions of 7 ft by 8 ft. Assume outdoor conditions of 90 F db and 50 percent relative humidity and that the blanket is at a temperature of 90 F. Estimate the time required for the blanket to become dry if the average heat-transfer coefficient on both sides of the blanket is 4 Btu/(hr-ft^2-F).

13-11. Redesign the air washer in Example 13-1 assuming counterflow of the air and water.

13-12. Determine the final state of the air, the cross-sectional area, and the height of a counterflow spray dehumidifier that operates as follows:

$$t_{l1} = 60 \text{ F} \qquad\qquad t_{a1} = 95 \text{ F}$$
$$t_{l2} = 50 \text{ F} \qquad\qquad t_{wb1} = 82 \text{ F}$$
$$G_a = 1200 \text{ lbm/(hr-ft}^2) \qquad\qquad \dot{Q}_a = 5000 \text{ cfm}$$
$$\dot{Q}_w = 55 \text{ gpm} \qquad\qquad h_a a_h = 60 \text{ Btu/(hr-F-ft}^3)$$
$$h_l a_h = 800 \text{ Btu/(hr-F-ft}^3)$$

13-13. A counterflow spray dehumidifier operates as stated below. Determine (a) the final state of the air, (b) the cross-sectional area, and (c) the height of the chamber.

$$t_{l1} = 16\text{ C} \qquad\qquad t_{a1} = 35\text{ C}$$
$$t_{l2} = 10\text{ C} \qquad\qquad t_{wb1} = 28\text{ C}$$
$$G_a = 1.63\text{ kg/(s-m}^2) \qquad \dot{Q}_a = 2.36\text{ m}^3/\text{s}$$
$$\dot{Q}_w = 3.5 \times 10^{-3}\text{ m}^3/\text{s} \qquad h_a a_h = 1040\text{ W/(C-m}^3)$$
$$h_l a_h = 13850\text{ W/(C-m}^3)$$

13-14. Solve Problem 13-12 assuming parallel flow of the air and water spray.

13-15. Solve Problem 13-13 assuming parallel flow of the air and water.

13-16. A parallel-flow air washer is to be used as an evaporative cooler. Air will enter at 100 F (38 C) dry bulb and 62 F (17 C) wet bulb and leave at 75 F (24 C) dry bulb. Other operating conditions are as follows:

$$t_{l1} = 80\,\text{F}\,(27\text{ C})$$
$$G_a = 1000\text{ lbm/(hr-ft}^2)\,[1.36\text{ kg/(m}^2\text{-s)}]$$
$$h_l a_h = 700\text{ Btu/(hr-ft}^3\text{-F)}\,[13\text{ kW/(m}^3\text{-C)}]$$
$$\dot{Q}_a = 4000\text{ ft}^3/\text{min}\,(1.9\text{ m}^3/\text{s})$$
$$h_a a_h = 55\text{ Btu(hr-ft}^3\text{-F)}\,[1\text{ kW/(m}^3\text{-C)}]$$
$$G_l c_l/G_a = 0.7\text{ Btu/(lbm-F)}\,[2.93\text{ kJ/(kg-C)}]$$

Determine the final state of the air, and the chamber cross-sectional area and length.

13-17. A counterflow cooling tower cools water from 104 to 85 F when the outside air has a wet bulb temperature of 76 F. The water flow rate is 2000 gpm and the air flow rate is 210,000 cfm. Calculate the transfer units for the tower.

13-18. A counterflow cooling tower cools water from 44 to 30 C. The outdoor air has a wet bulb temperature of 22 C. Water flows at the rate of 0.32 m³/s and the water-to-air mass flow ratio is 1.0. Estimate the transfer units for the tower.

13-19. Estimate the tower dimensions for Problem 13-17. The air mass velocity may be assumed to be 1800 lbma/(hr-ft²) and the overall mass-transfer coefficient per unit volume $U_i a_m$ is about 125 lbm/(hr-ft³).

13-20. Estimate the tower dimensions for Problem 13-18. Assume a mass velocity for the air of 2.7 kg/(s-m³).

13-21. Use Fig. 13-9 and Table 13-2 to select a suitable tower(s) for the conditions of Problem 13-17.

13-22. Complete Example 13-2 to find the height of the chamber.

13-23. Repeat Example 13-3 changing the entering air wet bulb temperature to 79 F. Compare the transfer units with those of Table 13-1. How will this affect the tower dimensions?

13-24. Repeat Example 13-3, changing the entering water temperature to 105 F. Compare the transfer units with those of Table 13-1. How will this affect the tower dimensions?

13-25. The condensers for a centrifugal chiller plant require 200 gpm with water entering at 85 F and leaving at 100 F, the outdoor ambient air wet bulb temperature is 76 F. (a) Select a suitable cooling tower using Fig. 13-9 and Table 13-2 and (b) compute the cooling range, approach, and the tower capacity.

Chapter 14

Extended Surface Heat Exchangers

The term *heat exchanger* is usually applied to a device in which two fluid streams separated by a solid surface exchange heat energy. These devices may take many forms. However, ordinary metal tubes are the main components of many types. The heat exchanger is the most widely used device in HVAC applications. Although it is only a part of the overall system, a heat exchanger is required in every heating and cooling design problem that the engineer solves.

The major applications in the HVAC field are as follows: refrigerant-to-water in the case of chillers and water-cooled condensers where shell and tube configurations are often used; water-to-air and refrigerant-to-air where finned tubes are often used; and air-to-air in the case of heat recovery applications where plate-fin type surfaces or rotating heat exchangers may be used.

The heat-transfer surface may operate under conditions where only sensible heat transfer occurs or where latent and sensible heat transfer occur simultaneously as with dehumidifying coils. Both of these cases are treated in this chapter. Pressure loss is an important consideration in the design of heat exchangers; this will be considered in conjunction with the heat transfer in the discussions that follow.

In general, complications arise when one attempts to describe the rate of heat transfer from one fluid to the other in a heat exchanger. For practical purposes

$$\dot{q} = UA\Delta t_m \tag{14-1}$$

where U is the familiar overall heat-transfer coefficient and A is the surface area associated with U. The mean temperature difference between the streams Δt_m must be used because Δt varies continuously throughout the heat exchanger. The overall heat-transfer coefficient U is also a variable resulting from changing physical properties and hydrodynamic characteristics from one part of the exchanger to another. In addition, the calculation of U is difficult and subject to large errors in some cases.

The use of Eq. 14-1 requires the assumption of an average value for U and the determination of a suitable mean temperature difference. An alternate approach to heat-exchanger problems involves the use of an average value of U and the relations between effectiveness ε and the *number of transfer units* NTU.

14-1   THE LMTD METHOD

With suitable assumptions it is possible to derive an expression for the mean temperature difference required in Eq. 14-1 for parallel and counterflow. The assumptions are as follows:

1. The overall heat-transfer coefficient U, the mass flow rates $\dot{m}_c$ and $\dot{m}_h$, and the specific heats c_c and c_h are all constants where the subscripts c and h refer to the cold and hot streams.
2. There is no heat loss or gain external to the heat exchanger and there is no axial conduction in the heat exchanger.
3. A single bulk temperature applies to each stream at a given cross section.

Figures 14-1 and 14-2 show counterflow and parallel flow heat exchangers that represent the simplest types. The subscripts i and o refer to inlet and outlet, respectively. For both counterflow and parallel flow the appropriate mean temperature is given by (1)

$$\Delta t_m = \frac{\Delta t_1 - \Delta t_2}{\ln(\Delta t_1/\Delta t_2)} \tag{14-2}$$

where Δt_2 and Δt_1 are defined in Figs. 14-1 and 14-2. This particular mean temperature difference is the *log mean temperature difference* and is designated LMTD.

Complex Flow Patterns

In many cases the flow paths in the heat exchanger are quite complex and in some cases expressions may be developed for Δt_m; however, they are generally so complicated that charts have been developed to replace the equations. The concept of a correction factor F is used where

$$\dot{q} = UA(F)(\Delta t_m) \tag{14-3}$$

where Δt_m is computed in the same manner as the LMTD for an equivalent counterflow exchanger. Figures 14-3 and 14-4 show charts for two common flow configurations encountered in HVAC applications. The shell and tube configuration shown in Fig. 14-3 is often used for chillers and water-cooled condensers, whereas the cross-flow configuration occurs with many heating and cooling applications where air flows normal to a bank of finned tubes. The parameters P and R are defined as

$$P = \frac{t_{co} - t_{ci}}{t_{hi} - t_{ci}} \quad \text{and} \quad R = \frac{t_{hi} - t_{ho}}{t_{co} - t_{ci}} \tag{14-4}$$

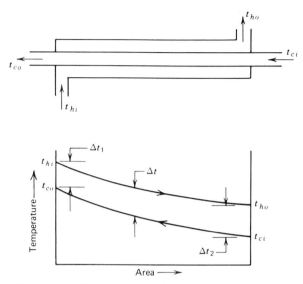

Figure 14-1 Counterflow heat exchanger.

P may be thought of as an effectiveness of the heat exchanger when $(\dot{m}c_p)_c < (\dot{m}c_p)_h$ where $(\dot{m}c_p)$ is the *fluid capacity rate C*. If the numerator and denominator of P are multiplied by C_c, P becomes the ratio of the actual heat transfer to the cold fluid to the heat transfer if the cold fluid is heated to the temperature of the entering hot fluid. P is thus the actual heat transfer divided by the theoretical maximum heat transfer to the cold fluid. When $C_h < C_c$, P is the effectiveness divided by R. It may also be shown that $R = C_c/C_h$.

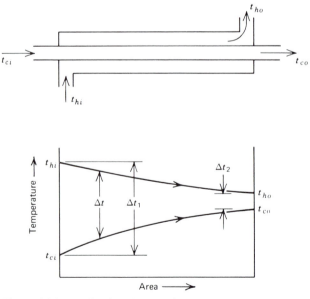

Figure 14-2 Parallel flow heat exchanger.

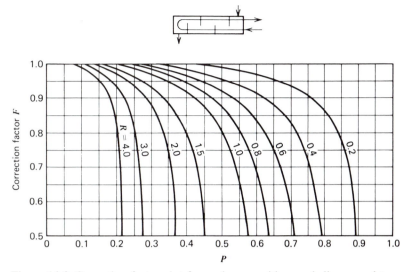

Figure 14-3 Correction factor plot for exchanger with one shell pass and two, four, or any multiple of tube passes.

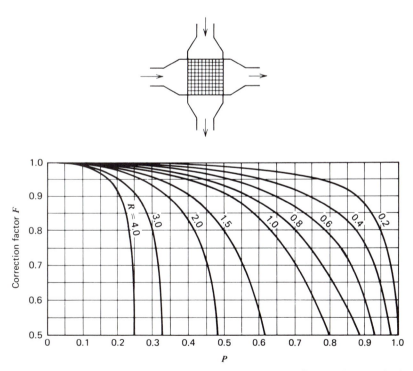

Figure 14-4 Correction factor plot for single-pass cross-flow exchanger, both fluids unmixed.

Calculations involving the LMTD are straightforward when the fluid inlet and outlet temperatures are known. If three of the temperatures are known, together with the fluid capacity rates of the streams, the fourth temperature may be calculated by a simple energy balance. When two of the temperatures are unknown, however, trial-and-error procedures are required because of the form of the equation for the LMTD, Eq. 14-2. In the design of heat exchangers this is often true.

14-2 THE NTU METHOD

The NTU method has the advantage of eliminating the trial-and-error procedure of the LMTD method for many practical problems when only the inlet fluid temperatures are known. NTU stands for the "Number of Transfer Units."

Heat exchanger effectiveness was mentioned in the previous section as

$$\varepsilon = \frac{\text{Actual heat-transfer rate}}{\text{Maximum possible heat-transfer rate}}$$

The actual heat-transfer rate is given by

$$\dot{q} = C_h(t_{hi} - t_{ho}) = C_c(t_{co} - t_{ci}) \tag{14-5}$$

The maximum possible heat-transfer rate is expressed by

$$\dot{q}_{max} = C_{min}(t_{hi} - t_{ci}) \tag{14-6}$$

This is true because the maximum heat transfer would occur if one of the fluids were to undergo a temperature change equal to the maximum in the heat exchanger, $(t_{hi} - t_{ci})$. The fluid experiencing the maximum temperature change must be the one with the minimum value of C to satisfy the energy balance.

The fluid with the minimum value of C may be the hot or the cold fluid. For $C_h = C_{min}$, using Eqs. 14-5 and 14-6

$$\varepsilon = \frac{\dot{q}}{\dot{q}_{max}} = \frac{C_h(t_{hi} - t_{ho})}{C_{min}(t_{hi} - t_{ci})} = \frac{(t_{hi} - t_{ho})}{(t_{hi} - t_{ci})} \tag{14-7a}$$

For $C_c = C_{min}$

$$\varepsilon = \frac{\dot{q}}{\dot{q}_{max}} = \frac{C_c(t_{co} - t_{ci})}{C_{min}(t_{hi} - t_{ci})} = \frac{(t_{co} - t_{ci})}{(t_{hi} - t_{ci})} \tag{14-8}$$

It is therefore necessary to have two expressions for the effectiveness (Eqs. 14-7 and 14-8). When effectiveness is known, the outlet temperature may be easily computed. For example, when $C_h < C_c$

$$t_{ho} = \varepsilon(t_{ci} - t_{hi}) + t_{hi} \tag{14-7b}$$

Also

$$t_{co} = \frac{\dot{q}}{C_c} + t_{ci} = \frac{C_h}{C_c}(t_{hi} - t_{ho}) + t_{ci} \qquad \text{(14-9a)}$$

or

$$t_{co} = \frac{C_h}{C_c}\varepsilon(t_{hi} - t_{ci}) + t_{ci} \qquad \text{(14-9b)}$$

Expressions for ε for several flow configurations are shown in Table 14-1. Figures 14-5, 14-6, and 14-7 show graphical representations of the effectiveness for three common cases. The NTU parameter is defined as UA/C_{min} and may be thought of as a heat-transfer size factor. It may also be observed that flow configuration is unimportant when $C_{min}/C_{max} = 0$. This corresponds to the situation of one fluid undergoing a phase change where c_p may be thought of as being infinite. Evaporating or condensing refrigerants as well as condensing water vapor are examples where $C_{min}/C_{max} = 0$.

The NTU method has gained greatest acceptance in connection with design of *compact heat exchangers* where a large surface area per unit volume exists. An arbitrary definition has been proposed that states that the ratio of surface area to volume in a compact exchanger is greater than 200 ft^2/ft^3 or 656 m^2/m^3. The common finned-tube exchangers used extensively in HVAC systems generally fall into this category. Compact heat exchangers are used in applications where at least one fluid is a gas. Because the thermal resistance associated with the gas film is high compared with a

Table 14-1 Thermal Effectiveness of Heat Exchangers with Various Flow Arrangements

Parallel flow:	$\varepsilon = \dfrac{1 - \exp[-\text{NTU}(1 + C)]}{1 + C}$
Counterflow:	$\varepsilon = \dfrac{1 - \exp[-\text{NTU}(1 - C)]}{1 - C\exp[-\text{NTU}(1 - C)]}$
Cross flow (both streams are unmixed):[a]	$\varepsilon = 1 - \exp\left\{\dfrac{1}{C\eta}[\exp((-\text{NTU})(C)(\eta)) - 1]\right\}$ where $\eta = \text{NTU}^{-0.22}$
Cross flow (both streams are mixed):	$\varepsilon = \text{NTU}\left\{\dfrac{\text{NTU}}{1 - \exp(-\text{NTU})} + \dfrac{(\text{NTU})(C)}{1 - \exp[-(\text{NTU})(C)]} - 1\right\}^{-1}$
Cross flow (stream C_{min} is unmixed):	$\varepsilon = \dfrac{1}{C}\{1 - \exp[-C(1 - \exp(-\text{NTU}))]\}$
Cross flow (stream C_{max} is unmixed):	$\varepsilon = 1 - \exp\left\{-\dfrac{1}{C}[1 - \exp(-(\text{NTU})(C))]\right\}$
1–2 Parallel counterflow:	$\varepsilon = 2\left\{1 + C + \dfrac{1 + \exp[-\text{NTU}(1 + C^2)^{1/2}]}{1 - \exp[-\text{NTU}(1 + C^2)^{1/2}]}(1 + C^2)^{1/2}\right\}^{-1}$ where $\text{NTU} = (UA/C_{min})$, and $C = C_{min}/C_{max}$

[a]This is an approximate expression.

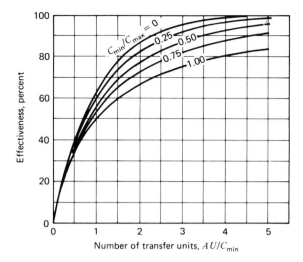

Figure 14-5 Effectiveness for counterflow exchanger.

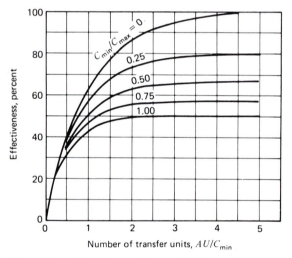

Figure 14-6 Effectiveness for parallel flow exchanger.

fluid such as water, fins are used to increase the heat-transfer area and decrease the thermal resistance.

14-3 HEAT TRANSFER—SINGLE-COMPONENT FLUIDS

The heat transfer rate from one fluid to the other in a heat exchanger is expressed by Eq. 14-3:

$$\dot{q} = UA(F)\,\Delta t_m \qquad \textbf{(14-3)}$$

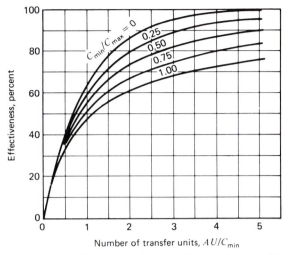

Figure 14-7 Effectiveness for cross-flow exchanger with
fluids unmixed.

when the LMTD method is used. When the NTU method is utilized, the heat-transfer
rate is generally computed from the temperature change for either fluid. For example

$$\dot{q} = (\dot{m}c_p)_h(t_{hi} - t_{ho}) \tag{14-10}$$

The temperature change is obtained from the effectiveness, which depends on NTU,
where

$$\text{NTU} = \frac{UA}{(\dot{m}c_p)_{min}} \tag{14-11}$$

It is evident that an average value of the overall coefficient U must be known for both
design methods.

The concept of overall thermal resistance and the overall heat-transfer coefficient was
discussed in Chapter 5. The general procedure is the same for heat exchangers. For a
simple heat exchanger without fins the overall coefficient U is given by

$$\frac{1}{UA} = \frac{1}{h_o A_o} + \frac{\Delta x}{k A_m} + \frac{1}{h_i A_i} + \frac{R_{fi}}{A_i} + \frac{R_{fo}}{A_o} \tag{14-12a}$$

where

h_o = heat-transfer coefficient on the outside, Btu/(hr-ft²-F) or W/(m²-C)
h_i = heat-transfer coefficient on the inside, Btu/(hr-ft²-F) or W/(m²-C)
Δx = thickness of the separating wall, ft or m
k = thermal conductivity of the separating wall, (Btu-ft)/
    (ft²-hr-F) or (W-m)/(m²-C)
A = area, ft² or m² where o, m, and i refer to outside, mean, and inside, respectively
R_f = fouling factor, (hr-ft²-F)/Btu or (m²-C)/W

In general the areas A_o, A_m, A_i are not equal and U may be referenced to any one of the three. Let $A = A_o$; then

$$\frac{1}{U_o} = \frac{1}{h_o} + \frac{\Delta x}{k(A_m/A_o)} + \frac{1}{h_i(A_i/A_o)} + \frac{R_{fi}}{(A_i/A_o)} + R_{fo} \qquad (14\text{-}12\text{b})$$

Fin Efficiency

Many of the heat exchangers used in HVAC systems have fins on one or both sides. Because the fins do not have a uniform temperature, the fin efficiency η is used to describe the heat-transfer rate.

$$\eta = \frac{\text{Actual heat transfer}}{(\text{Heat transfer with fin all at the base temperature } t_b)}$$

Figure 14-8 shows a simple fin. Since the base on which the fin is mounted also transfers heat, another parameter similar to fin efficiency is defined, called the surface effectiveness η_s:

$$\eta_s = \frac{\text{Actual heat transfer for fin and base}}{(\text{Heat transfer for fin and base when the fin is at the base temperature } t_b)}$$

If we assume that h is uniform over the fin and base surface, the actual heat-transfer rate is then given by

$$\dot{q} = hA\eta_s(t_b - t_\infty) \qquad (14\text{-}13)$$

where A is the total surface area of the fin and base. This may be written

$$\eta_s = \frac{\dot{q}}{hA(t_b - t_\infty)} = \frac{hA_b(t_b - t_\infty) + hA_f\eta(t_b - t_\infty)}{hA(t_b - t_\infty)} \qquad (14\text{-}14)$$

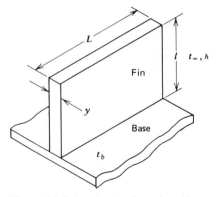

Figure 14-8 A simple fin of uniform cross section.

where $A = A_b + A_f$. Assuming that h is constant over the fin and base, we see that

$$\eta_s = \frac{A_b + \eta A_f}{A} = 1 - \frac{A_f}{A}(1 - \eta) \tag{14-15}$$

The thermal resistance is given by

$$R' = \frac{1}{hA\eta_s} \tag{14-16}$$

For a case where both sides of the heat exchanger have fins, the overall coefficient U is, assuming no fouling

$$\frac{1}{UA} = \frac{1}{h_o A_o \eta_{so}} + \frac{\Delta x}{k A_m} + \frac{1}{h_i A_i \eta_{si}} \tag{14-17}$$

If $A = A_o$

$$\frac{1}{U_o} = \frac{1}{h_o \eta_{so}} + \frac{\Delta x}{k(A_m/A_o)} + \frac{1}{h_i \eta_{si}(A_i/A_o)} \tag{14-18}$$

The second term on the right-hand side of Eqs. 14-17 and 14-18 represents the thermal resistance of the base and is often negligible.

Extended or finned surfaces may take on many forms ranging from the simple plate of uniform cross section shown in Fig. 14-8 to complex patterns attached to tubes. Several common configurations are considered in this section, starting with the fin of uniform cross section. The heat transfer rate for the fin may be shown to be (1)

$$\dot{q} = (t_b - t_\infty)mkA_c \tanh(ml) \tag{14-19}$$

when heat transfer from the tip is zero (1). The parameter m is given by

$$m = \left[\frac{hP}{kA_c}\right]^{1/2} \tag{14-20}$$

where

k = thermal conductivity of the fin material, (Btu-ft)/(ft^2-hr-F) or (W-m)/(m^2-C)
P = perimeter of the fin $2(L + y)$, ft or m
A_c = cross-sectional area of the fin $(L)(Y)$, ft^2 or m^2
h = heat-transfer coefficient, Btu/(hr-ft^2-F) or W/(m^2-C)

If we use the definition of fin efficiency given,

$$\eta = \frac{(t_b - t_\infty)mkA_c \tanh(ml)}{hA(t_b - t_\infty)} \tag{14-21}$$

or

$$\eta = \frac{\tanh(ml)}{(ml)} \tag{14-22}$$

The heat-transfer coefficient used to define m is assumed to be a constant. In a practical heat exchanger h will vary over the surface of the fins and will probably change between the inlet and outlet of the exchanger. The only practical solution is to use an average value of h for the complete surface. Equation 14-22 may be applied to a surface such as that shown in Fig. 14-25.

Most fins are very thin and $L \gg y$. In this case the parameter m defined by Eq. 14-20 may be simplified by setting $P = 2L$. Then

$$m = \left[\frac{2hL}{kLy}\right]^{1/2} = \left[\frac{2h}{ky}\right]^{1/2} \tag{14-23}$$

This approximation is often applied without explanation.

Finned tube heat exchangers are very popular for water-to-air or refrigerant-to-air applications. Figure 14-9 shows a sketch of a tube with circular fins. The diagram is somewhat idealized, since in practice the fin is usually wound on the tube in a helix from one continuous strip of material. A typical circular finned-tube water coil is shown in Fig. 14-10. Typically the fin will be quite thin. In the case of the circular fin the solution for the fin efficiency is very complex and is not generally used for practical problems; however, Fig. 14-11 shows a plot of the solution. An approximate but quite accurate method of predicting η for a circular fin has been developed by Schmidt (2). The method is largely empirical but has many advantages when an analytical expression is required. The method is summarized as follows:

$$\eta = \frac{\tanh(mr\phi)}{(mr\phi)} \tag{14-24}$$

where m is defined by Eq. 14-23 and

$$\phi = \left(\frac{R}{r} - 1\right)[1 + 0.35 \ln(R/r)] \tag{14-25}$$

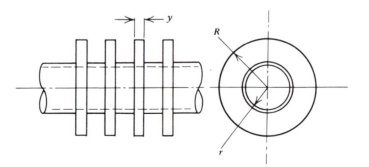

Figure 14-9 Tube with circular fins.

Figure 14-10 Circular finned-tube water coil. (Courtesy Thermal Corporation, Houston, TX.)

When R/r is between 1.0 and 8 and η falls between 0.5 and 1.0, the error is less than one percent of the value of the fin efficiency taken from Fig. 14-11.

Continuous-plate fins are also used extensively in finned-tube heat exchangers. In this case each fin extends from tube to tube. Figure 14-12 shows such an arrangement. It is not possible to obtain a closed analytical solution for this type of fin, and approximate methods are necessary. Consider the rectangular tube array of Fig. 14-13 with continuous-plate fins. When it is assumed that the heat-transfer coefficient is constant over the fin surface, an imaginary rectangular fin may be defined as shown. The outline of the fin is an equipotential line where the temperature gradient is zero. The problem is then to find η for a rectangular fin. Zabronsky (3) has suggested that a circular fin of equal area be substituted for purposes of calculating η; however, Carrier and Anderson (4) have shown that the efficiency of a circular fin of equal area is not accurate; they recommend the sector method. Rich (5) developed charts shown in reference 6 to facilitate use of the sector method. Schmidt (2) describes an approach to this problem that is nearly as accurate as the sector method and has the advantage of simplicity. Again the procedure is empirical; however, Schmidt tested the method statistically using maximum and minimum values of η that must bracket the actual fin efficiency. The method is based on the selection of a circular fin with a radius R_e that has the same fin efficiency as the rectangular fin. After R_e is determined, Eq. 14-24 is used for the calculation of η. For the rectangular fin

$$\frac{R_e}{r} = 1.28\psi(\beta - 0.2)^{1/2} \tag{14-26}$$

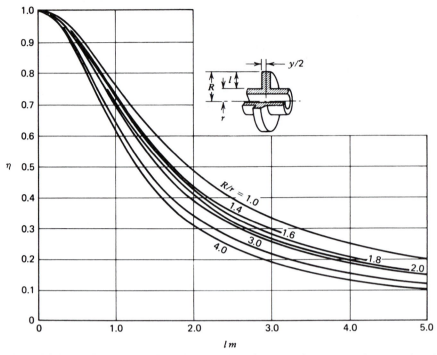

Figure 14-11 Performance of circumferential fins of rectangular cross section. (Reprinted by permission from *ASME Transactions,* Vol. 67, 1945.)

where

$$\psi = \frac{M}{r} \quad \text{and} \quad \beta = \frac{L}{M}$$

M and L are defined in Fig. 14-13 where L is always selected to be greater than or equal to M. In other words $\beta \geq 1$. The parameter ϕ given by Eq. 14-25 is computed using R_e instead of R.

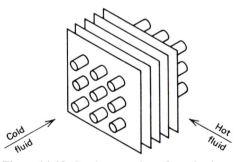

Figure 14-12 Continuous plate–fin–tube heat exchanger.

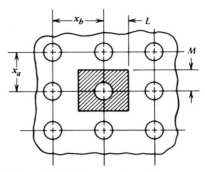

Figure 14-13 Rectangular tube array.

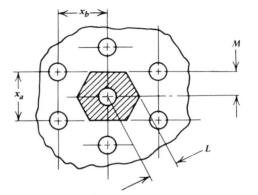

Figure 14-14 Hexangular tube array.

Figure 14-14 shows a triangular tube layout with continuous-plate fins. Here a hexangular fin results, which may be analyzed by the sector method (6). Schmidt (2) also analyzed this result and gives the following empirical relation, which is similar to Eq. 14-26:

$$\frac{R_e}{r} = 1.27\psi(\beta - 0.3)^{1/2} \tag{14-27}$$

where

$$\psi = \frac{M}{r} \quad \text{and} \quad \beta = \frac{L}{M}$$

M and L are defined in Fig. 14-14 where $L \geq M$. Equations 14-24 and 14-25 are used to compute η.

Special types of fins are sometimes used such as spines or fins of nonuniform cross section. Reference 6 contains data pertaining to these surfaces.

In the foregoing discussion we have assumed that the fins are rigidly attached to the base material so that zero thermal contact resistance exists. This may not always be true, particularly for plate–fin–tube surfaces. Eckels (7) has developed an empirical relation to predict the unit contact resistance for plate–fin–tube surfaces as follows:

$$R_{ct} = C \left[\frac{D_t \left(\frac{s}{y} - 1 \right)^2}{y} \right]^{0.6422} \tag{14-28}$$

where

R_{ct} = unit contact resistance, (hr-ft^2-F)/Btu or (m^2-C)/W
C = a constant, 2.222×10^{-6} for English units and 3.913×10^{-7} for SI units
D_t = outside tube diameter, in. or m
s = fin spacing, in. or m
y = fin thickness, in. or m

This unit contact resistance is associated with the outside tube area and added to Eq. 14-17. Because this contact resistance is undesirable as well as difficult to predict, every effort should be made to eliminate it in the manufacture of the heat exchanger. If tests are made for a surface, the contact resistance is usually reflected in the heat-transfer coefficients obtained.

14-4 TRANSPORT COEFFICIENTS INSIDE TUBES

Most HVAC heat exchanger applications of flow inside tubes and passages involve water, water vapor, and boiling or condensing refrigerants. The smooth copper tube is by far the most common geometry with these fluids. Forced convection turbulent flow is the most important mode; however, laminar flow sometimes occurs.

Turbulent Flow of Liquids Inside Tubes

Probably the most widely used heat-transfer correlation for this common case is the Dittus-Boelter equation (1)

$$\frac{\bar{h}D}{k} = 0.023(\mathrm{Re}_D)^{0.8}(\mathrm{Pr})^n \tag{14-29}$$

where

$$n = 0.4 \qquad t_{\mathrm{wall}} > t_{\mathrm{bulk}}$$
$$n = 0.3 \qquad t_{\mathrm{wall}} < t_{\mathrm{bulk}}$$

Equation 14-29 applies under conditions of $\mathrm{Re}_D > 10,000$, $0.7 < \mathrm{Pr} < 100$, and $L/D > 60$. All fluid properties should be evaluated at the arithmetic mean bulk temperature of the fluid. Appendix B gives the thermophysical properties required in Eq. 14-29 for some common liquids and gases. Reference 6 gives other, similar correlations for special conditions. Equation 14-29 may be used for annular or noncircular cross sections for approximate calculations. In this case the tube diameter D is replaced by the hydraulic diameter D_h.

$$D_h = \frac{4(\text{Cross-sectional area})}{(\text{Wetted perimeter})} \qquad \text{(14-30)}$$

Reference 8 gives extensive data for noncircular flow channels when more accurate values are required.

Pressure drop for flow of liquids inside pipes and tubes was discussed in Chapter 10. The same procedure applies to heat-exchanger tubes; we must still take into account the considerable increase in equivalent length caused by the many U-turns, tube inlets and exits, and the headers required in most heat exchangers.

Laminar Flow of Liquids Inside Tubes

The recommended correlation for predicting the average film coefficient in laminar flow in tubes is

$$\frac{\bar{h}D}{k} = 1.86 \left[\text{Re}_D \text{Pr} \frac{D}{L} \right]^{1/3} \left(\frac{\mu}{\mu_s} \right)^{0.14} \qquad \text{(14-31)}$$

When the term in brackets is less than about 20, Eq. 14-31 becomes invalid; however, this will not occur for most heat-exchanger applications. Properties should be evaluated at the arithmetic mean bulk temperature except for μ_s, which is evaluated at the wall temperature.

A word of caution is appropriate concerning the transition from laminar to turbulent flow. This region is defined approximately by $2000 < \text{Re}_D < 10{,}000$. Prediction of heat transfer and friction coefficients is uncertain during transition. The usual practice is to avoid the region by proper selection of tube size and flow rate.

Ghajar and co-workers (9,10) have investigated heat transfer and pressure loss in the transition region and present correlations for this case.

Pressure drop is computed as described earlier for turbulent flow in tubes and in Chapter 10. For laminar flow the friction factor (Moody) is given by

$$f = \frac{64}{\text{Re}_D} \qquad \text{(14-32)}$$

Ethylene Glycol–Water Solutions

In many systems it is necessary to add ethylene glycol to the water to prevent freezing and consequent damage to the heat exchangers and other components. The effect of the glycol on flow friction was discussed in Chapter 10, and it was shown that the lost head is generally increased when a glycol–water solution is used. The heat transfer is also adversely affected. Figures 14-15 and 14-16 give the specific heat and thermal conductivity of ethylene glycol solutions as a function of temperature and concentration. Similar data for specific gravity and viscosity are given in Chapter 10. It is very important to anticipate the use of glycol solutions during the design phase of a project

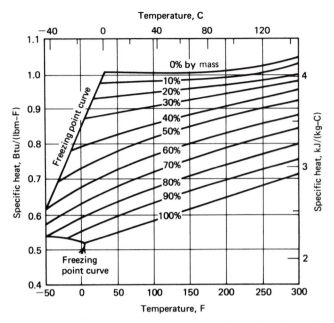

Figure 14-15 Specific heat of aqueous solutions of ethylene glycol. (Reprinted by permission from *ASHRAE Handbook, Fundamentals Volume,* 1989.)

because the heat-transfer coefficient using a 30 percent glycol solution may be as much as 40 percent less than the coefficient using pure water. This is mainly because of the lower thermal conductivity and specific heat of the glycol solution.

Condensation and Evaporation Inside Horizontal Tubes

The prediction of heat transfer and pressure drop in two-phase flow is much more uncertain than with a single-phase flow. The mixture of vapor and liquid can vary considerably in composition and hydrodynamic behavior, and it is generally not possible to describe all conditions with one relation. Two-phase flow inside horizontal tubes is the most common situation in HVAC systems, and one or two correlations are presented for this case.

The following relations from reference 6 apply to film condensation, the dominant mode

$$\frac{\bar{h}D}{k_l} = 13.8(\mathrm{Pr})_l^{1/3}\left(\frac{i_{fg}}{c_{pl}\,\Delta t}\right)^{1/6}\left[\frac{DG_v}{\mu_l}\left(\frac{\rho_l}{\rho_v}\right)^{1/2}\right]^{0.2} \tag{14-33}$$

where

$$\frac{DG}{\mu_l} < 5000 \quad \text{and} \quad 1000 < \frac{DG_v}{\mu_l}\left(\frac{\rho_l}{\rho_v}\right)^{1/2} < 20,000$$

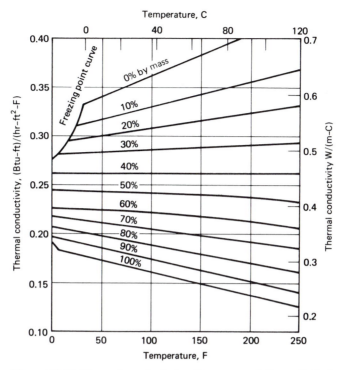

Figure 14-16 Thermal conductivity of aqueous solutions of ethylene glycol. (Reprinted by permission from *ASHRAE Handbook, Fundamentals Volume, 1989.*)

The subscripts l and v refer to liquid and vapor, respectively, and Δt is the difference between the fluid saturation temperature and the wall surface temperature. When

$$20{,}000 < \frac{DG_v}{\mu_l}\left(\frac{\rho_l}{\rho_v}\right)^{1/2} < 100{,}000$$

$$\frac{\overline{h}D}{k_l} = 0.1(\mathrm{Pr})_l^{1/3}\left(\frac{i_{fg}}{c_{pl}\Delta t}\right)^{1/6}\left[\frac{DG_v}{\mu_l}\left(\frac{\rho_l}{\rho_v}\right)^{1/2}\right]^{2/3} \tag{14-34}$$

Equations 14-33 and 14-34 are for condensing saturated vapor; however, little error is introduced for superheated vapor when the wall temperature is below the saturation temperature and h is calculated for saturated vapor. Appendices A and B give the required properties.

The average heat-transfer coefficients for evaporating R-12 and R-22 may be estimated from the following relation from reference 6:

$$\frac{\overline{h}D}{k_l} = C_1\left[\left(\frac{GD}{\mu_l}\right)^2\left(\frac{J\,\Delta x\,i_{fg}g_c}{Lg}\right)\right]^n \tag{14-35}$$

where

J = Joules equivalent, 778 (ft-lbf)/Btu or 1 for SI units

Δx = change in quality of the refrigerant, mass of vapor per unit mass of the mixture

i_{fg} = enthalpy of vaporization, Btu/lbm or J/kg

L = length of the tube, ft or m

C_1 = constant = 9×10^{-4} when $x_e < 0.9$, and 8.2×10^{-3} when $x_e \geq 1.0$ (x_e is the quality of the refrigerant leaving the tube)

n = constant = 0.5   when $x_e < 0.9$,   and 0.4 when $x_e \geq 1.0$

The correlation was obtained from tests made using copper tubes having diameters of 0.47 and 0.71 in. and lengths from 13 to 31 ft. Evaporating temperatures varied from -4 to 32 F. Equation 14-35 is sufficient for most HVAC applications where Appendices A and B give the required properties.

The pressure loss that occurs with a gas–liquid flow is of interest. Experience has shown that pressure losses in two-phase flow are usually much higher than would occur for either phase flowing along at the same mass rate. As in any flow, the total pressure loss along a tube depends on three factors: (1) friction, due to viscosity, (2) change of elevation, and (3) acceleration of the fluid.

Friction is present in any flow situation, although in some cases it may contribute less than the other two factors. In horizontal flow the change in elevation is zero, and there would be no pressure drop due to this factor. Where there is a small change in gas density or little evaporation occurring, the pressure drop due to acceleration would usually be small. In flow with large changes of density or where evaporation is present, however, the acceleration pressure drop may be very significant.

There has been a variety of schemes proposed for predicting the pressure drop in two-phase flow. Some investigators have attempted to develop theories and equations valid only for a particular flow pattern. These have the disadvantage of requiring that the type of flow pattern be known, and they lead to complexities in those situations where flow patterns change along the flow. Other investigators have attempted to develop single theories valid over several other regimes. These attempts have not been completely successful.

A widely used method of predicting pressure drop in two-phase flow was developed by Lockhart and Martinelli (27). Their method, which is essentially empirical, is based on measurements of isothermal flow of air and several liquids under essentially incompressible conditions. *The pressure drop predicted is that due to friction only.*

Extensive work has been devoted to the two-phase pressure-loss problem, but available methods remain very complex and impractical for general use. Therefore, the manufacturers of coils have resorted to experimental data for specific coils and refrigerants. Figure 14-17 is an example of such data for $\frac{1}{2}$-inch O.D. tubes. Trial and error may be required to obtain suitable circuiting of the coil. Circuiting will be discussed later.

14-5   TRANSPORT COEFFICIENTS OUTSIDE TUBES AND COMPACT SURFACES

Air is the most common flow medium in this case except for shell and tube evaporators and condensers where heat is transferred between a refrigerant inside the tubes to water outside the tubes. Water-to-water applications may also occur. In the case of bare tube bundles, the flow may be across the tubes, parallel to the tubes, or a combination of the

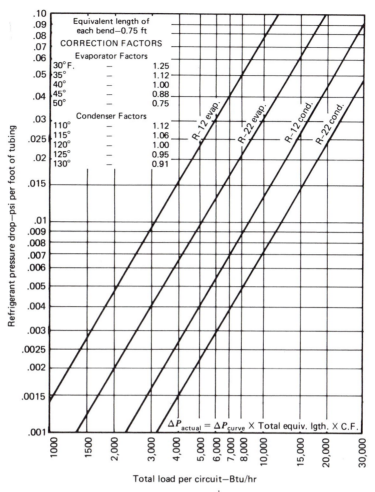

Figure 14-17 Refrigerant pressure loss in $\frac{1}{2}$ in. O.D. tubes.

two when baffles are used. Compact surfaces such as finned tubes or plate fins will usually have air flowing parallel to the fins and normal to the tubes.

Bare Tubes with Parallel Flow

Kays and Crawford (12) discuss both laminar and turbulent heat transfer with flow parallel to the tubes and present results for constant heat flux per unit length of tube for fully developed velocity and temperature profiles. References 13 and 14 also pertain to this problem as well as pressure loss with flow parallel to tubes.

Bare Tubes with a Combination of Parallel and Cross Flow

This flow pattern occurs when baffles are used in a tube and shell exchanger. Empirical data are usually employed because of the complex nature of the flow pattern. The most successful of these methods developed from the work of Tinker and was clarified by Devore (15); the shell side flow was split into the basic components of pure cross-flow, baffle-to-baffle leakage flow, tube-to-baffle leakage flow, and bundle-to-shell bypass flow. Tinker made gross simplifications to describe the individual flow streams and developed a set of equations that could be solved simultaneously. Tinker's method was utilized in part to develop the Delaware method reported by Bell (16). This method avoided some of the complexities of Tinker's method by assuming independent correction factors for the individual streams. A study by Palen and Taborek (17) using extensive data banks, and the digital computer has led to a method that improves on the older ones.

Bare Tubes in Cross Flow

The most common application of bare tubes in pure cross flow involves air. Although this application is rapidly going out of style in favor of finned tubes, considerable data are available for tubes in cross flow, as shown in Fig. 14-18 and reference 8. The manner of presentation is quite typical of that used for all types of compact heat-exchanger surfaces where the *j*-factor, introduced in Chapter 13

$$j = \frac{\bar{h}}{Gc_p} \text{Pr}^{2/3} \tag{14-36}$$

and the Fanning friction factor f are plotted versus the Reynolds number

$$\text{Re} = \frac{GD_h}{\mu} \tag{14-37}$$

The number of rows of tubes in the flow direction has an effect on the *j*-factor and the heat-transfer coefficient h. The data of Fig. 14-18 are applicable to an exchanger with four rows of tubes (18). For bare tubes in cross-flow, the relation between the heat-transfer coefficient for a finite number of tube rows N and that for an infinite number of tube rows is given approximately by

$$\frac{\bar{h}}{h_\infty} = 1 - 0.32e^{-0.15N_r} \tag{14-38}$$

when $2 < N_r < 10$. One might expect the friction factor to also depend on the number of tube rows; however, this does not seem to hold true. The assumption is that since a contraction and expansion occur for each row, the friction factor is the same for each row.

The mechanical energy equation, Eq. 10-1d, with the elevation and work terms zero, expresses the lost head for a bank of tubes oriented as shown in Fig. 14-19:

$$\frac{(P_{01} - P_{02})g_c}{\rho_m g} = l_h \tag{14-39}$$

where l_h is made up losses resulting from a change in momentum, friction, and entrance

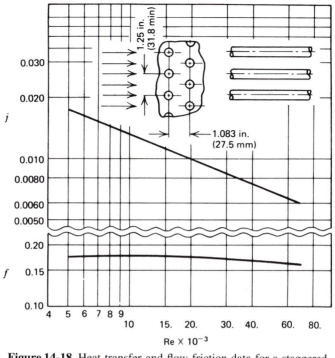

Figure 14-18 Heat-transfer and flow-friction data for a staggered
tube bank, 4 rows of tubes. (Reprinted by permission
from *ASHRAE Transactions,* Vol. 79, Part II, 1973.)

and exit contraction and expansion losses. Integration of the momentum equation
through the heat exchanger core, Fig. 14-19, yields (8)

$$\frac{\Delta P_0 g_c}{\rho_m g} = \frac{G_c^2}{2 g \rho_m \rho_1} \left[(K_i + 1 - \sigma^2) \right.$$

$$\left. + 2 \left(\frac{\rho_1}{\rho_2} - 1 \right) + f \frac{A}{A_c} \frac{\rho_1}{\rho_m} - (1 - \sigma^2 - K_e) \frac{\rho_1}{\rho_2} \right] \quad \textbf{(14-40)}$$

where K_i and K_e are entrance and exit loss coefficients that will be discussed in the next

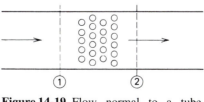

Figure 14-19 Flow normal to a tube
bank.

section and σ is the ratio of minimum flow-to-frontal area of the exchanger. It may be shown that

$$\frac{A}{A_c} = \frac{4L}{D_h} \tag{14-41}$$

which is a result of the hydraulic diameter concept.

$$A = \text{total heat transfer area, ft}^2 \text{ or m}^2$$

$$A_c = \text{flow cross-sectional area, ft}^2 \text{ or m}^2$$

Referring to Eq. 14-40 the first and last terms in the brackets account for entrance and exit losses, whereas the second and third terms account for flow acceleration and friction, respectively. In the case of tube bundles, the entrance and exit effects are included in the friction term, that is, $K_i = K_e = 0$. Equation 14-40 then becomes

$$l_h = \frac{G_c^2}{2g\rho_m\rho_1}\left[(1 + \sigma^2)\left(\frac{\rho_1}{\rho_2} - 1\right) + f\frac{A}{A_c}\frac{\rho_1}{\rho_m}\right] \tag{14-42}$$

G_c is based on the minimum flow area and ρ_m is the mean density between inlet and outlet given by

$$\rho_m = \frac{1}{A}\int_A \rho \, dA \tag{14-43}$$

Equation 14-43 is difficult to evaluate. An arithmetic average is usually a good approximation except for parallel flow:

$$\rho_m \approx \frac{\rho_1 + \rho_2}{2} \tag{14-44}$$

A useful nondimensional form of Eq. 14-42 is given by

$$\frac{\Delta P_0}{P_{01}} = \frac{G_c^2}{2g_c\rho_1 P_{01}}\left[(1 + \sigma^2)\left(\frac{\rho_1}{\rho_2} - 1\right) + f\frac{A}{A_c}\frac{\rho_1}{\rho_m}\right] \tag{14-45}$$

where ΔP_0 and P_{01} have units of lbf/ft^2 or Pa. Equations 14-42 and 14-45 are also valid for finned tubes or any other surface that does not have abrupt contractions or expansions.

Finned-tube Heat-transfer Surfaces

Heat-transfer surfaces may take many forms; however, finned tubes are the most popular in HVAC applications. The manner in which data are presented is the same as that shown for bare tubes in Fig. 14-18 and the lost head may be computed using Eq. 14-45. Rich (18, 19) has studied the effect of both fin spacing and tube rows for the plate–

fin–tube geometry. Both the j-factor and friction factor decrease as the fin spacing is decreased. The decrease in j-factor was about 50 percent and the decrease in friction factor was about 75 percent as fin pitch was increased from 3 to 20 over the Reynolds number ($G_c D_h/\mu$) range of 500 to 1500. Figure 14-20 shows how the data correlated when the Reynolds number was based on the tube row spacing χ_b. For a given fin pitch it was found that the j-factors decreased as the number of tube rows was increased from 1 to 6 in the useful Reynolds number range. This is contrary to the behavior of bare tubes and results from the difference in the flow fields in each case. The one-row coil had significantly higher j-factors than the others, as was known from previous work done by Shepherd (20). Figure 14-21 shows the j-factor data for the coils with various numbers of tube rows. Note that the Reynolds number is based on the tube row spacing. The combination of Figs. 14-20 and 14-21 therefore gives performance data for all heat exchangers of this one tube diameter and tube pattern with variable fin pitch and number of tube rows. Other surfaces with tube diameters and patterns in the same range will behave similarly. The study of tube-row effect (18) also showed that all rows in a plate–fin–tube coil do not have the same heat-transfer rate (Fig. 14-22). The j-factors are less for each successive row in the useful (low) Reynolds number range.

The friction factors behave in a manner similar to that discussed before for bare tube banks; therefore, it is assumed that there is no tube-row effect.

Caution should be exercised in using published data for plate–fin–tube heat exchangers, especially if the number of rows is not given.

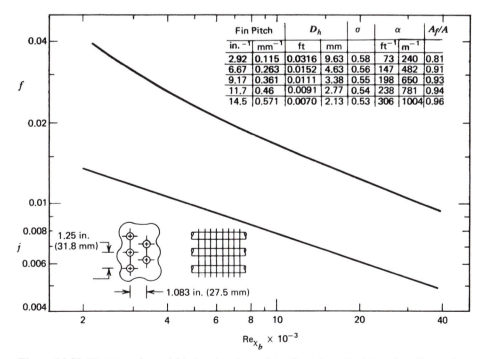

Figure 14-20   Heat-transfer and friction data for a plate–fin–tube coil with various fin spacings and 5 rows of tubes. (Reprinted by permission from *ASHRAE Transactions,* Vol. 79, Part II, 1973.)

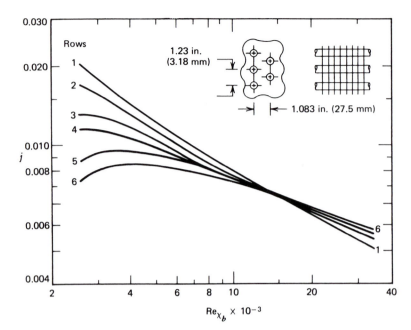

Figure 14-21 Heat-transfer data for plate–fin–tube coils with various numbers of tube rows. (Reprinted by permission from *ASHRAE Transactions,* Vol. 81, Part I, 1975.)

Research by McQuiston (21, 22) has resulted in the correlation of plate–fin–tube transport data that account for geometric variables as well as hydrodynamic effects. Figure 14-23 shows *j*-factors plotted versus the parameter *JP*, which is defined as

$$JP = \text{Re}_D^{-0.4} \left(\frac{A}{A_t} \right)^{-0.15} \tag{14-46}$$

where

$$\text{Re}_D = \frac{G_c D}{\mu} \tag{14-47}$$

and

$$\frac{A}{A_t} = \frac{4}{\pi} \frac{\chi_b}{D_h} \frac{\chi_a}{D} \sigma \tag{14-48}$$

In this case the Reynolds number is based on the outside tube diameter and A/A_t is the ratio of the total heat-transfer area to the area of the bare tubes without fins. Note that A/A_t becomes 1.0 for a bare tube bank and the correlation takes a familiar form. The tube-row effect is not accounted for in Fig. 14-23 and must be done separately using Fig. 14-21, which is described approximately by

$$j_n/j_1 = 1 - 1280 \, N_r \text{Re}_{\chi_b}^{-1.2} \tag{14-49}$$

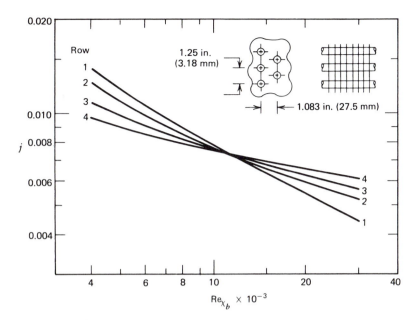

Tube outside diameter, 0.521 in. or 13.2 mm
Fin pitch, 14.5 fins/in. or 0.571 fins/mm
Fin thickness, 0.006 in. or 0.152 mm
Free flow area/frontal area, 0.536

Figure 14-22 Heat-transfer data for each row of a 4-row plate–fin–tube coil. (Reprinted by permission from *ASHRAE Transactions,* Vol. 81, Part I, 1975.)

where the subscripts n and 1 pertain to the number of tube rows. Because Fig. 14-23 is for 4 rows of tubes, it is more convenient to write

$$j_n/j_4 = \frac{1 - 1280\, N_r \mathrm{Re}_{\chi_b}^{-1.2}}{1 - 5120\, \mathrm{Re}_{\chi_b}^{-1.2}} \tag{14-50}$$

where j_4 is read from Fig. 14-23.

Generalized correlation of friction data is more involved than that for heat-transfer data. Figure 14-24 shows such a correlation using a parameter FP defined as

$$FP = \mathrm{Re}_D^{-0.25} \left(\frac{D}{D^*}\right)^{0.25} \left[\frac{\chi_a - D}{4(s - y)}\right]^{-0.4} \left[\frac{\chi_a}{D^*} - 1\right]^{-0.5} \tag{14-51}$$

where D^* is a hydraulic diameter defined by

$$D^* = \frac{D\left(\dfrac{A}{A_t}\right)}{1 + \dfrac{(\chi_a - D)}{s}} \tag{14-52}$$

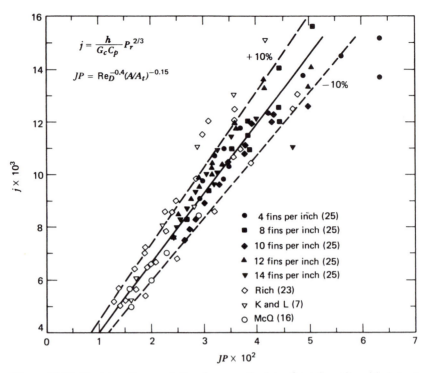

Figure 14-23 Heat-transfer correlation for smooth plate–fin–tube coils with 4 rows of tubes. (Reprinted by permission from *ASHRAE Transactions*, Vol. 84, Part I, 1978.)

The correlating parameters of Eqs. 14-51 and 15-52 have evolved over a long period of time from observations of experimental data. The friction data scatter more than the heat-transfer data of Fig. 14-23, which are typical. Note that the data from reference 21 are much more consistent than some of the other data that date back more than 20 years.

The data presentations of Figs. 14-23 and 14-24 have the advantage of generality and are also adaptable to the situation where moisture is condensing on the surface. This will be discussed later in this chapter. These same types of correlations may be used for other types of finned surfaces such as circular and wavy fins.

Plate–Fin Heat-Transfer Surfaces

Figure 14-25 illustrates the plate–fin heat-transfer surface. The fins may have several variations such as louvers, strips, or waves. Plain smooth fins are generally not used because of the low heat-transfer coefficients that arise when the flow length becomes long. The types mentioned earlier disturb the boundary layer so that the length does not influence the heat-transfer or flow-friction coefficients. Figure 14-25 is an example of data for a louvered plate–fin surface.

In computing the lost head for these surfaces, one must consider the entrance and exit losses resulting from abrupt contraction and expansion. The entrance and exit losses are

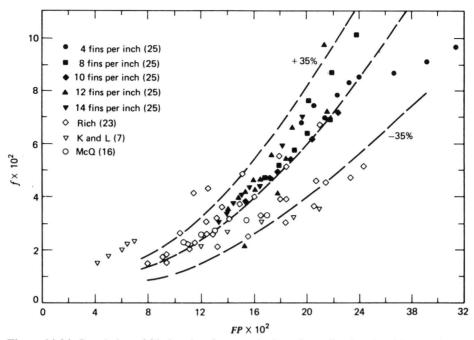

Figure 14-24 Correlation of friction data for smooth plate–fin–coils. (Reprinted by permission from *ASHRAE Transactions,* Vol. 84, Part I, 1978.)

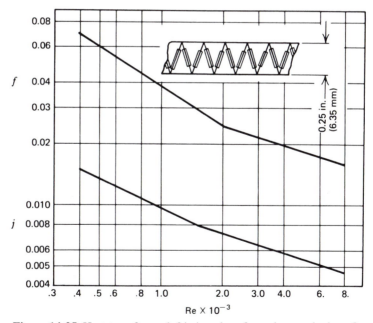

Figure 14-25 Heat-transfer and friction data for a louvered plate–fin surface.

expressed in terms of a loss coefficient K and the velocity head inside the heat exchanger core. Thus for the entrance

$$\Delta P_{0i} = K_i \frac{G_c^2}{2\rho_i g_c} \qquad (14\text{-}53)$$

and for the exit

$$\Delta P_{0e} = K_e \frac{G_c^2}{2\rho_e g_c} \qquad (14\text{-}54)$$

Equations 14-53 and 14-54 are included in Eq. 14-40:

$$\frac{\Delta P_0}{P_{01}} = \frac{G_c^2}{2g_c P_{01}\rho_1}\left[(K_i + 1 - \sigma^2)\right.$$
$$\left. + 2\left(\frac{\rho_1}{\rho_2} - 1\right) + f\frac{A}{A_c}\frac{\rho_1}{\rho_m} - (1 - \sigma^2 - K_e)\frac{\rho_1}{\rho_2}\right] \qquad (14\text{-}55)$$

The entrance and exit loss coefficients depend on the type of surface, the contraction ratio, and the Reynolds number $G_c D_h/\mu$. The degree to which the velocity profile has developed is also important. Reference 8 gives entrance and exit loss coefficients that apply to surfaces such as that shown in Fig. 14-25. The loss coefficients are a function of Reynolds number. However, most plate–fin surfaces have flow interruptions that cause continual redevelopment of the boundary layer, which is equivalent to a very high

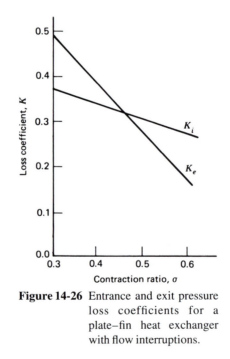

Figure 14-26 Entrance and exit pressure loss coefficients for a plate–fin heat exchanger with flow interruptions.

Reynolds number condition. Figure 14-26 gives loss coefficients applicable to plate–fin type surfaces with flow interruptions such as that of Fig. 14-25.

14-6   DESIGN PROCEDURES FOR SENSIBLE HEAT TRANSFER

It is difficult to devise one procedure for designing all heat exchangers because the given parameters vary from situation to situation. All of the terminal temperatures may be known or only the inlet temperatures may be given. The mass flow rates may be fixed in some cases and a variable in others. Usually the surface area is not given.

Earlier in the chapter the LMTD and NTU methods were described as the two general heat-exchanger design procedures. Either method may be used, but the NTU method has certain advantages. Consider the design problem where t_{hi}, t_{ho}, t_{ci}, $\dot{m}_c$, and $\dot{m}_h$ are known and the surface area A is to be determined. With either approach the heat-transfer coefficients must be determined as previously discussed so that the overall coefficient U can be computed. The NTU approach then proceeds as follows:

1. Compute the effectiveness ε and C_{min}/C_{max} from the given data.
2. Determine the NTU for the particular flow arrangement from the ε-NTU curve, such as Fig. 14-5 or Table 14-1.
3. Compute A from $A = \text{NTU}(C_{min}/U)$.

The LMTD approach is as follows:

1. Compute P and R from the given terminal temperatures.
2. Determine the correction factor F from the appropriate curve, such as Fig. 14-3.
3. Calculate the LMTD for an equivalent counterflow exchanger using Eq. 14-2.
4. Calculate A from $A = \dot{q}/U(F)(\text{LMTD})$ where

$$\dot{q} = C_c(t_{co} - t_{ci}) = C_h(t_{hi} - t_{ho})$$

The NTU approach requires somewhat less effort in this case.

Consider the design problem where A, U, $\dot{m}_c$, $\dot{m}_h$, t_{hi}, and t_{ci} are given, and it is necessary to find the outlet temperatures t_{ho}, t_{co}. The NTU approach is as follows:

1. Calculate the NTU $= UA/C_{min}$ from given data.
2. Find ε from the appropriate curve for the flow arrangement using NTU and C_{min}/C_{max}.
3. Compute one outlet temperature from Eq. 14-7 or 14-8.
4. Compute the other outlet temperature from

$$\dot{q} = C_c(t_{co} - t_{ci}) = C_h(t_{hi} - t_{ho})$$

The LMTD approach requires iteration as follows:

1. Calculate R from $R = C_c/C_h$.
2. Assume one outlet temperature in order to compute P (first approximation).
3. Find F from the appropriate curve (first approximation).
4. Evaluate LMTD (first approximation).
5. Determine $\dot{q} = UAF(\text{LMTD})$ (first approximation).
6. Calculate outlet temperature to compare with the assumption of step 2.
7. Repeat steps 2 through 6 until satisfactory agreement is obtained.

It is obvious that the NTU method is much more straightforward in this later problem. A complete design problem is illustrated in the following example.

EXAMPLE 14-1

Design a water-to-air heating coil of the continuous plate–fin–tube type. The required duty for the coil is as follows:

Heat outdoor air from 50 to about 100 F
Air flow rate of 2000 cfm
Entering water temperature, 150 F
Leaving water temperature, 140 F
Air face velocity should not exceed 1000 fpm
Water-side head loss should not exceed 10 ft wg
Water connections must be on the same end of the coil
Air-side pressure drop should not exceed 1.2 in. wg

SOLUTION

Figure 14-27 is a schematic of a typical water-to-air heating coil that has multiple rows of tubes. Although the water may be routed through the tubes in many different ways, the circuiting is usually such that counterflow will be approached as shown. Counterflow can usually be assumed when three or more rows are used. Because the water inlet and outlet connections must be on the same end of the coil in this case, a multiple of two rows is used if possible.

The first step in the solution will be to compute the overall heat-transfer coefficient U, which will be based on the air-side area. Equation 14-18 applies where η_{si} is equal to one and the wall thermal resistance is negligible.

$$\frac{1}{U_o} = \frac{1}{h_o \eta_{so}} + \frac{1}{h_i (A_i/A_o)}$$

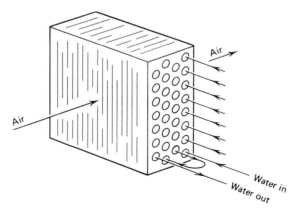

Figure 14-27 A typical heating coil circuited to approach counterflow.

The subscript o refers to the air side and the i refers to the water side. Equation 14-29 will be used to find the coefficient h_i assuming a water velocity of 4 ft/sec. Experience has shown that velocities greater than 5 ft/sec or 1.5 m/s result in very high lost head. Since at this point the tube diameter must be established, a surface geometry must be selected.

One standard plate–fin–tube surface uses $\frac{1}{2}$-in. tubes in a triangular layout as shown in Fig. 14-14 with χ_a of 1.25 in. and χ_b of 1.083 in. The fin pitch is 8 fins/in. and the fin thickness is 0.006 in. As a result of fabrication of the coil, the final tube outside diameter is 0.525 in. with a wall thickness of 0.015 in. Other geometric data will be given as required and the j-factor and friction factor will be obtained from Figs. 14-23 and 14-24. The Reynolds number based on the tube inside diameter is then

$$\mathrm{Re}_D = \frac{\rho \bar{V} D}{\mu} = \frac{61.5(4)(0.4831/12)}{(1.04/3600)} = 34{,}275$$

where ρ and μ are evaluated at 145 F. The Prandtl number is

$$\mathrm{Pr} = \frac{\mu c_p}{k} = \frac{(1.04)(1.0)}{(0.38)} = 2.74$$

Then using Eq. 14-29

$$\bar{h}_i = 0.023 \frac{k}{D} (\mathrm{Re}_D)^{0.8} (\mathrm{Pr})^{0.3}$$

$$\bar{h}_i = 0.023 \frac{0.38}{(0.483/12)} (34{,}275)^{0.8} (2.74)^{0.3}$$

$$\bar{h}_i = 1250 \ \mathrm{Btu/(hr\text{-}ft^2\text{-}F)}$$

where the exponent on the Prandtl number is for $t_{\mathrm{wall}} < t_{\mathrm{bulk}}$ and L/D has been assumed to be larger than 60.

To compute the air-side heat-transfer coefficient it is necessary to know the air velocity or air mass velocity inside the core. Because the coil face velocity cannot exceed 1000 ft/min, a face velocity of 900 ft/min will be assumed. Then

$$\dot{m}_a = G_{fr} A_{fr} = G_c A_c$$

and

$$G_c = G_{fr} \frac{A_{fr}}{A_c} = \frac{G_{fr}}{\sigma}$$

where the subscript fr refers to the face of the coil and c refers to the minimum flow area inside the coil. The ratio of minimum flow-to-frontal area for this case has been calculated to be 0.555.

$$G_{fr} = \rho_{fr}\overline{V}_{fr} = \frac{14.7(144)(900)60}{53.35(510)} = 4200 \text{ lbm/(hr-ft}^2)$$

and

$$G_c = \frac{4200}{(0.555)} = 7569 \text{ lbm/(hr-ft}^2)$$

The j-factor correlation of Fig. 14-23 is based on the parameter JP, which is defined by Eq. 14-46. The Reynolds number is then

$$\text{Re}_D = \frac{G_c D}{\mu} = \frac{7569(0.525/12)}{0.044} = 7578$$

and the parameter A/A_t defined by Eq. 14-48 is

$$A/A_t = \frac{4(1.083)1.25(0.555)}{\pi(0.01312)12(0.525)} = 11.6$$

where the hydraulic diameter is another known dimension of the coil. The parameter JP is

$$JP = (7578)^{-0.4}(11.6)^{-0.15} = 0.0194$$

The j-factor is now read from Fig. 14-23 as 0.0066. Then

$$\text{StPr}^{2/3} = \left(\frac{\overline{h}_o}{G_c c_p}\right)\left(\frac{\mu c_p}{k}\right)^{2/3} = 0.0066$$

or

$$\overline{h}_o = 0.0066(7569)(0.24)(0.71)^{-2/3} = 15.1 \text{ Btu/(hr-ft}^2\text{-F)}$$

The next step is to compute the fin efficiency and the surface effectiveness. Equations 14-23, 14-24, 14-25, and 14-27 will be used. The equivalent fin radius R_e is first computed from Eq. 14-27. The dimensions L and M are found as follows by referring to Fig. 14-14.

$$\text{Dim}_1 = \frac{X_a}{2} = \frac{1.25}{2} = 0.625 \text{ in.}$$

$$\text{Dim}_2 = \frac{[(X_a/2)^2 + X_b^2]^{1/2}}{2}$$

$$= \frac{[(0.625)^2 + (1.083)^2]^{1/2}}{2} = 0.625 \text{ in.}$$

Because Dim_1 is equal to Dim_2 in this case

$$L = M = 0.625 \text{ in.}$$

Then

$$\psi = \frac{M}{r} = \frac{0.625}{(0.525/2)} = 2.38$$

$$\beta = \frac{L}{M} = \frac{0.625}{0.625} = 1.0$$

and

$$\frac{R_e}{r} = 1.27(2.38)(1.0 - 0.3)^{1/2} = 2.53$$

From Eq. 14-25

$$\phi = (2.53 - 1)(1 + 0.35 \ln 2.53) = 2.03$$

and using Eq. 14-23

$$m = \left[\frac{2(15.1)}{100(0.006/12)} \right]^{1/2} = 24.6 \text{ ft}^{-1}$$

where the thermal conductivity k of the fin material has been assumed equal to 100 (Btu-ft)/(ft^2-hr-F), which is typical of aluminum fins. Then from Eq. 14-24,

$$\eta = \frac{\tanh\left[(24.6)(0.525/24)(2.03)\right]}{(0.525/24)(2.03)(24.6)} = 0.73$$

The surface effectiveness η_{so} is then computed using Eq. 14-15 where A_f/A is 0.919.

$$\eta_{so} = 1 - 0.919(1 - 0.73) = 0.75$$

The ratio of the water-side to air-side heat-transfer areas must finally be determined. The ratio of the total air-side heat-transfer area to the total volume (A_o/V) or α is given as 170 ft^{-1}. The ratio of the water-side heat-transfer area to the total volume (A_i/V) is closely approximated by

$$\frac{A_i}{V} = \frac{D_i \pi}{X_a X_b}$$

$$\frac{A_i}{A_o} = \left(\frac{A_i}{V}\right) \Big/ \left(\frac{A_o}{V}\right) = \frac{\pi D_i}{X_a X_b \alpha} \tag{14-56}$$

$$\frac{A_i}{A_o} = \frac{\pi(0.483/12)}{(1.25/12)(1.083/12)(170)} = 0.079$$

The overall coefficient U is then given by

$$\frac{1}{U_o} = \frac{1}{15.1(0.75)} + \frac{1}{1248(0.079)} = 0.098$$

and

$$U_o = 10.2 \text{ Btu/(hr-ft}^2\text{-F)}$$

The fluid capacity rates will now be computed. For the air

$$\dot{m} = \rho\dot{Q} = \frac{14.7(144)}{53.35(510)}(2000)(60) = 9{,}336 \text{ lbm/hr}$$

and

$$C_{air} = C_c = 0.24(9336) = 2241 \text{ Btu/(hr-F)}$$

For the water

$$\dot{q} = C_w(t_{wi} - t_{wo}) = C_{air}(t_{ao} - t_{ai})$$

and

$$C_w = C_h = C_{air}\frac{(t_{ao} - t_{ai})}{(t_{wi} - t_{wo})}$$

$$C_w = 2241\frac{(100 - 50)}{(150 - 140)} = 11{,}205 \text{ Btu/(hr-F)}$$

Since $C_w > C_{air}$, $C_{air} = C_{min} = C_c$, $C_w = C_h = C_{max}$, and

$$\frac{C_{min}}{C_{max}} = \frac{2241}{11{,}205} = 0.20$$

The effectiveness ε is given by Eq. 14-8:

$$\varepsilon = \frac{t_{co} - t_{ci}}{t_{hi} - t_{ci}} = \frac{100 - 50}{150 - 50} = 0.50$$

Now assuming that the flow arrangement is counterflow, the NTU is read from Fig. 14-5 at $\varepsilon = 0.5$ and $C_{min}/C_{max} = 0.2$ as 0.74. Then

$$\text{NTU} = \frac{U_o A_o}{C_{min}}$$

$$A_o = \frac{0.74(2241)}{10.2} = 163 \text{ ft}^2$$

The total volume of the heat exchanger is given by

$$V = \frac{A_o}{\alpha} = \frac{163}{170} = 0.96 \text{ ft}^3$$

Since a face velocity of 900 ft/min was assumed, the face area is

$$A_{fr} = \frac{\dot{Q}}{V_{fr}} = \frac{2000}{900} = 2.22 \text{ ft}^2$$

and the depth is

$$L = \frac{V}{A_{fr}} = 0.96/2.22 = 0.43 \text{ ft} = 5.18 \text{ in.}$$

The number of rows of tubes N_r will then be

$$N_r = \frac{L}{X_b} = 5.18/1.083 = 4.78$$

Since N_r must be an integer and a multiple of two for the flow arrangement of Fig. 14-27, six rows must be used. This will overdesign the heat exchanger. Another possibility is to use five rows with a different circuiting arrangement so that the water connections are on the same end of the coil. This will be considered later when the lost head on the water side is computed.

The lost head on the air side of the exchanger is given by Eq. 14-45, where the ratio A/A_c is given by

$$\frac{A}{A_c} = \frac{(\alpha V)}{(\sigma A_{fr})} = \frac{170(0.96)}{0.555(2.22)} = 132$$

The mass velocity G_c was previously computed as 7569 lbm/(hr-ft²) and the mean density ρ_m is approximately

$$\rho_m = \frac{P}{2R} \left(\frac{1}{T_{ci}} + \frac{1}{T_{co}} \right)$$

$$\rho_m = \frac{14.7(144)}{2(53.35)} \left[\frac{1}{510} + \frac{1}{560} \right] = 0.074 \text{ lbm/ft}^3$$

The friction factor is read from Fig. 14-24 with FP computed from Eq. 14-51. Using Eq. 14-52

$$\frac{D^*}{D} = \frac{11.6}{1 + (1.25 - 0.525)/0.125} = 1.71$$

and

$$FP = (7578)^{-0.25}(1.71)^{-0.25} \left[\frac{1.25 - 0.525}{4(0.125 - 0.006)} \right]^{-0.4} \left[\frac{1.25}{0.898} - 1 \right]^{-0.5}$$

$$= 0.130$$

then from Fig. 14-24, $f = 0.027$,

$$\Delta P_0 = \frac{(7569)^2}{2(0.078)(32.2)(3600)^2} \left\{ [1 + (0.555)^2] \left(\frac{0.078}{0.071} - 1 \right) \right.$$

$$\left. + 0.027(132) \left(\frac{0.078}{0.074} \right) \right\}$$

$$\Delta P_0 = 3.42 \ \text{lbf/ft}^2$$

or

$$\Delta P_0 = \frac{3.42}{62.4} = 0.055 \ \text{ft wg} = 0.66 \ \text{in. wg}$$

The lost head on the tube side of the exchanger must now be computed. Recall that a velocity of 4 ft/sec was assumed to compute the heat-transfer coefficient h_i. It has also been determined that at least five rows of tubes are required and the water connections must be on the same end of the exchanger. Therefore, consider the arrangement shown in Fig. 14-28. If we use two passes per row of tubes, the water enters and leaves the same end of the coil. For the coil shown there are five separate water circuits.

The flow cross-sectional area for the water may be determined from the fluid capacity rate for the water and the continuity equation:

$$\dot{m}_w = \overline{V} A \rho = \frac{C_h}{c_p}$$

and

$$A = \frac{C_h}{\overline{V} \rho c_p} = \left(\frac{11,205}{3600} \right) \frac{1}{(4)(61.5)(1.0)}$$

$$A = 0.01265 \ \text{ft}^2$$

For N number of tubes

$$A = N \frac{\pi}{4} D_i^2$$

and

$$N = \frac{4A}{\pi D_i^2} = \frac{4(0.01265)(144)}{\pi(0.483)^2} = 9.94$$

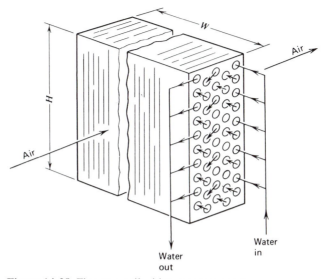

Figure 14-28 Five-row coil with two passes per row.

Since N must be an integer, 10 tubes are required and the water velocity is reduced somewhat. This reduction in velocity will not significantly reduce the heat-transfer coefficient h_i. To adapt to the flow arrangement of Fig. 14-28, a coil that is 20 tubes high must be used. Then the height H becomes

$$H = 20X_a = 20(1.25) = 25 \text{ in.}$$

The frontal area A_{fr} was previously found to be 2.22 ft^2. Then the width W is

$$W = \frac{A_{fr}}{H} = \frac{2.22}{(25/144)} = 12.8 \text{ in.}$$

This arrangement will meet all of the design requirements; however, the shape of the coil, height of 25 in. and width of 12.8 in., may be unacceptable. If so, another alternative must be sought such as using six rows of tubes, placing the headers on opposite ends, or some other modification.

The lost head l_{fw} will be computed using Eq. 10-6. Lost head in the return bends will be accounted for by assuming a loss coefficient of 2 for each bend. The flow length L_w is

$$L_w = 2(5)(12.8/12) = 10.7 \text{ ft}$$

and the Moody friction factor is 0.023 from Fig. 10-1 at a Reynolds number of 34,275, which takes into account the lower water velocity. There are nine return bends in each circuit. Then

$$l_{fw} = 0.023 \frac{(10.7)}{(0.483/12)} \frac{(4)^2}{(64.4)} + 2(9)\frac{(4)^2}{(64.4)}$$

$$l_{fw} = 6 \text{ ft wg}$$

Assuming that the 5-row configuration is not satisfactory, reconsider the circuiting of Fig. 14-27 and a 6-row coil. The coil will then have 10 tubes per row and

$$H = 10X_a = 10(1.25) = 12.5 \text{ in.}$$

The width will then be

$$W = \frac{2.22}{(12.5/144)} = 25.6 \text{ in.}$$

and

$$L_w = 2(6)(25.6/12) = 25.6 \text{ ft}$$

Assuming that the friction factor is unchanged

$$l_{fw} = 0.023 \frac{(25.6)(4)^2}{(0.483/12)64.4} + 2(5)\frac{(4)^2}{64.4} = 6.2 \text{ ft wg}$$

The geometry of the 6-row coil is more reasonable. This value of the lost head does not include the losses in the inlet and outlet headers. Header losses may be substantial, depending on the design and fabrication, and may be equal to the losses in the tubes and return bends.

There are many different ways the heat-exchanger design problem may be posed. About the same amount of work is involved in every case, however. The previous example shows that the process is laborious and time consuming. Therefore, almost all manufacturers have devised computer programs that carry out the design process quickly and accurately. Because of the speed of a computer, a simulation rather than a design approach may be used where the performance of a given configuration is determined.

14-7   COMBINED HEAT AND MASS TRANSFER

When the heat exchanger surface in contact with moist air is at a temperature below the dewpoint temperature for the air, condensation of vapor will occur. Typically the air dry bulb temperature and the humidity ratio both decrease as the air flows through the exchanger. Therefore, sensible and latent heat transfer occur simultaneously. This process is similar to that occurring in the spray dehumidifier discussed in Chapter 13 and can be analyzed using the same procedure; however, this is not generally done.

The problem of cooling coil analysis and design is complicated by the uncertainty in

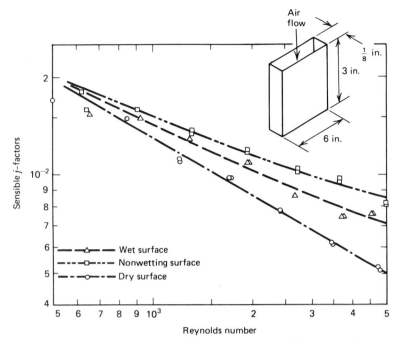

Figure 14-29 Sensible heat transfer j-factors for a parallel plate exchanger. (Reprinted by permission from *ASHRAE Transactions,* Vol. 82, Part II, 1976.)

determining the transport coefficients h, h_d, and f. It would be very convenient if heat transfer and friction data for dry heating coils, such as that shown in Figs. 14-23 and 14-24, could be used with the Colburn analogy of Eq. 13-20a to obtain the mass-transfer coefficients. But this approach is not reliable, and recent work (21, 22, 23, 24) has shown that the analogy does not always hold true. Figure 14-29 shows j-factors for a simple parallel plate exchanger that were obtained for different surface conditions. Although these particular j-factors are for the sensible heat transfer, the mass transfer j-factors and the friction factors exhibit the same behavior. Note that the dry surface j-factors fall below those obtained under dehumidifying conditions with the surface wet. The converging-diverging nature of the curves of Fig. 14-29 can be explained by the roughness introduced by the water on the surface and the nature of the boundary layers at different Reynolds numbers. The velocity, temperature, and concentration boundary layer thicknesses can all be approximated by

$$\frac{\delta}{x} = \frac{5}{\sqrt{Re_x}}$$ (14-57)

where

δ = boundary layer thickness
x = distance from inlet, measured in the same units as δ
Re_x = Reynolds number based on x

Equation 14-57 shows that at low Reynolds numbers the boundary layer grows quickly; the droplets are soon covered and have little effect on the flow field. As the Reynolds number is increased, the boundary layer becomes thin and more of the total flow field is exposed to the droplets. The roughness caused by the droplets induces mixing and larger j-factors. The data of Fig. 14-29 cannot be applied to all surfaces because the length of the flow channel is also an important variable. It seems certain, however, that the water collecting on the surface is responsible for breakdown of the j-factor analogy. The j-factor analogy is approximately true when the surface conditions are identical (24). That is, when the surface is wetted, the sensible and mass transfer j-factors are in close agreement. This is of little use, however, because a wet test must be made to obtain this information. Under some conditions it is possible to obtain a film of condensate on the surface instead of droplets. For example, aluminum when thoroughly degreased and cleaned with a harsh detergent in hot water experiences filmwise condensation (21). Figure 14-30 shows j-factor and friction data for a plate–fin–tube surface under dry conditions and with filmwise and dropwise condensation. The trends are the same as those shown in Fig. 14-29. The friction factors are influenced by the water on the surface over the complete Reynolds number range, whereas the j-factors are affected only at the higher Reynolds numbers. The data shown correspond to face velocities of 200 to 800 ft/min (1 to 4 m/s) with air at standard conditions. Although not shown, the mass transfer j-factors show the same trends and are in reasonable agreement with the wet surface j-factors shown in Fig. 14-30.

Recent research involving plate-fin-tube surfaces (22) has resulted in correlations that relate dry sensible j- and f-factors to those for wetted dehumidifying surfaces. Expressions were developed that modify the parameters JP and FP of Figs. 14-23 and 14-24 for wet surface conditions. In developing these functions, it was found that a Reynolds number based on fin spacing and the ratio of fin spacing to space between the fins were useful. For film-type condensation, the modifying functions are:

Sensible j-factor

$$J(s) = 0.84 + 4 \times 10^{-5}(\mathrm{Re}_s)^{1.25} \tag{14-58}$$

Total j-factor

$$J_t(s) = [0.95 + 4 \times 10^{-5}(\mathrm{Re}_s)^{1.25}] \left(\frac{s}{s - y}\right)^2 \tag{14-59}$$

Friction factor

$$F(s) = [1 + (\mathrm{Re}_s)^{-0.4}] \left(\frac{s}{s - y}\right)^{1.5} \tag{14-60}$$

For dehumidifying conditions, the abscissa of Fig. 14-23 is changed to $J(s)JP$ and $J_t(s)JP$. The abscissa of Fig. 14-24 is changed to $F(s)FP$.

Enthalpy Potential

The enthalpy potential was mentioned in Chapter 13 and will be more fully justified here. The heat transfer from moist air to a surface at a temperature below the air

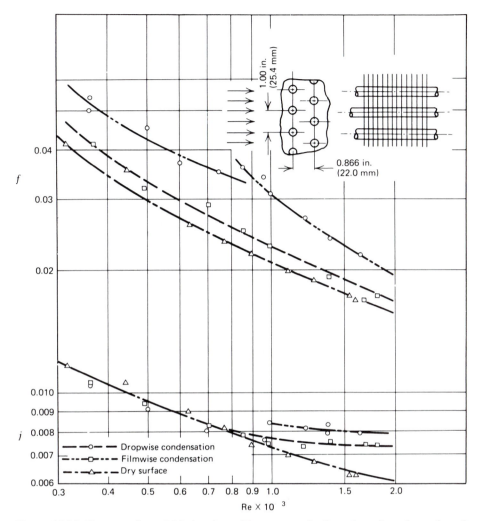

Figure 14-30 Heat-transfer and friction data with mass transfer for a plate–fin–tube surface, 4 rows of tubes.

dewpoint may be expressed as

$$\frac{\dot{q}}{A} = h(t_w - t_\infty) + h_d(W_w - W_\infty)i_{fg} \qquad \textbf{(14-61)}$$

Using the analogy of Eq. 13-20a with Le = 1 we see that

$$\frac{\dot{q}}{A} = h_d[c_{pa}(t_w - t_\infty) + (W_w - W_\infty)i_{fg}] \qquad \textbf{(14-62)}$$

The enthalpy of vaporization i_{fg} is evaluated at the wall temperature. Even though the Colburn analogy is not always precise, there is a proportionality between h and h_d, which is all that is required here. The enthalpy of the saturated moist air at the wall is given by

$$i_w = c_{pa}t_w + W_w(i_f + i_{fg})$$

(14-63)

where i_f and i_{fg} are evaluated at the wall temperature. For the moist air in the free stream

$$i_\infty = c_{pa}t_\infty + W_\infty[i_f + i_{fg} + c_{pv}(t_\infty - t_w)]$$

(14-64)

The temperature t_w should be the dewpoint temperature and i_f and i_{fg} should be evaluated at the dewpoint temperature. However, the errors tend to compensate and Eq. 14-64 is a very good approximation. The difference in enthalpy between the surface i_w and the free stream i_∞ is then

$$i_w - i_\infty = c_{pa}(t_w - t_\infty) + i_{fg}(W_w - W_\infty) + i_f(W_w - W_\infty) + W_\infty c_{pv}(t_w - t_\infty)$$ (14-65)

Comparison of Eqs. 14-62 and 14-65 then yields

$$\frac{\dot{q}}{A} = h_d[(i_w - i_\infty) - i_f(W_w - W_\infty) - W_\infty c_{pv}(t_w - t_\infty)]$$

(14-66)

The last two terms are typically about 0.5 percent of $(i_w - i_\infty)$ and can be neglected. Thus, the driving potential for simultaneous transfer of heat and mass is enthalpy to a close approximation, whereas temperature and concentration are the driving potentials for sensible heat and mass, respectively.

Equation 14-66 expresses the total heat transfer at a particular location in the heat exchanger; however, the moist air enthalpy at the surface i_w and in the free stream i_∞ vary throughout the exchanger, as shown in Fig. 14-31 for counterflow. In addition, most coils will have fins that must be accounted for. Then

$$\dot{q} = h_d A \eta_{ms} \Delta i_m$$

(14-67)

where η_{ms} is the surface effectiveness with combined heat and mass transfer and Δi_m is

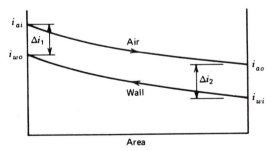

Figure 14-31 Enthalpy difference in a counterflow dehumidifying coil.

some mean enthalpy difference. With suitable assumptions it can be shown that Δi_m has the same form as the LMTD for counterflow:

$$\Delta i_m = \frac{\Delta i_1 - \Delta i_2}{\ln \dfrac{\Delta i_1}{\Delta i_2}} \tag{14-68}$$

This is true because i_w is directly proportional to t_c, the refrigerant temperature. Equation 14-67 expresses the total heat-transfer rate from the wall to the airstream where the wall temperature is not known explicitly. However, the heat-transfer rate from the refrigerant to the wall is given by

$$\dot{q} = h_i A_i (\Delta t_m)_i \tag{14-69}$$

where $(\Delta t_m)_i$ expresses the mean temperature difference between the refrigerant and the wall and the thermal resistance of the thin wall has been neglected. A simple iterative procedure is then necessary to solve Eqs. 14-67 and 14-69 for the total heat-transfer rate.

It was mentioned earlier in this chapter that the heat-transfer coefficient decreases from the inlet to the exit of the coil as shown in Fig. 14-22. This has a direct effect on total heat-transfer calculation because the coil surface temperature is higher than expected at the inlet due to a higher heat-transfer rate there. This should be taken into account because a portion of the coil near the air inlet may be at a temperature greater than the dewpoint with no mass transfer occurring.

The sensible heat transfer from the moist air to the refrigerant is computed for counterflow by Eq. 14-3:

$$\dot{q}_s = UA(\text{LMTD}) \tag{14-3}$$

where LMTD is given by Eq. 14-2 and U is given by Eq. 14-18 with η_s equal to η_{ms}.
The latent heat transfer is then easily computed from

$$\dot{q}_l = \dot{q} - \dot{q}_s \tag{14-70}$$

It is also true that

$$\dot{q}_l = \dot{m}_a (W_i - W_o) i_{fg} \tag{14-71}$$

Fin Efficiency with Mass Transfer

The fin efficiency with combined heat and mass transfer is lower than the value obtained with only sensible heat transfer. Although the basic definition is unchanged from that given in Section 14-3, the analysis is more complex and not exact. Threlkeld (25) has solved the problem; his solution, however, depends on knowing the water film thickness on the fin. This is a very nebulous quantity and impractical to use. An accepted method is outlined in reference 26 and is an adaptation of the work of Ware and Hacha (27) and others. Although this method appears to give good results for performance and rating purposes, it has some undesirable features since the coil surface temperature is assumed to be the only parameter affecting the fin efficiency regardless of the moist air conditions. Another disturbing feature is failure of the solution to reduce to the dry coil case when the surface and moist air conditions warrant this. These inconsistencies are troublesome when making general coil studies.

A fin of uniform cross section as shown in Fig. 14-8 has been analyzed by McQuiston (28). The method is approximate but reduces to the case of zero mass transfer and is adaptable to circular and plate–fin–tube surfaces. The analysis is outlined as follows: An energy balance on an elemental volume yields the following differential equation, assuming one-dimensional heat transfer and constant properties:

$$\frac{d^2t}{dx^2} = \frac{P}{kA_c}[h(t - t_\infty) + h_d i_{fg}(W - W_\infty)] \tag{14-72a}$$

where

t = temperature of the element, F or C
x = distance measured from base of fin, ft or m
P = circumference of the fin, ft or m
k = thermal conductivity of the fin material, (Btu-ft)/(hr-ft²-F) or(W-m)/(m²-C)
A_c = cross-sectional area of the fin, ft² or m²
h = convective heat-transfer coefficient, Btu/(hr-ft²-F) or W/(m²-C)
t_∞ = temperature of the air–vapor mixture flowing around the fin, F or C
h_d = convective mass-transfer coefficient, lbm/(ft²-hr) or kg/(m²-s)
i_{fg} = latent heat of vaporization of water, Btu/lbm or J/kg
W = humidity ratio of saturated air at temperature t, lbmw/lbma or kgw/kga
W_∞ = humidity ratio of the air–vapor mixture, lbmw/lbma or kgw/kga

The analogy of Eq. 13-20b will be used to obtain the mass-transfer coefficient h_d with Le = 1

$$h_d = \frac{h}{c_{pa}} \tag{14-73}$$

where

c_{pa} = specific heat capacity of the dry air, Btu/(lbm-F) or W/(kg-C)

As suggested, the coefficient h should be for a wet surface. Other correlations may also be used. Combining Eqs. 14-72a and 14-73 gives

$$\frac{d^2t}{dx^2} = \frac{hP}{kA_c}\left[(t - t_\infty) + \frac{i_{fg}}{c_{pa}}(W - W_\infty)\right] \tag{14-72b}$$

Now, if $(W - W_\infty)$ is simply related to $(t - t_\infty)$, Eq. 14-72b can be easily solved for the temperature distribution in the fin. To justify such a simplification consider the physical aspects of a typical cooling and dehumidifying coil.

Let the air–vapor mixture enter an exchanger at a fixed condition designated by point 1 on the psychrometric chart of Fig. 14-32. Consider an evaporator with a constant temperature refrigerant operating such that the moist air very near the wall is at a temperature designated by point $w2$. The humidity ratio of the leaving air W_2 will approach W_{w2}, as shown in Fig. 14-33. The process line 1–2 in Fig. 14-32 can be

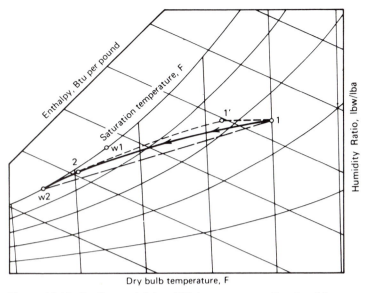

Figure 14-32 Cooling and dehumidifying processes. (Reprinted by permission from *ASHRAE Transactions*, Vol. 81, Part I, 1975.)

approximated by a straight line and a simple relationship between $(W - W_\infty)$ and $(t - t_\infty)$ exists. In fact,

$$(W_{w2} - W_1) = C(t_{w2} - t_1) \tag{14-74}$$

where C is a constant. An examination of data for many coils and various operating conditions shows that C will typically vary less than 10 percent from inlet to exit. It then seems reasonable to use an average value such as

$$C_{avg} = \frac{(C_1 + C_2)}{2} \tag{14-75}$$

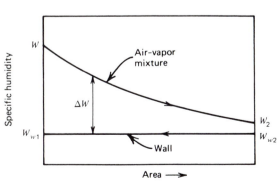

Figure 14-33 Specific humidity for a constant-temperature refrigerant.

Due to the shape of the saturation curve, precise location of the point $w2$ on Fig. 14-32 does not greatly affect the value of C for a particular coil condition. On the other hand, C is very sensitive to the location of point 1. For example, for $t_w = 45$ F and $t_1 = 80$ F, C varies from 1.4×10^{-4} to 0.0 as ϕ_1 varies from 50 to 29 percent. For the last condition there will be no condensation on the surface and only sensible heat transfer will occur.

When chilled water is used as a cooling medium in a counterflow arrangement, Fig. 14-34 applies. In this case the wall temperature is somewhat higher where the air enters the exchanger. Typical conditions for the moist air very near the wall are shown on Fig. 14-32 as points $w1$ and $w2$. Here the surface is completely wetted.

At the inlet to the exchanger the value of C is given by

$$C_1 = \frac{W_{w1} - W_1}{t_{w1} - t_1} \tag{14-76a}$$

whereas at the exit

$$C_2 = \frac{W_{w2} - W_2}{t_{w2} - t_2} \tag{14-76b}$$

Again C_1 is less than C_2 and an average value should be used. In most cases C will change less than 10 percent from inlet to outlet.

Process $1-1'-2$ (Fig. 14-32) approximates a situation where the coil is partially dry. In this case C_1 is zero until the air reaches the location in the coil where the surface temperature is below the air dewpoint temperature. C then increases to the value of C_2 at the exit. Again an average value of C may be used or C_2 may be used to obtain a conservative solution.

The differential equation describing the temperature distribution in a thin fin of uniform cross section thus becomes

$$\frac{d^2(t - t_\infty)}{dx^2} = M^2(t - t_\infty) \tag{14-77}$$

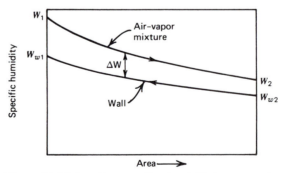

Figure 14-34 Specific humidity for chilled water as the refrigerant.

where

$$M^2 = \frac{hP}{kA_c}\left[1 + \frac{Ci_{fg}}{c_{pa}}\right] \qquad \textbf{(14-78a)}$$

When there is no condensation $C = 0$ and

$$M^2 = \frac{hP}{kA_c} = m^2 \qquad \textbf{(14-78b)}$$

The fin efficiency derived from the well-known solution of Eq. 14-77 where t_∞ and M are constants and the following boundary conditions are used:

$$x = 0 \qquad t = t_w \qquad \textbf{(14-79)}$$

$$x = l \qquad \frac{dt}{dx} = 0$$

becomes

$$\eta_m = \frac{\tanh(Ml)}{(Ml)} \qquad \textbf{(14-80)}$$

The approximation of Eq. 14-23 is also used here

$$\frac{hP}{kA_c} \approx \frac{2h}{ky} \qquad \textbf{(14-81)}$$

Equation 14-80 is identical in form to the equation for the fin efficiency with no mass transfer (Eq. 14-22).

The solution may be applied to circular fins on a tube (Eq. 14-24), or to the case of plate–fin–tube heat transfer surfaces (Eqs. 14-26 and 14-27). Figure 14-11 may also be used with m replaced by M.

The surface effectiveness has the same form as Eq. 14-15

$$\eta_{ms} = 1 - \frac{A_f}{A}(1 - \eta_m) \qquad \textbf{(14-82)}$$

The method presented is thought to be the most accurate available and is simple and straightforward to use. The method is readily adapted to the computer and is easy to use with hand calculations.

Transport Coefficients

The heat transfer and friction coefficients on the refrigerant side of the exchanger are determined by the methods discussed in Section 14-4. Chilled water and evaporating refrigerants are the usual cases.

The heat, mass, and friction coefficients on the air side of the exchanger should be obtained from correlations based on test data, if at all possible, since the analogy method is unreliable. The correlations of Figs. 14-23 and 14-24 as modified using $J(s)$, $J_i(s)$, and $F(s)$ (discussed at the beginning of this section) are recommended for plate–fin–tube coils. Other finned tube surfaces have similar behavior. For example, the dry-surface heat-transfer coefficients, for circular finned tubes in a staggered tube pattern, are well correlated by

$$j = 0.38\,JP \qquad (14\text{-}83)$$

and the friction factors are given by

$$f = 1.53\,(FP)^2 \qquad (14\text{-}84)$$

JP and FP may then be modified for a wet surface using Eqs. 14-58, 14-59, and 14-60. No information is available on the effect of tube rows on mass-transfer coefficients; however, it should be similar to that for sensible heat transfer for a dry surface (Fig. 14-21).

EXAMPLE 14-2

Estimate the heat, mass, and friction coefficients for a 4-row cooling coil that has the geometry of Fig. 14-20 with 12 fins per inch. The face velocity of the air is 600 ft/min and has an entering temperature of 80 F. The air leaves the coil at a temperature of 60 F. The air is at standard barometric pressure.

SOLUTION

The correlations of Figs. 14-23 and 14-24 will be used with the parameters JP, FP, $J(s)$, and $F(s)$ computed from Eqs. 14-46 through 14-60. The mass velocity is

$$G_c = \frac{60(600)14.7(144)}{0.54(53.35)540} = 4900\ \text{lbm/(hr-ft}^2)$$

The Reynolds number based on tube diameter is then

$$\frac{G_c D}{\mu} = \frac{4900(0.525/12)}{0.044} = 4870$$

where data from Fig. 14-20 and Table B-4a are used. To compute the parameter A/A, assume a 4-row coil, 1 ft in length, with 10 tubes in the face. This coil has a volume of

$$V = \frac{10(1.25)(12)4(1.083)}{1728} = 0.376\ \text{ft}^3$$

and a total of 40 tubes. The total outside surface area of the tubes is

$$A_t = 40\,\pi\,D\,L = 40\,\pi(0.525/12)(1) = 5.498\ \text{ft}^2$$

From Fig. 14-20, $A/V = \alpha = 238$ ft^2/ft^3; then

$$\frac{A}{A_t} = \frac{238(0.376)}{5.498} = 16.28$$

Using Eq. 14-46

$$JP = (4870)^{-0.4}(16.28)^{-0.15} = 0.022$$

The Reynolds number based on the fin spacing is

$$\text{Re}_s = \text{Re}_D \left(\frac{s}{D}\right) = 4870 \left(\frac{0.0833}{0.525}\right) = 773$$

Then

$$J(s) = 0.84 + (4 \times 10^{-5})(773)^{1.25} = 1.003$$

and

$$J_i(s) = [0.95 + (4 \times 10^{-5})(773)^{1.25}] \left[\frac{0.0833}{0.0833 - 0.006}\right]^2 = 1.293$$

Then

$$JP\, J(s) = 0.0220(1.003) = 0.022$$

and

$$JP\, J_i(s) = 0.0220(1.293) = 0.028$$

Using Fig. 14-23, $j = 0.0071$ and $j_i = 0.0088$
   From Eq. 14-36

$$h = \frac{j\, G_c\, c_p}{\text{Pr}^{2/3}} = \frac{0.0071(4900)0.24}{(0.7)^{2/3}} = 10.5 \text{ Btu/(hr-ft}^2\text{ F)}$$

and from Eq. 13-19a

$$h_d = \frac{j_i\, G_c}{\text{Sc}^{2/3}} = \frac{0.0088(4900)}{(0.6)^{2/3}} = 60.6 \text{ lbma/(ft}^2\text{-hr)}$$

The friction factor will be determined from Fig. 14-24. Using Eqs. 14-51 and 14-52

$$D^* = \frac{0.525(16.28)}{1 + (1.25 - 0.525)/0.0833} = 0.881$$

and

$$FP = (4870)^{-0.25} \left(\frac{0.525}{0.881}\right)^{0.25} \left[\frac{1.25 - 0.525}{4(0.0833 - 0.006)}\right]^{-0.4}$$

$$\times \left[\frac{1.25}{0.881} - 1\right] = 0.116$$

Using Eq. 14-60

$$F(s) = (1 + 773^{-0.4}) \left(\frac{0.0833}{0.0833 - 0.006}\right)^{1.5} = 1.20$$

Then

$$FP\, J(s) = 0.116(1.20) = 0.14$$

and from Fig. 14-24, $f = 0.031$.

Note that the presence of moisture on the surface increases h, h_d, and f.

Design Calculations

The design of a heat exchanger for simultaneous transfer of heat and mass is more complicated than for sensible heat transfer because the total energy transfer from the air has two components:

$$\dot{q} = \dot{m}_c c_{pc}(t_{co} - t_{ci}) = \dot{m}_a c_p(t_{ai} - t_{ao}) + \dot{m}_a i_{fg}(W_{ai} - W_{ao}) \qquad \textbf{(14-85)}$$

The simplest case occurs when the terminal conditions are all known because

$$\dot{q} = \dot{m}_a(i_{ai} - i_{ao}) = \dot{m}_c c_{pc}(t_{co} - t_{ci}) \qquad \textbf{(14-86)}$$

and

$$\dot{q}_s = \dot{m}_a c_p(t_{ai} - t_{ao}) = UA(\text{LMTD}) \qquad \textbf{(14-87)}$$

When only inlet conditions are known, a double-iteration loop is required and hand calculations become impractical. The digital computer may be used to great advantage in this case. Figure 14-35 shows a flow diagram for a successful computer program.

The method discussed in Chapter 13 for the spray dehumidifier may be used to predict performance of a dehumidifying coil when hand calculations are necessary. One modification is necessary in Eq. 13-28, where the liquid film coefficient h_l must be replaced by an overall coefficient defined by

$$\frac{1}{UA} = \frac{1}{h_c A_c} + \frac{\Delta x}{k A_m} \qquad \textbf{(14-88)}$$

The first term on the right is the refrigerant-side film resistance and the second term is the wall thermal resistance. Otherwise the procedure is identical. The results will be approximate, however.

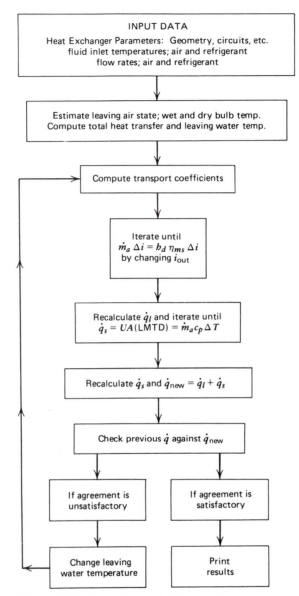

Figure 14-35 Flow diagram for heat-exchanger computer
program.

The American Refrigeration Institute has a certification program for heating and cooling coils (26) that embodies a calculation procedure for dehumidifying coils. The standard is intended to standardize the rating of coils by setting up detailed procedures to calculate performance. The method has many disadvantages when used for coil design, however.

REFERENCES

1. J. D. Parker and F. C. McQuiston, *Introduction to Fluid Mechanics and Heat Transfer,* 2nd ed., Kendall-Hunt, Dubuque, IA, 1988.

2. T. E. Schmidt, "La Production Calorifique des Surfaces Munies D'ailettes," *Annexe Du Bulletin De L'Institut International Du Froid,* Annexe G-5, 1945–46.

3. H. Zabronsky, "Temperature Distribution and Efficiency of a Heat Exchanger Using Square Fins on Round Tubes," *ASME Transactions, Journal of Applied Mechanics,* Vol. 77, December 1955.

4. W. H. Carrier and S. W. Anderson, "The Resistance to Heat Flow Through Finned Tubing," *Heating, Piping, and Air Conditioning,* May 1944.

5. D. G. Rich, "The Efficiency and Thermal Resistance of Annular and Rectangular Fins," Presented at the 1966 International Heat Transfer Conference.

6. *ASHRAE Handbook, Fundamentals Volume,* American Society of Heating, Refrigerating and Air-Conditioning Engineers, Inc., Atlanta, GA, 1985.

7. R. W. Eckels, "Contact Conductance of Mechanically Expanded Plate Finned Tube Heat Exchangers," Westinghouse Research Laboratories, Scientific Paper No. 77-1E9-SURCO-P1, March 1977.

8. W. M. Kays and A. L. London, *Compact Heat Exchangers,* 2nd ed., McGraw-Hill, San Francisco, 1964.

9. A. J. Ghajar and K. F. Madon, "Pressure Drop Measurements in the Transition Region for a Circular Tube with Three Different Inlet Configurations," *Experimental Thermal and Fluid Science,* Vol. 5, No. 1, Jan. 1992.

10. A. J. Ghajar and L. M. Tam, "Correlations for Forced and Mixed Convection in Straight Duct Flows with Three Different Inlet Configurations," *Proceedings of the 28th ASME/ AICHE National Heat Transfer Conference,* San Diego, CA, August 1992.

11. R. W. Lockhart, and Martinelli, R. D., "Proposed Correlation of Data for Isothermal, Two-Phase, Two-Component Flow in Pipes," *Chemical Engineering Progress,* Vol. 45, No. 39, 1949.

12. W. M. Kays and M. E. Crawford, *Convective Heat and Mass Transfer,* 2nd ed., McGraw-Hill, San Francisco, 1980.

13. E. M. Sparrow, A. L. Loeffler, Jr., and H. A. Hubbard, *Transactions ASME, Journal of Heat Transfer,* November 1961.

14. R. G. Diessler and M. F. Taylor, *NACA TN 2351,* Washington, DC, 1955.

15. A. Devore, *Petroleum Refiner,* Vol. 40, No. 5, May 1961 and Vol. 41, No. 12, December 1962.

16. K. J. Bell, "Bulletin No. 5," University of Delaware Engineering Experimental Station, 1963.

17. J. W. Palen and Jerry Taborek, "Solution of Shell Side Flow Pressure Drop and Heat Transfer by Stream Analysis Method," Chemical Engineering Progress Symposium Series, Vol. 65, No. 92, 1969.

18. Donald G. Rich, "The Effect of the Number of Tube Rows on Heat Transfer Performance of Smooth Plate Fin-and-Tube Heat Exchangers," *ASHRAE Transactions,* Vol. 81, Part I, 1975.

19. D. G. Rich, "The Effect of Fin Spacing on the Heat Transfer and Friction Performance of Multirow Smooth Plate Fin-and-Tube Heat Exchangers," *ASHRAE Transactions,* Vol. 73, June 1973.

20. D. G. Shepherd, "Performance of One-Row Tube Coils with Thin-Plate Fins, Low Velocity Forced Connection," *Heating, Piping, and Air Conditioning,* April 1956.

21. F. C. McQuiston, "Heat, Mass, and Momentum Transfer Data for Five Plate-Fin-Tube Heat Transfer Surfaces," *ASHRAE Transactions,* Vol. 84, Part 1, 1978.

22. F. C. McQuiston, "Correlation of Heat, Mass, and Momentum Transport Coefficients for Plate-Fin-Tube Heat Transfer Surfaces," *ASHRAE Transactions,* Vol. 84, Part 1, 1978.

23. J. L. Guillory and F. C. McQuiston, "An Experimental Investigation of Air Dehumidification in a Parallel Plate Exchanger," *ASHRAE Transactions,* Vol. 29, June 1973.

24. F. C. McQuiston, "Heat, Mass, and Momentum Transfer in a Parallel Plate Dehumidifying Exchanger," *ASHRAE Transactions,* Vol. 82, Part II, 1976.

25. J. L. Threlkeld, *Thermal Environmental Engineering,* 2nd ed., Prentice-Hall, Englewood Cliffs, NJ, 1970.

26. *ARI Standard 410-91,* Air Conditioning and Refrigeration Institute, Arlington, VA, 1991.

27. C. D. Ware and T. H. Hacha, "Heat Transfer from Humid Air to Fin and Tube Extended Surface Cooling Coils," ASME Paper, No. 60-HT-17, August 1960.

28. F. C. McQuiston, "Fin Efficiency with Combined Heat and Mass Transfer," *ASHRAE Transactions,* Vol. 81, Part I, 1975.

29. F. C. McQuiston, "Heat Transfer and Flow Friction Data for Two Fin-Tube Surfaces," *ASME Journal of Heat Transfer,* Vol. 93, May 1971.

PROBLEMS

14-1. Five thousand cfm (2.36 m^3/s) of standard air are heated to 120 F (50 C) in a cross-flow heat exchanger. The heating medium is water at 200 F (95 C), which is cooled to 180 F (82 C). Determine (a) the LMTD and correction factor F, (b) the fluid capacity rates C_c and C_h, (c) the flow rate of the water in gpm (m^3/s), (d) the overall conductance UA, (e) the NTU based on C_{min}, and (f) the effectiveness ε.

14-2. A heat exchanger is to be designed to heat 4000 cfm (1.9 m^3/s) of air from 50 F (10 C) to 110 F (43 C) using hot water at 180 F (82 C) in a cross-flow arrangement. The flow rate of the hot water is 25 gpm (1.6 L/s). Assume the overall heat-transfer coefficient based on the air side is 10 Btu/(hr-ft^{2}F) [57 W/(m^2-C)] and determine the air-side surface area using (a) the LMTD method and (b) the NTU method.

14-3. An air-cooled condenser operates in cross flow with a constant refrigerant temperature of 125 F (52 C). The air enters at 95 F (35 C) at a flow rate of 3200 cfm (1.5 m^3/s). The air-side surface area is 300 ft^2 (28 m^2) and the overall heat-transfer coefficient is 10 Btu/(hr-ft^2-F) [57 W/(m^2-C)]. (a) What is the temperature of the air leaving the condenser? (b) What is the flow rate for refrigerant 22 (R-22), assuming a change in state from saturated vapor to saturated liquid?

14-4. A tube has circular fins as shown in Fig. 14-9; R is 1.0 in., whereas r is 0.5 in. The fin thickness y is 0.008 in. and the fin pitch is 10 fins/in. The average heat-transfer coefficient for the fin and tube is 10 Btu/(hr-ft^2-F), whereas the thermal conductivity k of the fin and tube material is 90 (Btu-ft)/(ft^2-hr-F). Determine (a) the fin efficiency η using Fig. 14-11, (b) the fin efficiency using Eq. 14-24, and (c) compare (a) and (b).

14-5. Find the surface effectiveness for the finned tube in Problem 14-4 assuming the ratio of fin area to total area is 0.9.

14-6. The finned tube of Problem 14-4 has a refrigerant flowing on the inside with an average heat-transfer coefficient of 200 Btu/(hr-ft^2-F). The tube wall has a thickness of 0.015 in. Compute the overall heat-transfer coefficient based on the air-side surface area. Assume that the ratio of air-side to inside surface area is 9.

14-7. Water flows through the finned tube in Problem 14-4. Find the overall heat-transfer coefficient assuming a water heat-transfer coefficient of 1100 Btu/(hr-ft^2-F).

14-8. A plate–fin heat exchanger has fins that are 6 mm high, 0.16 m thick, with a fin pitch of 0.47 fins/mm. The average heat-transfer coefficient is 57 W/(m^2-C) and the thermal conductivity of the fin material is 173 W/(m-C). Find the fin efficiency.

14-9. Assume that the ratio of fin area to total area in Problem 14-8 is 0.85. Find the surface effectiveness.

14-10. Refrigerant 12 flows on the opposite side of the surface in Problem 14-8. Determine the overall heat-transfer coefficient if the refrigerant heat-transfer coefficient is 1.4 kW/(m^2-C). Neglect the thermal resistance of the plate.

14-11. Find the fin efficiency for a plate–fin–tube surface like that shown in Fig. 14-14 using the data shown in Table 14-2.

14-12. Assume that water is flowing in the finned tubes of Problem 14-11 with a heat-transfer coefficient of 600 Btu/(hr-ft^2-F) [3.4 kW/(m^2-C)] and compute the overall heat-transfer coefficient. Assume that the ratio of fin area to total air-side area is 0.9 and the ratio of air-side area to inside area is 10.

14-13. Compute the contact resistance for the surface of Problem 14-11. Assume a fin pitch of 12 fins per in.

14-14. Determine the average heat-transfer coefficient for water flowing in a $\frac{1}{2}$-in. Type L copper tube. The water flows at a rate of $2\frac{1}{2}$ gpm (0.16 L/s) and undergoes a temperature change from 180 (82) to 150 F (66 C). The tube is over 4 ft (1.5 m) long.

14-15. Repeat Problem 14-14 for a 30 percent ethylene glycol solution.

14-16. Rework Problem 14-14 for 60 percent ethylene glycol solution.

14-17. Estimate the heat-transfer coefficient for water flowing with an average velocity of 4 ft/sec (1.2 m/s) in a long rectangular 3 × 1 in. (75 × 25 mm) channel. The bulk temperature of the water is 45 F (7 C). Assume (a) cooling and (b) heating.

14-18. Repeat Problem 14-17 for 30 percent ethylene glycol instead of water.

14-19. Coolant flows through a long 12-mm I.D. tube with an average velocity of 1.5 m/s. The coolant undergoes a temperature change from 40 to 50 C. Compute the average heat-transfer coefficient for (a) water and (b) 30 percent ethylene glycol solution.

14-20. Water flows through a 0.34-in. I.D. tube with an average velocity of $\frac{1}{2}$ ft/sec. The mean bulk temperature of the water is 45 F and the length of the tube is 10 ft. Estimate the average heat-transfer coefficient for (a) water and (b) ethylene glycol (30 percent solution).

Table 14-2 Data for Problem 14-11

	a	b	y	D	h	k
(a)	1.12 in.	1.35 in.	0.010 in.	0.64 in.	10 Btu/(hr-ft^2-F)	90 (Btu-ft)/(hr-ft^2-F)
(b)	25 mm	22 mm	0.18 mm	10 mm	68 W/(m^2-C)	170 (W-m)/(m^2-C)

14-21. Estimate the average heat-transfer coefficient for water flowing in 10-mm I.D. tubes at a velocity of 0.10 m/s. The length of the tubes is 3 m and the mean bulk temperature of the water is 40 C.

14-22. Estimate the average heat-transfer coefficient for condensing water vapor in $\frac{5}{8}$-in. O.D. tubes that have walls 0.018 in. thick. The vapor is saturated at 5 psia and flows at the rate of 1 lbm/hr at the tube inlet. The quality of the vapor is 10 percent when it exits from the tube.

14-23. Water vapor condenses in 15-mm I.D. tubes. The flow rate is 0.126×10^{-3} kg/s per tube at an absolute pressure of 35 kPa. The vapor leaving the tubes has a quality of 12 percent. Estimate the average heat-transfer coefficient.

14-24. Refrigerant 22 enters the $\frac{3}{8}$-in. O.D. tubes of an evaporator at the rate of 80 lbm/hr per tube. The refrigerant enters at 70 psia with a quality of 20 percent and leaves with 10 F of superheat. The effective tube length is 5 ft. Estimate the average heat-transfer coefficient.

14-25. Estimate the average heat-transfer coefficient for evaporating refrigerant 22 in 8.5-mm I.D. tubes that are 2 m in length. The mass velocity of the refrigerant is 200 kg/(s-m^2) and the pressure and quality at the inlet are 210 kPa and 30 percent, respectively. Saturated vapor exits from the tubes.

14-26. Refer to Problem 14-14 and compute the lost head for a circuit that consists of 6 tubes, 6 ft in length connected by 5 U-bends. The equivalent length of each U-bend is 1 ft.

14-27. Refer to Problem 14-20 and compute the lost head for a circuit made up of 10 tubes, 10 ft in length, connected by 9 U-bends. The U-bends have an equivalent length of 1.5 ft each.

14-28. A 10-ton cooling coil uses volatile refrigerant on the tube side. There are 10 tube circuits made up of 6 tubes, $\frac{1}{2}$ in. in diameter, connected by 5 U-bends. The evaporating temperature is 30 F. Estimate the pressure loss for each circuit for (a) refrigerant 12 and (b) refrigerant 22.

14-29. For the surface shown in Fig. 14-20 with 6.7 fins per in., the air mass velocity G_{fr} is 1800 lbm/(hr-ft^2) and the air is heated from 70 to 120 F. Compute the heat-transfer coefficient and friction factor using (a) Fig. 14-20, (b) Figs. 14-23 and 14-24.

14-30. For the surface of Fig. 14-20, with 0.263 fins per mm, the air mass velocity is 4.5 kg/(m^2-s) and the mean bulk temperature is 20 C. Estimate the heat-transfer coefficient and friction factor using (a) Fig. 14-20, (b) Figs. 14-23 and 14-24.

14-31. Compute the lost pressure for Problem 14-29b assuming standard atmospheric pressure at the inlet.

14-32. Compute the lost pressure for Problem 14-30b assuming standard atmospheric pressure at the inlet.

14-33. Find the heat-transfer coefficient and friction factor for the surface of Fig. 14-25 assuming a mass velocity of 2700 lbm/(hr-ft^2). The air enters at 75 F and leaves at 55 F.

14-34. Compute the lost pressure for Problem 14-33 assuming a pressure of 14.6 lbf/in.^{2} at the inlet and flow length of 4 in.

14-35. Design a condenser for the following given conditions:
Tube side: refrigerant 22, 2-phase, $t = 120$ F, $h_i = 200$ Btu/(hr-ft^2-F)
Finned side: $t_{in} = 100$ F, heat transfer rate $= 40,000$ Btu/hr, maximum head loss $= 0.3$ in. wg

Construction: all aluminum, $k = 100$ Btu/(hr-ft^2-F).

Use the surface of Fig. 14-20 with 8 fins per in. (a) Carry out the design by hand calculations. (b) Use a computer program to find a design.

14-36. Design a hot water heating coil for a capacity of 200,000 Btu/hr. Use the surface of Fig. 14-20 with two rows of tubes and 6.7 fins per in. Air enters the coil at 70 F and water enters at 150 F. Sketch the circuiting arrangement for the tube side. Compute the head loss for both the water and air. Make the width of the exchanger about twice the height. (a) Use hand calculation method. (b) Use a computer program.

14-37. Air enters a fin–tube heat exchanger described by Fig. 14-20 (6.7 fins per in.) at the rate of 2000 cfm at 14.696 psia and with a face velocity of 1000 ft/min. The entering and leaving air temperatures are 40 and 110 F, respectively. Hot water flows through the tubes with an average velocity of 3.5 ft/sec, entering at 200 F and leaving at 190 F. Assume counterflow and use the NTU method or a computer program to determine (a) the heat-transfer area required and (b) the heat-exchanger dimensions and the number of tube rows in the air-flow direction. (c) Compute the lost head on the air side in in. wg, and (d) select circuiting for the tubes and compute the head loss for the water side.

14-38. The surface of Problem 14-29 has 8 rows of tubes. Compute the heat-transfer coefficient.

14-39. The surface of Problem 14-30 has 6 rows of tubes. Compute the heat-transfer coefficient.

14-40. The surface of Fig. 14-20 is to be sized for a condenser in an air-conditioning unit. The condenser must reject 4.4 kW. Assume refrigerant 22 enters the condenser at 51 C as saturated vapor; it leaves as saturated liquid at the same temperature. Air enters the condenser at 35 C. Use a 5-row coil with aluminum fins with 0.571 fins per mm. Compute the air-side head loss in Pa. Use hand or computer methods to solve.

14-41. Compute the surface effectiveness with dehumidification for the surface of Fig. 14-20 with a fin pitch of 11.7 fins per inch. The face velocity of the entering air is 500 ft/min, 85 F dry bulb and 70 F wet bulb. The surface temperatures where the air enters and leaves are 55 F and 45 F, respectively.

14-42. Air enters a finned coil at 80 F db, 67 F wb, and standard atmospheric pressure. The chilled water has a temperature of 50 F at this position in the coil. Will moisture condense from the air for the following assumed conditions? Copper tubes with wall thickness of 0.018 in. and thermal conductivity of 190 Btu/(ft-hr-F). The heat-transfer coefficient on the inside of the tubes is 1000 Btu/(hr-ft^2-F). The product of total heat-transfer coefficient and surface effectiveness ($h_d \eta_o$) is 60 lbma/(ft^2-hr) and the ratio of total surface area to inside tube surface area is 12.

14-43. Consider a cooling coil with air entering at 27 C db, 19 C wb, and standard atmospheric pressure. The coil is all-aluminum construction with tube wall thickness of 0.50 mm and thermal conductivity of 58 W/(m-C). The heat-transfer coefficient on the inside of the tubes is 53 W/(m^2-C). The product of total heat-transfer coefficient and surface effectiveness ($h_d \eta_o$) is 2.5 kg/(m^2-hr). The ratio of total surface area to inside tube surface area is 14. Will condensation occur on the surface at this location if the chilled water has a temperature of 14.3 C at this location in the coil?

14-44. A heat exchanger surface like Fig. 14-20 has a face area of 10 ft^2, 5 rows of tubes, and aluminum fins (11.7 fins/in.). Moist air enters the exchanger at 80 F dry bulb and 68 F wet bulb, at standard atmospheric pressure and with a face velocity of 650 ft/min. Use a coil simulation program to compute the (a) leaving air conditions, (b) total heat-transfer rate, (c) sensible heat-transfer rate, and (d) pressure loss for the air. The refrigerant is water that undergoes a temperature change from 45 to 50 F in a counterflow arrangement.

14-45. A heat exchanger made like the surface of Fig. 14-20 has a face area of 1 m^2 and 4 rows of tubes and aluminum fins (0.461 fins/mm). Air enters the coil at 26 C dry bulb and 20 C wet bulb with a face velocity of 3.5 m/s and a pressure of 100 kPa absolute pressure. Use a coil simulation program to determine the (a) leaving air conditions, (b) total heat-transfer rate, (c) sensible heat-transfer rate, and (d) pressure loss for the air. The refrigerant is water that undergoes a change in temperature from 7 to 10 C in counterflow.

14-46. Suppose the coil of Example 14-2 had 6 rows of tubes instead of 4 and estimate the heat, mass, and friction coefficients.

14-47. Use the method described in Chapter 13 for the spray dehumidifier to solve Problem 14-44.

14-48. Use the method described in Chapter 13 for the spray dehumidifier to solve Problem 14-45.

14-49. Rework Problem 14-44 assuming barometric pressure at 5000 ft elevation.

14-50. Rework Problem 14-45 assuming barometric pressure at 1500 m.

Chapter 15

Refrigeration

Almost every HVAC system that provides a cooling effect depends on a refrigeration system to provide either a cold liquid such as water or brine or a direct removal of sensible and latent heat from an air stream. Refrigeration systems in some cases also provide heating for HVAC systems and are commonly called *heat pumps* (see Chapter 2). Refrigeration is a specialization separate from HVAC system design, and few engineers are expert in both areas. Because both the control and performance of HVAC systems are significantly affected by the performance of the refrigeration system, those who are involved with HVAC systems need to have some basic knowledge of refrigeration. This understanding helps HVAC engineers to make intelligent selection of refrigeration equipment and allows them to properly fit this equipment into the overall system. ASHRAE Standard 15, "Safety Code for Mechanical Refrigeration" (1) establishes reasonable safeguards of life, limb, health, and property; defines practices that are consistent with safety; and prescribes safety standards for the industry. Engineers designing HVAC systems should become familiar with these requirements.

Concerns about depletion of atmospheric ozone and about global warming have focused attention on the release into the atmosphere of certain refrigerants widely used in the HVAC field. Most of the major industrial countries have agreed on schedules to stop or reduce the production of those refrigerants thought to be particularly harmful to the earth's atmosphere. Restrictive regulations on refrigerant production, recovery, and release have been issued in the United States and other countries, and taxes have been assessed to discourage the use of some of these refrigerants. Compressor, refrigerant, and refrigerating system manufacturers are attempting to find substitutes for the refrigerants that must be replaced and to design equipment that will function acceptably with these new substitutes. A renewed interest in ammonia systems has also occurred because of the use of an environmentally acceptable refrigerant.

15-1 THE PERFORMANCE OF REFRIGERATION SYSTEMS

The instantaneous performance of any refrigerating system when used for cooling is expressed in terms of the cooling coefficient of performance or COP_c.

$$COP_c = \frac{\text{Useful refrigerating effect}}{\text{Net energy input}} \tag{15-1}$$

COP_c is a dimensionless quantity, expressible as a pure number. In the United States the performance of HVAC systems is often given in dimensional terms, Btu/(W-hr), as an *energy efficiency ratio* or EER. Since 3.412 Btu = 1.0 W-hr, an EER rating of 10.0 would be equivalent to a COP_c of 10.0/3.412 or 2.93. The anticipated performance of a refrigerating device, over an average season, is sometimes called the *seasonal energy efficiency ratio* or SEER, also expressed in Btu/(W-hr).

Refrigerating systems used for heating, which are usually referred to as heat pumps may be evaluated in terms of a heating coefficient of performance or COP_h. In this case it is the useful heating effect that is considered.

$$COP_h = \frac{\text{Useful heating effect}}{\text{Net energy input}} \qquad (15\text{-}2)$$

Heat pumps may also be rated in terms of energy efficiency ratios and seasonal energy efficiency ratios by expressing Eq. 15-2 in Btu/(W-hr).

It is sometimes convenient to rate refrigerating systems or heat pumps in terms of the power requirements per ton. For typical power units this is usually given as

$$\text{hp/ton} = \frac{(12{,}000) \text{ Btu/ton-hr}}{(\text{COP})(2545) \text{ Btu/hp-hr}} = \frac{4.72}{\text{COP}} \qquad (15\text{-}3)$$

$$\text{kW/ton} = \frac{(12{,}000) \text{ Btu/ton-hr}}{(\text{COP})(3412) \text{ Btu/kW-hr}} = \frac{3.52}{\text{COP}} \qquad (15\text{-}4)$$

In studying the performance of refrigeration cycles the concept of the thermodynamically *reversible cycle* is useful. Two important characteristics of a reversible cycle are as follows:

1. No refrigeration cycle can have a higher coefficient of performance than that of a reversible cycle operating between the same source and sink temperature.
2. All reversible refrigeration cycles operated between the same source and sink temperatures have identical coefficients of performance.

The most convenient reversible cycle to use as an ideal refrigerating cycle is the Carnot refrigeration cycle, consisting of two isothermal processes and two adiabatic processes. Such a cycle is shown in Fig. 15-1 on temperature–entropy coordinates. Because of characteristic 2 listed above, no particular working medium needs to be specified. Notice that in the Carnot refrigeration cycle all of the heat is absorbed at the lower (evaporator) temperature T_e in process 2–3 at constant temperature and all of the heat is rejected at a constant higher (condenser) temperature T_c in process 4–1. It can be easily shown that the COP_c for this case is a function only of these two temperatures, expressed in absolute degrees (K or R):

$$\text{Carnot cycle } COP_c = \frac{T_e}{T_c - T_e} \qquad (15\text{-}5)$$

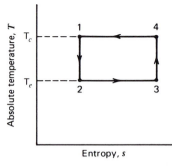

Figure 15-1 The Carnot refrigeration cycle.

For heating (as with a heat pump) the cycle would look the same as in Fig. 15-1 but
the objective would be different and the heating coefficient of performance would be

$$\text{Carnot cycle COP}_h = \frac{T_c}{T_c - T_e} \tag{15-6}$$

Equations 15-5 and 15-6 are valid only for cycles where heat is received by the
refrigerating device at a constant temperature T_e and rejected at a constant temperature
T_c. These equations give an upper limit to strive for; for a variety of reasons actual
devices typically give much lower coefficients of performance. No practical refrigeration device has been developed to operate on the Carnot cycle, but it is quite beneficial
to use the concept as a standard of perfection and as a guide in the design of real cycles.

The *refrigerating efficiency* η is the ratio of the coefficient of performance of a cycle
or system to that of an ideal cycle or system

$$\eta = \frac{\text{COP}_c}{\text{COP}_{\text{ideal}}} \tag{15-7}$$

15-2 THE THEORETICAL SINGLE-STAGE COMPRESSION CYCLE

Of the three basic refrigeration methods—vapor compression, absorption, and thermo-
electric—the most common method in the HVAC industry is vapor compression, with
absorption a distance second choice. These methods have many complex variations but
only the basic compression and absorption cycles will be discussed in this chapter. The
theoretical single-stage compression cycle will be discussed first.

A schematic of the theoretical single-stage vapor compression cycle, the simplest
approximation to real, practical operating cycles is shown in Fig. 15-2. The thermodynamic processes for this theoretical cycle are shown on both temperature–entropy and
pressure–enthalpy diagrams in that same figure. Although practical and simple, this
cycle has two features that keep it from having as high a coefficient of performance as
the Carnot cycle discussed earlier. The first is the fact that the flow through the
expansion valve, process 1–2 is an irreversible throttling process where an opportunity
to produce useful work is lost. The second nonideal feature is that heat rejection, process
4–1, does not occur at constant temperature. We will see in Section 15-4 that even this
nonideal cycle is not a practical reality. But it is simple, it does show useful trends, and

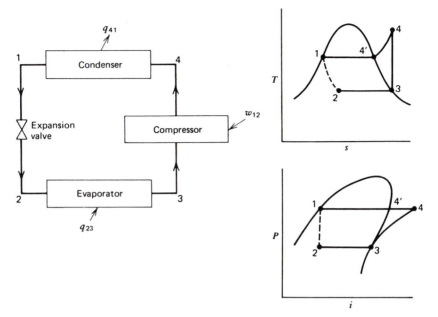

Figure 15-2 The theoretical single-stage cycle.

it can be modified to closely approximate real systems. For these reasons the cycle shown in Fig. 15-2 is a good model for understanding the basic features of vapor compression systems.

Refrigerant entering the compressor is assumed to be dry saturated vapor at the evaporator pressure. This is a convenient place to begin an analysis because we can easily determine all fluid properties here. The compression process 3–4 is assumed to be reversible and adiabatic and, therefore, isentropic, and is continued until the condenser pressure is reached. Point 4 is obviously in the superheated vapor region. The process 4–1 is carried out at constant pressure with the temperature of the vapor decreasing until the saturated vapor condition is reached at 4′; then the process is both at constant temperature and constant pressure during the condensation from 4′ to 1. At point 1 the refrigerant leaves the condenser as a saturated liquid. It is then expanded through a throttling valve where partial evaporation occurs as the pressure drops across the valve. The throttling process 1–2 is irreversible with an increase in entropy occurring. For this reason the process is shown as a dashed line in Fig. 15-2. For a throttling process the enthalpy at exit and at inlet are equal. To determine the coefficient of performance of this cycle the useful refrigerating effect and the net energy input must be determined. For steady flow of one pound of refrigerant:

$$i_3 - i_2 = i_3 - i_1$$

$$\text{Useful refrigerating effect} = q_{23} = i_3 - i_2 = i_3 - i_1$$

$$\text{Net energy input} = w_{34} = i_4 - i_3$$

$$\text{COP} = \frac{i_3 - i_1}{i_4 - i_3}$$

EXAMPLE 15-1

For a theoretical single-stage cycle operating with a condenser pressure of 250 psia and an evaporator pressure of 40 psia and using refrigerant 22 as the working fluid determine the (a) COP, (b) temperature at exit to compressor, (c) temperature in condenser, and (d) temperature in evaporator.

SOLUTION

Use the pressure-enthalpy diagram for refrigerant 22, Chart 5.

$$\text{(a) COP} = \frac{i_3 - i_1}{i_4 - i_3} = \frac{104.6 - 43.1}{125.4 - 104.6} = \frac{61.5}{20.8} = 2.96$$

(b) $t_4 = 168\,\text{F}$

(c) $t_1 = 116\,\text{F}$

(d) $t_2 = 2\,\text{F}$

The coefficient of performance is closely related to the evaporating and condensing temperatures. For the highest COP the cycle should be operated at the lowest possible condensing temperature and the highest possible evaporating temperature.

EXAMPLE 15-2

Compute the COP and refrigerating efficiency for a theoretical single-stage cycle operating with a condenser pressure of 1219 kPa and an evaporator pressure of 204 kPa using R-12 as the refrigerant. The work for the cycle is 31 kJ/kg of refrigerant.

SOLUTION

From Fig. 15-2, the COP may be expressed as

$$\text{COP} = \frac{i_3 - i_1}{i_4 - i_3} = \frac{i_3 - i_1}{w_{34}}$$

And from Table A-3*b*

$$i_3 = 346.3\,\text{kJ/kg} \qquad i_1 = 248.9\,\text{kJ/kg}$$

Then

$$\text{COP} = \frac{346.3 - 248.9}{31} = 3.14$$

A Carnot cycle operating between the saturation temperatures corresponding to the given condenser and evaporator pressures has a COP given by Eq. 15-5 where from

Table A-3*b*

$$T_e = -12 + 273 = 261 \text{ K}$$

$$T_c = 50 + 273 = 323 \text{ K}$$

$$\text{COP}_c = \frac{261}{323 - 261} = 4.2$$

The refrigerating efficiency is then

$$\eta_R = \frac{\text{COP}}{\text{COP}_c} = \frac{3.14}{4.2} = 0.75 \text{ or } 75 \text{ percent}$$

15-3   REFRIGERANTS

The fluid used for energy exchanges in a refrigerating or heat pump system is called the *refrigerant*. The refrigerant usually absorbs heat while undergoing a phase change (in the evaporator) and then is compressed to a higher pressure and a higher temperature, allowing it to transfer that energy (in the condenser) directly or indirectly to the atmosphere or to a medium being purposefully heated. A refrigerant's suitability for a given application depends on many factors including its thermodynamic, physical, and chemical properties, and its safety. The significance or relative importance of each characteristic varies from one application to the next, and there is no such thing as an ideal refrigerant for all applications. Some of the characteristics of general importance are as follows:

Thermodynamic Characteristics

1. *High latent enthalpy of vaporization.* This means a large refrigerating effect per unit mass of the refrigerant circulated. In small-capacity systems, too low a flow rate may actually lead to problems.
2. *Low freezing temperature.* The refrigerant must not solidify during normal operating conditions.
3. *Relatively high critical temperatures.* Large amounts of power would otherwise be required for compression.
4. *Positive evaporating pressure.* Pressure in the evaporator should be above atmospheric to prevent air from leaking into the system.
5. *Relatively low condensing pressure.* Otherwise expensive piping and equipment will be required.

Physical and Chemical Characteristics

1. *High dielectric strength of vapor.* This permits the use in hermetically sealed compressors where vapor may come in contact with motor windings.
2. *Good heat-transfer characteristics.* Thermophysical properties should be such that high heat-transfer coefficients can be obtained. This covers properties such as density, specific heat, thermal conductivity, and viscosity.

3. *Satisfactory oil solubility.* Oil may be dissolved in the refrigerant or refrigerant in the oil. This can affect lubrication and heat-transfer characteristics and lead to oil logging in the evaporator. A system must be designed with the oil solubility characteristics in mind.

4. *Low water solubility.* Water in a refrigerant can lead either to freeze-up in the expansion devices or corrosion.

5. *Inertness and stability.* The refrigerant must not react with materials that will contact it and its own chemical makeup must not change with time.

Safety

1. *Nonflammability.* The refrigerant should not burn or support combustion when mixed with air.

2. *Nontoxicity.* The refrigerant should not be harmful to humans directly or indirectly through foodstuffs.

3. *Nonirritability.* The refrigerant should not irritate humans (eyes, nose, lungs, or skin).

Effect on the Environment

1. *Ozone depletion potential.* The refrigerant's potential to deplete the ozone in the upper atmosphere should be low.

2. *Global warming potential.* The refrigerant's potential to persist in the upper atmosphere and to trap the radiation emitted by the earth (the greenhouse effect) should be low.

In addition to these characteristics, the refrigerant should be of relative low cost and be easy to detect in case of leaks.

Concerns about the greenhouse effect (atmospheric warming), the depletion of the ozone layer at high altitudes, and the government regulations that have and that will come from those concerns have the full attention of the HVAC industry. A major level of interest and activity is in the development and/or use of refrigerants with low ozone depletion potenital (ODP) and low greenhouse warming potential (GWP). Additionally, the concerns have caused the development of a strong emphasis on refrigerant recovery methods and on reduction of refrigerant leakage.

The safety of all types of products has received increased emphasis lately, and refrigerants are no exception. Figure 15-3 shows refrigerant safety group classifications, consisting of two alphanumeric characters (2). The capital letter indicates the level of toxicity and the Arabic numeral denotes the level of flammability.

Refrigerants generally are from one of four classes of elements or compounds: halocarbons, hydrocarbons, organic, and inorganic. Table 15-1 lists several of the more important refrigerants. A more complete list may be found in references 2 and 3. The halocarbons, a popular class of refrigerants in use for over half a century, were initially introduced by the du Pont Company. Now produced by several manufacturers, these products now make up the bulk of the refrigerants used in the HVAC industry. Many of these refrigerants are still referred to as ''Freons,'' the trade name given to them by du Pont, but because companies manufacture and sell them under different brand names it is proper to refer to them, and to all refrigerants, by the generic numbers designated by

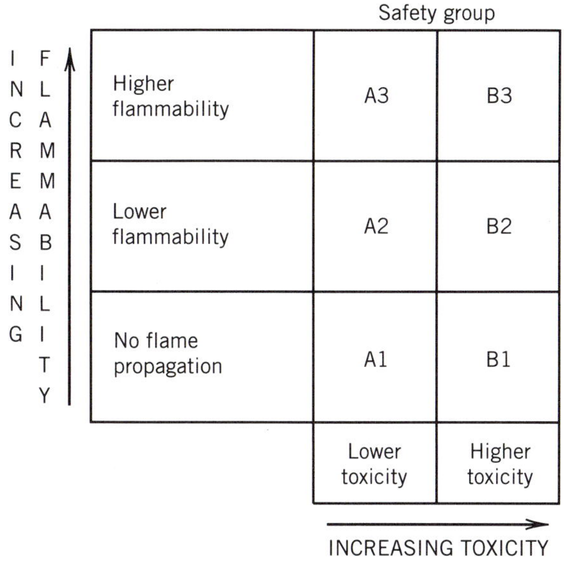

Safety group

Figure 15-3 Refrigerant safety group classification (2).

ASHRAE. These designations, along with refrigerant safety group classifications, have been given by ASHRAE Standard 34 (2).

According to Standard 34 (2), when the generic number designations appear in technical publications, or equipment nameplates, and in specifications, they should be preceded by the capital letter R, the word *Refrigerant,* or the word *Refrigerants.* The manufacturer's trademark or tradename may be used as an alternative prefix in technical publications but not on equipment nameplates or in specifications. Composition-designating prefixes, to be described below, should be used only in nontechnical publications in which the potential for ozone depletion is pertinent.

The chemical composition of these halocarbon and hydrocarbon refrigerants in the methane, ethane, propane, and cyclobutane series and the molecular structure of the methane, ethane, and most of the propane series are explicitly determined without ambiguity from the number designation. The converse is also true. The following list summarizes major points of the ASHRAE designation system (2):

1. The first digit on the right is the number of fluorine (F) atoms in the compound.
2. The second digit from the right is one more than the number of hydrogen (H) atoms in the compound.
3. The third digit from the right is one less than the number of carbon (C) atoms in the compound. When this digit is zero, it is omitted from the number.
4. Blends are designated by their respective refrigerant numbers and weight proportions, named in the order of increasing normal boiling points of the components, for example R-22/12(90/10).
5. Zeotropic blends that have been commercialized are assigned an identifying number in the 400 series accompanied by the weight proportions of the components, for example R-400(90/10) for mixtures of R-12 and R-114.

Table 15-1 Properties of Selected Refrigerants

Refrigerant Number	Chemical Name	Chemical Formula	Molecular Mass	Normal Boiling Point C	Normal Boiling Point F	Safety Group
	Methane Series					
11	trichlorofluoromethane	CCl_3F	137.4	24	75	A1
12	dichlorodifluoromethane	CCl_2F_2	120.9	−30	−22	A1
13	chlorotrifluoromethane	$CClF_3$	104.5	−81	−115	A1
14	carbon tetrafluoride	CF_4	88.0	−128	−198	A1
21	dichlorofluoromethane	$CHCl_2F$	102.9	9	48	B1
22	chlorodifluoromethane	$CHClF_2$	86.5	−41	−41	A1
23	trifluoromethane	CHF_3	70.0	−82	−116	
50	methane	CH_4	16.0	−161	−259	A3
	Ethane Series					
114	1,2-dichlorotetrafluoroethane	$CClF_2CClF_2$	170.9	4	38	A1
123	2,2-dichloro-1,1,1-trifluoroethane	$CHCl_2CF_3$	153.0	27	81	B1
124	2-chloro-1,1,1,2-tetrafluoroethane	$CHClFCF_3$	136.5	−12	10	

125	pentafluoroethane	CHF_2CF_3	120.0	−49	−56	
134a	1,1,1,2-tetrafluoroethane	CH_2FCF_3	102.0	−26	−15	A1
143a	1,1,1-trifluoroethane	CH_3CF_3	84.0	−47	−53	A2
152a	1,1-difluoroethane	CH_3CHF_2	66.0	−25	−13	A3
170	ethane	CH_3CH_3	30.0	−89	−128	A3
	Propane Series					
290	propane	$CH_3CH_2CH_3$	44.0	−42	−44	A3
	Inorganic Compounds					
717	ammonia	NH_3	17.0	−33	−28	B2
718	water	H_2O	18.0	100	212	A1
744	carbon dioxide	CO_2	44.0	−78[a]	−109[a]	A1
764	sulfur dioxide	SO_2	64.1	−10	14	B1
	Zeotropes					
400	R-12/114 (must be specified)	none	none			A1/A1
	Azeotropes					
502	R-22/115 (48.8/51.2)	19 66	112.0	−45	−49	A1

[a]Sublimes

6. Azeotropes that have been commercialized are assigned and identifying number in the 500 series with no composition shown.
7. Organic refrigerants are assigned serial numbers in the 600 series.
8. Inorganic compounds are designated by adding 700 to their molecular mass; for example, water is 718. When two or more compounds have the same molecular mass, uppercase letters are assigned to distinguish them.
9. The letter C is used before number designations to identify cyclic derivatives. Lowercase letters are appended after numbers to distinguish isomers— refrigerants with the same chemical composition but with differing molecular structures.

The present concern with refrigerants' effect on the environment is primarily due to the release of chlorine (C1) in the upper atmosphere, which is believed to react with the ozone (O_3). Reduction of ozone levels in the upper atmosphere can lead to reduced screening of harmful ultraviolet rays from the sun. This in turn is thought to increase the risk of skin cancers and perhaps to lead to other damaging effects on living material. Several of the most popular refrigerants not only contain chlorine but, as good refrigerants, are also highly stable. This stability results in released refrigerants being able to persist until they reach the upper atmosphere, where sunlight promotes the reaction with the ozone. Less stable materials containing chlorine are not as likely to reach the high altitudes and to cause the damage to the ozone layer.

The refrigerants of primary concern are called *chlorofluorocarbons* or CFCs. These include R-11, R-12, R-113, R-114, and R-115. In discussion of their effect on the environment, it is accepted practice to refer to them with the prefix CFC; thus CFC-11, CFC-12, and so on, would be proper designations in such cases.

Another group of refrigerants contain chlorine, but because they also retain a hydrogen molecule in their molecular structure they are less persistent than the CFCs and thus represent a decreased threat to the ozone layer. These are called *hydrochloro-fluorocarbons* or HCFCs. In this group are R-22, R-123, R-124, R-141b, and R-142b. It appears that regulations and restrictions on the use of the HCFCs will continue to be put forth, although they may come about at a slower pace than for the CFCs.

The group of fluorocarbons that appear to be the least harmful to the ozone layer are the *hydrofluorocarbons* or HFCs, which contain no chlorine. In this group are R-125, R-134a, R-143a, and R-152a.

Unfortunately there do not appear to be any good "drop in" substitutes for the refrigerants of concern. Variations in boiling points and refrigerating efficiencies cause some of the concern, but the bigger problems could be in safety and in lubricant and elastomer compatability. R-134a appears to be a good candidate as a substitute for R-12, widely used in automotive and commercial refrigeration applications. R-123 has been promoted as a candidate to replace R-11, a very common refrigerant for systems with centrifugal compressors. Mixtures of two or more refrigerant types, including both zeotropes and azeotropes, are continuing to be tested and may be useful in some instances.

The limitations on possible use of a halocarbon to solve the refrigerant problem is illustrated by Figure 15-4. There the possible structures in the methane series and the ethane series are shown with increasing numbers of halogen atoms (decreasing number of hydrogen molecules) from top to bottom in each case. Fully halogenated molecules, which contain no hydrogen and which are considered to be environmentally unaccept-

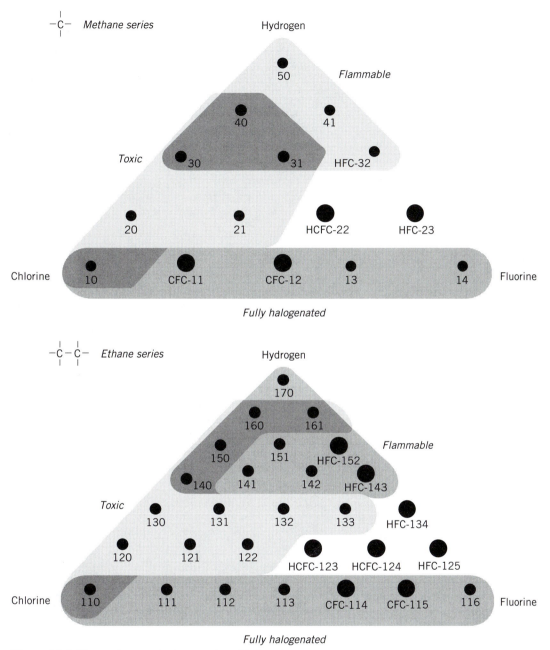

Figure 15-4 The methane and ethane series of halocarbons.

able, exist in the bottom row in both series. Molecules to the left have mostly chlorine atoms, and molecules to the right have mostly fluorine atoms.

It can be seen in the methane series that the upper six compounds are flammable, six of the leftmost compounds are toxic, and the five in the bottom row are fully halogenated. Only two compounds in that entire series, R-22 (an HCFC) and R-23 (an

HFC), do not have these limitations. R-23 is not widely used as a refrigerant owing to its thermodynamic properties. Unfortunately R-22, the most popular refrigerant in residential and light commercial air conditioning, is considered to have some environmental impact and may undergo future government restrictions in manufacture. R-22's replacement in heat pump applications appears to be one of the more difficult problems to solve.

In the ethane series R-123, R-124, R-125, and R-134 are the only partially halogenated compounds free from toxicity and flammability problems. Of these R-123 and R-134a have created the most interest, with production of R-134a already underway and promising to be one of the more popular future refrigerants. R-123 is an HCFC and may face future regulatory restrictions. The thermodynamic properties of R-124 and R-125 make them of less interest in HVAC applications.

Compounds in the more complex propane and ether series are being evaluated as potential replacements for some present refrigerants.

EXAMPLE 15-3

(a) Determine the ASHRAE number designation for dichlorotetrafluoroethane, $CClF_2CClF_2$. (b) Give the chemical composition of R-11.

SOLUTION

(a) There are four fluorine atoms, no hydrogen atoms, and two carbon atoms per molecule:

$$(2 - 1)(0 + 1)(4)$$

Thus the ASHRAE Designation is 114. (b). R-11 should have one fluorine atom, no hydrogen atoms, and one carbon atom. The one carbon atom indicates a methane, and since there is only one fluorine atom and no hydrogen atoms, there must be three chlorine atoms to complete the structure. Thus R-11 is trichloromonofluoromethane or CCl_3F.

15-4   COMPRESSION CYCLE COMPONENTS

The four essential parts of the vapor compression refrigeration cycle are shown in Fig. 15-2, the compressor, evaporator, condenser, and expansion valve. The compressor, which is the most mechanically complex of these components, is often the most expensive single item in the system. A more complete discussion of compressors can be found in the *ASHRAE Handbook, HVAC Systems and Equipment Volume* (4).

Compressors may be of two basic types, positive displacement and dynamic. *Positive displacement compressors* increase the pressure of the refrigerant vapor by reducing the volume. Examples are the reciprocating, rotary (rolling piston, rotary vane, single screw, and twin screw), scroll, and trochoidal. *Dynamic compressors* increase the pressure of refrigerant vapor by a continuous transfer of angular momentum to the vapor from the rotating member followed by a conversion of this momentum into a pressure rise. The centrifugal compressor is of this type.

Reciprocating Compressors

Most reciprocating compressors are single-acting, using pistons driven directly through a pin and connecting rod from the crankshaft. Double-acting compressors are not extensively used.

The halocarbon compressor, the most widely used, is manufactured in three types of design: (1) open, (2) semi- or bolted hermetic, and (3) the welded-shell hermetic. Ammonia compressors are manufactured only in the open-type design. In the open-type compressor the shaft extends through a seal in the crankcase to an external drive.

In hermetic compressors the motor and compressor are contained within the same pressure vessel with the motor shaft as part of the compressor crankshaft, and with the motor in contact with the refrigerant (Fig. 15-5a). A semihermetic, accessible, or serviceable hermetic compressor is of bolted construction capable of field repair (Fig. 15-5b). A welded-shell (sealed) hermetic compressor is one in which the motor-compressor is mounted inside a steel shell that in turn is sealed by welding.

The idealized reciprocating compressor is assumed to operate in a reversible adiabatic manner; pressure losses in the valves, intake, and exhaust manifolds are neglected. We will now consider the effect of some of the unavoidable realities of the compressor.

Figure 15-6 shows a schematic indicator diagram for a reciprocating compressor. The pressure in the cylinder is assumed constant during the exhaust process $P_c = P_d$ and during the intake process P_a and P_b; however, the pressure loss in the valves is to be considered. The gas remaining in the clearance volume V_d expands in a polytropic process from state d to state a. State b is generally different from that at a because of the mixing of the expanded clearance volume vapor and the intake vapor. The vapor is

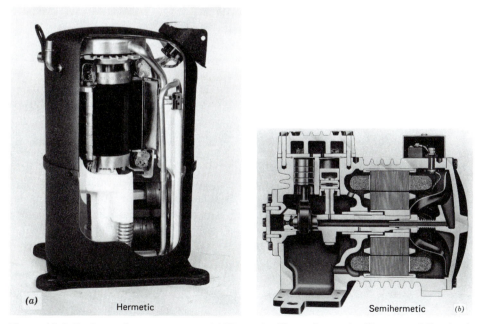

Figure 15-5 Reciprocating compressors. (a) Hermetic, (b) semihermetic. (Courtesy of Copeland Corporation, Sidney, OH.)

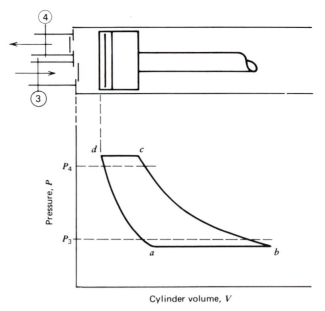

Figure 15-6 Schematic indicator diagram for a reciprocating compressor.

compressed from state b to state c in a polytropic process. Heat transfer may occur during the exhaust process c–d. Figure 15-7 is a pressure-specific volume diagram for the compressor of Fig. 15-6. In the state diagram of Fig. 15-7 it is assumed that

1. The same polytropic exponent n applies to the compression process b–c and the expansion process d–a.
2. Heat transfer during the exhaust process is negligible; therefore, states c and d are identical.
3. The state of the mixture of reexpanded clearance volume gas a and the intake gas 3 is the same as state b and is designated as a'. This is actually a result of assumptions 1 and 2 above. Note that the work required to compress the clearance volume vapor is just balanced by the work done in the expansion of the clearance volume vapor.

The *volumetric efficiency* η_v is defined as the ratio of the actual mass of the vapor compressed to the mass of vapor that could be compressed if the intake volume equaled the piston displacement and the state of the vapor at the beginning of the compression were equal to state 3 of Fig. 15-7. It may be shown that

$$\eta_v = \left[1 + C - C \left(\frac{P_c}{P_b} \right)^{1/n} \right] \frac{v_3}{v_b} \tag{15-8}$$

The *clearance factor* C is defined as the ratio of the clearance volume to the piston

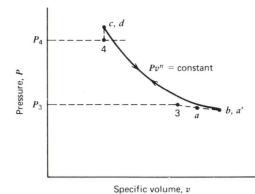

Figure 15-7 Pressure-specific volume diagram for compressor of Fig. 15-6.

displacement or

$$C = \frac{V_d}{V_b - V_d} \tag{15-9}$$

The three main factors that influence the volumetric efficiency are accounted for in Eq. 15-8. The term within the brackets describes the effect of the clearance volume and reexpansion of the clearance volume vapor, whereas the term v_3/v_b accounts for the pressure drop and heating of the intake vapor. Obviously, it would be desirable for the clearance volume to be zero, but this is not possible owing to mechanical constraints. The valve plate in the compressor must be thick enough to withstand the forces of high pressure, which causes the exhaust valve ports to have considerable volume. There must also be space for the intake valve to open. Compressor designers are constantly seeking ways to reduce the clearance volume.

Volumetric efficiency may also be expressed as

$$\eta_v = \frac{\dot{m} v_3}{PD} \tag{15-10}$$

where PD is the piston displacement in volume per unit of time. Then combining Eqs. 15-8 and 15-10, the mass flow rate of refrigerant is given by

$$\dot{m} = \left[1 + C - C \left(\frac{P_c}{P_b} \right)^{1/n} \right] \frac{PD}{v_b} \tag{15-11}$$

The purpose of the compressor is to move refrigerant through the system as efficiently as possible. The important effect of the clearance volume is especially evident in Eq. 15-11.

Although the polytropic exponent n must be determined experimentally, it may be approximated by the isentropic exponent k when other data are not available. Typical values are $k = 1.13$ for R-12 and 1.16 for R-22.

An expression for the compressor work may be derived subject to the same assumptions used to obtain volumetric efficiency:

$$w = \frac{n}{(n-1)} P_b v_b \left[\left(\frac{P_c}{P_b} \right)^{(n-1)/n} - 1 \right] \tag{15-12}$$

The power requirement for the compressor is given by

$$\dot{W} = \frac{\dot{m}w}{\eta_m} \tag{15-13}$$

where η_m is the compressor mechanical efficiency. Equations 15-8 through 15-13 are not intended to predict the actual performance of real compressors but rather to show the relationships among the important variables. Figure 15-8 shows representative performance data for a hermetic compressor operating with refrigerant 22. Manufacturers furnish performance data in this form or as a table.

EXAMPLE 15-4

Refrigerant 12 vapor enters the suction header of a single-stage reciprocating compressor at 45 psia and 40 F. The discharge pressure is 200 psia. Pressure drop in the suction valve is 2 psi and the pressure loss in the discharge valve is 4 psi. The vapor is superheated 12 F during the intake stroke. The clearance volume is 5 percent of piston displacement. Determine the (a) volumetric efficiency, (b) compressor pumping capacity if the piston displacement is 10 in.^{3} and the crankshaft rotates at 1725 rpm, and (c) shaft horsepower required for a mechanical efficiency of 70 percent.

SOLUTION

The $P-v$ diagram of Fig. 15-7 will be used to aid in the problem solution. From the given data

$$P_4 = 200 \text{ psia}; \qquad P_c = P_d = P_4 + 4 = 204 \text{ psia}$$

$$P_3 = 45 \text{ psia}; \qquad P_b = P_3 - 2 = 43 \text{ psia}$$

From Chart 4, $v_3 = 0.92$ ft^3/lbm and at point b, $t_b = 52$ F. $P_b = 43$ psia and $v_b = 0.98$ ft^3/lbm. The clearance factor is given as 0.05, and it will be assumed that $n = k = 1.13$. Then using Eq. 15-8

$$\eta_v = \left[1 + 0.05 - (0.05) \left(\frac{204}{43} \right)^{1/1.13} \right] \frac{0.92}{0.98} = 0.8$$

Now from Eq. 15-10

$$\dot{m} = \frac{\eta_v(PD)}{v_3} = \frac{(0.8)}{(0.92)} \frac{(10)(1725)}{(1728)} = 8.68 \text{ lbm/min}$$

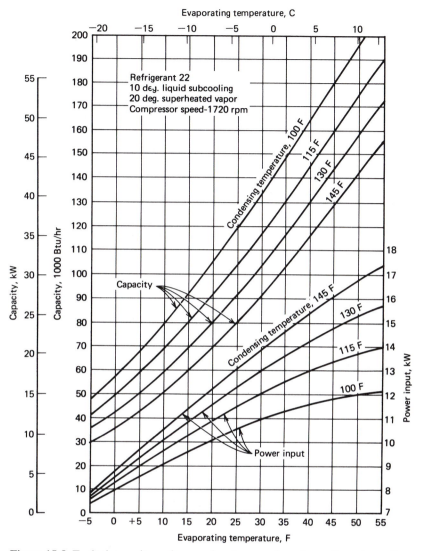

Figure 15-8 Typical capacity and power input curves for a hermetic reciprocating compressor. (Reprinted by permission from *ASHRAE Handbook, Equipment Volume*, 1975.)

The shaft power required will be computed using Eqs. 15-12 and 15-13

$$\dot{W} = \frac{\dot{m}w}{\eta_m} = \left(\frac{\dot{m}}{\eta_m}\right)\frac{n}{(n-1)}P_b v_b \left[\left(\frac{P_c}{P_b}\right)^{(n-)1/n} - 1\right]$$

$$\dot{W} = \left(\frac{8.68}{0.7}\right)\frac{1.13}{(1.13-1)}(43)(144)0.98\left[\left(\frac{204}{43}\right)^{(1.13-1)/1.13} - 1\right]$$

$$\dot{W} = 128,298 \text{ (ft-lbf)/min} = 3.9 \text{ hp}$$

Rotary Compressors

Rotary compressors are characterized by their circular or rotary motion as opposed to reciprocating motion. Their positive displacement compression process is nonreversing and either continuous or cyclical depending on the mechanism employed. Most are direct-drive machines.

Figures 15-9 and 15-10 show two common types of rotary compressors: the rolling piston type and the rotating vane type. These two machines are very similar with respect to size, weight, thermodynamic performance, field of greatest applications, range of sizes, durability, and sound level.

The rotary compressor performance is characterized by high volumetric efficiency because of its small clearance volume, and correspondingly low reexpansion loss. Rotary vane compressors have a low weight-to-displacement ratio, which, in combination with small size, makes them suited to transport applications. Figure 15-11 shows performance typical of rolling-piston compressors.

Single-screw Compressors

The single-screw compressor consists of a single cylindrical main rotor that works with a pair of gaterotors. The rotors may vary considerably in geometry. These are most often used in the liquid injection mode, where sufficient liquid cools and seals the compressor. These compressors can operate with pressure ratios above 20 in single stage, and are available in capacities from 20 to 1300 tons (4).

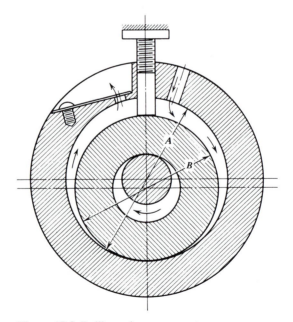

Figure 15-9 Rolling piston-type rotary compressor. (Reprinted by permission from *ASHRAE Handbook, Equipment Volume,* 1975.)

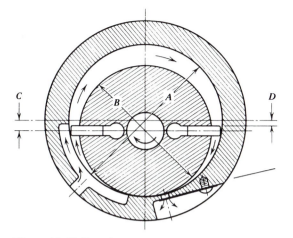

Figure 15-10 Rotating vane-type rotary compressor.
(Reprinted by permission from *ASHRAE*
Handbook, Equipment Volume, 1975.)

Double Helical Rotary (Twin-screw) Compressor

The double helical rotary or twin-screw compressor, like the single-screw compressor, belongs to the broad class of positive displacement compressors. Although it was developed in the early 1930s and has been used in the gas and process industries, it was first introduced to the refrigeration industy in the late 1950s.

The machine essentially consists of two mating helically grooved rotors, a male (lobes) and a female (gullies), in a stationary housing with suitable inlet and outlet gas ports (Fig. 15-12). The flow of gas in the rotors is both radial and axial. With a four-lobe male rotor rotating at 3600 rpm, the six-lobe female rotor follows at $4/6 \times (3600)$ or 2400 rpm. The female rotor can be driven by the male rotor.

Compression in this machine is obtained by direct volume reduction with pure rotary motion. The four basic continuous phases of the working cycle are

1. *Suction.* A pair of lobes unmesh on the inlet port side and gas flows in the increasing volume formed between the lobes and the housing until the lobes are completely unmeshed.
2. *Transfer.* The trapped pocket isolated from the inlet and outlet is moved circumferentially at constant suction pressure.
3. *Compression.* When remeshing starts at the inlet end, the trapped volume is reduced and the charge is gradually moved helically and compressed simultaneously toward the discharge end as the lobes' mesh point moves along axially.
4. *Discharge* starts when the compressed volume has been moved to the axial ports on the discharge end of the machine and continues until all the trapped gas is completely squeezed out.

During the remeshing period of compression and discharge, a fresh charge is drawn through the inlet on the opposite side of the meshing point. With four male lobes rotating at 3600 rpm, four interlobe volumes are filled and discharged per revolution, providing 14,400 discharges per minute. Since the intake and discharge cycles overlap

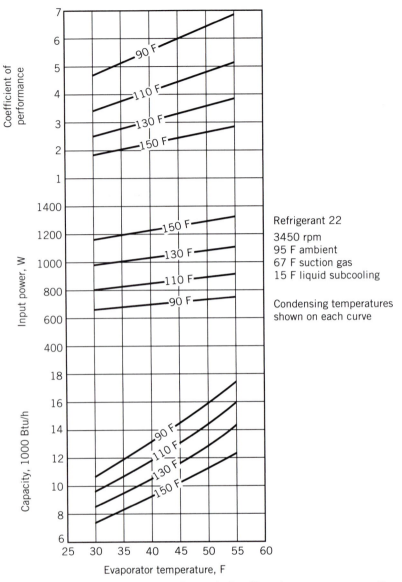

Figure 15-11 Performance curves for typical rolling-piston compressor. (Reprinted by permission from *ASHRAE Handbook, HVAC Systems and Equipment Volume,* 1992.)

effectively, a smooth continuous flow of gas results. Typical performance levels for oil-flooded, twin-screw compressors are given in Figure 15-13.

Orbital Compressors

One type of orbital compressor is the scroll compressor, a rotary motion, positive displacement machine that uses two interfitting, spiral-shaped scroll members to

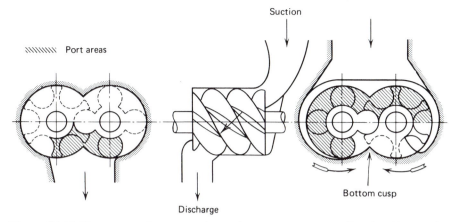

Figure 15-12 The compression process in a twin-screw compressor. (Reprinted by permission from *ASHRAE Handbook, HVAC Systems and Equipment Volume,* 1992.)

compress the refrigerant vapor. Capacities range from approximately 1 to 15 tons (3.5 to 53 kW) with current use in residential and commerical air-conditioning (including heat pumps) and automotive applications. Recent advances in manufacturing technology have permitted the close machining tolerances necessary in the scroll members for effective operation. Advantages include low noise as well as high efficiency. Disadvantages include an incompatibility with solid contaminants and a poor performance at low suction pressures. Good lubrication is critical.

Typically scroll members are a geometrically matched pair assembled 180 degrees out of phase. Each scroll member is bound by a base plate and is open on the other end of the vane. The two scrolls are fitted together so as to form pockets between the two base plates and the various lines of contact between the vanes. One scroll is held fixed and the other moves in an orbital path so that the contact between the flanks of the scrolls is maintained and moves progressively inward (Fig. 15-14). The pockets formed between the scrolls are reduced in size as the gas is moved inward toward the discharge port. Capacity is controlled either by using variable speed drive or by opening and closing porting holes between the suction side and the compression chambers. Most scroll compressors are of the hermetic type.

Trochoidal compressors are small, rotary, orbital, positive displacement devices that can run at speeds up to 9000 rpm. They are manufactured in a variety of configurations, including the Wankel design. Wankel solved earlier sealing problems and produced a trochoidal compressor with a three-sided epitrochoidal piston (motor) and two-envelope cylinder (casing) in capacities up to 2 tons (7 kW).

Centrifugal Compressors

Centrifugal compressors or turbocompressors are members of a family of turbomachines that includes fans, propellors, and turbines. Such machines are characterized by a continuous exchange of angular momentum between a rotating mechanical element and a steadily flowing fluid. Because their flows are continuous, turbomachines have greater volumetric capacities, size-for-size, than do positive displacement devices. For effec-

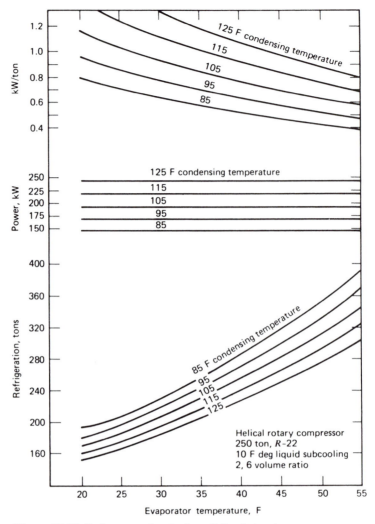

Figure 15-13 Performance levels for oil-flooded twin-screw compressors. (Reprinted by permission from *ASHRAE Handbook, Equipment Volume*, 1975.)

tive momentum exchange, their rotative speeds must be higher, but little vibration or wear results because of the steadiness of the motion and the absence of contacting parts.

Centrifugal compressors are used in a wide variety of refrigeration and air-conditioning applications. As many as eight or nine stages can be installed in a single casing. Almost any refrigerant can be used.

Loading the Compressor

The performance diagrams of Figs. 15-8, 15-11, and 15-13 are valuable aids in designing and analyzing refrigeration systems. The compressor capacity is given in terms of the load on the evaporator, the evaporator temperature, and the condensing

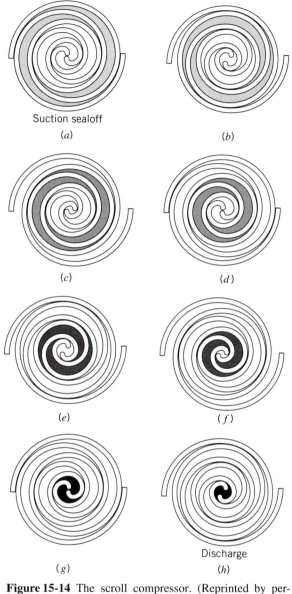

Suction sealoff
(a)   (b)

(c)   (d)

(e)   (f)

Discharge
(g)   (h)

Figure 15-14 The scroll compressor. (Reprinted by permission from *ASHRAE Handbook, HVAC Systems and Equipment Volume*, 1992.)

temperature. This makes it possible to plot curves representing the capacity of the evaporator and condenser on the same chart. Figure 15-15 is such a plot for a semihermetic reciprocating compressor at a condensing temperature of 130 F. The evaporator curve is for an evaporator at fixed air flow rate and entering air temperatures. The system must operate where the two curves cross. The curves representing the power input to the compressor and heat rejected in the condenser are also plotted.

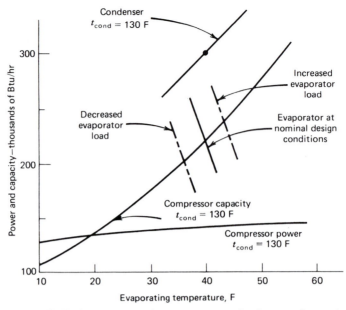

Figure 15-15 Compressor and evaporator capacity for a semihermetic reciprocating compressor.

If the load on the evaporator changes as the result of a change in air flow rate or temperature, the evaporator curve will move to the right or left as shown in Fig. 15-15, depending on whether the load decreases or increases. In all probability the condensing temperature will also change as the load changes, but we will assume it to be constant for this discussion. The important point is that the curves for the evaporator and compressor define the operating condition for the compressor. Compressor power decreases as the evaporating temperature decreases and vice versa. Therefore, the evaporating temperature is an indication of compressor load. A system may be checked in the field by measuring the compressor suction pressure or power. Suction pressure is approximately equal to evaporator pressure, which is directly related to evaporator temperature. For example, a low pressure indicates that the compressor is not loaded. This is an indication that the evaporator is not transferring the expected quantity of heat to the refrigerant. This may be due to any number of things, such as decreased air flow over the evaporator due to clogged filters or a loose fan belt, decreased air temperatures, and so on. The opposite case could also occur, where the compressor is overloaded, as indicated by an increased suction pressure and evaporator temperature.

Control

A compressor will be at partial load much of the time. With small systems, the compressor can be started and stopped by thermostat action when the demand for cooling is satisfied. However, it is not desirable to start and stop larger compressors at frequent intervals. In this case a technique is employed to reduce the compressor capacity referred to as *compressor unloading*. This may be done by reducing the compressor speed, holding the intake valves open, or increasing the clearance volume.

Let us consider the reciprocating compressor. Variable-speed compressors are available; however, unloading by holding the intake valves open is the most widely used method. Large compressors have two or more cylinders, and each cylinder can be unloaded independent of the others. For example, if one of two cylinders is unloaded, the capacity of the compressor would be decreased by about one-half. This control action is initiated by sensing the suction pressure and carried out by an electric solenoid. Capacity may also be controlled by varying the clearance volume.

As the system becomes larger, two or more compressors may be used, each with capacity control, for further flexibility in matching the refrigeration capacity to the cooling load.

Condensers and Evaporators

Aside from the compressors, the heat exchangers (condensers and evaporators) are major cost items in a refrigerating system and very often take up the most space. Their proper configuration and capacity are critical to proper performance. Heat exchangers have been discussed in some detail in Chapter 14 in regard to their heat exchange and pressure drop characteristics.

Condensers

The condenser is a heat exchanger that usually rejects all the heat from the refrigeration system. This includes not only the heat absorbed by the evaporator but also the energy input to the compressor. The condenser accepts hot, high-pressure refrigerant, usually a superheated gas, from the compressor and rejects heat from the gas to some cooler substance, usually air or water. As energy is removed from the gas it condenses, and this condensate is drained so that it may continue its path back through the expansion valve or capillary to the evaporator.

Condensers may be water cooled, air cooled, or evaporative (air and water cooled). Important design criteria for condensers is given in the *ASHRAE Handbook, HVAC Systems and Equipment Volume* (4). ARI Standard 450 gives industry criteria for standard equipment, standard safety provisions, marking, fouling factors, and recommended rating points for water-cooled condensers (5). Methods of testing for rating water-cooled refrigerant condensers is given in ASHRAE Standard 22 (6). Rating of remote mechanical-draft air-cooled refrigerant condensers are covered by ARI Standard 460 (7) and ASHRAE Standard 20 (8). Design criteria, use of materials, and testing for safety are given in ANSI/ASHRAE Standard 15 (1). As pressure vessels most condensers must meet the requirements of national, state, and local codes, which are often patterned from the ASME Boiler and Pressure Vessel Code, Unfired Pressure Vessels, Section VIII.

Evaporators

Where an expansion valve or capillary tube is used, the refrigerant is usually received in the evaporator in a two-phase, low-temperature condition, having partially evaporated in the throttling process. Energy is picked up in the evaporator by heat transfer from a medium at some slightly higher temperature, causing the refrigerant to evaporate. Most evaporators are designed and controlled to bring the refrigerant to a small degree of superheat as it leaves the evaporator so as to protect the compressor downstream from the damaging effects of liquid. The medium transferring heat to the evaporator may be the air stream to be cooled (*direct-expansion* or *DX coils*) or may be water or brine, as in the case of chillers. Where a small temperature difference is required between the refrigerant and the medium being cooled, a flooded evaporator is sometimes used. In this case the evaporator coil is supplied refrigerant liquid, which circulates from the coil to a surge tank, where refrigerant vapor is drawn into the compressor suction line. Circulation between the surge tank and the evaporator coil may be controlled by a thermosyphon effect or by forced pump circulation.

Refrigeration Control Valves

The most important component in the vapor compression cycle from the standpoint of system control is the expansion valve. Proper manipulation of the expansion valve either manually or automatically can go a long way toward optimizing system performance; other devices are sometimes required, however, to control evaporator temperature or compressor load. The most common control valves are described in the following paragraphs.

Thermostatic Expansion Valve

The thermostatic expansion valve positions the valve spool to admit the refrigerant as required by evaporator load (Fig. 15-16). A bulb containing a small refrigerant charge is connected by a small tube to the chamber above the valve diaphragm. The bulb is

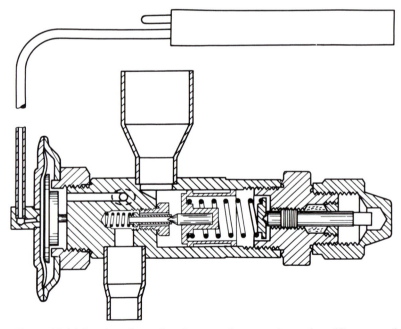

Figure 15-16 Interior view of a thermostatic expansion valve. (Courtesy of Sporlan Valve Company, St. Louis, MO.)

clamped to the refrigerant line where refrigerant leaves the evaporator. Spring pressure tends to close the valve, whereas bulb pressure tends to open it. The bulb is essentially a temperature-sensing element, and several degrees superheat of the refrigerant leaving the evaporator are required before the expansion valve will open. As the load on the evaporator increases, superheat will increase and the valve opens to allow a larger refrigerant flow rate. As the load decreases, the valve will close, maintaining a super-heated vapor at the evaporator exit. This action protects the compressor from liquid slugging and decreases the compressor load as evaporator load decreases. Most thermo-static expansion valves are adjustable. About 5 to 10 F (3 to 5 C) superheat is usually maintained.

It is possible for the thermostatic expansion valve to overload the compressor as evaporator load continues to increase. To help prevent this condition a liquid–vapor charge may be used in the bulb. The charge is such that all the liquid is evaporated at some predetermined temperature. Very little bulb pressure rise will result with a temperature rise above this point, and the valve will not open significantly with further increase in evaporator load. Thus, the load on the compressor will be limited.

When an evaporator is subjected to a large variation in load, the pressure drop of the refrigerant as it flows through the coil will vary. The refrigerant saturation temperature will therefore vary at the coil exit, and different amounts of superheat will result at light and heavy loads. This condition is controlled by the use of an external equalizer (Fig. 15-17). The valve then senses pressure at about the same point as it senses temperature, and the superheat will be correct for whatever pressure exists in the suction line. The external equalizer must be used when one expansion valve feeds parallel circuits, as shown in Fig. 15-17.

Any number of evaporators may be operated in parallel on the same compressor where a thermostatic expansion valve controls the feed to each one. Each valve controls the flow as required by that evaporator.

The thermostatic expansion valve must be selected to match the refrigeration load, the refrigerant type, the pressure difference across the valve, and the refrigerant temperature. The valves may be adjusted but do not have sufficient range to work well with different refrigerants.

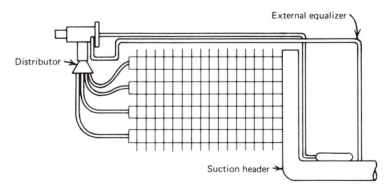

Figure 15-17 External equalizer with refrigerant distributor feeding parallel circuits.

Capillary Tube

The capillary tube is not a valve but effectively replaces the expansion valve in many applications. It is a long thin tube placed between the condenser and the evaporator. The small diameter and relatively large length of the tube produces a large pressure drop. Effective control of the system results because the tube allows the flow of liquid more readily than vapor. Although the capillary tube operates most efficiently at one particular set of conditions, there is only a slight loss in efficiency at off-design conditions in small systems.

The main advantage of the capillary tube is its simplicity and low cost. Being a passive device it is not subject to wear. However, the very small bore of the tube is subject to plugging if precautions are not taken to maintain a clean system. Moisture can also cause plugging due to ice formation.

Subcooled liquid refrigerant enters a properly sized capillary tube as shown in Fig. 15-18. Pressure loss because of friction causes the refrigerant to reach saturation at some point in the tube. At that point pressure loss becomes much more pronounced as vapor forms and the refrigerant temperature decreases rapidly to the tube exit. Critical flow conditions may occur at the tube exit with a pronounced pressure loss at that point. The amount of refrigerant flowing through the tube depends on the overall pressure difference between the condenser and the evaporator and the saturation pressure of the refrigerant entering the tube. Therefore, a change in evaporator load or environmental conditions that increase the pressure difference or decrease the saturation pressure will cause an increase in refrigerant flow rate. Fortunately an increase in load increases pressure difference, and a decrease in condensing medium temperature decreases the pressure difference. This partially explains the favorable control characteristics of the capillary tube.

Although there are calculation procedures for the sizing of capillary tubes, the final selection must be made by trial and error. The equipment volume of the *ASHRAE Handbook* (9) contains charts for preliminary selection of capillary tubes.

The charging of capillary tube systems is rather critical. This is because part of the passive control of the system results from the flooding of the condenser at part-load conditions, which acts to decrease the condenser size, decrease subcooling, and increase the saturation pressure at the capillary inlet. If other than the proper charge is used, this

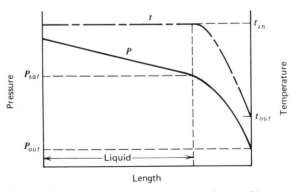

Figure 15-18 Pressure and temperature of the refrigerant in a capillary tube.

important effect is lost. Too much refrigerant can also cause compressor flooding during the off cycle and result in valve damage on startup.

Evaporator Pressure Regulator

The purpose of an evaporator pressure regulator is to maintain a relatively constant minimum pressure in the evaporator. Because most of the evaporator surface is subjected to two-phase refrigerant, a constant minimum temperature will also be maintained. Evaporator pressure is sensed internally and balanced by a spring-loaded diaphragm. When evaporator pressure falls below a set value, the valve will close, restricting the flow of refrigerant so that evaporator pressure will rise.

The main applications of evaporator pressure regulators are to set a minimum evaporator temperature to permit the use of different evaporators at different pressures on the same compressor.

Suction Pressure Regulator

The suction pressure regulator has the function of limiting the maximum pressure at the compressor suction. Since the compressor load is determined by suction pressure, the suction pressure regulator is a load-limiting device. This valve functions very much like the evaporator pressure regulator except that it senses compressor suction pressure. The suction pressure regulator also reduces the compressor load during the startup period because the valve will remain closed until suction pressure is reduced to a set pressure.

15-5   THE REAL SINGLE-STAGE CYCLE

Figure 15-19 illustrates a practical single-stage cycle schematically. The practical aspects of the compressor were discussed in a previous section and are shown again as state points 3 and 4. Other factors cause the complete cycle to deviate from the theoretical. These are the pressure losses in all connecting tubing that increase the power requirement for the cycle and the heat transfer to and from the various components. Heat transfer will generally be away from the system on the high-pressure side of the cycle and will improve cycle performance. Exposed surfaces on the low-pressure side of the cycle will generally be at a lower temperature than the environment, and any heat transfer will generally degrade cycle performance. As shown in Fig. 15-19, $\dot{q}_{3'3}$ increases compressor load and $\dot{q}_{44'}$ decreases compressor load.

It should be mentioned that the condenser and evaporator heat-transfer surfaces have pressure losses that also contribute to the overall compressor power requirement. The following example illustrates a complete cycle analysis.

EXAMPLE 15-5

A 10-ton refrigeration unit operates on R-22, has a single-stage reciprocating compressor with a clearance volume of 3 percent, and operates at 1725 rpm. The refrigerant leaves the evaporator at 70 psia and 40 F. Pressure loss in the evaporator is 5 psi. Pressure loss in the suction valve is 2 psi, and there is 20 F superheat at the beginning of the compression stroke. The compressor discharge pressure is 300 psia, a pressure loss

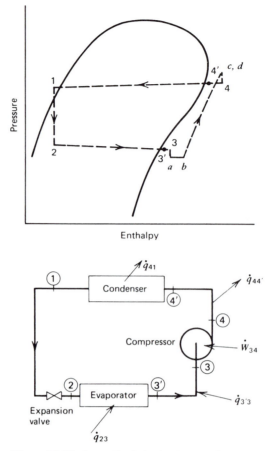

Figure 15-19 A practical single-stage cycle.

of 5 psi exists in the discharge valve, and the vapor is desuperheated 10 F between the compressor discharge and the condenser inlet. Neglect pressure loss in tubing between the compressor and condenser, but the condenser pressure loss is 5 psi. The liquid has a temperature of 110 F at the expansion valve. Determine the following: (a) compressor volumetric efficiency, (b) compressor piston displacement, (c) shaft power input with mechanical efficiency of 75 percent, (d) heat rejected in the condenser, and (e) power input to compressor per ton of cooling effect.

SOLUTION

Most of the state points as shown in Fig. 15-19 may be located from the given data.

$$P_1 = 290 \text{ psia}: \qquad T_1 = 110 \text{ F}; \qquad i_2 = i_1$$

$$P_{3'} = 70 \text{ psia}; \qquad t_{3'} = 40 \text{ F}; \qquad P_2 = P_{3'} + 5 = 75 \text{ psia}$$

$$P_a = P_b = P_{3'} - 2 = 68 \text{ psia}; \qquad t_b = 50 \text{ F}$$

$$P_c = P_d = 300 \text{ psia}; \qquad P_4 = P_{4'} = P_c - 5 = 295 \text{ psia}$$

States *c*, *d*, and 4 may be completely determined following the compressor and evaporator analysis. From Chart 5, $v_3 = 0.78$ ft³/lbm and $v_b = 0.81$ ft³/lbm. Assuming that $n = k = 1.16$, the volumetric efficiency may be computed from Eq. 15-8 as

$$\eta_v = \left[1 + 0.03 - (0.03) \left(\frac{300}{68} \right)^{1/1.16} \right] \frac{0.78}{0.81} = 0.89$$

For the evaporator

$$\dot{q}_e = \dot{m}(i_{3'} - i_2) = \dot{m}(i_{3'} - i_1)$$

or

$$\dot{m} = \frac{\dot{q}_e}{(i_{3'} - i_1)}$$

From Chart 5, $i_{3'} = 109$ Btu/lbm, $i_1 = 41$ Btu/lbm

$$\dot{m} = \frac{10(12,000)}{(109 - 41)} = 1765 \text{ lbm/hr}$$

From Eq. 15-10

$$\eta_v = \frac{\dot{m} v_3}{PD} \quad \text{or} \quad PD = \frac{\dot{m} v_3}{\eta_v}$$

$$PD = \frac{1765(0.78)}{0.89} = 1547 \text{ ft}^3/\text{hr}$$

or

$$\frac{PD}{\text{cycle}} = 25.8 \text{ in.}^3$$

The work per pound of refrigerant may be computed from Eq. 15-12:

$$w = \frac{n}{n-1} P_b v_b \left[\left(\frac{P_c}{P_b} \right)^{(n-1)/n} - 1 \right]$$

$$w = \frac{1.16}{(1.16 - 1)} 68(144)(0.81) \left[\left(\frac{300}{68} \right)^{(1.16 - 1)/1.16} - 1 \right]$$

$$w = 13,064 \text{ (ft-lbf)/lbm}$$

$$\dot{W} = \dot{m} w = \frac{13,064(1765)}{778} = 29,638 \text{ Btu/hr}$$

$$\dot{W}_{sh} = \frac{\dot{W}}{\eta_m} = 39,517 \text{ Btu/hr} = 15.5 \text{ hp}$$

The heat rejected in the condenser is given by

$$\dot{q}_c = \dot{m}(i_1 - i_{4'})$$

However, state 4' has not been completely determined. Since the polytropic exponent was assumed equal to k, the isentropic compression process follows a line of constant entropy on Chart 5. Then $t_c = t_d = 188\,\mathrm{F}$, $t_4 \approx 185\,\mathrm{F}$, and $t_{4'} = t_4 - 10 = 175\,\mathrm{F}$. Then $i_{4'} = 125$ Btu/lbm.

$$\dot{q}_c = 1765(41 - 125) = -148,000 \text{ Btu/hr}$$

The total power input to the compressor per ton of cooling effect is

$$\frac{\mathrm{hp}}{\mathrm{ton}} = \frac{15.5}{10} = 1.55$$

Refrigerant Piping

The sizing of the piping for a refrigeration system is a very important step in the overall design process. Although the process is quite similar to that outlined in Chapter 10 for liquids, one other criterion must be considered. That is the need to ensure transport of oil from the system back to the compressor. Most of the oil resides in the compressor crankcase, but a small amount may travel out into the system, especially during startup. Where liquid is present between the condenser and the evaporator, the oil moves along with the refrigerant quite readily. However, where only vapor is flowing, movement of the oil is dependent on the refrigerant velocity. This is especially true where the vapor is traveling vertically upward.

Refrigerant pipe sizing has become a science to itself, and many manufacturers publish procedures for use of their distributors. Reference 10 should be consulted for further details.

System Control

Control of the reciprocating compressor in response to the load was discussed in Section 15-4. Although this is probably the most important aspect of controlling the complete system, some other factors should be mentioned. The compressor may be gradually unloaded but eventually it will have to be stopped when the load is zero. This may be done by simply turning off the power; however, with large reciprocating compressors, this will lead to problems on startup due to possible flooding of the compressor by liquid refrigerant. Therefore, shutdown is usually done on a *pump down cycle.*

Referring to Figure 15-20, when the thermostat is satisfied the solenoid valve closes, stopping the flow of refrigerant to the expansion valve. The compressor continues to run, reducing the pressure in the system between the solenoid valve and the suction side of the compressor, while compressed hot vapor is condensed and stored in the receiver. When the suction pressure reaches the set point of the low-pressure sensor, the controller disconnects the power to the compressor. Later, when the thermostat calls for

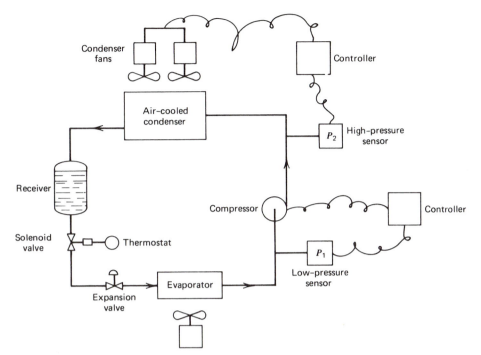

Figure 15-20 Refrigeration system pump-down and condenser control.

cooling, the solenoid valve opens and refrigerant flows toward the compressor with a consequent rise in pressure. When the pressure reaches the low-pressure switch set point, the controller starts the compressor. During the off cycle there is only a small amount of refrigerant in the system on the suction side of the compressor, and since the low-pressure switch is set for a relatively low pressure, the compressor starts under a relatively light load. After it starts, the unloading system takes over control of the compressor.

Another aspect of system control relates to the air-cooled condenser. The condenser fan is controlled to start and stop as the compressor starts and stops. This is usually adequate for small systems at normal outdoor temperatures during the summer. However, when the outdoor ambient becomes low, the condensing pressure may be so low as to cause operational problems. Therefore, a system designed for low ambient operation must be provided with *head pressure control*. This simply reduces the flow of air over the condenser in response to the compressor discharge pressure. Inlet dampers may be used, which are modulated to maintain some minimum head pressure. Another approach is to use two or more condenser fans where one or more of the fans may be cycled off to maintain the head pressure. Either of these methods may be controlled as shown in Fig. 15-20.

The reader is referred to reference 9 for information relating to the control of large refrigerating systems using centrifugal or absorption machines.

The following example illustrates the use of the compressor performance curves shown in Figs. 15-8, 15-11 and 15-13.

EXAMPLE 15-6

Construct the pressure–enthalpy diagram for a system using the compressor of Fig. 15-8 with an evaporating temperature of 35 F and a condensing temperature of 115 F. Subcooling and superheat are as specified in Fig. 15-8. Neglect system pressure losses and piping heat losses and gains.

SOLUTION

Chart 5 is used to obtain thermodynamic properties and to locate the isotherms for 35 F and 115 F, respectively, as shown in Fig. 15-21. The subcooling is 10 F; therefore, state 1 is determined by the 115 F and 105 F isotherms at the condenser pressure. The superheat is 20 F and state 3 is located on the 55 F isotherm at the evaporator pressure. As a result of the throttling process, state 2 is located on the 35 F isotherm where $i_2 = i_1$. Since the data of Fig. 15-8 are for a real compressor, the process from 3 to 4 is not isentropic and some other method will have to be used to locate state 4. This can be done by computing the heat transfer to the refrigerant in process 2–3 and the work done on the refrigerant in process 3–4:

$$\dot{q}_{23} = \dot{m}(i_3 - i_2)$$

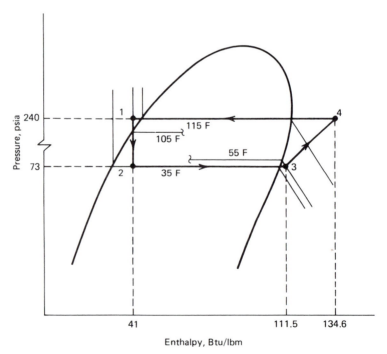

Figure 15-21 Pressure–enthalpy diagram for Example 15-6.

and

$$\dot{m} = \frac{\dot{q}_{23}}{(i_3 - i_2)}$$

From Chart 5 $i_1 = i_2 = 41$ Btu/lbm, $i_3 = 111.5$ Btu/lbm, and $\dot{q}_{23} = 130,000$ Btu/hr from Fig. 15-8. Then

$$\dot{m} = \frac{130,000}{(111.5 - 41)} = 1844 \text{ lbm/hr}$$

The compressor power is 12.5 kW or 42,650 Btu/hr from Fig. 15-8. The heat rejected from the refrigerant in the condenser is then

$$\dot{q}_{41} = \dot{q}_{23} + \dot{w}_{34} = 130,000 + 42,650 = 172,650 \text{ Btu/hr}$$

Also

$$\dot{q}_{41} = \dot{m}(i_4 - i_1)$$

and

$$i_4 = i_1 + \frac{\dot{q}_{41}}{\dot{m}} = 41 + \left(\frac{172,650}{1844}\right) = 134.6 \text{ Btu/lbm}$$

The complete cycle is thus determined in Fig. 15-21. Note an increase in entropy for the compressor of about 0.016 Btu/(lbm-R).

15-6   ABSORPTION REFRIGERATION

The mechanical vapor compression refrigeration system described in the previous sections is an efficient and practical method. However, the required energy input is shaft work or power, which is high-grade energy and expensive. The relatively large amount of work required is because of the compression of vapor that has a large volume.

It is possible to replace the vapor-compression process with a series of processes where the refrigerant vapor is absorbed by a liquid and then the liquid solution is pumped to the higher pressure. Although the shaft work required is much less, large quantities of heat must be transferred to the system. Figure 15-22 illustrates the essential features of the absorption cycle. Note that the compressor of the vapor compression cycle has been replaced by the generator, absorber, and pump, and that the solution of refrigerant and absorbent circulates through these elements. The refrigerant alone flows through the condenser, expansion valve, and evaporator. Heat is transferred to the solution in the generator, and the refrigerant vapor is separated from the solution, whereas the opposite process occurs in the absorber. The ease with which these processes may be carried out in practice depends on the refrigerant–absorbent pair and the many refinements to the cycle required in actual practice. These factors will be discussed later. The overall performance of the absorption cycle in terms of refrigerating effect per unit of energy input is generally poor; however, waste heat such as that

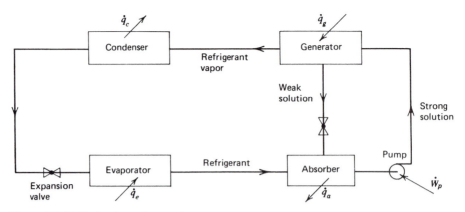

Figure 15-22 Basic absorption cycle.

rejected from a power plant can often be used to achieve better overall energy conservation.

Properties and Processes of Homogeneous Binary Mixtures

A homogeneous mixture is uniform in composition and cannot be separated into its constituents by pure mechanical means. The properties such as pressure, density, and temperature are uniform throughout the mixture. The thermodynamics of binary mixtures is extensively covered by Bosnjakovic (11).

The thermodynamic state of a mixture cannot be specified by two independent properties as may be done with a pure substance. The composition of the mixture as described by the *concentration x*, the ratio of the mass of one constituent to the mass of the mixture, is required in addition to two other independent properties such as pressure and temperature.

The miscibility of a mixture is an especially important characteristic for an absorption system. A mixture is miscible if it does not separate after mixing. A miscible mixture is homogeneous. Some mixtures may not be miscible under all conditions, with temperature being the main property influencing miscibility. Oil–refrigerant 22 is miscible at higher temperatures but nonmiscible at low temperatures. Binary mixtures for an absorption refrigeration cycle must be completely miscible in both the liquid and vapor phases with no miscibility gaps.

The behavior of a binary mixture in the vicinity of the saturation region is important to absorption refrigeration. The use of imaginary experiments and a temperature versus concentration diagram is helpful to understand this behavior. Figure 15-23*a* shows a piston–cylinder arrangement containing a binary mixture in the liquid phase with a concentration of material B, x_1. The piston has a fixed mass and is frictionless, so that the pressure of the mixture is always constant. Figure 15-23*c* is the t–x diagram for the mixture with the initial state shown as point 1. As heat is slowly transferred to the mixture, the temperature will rise, and at point 2 vapor will begin to form and collect under the piston as shown in Fig. 15-23*b*. If the experiment is stopped at some point above this temperature and the concentrations of the vapor and liquid determined, a rather surprising result is observed. For example, the state of the liquid is at point 3 while the state of the vapor is at point 4 and the concentration of material B in the vapor

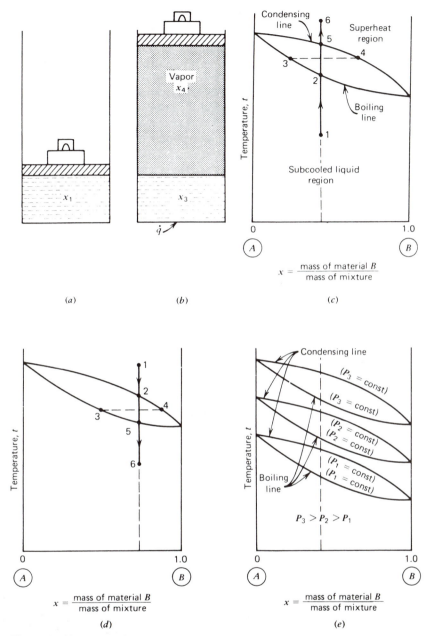

Figure 15-23 Evaporation and condensation characteristics for a homogeneous binary
mixture. (*a*) Subcooled liquid, (*b*) Two-phase mixture, (*c*) *t–x* diagram,
(*d*) *t–x* diagram, (*e*) *t–x* diagram

x_4 is larger than in the liquid x_3. We shall see later that this phenomenon is quite helpful
in the absorption cycle.

Continued heating of the mixture will gradually vaporize all the liquid (state point 5)
with concentration $x_5 = x_1$. Further heating will superheat the vapor to point 6. When
the superheated vapor is cooled at constant pressure, the entire process will be reversed,

as shown in Fig. 15-23*d*. Additional experiments carried out at different concentrations but at the same pressure establish the boiling and condensing lines shown in Figs. 15-23*c* and 15-23*d*. If the pressure is changed and the experiments repeated at various concentrations, the boiling and condensing lines will be displaced as shown in Fig. 15-23*e*. Unlike a pure substance, binary mixtures do not have a single saturation temperature for each pressure, because the saturation temperatures depend on the concentration.

A more useful representation of the properties of a binary mixture is the enthalpy–concentration diagram. Figure 15-24 is a schematic i–x diagram including the liquid and vapor regions for a homogeneous binary mixture. The condensing and boiling lines for a given pressure do not converge at $x = 0$ and $x = 1$ but are separated by a distance proportional to the enthalpy of vaporization of each constituent. Lines of constant temperature are shown in the liquid and vapor regions but not in the saturation region. When the boiling and condensing lines for more than one pressure are shown on one chart, it becomes difficult to distinguish the proper isotherms for each pressure in the saturation region. However, these isotherms may be located as needed.

Chart 2 is an enthalpy–concentration diagram for an ammonia–water mixture. Chart 2 covers the subcooled and saturation region but does not have temperature lines in the vapor region. Lines of constant temperature in the saturation region are located using the equilibrium construction lines. A vertical line is drawn upward from the saturated

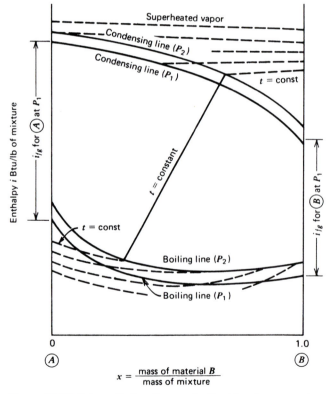

Figure 15-24 Schematic i–x diagram for liquid and vapor regions for a homogeneous binary mixture.

liquid state to the equilibrium construction line for the same pressure. From that point a horizontal line is drawn to the saturated vapor line for the same pressure. This intersection gives the saturated vapor-state point. The isotherm for the given temperature is the straight line connecting the saturated liquid and vapor states.

The most common thermodynamic processes involved in industrial systems and absorption refrigeration are described. These are the adiabatic and nonadiabatic mixing of two streams, heating and cooling including vaporization and condensation, and throttling. An understanding of these processes will help in understanding the absorption cycle. The i–x diagram will be quite useful in the solution of these problems.

Adiabatic Mixing of Two Streams

Figure 15-25 shows a mixing chamber where two binary streams of different concentration and enthalpy mix in a steady-flow process. Determination of the state of the stream leaving the mixing chamber involves energy and mass balances on the control volume defined by the mixing chamber. The energy balance is

$$i_1 \dot{m}_1 + i_2 \dot{m}_2 = i_3 \dot{m}_3 \qquad (15\text{-}14)$$

while the overall mass balance is given by

$$\dot{m}_1 + \dot{m}_2 = \dot{m}_3 \qquad (15\text{-}15)$$

and the mass balance for one constituent is

$$\dot{m}_1 x_1 + \dot{m}_2 x_2 = \dot{m}_3 x_3 \qquad (15\text{-}16)$$

Elimination of $\dot{m}_3$ from Eqs. 15-14, 15-15, and 15-16 gives

$$\frac{\dot{m}_1}{\dot{m}_2} = \frac{i_2 - i_3}{i_3 - i_1} = \frac{x_2 - x_3}{x_3 - x_1} \qquad (15\text{-}17)$$

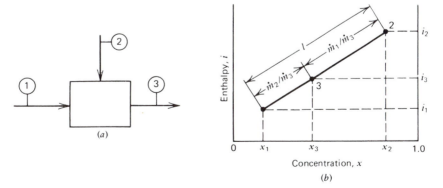

Figure 15-25   Steady-flow adiabatic mixing process.

Equation 15-17 defines a straight line on the i–x diagram as shown in Fig. 15-25, and state 3 must lie on this line. It may be shown that

$$x_3 = x_1 + \frac{\dot{m}_2}{\dot{m}_3}(x_2 - x_1) \tag{15-18}$$

$$i_3 = i_1 + \frac{\dot{m}_2}{\dot{m}_3}(i_2 - i_1) \tag{15-19}$$

The i–x diagram may be used to advantage to solve mixing problems, but the procedure is somewhat involved when the final state is in the mixture region.

EXAMPLE 15-7

Twenty lbm per minute of liquid water–ammonia solution at 150 psia, 220 F, and concentration of 0.25 lbm ammonia per lbm of solution is mixed in a steady-flow adiabatic process with 10 lbm per minute of saturated water–ammonia solution at 150 psia and 100 F. Determine the enthalpy, concentration, and temperature of the mixture.

SOLUTION

Chart 2 will be used to carry out the solution and Fig. 15-26 shows the procedure. State 1 is a subcooled condition located on the diagram at $t = 220$ F and $x = 0.25$. State 2 is a saturated liquid and is located at the intersection of the 150 psia boiling line and the 100

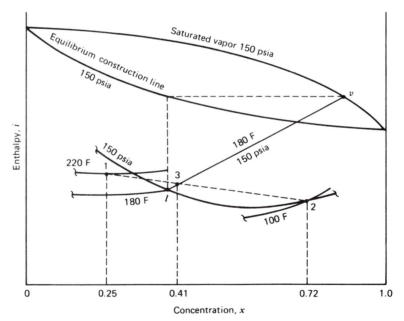

Figure 15-26 Schematic i–x diagram for Example 15-7.

F temperature line. States 1 and 2 are joined by a straight line. State 3 is located on the connecting line and is determined by the concentration x_3 or enthalpy i_3. Using Eq. 15-18, we get

$$x_3 = x_1 + \frac{\dot{m}_2}{\dot{m}_3}(x_2 - x_1) = 0.25 + \frac{10}{30}(0.72 - 0.25)$$

$$x_3 = 0.41 \text{ lbm ammonia/lbm mixture}$$

Because the resulting mixture state lies within the saturation region, it is a mixture of vapor and liquid. To determine the temperature t_3 a graphical trial-and-error procedure shown in Fig. 15-26 is used. The fractional proportions of liquid and vapor in the mixture may also be determined from Chart 2 in a graphical manner.

$$\frac{\dot{m}_v}{\dot{m}_3} = \frac{\overline{l3}}{\overline{lv}} = 0.038 \tag{15-20}$$

Therefore, the mixture is 3.8 percent vapor and 96.2 percent liquid.

Mixing of Two Streams with Heat Exchange

This type of process is quite common and occurs in the absorber of the absorption refrigeration device. In this case shown in Fig. 15-27 the energy balance becomes

$$\dot{m}_1 i_1 + \dot{m}_2 i_2 = \dot{m}_3 i_3 + \dot{q} \tag{15-21}$$

The mass balance equations are identical to those for adiabatic mixing

$$\dot{m}_1 + \dot{m}_2 = \dot{m}_3 \tag{15-15}$$

$$\dot{m}_1 x_1 + \dot{m}_2 x_2 = \dot{m}_3 x_3 \tag{15-16}$$

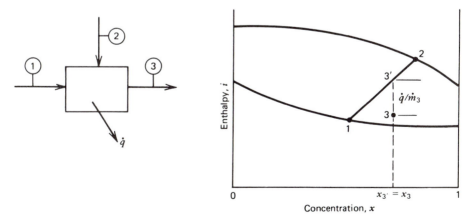

Figure 15-27 Steady-flow mixing of two streams with heat transfer.

The equation for the concentration x_3 is the same as Eq. 15-18, however, the enthalpy i_3 is given by

$$i_3 = i_1 + \frac{\dot{m}_2}{\dot{m}_3}(i_2 - i_1) - \frac{\dot{q}}{\dot{m}_3} \tag{15-22}$$

Equation 15-22 differs from Eq. 15-16 only in the last term. The significance of this is shown in the i–x diagram of Fig. 15-27. Point 3′ represents a state that would occur with adiabatic mixing. Point 3 is located a distance $\dot{q}/\dot{m}_3$ directly below point 3′ because $x_{3'} = x_3$ and heat is removed. If heat were added, then point 3 would be above point 3′.

EXAMPLE 15-8

Five lbm per minute of saturated liquid aqua–ammonia at 100 psia and 220 F are mixed with 10 lbm/min of saturated vapor aqua–ammonia at 100 psia and 220 F. Heat transfer from the mixture is 4200 Btu/min. Find the enthalpy, temperature, and concentration of the final mixture.

SOLUTION

The concentration of the mixture x_3 is given by Eq. 15-18. Using Chart 2, we get

$$x_3 = 0.214 + \frac{10}{15}(0.845 - 0.214) = 0.635 \frac{\text{lbm ammonia}}{\text{lbm of mixture}}$$

The total mass flow rate $\dot{m}_3$ is the sum of the two inlet flow rates and

$$\frac{\dot{q}}{\dot{m}_3} = \frac{4200}{15} = 280 \text{ Btu/lbm}$$

Then using Chart 2, point 3 is located at $x_3 = x_{3'} = 0.635$ and

$$i_3 = i_{3'} - \frac{\dot{q}}{\dot{m}_3} = 578 - 280 = 298 \text{ Btu/lbm}$$

Using a straightedge and the equilibrium construction line, we find that the final temperature t_3 is about 148 F.

Heating and Cooling Processes

The absorption refrigeration cycle requires vaporization and condensation processes. For ammonia–water systems rectification is necessary to produce high-purity ammonia vapor. This is done by alternate heating and cooling processes. Figure 15-28 shows a simplified arrangement to accomplish this. To better visualize the processes, the liquid and vapor phases are separated following each heat exchanger.

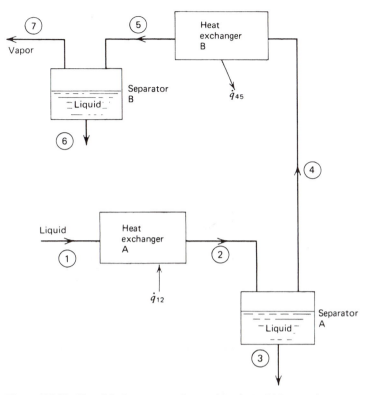

Figure 15-28 Simplified apparatus for rectification of binary mixture.

For heat exchanger A

$$\dot{q}_{12} = \dot{m}_1(i_2 - i_1)$$

$$\dot{m}_1 = \dot{m}_2$$

$$x_1 = x_2$$

For separator A,

$$\dot{m}_2 i_2 = \dot{m}_3 i_3 + \dot{m}_4 i_4$$

$$\dot{m}_2 = \dot{m}_3 + \dot{m}_4$$

$$\dot{m}_2 x_2 = \dot{m}_3 x_3 + \dot{m}_4 x_4$$

Combination of the foregoing energy- and mass-balance equations yields

$$\frac{\dot{m}_3}{\dot{m}_2} = \frac{x_4 - x_2}{x_4 - x_3} = \frac{i_4 - i_2}{i_4 - i_3} \qquad \textbf{(15-23)}$$

$$\frac{\dot{m}_4}{\dot{m}_2} = \frac{x_2 - x_3}{x_4 - x_3} = \frac{i_2 - i_3}{i_4 - i_3} \qquad \textbf{(15-24)}$$

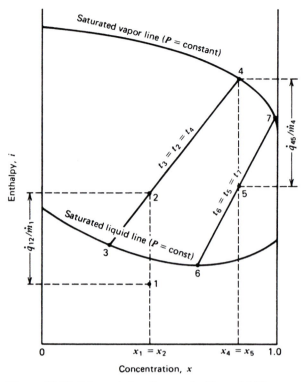

Figure 15-29 Schematic $i-x$ diagram for Figure 15-27.

Figure 15-29 shows the state points 1, 2, 3, and 4 on the $i-x$ diagram. The heat transfer $\dot{q}_{12}/\dot{m}_1$ may be determined graphically as shown, and the fractional components for the separator may be determined directly from the diagram.

$$\frac{\dot{m}_3}{\dot{m}_2} = \frac{\overline{24}}{\overline{34}} \quad \text{and} \quad \frac{\dot{m}_4}{\dot{m}_2} = \frac{\overline{32}}{\overline{34}}$$

The symbol $\overline{24}$ refers to the length of the line connecting points 2 and 4. Heat exchanger *B* and separator *B* may be analyzed in exactly the same way. The state points are shown in Fig. 15-29. Notice that the vapor at state 7 is almost 100 percent ammonia. This is necessary for the aqua-ammonia absorption cycle to operate efficiently. In actual practice the simple arrangement of Fig. 15-28 is inadequate for the separation of a binary mixture, and a rectifying column must be introduced between the two heat exchangers. Heat exchanger *A* is called the generator and heat exchanger *B* is referred to as the *dephlegmator*. Analysis of the dephlegmator is somewhat involved and the reader is referred to reference 12.

Throttling Process

The throttling process occurs in most refrigeration cycles. A throttling valve is shown schematically in Fig. 15-30a. Although evaporation occurs in throttling, and the

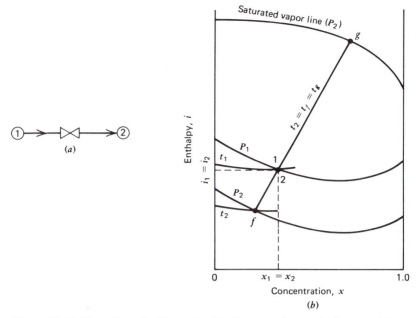

Figure 15-30 Throttling of a binary liquid mixture under steady-flow conditions. (*a*) Throttle valve, (*b*) *i–x* diagram.

temperature of the mixture changes, an energy balance gives $i_2 = i_1$ and the concentration remains constant $x_2 = x_1$. The state points 1 and 2 are identical on the *i–x* diagram of Fig. 15-30*b*; however, it should be noted that state 1 is at pressure P_1 and state 2 is at a pressure P_2. The line $\overline{f2g}$ is located by trial and error using a straightedge and the equilibrium construction line. The temperature t_2 will generally be less than t_1. The fractional components of liquid and vapor may be determined from the line segment ratios of $\overline{f2g}$.

EXAMPLE 15-9

Saturated aqua-ammonia at 100 psia, 240 F, with a concentration of 0.3 lbm ammonia/lbm of mixture is throttled in a steady-flow process to 30 psia. Find the temperature and fractions of liquid and vapor at state 2.

SOLUTION

State 1 is first located using Chart 2 and is shown schematically in Fig. 15-31. The isotherm for 240 F has been located using a straightedge. In this case state 1 is a two-phase mixture. State 2 is located at the same point as state 1 on the diagram. The isotherm that passes through point 2 for a pressure of 30 psia is then located by trial and error and is found to be 180 F as shown in Fig. 15-31. The fraction of vapor at state 2 is

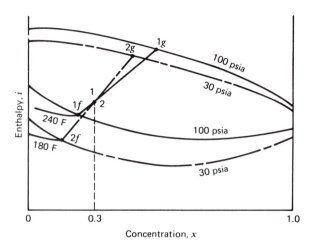

Figure 15-31 Schematic of throttling process of Example 15-9.

given by

$$\frac{\dot{m}_v}{\dot{m}} = \frac{\overline{22f}}{\overline{2f2g}} = \frac{2.2 \text{ in.}}{7.5 \text{ in.}} = 0.289 \frac{\text{lbm vapor}}{\text{lbm mixture}}$$

$$\frac{\dot{m}_f}{\dot{m}} = 1 - 0.289 = 0.711 \frac{\text{lbm liquid}}{\text{lbm mixture}}$$

15-7 THE THEORETICAL ABSORPTION REFRIGERATION SYSTEM

Figure 15-32 shows a schematic arrangement of components for a simple theoretical absorption cycle. For simplicity we shall assume that the absorbent does not vaporize in the generator; thus only the refrigerant flows through the condenser, expansion valve, and evaporator. The vapor leaving the evaporator is absorbed by the weak liquid in the absorber as heat is transferred from the mixture. The refrigerant-enriched solution is then pumped to the pressure level in the generator, where refrigerant vapor is driven off by heat transfer to the solution while the weak solution returns to the absorber by way of the intercooler.

An ideal system would be completely reversible; however, this is not theoretically possible with the system of Fig. 15-32. The Carnot cycle meets this requirement for a vapor-compression system. The maximum attainable coefficient of performance for the absorption cycle may be determined following a method developed by Bosnjakovic (11).

By the first law of thermodynamics

$$\dot{q}_a + \dot{q}_c = \dot{q}_e + \dot{q}_g + \dot{W}_p \tag{15-25}$$

where the waste heat is

$$\dot{q}_o = \dot{q}_a + \dot{q}_c \tag{15-26}$$

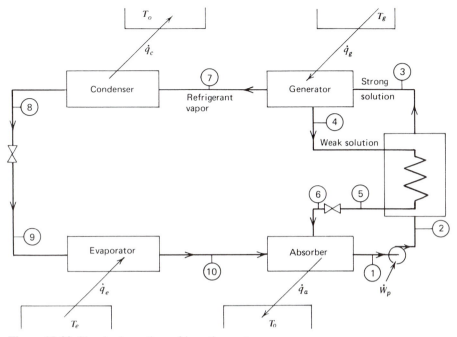

Figure 15-32 Simple absorption refrigeration system.

It will be assumed that the environment temperature T_o, the generator heating medium temperature T_g, and the refrigerated substance temperature T_e are all constant, absolute temperatures. The second law of thermodynamics requires that the net change in entropy for the system plus the surroundings must be greater than or equal to zero. Since the working fluid undergoes a cycle, its change in entropy is zero. Therefore,

$$\Delta S_{\text{total}} = \Delta S_g + \Delta S_e + \Delta S_o \geq 0 \qquad \textbf{(15-27)}$$

Because the reservoirs are internally reversible, their entropy changes may be computed as

$$\Delta S_g = -\frac{\dot{q}_g}{T_g} \qquad \Delta S_e = -\frac{\dot{q}_e}{T_e} \qquad \Delta S_o = \frac{\dot{q}_o}{T_o} \qquad \textbf{(15-28)}$$

and

$$\Delta S_{\text{total}} = -\frac{\dot{q}_g}{T_g} - \frac{\dot{q}_e}{T_e} + \frac{\dot{q}_o}{T_o} \geq 0 \qquad \textbf{(15-29)}$$

By combining Eqs. 15-25 and 15-29 it may be shown that

$$\dot{q}_g \frac{T_g - T_o}{T_g} \geq \dot{q}_e \frac{T_o - T_e}{T_e} - \dot{W}_p \qquad \textbf{(15-30)}$$

If the pump power $\dot{W}_p$ is neglected, Eq. 15-30 may be rearranged to give

$$\text{COP} = \frac{\dot{q}_e}{\dot{q}_g} \leq \frac{T_e(T_g - T_o)}{T_g(T_o - T_e)} \tag{15-31}$$

and when all the processes are reversible

$$(\text{COP})_{\text{max}} = \frac{T_e(T_g - T_o)}{T_g(T_o - T_e)} \tag{15-32}$$

Equation 15-32 shows that the maximum coefficient of performance for an absorption cycle is equal to the COP for a Carnot cycle operating between temperatures T_e and T_o multiplied by the efficiency of a Carnot engine operating between temperatures T_g and T_o. It is also shown that for a given environment temperature T_o, the COP will increase with an increase in T_g or T_e. Unfortunately, practical absorption cycles have COPs much less than that given by Eq. 15-32.

15-8   THE AQUA–AMMONIA ABSORPTION SYSTEM

The aqua–ammonia system is one of the oldest absorption refrigeration cycles. The ammonia is the refrigerant and the water is the absorbent. Because both the water and ammonia are volatile, the generator of the simple cycle must be replaced by the

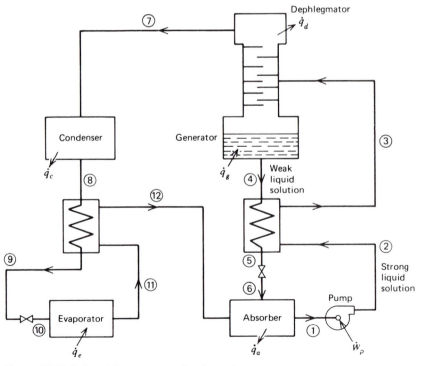

Figure 15-33 Industrial aqua-ammonia absorption system.

combination of generator, rectifying column, and dephlegmator as shown in Fig. 15-33. This is necessary to separate almost all of the water vapor from the ammonia vapor. Note that one additional heat exchanger has also been added. A complete cycle will now be analyzed making use of the basic processes previously described.

EXAMPLE 15-10

Consider the cycle of Fig. 15-33 and the following given data: condensing pressure, 200 psia; evaporating pressure, 30 psia; generator temperature, 240 F; temperature of vapor leaving dephlegmator, 130 F; and temperature of the strong solution entering the rectifying column, 200 F. The heat exchanger lowers the temperature of the liquid leaving the condenser 10 F.

States 1, 3, 4, 7, 8, and 12 are saturated. Pressure drop in components and connecting lines is negligible. The system produces 100 tons of refrigeration. Determine (a) properties P, t, x, and i for all state points of the system, (b) mass flow rate for all parts of the system, (c) power required for the pump assuming 75 percent mechanical efficiency, (d) system coefficient of performance, and (e) system refrigerating efficiency.

SOLUTION

(a) Table 15-2 is a tabulation of thermodynamic properties and flow rates for the problem. Figure 15-34 is a schematic i–x diagram showing the state points.

The given data establish all the pressures, and the temperatures t_3, t_4, and t_7 are given. States 3, 4, and 7 are saturated states and can be located on Chart 2 and the concentration and enthalpy values read. States 7 through 12 have the same concentration because no mixing occurs. It is also true that states 1, 2, and 3 have the same concentration and $x_4 = x_5 = x_6$ as well. Now states 1, 8, and 12 are saturation states and may be located on Chart 2 making use of the known pressures and concentrations and i_1, i_8, and i_{12} read. Since $t_9 = t_8 - 10$, state 9 is determined from t_9 and x_9 and i_9 are read from the chart.

Table 15-2 Properties and Flow Rates for Example 15-10

State	P psia	t F	x lbm NH$_3$/lbm	i Btu/lbm	$\dot{m}$ lbm/min
1	30	79	0.408	− 25	262.7
2	200	79	0.408	− 25.4	262.7
3	200	200	0.408	109	262.7
4	200	240	0.298	159	221.3
5	200	97	0.298	0	221.3
6	30	97	0.298	0	221.3
7	200	130	0.996	650	41.4
8	200	97	0.996	148	41.4
9	200	86	0.996	137	41.4
10	30	0	0.996	137	41.4
11	30	40	0.996	620	41.4
12	30	57	0.996	631	41.4

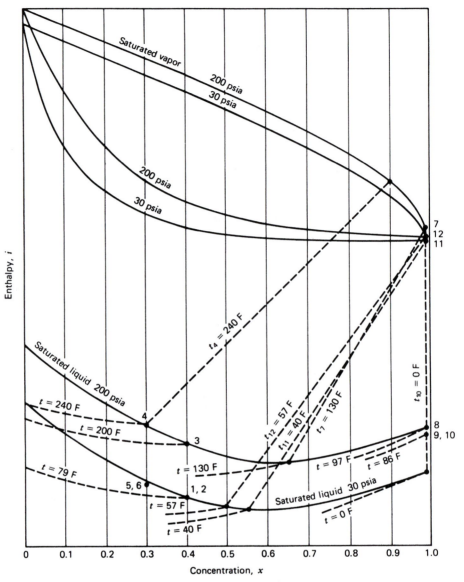

Figure 15-34 Schematic i–x diagram for Example 15-10.

(b) An energy balance between points 8 and 12 in Fig. 15-34 yields

$$\dot{m}_8 i_8 + \dot{q}_e = \dot{m}_{12} i_{12} = \dot{m}_8 i_{12}$$

$$\dot{m}_8 = \frac{\dot{q}_e}{i_{12} - i_8} = \frac{(100)(200)}{(631 - 148)} = 41.4 \text{ lbm/min}$$

From Fig. 15-33 it is obvious that $\dot{m}_7 = \dot{m}_8 = \dot{m}_9 = \dot{m}_{10} = \dot{m}_{11} = \dot{m}_{12}$.

Mass balance on the absorber gives

$$\dot{m}_{12} + \dot{m}_6 = \dot{m}_1$$

$$\dot{m}_{12}x_{12} + \dot{m}_6 x_6 = \dot{m}_1 x_1$$

and

$$\dot{m}_6 = \dot{m}_{12}\frac{(x_{12} - x_1)}{(x_1 - x_6)} = 41.4\frac{(0.996 - 0.408)}{(0.408 - 0.298)}$$

$$\dot{m}_6 = 221.3 \text{ lbm/min}$$

$$\dot{m}_1 = \dot{m}_{12} + \dot{m}_6 = 221.3 + 41.4 = 262.7 \text{ lbm/min}$$

From Fig. 15-34, $\dot{m}_1 = \dot{m}_2 = \dot{m}_3$ and $\dot{m}_4 = \dot{m}_5 = \dot{m}_6$.
   (c) The pump work may be expressed as

$$w_{12} = i_1 - i_2 = \frac{(P_1 - P_2)v_1}{J}$$

where $J = 778$ (ft-lbf)/Btu. The specific volume v_1 is an empirically determined quantity and is given in reference 10. For $x_1 = 0.408$ and $t_1 = 79$ F, $v_1 = 0.0187$ ft^3/lbm. Then

$$w_{12} = \frac{144(30 - 200)(0.0187)}{778}$$

$$w_{12} = -0.588 \text{ Btu/lbm}$$

$$\dot{W}_p = w_{12}\dot{m}_1 = (-0.588)(262.7) = -154.6 \text{ Btu/min}$$

$$i_2 = i_1 - w_{12} = -25 - (-0.588) = -24.4 \text{ Btu/lbm}$$

The enthalpy, pressure, and concentration at state 2 then establish t_2.
   The energy balance on the solution heat exchanger yields

$$\dot{m}_2 i_2 + \dot{m}_4 i_4 = \dot{m}_3 i_3 + \dot{m}_5 i_5$$

$$\dot{m}_4 = \dot{m}_5 \quad \text{and} \quad \dot{m}_2 = \dot{m}_3$$

Then

$$i_5 = i_4 - \frac{\dot{m}_2}{\dot{m}_4}(i_3 - i_2) = 159 - \frac{262.7}{221.3}[109 - (-24.4)]$$

$$i_5 = 0.64 \approx 0 \text{ Btu/lbm}$$

The temperature t_5 is then read from Chart 2 as a function of P_5, x_5, and i_5. The enthalpy $i_6 = i_5$ are coincident on Chart 2; however, state 5 is at 200 psia, whereas state 6 is at 30 psia. Since state 6 is subcooled, $t_6 = t_5$.

States 9 and 10 are coincident on Chart 2 since the process is one of throttling. Temperature t_{10} may be found using the method described in Example 15-8.

The energy balance on the evaporator gives

$$\dot{m}_{10}i_{10} + \dot{q}_e = \dot{m}_{11}i_{11}$$

$$\dot{m}_{10} = \dot{m}_{11}$$

$$i_{11} = \frac{\dot{q}_e}{\dot{m}_{10}} + i_{10}$$

$$i_{11} = \frac{(200)(100)}{41.4} + 137 = 620 \text{ Btu/lbm}$$

State 11 is a mixture of liquid and vapor, and its temperature is found from Chart 2 to be 40 F.

The pump horsepower requirement may be computed from $\dot{W}_p$ above.

$$\text{Power} = \frac{\dot{W}_p}{\eta_m} = \frac{154.6(778)}{0.75(33,000)} = 4.86 \text{ hp}$$

(d) To compute the coefficient of performance it is necessary to establish the generator heat-transfer rate. However, this cannot be done without an analysis of the complete generator–rectifier column dephlegmator unit. To make a reasonable analysis, details of the column design and experimental data are required. For purposes of simplicity it will be assumed that the generator heat-transfer rate is 115 percent of the net heat transfer for the complete rectifying unit. An energy balance gives

$$\dot{m}_3 i_3 + \dot{q}_g = \dot{m}_4 i_4 + \dot{m}_7 i_7 + \dot{q}_d$$

$$\dot{q}_{\text{net}} = (\dot{q}_g - \dot{q}_d) = \dot{m}_4 i_4 + \dot{m}_7 i_7 - \dot{m}_3 i_3$$

$$= 221.3(159) + 41.4(650) - 262.7(109)$$

$$\dot{q}_{\text{net}} = 33,462.4 \text{ Btu/min}$$

Then

$$\dot{q}_g = 1.15 \, \dot{q}_{\text{net}} = 38,480 \text{ Btu/min}$$

$$\dot{q}_d = 38,480 - 33,462 = 5020 \text{ Btu/min}$$

Neglecting the pump work

$$\text{COP} = \frac{\dot{q}_e}{\dot{q}_g} = \frac{100(200)}{38,480} = 0.52$$

(e) The maximum COP may be calculated from Eq. 15-32. It will be assumed that t_o = 80 F, t_g = 240 F, and t_e = 0 F were given.

$$(COP)_{max} = \frac{460(700 - 540)}{700(540 - 460)} = 1.31$$

and

$$\eta_R = \frac{COP}{(COP)_{max}} = \frac{0.52}{1.31} = 0.40$$

REFERENCES

1. ASHRAE Standard 15-89, "Safety Code for Mechanical Refrigeration," American Society of Heating, Refrigerating and Air-Conditioning Enginers, Inc., Atlanta, GA, 1989.
2. ASHRAE Standard 39-42, "Number Designation and Safety Classification of Refrigerants," American Society of Heating, Refrigerating and Air-Conditioning Engineers, Inc., Atlanta, GA, 1992.
3. *ASHRAE Handbook, Refrigeration Volume*, American Society of Heating, Refrigerating and Air-Conditioning Engineers, Inc., Atlanta, GA, 1990.
4. *ASHRAE Handbook, HVAC Systems and Equipment Volume,* American Society of Heating, Refrigerating and Air-Conditioning Engineers, Inc., Atlanta, GA, 1992.
5. ARI Standard 450-87, "Water-Cooled Refrigerant Condensers, Remote Type," Air Conditioning and Refrigeration Institute, Arlington, VA, 1987.
6. ASHRAE Standard 22-92, "Methods of Testing for Rating Water-Cooled Refrigerant Condensers," American Society of Heating, Refrigerating and Air-Conditioning Engineers, Inc., Atlanta, GA, 1992.
7. ARI Standard 460-87, "Remote Mechanical Draft Air-Cooled Refrigerant Condensers," Air Conditioning and Refrigeration Institute, Arlington, VA, 1987.
8. ASHRAE Standard 20-70, "Methods of Testing for Rating Remote Mechanical-Draft Air-Cooled Refrigerant Condensers," American Society of Heating, Refrigerating and Air-Conditioning Engineers, Inc., Atlanta, GA, 1970.
9. *ASHRAE Handbook, Equipment Volume*, American Society of Heating, Refrigerating and Air-Conditioning Engineers, Inc., Atlanta, GA, 1988.
10. *ASHRAE Handbook, Fundamentals Volume*, American Society of Heating, Refrigerating and Air-Conditioning Engineers, Inc., Atlanta, GA, 1989.
11. Fran Bosnjakovic, *Technical Thermodynamics*, translated by Perry L. Blackshear, Jr., Holt, Rinehart, and Winston, New York, 1965.
12. James L. Threlkeld, *Thermal Environmental Engineering*, 2nd ed., Prentice-Hall, Englewood Cliffs, NJ, 1970.

PROBLEMS

15-1. An ideal single-stage vapor compression refrigeration cycle uses R-22 as the working fluid. The condensing temperature is 110 F (43 C) and the evaporating temperature is 40 F (4.5 C). The system produces 10 tons (35.2 kW) of cooling effect. Determine the (a) coefficient of performance, (b) refrigerating efficiency, (c) hp/ton (kW input/kW of refrigeration), (d) mass flow rate of the refrigerant, (e) theoretical input to the compressor in hp (kW), and (f) theoretical piston displacement of the compressor in cfm (m³/s).

15-2. A vapor compression refrigeration cycle uses R-22 and follows the theoretical single-stage cycle. The condensing temperature is 48 C and the evaporating temperature is − 18 C. Power input to the cycle is 2.5 kW and the mass flow rate of refrigerant is 0.05 kg/s. Determine (a) heat rejected from the condenser, (b) the coefficient of performance, (c) the enthalpy at the compressor exit, and (d) the refrigerating efficiency.

15-3. An R-12 system is arranged as shown in Fig. 15-35. Compression is isentropic. Assume frictionless flow. Find the system hp/ton.

15-4. Consider a reciprocating compressor operating with R-12. Refrigerant enters the cylinder at 20 psia (138 kPa) and 20 F (−7 C), but leaves the evaporator saturated at 0.5 F (− 18 C). The vapor is discharged from the cylinder at 180 psia (1.24 MPa). Compute the volumetric efficiency for (a) a clearance factor of 0.03; (b) a clearance factor of 0.15; (c) compare the mass flow rates for parts (a) and (b); and (d) compare the power input to the compressor in parts (a) and (b).

15-5. Consider a 4-cylinder, 3-in. bore by 4-in. stroke, 800 rpm, single-acting compressor for use with R-12. The proposed operating condition for the compressor is 100 F condensing temperature and 40 F evaporating temperature. It is estimated that the refrigerant will enter the expansion valve as a saturated liquid, that vapor will leave the evaporator at a temperature of 45 F, and that vapor will enter the compressor at a temperature of 55 F. Assume a compressor volumetric efficiency of 70 percent and frictionless flow. Calculate the refrigeration capacity in tons.

15-6. Consider the compressor of Fig. 15-8. (a) Construct the pressure-enthalpy diagram for a condensing temperature of 130 F (54 C) and an evaporating temperature of 45 F (7 C). (b) What is the heat-transfer rate in the evaporator and the power input? (c) Suppose that the load on the evaporator decreases to 110,000 Btu/hr (32 kW) and find the evaporating temperature and power input. Assume that the condensing temperature remains constant.

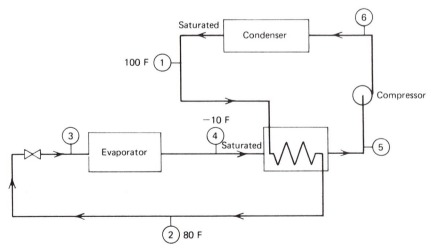

Figure 15-35 Schematic for Problem 15-3.

15-7. Refer to Problem 15-6a and sketch the capacity and power curves for 130 F (54 C) from Fig. 15-8. (a) Sketch the capacity curve for the evaporator assuming its capacity is proportional to the evaporating temperature in the ratio 4000 Btu/(hr-F) (2.1 kW/C). (b) The evaporator load decreases to 130,000 Btu/hr (38 kW) while the condensing temperature decreases to 115 F (46 C). Sketch the new evaporator-capacity curve. (c) Suppose in part (a) that the evaporator operating conditions remain fixed but the condensing temperature increases to 145 F (63 C). What will the capacity and evaporating temperature be?

15-8. A refrigeration system using the compressor described in Fig. 15-8 is designed to operate with a condensing temperature of 115 F with an evaporating temperature of 50 F when the outdoor ambient is 95 F. After the system is put into operation, the evaporator pressure is measured to be 69 psia and the power to the compressor is 12 kW. Superheat and subcooling are assumed to be as given in Fig. 15-8 and the ambient is about 95 F. (a) Estimate the condensing and evaporating temperatures, (b) compare the actual and expected performance, and (c) suggest what might be done to obtain the design conditions.

15-9. Saturated R-22 at 45 F (7 C) enters the compressor of a single-stage system. Discharge pressure is 275 psia (1.90 MPa). Suction valve pressure drop is 2 psi (13.8 kPa). Discharge valve pressure drop is 4 psi (27.6 kPa). Assume the vapor is superheated 10 F (5.6 C) in the cylinder during the intake stroke. Piston clearance is 5 percent. Determine the (a) volumetric efficiency, (b) pumping capacity in lbm/min (kg/s) for 20 ft^3/min (9.44 l/s) piston displacement, and (c) horsepower (kW) requirement if mechanical efficiency is 80 percent.

15-10. Consider the single-stage vapor-compression cycle shown in Fig. 15-36. Design conditions using R-12 are:

$$\dot{q}_L = 30{,}000 \text{ Btu/hr} \qquad P_3 = 200 \text{ psia}$$

$$P_1 = 60 \text{ psia saturated} \qquad P_3 - P_4 = 2 \text{ psi}$$

$$P_2 = 55 \text{ psia} \qquad C = 0.04$$

$$T_2 = 60 \text{ F} \qquad \eta_m = 0.90$$

$$PD = 9.4 \text{ cfm}$$

(a) Determine $\dot{W}$, $\dot{q}_H$, $\dot{m}_{12}$, and sketch the cycle on a P–i diagram. If the load $\dot{q}_L$ decreases to 24,000 Btu/hr and the system comes to equilibrium with $P_2 = 50$ psia and $T_2 = 50$ F, (b) determine $\dot{W}$, $\dot{q}_H$, $\dot{m}$ and locate the cycle on a P–i diagram.

15-11. Consider an ordinary single-stage vapor-compression air-conditioning system. Because of clogged filters the air flow over the evaporator is gradually reduced to a very low level. Explain how the evaporator and compressor will be affected if the system continues to operate.

15-12. A vapor-compression cycle is subject to short periods of very light load; it is not practical to shut the system down. During these periods of light load, moisture condenses from the air flowing over the evaporator and freezes. Suggest a modification to the system to prevent this condition.

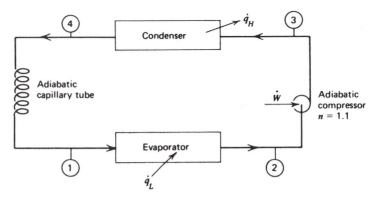

Figure 15-36 Schematic for Problem 15-10.

15-13. A vapor-compression cycle is subject to occasional overload that leads to the tripping of circuit breakers. Explain how the system can be modified to prevent compressor overload without shutting the system off.

15-14. A saturated liquid aqua–ammonia solution at 220 F and 200 psia is throttled to a pressure of 10 psia. Find (a) the temperature after the throttling process, and (b) the relative portions of liquid and vapor in the mixture after throttling.

15-15. A solution of ammonia and water at 180 F, 100 psia, and with a concentration of 0.25 lbm ammonia per lbm of solution is heated at constant pressure to a temperature of 280 F. The vapor is then separated from the liquid and cooled to a saturated liquid at 100 psia. What are the temperature and concentration of the saturated liquid?

15-16. It is proposed to use hot water at 180 F (82 C) from a solar collector system to operate a simple absorption cycle. Compute the maximum possible coefficient of performance assuming an environment temperature of 100 F (38 C) and an air-cooled evaporator with the air temperature at 75 F (24 C).

APPENDIX A

Thermodynamic Properties

Table A-1a Water at Saturation—English Units

Fahrenheit Temperature	Pressure P psia	Specific Volume ft³/lbm		Enthalpy Btu/lbm			Entropy Btu/(lbm-R)		
		Saturated Solid v_i	Saturated Vapor $v_g \times 10^{-3}$	Saturated Solid i_i	Sublimation i_{ig}	Saturated Vapor i_g	Saturated Solid s_i	Sublimation s_{ig}	Saturated Vapor s_g
32.018	0.0887	0.01747	3.302	−143.34	1218.7	1075.4	−0.292	2.479	2.187
30	0.0808	0.01747	3.607	−144.35	1218.9	1074.5	−0.294	2.489	2.195
25	0.0641	0.01746	4.506	−146.84	1219.1	1072.3	−0.299	2.515	2.216
20	0.0505	0.01745	5.655	−149.31	1219.4	1070.1	−0.304	2.542	2.238
15	0.0396	0.01745	7.13	−151.75	1219.7	1067.9	−0.309	2.569	2.260
10	0.0309	0.01744	9.04	−154.17	1219.9	1065.7	−0.314	2.597	2.283
5	0.0240	0.01743	11.52	−156.56	1220.1	1063.5	−0.320	2.626	2.306
0	0.0185	0.01743	14.77	−158.93	1220.2	1061.2	−0.325	2.655	2.330

Fahrenheit Temperature	Pressure P psia	Specific Volume ft³/lbm		Enthalpy Btu/lbm			Entropy Btu/(lbm-R)		
		Saturated Liquid v_f	Saturated Vapor v_g	Saturated Liquid i_f	Evaporation i_{fg}	Saturated Vapor i_g	Saturated Liquid s_f	Evaporation s_{fg}	Saturated Vapor s_g
32.018	0.08866	0.016022	3302	0.01	1075.4	1075.4	0.00000	2.1869	2.1869
36	0.10397	0.016021	2839	4.00	1073.1	1077.1	0.00810	2.1648	2.1729
40	0.12166	0.016020	2445	8.02	1070.9	1078.9	0.01617	2.1430	2.1592
44	0.14196	0.016021	2112	12.03	1068.6	1080.7	0.02418	2.1215	2.1457
48	0.16520	0.016023	1829	16.05	1066.4	1082.4	0.03212	2.1003	2.1324
52	0.19173	0.016026	1588.6	20.07	1064.1	1084.2	0.04000	2.0795	2.1195
56	0.2219	0.016030	1382.9	24.08	1061.9	1085.9	0.04781	2.0590	2.1068
60	0.2563	0.016035	1206.9	28.08	1059.6	1087.7	0.05555	2.0388	2.0943
64	0.2952	0.016040	1055.8	32.09	1057.3	1089.4	0.06323	2.0189	2.0821
68	0.3391	0.016047	925.8	36.09	1055.1	1091.2	0.07084	1.9993	2.0701
72	0.3887	0.016055	813.7	40.09	1052.8	1092.9	0.07839	1.9800	2.0584
76	0.4446	0.016063	716.8	44.09	1050.6	1094.7	0.08589	1.9610	2.0469
80	0.5073	0.016073	632.8	48.09	1048.3	1096.4	0.09332	1.9423	2.0356
84	0.5776	0.016083	559.8	52.08	1046.0	1098.1	0.10069	1.9238	2.0245

88	0.6562	0.016094	496.3	56.07	1043.8	1099.9	0.10801	1.9056	2.0136
92	0.7439	0.016105	440.9	60.06	1041.5	1101.6	0.11527	1.8877	2.0030
96	0.8416	0.016117	392.4	64.06	1039.2	1103.3	0.12248	1.8700	1.9925
100	0.9503	0.016130	350.0	68.05	1037.0	1105.0	0.12963	1.8526	1.9822
104	1.0708	0.016144	312.8	72.04	1034.7	1106.7	0.13674	1.8354	1.9722
108	1.2044	0.016158	280.0	76.03	1032.4	1108.4	0.14379	1.8185	1.9623
112	1.3520	0.016173	251.1	80.02	1030.1	1110.2	0.15079	1.8017	1.9525
116	1.5150	0.016189	225.6	84.01	1027.8	1111.9	0.15775	1.7852	1.9430
120	1.6945	0.016205	203.0	88.00	1025.5	1113.5	0.16465	1.7690	1.9336
124	1.8921	0.016221	183.1	91.99	1023.2	1115.2	0.17152	1.7529	1.9244
128	2.1090	0.016239	165.30	95.98	1020.9	1116.9	0.17833	1.7370	1.9154
132	2.347	0.016256	149.51	99.98	1018.6	1118.6	0.18510	1.7214	1.9065
136	2.607	0.016275	135.44	103.97	1016.3	1120.3	0.19183	1.7059	1.8978
140	2.892	0.016293	122.88	107.96	1014.0	1121.9	0.19851	1.6907	1.8892
144	3.203	0.016313	111.66	111.96	1011.6	1123.6	0.20515	1.6756	1.8807
148	3.541	0.016333	101.61	115.96	1009.3	1125.2	0.21175	1.6607	1.8725
152	3.910	0.016353	92.60	119.96	1006.9	1126.9	0.21831	1.6460	1.8643
156	4.310	0.016374	84.51	123.96	1004.6	1128.5	0.22482	1.6315	1.8563
160	4.745	0.016395	77.23	127.96	1002.2	1130.1	0.23130	1.6171	1.8484
164	5.216	0.016417	70.67	131.96	999.8	1131.8	0.23774	1.6029	1.8407
168	5.726	0.016439	64.76	135.97	997.4	1133.4	0.24414	1.5889	1.8330
172	6.277	0.016462	59.42	139.97	995.0	1135.0	0.25050	1.5750	1.8255
176	6.873	0.016485	54.58	143.98	992.6	1136.6	0.25682	1.5613	1.8181
180	7.515	0.016509	50.20	147.99	990.2	1138.2	0.26311	1.5478	1.8109
184	8.206	0.016533	46.23	152.01	987.7	1139.7	0.26936	1.5344	1.8037
188	8.951	0.016558	42.62	156.02	985.3	1141.3	0.27557	1.5211	1.7967
192	9.750	0.016583	39.34	160.04	982.8	1142.9	0.28175	1.5080	1.7898
196	10.609	0.016608	36.36	164.05	980.4	1144.4	0.28789	1.4951	1.7830
200	11.529	0.016634	33.63	168.07	977.9	1145.9	0.29400	1.4822	1.7762
204	12.515	0.016661	31.15	172.10	975.4	1147.5	0.30008	1.4695	1.7696
208	13.570	0.016688	28.88	176.13	972.9	1149.0	0.30612	1.4570	1.7631
212	14.698	0.016716	26.80	180.16	970.3	1150.5	0.31213	1.4446	1.7567
216	15.903	0.016743	24.90	184.18	967.8	1152.0	0.31811	1.4322	1.7504
220	17.188	0.016772	23.15	188.22	965.3	1153.5	0.32406	1.4201	1.7441
224	18.557	0.016801	21.55	192.26	962.7	1154.9	0.32998	1.4080	1.7380
228	20.015	0.016830	20.08	196.30	960.1	1156.4	0.33586	1.3961	1.7319

Table A-1a Water at Saturation—English Units (*continued*)

Fahrenheit Temperature	Pressure P psia	Specific Volume ft³/lbm		Enthalpy Btu/lbm			Entropy Btu/(lbm-R)		
		Saturated Liquid v_f	Saturated Vapor v_g	Saturated Liquid i_f	Evaporation i_{fg}	Saturated Vapor i_g	Saturated Liquid s_f	Evaporation s_{fg}	Saturated Vapor s_g
232	21.57	0.016860	18.723	200.34	957.5	1157.9	0.34172	1.3842	1.7260
236	23.22	0.016891	17.476	204.39	954.9	1159.3	0.34755	1.3725	1.7201
240	24.97	0.016922	16.327	208.44	952.3	1160.7	0.35335	1.3609	1.7143
244	26.82	0.016953	15.267	212.49	949.6	1162.1	0.35912	1.3494	1.7085
248	28.79	0.016985	14.287	216.55	947.0	1163.5	0.36486	1.3380	1.7029
248	28.79	0.016985	14.287	216.55	947.0	1163.5	0.36486	1.3380	1.7029
252	30.88	0.017017	13.382	220.62	944.3	1164.9	0.37058	1.3267	1.6973

Source: Abridged from *Steam Tables* (*English Units*), by Joseph H. Keenan, F. G. Keyes, P. G. Hill, and J. G. Moore. Copyright © 1969 by John Wiley & Sons, Inc., New York. Reprinted by permission of John Wiley & Sons, Inc.

Table A-1b Water at Saturation—SI Units

Temperature C	Temperature K	Pressure Pa $P \times 10^{-5}$	Specific Volume m³/kg Saturated Solid $v_i \times 10^3$	Specific Volume m³/kg Saturated Vapor v_g	Enthalpy kJ/kg Saturated Solid i_i	Enthalpy kJ/kg Sublimation i_{ig}	Enthalpy kJ/kg Saturated Vapor i_g	Entropy kJ/(kg-K) Saturated Solid s_i	Entropy kJ/(kg-K) Sublimation s_{ig}	Entropy kJ/(kg-K) Saturated Vapor s_g
0.01	273.16	0.006113	1.0908	2.061	−333.40	2834.8	2501.4	−1.221	10.378	9.156
−2	271.15	0.005176	1.0904	2.417	−337.62	2835.3	2497.7	−1.237	10.456	9.219
−4	269.15	0.004375	1.0901	2.838	−341.78	2835.7	2494.0	−1.253	10.536	9.283
−6	267.15	0.003689	1.0898	3.342	−345.91	2836.2	2490.3	−1.268	10.616	9.348
−8	265.15	0.003102	1.0894	3.944	−350.02	2836.6	2486.6	−1.284	10.698	9.414
−10	263.15	0.002602	1.0891	4.667	−354.09	2837.0	2482.9	−1.299	10.781	9.481
−12	261.15	0.002176	1.0888	5.537	−358.14	2837.3	2479.2	−1.315	10.865	9.550
−14	259.15	0.001815	1.0884	6.588	−362.15	2837.6	2475.5	−1.331	10.950	9.619
−16	257.15	0.001510	1.0881	7.860	−366.14	2837.9	2471.8	−1.346	11.036	9.690

Temperature C	Temperature K	Pressure Pa $P \times 10^{-5}$	Specific Volume m³/kg Saturated Liquid $v_f \times 10^3$	Specific Volume m³/kg Saturated Vapor $v_g \times 10^3$	Enthalpy kJ/kg Saturated Liquid i_f	Enthalpy kJ/kg Evaporation i_{fg}	Enthalpy kJ/kg Saturated Vapor i_g	Entropy kJ/(kg-K) Saturated Liquid s_f	Entropy kJ/(kg-K) Evaporation s_{fg}	Entropy kJ/(kg-K) Saturated Vapor s_g
0.01	273.16	0.006113	1.0002	206 136	0.01	2501.3	2501.4	0.0000	9.1562	9.1562
2	275.15	0.007056	1.0001	179 889	8.37	2496.7	2505.0	0.0305	9.0730	9.1035
4	277.15	0.008131	1.0001	157 232	16.78	2491.9	2508.7	0.0610	8.9904	9.0514
6	279.15	0.009349	1.0001	137 734	25.20	2487.2	2512.4	0.0912	8.9090	9.0003
8	281.15	0.010724	1.0002	120 917	33.60	2482.5	2516.1	0.1212	8.8289	8.9501
10	283.15	0.012276	1.0004	106 379	42.01	2477.7	2519.8	0.1510	8.7498	8.9008
12	285.15	0.014022	1.0005	93 784	50.41	2473.0	2523.4	0.1806	8.6718	8.8524
14	287.15	0.015983	1.0008	82 848	58.80	2468.3	2527.1	0.2099	8.5949	8.8048
16	289.15	0.018181	1.0011	73 333	67.19	2463.6	2530.8	0.2390	8.5191	8.7582
18	291.15	0.020640	1.0014	65 038	75.58	2458.8	2534.4	0.2679	8.4438	8.7123
20	293.15	0.02339	1.0018	57 791	83.96	2454.1	2538.1	0.2966	8.3706	8.6672
22	295.15	0.02645	1.0022	51 447	92.33	2449.4	2541.7	0.3251	8.2979	8.6229
24	297.15	0.02985	1.0027	45 883	100.70	2444.7	2545.4	0.3534	8.2261	8.5794

Table A-1b Water at Saturation—SI Units (*continued*)

Temperature		Pressure Pa $P \times 10^{-5}$	Specific Volume m³/kg		Enthalpy kJ/kg			Entropy kJ/(kg-K)		
C	K		Saturated Liquid $v_f \times 10^3$	Saturated Vapor v_g	Saturated Liquid i_f	Sublimation i_{fg}	Saturated Vapor i_g	Saturated Liquid s_f	Sublimation s_{fg}	Saturated Vapor s_g
26	299.15	0.03363	1.0032	40 994	109.07	2439.9	2549.0	0.3814	8.1552	8.5367
28	301.15	0.03782	1.0037	36 690	117.43	2435.2	2552.6	0.4093	8.0854	8.4946
30	303.15	0.04246	1.0043	32 894	125.79	2430.5	2556.3	0.4369	8.0164	8.4533
32	305.15	0.04759	1.0050	29 540	134.15	2425.7	2559.9	0.4644	7.9483	8.4127
34	307.15	0.05324	1.0056	26 571	142.50	2421.0	2563.5	0.4917	7.8811	8.3728
36	309.15	0.05947	1.0063	23 940	150.86	2416.2	2567.1	0.5188	7.8147	8.3336
38	311.15	0.06632	1.0071	21 602	159.21	2411.5	2570.7	0.5458	7.7492	8.2950
40	313.15	0.07384	1.0078	19 523	167.57	2406.7	2574.3	0.5725	7.6845	8.2570
42	315.15	0.08208	1.0086	17 671	175.91	2401.9	2577.9	0.5991	7.6206	8.2197
44	317.15	0.09111	1.0095	16 018	184.27	2397.2	2581.4	0.6255	7.5574	8.1829
46	319.15	0.10098	1.0103	14 540	192.62	2392.4	2585.0	0.6518	7.4950	8.1468
48	321.15	0.11175	1.0112	13 218	200.97	2387.6	2588.5	0.6779	7.4334	8.1113
50	323.15	0.12349	1.0121	12 032	209.33	2382.7	2592.1	0.7038	7.3725	8.0763
52	325.15	0.13628	1.0131	10 968	217.69	2377.9	2595.6	0.7296	7.3123	8.0419
54	327.15	0.15019	1.0141	10 011	226.04	2373.1	2599.1	0.7552	7.2528	8.0080
56	329.15	0.16529	1.0151	9149	234.41	2368.2	2602.6	0.7807	7.1940	7.9747
58	331.15	0.18166	1.0161	8372	242.77	2363.4	2606.1	0.8060	7.1359	7.9419
60	333.15	0.19940	1.0172	7671	251.13	2358.5	2609.6	0.8312	7.0784	7.9096
62	335.15	0.21860	1.0182	7037	259.49	2353.6	2613.1	0.8562	7.0216	7.8778
64	337.15	0.23934	1.0194	6463	267.86	2348.7	2616.5	0.8811	6.9654	7.8465
66	339.15	0.2617	1.0205	5943	276.23	2343.7	2620.0	0.9058	6.9098	7.8156
68	341.15	0.2859	1.0217	5471	284.61	2338.8	2623.4	0.9304	6.8548	7.7852
70	343.15	0.3119	1.0228	5042	292.98	2333.8	2626.8	0.9549	6.8004	7.7553
72	345.15	0.3399	1.0240	4652	301.36	2328.9	2630.2	0.9792	6.7466	7.7258
74	347.15	0.3699	1.0253	4297	309.74	2323.9	2633.6	1.0034	6.6934	7.6968
76	349.15	0.4022	1.0265	3973	318.13	2318.9	2637.0	1.0275	6.6407	7.6682
78	351.15	0.4368	1.0278	3677	326.51	2313.8	2640.3	1.0515	6.5885	7.6400
80	353.15	0.4739	1.0291	3407	334.91	2308.8	2643.7	1.0753	6.5369	7.6122
82	355.15	0.5136	1.0305	3160	343.30	2303.7	2647.0	1.0990	6.4858	7.5848

84	357.15	0.5560	1.0318	2934	351.70	2298.6	2650.3	1.1225	6.4353	7.5578
86	359.15	0.6014	1.0332	2726	360.10	2293.5	2653.6	1.1460	6.3852	7.5312
88	361.15	0.6498	1.0346	2536	368.51	2288.3	2656.6	1.1693	6.3356	7.5050
90	363.15	0.7014	1.0360	2361	376.92	2283.2	2660.1	1.1925	6.2866	7.4791
92	365.15	0.7564	1.0375	2200	385.33	2278.0	2663.3	1.2156	6.2379	7.4536
94	367.15	0.8149	1.0389	2052	393.75	2272.8	2666.5	1.2386	6.1898	7.4284
96	369.15	0.8771	1.0404	1915.0	402.17	2267.6	2669.7	1.2615	6.1421	7.4036
98	371.15	0.9433	1.0420	1789.1	410.61	2262.3	2672.9	1.2842	6.0948	7.3791
100	373.15	1.0135	1.0435	1672.9	419.04	2257.0	2676.1	1.3069	6.0480	7.3549
102	375.15	1.0880	1.0451	1565.5	427.48	2251.7	2679.2	1.3294	6.0016	7.3310
104	377.15	1.1669	1.0467	1466.2	435.92	2246.4	2682.3	1.3518	5.9557	7.3075
106	379.15	1.2506	1.0483	1374.3	444.37	2241.0	2685.4	1.3741	5.9101	7.2843
108	381.15	1.3391	1.0499	1289.1	452.83	2235.6	2688.4	1.3964	5.8650	7.2613
110	383.15	1.4327	1.0516	1210.2	461.30	2230.2	2691.5	1.4185	5.8202	7.2387
112	385.15	1.5317	1.0533	1136.9	469.76	2224.7	2694.5	1.4405	5.7758	7.2163
114	387.15	1.6362	1.0550	1068.8	478.24	2219.2	2697.5	1.4624	5.7318	7.1942
116	389.15	1.7465	1.0568	1005.5	486.72	2213.7	2700.5	1.4842	5.6882	7.1724
118	391.15	1.8627	1.0585	946.7	495.21	2208.2	2703.4	1.5060	5.6449	7.1509
120	393.15	1.9853	1.0603	891.9	503.71	2202.6	2706.3	1.5276	5.6020	7.1296
122	395.15	2.1143	1.0622	840.8	512.21	2197.0	2709.2	1.5492	5.5594	7.1086

Source: Abridged from, *Steam Tables (International Edition—Metric Units),* by Joseph H. Keenan, F. G. Keyes, P. G. Hill, and J. G. Moore. Copyright © 1969 by John Wiley & Sons, Inc., New York. Reprinted by permission of John Wiley & Sons, Inc.

Table A-2a Moist Air, Standard Atmospheric Pressure, 14.696 psi—English Units

Temp. F	Humidity Ratio lb_w/lb_a W_s	Volume ft³/lb dry air v_a	Volume v_{as}	Volume v_s	Enthalpy Btu/lb dry air i_a	Enthalpy i_{as}	Enthalpy i_s	Entropy Btu/(lb dry air·F) s_a	Entropy s_{as}	Entropy s_s	Condensed Water Enthalpy Btu/lb i_w	Condensed Water Entropy Btu/(lb·F) s_w	Vapor Press. in Hg p_s	Temp. F
0	0.0007875	11.579	0.015	11.594	0.0	0.835	0.835	0.00000	0.00192	0.00192	−158.89	−0.3243	0.037671	0
5	0.0010207	11.706	0.019	11.725	1.201	1.085	2.286	0.00260	0.00247	0.00506	−156.52	−0.3192	0.048814	5
10	0.0013158	11.832	0.025	11.857	2.402	1.402	3.804	0.00517	0.00315	0.00832	−154.13	−0.3141	0.062901	10
15	0.0016874	11.959	0.032	11.991	3.603	1.801	5.404	0.00771	0.00400	0.01171	−151.71	−0.3089	0.080623	15
20	0.0021531	12.085	0.042	12.127	4.804	2.303	7.107	0.01023	0.00505	0.01528	−149.27	−0.3038	0.102798	20
25	0.0027339	12.212	0.054	12.265	6.005	2.930	8.935	0.01272	0.00636	0.01908	−146.80	−0.2987	0.130413	25
30	0.0034552	12.338	0.068	12.406	7.206	3.711	10.917	0.01519	0.00796	0.02315	−144.31	−0.2936	0.164631	30
32	0.0037895	12.389	0.075	12.464	7.687	4.073	11.760	0.01617	0.00870	0.02487	−143.30	−0.2915	0.180479	32
36	0.004452	12.490	0.089	12.579	8.648	4.793	13.441	0.01811	0.01016	0.02827	4.05	0.0081	0.21181	36
40	0.005216	12.591	0.105	12.696	9.609	5.624	15.233	0.02004	0.01183	0.03187	8.07	0.0162	0.24784	40
44	0.006094	12.692	0.124	12.816	10.570	6.582	17.152	0.02196	0.01374	0.03570	12.09	0.0242	0.28918	44
48	0.007103	12.793	0.146	12.939	11.531	7.684	19.215	0.02386	0.01592	0.03978	16.10	0.0321	0.33651	48
52	0.008259	12.894	0.171	13.065	12.492	8.949	21.441	0.02575	0.01840	0.04415	20.11	0.0400	0.39054	52
56	0.009580	12.995	0.200	13.195	13.453	10.397	23.850	0.02762	0.02122	0.04884	24.11	0.0478	0.45205	56
60	0.011087	13.096	0.233	13.329	14.415	12.052	26.467	0.02947	0.02442	0.05389	28.11	0.0555	0.52193	60
64	0.012805	13.198	0.271	13.468	15.376	13.942	29.318	0.03132	0.02804	0.05936	32.11	0.0632	0.60113	64
68	0.014758	13.299	0.315	13.613	16.337	16.094	32.431	0.03315	0.03214	0.06529	36.11	0.0708	0.69065	68
72	0.016976	13.400	0.365	13.764	17.299	18.543	35.841	0.03496	0.03677	0.07173	40.11	0.0783	0.79167	72
76	0.019491	13.501	0.422	13.923	18.260	21.323	39.583	0.03676	0.04199	0.07875	44.10	0.0858	0.90533	76
80	0.022340	13.602	0.487	14.089	19.222	24.479	43.701	0.03855	0.04787	0.08642	48.10	0.0933	1.03302	80
84	0.025563	13.703	0.561	14.264	20.183	28.055	48.238	0.04033	0.05448	0.09481	52.09	0.1006	1.17608	84

Temp														Temp
88	0.029208	13.804	0.646	14.450	21.145	32.105	53.250	0.04209	0.06192	0.10401	56.09	0.1080	1.33613	88
92	0.033323	13.905	0.742	14.647	22.107	36.687	58.794	0.04384	0.07028	0.11412	60.08	0.1152	1.51471	92
96	0.037972	14.006	0.852	14.858	23.069	41.871	64.940	0.04558	0.07968	0.12525	64.07	0.1224	1.71372	96
100	0.043219	14.107	0.976	15.084	24.031	47.730	71.761	0.04730	0.09022	0.13752	68.07	0.1296	1.93492	100
104	0.049140	14.208	1.118	15.326	24.993	54.354	79.346	0.04901	0.10206	0.15108	72.06	0.1367	2.18037	104
108	0.055826	14.309	1.279	15.588	25.955	61.844	87.799	0.05071	0.11537	0.16608	76.05	0.1438	2.45232	108
112	0.063378	14.411	1.462	15.872	26.917	70.319	97.237	0.05240	0.13032	0.18272	80.05	0.1508	2.75310	112
116	0.071908	14.512	1.670	16.181	27.879	79.906	107.786	0.05408	0.14713	0.20121	84.04	0.1577	3.08488	116
120	0.081560	14.613	1.906	16.519	28.842	90.770	119.612	0.05575	0.16605	0.22180	88.04	0.1647	3.45052	120
124	0.092500	14.714	2.176	16.890	29.805	103.102	132.907	0.05740	0.18739	0.24480	92.03	0.1715	3.85298	124
128	0.104910	14.815	2.485	17.299	30.767	117.111	147.878	0.05905	0.21149	0.27054	96.03	0.1783	4.29477	128
132	0.119023	14.916	2.837	17.753	31.730	133.066	164.796	0.06068	0.23876	0.29944	100.02	0.1851	4.77919	132
136	0.135124	15.017	3.242	18.259	32.693	151.294	183.987	0.06230	0.26973	0.33203	104.02	0.1919	5.30973	136
140	0.153538	15.118	3.708	18.825	33.656	172.168	205.824	0.06391	0.30498	0.36890	108.02	0.1985	5.88945	140
144	0.174694	15.219	4.245	19.464	34.620	196.183	230.802	0.06551	0.34530	0.41081	112.02	0.2052	6.52241	144
148	0.199110	15.320	4.869	20.189	35.583	223.932	259.514	0.06710	0.39160	0.45871	116.02	0.2118	7.21239	148
152	0.227429	15.421	5.596	21.017	36.546	256.158	292.705	0.06868	0.44507	0.51375	120.02	0.2184	7.96306	152
156	0.260512	15.522	6.450	21.972	37.510	293.849	331.359	0.07025	0.50723	0.57749	124.02	0.2249	8.77915	156
160	0.29945	15.623	7.459	23.082	38.474	338.263	376.737	0.07181	0.58007	0.65188	128.02	0.2314	9.6648	160
164	0.34572	15.724	8.664	24.388	39.438	391.095	430.533	0.07337	0.66622	0.73959	132.03	0.2378	10.6250	164
168	0.40131	15.825	10.117	25.942	40.402	454.630	495.032	0.07491	0.76925	0.84415	136.03	0.2442	11.6641	168
172	0.46905	15.926	11.894	27.820	41.366	532.138	573.504	0.07644	0.89423	0.97067	140.04	0.2506	12.7880	172
176	0.55294	16.027	14.103	30.130	42.331	628.197	670.528	0.07796	1.04828	1.12624	144.05	0.2569	14.0010	176
180	0.65911	16.128	16.909	33.037	43.295	749.871	793.166	0.07947	1.24236	1.32183	148.06	0.2632	15.3097	180
184	0.79703	16.229	20.564	36.793	44.260	908.061	952.321	0.08098	1.49332	1.57430	152.07	0.2694	16.7190	184
188	0.98272	16.330	25.498	41.828	45.225	1121.174	1166.399	0.08247	1.82963	1.91210	156.08	0.2756	18.2357	188
192	1.24471	16.431	32.477	48.908	46.190	1422.047	1468.238	0.08396	2.30193	2.38589	160.10	0.2818	19.8652	192
196	1.64070	16.532	43.046	59.578	47.155	1877.032	1924.188	0.08543	3.01244	3.09787	164.12	0.2880	21.6152	196
200	2.30454	16.633	60.793	77.426	48.121	2640.084	2688.205	0.08690	4.19787	4.28477	168.13	0.2941	23.4906	200

Source: Abridged by permission from *ASHRAE Handbook, Fundamentals Volume*, 1985.

Table A-2b Moist Air, Standard Atmospheric Pressure, 101.325 kPa—SI Units

Temp. C	Humidity Ratio kg_w/kg_a W_s	Volume m³/kg dry air v_a	v_{as}	v_s	Enthalpy kJ/kg dry air i_a	i_{as}	i_s	Entropy kJ/(kg dry air·K) s_a	s_{as}	s_s	Condensed Water Enthalpy kJ/kg i_w	Entropy kJ/(kg·K) s_w	Vapor Press. kPa p_s	Temp. C
−18	0.0007711	0.7222	0.0009	0.7231	−18.103	1.902	−16.201	−0.0686	0.0079	−0.0607	−370.02	−1.3596	0.12492	−18
−16	0.0009303	0.7279	0.0011	0.7290	−16.092	2.299	−13.793	−0.0607	0.0094	−0.0513	−366.06	−1.3441	0.15068	−16
−14	0.0011191	0.7336	0.0013	0.7349	−14.080	2.769	−11.311	−0.0529	0.0113	−0.0416	−362.07	−1.3287	0.18122	−14
−12	0.0013425	0.7393	0.0016	0.7409	−12.069	3.327	−8.742	−0.0452	0.0134	−0.0318	−358.06	−1.3132	0.21732	−12
−10	0.0016062	0.7450	0.0019	0.7469	−10.057	3.986	−6.072	−0.0375	0.0160	−0.0215	−354.01	−1.2978	0.25991	−10
− 8	0.0019166	0.7507	0.0023	0.7530	−8.046	4.764	−3.283	−0.0299	0.0189	−0.0110	−349.93	−1.2824	0.30999	− 8
− 6	0.0022811	0.7563	0.0028	0.7591	−6.035	5.677	−0.357	−0.0223	0.0224	0.0000	−345.82	−1.2669	0.36874	− 6
− 4	0.0027081	0.7620	0.0033	0.7653	−4.023	6.751	2.728	−0.0148	0.0264	0.0115	−341.69	−1.2515	0.43748	− 4
− 2	0.0032074	0.7677	0.0039	0.7717	−2.011	8.007	5.995	−0.0074	0.0310	0.0236	−337.52	−1.2361	0.51773	− 2
0	0.0037895	0.7734	0.0047	0.7781	−0.000	9.473	9.473	0.0000	0.0364	0.0364	−333.32	−1.2206	0.61117	0
2	0.004381	0.7791	0.0055	0.7845	2.012	10.970	12.982	0.0073	0.0419	0.0492	8.49	0.0306	0.7060	2
4	0.005054	0.7848	0.0064	0.7911	4.024	12.672	16.696	0.0146	0.0480	0.0627	16.91	0.0611	0.8135	4
6	0.005818	0.7904	0.0074	0.7978	6.036	14.608	20.644	0.0219	0.0550	0.0769	25.32	0.0913	0.9353	6
8	0.006683	0.7961	0.0085	0.8046	8.047	16.805	24.852	0.0290	0.0628	0.0919	33.72	0.1213	1.0729	8
10	0.007661	0.8018	0.0098	0.8116	10.059	19.293	29.352	0.0362	0.0717	0.1078	42.11	0.1511	1.2280	10
12	0.008766	0.8075	0.0113	0.8188	12.071	22.107	34.179	0.0433	0.0816	0.1248	50.50	0.1806	1.4026	12
14	0.010012	0.8132	0.0131	0.8262	14.084	25.286	39.370	0.0503	0.0927	0.1430	58.88	0.2099	1.5987	14
16	0.011413	0.8188	0.0150	0.8338	16.096	28.867	44.963	0.0573	0.1051	0.1624	67.26	0.2389	1.8185	16
18	0.012989	0.8245	0.0172	0.8417	18.108	32.900	51.008	0.0642	0.1190	0.1832	75.63	0.2678	2.0643	18
20	0.014758	0.8302	0.0196	0.8498	20.121	37.434	57.555	0.0711	0.1346	0.2057	84.00	0.2965	2.3389	20
22	0.016741	0.8359	0.0224	0.8583	22.133	42.527	64.660	0.0779	0.1519	0.2298	92.36	0.3249	2.6448	22
24	0.018963	0.8416	0.0256	0.8671	24.146	48.239	72.385	0.0847	0.1712	0.2559	100.73	0.3531	2.9852	24
26	0.021448	0.8472	0.0291	0.8764	26.159	54.638	80.798	0.0915	0.1927	0.2842	109.09	0.3812	3.3633	26
28	0.024226	0.8529	0.0331	0.8860	28.172	61.804	89.976	0.0982	0.2166	0.3148	117.45	0.4090	3.7823	28

30	0.027329	0.8586	0.0376	0.8962	30.185	69.820	100.006	0.1048	0.2432	0.3481	125.81	0.4367	4.2462	30
32	0.030793	0.8643	0.0426	0.9069	32.198	78.780	110.979	0.1115	0.2728	0.3842	134.17	0.4642	4.7586	32
34	0.034660	0.8700	0.0483	0.9183	34.212	88.799	123.011	0.1180	0.3056	0.4236	142.53	0.4915	5.3242	34
36	0.038971	0.8756	0.0546	0.9303	36.226	99.983	136.209	0.1246	0.3420	0.4666	150.89	0.5186	5.9468	36
38	0.043778	0.8813	0.0618	0.9431	38.239	112.474	150.713	0.1311	0.3824	0.5135	159.25	0.5456	6.6315	38
40	0.049141	0.8870	0.0698	0.9568	40.253	126.430	166.683	0.1375	0.4273	0.5649	167.61	0.5724	7.3838	40
42	0.055119	0.8927	0.0788	0.9714	42.268	142.007	184.275	0.1439	0.4771	0.6211	175.97	0.5990	8.2081	42
44	0.061791	0.8983	0.0888	0.9872	44.282	159.417	203.699	0.1503	0.5325	0.6828	184.33	0.6254	9.1110	44
46	0.069239	0.9040	0.1002	1.0042	46.296	178.882	225.179	0.1566	0.5940	0.7507	192.69	0.6517	10.0982	46
48	0.077556	0.9097	0.1129	1.0226	48.311	200.644	248.955	0.1629	0.6624	0.8253	201.06	0.6778	11.1754	48
50	0.086858	0.9154	0.1272	1.0425	50.326	225.019	275.345	0.1692	0.7385	0.9077	209.42	0.7038	12.3503	50
52	0.097272	0.9211	0.1433	1.0643	52.341	252.340	304.682	0.1754	0.8234	0.9988	217.78	0.7296	13.6293	52
54	0.108954	0.9267	0.1614	1.0882	54.357	283.031	337.388	0.1816	0.9182	1.0998	226.15	0.7552	15.0205	54
56	0.122077	0.9324	0.1819	1.1143	56.373	317.549	373.922	0.1877	1.0243	1.2120	234.52	0.7807	16.5311	56
58	0.136851	0.9381	0.2051	1.1432	58.389	356.461	414.850	0.1938	1.1432	1.3370	242.88	0.8061	18.1691	58
60	0.15354	0.9438	0.2315	1.1752	60.405	400.458	460.863	0.1999	1.2769	1.4768	251.25	0.8313	19.9439	60
62	0.17244	0.9494	0.2614	1.2109	62.421	450.377	512.798	0.2059	1.4278	1.6337	259.62	0.8563	21.8651	62
64	0.19393	0.9551	0.2957	1.2508	64.438	507.177	571.615	0.2119	1.5985	1.8105	268.00	0.8812	23.9405	64
66	0.21848	0.9608	0.3350	1.2958	66.455	572.116	638.571	0.2179	1.7927	2.0106	276.37	0.9060	26.1810	66
68	0.24664	0.9665	0.3803	1.3467	68.472	646.724	715.196	0.2238	2.0147	2.2385	284.75	0.9306	28.5967	68
70	0.27916	0.9721	0.4328	1.4049	70.489	732.959	803.448	0.2297	2.2699	2.4996	293.13	0.9551	31.1986	70
72	0.31698	0.9778	0.4941	1.4719	72.507	833.335	905.842	0.2356	2.5655	2.8010	301.51	0.9794	33.9983	72
74	0.36130	0.9835	0.5662	1.5497	74.525	951.077	1025.603	0.2414	2.9104	3.1518	309.89	1.0037	37.0063	74
76	0.41377	0.9892	0.6519	1.6411	76.543	1090.628	1167.172	0.2472	3.3171	3.5644	318.28	1.0278	40.2369	76
78	0.47663	0.9948	0.7550	1.7498	78.562	1257.921	1336.483	0.2530	3.8023	4.0553	326.67	1.0517	43.7020	78
80	0.55295	1.0005	0.8805	1.8810	80.581	1461.200	1541.781	0.2587	4.3890	4.6477	335.06	1.0755	47.4135	80
82	0.64724	1.0062	1.0360	2.0422	82.600	1712.547	1795.148	0.2644	5.1108	5.3753	343.45	1.0993	51.3860	82
84	0.76624	1.0119	1.2328	2.2446	84.620	2029.983	2114.603	0.2701	6.0181	6.2882	351.85	1.1228	55.6337	84
86	0.92062	1.0175	1.4887	2.5062	86.640	2442.036	2528.677	0.2757	7.1901	7.4658	360.25	1.1463	60.1727	86
88	1.12800	1.0232	1.8333	2.8565	88.661	2995.890	3084.551	0.2813	8.7580	9.0393	368.65	1.1696	65.0166	88
90	1.42031	1.0289	2.3199	3.3488	90.681	3776.918	3867.599	0.2869	10.9586	11.2455	377.06	1.1928	70.1817	90

Source: Abridged by permission from *ASHRAE Handbook, Fundamentals Volume*, 1985.

Table A-3a Refrigerant 12 Saturation Properties—English Units

Fahrenheit Temperature	Pressure psia P	Volume ft³/lbm Liquid v_f	Volume ft³/lbm Vapor v_g	Enthalpy Btu/lbm Liquid i_f	Enthalpy Btu/lbm Latent i_{fg}	Enthalpy Btu/lbm Vapor i_g	Entropy Btu/(lbm-R) Liquid s_f	Entropy Btu/(lbm-R) Vapor s_g	Fahrenheit Temperature
−40	9.3076	0.010564	3.8750	0	72.913	72.913	0	0.17373	−40
−36	10.320	0.010607	3.5198	0.8434	72.511	73.354	0.001995	0.17313	−36
−32	11.417	0.010651	3.2035	1.6887	72.106	73.795	0.003976	0.17257	−32
−28	12.604	0.010696	2.9214	2.5358	71.698	74.234	0.005942	0.17203	−28
−24	13.886	0.010741	2.6691	3.3848	71.288	74.673	0.007894	0.17151	−24
−20	15.267	0.010788	2.4429	4.2357	70.874	75.110	0.009831	0.17102	−20
−16	16.753	0.010834	2.2399	5.0885	70.456	75.545	0.011755	0.17055	−16
−12	18.348	0.010882	2.0572	5.9434	70.036	75.979	0.013666	0.17010	−12
−8	20.059	0.010931	1.8924	6.8003	69.611	76.411	0.015564	0.16967	−8
−4	21.891	0.010980	1.7436	7.6594	69.183	76.842	0.017449	0.16927	−4
0	23.849	0.011030	1.6089	8.5207	68.750	77.271	0.019323	0.16888	0
4	25.939	0.011082	1.4867	9.3843	68.314	77.698	0.021184	0.16851	4
8	28.167	0.011134	1.3758	10.250	67.873	78.123	0.023033	0.16815	8
12	30.539	0.011187	1.2748	11.118	67.428	78.546	0.024871	0.16782	12
16	33.060	0.011241	1.128	11.989	66.977	78.966	0.026699	0.16750	16
20	35.736	0.011296	1.0988	12.863	66.522	79.385	0.028515	0.16719	20
24	38.574	0.011352	1.0220	13.739	66.061	79.800	0.030322	0.16690	24
28	41.580	0.011409	0.95173	14.618	65.596	80.214	0.032118	0.16662	28
32	44.760	0.011468	0.88725	15.500	65.124	80.624	0.033905	0.16635	32
36	48.120	0.011527	0.82803	16.384	64.647	81.031	0.035683	0.16610	36
40	51.667	0.011588	0.77357	17.273	64.163	81.436	0.037453	0.16586	40
44	55.407	0.011650	0.72341	18.164	63.673	81.837	0.039213	0.16562	44
48	59.347	0.011714	0.67715	19.059	63.177	82.236	0.040966	0.16540	48
52	63.494	0.011779	0.63444	19.957	62.673	82.630	0.042711	0.16519	52
56	67.853	0.011845	0.59495	20.859	62.162	83.021	0.044449	0.16499	56

60	72.433	0.011913	0.55839	21.766	61.643	83.409	0.046180	0.16479	60
64	77.239	0.011982	0.52450	22.676	61.116	83.792	0.047905	0.16460	64
68	82.279	0.012053	0.49305	23.591	60.580	84.171	0.049624	0.16442	68
72	87.559	0.012126	0.46383	24.511	60.035	84.546	0.051338	0.16425	72
76	93.087	0.012201	0.43666	25.435	59.481	84.916	0.053047	0.16408	76
80	98.870	0.012277	0.41135	26.365	58.917	85.282	0.054751	0.16392	80
84	104.92	0.012356	0.38776	27.300	58.343	85.643	0.056452	0.16376	84
88	111.23	0.012437	0.36575	28.241	57.757	85.998	0.058149	0.16360	88
92	117.82	0.012520	0.34518	29.187	57.161	86.348	0.059844	0.16345	92
96	124.70	0.012605	0.32594	30.140	56.551	86.691	0.061536	0.16330	96
100	131.86	0.012693	0.30794	31.100	55.929	87.029	0.632273	0.16315	100
104	139.33	0.012783	0.29106	32.067	55.293	87.360	0.064916	0.16301	104
108	147.11	0.012876	0.27524	33.041	54.643	87.684	0.066606	0.16286	108
112	155.19	0.012972	0.26037	34.023	53.978	88.001	0.068296	0.16271	112
116	163.61	0.013072	0.24641	35.014	53.296	88.310	0.069987	0.16256	116
120	172.35	0.013174	0.23326	36.013	52.597	88.610	0.071680	0.16241	120
124	181.43	0.013280	0.22089	37.021	51.881	88.902	0.073376	0.16226	124
128	190.86	0.013390	0.20922	38.040	51.144	89.184	0.075075	0.16210	128
132	200.64	0.013504	0.19821	39.069	50.387	89.456	0.076779	0.16194	132
136	210.79	0.013623	0.18782	40.110	49.608	89.718	0.078489	0.16177	136
140	221.32	0.013746	0.17799	41.162	48.805	89.967	0.080205	0.16159	140
144	232.22	0.013874	0.16868	42.227	47.977	90.204	0.081928	0.16140	144
148	243.51	0.014008	0.15987	43.306	47.122	90.428	0.083661	0.16120	148
152	255.20	0.014148	0.15151	44.399	46.238	90.637	0.085404	0.16099	152
156	267.30	0.014925	0.14358	45.508	45.322	90.830	0.087159	0.16077	156
160	279.82	0.014449	0.13604	46.633	44.373	91.006	0.088927	0.16053	160
164	292.77	0.014611	0.12886	47.777	43.386	91.163	0.090710	0.16027	164
168	306.15	0.014782	0.12202	48.939	42.360	91.299	0.092511	0.16000	168
172	319.97	0.014963	0.11550	50.123	41.290	91.413	0.094330	0.15969	172

Table A-3b Refrigerant 12 Saturation Properties—SI Units

Celsius Temperature	Pressure Pa $P \times 10^{-5}$	Enthalpy kJ/kg i_f	Enthalpy kJ/kg i_{fg}	Enthalpy kJ/kg i_g	Entropy kJ/(kg-K) s_f	Entropy kJ/(kg-K) s_g	Specific Volume m³/kg $v_f \times 10^3$	Specific Volume m³/kg $v_g \times 10^3$
-40	0.6412	163.949	169.601	333.551	0.85804	1.58548	0.69547	242.094
-35	0.8065	168.369	167.489	335.858	0.87675	1.58004	0.66561	195.536
-30	1.0034	172.811	165.342	338.153	0.89516	1.57516	0.67198	159.487
-28	1.0919	174.595	164.472	339.067	0.90243	1.57334	0.67459	147.373
-26	1.1864	176.381	163.596	339.978	0.90967	1.57160	0.67724	136.374
-24	1.2871	178.172	162.713	340.885	0.91686	1.56993	0.67993	126.364
-22	1.3943	179.966	161.824	341.790	0.92400	1.56833	0.68267	117.243
-20	1.5083	181.765	160.927	342.692	0.93110	1.56680	0.68544	108.915
-18	1.6294	183.568	160.022	343.590	0.93816	1.56533	0.68827	101.303
-16	1.7577	183.375	159.110	344.485	0.94518	1.56393	0.69113	94.3375
-14	1.8938	187.186	158.189	345.375	0.95216	1.56258	0.69404	87.9461
-12	2.0377	189.003	157.259	346.262	0.95911	1.56129	0.69701	82.0803
-10	2.1898	190.823	156.322	347.144	0.96601	1.56005	0.70002	76.6898
-8	2.3505	192.648	155.374	348.022	0.97288	1.55886	0.70308	71.7262
-6	2.5199	194.478	154.418	348.896	0.97971	1.55773	0.70619	67.1525
-4	2.6986	196.313	153.452	249.765	0.98651	1.55664	0.70936	62.9298
-2	2.8866	198.154	152.475	350.629	0.99327	1.55560	0.71259	59.0285
0	3.0842	200.000	151.488	351.488	1.00000	1.55460	0.71587	55.4217
2	3.2922	201.853	150.489	352.342	1.00670	1.55364	0.71922	52.0763
4	3.5104	203.711	149.478	353.189	1.01338	1.55272	0.72262	48.9758
6	3.7394	205.574	148.457	354.031	1.02002	1.55184	0.72609	46.0982
8	3.9794	207.445	147.422	354.867	1.02663	1.55098	0.72963	43.4225
10	4.2306	209.322	146.375	355.697	1.03322	1.55017	0.73324	40.9363
12	4.4939	211.207	145.313	356.520	1.03978	1.54939	0.73692	38.6171
14	4.7689	213.099	144.237	357.336	1.04632	1.54863	0.74068	36.4570
16	5.0564	214.998	143.147	358.145	1.05284	1.54790	0.74451	34.4409

18	5.3567	216.905	142.041	358.947	1.05934	1.54720	0.74843	32.5569
20	5.6703	218.821	140.919	359.740	1.06581	1.54652	0.75243	30.7944
22	5.9969	220.745	139.781	360.526	1.07227	1.54586	0.75652	29.1470
24	6.3374	222.678	138.625	361.303	1.07871	1.54523	0.76069	27.6030
26	6.6922	224.621	137.450	362.071	1.08514	1.54461	0.76497	25.1552
28	7.0615	226.574	136.256	362.829	1.09155	1.54400	0.76935	24.7958
30	7.4455	228.537	135.043	363.579	1.09795	1.54341	0.77383	23.5200
32	7.8451	230.511	133.806	364.317	1.10433	1.54283	0.77842	22.3192
34	8.2598	232.495	132.551	365.046	1.11071	1.54226	0.78312	21.1912
36	8.6909	234.493	131.270	365.763	1.11708	1.54170	0.78794	20.1276
38	9.1384	236.503	129.965	366.468	1.12345	1.54114	0.79290	19.1251
40	9.6026	238.526	128.635	367.161	1.12981	1.54059	0.79798	18.1799
42	10.084	240.561	127.280	367.841	1.13617	1.54004	0.80321	17.2884
44	10.583	242.612	125.895	368.507	1.14253	1.53949	0.80858	16.4449
46	11.099	244.678	124.481	369.159	1.14889	1.53893	0.81411	15.6479
48	11.634	246.758	123.039	369.797	1.15525	1.53837	0.81981	14.8945
50	12.188	248.855	121.562	370.417	1.16162	1.53780	0.82568	14.1803
52	12.761	250.969	120.053	371.022	1.16799	1.53722	0.83173	13.5040
54	13.353	253.101	118.507	371.608	1.17438	1.53662	0.83799	12.8623
56	13.966	255.252	116.924	372.176	1.18078	1.53601	0.84446	12.2532
58	14.600	257.422	115.301	372.723	1.18719	1.53537	0.85115	11.6747
60	15.253	259.611	113.639	373.250	1.19361	1.53472	0.85808	11.1253
62	15.928	261.819	111.937	373.756	1.20005	1.53404	0.86528	10.6033
64	16.625	264.051	110.187	374.239	1.20650	1.53332	0.87275	10.1061
66	17.344	266.302	108.394	374.696	1.21297	1.53258	0.88053	9.63298
68	18.086	268.577	106.552	375.129	1.21947	1.53180	0.88863	9.18216
70	18.851	270.873	104.662	373.534	1.22598	1.53098	0.89709	8.75265
75	20.867	276.701	99.726	376.428	1.24231	1.52876	0.92001	7.76538
80	23.038	282.647	94.493	377.140	1.25869	1.52626	0.94603	6.88968

Source: Abstracted by permission from "Thermodynamic Table for Refrigerant 12 in SI-Units" International Institute of Refrigeration, Paris.

Table A-4a Refrigerant 22 Saturation Properties—English Units

Fahrenheit Temperature	Pressure psia P	Volume ft³/lbm Liquid v_f	Volume ft³/lbm Vapor v_g	Enthalpy Btu/lbm Liquid i_f	Enthalpy Btu/lbm Latent i_{fg}	Enthalpy Btu/lbm Vapor i_g	Entropy Btu/(lbm-R) Liquid s_f	Entropy Btu/(lbm-R) Vapor s_g	Fahrenheit Temperature
−40	15.222	0.011363	3.2957	0.000	100.257	100.257	0.00000	0.23888	−40
−36	16.859	0.011415	2.9954	1.014	99.682	100.696	0.00240	0.23767	−36
−32	18.633	0.011469	2.7278	2.035	99.097	101.132	0.00479	0.23649	−32
−28	20.549	0.011523	2.4887	3.061	98.503	101.564	0.00716	0.23534	−28
−24	22.617	0.011578	2.2746	4.093	97.899	101.992	0.00953	0.23423	−24
−20	24.845	0.011634	2.0826	5.131	97.285	102.415	0.01189	0.23315	−20
−16	27.239	0.011691	1.9099	6.175	96.660	102.835	0.01425	0.23210	−16
−12	29.809	0.011749	1.7544	7.224	96.025	103.250	0.01659	0.23108	−12
−8	32.563	0.011808	1.6141	8.280	95.380	103.660	0.01892	0.23008	−8
−4	35.509	0.011868	1.4872	9.341	94.724	104.065	0.02125	0.22912	−4
0	38.657	0.011930	1.3723	10.409	94.056	104.465	0.02357	0.22817	0
4	42.014	0.011992	1.2680	11.483	93.378	104.860	0.02587	0.22725	4
8	45.591	0.012056	1.1732	12.562	92.688	105.250	0.02818	0.22636	8
12	49.396	0.012121	1.0869	13.648	91.986	105.633	0.03047	0.22548	12
16	53.438	0.012188	1.0082	14.739	91.272	106.011	0.03275	0.22463	16
20	57.727	0.012255	0.93631	15.837	90.545	106.383	0.03503	0.22379	20
24	62.272	0.012325	0.87055	16.942	89.807	106.748	0.03730	0.22297	24
28	67.083	0.012395	0.81031	18.052	89.055	107.107	0.03958	0.22217	28
32	72.169	0.012468	0.75503	19.169	88.290	107.459	0.04182	0.22139	32
36	77.540	0.012542	0.70425	20.292	87.512	107.804	0.04407	0.22062	36
40	83.206	0.012618	0.65753	21.422	86.720	108.142	0.04632	0.21986	40
44	89.177	0.012695	0.61448	22.558	85.914	108.472	0.04855	0.21912	44
48	95.463	0.012775	0.57467	23.701	85.094	108.795	0.05079	0.21839	48
52	102.07	0.012856	0.53808	24.851	84.258	109.109	0.05301	0.21768	52
56	109.02	0.012940	0.50414	26.008	83.407	109.415	0.05523	0.21697	56

60	116.31	0.013025	0.47272	27.172	82.540	109.712	0.05745	0.21627	60
64	123.96	0.013114	0.44358	28.344	81.656	110.000	0.05966	0.21558	64
68	131.97	0.013204	0.41653	29.523	80.755	110.278	0.06186	0.21490	68
72	140.37	0.013297	0.39139	30.170	79.836	110.547	0.06406	0.21422	72
76	149.15	0.013393	0.36800	31.906	78.899	110.805	0.06626	0.21355	76
80	158.33	0.013492	0.34621	33.109	77.943	111.052	0.06846	0.21288	80
84	167.92	0.013594	0.32588	34.322	76.966	111.288	0.07065	0.21222	84
88	177.93	0.013700	0.30690	35.543	75.968	111.512	0.07285	0.21155	88
92	188.37	0.013809	0.28917	36.774	75.949	111.723	0.07504	0.21089	92
96	199.26	0.013921	0.27257	38.016	73.905	111.921	0.07723	0.21023	96
100	210.60	0.014038	0.25702	39.267	72.838	112.105	0.07942	0.20956	100
104	222.42	0.014159	0.24244	40.530	71.744	112.274	0.08161	0.20889	104
108	234.71	0.014285	0.22875	41.804	70.623	112.427	0.08381	0.20821	108
112	247.50	0.014416	0.21589	43.091	69.473	112.564	0.08601	0.20753	112
116	260.79	0.014552	0.20378	44.391	68.291	112.682	0.08821	0.20684	116
120	274.60	0.014694	0.19238	45.705	67.077	112.782	0.09042	0.20613	120
124	288.95	0.014843	0.18163	47.034	65.826	112.860	0.09264	0.20542	124
128	303.84	0.014999	0.17147	48.380	64.537	112.917	0.09487	0.20468	128
132	319.29	0.015163	0.16187	49.743	63.206	112.949	0.09711	0.20393	132
136	335.32	0.015336	0.15279	51.125	61.829	112.954	0.09936	0.20315	136
140	351.94	0.015518	0.14418	52.528	60.403	112.931	0.10163	0.20235	140
144	369.17	0.015712	0.13600	53.955	59.922	112.877	0.10391	0.20152	144
148	387.03	0.015917	0.12823	55.406	57.380	112.787	0.10622	0.20065	148
152	405.52	0.016137	0.12083	56.887	55.887	112.658	0.10856	0.19974	152
156	424.68	0.016372	0.11376	58.399	54.087	112.485	0.11093	0.19878	156
160	444.53	0.016627	0.10710	59.948	52.316	112.263	0.11334	0.19776	160
164	465.07	0.016902	0.10054	61.538	50.446	111.984	0.11580	0.19688	164
168	486.34	0.017204	0.094309	63.178	48.461	111.639	0.11831	0.19552	168
172	508.37	0.017538	0.088299	64.875	46.340	111.216	0.12089	0.19425	172

Source: Copyright (1964) by the Du Pont Company. Reprinted by permission.

Table A-4b Refrigerant 22 Saturation Properties—SI Units

Celsius Temperature	Pressure Pa	Enthalpy kJ/kg			Entropy kJ/(kg-K)		Specific Volume m³/kg	
	$P \times 10^{-5}$	i_f	i_{fg}	i_g	s_f	s_g	$v_f \times 10^3$	$v_g \times 10^3$
-40	1.0490	155.624	233.204	388.828	0.82489	1.82512	0.70935	205.841
-35	1.3162	160.923	230.162	391.085	0.84742	1.81388	0.71679	166.470
-30	1.6340	166.291	227.008	393.299	0.86975	1.80337	0.72451	135.907
-28	1.7768	168.458	225.712	394.170	0.87863	1.79934	0.72768	125.613
-26	1.9291	170.636	224.398	395.034	0.88748	1.79542	0.73091	116.262
-24	2.0912	172.826	223.064	395.890	0.89630	1.79160	0.73418	107.747
-22	2.2639	175.027	221.709	396.736	0.90509	1.78786	0.37751	99.9741
-20	2.4472	177.239	220.335	397.574	0.91385	1.78423	0.74090	92.8825
-18	2.6418	179.463	218.939	398.402	0.92259	1.78066	0.74435	86.3881
-16	2.8481	181.698	217.522	399.220	0.93129	1.77719	0.74785	80.4427
-14	3.0666	183.945	216.083	400.028	0.93997	1.77378	0.75142	74.9848
-12	3.2976	186.204	214.622	400.826	0.94862	1.77046	0.75505	69.9736
-10	3.5416	188.473	213.140	401.613	0.95725	1.76720	0.75875	65.3646
-8	3.7992	190.755	211.634	402.388	0.96585	1.76401	0.76251	61.1184
-6	4.0708	193.049	210.104	403.153	0.97442	1.76089	0.76635	57.2017
-4	4.3568	195.354	208.551	403.905	0.98297	1.75782	0.77026	53.5871
-2	4.6577	197.671	206.974	404.645	0.99150	1.75482	0.77425	50.2452
0	4.9740	200.000	205.372	405.373	1.00000	1.75187	0.77832	47.1523
2	5.3065	202.343	203.744	406.087	1.00848	1.74896	0.78248	44.2848
4	5.6549	204.690	202.092	406.788	1.01694	1.74612	0.78672	41.6278
6	6.0207	207.062	200.412	407.474	1.02537	1.74331	0.79105	39.1576
8	6.4036	209.441	198.706	408.147	1.03378	1.74055	0.79548	36.8624
10	6.8045	211.834	196.971	408.804	1.04218	1.73782	0.80001	34.7261
12	7.2239	214.240	195.207	409.447	1.05056	1.73514	0.80464	32.7357
14	7.6626	216.660	193.412	410.072	1.05892	1.73248	0.80937	30.8778
16	8.1203	219.092	191.589	410.681	1.06726	1.72985	0.81423	29.1453

18	8.5980	221.539	189.735	411.274	1.07559	1.72726	0.81920	27.5268
20	9.0967	224.002	187.846	411.847	1.08390	1.72468	0.82430	26.0109
22	9.6158	226.477	185.926	412.403	1.09220	1.72214	0.82953	24.5937
24	10.157	228.970	183.907	412.939	1.10049	1.71960	0.83489	23.2643
26	10.720	231.477	181.978	413.455	1.10876	1.71708	0.84041	22.0177
28	11.306	234.001	179.949	413.950	1.11703	1.71457	0.84608	20.8476
30	11.915	236.543	177.881	414.424	1.12530	1.71207	0.85191	19.7478
32	12.549	239.102	175.771	414.873	1.13356	1.70957	0.85791	18.7129
34	13.206	241.678	173.621	415.300	1.14181	1.70708	0.86410	17.7397
36	13.889	244.276	171.424	415.700	1.15007	1.70457	0.87049	16.8217
38	14.597	246.892	169.182	416.074	1.15833	1.70206	0.87708	15.9564
40	15.331	249.530	166.890	416.419	1.16659	1.69953	0.88390	15.1396
42	16.092	252.190	164.546	416.735	1.17487	1.69699	0.89095	14.3677
44	16.881	254.873	162.148	417.020	1.18315	1.69442	0.89825	13.6380
46	17.698	257.581	159.690	417.271	1.19145	1.69181	0.90583	12.9471
48	18.543	260.314	157.173	417.487	1.19977	1.68918	0.91371	12.2929
50	19.418	263.073	154.594	417.667	1.20811	1.68650	0.92190	11.6731
52	20.323	265.866	151.938	417.804	1.21648	1.68377	0.93044	11.0839
54	21.258	268.686	149.214	417.900	1.22488	1.68099	0.93936	10.5251
56	22.227	271.546	146.400	417.946	1.23334	1.67812	0.94869	9.99235
58	23.227	274.441	143.502	417.943	1.24183	1.67518	0.95847	9.48558
60	24.260	277.373	140.514	417.887	1.25038	1.67215	0.96874	9.00332
62	25.326	280.348	137.423	417.771	1.25899	1.66903	0.97957	8.54301
64	26.428	283.373	134.216	417.589	1.26768	1.66577	0.99101	8.10286
66	27.566	286.451	130.883	417.334	1.27647	1.66238	1.00314	7.68155
68	28.740	289.585	127.417	417.002	1.28535	1.65884	1.01604	7.27841
70	29.952	292.786	123.795	416.581	1.29436	1.65512	1.02983	6.89127
75	33.153	301.133	113.925	415.058	1.31757	1.64480	1.06911	5.98528
80	36.614	310.139	102.511	412.650	1.34222	1.63249	1.11803	5.15095

Source: Abstracted by permission from "Thermodynamic Table for Refrigerant R22 in SI-Units," International Institute of Refrigeration, Paris.

Thermophysical Properties

Table B-1a Refrigerant 12—English Units

Fahrenheit Temperature	Viscosity, μ lbm/(ft-hr)			Thermal Conductivity, k Btu/(hr-ft-F)			Specific Heat, c_p Btu/(lbm-F)			Fahrenheit Temperature
	Saturated Liquid	Saturated Vapor	Gas $P = 1$ atm	Saturated Liquid	Saturated Vapor	Gas $P = 1$ atm	Saturated Liquid	Saturated Vapor	Gas $P = 0$	
−140	2.44			0.0655			0.199		0.1085	−140
−120	1.95			0.0631			0.202		0.1233	−120
−100	1.593			0.0608			0.204		0.1160	−100
−80	1.331			0.0585			0.207		0.1196	−80
−60	1.132			0.0561			0.209	0.126	0.1230	−60
−40	0.978			0.0538			0.212	0.133	0.1264	−40
−20	0.856	0.0246	0.0246	0.0514	0.0040	0.00400	0.214	0.139	0.1296	−20
0	0.758	0.0262	0.0258	0.0490	0.0043	0.00431	0.217	0.145	0.1327	0
20	0.679	0.0276	0.0269	0.0467	0.0046	0.00463	0.220	0.150	0.1356	20
40	0.613	0.0287	0.0280	0.0443	0.0050	0.00495	0.224	0.157	0.1385	40
60	0.557	0.0297	0.0290	0.0420	0.0053	0.00528	0.229	0.164	0.1413	60
80	0.511	0.0308	0.0301	0.0397	0.0056	0.00561	0.234	0.174	0.1439	80
100	0.471	0.0320	0.0312	0.0373	0.0060	0.00594	0.240	0.185	0.1465	100
120	0.436	0.0335	0.0322	0.0350	0.0064	0.00627	0.251	0.199	0.1490	120
140	0.403	0.0354	0.0332	0.0326	0.0068	0.00660	0.266	0.216	0.1513	140
160	0.366	0.0380	0.0342	0.0302	0.0072	0.00694	0.288	0.235	0.1536	160
180	0.325	0.0412	0.0352	0.0276	0.0076	0.00728	0.317	0.290	0.1558	180
200	0.270	0.0453	0.0362	0.0246	0.0083	0.00763	0.356	0.362	0.1579	200
220	0.198	0.051	0.0372	0.0204	0.0093	0.00798	0.406		0.1599	220
230	0.148	0.060	0.0377	0.0161	0.0107	0.00816			0.1609	230
234	0.075	0.075	0.0379	0.0130	0.130	0.00823			0.1612	234
240			0.0382			0.00834			0.1618	240
260			0.0392			0.00871			0.1637	260
280			0.0401			0.00908			0.1654	280
300			0.0410			0.00945			0.1671	300
320			0.0420			0.00982			0.1687	320
340			0.0429			0.0101			0.1703	340
360			0.0438			0.0105			0.1718	360
380			0.0447			0.0108			0.1732	380
400			0.0456			0.0112			0.1746	400

Source: Reprinted by permission from *ASHRAE Handbook, Fundamentals Volume,* 1972.

Table B-1b Refrigerant 12—SI Units

Celsius Temperature	Viscosity, $\mu \times 10^3$ (N-s/m²)			Thermal Conductivity, k W/(m-K)			Specific Heat, c_p kJ/(kg-K)			Kelvin Temperature
	Saturated Liquid	Saturated Vapor	Gas $P=101.33$ kPa	Saturated Liquid	Saturated Vapor	Gas $P=101.33$ kPa	Saturated Liquid	Saturated Vapor	Gas $P=0$	
−95	1.01			0.113			0.833		0.4543	178
−84	0.806			0.109			0.846		0.4702	189
−73	0.659			0.105			0.854		0.4857	200
−62	0.550			0.101			0.867		0.5007	211
−51	0.468			0.097			0.875	0.528	0.5150	222
−40	0.404			0.093			0.888	0.557	0.5292	233
−29	0.354	0.0102	0.0102	0.089	0.0069	0.0069	0.896	0.582	0.5426	244
−18	0.313	0.0108	0.0107	0.085	0.0074	0.0075	0.909	0.607	0.5556	255
−7	0.281	0.0114	0.0111	0.081	0.0080	0.0080	0.921	0.628	0.5677	266
4	0.253	0.0119	0.0116	0.077	0.0087	0.0086	0.938	0.657	0.5799	277
16	0.230	0.0123	0.0120	0.073	0.0920	0.0091	0.959	0.687	0.5916	289
27	0.211	0.0127	0.0124	0.069	0.0097	0.0097	0.980	0.729	0.6025	300
38	0.195	0.0132	0.0129	0.065	0.0104	0.0103	1.00	0.775	0.6113	311
49	0.180	0.0138	0.0133	0.061	0.0111	0.0109	1.05	0.833	0.6238	322
60	0.167	0.0146	0.0137	0.056	0.0118	0.0114	1.11	0.904	0.6335	333
71	0.151	0.0157	0.0141	0.052	0.0125	0.0120	1.21	0.984	0.6431	344
82	0.134	0.0170	0.0146	0.048	0.0132	0.0126	1.33	1.21	0.6523	355
93	0.112	0.0187	0.0150	0.043	0.0173	0.0132	1.49	1.52	0.6611	366
104	0.082	0.021	0.0154	0.035	0.0161	0.0138	1.70		0.6695	377
110	0.061	0.025	0.0156	0.028	0.0185	0.0141			0.6737	383
112	0.031	0.031	0.0157	0.022	0.0225	0.0142			0.6749	385
116			0.0158			0.0144			0.6774	389
127			0.0162			0.0151			0.6854	400
138			0.0166			0.0157			0.6925	411
149			0.0169			0.0164			0.6996	422
160			0.0174			0.0170			0.7063	433
171			0.0177			0.0175			0.7130	444
183			0.0181			0.0182			0.7193	456
194			0.0185			0.0187			0.7252	467
205			0.0189			0.0194			0.7310	478

Source: Adapted by permission from *ASHRAE Handbook, Fundamentals Volume,* 1972.

Table B-2a Refrigerant 22—English Units

Fahrenheit Temperature	Viscosity, μ lbm/(ft-hr)			Thermal Conductivity, k Btu/(hr-ft-F)			Specific Heat, c_p Btu/(lbm-F)			Fahrenheit Temperature
	Saturated Liquid	Saturated Vapor	Gas $P = 1$ atm	Saturated Liquid	Saturated Vapor	Gas $P = 1$ atm	Saturated Liquid	Saturated Vapor	Gas $P = 0$	
−100	1.153			0.0789			0.255		0.1260	−100
−80	1.002			0.0757			0.256		0.1292	−80
−60	0.883			0.0725			0.259	0.139	0.1324	−60
−40	0.788	0.0242	0.0242	0.0693	0.0040	0.00404	0.262	0.146	0.1356	−40
−20	0.710	0.0254	0.0254	0.0661	0.0044	0.00443	0.266	0.152	0.1388	−20
0	0.646	0.0266	0.0265	0.0630	0.0048	0.00481	0.271	0.158	0.1420	0
20	0.592	0.0279	0.0276	0.0598	0.0052	0.00520	0.276	0.165	0.1452	20
40	0.546	0.0292	0.0288	0.0566	0.0056	0.00558	0.283	0.175	0.1484	40
60	0.507	0.0306	0.0299	0.0534	0.0060	0.00597	0.291	0.187	0.1515	60
80	0.474	0.0321	0.0310	0.0502	0.0064	0.00635	0.300	0.204	0.1546	80
100	0.444	0.0339	0.0321	0.0471	0.0068	0.00674	0.313	0.226	0.1577	100
120	0.422	0.0360	0.0333	0.0439	0.0072	0.00712	0.332	0.253	0.1608	120
140	0.387	0.0383	0.0344	0.0407	0.0077	0.00751	0.357	0.288	0.1638	140
160	0.340	0.0411	0.0354	0.0371	0.0084	0.00790	0.390	0.332	0.1668	160
180	0.285	0.045	0.0365	0.0318	0.0105	0.00828	0.433		0.1697	180
190	0.244	0.049	0.0371	0.0288	0.0119	0.00848			0.1712	190
200	0.182	0.058	0.0376	0.0238	0.0140	0.00867			0.1726	200
205	0.074	0.074	0.0379	0.0177	0.0177	0.00876			0.1733	205
220			0.0387			0.00905			0.1754	220
240			0.0398			0.00944			0.1782	240
260			0.0408			0.00982			0.1810	260
280			0.0419			0.01021			0.1836	280
300			0.0429			0.01059			0.1863	300
320			0.0439			0.0110			0.1888	320
340			0.0450			0.0114			0.1913	340
360			0.0460			0.0118			0.1937	360
380			0.0470			0.0121			0.1960	380
400			0.0480			0.0125			0.1983	400
420			0.0490			0.0129			0.2005	420
440			0.0500			0.0133			0.2026	440

Source: Reprinted by permission from *ASHRAE Handbook, Fundamentals Volume,* 1972.

Table B-2b Refrigerant 22—SI Units

Celsius Temperature	Viscosity, $\mu \times 10^3$ (N-s)/m² Saturated Liquid	Saturated Vapor	Gas P = 101.33 kPa	Thermal Conductivity, k W/(m-K) Saturated Liquid	Saturated Vapor	Gas P = 101.33 kPa	Specific Heat, c_p kJ/(kg-K) Saturated Liquid	Saturated Vapor	Gas P = 0	Kelvin Temperature
− 73	0.4766		0.0100	0.137			1.07		0.5275	200
− 62	0.4142		0.0105	0.131			1.07		0.5409	211
− 51	0.365		0.0110	0.125			1.08	0.582	0.5543	222
− 40	0.326	0.0100	0.0114	0.120	0.0069	0.00699	1.10	0.611	0.5677	233
− 29	0.294	0.0105	0.0119	0.114	0.0076	0.00767	1.11	0.636	0.5811	244
− 18	0.267	0.0110	0.0124	0.109	0.0083	0.00832	1.13	0.662	0.5945	255
− 7	0.245	0.0115	0.0128	0.103	0.0090	0.00900	1.16	0.691	0.6079	266
4	0.226	0.0121	0.0133	0.0980	0.0097	0.00966	1.18	0.733	0.6213	277
16	0.210	0.0126	0.0138	0.0924	0.010	0.0103	1.22	0.783	0.6343	289
27	0.196	0.0133	0.0142	0.0869	0.011	0.0110	1.26	0.854	0.6473	300
38	0.184	0.0140	0.0146	0.0815	0.012	0.0117	1.31	0.946	0.6603	311
49	0.174	0.0149	0.0151	0.0760	0.012	0.0123	1.39	1.06	0.6732	322
60	0.160	0.0158	0.0153	0.0704	0.013	0.0130	1.49	1.21	0.6858	333
71	0.141	0.0170	0.0155	0.0642	0.015	0.0137	1.63	1.39	0.6984	344
82	0.118	0.019	0.0157	0.0550	0.018	0.0143	1.81		0.7105	355
88	0.101	0.020	0.0160	0.0498	0.021	0.0147			0.7168	361
94	0.075	0.024	0.0165	0.0412	0.024	0.0150			0.7226	367
96	0.031	0.031	0.0169	0.0306	0.031	0.0152			0.7256	369
105			0.0173			0.0157			0.7344	378
116			0.0177			0.0163			0.7461	389
127			0.0181			0.0170			0.7578	400
138			0.0186			0.01767			0.7687	411
149			0.0190			0.01833			0.7800	422
160			0.0194			0.0190			0.7905	433
171			0.0198			0.0197			0.8009	444
183			0.0203			0.0204			0.8110	456
194			0.0207			0.0209			0.8206	467
205						0.0216			0.8302	478
216						0.0223			0.8395	489
227						0.0230			0.8482	500

Source: Adapted by permission from *ASHRAE Handbook. Fundamentals Volume, 1972.*

Table B-3a Water–English Units

Fahrenheit Temperature	Viscosity, μ lbm/(ft-hr)			Thermal Conductivity, k Btu/(hr-ft-F)			Specific Heat, c_p Btu/(lbm-F)				Fahrenheit Temperature
	Saturated Liquid	Saturated Vapor	Gas $P = 1$ atm	Saturated Liquid	Saturated Vapor	Gas $P = 1$ atm	Saturated Liquid	Saturated Vapor	Gas $P = 0$	Gas $P = 1$ atm	
32	4.23	0.0192		0.329	0.0100		1.007		0.4438		32
40	3.65	0.0197		0.334	0.0103		1.005		0.4442		40
50	3.09	0.0202		0.339	0.0105		1.003	0.4446	0.4445		50
60	2.64	0.0207		0.345	0.0107		1.001	0.4454	0.4448		60
70	2.29	0.0213		0.350	0.0109		1.000	0.4464	0.4452		70
80	2.01	0.0218		0.354	0.0112		0.999	0.4475	0.4456		80
90	1.77	0.0224		0.359	0.0114		0.998	0.4488	0.4459		90
100	1.58	0.0229		0.363	0.0116		0.998	0.4522	0.4463		100
120	1.29	0.0240		0.371	0.0121		0.998	0.4567	0.4472		120
140	1.080	0.0251		0.378	0.0125		1.000	0.4626	0.4480		140
160	0.927	0.0261		0.383	0.0130		1.001	0.4699	0.4490		160
180	0.804	0.0272		0.388	0.0135		1.003	0.4791	0.4500		180
200	0.708	0.0283		0.392	0.0140		1.006	0.4859	0.4510		200
212	0.660	0.0289	0.0289	0.394	0.0143	0.0143	1.008	0.4902	0.4516	0.4892	212
220	0.631	0.0292	0.0294	0.395	0.0145	0.0144	1.009	0.5019	0.4521	0.4867	220
240	0.567	0.0302	0.0305	0.397	0.0151	0.0147	1.013	0.5157	0.4532	0.4819	240
260	0.515	0.0312	0.0315	0.398	0.0158	0.0153	1.017	0.5319	0.4544	0.4780	260
280	0.471	0.0322	0.0326	0.398	0.0165	0.0158	1.022	0.5508	0.4556	0.4750	280
300	0.433	0.0331	0.0337	0.397	0.0172	0.0164	1.029	0.573	0.4569	0.4729	300
320	0.401	0.0341	0.0348	0.396	0.0181	0.0170	1.040	0.598	0.4582	0.4714	320
340	0.374	0.0350	0.0359	0.394	0.0190	0.0175	1.053	0.626	0.4595	0.4706	340
360	0.350	0.0359	0.0369	0.391	0.0199	0.0181	1.066	0.695	0.4609	0.4703	360
400	0.311	0.0377	0.0391	0.383	0.0222	0.0194	1.085	0.968	0.4638	0.4709	400
500	0.248	0.0427	0.0445	0.349	0.0306	0.0226	1.180	1.71	0.4715	0.4745	500
600	0.197	0.0504	0.0499	0.296	0.0486	0.0262	1.528		0.4798	0.4817	600
700			0.0553	0.188	0.1203	0.0298			0.4885	0.4895	700
706	0.101	0.101	0.0556	0.139	0.139	0.0300			0.4890	0.4900	706
750			0.0580			0.0317			0.4929	0.4935	750
800			0.0607			0.0337			0.4973	0.4977	800
1000			0.0715			0.0418			0.5150	0.5152	1000

Source: Reprinted by permission from *ASHRAE Handbook, Fundamentals Volume*, 1972.

Table B-3b Water—SI Units

Celsius Temperature	Viscosity, $\mu \times 10^3$ (N·s/m²)			Thermal Conductivity, k W/(m·K)			Specific Heat, c_p kJ/(kg·K)				Kelvin Temperature
	Saturated Liquid	Saturated Vapor	Gas P = 101.33 kPa	Saturated Liquid	Saturated Vapor	Gas P = 101.33 kPa	Saturated Liquid	Saturated Vapor	Gas P = 0	Gas P = 101.33 kPa	
0	1.75	0.00794		0.569	0.0173		4.216		1.858		273
4	1.51	0.00814		0.578	0.0178		4.208		1.860		277
10	1.28	0.00835		0.587	0.0182		4.199		1.861		283
16	1.09	0.00856		0.597	0.0185		4.191	1.861	1.862		289
21	0.947	0.00880		0.606	0.0189		4.187	1.865	1.864		294
27	0.831	0.00901		0.613	0.0194		4.183	1.869	1.866		300
32	0.732	0.00926		0.621	0.0197		4.178	1.874	1.867		305
38	0.653	0.00947		0.628	0.0201		4.178	1.879	1.869		311
49	0.533	0.00992		0.642	0.0209		4.178	1.893	1.872		322
60	0.446	0.0104		0.654	0.0216		4.187	1.912	1.876		333
71	0.383	0.0108		0.663	0.0225		4.191	1.937	1.880		344
82	0.332	0.0112		0.672	0.0234		4.199	1.967	1.884		355
93	0.293	0.0117		0.678	0.0242		4.212	2.006	1.888		366
100	0.273	0.0119	0.0119	0.682	0.0247	0.0247	4.220	2.034	1.891	2.048	373
104	0.261	0.121	0.0122	0.684	0.0251	0.0249	4.224	2.052	1.893	2.038	377
116	0.234	0.0125	0.0126	0.687	0.0261	0.0254	4.241	2.101	1.897	2.018	389
127	0.213	0.0129	0.0130	0.689	0.0273	0.0265	4.258	2.159	1.902	2.001	400
138	0.195	0.0133	0.0135	0.689	0.0286	0.0273	4.280	2.227	1.908	1.989	411
149	0.179	0.0137	0.0139	0.687	0.0298	0.0284	4.308	2.306	1.913	1.980	422
160	0.166	0.0141	0.0144	0.685	0.0313	0.0294	4.354	2.399	1.918	1.974	433
171	0.155	0.0145	0.0148	0.682	0.0329	0.0303	4.409	2.504	1.924	1.970	444
183	0.145	0.0148	0.0153	0.677	0.0344	0.0313	4.463	2.621	1.930	1.969	456
205	0.129	0.0156	0.0162	0.663	0.0384	0.0336	4.543	2.910	1.942	1.972	478
260	0.103	0.0177	0.0184	0.604	0.0530	0.0391	4.940	4.053	1.974	1.987	533
316	0.0814	0.208	0.0206	0.512	0.0841	0.0453	6.397	7.159	2.009	2.017	589
371			0.0229	0.325	0.208	0.0516			2.045	2.049	644
375	0.0418	0.0418	0.0230	0.241	0.241	0.0519			2.047	2.052	648
399			0.0240			0.0549			2.064	2.066	672
427			0.0251			0.0583			2.082	2.084	700
538			0.0296			0.0723			2.156	2.157	811

Source: Adapted by permission from *ASHRAE Handbook, Fundamentals Volume*, 1972.

Table B-4a Air—English Units

Fahrenheit Temperature	Viscosity, μ lbm/(ft-hr) Saturated Liquid	Saturated Vapor	Gas $P = 1$ atm	Thermal Conductivity, k Btu/(hr-ft-F) Saturated Liquid	Saturated Vapor	Gas $P = 1$ atm	Specific Heat, c_p Btu/(lbm-F) Saturated Liquid	Saturated Vapor	Gas $P = 0$	Gas $P = 1$ atm	Fahrenheit Temperature
-352	0.7865			0.1040	0.00312						-352
-334	0.5348			0.0942	0.00370						-334
-316	0.3993	0.0133		0.0838	0.00433		0.4690		0.2394		-316
-298	0.3194	0.0157	0.0154	0.0740	0.00497	0.00480	0.4962	0.2676	0.2394		-298
-280	0.2664	0.0182	0.0171	0.0636	0.00584	0.00532	0.5268	0.2891	0.2394	0.2456	-280
-262	0.2297	0.0208	0.0188	0.0537	0.00705	0.00589	0.5784	0.3440	0.2394	0.2442	-262
-244	0.1815	0.0247	0.0204	0.0439	0.00890	0.00641	0.6689	0.4778	0.2394	0.2430	-244
-226	0.1016	0.0346	0.0220	0.0312	0.01213	0.00693			0.2394	0.2422	-226
-220.6	0.05009	0.05009	0.0225	0.01964	0.01964	0.00711			0.2394	0.2420	-220.6
-208			0.0236			0.00745			0.2394	0.2418	-208
-190			0.0251			0.00797			0.2394	0.2415	-190
-172			0.0266			0.00844			0.2394	0.2411	-172
-136			0.0295			0.00948			0.2394	0.2406	-136
-100			0.0324			0.0105			0.2394	0.2403	-100
-64			0.0351			0.0114			0.2396	0.2403	-64
-28			0.0375			0.0124			0.2396	0.2401	-28
8			0.0402			0.0134			0.2396	0.2401	8
44			0.0426			0.0142			0.2399	0.2403	44
80			0.0448			0.0151			0.2401	0.2403	80
116			0.0472			0.0160			0.2403	0.2406	116
152			0.0494			0.0168			0.2406	0.2408	152
188			0.0516			0.0176			0.2411	0.2413	188
224			0.0535			0.0183			0.2415	0.2418	224
260			0.0554			0.0191			0.2420	0.2422	260
296			0.0576			0.0199			0.2427	0.2430	296
332			0.0593			0.0206			0.2434	0.2437	332
368			0.0612			0.0214			0.2442	0.2444	368
404			0.0632			0.0221			0.2449	0.2451	404
440			0.0649			0.0228			0.2458	0.2461	440
620			0.0733			0.0264			0.2511	0.2513	620

Source: Adapted by permission from *ASHRAE Handbook, Fundamentals Volume,* 1972.

Table B-4b Air—SI Units

Celsius Temperature	Viscosity, $\mu \times 10^3$ (N-s)/m²			Thermal Conductivity, k W/(m-K)			Specific Heat, c_p kJ/(kg-K)				Kelvin Temperature
	Saturated Liquid	Saturated Vapor	Gas $P = 101.33$ kPa	Saturated Liquid	Saturated Vapor	Gas $P = 101.33$ kPa	Saturated Liquid	Saturated Vapor	Gas $P = 0$	Gas $P = 101.33$ kPa	
−213	0.325			0.180	0.0054						60
−203	0.221			0.163	0.0064						70
−193	0.165	0.0055		0.145	0.0075		1.963		1.002		80
−183	0.132	0.0065	0.00635	0.128	0.0086	0.0083	2.077	1.12	1.002		90
−173	0.1101	0.0075	0.00706	0.110	0.0101	0.0092	2.205	1.21	1.002	1.028	100
−163	0.0949	0.0086	0.00775	0.093	0.0122	0.0102	2.421	1.44	1.002	1.022	110
−153	0.0750	0.0102	0.00843	0.076	0.0154	0.0111	2.80	2.00	1.002	1.017	120
−143	0.0420	0.0143	0.00909	0.054	0.021	0.0120			1.002	1.014	130
−140	0.0207	0.0207	0.00929	0.034	0.034	0.0123			1.002	1.013	133
−133			0.00974			0.0129			1.002	1.012	140
−123			0.01038			0.0138			1.002	1.011	150
−113			0.0110			0.0146			1.002	1.009	160
−93			0.0122			0.0164			1.002	1.007	180
−73			0.0134			0.0181			1.002	1.006	200
−53			0.0145			0.0198			1.003	1.006	220
−33			0.0155			0.0215			1.003	1.005	240
−13			0.0166			0.0231			1.003	1.005	260
7			0.0176			0.0246			1.004	1.006	280
27			0.0185			0.0261			1.005	1.006	300
47			0.0195			0.0276			1.006	1.007	320
67			0.0204			0.0290			1.007	1.008	340
87			0.0213			0.0304			1.009	1.010	360
107			0.0221			0.0317			1.011	1.012	380
127			0.0229			0.0331			1.013	1.014	400
147			0.0238			0.0344			1.016	1.017	420
167			0.0245			0.0357			1.019	1.020	440
187			0.0253			0.0370			1.022	1.023	460
207			0.0261			0.0383			1.025	1.026	480
227			0.0268			0.0395			1.029	1.030	500
327			0.0303			0.0456			1.051	1.052	600

Source: Reprinted by permission from *ASHRAE Handbook, Fundamentals Volume,* 1972.

APPENDIX C

Weather Data

Table C-1 Climatic Conditions for the United States and Canada

State and Station	Latitude, Degree and Minutes	Longitude, Degree and Minutes	Elevation		Winter				Summer					
			ft	m	99 Percent		$97\frac{1}{2}$ Percent		Dry Bulb $2\frac{1}{2}$ Percent		Wet Bulb $2\frac{1}{2}$ Percent		Daily Range	
					F	C	F	C	F	C	F	C	F	C
Alabama, Birmingham AP	33 3	86 5	610	186	17	-8	21	-6	94	34	75	24	21	12
Alaska, Anchorage AP	61 1	150 0	90	27	-23	-31	-18	-28	68	20	58	14	15	8
Arizona, Tucson AP	32 1	111 0	2584	788	28	-2	32	0	102	39	66	19	26	14
Arkansas, Little Rock AP	34 4	92 1	257	78.3	15	-9	20	-7	96	36	77	25	22	12
California, San Francisco AP	37 4	122 2	8	2.4	35	2	38	3	77	25	63	17	20	11
Colorado, Denver AP	39 5	104 5	5283	1610	-5	-21	1	-17	91	33	59	15	28	16
Connecticut, Bridgeport AP	41 1	73 1	7	2.1	-6	-21	9	-13	84	29	71	22	18	10
Delaware, Wilmington AP	39 4	75 3	78	24	10	-12	14	-10	89	32	74	23	20	11
District of Columbia, Washington National	38 5	77 0	14	4.3	14	-10	17	-8	91	33	74	23	18	10
Florida, Tallahassee AP	30 2	84 2	58	18	27	-3	30	-1	92	33	76	24	19	11
Georgia, Atlanta AP	33 4	84 3	1005	306	17	-8	22	-6	92	33	74	23	19	11
Hawaii, Honolulu AP	21 2	158 0	7	2.1	62	17	63	17	86	30	73	23	12	7
Idaho, Boise AP	43 3	116 1	2842	866	3	-16	10	-12	94	34	64	18	31	17
Illinois, Chicago O'Hare AP	42 0	87 5	658	201	-8	-22	-4	-20	89	32	74	23	20	11
Indiana, Indianapolis AP	39 4	86 2	793	242	-2	-19	2	-17	90	32	74	23	22	12
Iowa, Sioux City AP	42 2	96 2	1095	334	-11	-24	-7	-22	92	33	74	23	24	13
Kansas, Wichita AP	37 4	97 3	1321	403	3	-16	7	-14	98	37	73	23	23	13
Kentucky, Louisville AP	38 1	85 4	474	144	5	-15	10	-12	93	34	75	24	23	13
Louisiana, Shreveport AP	32 3	93 5	252	76.8	20	-7	25	-4	96	36	76	24	20	11
Maine, Caribou AP	46 5	68 0	624	190	-18	-28	-13	-25	81	27	67	19	21	12
Maryland, Baltimore AP	39 1	76 4	146	44.5	10	-12	13	-11	91	33	75	24	21	12
Massachusetts, Boston AP	42 2	71 0	15	4.6	-6	-14	9	-13	88	31	71	22	16	9
Michigan, Lansing AP	42 5	84 4	852	260	-3	-19	1	-17	87	31	72	22	24	13
Minnesota, Minneapolis/St. Paul	44 5	93 1	822	251	-16	-27	-12	-24	89	32	73	23	22	12
Mississippi, Jackson AP	32 2	90 1	330	101	21	-6	25	-4	95	35	76	24	21	12
Missouri, Kansas City AP	39 1	94 4	742	226	2	-17	6	-14	96	36	74	23	20	11

TABLE C.

Location														
Montana, Billings AP	45 4	108 3	3567	1087	-15	-26	-10	-23	91	33	64	18	31	17
Nebraska, Lincoln CO	40 5	96 5	1150	351	-5	-21	-2	-19	95	35	74	23	24	13
Nevada, Las Vegas AP	36 1	115 1	2162	659	25	-4	28	-12	106	41	65	18	30	17
New Hampshire, Concord	43 1	71 3	339	103	-8	-22	-3	-19	87	31	70	21	26	14
New Jersey, Atlantic City CO	39 3	74 3	11	3.4	10	-12	13	-11	89	32	74	23	18	10
New Mexico, Albuquerque AP	35 0	106 4	5310	1618	12	-11	16	-9	94	34	61	16	30	17
New York, Syracuse AP	43 1	76 1	424	129	-3	-19	2	-17	87	31	71	22	20	11
North Carolina, Charlotte AP	35 1	81 0	735	224	18	-8	22	-6	93	34	74	23	20	11
North Dakota, Bismarck AP	46 5	100 5	1647	502	-23	-31	-19	-28	91	33	68	20	27	15
Ohio, Cleveland AP	41 2	81 5	777	237	1	-17	5	-15	88	31	72	22	22	12
Oklahoma, Stillwater	36 1	97 1	884	269	8	-13	13	-11	96	36	74	23	24	13
Oregon, Pendleton AP	45 4	118 5	1492	455	-2	-19	5	-15	93	34	64	18	29	16
Pennsylvania, Pittsburgh AP	40 3	80 0	1137	347	1	-17	5	-15	86	30	71	22	22	12
Rhode Island, Providence AP	41 4	71 3	55	17	5	-15	9	-13	86	30	72	22	19	11
South Carolina, Charleston AFB	32 5	80 0	41	12	24	-4	27	-3	91	33	78	26	18	10
South Dakota, Rapid City	44 0	103 0	3165	965	-11	-24	-7	-22	92	33	65	18	28	16
Tennessee, Memphis AP	35 0	90 0	263	80.2	13	-11	18	-8	95	35	76	24	21	12
Texas, Dallas AP	32 5	96 5	481	147	18	-8	22	-6	100	38	75	24	20	11
Utah, Salt Lake City	40 5	112 0	4220	1286	3	-16	8	-13	95	35	62	17	32	18
Vermont, Burlington AP	44 3	73 1	331	101	-12	-24	-7	-22	85	29	70	21	23	13
Virginia, Norfolk AP	36 5	76 1	26	7.9	20	-7	22	-6	91	33	76	24	18	10
Washington, Spokane AP	47 4	117 3	2357	718	-6	-21	2	-17	90	32	63	17	28	16
West Virginia, Charleston AP	38 2	81 4	939	286	7	-14	11	-12	90	32	73	23	20	11
Wisconsin, Milwaukee AP	43 0	87 5	672	205	-8	-22	-4	-20	87	31	73	23	21	12
Wyoming, Casper AP	42 5	106 3	5319	1621	-11	-24	-5	-21	90	32	57	14	31	17
Alberta, Calgary AP	51 1	114 1	3540	1079	-27	-33	-23	-31	81	27	61	16	25	14
British Columbia, Vancouver AP	49 1	123 1	16	4.9	15	-9	19	-7	77	25	66	19	17	9
Manitoba, Winnipeg AP	49 5	97 1	786	240	-30	-34	-27	-33	86	30	71	22	22	12
New Brunswick, Fredericton	45 5	66 3	74	23	-16	-27	-11	-24	85	29	69	21	23	13
Nova Scotia, Halifax AP	44 4	63 3	136	41.5	1	-17	5	-15	76	24	65	18	16	9
Ontario, Ottawa AP	45 2	76 0	339	103	-17	-27	-13	-25	87	31	71	22	21	12
Quebec, Montreal AP	45 3	74 0	98	30	-16	-27	-10	-23	85	29	72	22	18	10
Saskatchewan, Regina AP	50 3	105 0	1884	574	-33	-36	-29	-34	88	31	68	20	27	15

Source: Abridged by permission from *ASHRAE Handbook, Fundamentals Volume, 1977.*

TABLE C.

Table C-2 Average Degree Days[a] for Cities in the United States and Canada

State and Station	Average Winter Temperature F	Average Winter Temperature C	July	August	September	October	November	December	January	February	March	April	May	June	Yearly Total
Alabama, Birmingham	54.2	12.7	0	0	6	93	363	555	592	462	363	108	9	0	2551
Alaska, Anchorage	23.0	5.0	245	291	516	930	1284	1572	1631	1316	1293	879	592	315	10864
Arizona, Tucson	58.1	14.8	0	0	0	25	231	406	471	344	242	75	6	0	1800
Arkansas, Little Rock	50.5	10.6	0	0	9	127	465	716	756	577	434	126	9	0	3219
California, San Francisco	53.4	12.2	82	78	60	143	306	462	508	395	363	279	214	126	3015
Colorado, Denver	37.6	3.44	6	9	117	428	819	1035	1132	938	887	558	288	66	6283
Connecticut, Bridgeport	39.9	4.72	0	0	66	307	615	986	1079	966	853	510	208	27	5617
Delaware, Wilmington	42.5	6.17	0	0	51	270	588	927	980	874	735	387	112	6	4930
District of Columbia, Washington	45.7	7.94	0	0	33	217	519	834	871	762	626	288	74	0	4224
Florida, Tallahassee	60.1	15.9	0	0	0	28	198	360	375	286	202	86	0	0	1485
Georgia, Atlanta	51.7	11.28	0	0	18	124	417	648	636	518	428	147	25	0	2961
Hawaii, Honolulu	74.2	23.8	0	0	0	0	0	0	0	0	0	0	0	0	0
Idaho, Boise	39.7	4.61	0	0	132	415	792	1017	1113	854	722	438	245	81	5809
Illinois, Chicago	35.8	2.44	0	12	117	381	807	1166	1265	1086	939	534	260	72	6639
Indiana, Indianapolis	39.6	4.56	0	0	90	316	723	1051	1113	949	809	432	177	39	5699
Iowa, Sioux City	43.0	1.10	0	9	108	369	867	1240	1435	1198	989	483	214	39	6951
Kansas, Wichita	44.2	7.11	0	0	33	229	618	905	1023	804	645	270	87	6	4620
Kentucky, Louisville	44.0	6.70	0	0	54	248	609	890	930	818	682	315	105	9	4660
Louisiana, Shreveport	56.2	13.8	0	0	0	47	297	477	552	426	304	81	0	0	2184
Maine, Caribou	24.4	−3.89	78	115	336	682	1044	1535	1690	1470	1308	858	468	183	9767
Maryland, Baltimore	43.7	6.83	0	0	48	264	585	905	936	820	679	327	90	0	4654
Massachusetts, Boston	40.0	4.40	0	9	60	316	603	983	1088	972	846	513	208	36	5634
Michigan, Lansing	34.8	1.89	6	22	138	431	813	1163	1262	1142	1011	579	273	69	6909
Minnesota, Minneapolis	28.3	−1.72	22	31	189	505	1014	1454	1631	1380	1166	621	288	81	8382
Mississippi, Jackson	55.7	13.5	0	0	0	65	315	502	546	414	310	87	0	0	2239
Missouri, Kansas City	43.9	6.94	0	0	39	220	612	905	1032	818	682	294	109	0	4711
Montana, Billings	34.5	1.72	6	15	186	487	897	1135	1296	1100	970	570	285	102	7049
Nebraska, Lincoln	38.8	4.11	0	6	75	301	726	1066	1237	1016	834	402	171	30	5864

Location	Avg °F	Avg °C													Annual
Nevada, Las Vegas	53.5	12.28		0	0	78	387	617	688	487	335	111	6	0	2709
New Hampshire, Concord	33.0	0.60	6	50	177	505	822	1240	1358	1184	1032	636	298	75	7383
New Jersey, Atlantic City	43.2	6.56	0	0	39	251	549	880	936	848	741	420	133	15	4812
New Mexico, Albuquerque	45.0	7.20	0	0	12	229	642	868	930	703	595	288	81	0	4348
New York, Syracuse	35.2	2.11	6	28	132	415	744	1153	1271	1140	1004	570	248	45	6756
North Carolina, Charlotte	50.4	10.56	0	0	6	124	438	691	691	582	481	156	22	0	3191
North Dakota, Bismarck	26.6	−2.67	34	28	222	577	1083	1463	1708	1442	1203	645	329	117	8851
Ohio, Cleveland	37.2	3.22	9	25	105	384	738	1088	1159	1047	918	552	260	66	6351
Oklahoma, Stillwater	48.3	9.39	0	0	15	164	498	766	868	664	527	189	34	0	3725
Oregon, Pendleton	42.6	6.22	0	0	111	350	711	884	1017	773	617	396	205	63	5127
Pennsylvania, Pittsburgh	38.4	3.89	0	9	105	375	726	1063	1119	1002	874	480	195	39	5987
Rhode Island, Providence	38.8	4.11	0	16	96	372	660	1023	1110	988	868	534	236	51	5954
South Carolina, Charleston	56.4	13.9		0	0	59	282	471	487	389	291	54	0	0	2033
South Dakota, Rapid City	33.4	1.11	22	12	165	481	897	1172	1333	1145	1051	615	326	126	7345
Tennessee, Memphis	50.5	10.6	0	0	18	130	447	698	729	585	456	147	22	0	3232
Texas, Dallas	55.3	13.3		0	0	62	321	524	601	440	319	90	6	0	2363
Utah, Salt Lake City	38.4	3.89		0	81	419	849	1082	1172	910	763	459	233	84	6052
Vermont, Burlington	29.4	−1.11	28	65	207	539	891	1349	1513	1333	1187	714	353	90	8269
Virginia, Norfolk	49.2	9.89		0	0	136	408	698	738	655	533	216	37	0	3421
Washington, Spokane	36.5	2.83	9	25	168	493	879	1082	1231	980	834	531	288	135	6655
West Virginia, Charleston	44.8	7.44	0	0	63	254	591	865	880	770	648	300	96	9	4476
Wisconsin, Milwaukee	32.6	0.667	43	47	174	471	876	1252	1376	1193	1054	642	372	135	7635
Wyoming, Casper	33.4	1.11	6	16	192	524	942	1169	1290	1084	1020	657	381	129	7410
Alberta, Calgary	—	—	109	186	402	719	1110	1389	1575	1379	1268	798	477	291	9703
British Columbia, Vancouver	—	—	81	87	219	456	657	787	862	723	676	501	310	156	5515
Manitoba, Winnipeg	—	—	38	71	322	683	1251	1757	2008	1719	1465	813	405	147	10679
New Brunswick, Fredericton	—	—	78	68	234	592	915	1392	1541	1379	1172	753	406	141	8671
Nova Scotia, Halifax	—	—	58	51	180	457	710	1074	1213	1122	1030	742	487	237	7361
Ontario, Ottawa	—	—	25	81	222	567	936	1469	1624	1441	1231	708	341	90	8735
Quebec, Montreal	—	—	9	43	165	521	882	1392	1566	1381	1175	684	316	69	8203
Saskatchewan, Regina	—	—	78	93	360	741	1284	1711	1965	1687	1473	804	409	201	10806

[a] Based on degrees F, quantities may be converted to degree days based on degrees C by multiplying by 5/9. This assumes 18 C corresponds to 65 F.

Source: Abridged by permission from *ASHRAE Handbook, Systems Volume,* 1980.

Table C-3 Annual Bin Weather Data for Oklahoma City, Oklahoma, 35° 24' N 97° 36' W, 1285 Ft Elevation

Hours of Dry-bulb Occurrence: MCWB

Temperature Bin-F	Time Group						Total
	1–4	5–8	9–12	13–16	17–20	21–24	
100/104	0: 0	0: 0	0: 0	2: 77	0: 0	0: 0	2: 77
95/ 99	0: 0	0: 0	5: 74	70: 76	29: 76	0: 0	104: 76
90/ 94	0: 0	0: 0	55: 76	153: 74	88: 74	0: 0	296: 74
85/ 89	2: 73	0: 0	116: 74	145: 72	120: 73	24: 74	407: 73
80/ 84	20: 72	33: 73	148: 72	153: 68	168: 71	96: 72	618: 71
75/ 79	121: 71	93: 71	132: 68	115: 65	144: 67	171: 70	776: 69
70/ 74	229: 68	221: 69	138: 65	118: 58	117: 63	186: 68	1009: 66
65/ 69	161: 64	161: 65	98: 58	98: 54	93: 56	136: 62	747: 61
60/ 64	120: 58	99: 58	95: 53	135: 51	96: 51	97: 55	642: 54
55/ 59	87: 53	104: 53	105: 50	108: 47	116: 49	81: 51	601: 50
50/ 54	96: 48	103: 49	137: 46	94: 45	133: 45	121: 47	684: 46
45/ 49	98: 43	96: 43	100: 42	66: 40	87: 41	122: 42	569: 42
40/ 44	150: 39	121: 39	98: 38	67: 37	91: 38	140: 38	667: 38
35/ 39	144: 35	153: 35	89: 34	54: 34	76: 34	105: 34	621: 35
30/ 34	107: 31	140: 30	74: 30	40: 30	50: 30	93: 30	504: 30
25/ 29	63: 25	51: 26	27: 26	24: 24	24: 25	40: 25	229: 25
20/ 24	36: 20	41: 21	23: 20	17: 19	23: 19	31: 21	171: 20
15/ 19	19: 16	37: 15	17: 16	1: 12	5: 14	16: 15	95: 15
10/ 14	7: 10	7: 11	3: 9	0: 0	0: 0	1: 11	18: 10

Source: Reprinted by permission from *Bin and Degree Hour Weather Data for Simplified Energy Calculations*, ASHRAE, Inc., Atlanta, GA, 1986.

Table C-4 Annual Bin Weather Data for Chicago, Illinois, 41° 47' N 87° 45' W, 607 Ft Elevation

Temperature Bin-F	Time Group						
	1–4	5–8	9–12	13–16	17–20	21–24	Total
	Hours of Dry-bulb Occurrence: MCWB						
90/ 94	0: 0	0: 0	15: 74	64: 73	18: 71	0: 0	97: 73
85/ 89	0: 0	0: 0	64: 71	113: 68	45: 70	0: 0	222: 69
80/ 84	0: 0	7: 71	131: 67	113: 65	90: 67	21: 70	362: 67
75/ 79	25: 70	51: 68	114: 64	116: 64	125: 65	81: 68	512: 65
70/ 74	120: 66	125: 65	131: 62	111: 62	151: 63	167: 65	805: 64
65/ 69	156: 62	155: 62	88: 60	65: 58	80: 59	143: 60	687: 61
60/ 64	159: 57	130: 57	78: 54	73: 53	72: 55	103: 57	615: 56
55/ 59	111: 52	112: 52	93: 49	109: 49	96: 51	101: 52	622: 51
50/ 54	112: 48	92: 48	88: 46	89: 47	100: 47	104: 47	585: 47
45/ 49	92: 44	102: 43	97: 42	96: 40	83: 42	107: 43	577: 42
40/ 44	120: 38	104: 39	72: 37	94: 37	131: 37	115: 38	636: 38
35/ 39	106: 34	105: 34	133: 33	134: 33	131: 33	111: 34	720: 33
30/ 34	182: 30	182: 30	156: 29	129: 29	135: 30	173: 30	957: 30
25/ 29	93: 25	98: 25	66: 25	68: 24	94: 25	92: 25	511: 25
20/ 24	75: 21	75: 21	60: 20	40: 20	44: 21	60: 21	354: 21
15/ 19	63: 16	56: 16	30: 16	21: 15	33: 15	40: 16	243: 16
10/ 14	25: 11	32: 11	25: 10	13: 10	11: 11	19: 11	125: 11
5/ 9	5: 7	14: 6	11: 5	8: 6	16: 6	12: 5	66: 6
0/ 4	15: 1	16: 0	7: 0	4: 1	5: 0	11: 1	58: 0
−5/ −1	1: −4	4: −7	1: −6	0: 0	0: 0	0: 0	6: −6

Source: Reprinted by permission from *Bin and Degree Hour Weather Data for Simplified Energy Calculations,* ASHRAE, Inc., Atlanta, GA, 1986.

Table C-5 Annual Bin Weather Data for Denver, Colorado, 39° 45′ N 104° 52′ W 5283 Ft Elevation

Temperature Bin-F	Time Group						Total
	1–4	5–8	9–12	13–16	17–20	21–24	
	Hours of Dry-bulb Occurrence: MCWB						
95/ 99	0: 0	0: 0	0: 0	3: 60	0: 0	0: 0	3: 60
90/ 94	0: 0	0: 0	15: 58	88: 59	15: 58	0: 0	118: 59
85/ 89	0: 0	0: 0	72: 59	112: 58	51: 58	0: 0	235: 58
80/ 84	0: 0	1: 57	107: 58	146: 56	91: 57	3: 58	348: 57
75/ 79	0: 0	16: 58	124: 56	100: 54	120: 56	30: 57	390: 55
70/ 74	12: 55	49: 56	121: 54	95: 52	112: 53	83: 56	472: 54
65/ 69	90: 54	99: 54	103: 51	136: 49	129: 51	140: 54	697: 52
60/ 64	129: 52	118: 52	109: 48	106: 46	108: 50	129: 51	699: 50
55/ 59	133: 50	137: 49	120: 44	119: 43	106: 46	147: 47	762: 47
50/ 54	154: 44	128: 44	135: 41	138: 40	98: 41	130: 43	783: 42
45/ 49	137: 41	139: 40	118: 38	84: 38	118: 38	120: 40	716: 39
40/ 44	170: 36	138: 36	83: 34	81: 33	84: 35	109: 36	665: 35
35/ 39	148: 32	135: 32	127: 31	103: 31	120: 32	125: 32	758: 32
30/ 34	132: 29	145: 29	96: 28	71: 28	122: 28	147: 28	713: 28
25/ 29	148: 24	120: 24	59: 24	36: 25	79: 25	123: 24	565: 24
20/ 24	98: 20	116: 20	28: 19	14: 20	51: 20	92: 20	399: 20
15/ 19	35: 16	49: 15	17: 15	14: 14	20: 16	29: 16	164: 15
10/ 14	29: 10	21: 10	8: 9	9: 10	15: 10	24: 11	106: 10
5/ 9	19: 6	19: 6	10: 5	3: 4	7: 6	7: 5	65: 6
0/ 4	20: −1	23: 0	7: 0	2: 2	11: 0	17: −1	80: 0
− 5/ − 1	6: − 6	7: − 6	1: − 4	0: 0	3: − 6	5: − 5	22: − 6

Source: Reprinted by permission from *Bin and Degree Hour Weather Data for Simplified Energy Calculations,* ASHRAE, Inc., Atlanta, GA, 1986.

Table C-6 Annual Bin Weather Data for Washington, D.C., 38° 51′ N 77° 02′ W, 14 Ft Elevation

Temperature Bin-F	Time Group						
	1–4	5–8	9–12	13–16	17–20	21–24	Total
	Hours of Dry-bulb Occurrence: MCWB						
90/ 94	0: 0	0: 0	11: 76	60: 77	15: 77	0: 0	86: 77
85/ 89	0: 0	0: 0	54: 75	142: 72	84: 74	0: 0	280: 73
80/ 84	2: 74	11: 75	140: 72	172: 70	155: 71	51: 74	531: 71
75/ 79	89: 73	88: 73	166: 68	168: 64	169: 67	145: 71	825: 69
70/ 74	186: 69	188: 68	164: 64	135: 61	173: 63	198: 67	1044: 66
65/ 69	178: 64	156: 63	119: 59	81: 56	102: 59	178: 62	814: 61
60/ 64	133: 58	119: 58	86: 54	103: 51	73: 53	102: 57	616: 56
55/ 59	106: 53	137: 53	113: 50	97: 47	90: 48	75: 51	618: 51
50/ 54	99: 48	90: 48	95: 46	99: 44	129: 45	111: 47	623: 46
45/ 49	128: 43	110: 43	128: 41	108: 40	102: 41	135: 42	711: 41
40/ 44	137: 38	128: 38	99: 36	113: 36	130: 36	135: 38	742: 37
35/ 39	148: 33	142: 34	134: 32	107: 31	113: 32	137: 33	781: 33
30/ 34	138: 29	137: 29	97: 28	47: 29	84: 28	114: 28	617: 28
25/ 29	62: 25	96: 25	35: 24	19: 24	25: 24	52: 24	289: 24
20/ 24	39: 20	41: 20	8: 18	5: 19	10: 19	13: 19	116: 20
15/ 19	6: 15	5: 14	10: 13	4: 15	4: 15	6: 16	35: 15
10/ 14	9: 10	12: 11	1: 12	0: 0	2: 11	8: 10	32: 10

Source: Reprinted by permission from *Bin and Degree Hour Weather Data for Simplified Energy Calculations*, ASHRAE, Inc., Atlanta, GA, 1986.

APPENDIX D

Pipe and Tube Data

Table D-1 Steel Pipe Dimensions—English and SI Units

Nominal Pipe Size in.	Schedule Number	Diameter O.D. in.	O.D. mm	I.D. in.	I.D. mm	Wall Thickness in.	Wall Thickness mm	Inside Cross-sectional Area ft²	Area m² × 10³
$\frac{1}{4}$	40	0.540	13.7	0.364	9.25	0.088	2.23	0.00072	0.067
	80			0.302	7.67	0.119	3.02	0.00050	0.046
$\frac{3}{8}$	40	0.675	17.1	0.493	12.5	0.091	2.31	0.00133	0.124
	80			0.423	10.7	0.126	3.20	0.00098	0.091
$\frac{1}{2}$	40	0.840	21.3	0.622	15.8	0.109	2.77	0.00211	0.196
	80			0.546	13.9	0.147	3.73	0.00163	0.151
$\frac{3}{4}$	40	1.050	26.7	0.824	20.9	0.113	2.87	0.00371	0.345
	80			0.742	18.8	0.154	3.91	0.00300	0.279
1	40	1.315	33.4	1.049	26.6	0.133	3.38	0.00600	0.557
	80			0.957	24.3	0.179	4.55	0.00499	0.464
$1\frac{1}{2}$	40	1.900	48.3	1.610	40.9	0.145	3.68	0.01414	1.314
	80			1.500	38.1	0.200	5.08	0.01225	1.138
2	40	2.375	60.3	2.067	52.5	0.154	3.91	0.02330	2.165
	80			1.939	49.3	0.218	5.54	0.02050	1.905
$2\frac{1}{2}$	40	2.875	73.0	2.469	62.7	0.203	5.16	0.03322	3.086
	80			2.323	59.0	0.276	7.01	0.02942	2.733
3	40	3.500	88.9	3.068	77.9	0.216	5.49	0.05130	4.766
	80			2.900	73.7	0.300	7.62	0.04587	4.262
4	40	4.500	114.3	4.026	102.3	0.237	6.02	0.08840	8.213
	80			3.826	97.2	0.337	8.56	0.07986	7.419
5	40	5.563	141.3	5.047	128.1	0.258	6.55	0.1390	12.91
	80			4.813	122.3	0.375	9.53	0.1263	11.73
6	40	6.625	168.3	6.065	154.1	0.280	7.11	0.2006	18.64
	80			5.761	146.3	0.432	11.0	0.1810	16.82
8	40	8.625	219.1	7.981	202.7	0.322	8.18	0.3474	32.28
	80			7.625	193.7	0.500	12.7	0.3171	29.46
10	40	10.75	273.1	10.020	254.5	0.365	9.27	0.5475	50.86
	80			9.750	247.7	0.500	12.7	0.5185	48.17

Source: Adapted from A.S.A. Standards B36.10.

Table D-2 Type L Copper Tube Dimensions—English and SI Units

Nominal Tube Size in.	Diameter O.D. in.	O.D. mm	I.D. in.	I.D. mm	Wall Thickness in.	Wall Thickness mm	Inside Cross-sectional Area in.²	Inside Cross-sectional Area m² × 10³
$\frac{1}{4}$	0.375	9.53	0.315	8.00	0.030	0.762	0.0779	0.0503
$\frac{3}{8}$	0.500	12.7	0.43	11	0.035	0.889	0.145	0.094
$\frac{1}{2}$	0.625	15.9	0.545	13.8	0.040	1.02	0.233	0.150
$\frac{5}{8}$	0.750	19.1	0.666	16.9	0.042	1.07	0.348	0.225
$\frac{3}{4}$	0.875	22.2	0.785	19.9	0.045	1.14	0.484	0.312
1	1.125	28.58	1.025	26.04	0.050	1.27	0.825	0.532
$1\frac{1}{4}$	1.375	34.93	1.265	32.13	0.055	1.40	1.26	0.813
$1\frac{1}{2}$	1.625	41.28	1.505	38.23	0.060	1.52	1.78	1.15
2	2.125	53.98	1.985	50.42	0.070	1.78	3.10	2.00
$2\frac{1}{2}$	2.625	66.68	2.465	62.61	0.080	2.03	4.77	3.08
3	3.125	79.38	2.945	74.80	0.090	2.29	6.81	4.39
$3\frac{1}{2}$	3.625	92.08	3.425	87.00	0.100	2.54	9.21	5.94
4	4.125	104.8	3.905	99.19	0.110	2.79	12.0	7.74

Source: Based on ASTM B-88.

APPENDIX E

Useful Data

Length
1 ft = 30.48 cm
1 in. = 2.54 cm
1 m = 39.37 in.
1 micron = 10^{-6} m = 3.281×10^{-6} ft
1 mile = 5280 ft

Area
1 m^2 = 1550.1472 in.^{2}
1 m^2 = 10.76392 ft^2

Volume
1 ft^3 = 7.48 U.S. gallons = 1728 in.^{3}
1 m^3 = 6.1×10^4 in.^{3}
1 m^3 = 35.3147 ft^3
1 m^3 = 264.154 U.S. gallons

Mass
1 kg = 2.20462 lbm
1 lbm = 7000 grains = 453.5924 g

Force
1 N = 0.224809 lbf
1 lbf = 4.44822 N

Energy
1 Btu = 778.28 ft-lbf
1 kilocalorie = 10^3 calories = 3.968 Btu
1 J = 9.48×10^{-4} Btu = 0.73756 ft-lbf
1 kW-hr = 3412 Btu = 2.6562×10^6 ft-lbf

Table E-1 Conversion Factors (*Continued*)

Power

1 hp = 33,000 (ft-lbf)/min
1 hp = 745.7 W
1 W = 3.412 Btu/hr
1 W = 0.001341 hp
1 W = 0.0002843 tons of refrigeration

Pressure

1 atm = 14.6959 psia = 2116 lbf/ft^2 = 101325 N/m^2
1 in. wg = 249.08 Pa
1 in. of mercury = 3376.85 Pa
1 lbf/in.^{2} = 6894.76 Pa
1 Pa = 1 N/m^2 = 1.4504 × 10^{-4} lbf/in.^{2}

Temperature

1 degree R difference = 1 degree F difference = $\frac{5}{9}$ degree C difference = $\frac{5}{9}$ degree K difference
degree F = $\frac{9}{5}$ (degree C) + 32
degree C = $\frac{5}{9}$ (degree F − 32)

Thermal Conductivity

$$1 \frac{Btu}{hr\text{-}ft\text{-}F} = 0.004134 \frac{calorie}{s\text{-}cm\text{-}C} = 1.7307 \frac{W}{m\text{-}C}$$

$$1 \frac{W}{m\text{-}C} = 0.5778 \frac{Btu}{hr\text{-}ft\text{-}F}$$

$$1 \frac{Btu\text{-}in.}{hr\text{-}ft^2\text{-}F} = 0.1442 \frac{W}{m\text{-}C}$$

$$1 \frac{W}{m\text{-}C} = 6.933 \frac{Btu\text{-}in.}{hr.\text{-}ft^2\text{-}F}$$

Heat Transfer Coefficient

$$1 \frac{Btu}{hr\text{-}ft^2\text{-}F} = 5.678 \frac{W}{m^2\text{-}C}$$

$$1 \frac{W}{m^2\text{-}C} = 0.1761 \frac{Btu}{hr\text{-}ft^2\text{-}F}$$

Viscosity, Absolute

1 poise = 100 centipoises

$$1 \frac{lbm}{sec\text{-}ft} = 1490 \text{ centipoises} = 1.49 \frac{N\text{-}s}{m^2}$$

$$1 \frac{lbf\text{-}sec}{ft^2} = 47,800 \text{ centipoises}$$

$$1 \text{ centipose} = 0.001 \frac{N\text{-}s}{m^2}$$

Table E-1 Conversion Factors (*Continued*)

Viscosity, Kinematic
1 ft²/sec = 0.0929 m²/s
1 m²/s = 10.764 ft²/sec

Specific Heat

$$1\frac{\text{calorie}}{\text{gm-C}} = 1\frac{\text{Btu}}{\text{lbm-F}}$$

$$1\frac{\text{Btu}}{\text{lbm-F}} = 4186.8\frac{\text{J}}{\text{kg-C}}$$

$$1\frac{\text{kJ}}{\text{kg-C}} = 0.2388\frac{\text{Btu}}{\text{lbm-F}}$$

APPENDIX F

Symbols

Table F-1 Symbols Used in Control System Schematics

Centrifugal fan	Globe valve
Check valve	Humidstat, room
Coil for solenoid valve	Humidstat, duct
Contactor or motor starter; controller	Proportioning controller, solid-state
Control valve, three-way	Pump
Control valve, two-way	Receiver-controller
Damper motor	Relay
Motor starter	Reversing relay
Motor	Sequencing controller
Parallel-blade damper (for two-position control)	Smoke detector
	Static pressure controller

Table F-1 Symbols Used in Control System Schematics (*Continued*)

	Opposed-blade damper (for modulating control)		Thermal expansion valve, thermostatic expansion valve
	Pressure gauge		Thermometer, remote-bulb or insertion type
	Pressure regulator (pressure-reducing valve)		Thermostat or temperature sensor, insertion type
P	Pressure switch or sensor	T	Thermostat or temperature sensor, remote-bulb, duct or pipe, or insertion type
DP	Differential pressure sensor or controller		
		T	Thermostat, room
EP	Electric-pneumatic relay		
		VSC	Variable-speed controller
FS	Fire safety switch		
F	Flow switch		

Index